Reactions at Supports and Connections for a Three-Dimensional Structure

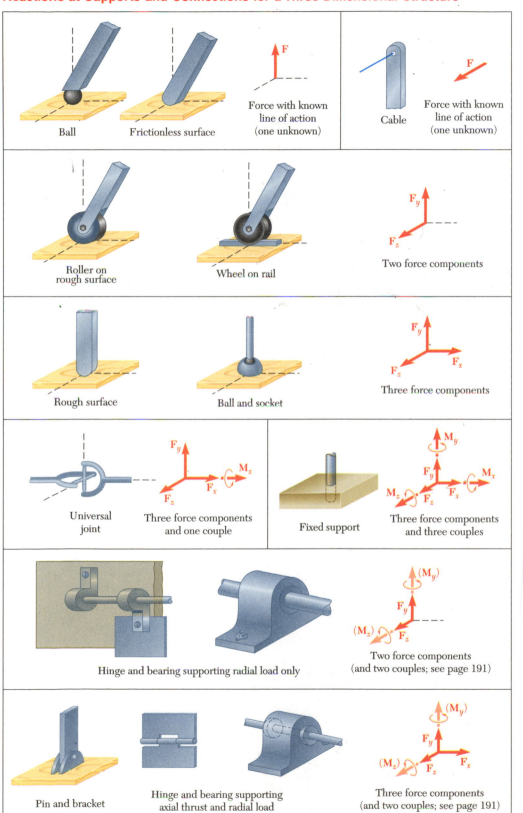

Support/Connection	Reaction
Ball / Frictionless surface	Force with known line of action (one unknown)
Cable	Force with known line of action (one unknown)
Roller on rough surface / Wheel on rail	Two force components
Rough surface / Ball and socket	Three force components
Universal joint	Three force components and one couple
Fixed support	Three force components and three couples
Hinge and bearing supporting radial load only	Two force components (and two couples; see page 191)
Pin and bracket / Hinge and bearing supporting axial thrust and radial load	Three force components (and two couples; see page 191)

Vector Mechanics
For Engineers

Statics

Eleventh Edition

Vector Mechanics For Engineers

Statics

Ferdinand P. Beer
Late of Lehigh University

E. Russell Johnston, Jr.
Late of University of Connecticut

David F. Mazurek
U.S. Coast Guard Academy

VECTOR MECHANICS FOR ENGINEERS: STATICS, ELEVENTH EDITION

Published by McGraw-Hill Education, 2 Penn Plaza, New York, NY 10121. Copyright © 2016 by
McGraw-Hill Education. All rights reserved. Printed in the United States of America. Previous editions
© 2013, 2010, and 2007. No part of this publication may be reproduced or distributed in any
form or by any means, or stored in a database or retrieval system, without the prior written consent
of McGraw-Hill Education, including, but not limited to, in any network or other electronic storage or
transmission, or broadcast for distance learning.

Some ancillaries, including electronic and print components, may not be available to customers outside the
United States.

This book is printed on acid-free paper.

1 2 3 4 5 6 7 8 9 0 DOW/DOW 1 0 9 8 7 6 5

ISBN 978-0-07-768730-4
MHID 0-07-768730-2

Managing Director: *Thomas Timp*
Global Brand Manager: *Raghothaman Srinivasan*
Director of Development: *Rose Koos*
Product Developer: *Robin Reed*
Brand Manager: *Thomas Scaife, Ph.D.*
Digital Product Analyst: *Dan Wallace*
Editorial Coordinator: *Samantha Donisi-Hamm*
Marketing Manager: *Nick McFadden*
LearnSmart Product Developer: *Joan Weber*
Content Production Manager: *Linda Avenarius*
Content Project Managers: *Jolynn Kilburg and Lora Neyens*
Buyer: *Laura Fuller*
Designer: *Matthew Backhaus*
Content Licensing Specialist (Image): *Carrie Burger*
Media Project Manager: *TK*
Typeface: *10/12 Times LT Std*
Printer: *R. R. Donnelley*

All credits appearing on page or at the end of the book are considered to be an extension of the copyright
page.

The photo on the cover shows One World Trade Center in New York City, the tallest skyscraper in the
Western Hemisphere. From its foundation to its structural components and mechanical systems, the design
and operation of the tower is based on the fundamentals of engineering mechanics.

Library of Congress Cataloging-in-Publication

Data On File

The Internet addresses listed in the text were accurate at the time of publication. The inclusion of a website
does not indicate an endorsement by the authors or McGraw-Hill Education, and McGraw-Hill Education
does not guarantee the accuracy of the information presented at these sites.

www.mhhe.com

About the Authors

Ferdinand P. Beer. Born in France and educated in France and Switzerland, Ferd received an M.S. degree from the Sorbonne and an Sc.D. degree in theoretical mechanics from the University of Geneva. He came to the United States after serving in the French army during the early part of World War II and taught for four years at Williams College in the Williams-MIT joint arts and engineering program. Following his service at Williams College, Ferd joined the faculty of Lehigh University where he taught for thirty-seven years. He held several positions, including University Distinguished Professor and chairman of the Department of Mechanical Engineering and Mechanics, and in 1995 Ferd was awarded an honorary Doctor of Engineering degree by Lehigh University.

E. Russell Johnston, Jr. Born in Philadelphia, Russ received a B.S. degree in civil engineering from the University of Delaware and an Sc.D. degree in the field of structural engineering from the Massachusetts Institute of Technology. He taught at Lehigh University and Worcester Polytechnic Institute before joining the faculty of the University of Connecticut where he held the position of chairman of the Department of Civil Engineering and taught for twenty-six years. In 1991 Russ received the Outstanding Civil Engineer Award from the Connecticut Section of the American Society of Civil Engineers.

David F. Mazurek. David holds a B.S. degree in ocean engineering and an M.S. degree in civil engineering from the Florida Institute of Technology and a Ph.D. degree in civil engineering from the University of Connecticut. He was employed by the Electric Boat Division of General Dynamics Corporation and taught at Lafayette College prior to joining the U.S. Coast Guard Academy, where he has been since 1990. He is a registered Professional Engineer in Connecticut and Pennsylvania, and has served on the American Railway Engineering & Maintenance-of-Way Association's Committee 15—Steel Structures since 1991. He is a Fellow of the American Society of Civil Engineers, and was elected to the Connecticut Academy of Science and Engineering in 2013. He was the 2014 recipient of both the Coast Guard Academy's Distinguished Faculty Award and its Center for Advanced Studies Excellence in Scholarship Award. Professional interests include bridge engineering, structural forensics, and blast-resistant design.

Brief Contents

Contents

1 Introduction   1

2 Statics of Particles   15

3 Rigid Bodies: Equivalent Systems of Forces   82

*Advanced or specialty topics

Preface

Objectives

NEW!

The 11th edition has undergone a complete rewrite to modernize and streamline the language throughout the text.

A primary objective in a first course in mechanics is to help develop a student's ability first to analyze problems in a simple and logical manner, and then to apply basic principles to its solution. A strong conceptual understanding of these basic mechanics principles is essential for successfully solving mechanics problems. We hope that this text, as well as the proceeding volume, *Vector Mechanics for Engineers: Dynamics,* will help instructors achieve these goals.†

General Approach

Vector analysis is introduced early in the text and is used in the presentation and discussion of the fundamental principles of mechanics. Vector methods are also used to solve many problems, particularly three-dimensional problems where these techniques result in a simpler and more concise solution. The emphasis in this text, however, remains on the correct understanding of the principles of mechanics and on their application to the solution of engineering problems, and vector analysis is presented chiefly as a convenient tool.‡

Practical Applications Are Introduced Early. One of the characteristics of the approach used in this book is that mechanics of *particles* is clearly separated from the mechanics of *rigid bodies.* This approach makes it possible to consider simple practical applications at an early stage and to postpone the introduction of the more difficult concepts. For example:

- In *Statics,* statics of particles is treated first (Chap. 2); after the rules of addition and subtraction of vectors are introduced, the principle of equilibrium of a particle is immediately applied to practical situations involving only concurrent forces. The statics of rigid bodies is considered in Chaps. 3 and 4. In Chap. 3, the vector and scalar products of two vectors are introduced and used to define the moment of a force about a point and about an axis. The presentation of these new concepts is followed by a thorough and rigorous discussion of equivalent systems of forces leading, in Chap. 4, to many practical applications involving the equilibrium of rigid bodies under general force systems.
- In *Dynamics,* the same division is observed. The basic concepts of force, mass, and acceleration, of work and energy, and of impulse and momentum are introduced and first applied to problems involving only particles. Thus, students can familiarize themselves with

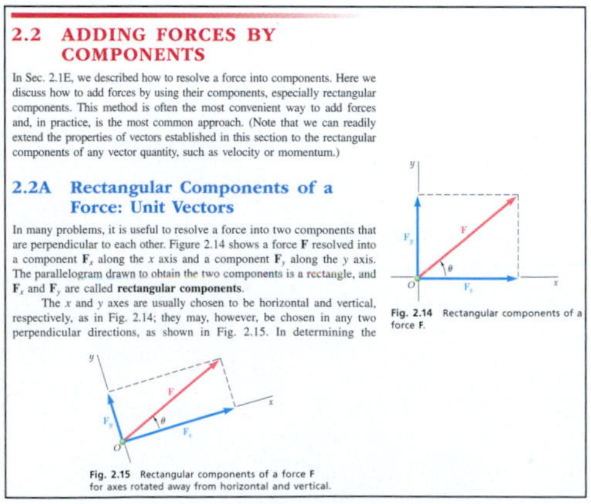

2.2 ADDING FORCES BY COMPONENTS

In Sec. 2.1E, we described how to resolve a force into components. Here we discuss how to add forces by using their components, especially rectangular components. This method is often the most convenient way to add forces and, in practice, is the most common approach. (Note that we can readily extend the properties of vectors established in this section to the rectangular components of any vector quantity, such as velocity or momentum.)

2.2A Rectangular Components of a Force: Unit Vectors

In many problems, it is useful to resolve a force into two components that are perpendicular to each other. Figure 2.14 shows a force **F** resolved into a component $\mathbf{F}_x$ along the x axis and a component $\mathbf{F}_y$ along the y axis. The parallelogram drawn to obtain the two components is a rectangle, and $\mathbf{F}_x$ and $\mathbf{F}_y$ are called **rectangular components**.

The x and y axes are usually chosen to be horizontal and vertical, respectively, as in Fig. 2.14; they may, however, be chosen in any two perpendicular directions, as shown in Fig. 2.15. In determining the

Fig. 2.14 Rectangular components of a force **F**.

Fig. 2.15 Rectangular components of a force **F** for axes rotated away from horizontal and vertical.

†Both texts also are available in a single volume, *Vector Mechanics for Engineers: Statics and Dynamics,* eleventh edition.

‡In a parallel text, *Mechanics for Engineers: Statics,* fifth edition, the use of vector algebra is limited to the addition and subtraction of vectors.

the three basic methods used in dynamics and learn their respective advantages before facing the difficulties associated with the motion of rigid bodies.

New Concepts Are Introduced in Simple Terms.
Since this text is designed for the first course in statics, new concepts are presented in simple terms and every step is explained in detail. On the other hand, by discussing the broader aspects of the problems considered, and by stressing methods of general applicability, a definite maturity of approach is achieved. For example, the concepts of partial constraints and statical indeterminacy are introduced early and are used throughout.

Fundamental Principles Are Placed in the Context of Simple Applications.
The fact that mechanics is essentially a *deductive* science based on a few fundamental principles is stressed. Derivations have been presented in their logical sequence and with all the rigor warranted at this level. However, the learning process being largely *inductive*, simple applications are considered first. For example:

- The statics of particles precedes the statics of rigid bodies, and problems involving internal forces are postponed until Chap. 6.
- In Chap. 4, equilibrium problems involving only coplanar forces are considered first and solved by ordinary algebra, while problems involving three-dimensional forces and requiring the full use of vector algebra are discussed in the second part of the chapter.

Systematic Problem-Solving Approach.
New to this edition of the text, all the sample problems are solved using the steps of *S*trategy, *M*odeling, *A*nalysis, and *R*eflect & *T*hink, or the "SMART" approach. This methodology is intended to give students confidence when approaching new problems, and students are encouraged to apply this approach in the solution of all assigned problems.

Free-Body Diagrams Are Used Both to Solve Equilibrium Problems and to Express the Equivalence of Force Systems.
Free-body diagrams are introduced early, and their importance is emphasized throughout the text. They are used not only to solve equilibrium problems but also to express the equivalence of two systems of forces or, more generally, of two systems of vectors. The advantage of this approach becomes apparent in the study of the dynamics of rigid bodies, where it is used to solve three-dimensional as well as two-dimensional problems. By placing the emphasis on "free-body-diagram equations" rather than on the standard algebraic equations of motion, a more intuitive and more complete understanding of the fundamental principles of dynamics can be achieved. This approach, which was first introduced in 1962 in the first edition of *Vector Mechanics for Engineers,* has now gained wide acceptance among mechanics teachers in this country. It is, therefore, used in preference to the method of dynamic equilibrium and to the equations of motion in the solution of all sample problems in this book.

4.1 EQUILIBRIUM IN TWO DIMENSIONS

In the first part of this chapter, we consider the equilibrium of two-dimensional structures; i.e., we assume that the structure being analyzed and the forces applied to it are contained in the same plane. Clearly, the reactions needed to maintain the structure in the same position are also contained in this plane.

4.1A Reactions for a Two-Dimensional Structure

The reactions exerted on a two-dimensional structure fall into three categories that correspond to three types of **supports** or **connections**.

1. **Reactions Equivalent to a Force with a Known Line of Action.** Supports and connections causing reactions of this type include *rollers, rockers, frictionless surfaces, short links and cables, collars on frictionless rods,* and *frictionless pins in slots.* Each of these supports and connections can prevent motion in one direction only. Figure 4.1 shows these supports and connections together with the reactions they produce. Each reaction involves *one unknown*—specifically, the magnitude of the reaction. In problem solving, you should denote this magnitude by an appropriate letter. The line of action of the reaction is known and should be indicated clearly in the free-body diagram.

   The sense of the reaction must be as shown in Fig. 4.1 for cases of a frictionless surface (toward the free body) or a cable (away from the free body). The reaction can be directed either way in the cases of double-track rollers, links, collars on rods, or pins in slots. Generally, we

NEW!

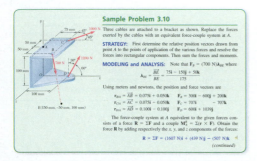

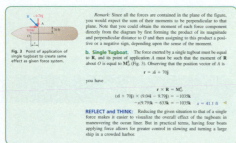

A Four-Color Presentation Uses Color to Distinguish Vectors.

Color has been used, not only to enhance the quality of the illustrations, but also to help students distinguish among the various types of vectors they will encounter. While there was no intention to "color code" this text, the same color is used in any given chapter to represent vectors of the same type. Throughout *Statics*, for example, red is used exclusively to represent forces and couples, while position vectors are shown in blue and dimensions in black. This makes it easier for the students to identify the forces acting on a given particle or rigid body and to follow the discussion of sample problems and other examples given in the text.

A Careful Balance Between SI and U.S. Customary Units Is Consistently Maintained.

Because of the current trend in the American government and industry to adopt the international system of units (SI metric units), the SI units most frequently used in mechanics are introduced in Chap. 1 and are used throughout the text. Approximately half of the sample problems and 60 percent of the homework problems are stated in these units, while the remainder are in U.S. customary units. The authors believe that this approach will best serve the need of students, who, as engineers, will have to be conversant with both systems of units.

It also should be recognized that using both SI and U.S. customary units entails more than the use of conversion factors. Since the SI system of units is an absolute system based on the units of time, length, and mass, whereas the U.S. customary system is a gravitational system based on the units of time, length, and force, different approaches are required for the solution of many problems. For example, when SI units are used, a body is generally specified by its mass expressed in kilograms; in most problems of statics it will be necessary to determine the weight of the body in newtons, and an additional calculation will be required for this purpose. On the other hand, when U.S. customary units are used, a body is specified by its weight in pounds and, in dynamics problems, an additional calculation will be required to determine its mass in slugs (or $lb \cdot s^2/ft$). The authors, therefore, believe that problem assignments should include both systems of units.

The *Instructor's and Solutions Manual* provides six different lists of assignments so that an equal number of problems stated in SI units and in U.S. customary units can be selected. If so desired, two complete lists of assignments can also be selected with up to 75 percent of the problems stated in SI units.

Optional Sections Offer Advanced or Specialty Topics.

A large number of optional sections have been included. These sections are indicated by asterisks and thus are easily distinguished from those which form the core of the basic statics course. They can be omitted without prejudice to the understanding of the rest of the text.

Among the topics covered in these additional sections are the reduction of a system of forces to a wrench, applications to hydrostatics, equilibrium of cables, products of inertia and Mohr's circle, the determination of the principal axes and the mass moments of inertia of a body of arbitrary shape, and the method of virtual work. The sections on the inertia properties of three-dimensional bodies are primarily intended for students who will later study in dynamics the three-dimensional motion of rigid bodies.

The material presented in the text and most of the problems require no previous mathematical knowledge beyond algebra, trigonometry, and elementary calculus; all the elements of vector algebra necessary to the understanding of the text are carefully presented in Chaps. 2 and 3. In general, a greater emphasis is placed on the correct understanding of the basic mathematical concepts involved than on the nimble manipulation of mathematical formulas. In this connection, it should be mentioned that the determination of the centroids of composite areas precedes the calculation of centroids by integration, thus making it possible to establish the concept of the moment of an area firmly before introducing the use of integration.

Guided Tour

Chapter Introduction. Each chapter begins with a list of learning objectives and an outline that previews chapter topics. An introductory section describes the material to be covered in simple terms, and how it will be applied to the solution of engineering problems.

Chapter Lessons. The body of the text is divided into sections, each consisting of one or more sub-sections, several sample problems, and a large number of end-of-section problems for students to solve. Each section corresponds to a well-defined topic and generally can be covered in one lesson. In a number of cases, however, the instructor will find it desirable to devote more than one lesson to a given topic. *The Instructor's and Solutions Manual* contains suggestions on the coverage of each lesson.

Concept Applications. Concept Applications are used within selected theory sections to amplify certain topics, and they are designed to reinforce the specific material being presented and facilitate its understanding.

Sample Problems. The Sample Problems are set up in much the same form that students will use when solving assigned problems, and they employ the SMART problem-solving methodology that students are encouraged to use in the solution of their assigned problems. They thus serve the double purpose of amplifying the text and demonstrating the type of neat and orderly work that students should cultivate in their own solutions. In addition, in-problem references and captions have been added to the sample problem figures for contextual linkage to the step-by-step solution.

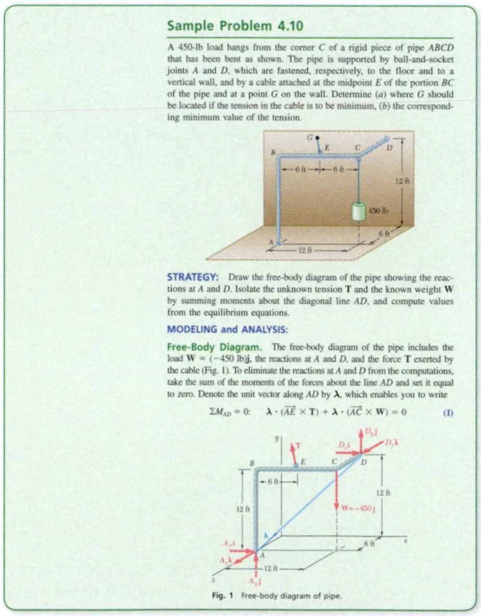

Sample Problem 4.10

A 450-lb load hangs from the corner C of a rigid piece of pipe $ABCD$ that has been bent as shown. The pipe is supported by ball-and-socket joints A and D, which are fastened, respectively, to the floor and to a vertical wall, and by a cable attached at the midpoint E of the portion BC of the pipe and at a point G on the wall. Determine (a) where G should be located if the tension in the cable is to be minimum, (b) the corresponding minimum value of the tension.

STRATEGY: Draw the free-body diagram of the pipe showing the reactions at A and D. Isolate the unknown tension **T** and the known weight **W** by summing moments about the diagonal line AD, and compute values from the equilibrium equations.

MODELING and ANALYSIS:

Free-Body Diagram. The free-body diagram of the pipe includes the load $\mathbf{W} = (-450 \text{ lb})\mathbf{j}$, the reactions at A and D, and the force **T** exerted by the cable (Fig. 1). To eliminate the reactions at A and D from the computations, take the sum of the moments of the forces about the line AD and set it equal to zero. Denote the unit vector along AD by $\boldsymbol{\lambda}$, which enables you to write

$$\Sigma M_{AD} = 0: \quad \boldsymbol{\lambda} \cdot (\overrightarrow{AE} \times \mathbf{T}) + \boldsymbol{\lambda} \cdot (\overrightarrow{AC} \times \mathbf{W}) = 0 \quad \text{(I)}$$

Fig. 1 Free-body diagram of pipe.

Solving Problems on Your Own. A section entitled *Solving Problems on Your Own* is included for each lesson, between the sample problems and the problems to be assigned. The purpose of these sections is to help students organize in their own minds the preceding theory of the text and the solution methods of the sample problems so that they can more successfully solve the homework problems. Also included in these sections are specific suggestions and strategies that will enable the students to more efficiently attack any assigned problems.

Homework Problem Sets. Most of the problems are of a practical nature and should appeal to engineering students. They are primarily designed, however, to illustrate the material presented in the text and to help students understand the principles of mechanics. The problems are grouped according to the portions of material they illustrate and, in general, are arranged in order of increasing difficulty. Problems requiring special attention are indicated by asterisks. Answers to 70 percent of the problems are given at the end of the book. Problems for which the answers are given are set in straight type in the text, while problems for which no answer is given are set in italic and red font color.

Over 300 of the homework problems in the text are new or revised.

Chapter Review and Summary. Each chapter ends with a review and summary of the material covered in that chapter. Marginal notes are used to help students organize their review work, and cross-references have been included to help them find the portions of material requiring their special attention.

Review Problems. A set of review problems is included at the end of each chapter. These problems provide students further opportunity to apply the most important concepts introduced in the chapter.

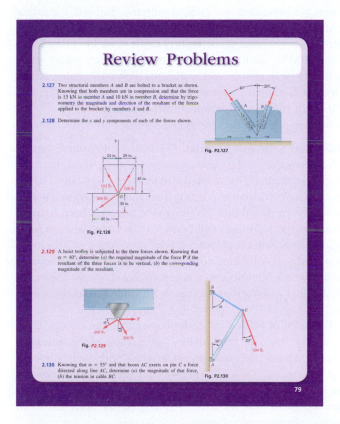

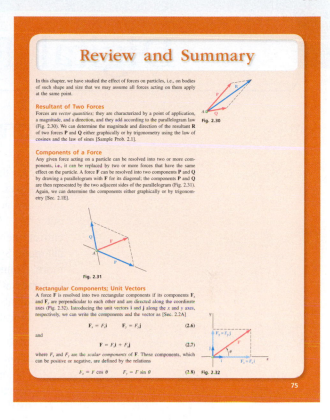

Computer Problems. Accessible through Connect are problem sets for each chapter that are designed to be solved with computational software. Many of these problems are relevant to the design process; they may involve the analysis of a structure for various configurations and loadings of the structure, or the determination of the equilibrium positions of a given mechanism that may require an iterative method of solution. Developing the algorithm required to solve a given mechanics problem will benefit the students in two different ways: (1) it will help them gain a better understanding of the mechanics principles involved; (2) it will provide them with an opportunity to apply their computer skills to the solution of a meaningful engineering problem.

Digital Resources

Connect® Engineering provides online presentation, assignment, and assessment solutions. It connects your students with the tools and resources they'll need to achieve success. With Connect Engineering you can deliver assignments, quizzes, and tests online. A robust set of questions and activities are presented and aligned with the textbook's learning outcomes. As an instructor, you can edit existing questions and author entirely new problems. Integrate grade reports easily with Learning Management Systems (LMS), such as WebCT and Blackboard—and much more. Connect Engineering also provides students with 24/7 online access to a media-rich eBook, allowing seamless integration of text, media, and assessments. **To learn more, visit connect.mheducation.com**

Find the following instructor resources available through Connect:

- **Instructor's and Solutions Manual.** *The Instructor's and Solutions Manual* that accompanies the eleventh edition features solutions to all end of chapter problems. This manual also features a number of tables designed to assist instructors in creating a schedule of assignments for their course. The various topics covered in the text have been listed in Table I and a suggested number of periods to be spent on each topic has been indicated. Table II prepares a brief description of all groups of problems and a classification of the problems in each group according to the units used. Sample lesson schedules are shown in Tables III, IV, and V, together with various alternative lists of assigned homework problems.
- **Lecture PowerPoint Slides** for each chapter that can be modified. These generally have an introductory application slide, animated worked-out problems that you can do in class with your students, concept questions, and "what-if?" questions at the end of the units.
- **Textbook images**
- **Computer Problem sets** for each chapter that are designed to be solved with computational software.
- **C.O.S.M.O.S.**, the Complete Online Solutions Manual Organization System that allows instructors to create custom homework, quizzes, and tests using end-of-chapter problems from the text.

 **LearnSmart** is available as an integrated feature of McGraw-Hill Connect. It is an adaptive learning system designed to help students learn faster, study more efficiently, and retain more knowledge for greater success. LearnSmart assesses a student's knowledge of course content through a series of adaptive questions. It pinpoints concepts the student does not understand and maps out a personalized study plan for success. This innovative study tool also has features that allow instructors to see exactly what students have accomplished and a built-in assessment tool for graded assignments.

SMARTBOOK™ SmartBook™ is the first and only adaptive reading experience available for the higher education market. Powered by an intelligent diagnostic and adaptive engine, SmartBook facilitates the reading process by identifying what content a student knows and doesn't know through adaptive assessments. As the student reads, the reading material constantly adapts to ensure the student is focused on the content he or she needs the most to close any knowledge gaps.

NEW!

Visit the following site for a demonstration of LearnSmart or SmartBook: www.learnsmartadvantage.com

CourseSmart. This text is offered through CourseSmart for both instructors and students. CourseSmart is an online browser where students can purchase access to this and other McGraw-Hill textbooks in a digital format. Through their browser, students can access the complete text online at almost half the cost of a traditional text. Purchasing the eTextbook also allows students to take advantage of CourseSmart's web tools for learning, which include full text search, notes and highlighting, and e-mail tools for sharing notes among classmates. To learn more about CourseSmart options, contact your sales representative or visit **www.coursesmart.com**.

Acknowledgments

A special thanks to Amy Mazurek who thoroughly checked the solutions and answers of all problems in this edition and then prepared the solutions for the *Instructor's and Solutions Manual*.

The authors thank the many companies that provided photographs for this edition.

We are pleased to acknowledge David Chelton, who carefully reviewed the entire text and provided many helpful suggestions for revising this edition.

The authors also thank the members of the staff at McGraw-Hill for their support and dedication during the preparation of this new edition. We particularly wish to acknowledge the contributions of Global Brand Manager Raghu Srinivasan, Brand Manager Thomas Scaife, Product Developers Robin Reed & Joan Weber, Content Project Manager Jolynn Kilburg, and Program Manager Lora Neyens.

David F. Mazurek

The authors gratefully acknowledge the many helpful comments and suggestions offered by focus group attendees and by users of the previous editions of *Vector Mechanics for Engineers:*

George Adams
Northeastern University

William Altenhof
University of Windsor

Sean B. Anderson
Boston University

Manohar Arora
Colorado School of Mines

Gilbert Baladi
Michigan State University

Brock E. Barry
United States Military

Francois Barthelat
McGill University

Oscar Barton, Jr
U.S. Naval Academy

M. Asghar Bhatti
University of Iowa

Shaohong Cheng
University of Windsor

Philip Datseris
University of Rhode Island

Daniel Dickrell, III
University of Florida

Timothy A. Doughty
University of Portland

Howard Epstein
University of Conneticut

Asad Esmaeily
Kansas State University,
Civil Engineering
Department

David Fleming
Florida Institute of
Technology

Ali Gordon
University of Central
Florida, Orlando

Jeff Hanson
Texas Tech University

David A. Jenkins
University of Florida

Shaofan Li
University of California,
Berkeley

Tom Mase
California Polytechnic
State University

Gregory Miller
University of Washington

William R. Murray
Cal Poly State University

Eric Musslman
University of Minnesota,
Duluth

Masoud Olia
Wentworth Institute of
Technology

Mark Olles
Rensselaer Polytechnic
Institute

Renee K. B. Petersen
Washington State
University

Carisa Ramming
Oklohoma State University

Amir G Rezaei
California State Polytechnic
University, Pomona

Martin Sadd
University of Rhode Island

Stefan Seelecke
North Carolina State
University

Yixin Shao
McGill University

Muhammad Sharif
The University of Alabama

Anthony Sinclair
University of Toronto

Lizhi Sun
University of California,
Irvine

Jeffrey Thomas
Northwestern University

Robert J. Witt
University of Wisconsin,
Madison

Jiashi Yang
University of Nebraska

Xiangwa Zeng
Case Western Reserve
University

List of Symbols

a	Constant; radius; distance		U	Work
$\mathbf{A}, \mathbf{B}, \mathbf{C}, \ldots$	Reactions at supports and connections		$\mathbf{V}$	Vector product; shearing force
$A, B, C, \ldots$	Points		V	Volume; potential energy; shear
A	Area		w	Load per unit length
b	Width; distance		$\mathbf{W}, W$	Weight; load
c	Constant		x, y, z	Rectangular coordinates; distances
C	Centroid		$\bar{x}, \bar{y}, \bar{z}$	Rectangular coordinates of centroid or center of gravity
d	Distance			
e	Base of natural logarithms		α, β, γ	Angles
$\mathbf{F}$	Force; friction force		γ	Specific weight
g	Acceleration of gravity		δ	Elongation
G	Center of gravity; constant of gravitation		$\delta\mathbf{r}$	Virtual displacement
h	Height; sag of cable		δU	Virtual work
$\mathbf{i}, \mathbf{j}, \mathbf{k}$	Unit vectors along coordinate axes		$\boldsymbol{\lambda}$	Unit vector along a line
$I, I_x, \ldots$	Moments of inertia		η	Efficiency
$\bar{I}$	Centroidal moment of inertia		θ	Angular coordinate; angle; polar coordinate
$I_{xy}, \ldots$	Products of inertia			
J	Polar moment of inertia		μ	Coefficient of friction
k	Spring constant		ρ	Density
k_x, k_y, k_O	Radii of gyration		ϕ	Angle of friction; angle
$\bar{k}$	Centroidal radius of gyration			
l	Length			
L	Length; span			
m	Mass			
$\mathbf{M}$	Couple; moment			
$\mathbf{M}_O$	Moment about point O			
$\mathbf{M}_O^R$	Moment resultant about point O			
M	Magnitude of couple or moment; mass of earth			
M_{OL}	Moment about axis OL			
$\mathbf{N}$	Normal component of reaction			
O	Origin of coordinates			
p	Pressure			
$\mathbf{P}$	Force; vector			
$\mathbf{Q}$	Force; vector			
$\mathbf{r}$	Position vector			
r	Radius; distance; polar coordinate			
$\mathbf{R}$	Resultant force; resultant vector; reaction			
R	Radius of earth			
$\mathbf{s}$	Position vector			
s	Length of arc; length of cable			
$\mathbf{S}$	Force; vector			
t	Thickness			
$\mathbf{T}$	Force			
T	Tension			

1

Introduction

The tallest skyscraper in the Western Hemisphere, One World Trade Center is a prominent feature of the New York City skyline. From its foundation to its structural components and mechanical systems, the design and operation of the tower is based on the fundamentals of engineering mechanics.

Objectives

- **Define** the science of mechanics and examine its fundamental principles.
- **Discuss** and compare the International System of Units and U.S. Customary Units.
- **Discuss** how to approach the solution of mechanics problems, and introduce the SMART problem-solving methodology.
- **Examine** factors that govern numerical accuracy in the solution of a mechanics problem.

1.1 What is Mechanics?

Mechanics is defined as the science that describes and predicts the conditions of rest or motion of bodies under the action of forces. It consists of the mechanics of *rigid bodies,* mechanics of *deformable bodies,* and mechanics of *fluids*.

The mechanics of rigid bodies is subdivided into **statics** and **dynamics**. Statics deals with bodies at rest; dynamics deals with bodies in motion. In this text, we assume bodies are perfectly rigid. In fact, actual structures and machines are never absolutely rigid; they deform under the loads to which they are subjected. However, because these deformations are usually small, they do not appreciably affect the conditions of equilibrium or the motion of the structure under consideration. They are important, though, as far as the resistance of the structure to failure is concerned. Deformations are studied in a course in mechanics of materials, which is part of the mechanics of deformable bodies. The third division of mechanics, the mechanics of fluids, is subdivided into the study of *incompressible fluids* and of *compressible fluids*. An important subdivision of the study of incompressible fluids is *hydraulics,* which deals with applications involving water.

Mechanics is a physical science, since it deals with the study of physical phenomena. However, some teachers associate mechanics with mathematics, whereas many others consider it as an engineering subject. Both these views are justified in part. Mechanics is the foundation of most engineering sciences and is an indispensable prerequisite to their study. However, it does not have the *empiricism* found in some engineering sciences, i.e., it does not rely on experience or observation alone. The rigor of mechanics and the emphasis it places on deductive reasoning makes it resemble mathematics. However, mechanics is not an *abstract* or even a *pure* science; it is an *applied* science.

The purpose of mechanics is to explain and predict physical phenomena and thus to lay the foundations for engineering applications. You need to know statics to determine how much force will be exerted on a point in a bridge design and whether the structure can withstand that force. Determining the force a dam needs to withstand from the water in a river requires statics. You need statics to calculate how much weight a crane can lift, how much force a locomotive needs to pull a freight train, or how much force a circuit board in a computer can withstand. The concepts of dynamics enable you to analyze the flight characteristics of a jet, design a building to resist earthquakes, and mitigate shock and vibration to passengers inside a vehicle. The concepts of dynamics enable you to calculate how much force you need to send a satellite into orbit, accelerate a 200,000-ton cruise ship, or design a toy truck that doesn't break. You will not learn how to do these things in this course, but the ideas and methods you learn here will be the underlying basis for the engineering applications you will learn in your work.

1.2 Fundamental Concepts and Principles

Although the study of mechanics goes back to the time of Aristotle (384–322 B.C.) and Archimedes (287–212 B.C.), not until Newton (1642–1727) did anyone develop a satisfactory formulation of its fundamental principles. These principles were later modified by d'Alembert, Lagrange, and Hamilton. Their validity remained unchallenged until Einstein formulated his **theory of relativity** (1905). Although its limitations have now been recognized, **newtonian mechanics** still remains the basis of today's engineering sciences.

The basic concepts used in mechanics are *space, time, mass,* and *force.* These concepts cannot be truly defined; they should be accepted on the basis of our intuition and experience and used as a mental frame of reference for our study of mechanics.

The concept of **space** is associated with the position of a point P. We can define the position of P by providing three lengths measured from a certain reference point, or *origin,* in three given directions. These lengths are known as the *coordinates* of P.

To define an event, it is not sufficient to indicate its position in space. We also need to specify the **time** of the event.

We use the concept of **mass** to characterize and compare bodies on the basis of certain fundamental mechanical experiments. Two bodies of the same mass, for example, are attracted by the earth in the same manner; they also offer the same resistance to a change in translational motion.

A **force** represents the action of one body on another. A force can be exerted by actual contact, like a push or a pull, or at a distance, as in the case of gravitational or magnetic forces. A force is characterized by its *point of application,* its *magnitude,* and its *direction;* a force is represented by a *vector* (Sec. 2.1B).

In newtonian mechanics, space, time, and mass are absolute concepts that are independent of each other. (This is not true in **relativistic mechanics**, where the duration of an event depends upon its position and the mass of a body varies with its velocity.) On the other hand, the concept of force is not independent of the other three. Indeed, one of the fundamental principles of newtonian mechanics listed below is that the resultant force acting on a body is related to the mass of the body and to the manner in which its velocity varies with time.

In this text, you will study the conditions of rest or motion of particles and rigid bodies in terms of the four basic concepts we have introduced. By **particle**, we mean a very small amount of matter, which we assume occupies a single point in space. A **rigid body** consists of a large number of particles occupying fixed positions with respect to one another. The study of the mechanics of particles is clearly a prerequisite to that of rigid bodies. Besides, we can use the results obtained for a particle directly in a large number of problems dealing with the conditions of rest or motion of actual bodies.

The study of elementary mechanics rests on six fundamental principles, based on experimental evidence.

- **The Parallelogram Law for the Addition of Forces.** Two forces acting on a particle may be replaced by a single force, called their *resultant,* obtained by drawing the diagonal of the parallelogram with sides equal to the given forces (Sec. 2.1A).
- **The Principle of Transmissibility.** The conditions of equilibrium or of motion of a rigid body remain unchanged if a force acting at a given point of the rigid body is replaced by a force of the same magnitude and same direction, but acting at a different point, provided that the two forces have the same line of action (Sec. 3.1B).
- **Newton's Three Laws of Motion.** Formulated by Sir Isaac Newton in the late seventeenth century, these laws can be stated as follows:

  FIRST LAW. If the resultant force acting on a particle is zero, the particle remains at rest (if originally at rest) or moves with constant speed in a straight line (if originally in motion) (Sec. 2.3B).

  SECOND LAW. If the resultant force acting on a particle is not zero, the particle has an acceleration proportional to the magnitude of the resultant and in the direction of this resultant force.

  As you will see in Sec. 12.1, this law can be stated as

  $$\mathbf{F} = m\mathbf{a} \tag{1.1}$$

  where $\mathbf{F}$, m, and $\mathbf{a}$ represent, respectively, the resultant force acting on the particle, the mass of the particle, and the acceleration of the particle expressed in a consistent system of units.

  THIRD LAW. The forces of action and reaction between bodies in contact have the same magnitude, same line of action, and opposite sense (Ch. 6, Introduction).

- **Newton's Law of Gravitation.** Two particles of mass M and m are mutually attracted with equal and opposite forces $\mathbf{F}$ and $-\mathbf{F}$ of magnitude F (Fig. 1.1), given by the formula

  $$F = G\frac{Mm}{r^2} \tag{1.2}$$

  where r = the distance between the two particles and G = a universal constant called the *constant of gravitation.* Newton's law of gravitation introduces the idea of an action exerted at a distance and extends the range of application of Newton's third law: the action $\mathbf{F}$ and the reaction $-\mathbf{F}$ in Fig. 1.1 are equal and opposite, and they have the same line of action.

A particular case of great importance is that of the attraction of the earth on a particle located on its surface. The force $\mathbf{F}$ exerted by the earth on the particle is defined as the **weight W** of the particle. Suppose we set

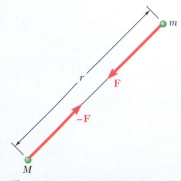

Fig. 1.1 From Newton's law of gravitation, two particles of masses M and m exert forces upon each other of equal magnitude, opposite direction, and the same line of action. This also illustrates Newton's third law of motion.

M equal to the mass of the earth, m equal to the mass of the particle, and r equal to the earth's radius R. Then introducing the constant

$$g = \frac{GM}{R^2} \tag{1.3}$$

we can express the magnitude W of the weight of a particle of mass m as[†]

$$W = mg \tag{1.4}$$

The value of R in formula (1.3) depends upon the elevation of the point considered; it also depends upon its latitude, since the earth is not truly spherical. The value of g therefore varies with the position of the point considered. However, as long as the point actually remains on the earth's surface, it is sufficiently accurate in most engineering computations to assume that g equals 9.81 m/s^2 or 32.2 ft/s^2.

The principles we have just listed will be introduced in the course of our study of mechanics as they are needed. The statics of particles carried out in Chap. 2 will be based on the parallelogram law of addition and on Newton's first law alone. We introduce the principle of transmissibility in Chap. 3 as we begin the study of the statics of rigid bodies, and we bring in Newton's third law in Chap. 6 as we analyze the forces exerted on each other by the various members forming a structure. We introduce Newton's second law and Newton's law of gravitation in dynamics. We will then show that Newton's first law is a particular case of Newton's second law (Sec. 12.1) and that the principle of transmissibility could be derived from the other principles and thus eliminated (Sec. 16.1D). In the meantime, however, Newton's first and third laws, the parallelogram law of addition, and the principle of transmissibility will provide us with the necessary and sufficient foundation for the entire study of the statics of particles, rigid bodies, and systems of rigid bodies.

As noted earlier, the six fundamental principles listed previously are based on experimental evidence. Except for Newton's first law and the principle of transmissibility, they are independent principles that cannot be derived mathematically from each other or from any other elementary physical principle. On these principles rests most of the intricate structure of newtonian mechanics. For more than two centuries, engineers have solved a tremendous number of problems dealing with the conditions of rest and motion of rigid bodies, deformable bodies, and fluids by applying these fundamental principles. Many of the solutions obtained could be checked experimentally, thus providing a further verification of the principles from which they were derived. Only in the twentieth century has Newton's mechanics found to be at fault, in the study of the motion of atoms and the motion of the planets, where it must be supplemented by the theory of relativity. On the human or engineering scale, however, where velocities are small compared with the speed of light, Newton's mechanics have yet to be disproved.

1.3 Systems of Units

Associated with the four fundamental concepts just discussed are the so-called *kinetic units*, i.e., the units of *length, time, mass,* and *force*. These units cannot be chosen independently if Eq. (1.1) is to be satisfied.

Photo 1.1 When in orbit of the earth, people and objects are said to be *weightless* even though the gravitational force acting is approximately 90% of that experienced on the surface of the earth. This apparent contradiction will be resolved in Chapter 12 when we apply Newton's second law to the motion of particles.

[†]A more accurate definition of the weight **W** should take into account the earth's rotation.

Three of the units may be defined arbitrarily; we refer to them as **basic units**. The fourth unit, however, must be chosen in accordance with Eq. (1.1) and is referred to as a **derived unit**. Kinetic units selected in this way are said to form a **consistent system of units**.

International System of Units (SI Units).[†]

In this system, which will be in universal use after the United States has completed its conversion to SI units, the base units are the units of length, mass, and time, and they are called, respectively, the **meter** (m), the **kilogram** (kg), and the **second** (s). All three are arbitrarily defined. The second was originally chosen to represent 1/86 400 of the mean solar day, but it is now defined as the duration of 9 192 631 770 cycles of the radiation corresponding to the transition between two levels of the fundamental state of the cesium-133 atom. The meter, originally defined as one ten-millionth of the distance from the equator to either pole, is now defined as 1 650 763.73 wavelengths of the orange-red light corresponding to a certain transition in an atom of krypton-86. (The newer definitions are much more precise and with today's modern instrumentation, are easier to verify as a standard.) The kilogram, which is approximately equal to the mass of 0.001 m³ of water, is defined as the mass of a platinum-iridium standard kept at the International Bureau of Weights and Measures at Sèvres, near Paris, France. The unit of force is a derived unit. It is called the **newton** (N) and is defined as the force that gives an acceleration of 1 m/s² to a body of mass 1 kg (Fig. 1.2). From Eq. (1.1), we have

$$1 \text{ N} = (1 \text{ kg})(1 \text{ m/s}^2) = 1 \text{ kg·m/s}^2 \tag{1.5}$$

The SI units are said to form an *absolute* system of units. This means that the three base units chosen are independent of the location where measurements are made. The meter, the kilogram, and the second may be used anywhere on the earth; they may even be used on another planet and still have the same significance.

The *weight* of a body, or the *force of gravity* exerted on that body, like any other force, should be expressed in newtons. From Eq. (1.4), it follows that the weight of a body of mass 1 kg (Fig. 1.3) is

$$W = mg$$
$$= (1 \text{ kg})(9.81 \text{ m/s}^2)$$
$$= 9.81 \text{ N}$$

Multiples and submultiples of the fundamental SI units are denoted through the use of the prefixes defined in Table 1.1. The multiples and submultiples of the units of length, mass, and force most frequently used in engineering are, respectively, the *kilometer* (km) and the *millimeter* (mm); the *megagram*[‡] (Mg) and the *gram* (g); and the *kilonewton* (kN). According to Table 1.1, we have

$$\begin{array}{ll} 1 \text{ km} = 1000 \text{ m} & 1 \text{ mm} = 0.001 \text{ m} \\ 1 \text{ Mg} = 1000 \text{ kg} & 1 \text{ g} = 0.001 \text{ kg} \\ 1 \text{ kN} = 1000 \text{ N} \end{array}$$

The conversion of these units into meters, kilograms, and newtons, respectively, can be effected by simply moving the decimal point three places

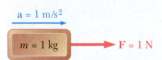

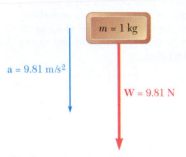

Fig. 1.2 A force of 1 newton applied to a body of mass 1 kg provides an acceleration of 1 m/s².

Fig. 1.3 A body of mass 1 kg experiencing an acceleration due to gravity of 9.81 m/s² has a weight of 9.81 N.

[†]SI stands for *Système International d'Unités* (French)
[‡]Also known as a *metric ton*.

Table 1.1   SI Prefixes

Multiplication Factor	Prefix[†]	Symbol
$1\ 000\ 000\ 000\ 000 = 10^{12}$	tera	T
$1\ 000\ 000\ 000 = 10^{9}$	giga	G
$1\ 000\ 000 = 10^{6}$	mega	M
$1\ 000 = 10^{3}$	kilo	k
$100 = 10^{2}$	hecto[‡]	h
$10 = 10^{1}$	deka[‡]	da
$0.1 = 10^{-1}$	deci[‡]	d
$0.01 = 10^{-2}$	centi[‡]	c
$0.001 = 10^{-3}$	milli	m
$0.000\ 001 = 10^{-6}$	micro	μ
$0.000\ 000\ 001 = 10^{-9}$	nano	n
$0.000\ 000\ 000\ 001 = 10^{-12}$	pico	p
$0.000\ 000\ 000\ 000\ 001 = 10^{-15}$	femto	f
$0.000\ 000\ 000\ 000\ 000\ 001 = 10^{-18}$	atto	a

[†]The first syllable of every prefix is accented, so that the prefix retains its identity. Thus, the preferred pronunciation of kilometer places the accent on the first syllable, not the second.

[‡]The use of these prefixes should be avoided, except for the measurement of areas and volumes and for the nontechnical use of centimeter, as for body and clothing measurements.

to the right or to the left. For example, to convert 3.82 km into meters, move the decimal point three places to the right:

$$3.82\text{ km} = 3820\text{ m}$$

Similarly, to convert 47.2 mm into meters, move the decimal point three places to the left:

$$47.2\text{ mm} = 0.0472\text{ m}$$

Using engineering notation, you can also write

$$3.82\text{ km} = 3.82 \times 10^{3}\text{ m}$$
$$47.2\text{ mm} = 47.2 \times 10^{-3}\text{ m}$$

The multiples of the unit of time are the *minute* (min) and the *hour* (h). Since 1 min = 60 s and 1 h = 60 min = 3600 s, these multiples cannot be converted as readily as the others.

By using the appropriate multiple or submultiple of a given unit, you can avoid writing very large or very small numbers. For example, it is usually simpler to write 427.2 km rather than 427 200 m and 2.16 mm rather than 0.002 16 m.[†]

Units of Area and Volume.
The unit of area is the *square meter* (m^2), which represents the area of a square of side 1 m; the unit of volume is the *cubic meter* (m^3), which is equal to the volume of a cube of side 1 m. In order to avoid exceedingly small or large numerical values when computing areas and volumes, we use systems of subunits obtained by respectively squaring and cubing not only the millimeter, but also two intermediate

[†]Note that when more than four digits appear on either side of the decimal point to express a quantity in SI units—as in 427 000 m or 0.002 16 m—use spaces, never commas, to separate the digits into groups of three. This practice avoids confusion with the comma used in place of a decimal point, which is the convention in many countries.

submultiples of the meter: the *decimeter* (dm) and the *centimeter* (cm). By definition,

$$1 \text{ dm} = 0.1 \text{ m} = 10^{-1} \text{ m}$$
$$1 \text{ cm} = 0.01 \text{ m} = 10^{-2} \text{ m}$$
$$1 \text{ mm} = 0.001 \text{ m} = 10^{-3} \text{ m}$$

Therefore, the submultiples of the unit of area are

$$1 \text{ dm}^2 = (1 \text{ dm})^2 = (10^{-1} \text{ m})^2 = 10^{-2} \text{ m}^2$$
$$1 \text{ cm}^2 = (1 \text{ cm})^2 = (10^{-2} \text{ m})^2 = 10^{-4} \text{ m}^2$$
$$1 \text{ mm}^2 = (1 \text{ mm})^2 = (10^{-3} \text{ m})^2 = 10^{-6} \text{ m}^2$$

Similarly, the submultiples of the unit of volume are

$$1 \text{ dm}^3 = (1 \text{ dm})^3 = (10^{-1} \text{ m})^3 = 10^{-3} \text{ m}^3$$
$$1 \text{ cm}^3 = (1 \text{ cm})^3 = (10^{-2} \text{ m})^3 = 10^{-6} \text{ m}^3$$
$$1 \text{ mm}^3 = (1 \text{ mm})^3 = (10^{-3} \text{ m})^3 = 10^{-9} \text{ m}^3$$

Note that when measuring the volume of a liquid, the cubic decimeter (dm^3) is usually referred to as a *liter* (L).

Table 1.2 shows other derived SI units used to measure the moment of a force, the work of a force, etc. Although we will introduce these units in later chapters as they are needed, we should note an important rule at

Table 1.2   Principal SI Units Used in Mechanics

Quantity	Unit	Symbol	Formula
Acceleration	Meter per second squared	. . .	m/s^2
Angle	Radian	rad	†
Angular acceleration	Radian per second squared	. . .	rad/s^2
Angular velocity	Radian per second	. . .	rad/s
Area	Square meter	. . .	m^2
Density	Kilogram per cubic meter	. . .	kg/m^3
Energy	Joule	J	N·m
Force	Newton	N	$kg \cdot m/s^2$
Frequency	Hertz	Hz	s^{-1}
Impulse	Newton-second	. . .	kg·m/s
Length	Meter	m	‡
Mass	Kilogram	kg	‡
Moment of a force	Newton-meter	. . .	N·m
Power	Watt	W	J/s
Pressure	Pascal	Pa	N/m^2
Stress	Pascal	Pa	N/m^2
Time	Second	s	‡
Velocity	Meter per second	. . .	m/s
Volume			
Solids	Cubic meter	. . .	m^3
Liquids	Liter	L	10^{-3} m^3
Work	Joule	J	N·m

†Supplementary unit (1 revolution = 2π rad = 360°).
‡Base unit.

this time: When a derived unit is obtained by dividing a base unit by another base unit, you may use a prefix in the numerator of the derived unit, but not in its denominator. For example, the constant k of a spring that stretches 20 mm under a load of 100 N is expressed as

$$k = \frac{100 \text{ N}}{20 \text{ mm}} = \frac{100 \text{ N}}{0.020 \text{ m}} = 5000 \text{ N/m} \text{ or } k = 5 \text{ kN/m}$$

but never as $k = 5$ N/mm.

U.S. Customary Units.

Most practicing American engineers still commonly use a system in which the base units are those of length, force, and time. These units are, respectively, the *foot* (ft), the *pound* (lb), and the *second* (s). The second is the same as the corresponding SI unit. The foot is defined as 0.3048 m. The pound is defined as the *weight* of a platinum standard, called the *standard pound,* which is kept at the National Institute of Standards and Technology outside Washington D.C., the mass of which is 0.453 592 43 kg. Since the weight of a body depends upon the earth's gravitational attraction, which varies with location, the standard pound should be placed at sea level and at a latitude of 45° to properly define a force of 1 lb. Clearly the U.S. customary units do not form an absolute system of units. Because they depend upon the gravitational attraction of the earth, they form a *gravitational* system of units.

Although the standard pound also serves as the unit of mass in commercial transactions in the United States, it cannot be used that way in engineering computations, because such a unit would not be consistent with the base units defined in the preceding paragraph. Indeed, when acted upon by a force of 1 lb—that is, when subjected to the force of gravity—the standard pound has the acceleration due to gravity, $g = 32.2$ ft/s² (Fig. 1.4), not the unit acceleration required by Eq. (1.1). The unit of mass consistent with the foot, the pound, and the second is the mass that receives an acceleration of 1 ft/s² when a force of 1 lb is applied to it (Fig. 1.5). This unit, sometimes called a *slug,* can be derived from the equation $F = ma$ after substituting 1 lb for F and 1 ft/s² for a. We have

$$F = ma \quad 1 \text{ lb} = (1 \text{ slug})(1 \text{ ft/s}^2)$$

This gives us

$$1 \text{ slug} = \frac{1 \text{ lb}}{1 \text{ ft/s}^2} = 1 \text{ lb·s}^2/\text{ft} \tag{1.6}$$

Comparing Figs. 1.4 and 1.5, we conclude that the slug is a mass 32.2 times larger than the mass of the standard pound.

The fact that, in the U.S. customary system of units, bodies are characterized by their weight in pounds rather than by their mass in slugs is convenient in the study of statics, where we constantly deal with weights and other forces and only seldom deal directly with masses. However, in the study of dynamics, where forces, masses, and accelerations are involved, the mass m of a body is expressed in slugs when its weight W is given in pounds. Recalling Eq. (1.4), we write

$$m = \frac{W}{g} \tag{1.7}$$

where g is the acceleration due to gravity ($g = 32.2$ ft/s²).

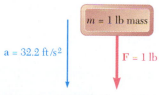

Fig. 1.4 A body of 1 pound mass acted upon by a force of 1 pound has an acceleration of 32.2 ft/s².

Fig. 1.5 A force of 1 pound applied to a body of mass 1 slug produces an acceleration of 1 ft/s².

Other U.S. customary units frequently encountered in engineering problems are the *mile* (mi), equal to 5280 ft; the *inch* (in.), equal to (1/12) ft; and the *kilopound* (kip), equal to 1000 lb. The *ton* is often used to represent a mass of 2000 lb but, like the pound, must be converted into slugs in engineering computations.

The conversion into feet, pounds, and seconds of quantities expressed in other U.S. customary units is generally more involved and requires greater attention than the corresponding operation in SI units. For example, suppose we are given the magnitude of a velocity $v = 30$ mi/h and want to convert it to ft/s. First we write

$$v = 30\frac{\text{mi}}{\text{h}}$$

Since we want to get rid of the unit miles and introduce instead the unit feet, we should multiply the right-hand member of the equation by an expression containing miles in the denominator and feet in the numerator. However, since we do not want to change the value of the right-hand side of the equation, the expression used should have a value equal to unity. The quotient (5280 ft)/(1 mi) is such an expression. Operating in a similar way to transform the unit hour into seconds, we have

$$v = \left(30\frac{\text{mi}}{\text{h}}\right)\left(\frac{5280 \text{ ft}}{1 \text{ mi}}\right)\left(\frac{1 \text{ h}}{3600 \text{ s}}\right)$$

Carrying out the numerical computations and canceling out units that appear in both the numerator and the denominator, we obtain

$$v = 44\frac{\text{ft}}{\text{s}} = 44 \text{ ft/s}$$

1.4   Converting Between Two Systems of Units

In many situations, an engineer might need to convert into SI units a numerical result obtained in U.S. customary units or vice versa. Because the unit of time is the same in both systems, only two kinetic base units need be converted. Thus, since all other kinetic units can be derived from these base units, only two conversion factors need be remembered.

Units of Length.   By definition, the U.S. customary unit of length is

$$1 \text{ ft} = 0.3048 \text{ m} \tag{1.8}$$

It follows that

$$1 \text{ mi} = 5280 \text{ ft} = 5280(0.3048 \text{ m}) = 1609 \text{ m}$$

or

$$1 \text{ mi} = 1.609 \text{ km} \tag{1.9}$$

Also,

$$1 \text{ in.} = \frac{1}{12} \text{ ft} = \frac{1}{12}(0.3048 \text{ m}) = 0.0254 \text{ m}$$

or

$$1 \text{ in.} = 25.4 \text{ mm} \qquad \textbf{(1.10)}$$

Units of Force.

Recall that the U.S. customary unit of force (pound) is defined as the weight of the standard pound (of mass 0.4536 kg) at sea level and at a latitude of 45° (where $g = 9.807 \text{ m/s}^2$). Then, using Eq. (1.4), we write

$$W = mg$$
$$1 \text{ lb} = (0.4536 \text{ kg})(9.807 \text{ m/s}^2) = 4.448 \text{ kg·m/s}^2$$

From Eq. (1.5), this reduces to

$$1 \text{ lb} = 4.448 \text{ N} \qquad \textbf{(1.11)}$$

Units of Mass.

The U.S. customary unit of mass (slug) is a derived unit. Thus, using Eqs. (1.6), (1.8), and (1.11), we have

$$1 \text{ slug} = 1 \text{ lb·s}^2/\text{ft} = \frac{1 \text{ lb}}{1 \text{ ft/s}^2} = \frac{4.448 \text{ N}}{0.3048 \text{ m/s}^2} = 14.59 \text{ N·s}^2/\text{m}$$

Again, from Eq. (1.5),

$$1 \text{ slug} = 1 \text{ lb·s}^2/\text{ft} = 14.59 \text{ kg} \qquad \textbf{(1.12)}$$

Although it cannot be used as a consistent unit of mass, recall that the mass of the standard pound is, by definition,

$$1 \text{ pound mass} = 0.4536 \text{ kg} \qquad \textbf{(1.13)}$$

We can use this constant to determine the *mass* in SI units (kilograms) of a body that has been characterized by its *weight* in U.S. customary units (pounds).

To convert a derived U.S. customary unit into SI units, simply multiply or divide by the appropriate conversion factors. For example, to convert the moment of a force that is measured as $M = 47 \text{ lb·in.}$ into SI units, use formulas (1.10) and (1.11) and write

$$M = 47 \text{ lb·in.} = 47(4.448 \text{ N})(25.4 \text{ mm})$$
$$= 5310 \text{ N·mm} = 5.31 \text{ N·m}$$

You can also use conversion factors to convert a numerical result obtained in SI units into U.S. customary units. For example, if the moment of a force is measured as $M = 40 \text{ N·m}$, follow the procedure at the end of Sec. 1.3 to write

$$M = 40 \text{ N·m} = (40 \text{ N·m})\left(\frac{1 \text{ lb}}{4.448 \text{ N}}\right)\left(\frac{1 \text{ ft}}{0.3048 \text{ m}}\right)$$

Carrying out the numerical computations and canceling out units that appear in both the numerator and the denominator, you obtain

$$M = 29.5 \text{ lb·ft}$$

The U.S. customary units most frequently used in mechanics are listed in Table 1.3 with their SI equivalents.

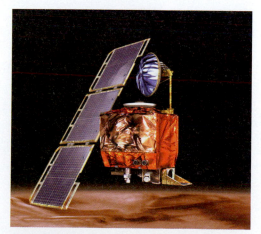

Photo 1.2 In 1999, The *Mars Climate Orbiter* entered orbit around Mars at too low an altitude and disintegrated. Investigation showed that the software on board the probe interpreted force instructions in newtons, but the software at mission control on the earth was generating those instructions in terms of pounds.

Table 1.3 U.S. Customary Units and Their SI Equivalents

Quantity	U.S. Customary Unit	SI Equivalent
Acceleration	ft/s^2	0.3048 m/s^2
	in./s^2	0.0254 m/s^2
Area	ft^2	0.0929 m^2
	in^2	645.2 mm^2
Energy	ft·lb	1.356 J
Force	kip	4.448 kN
	lb	4.448 N
	oz	0.2780 N
Impulse	lb·s	4.448 N·s
Length	ft	0.3048 m
	in.	25.40 mm
	mi	1.609 km
Mass	oz mass	28.35 g
	lb mass	0.4536 kg
	slug	14.59 kg
	ton	907.2 kg
Moment of a force	lb·ft	1.356 N·m
	lb·in.	0.1130 N·m
Moment of inertia		
Of an area	in^4	0.4162×10^6 mm^4
Of a mass	lb·ft·s^2	1.356 kg·m^2
Momentum	lb·s	4.448 kg·m/s
Power	ft·lb/s	1.356 W
	hp	745.7 W
Pressure or stress	lb/ft^2	47.88 Pa
	lb/in^2 (psi)	6.895 kPa
Velocity	ft/s	0.3048 m/s
	in./s	0.0254 m/s
	mi/h (mph)	0.4470 m/s
	mi/h (mph)	1.609 km/h
Volume	ft^3	0.02832 m^3
	in^3	16.39 cm^3
Liquids	gal	3.785 L
	qt	0.9464 L
Work	ft·lb	1.356 J

1.5 Method of Solving Problems

You should approach a problem in mechanics as you would approach an actual engineering situation. By drawing on your own experience and intuition about physical behavior, you will find it easier to understand and formulate the problem. Once you have clearly stated and understood the problem, however, there is no place in its solution for arbitrary methodologies.

The solution must be based on the six fundamental principles stated in Sec. 1.2 or on theorems derived from them.

Every step you take in the solution must be justified on this basis. Strict rules must be followed, which lead to the solution in an almost automatic fashion, leaving no room for your intuition or "feeling." After you have obtained an answer, you should check it. Here again, you may call upon your common sense and personal experience. If you are not completely satisfied with the result, you should carefully check your formulation of the problem, the validity of the methods used for its solution, and the accuracy of your computations.

In general, you can usually solve problems in several different ways; there is no one approach that works best for everybody. However, we have found that students often find it helpful to have a general set of guidelines to use for framing problems and planning solutions. In the Sample Problems throughout this text, we use a four-step method for approaching problems, which we refer to as the SMART methodology: **S**trategy, **M**odeling, **A**nalysis, and **R**eflect and **T**hink.

1. **Strategy.** The statement of a problem should be clear and precise, and it should contain the given data and indicate what information is required. The first step in solving the problem is to decide what concepts you have learned that apply to the given situation and to connect the data to the required information. It is often useful to work backward from the information you are trying to find: Ask yourself what quantities you need to know to obtain the answer, and if some of these quantities are unknown, how can you find them from the given data.

2. **Modeling.** The first step in modeling is to define the system; that is, clearly define what you are setting aside for analysis. After you have selected a system, draw a neat sketch showing all quantities involved with a separate diagram for each body in the problem. For equilibrium problems, indicate clearly the forces acting on each body along with any relevant geometrical data, such as lengths and angles. (These diagrams are known as **free-body diagrams** and are described in detail in Sec. 2.3C and the beginning of Ch. 4.)

3. **Analysis.** After you have drawn the appropriate diagrams, use the fundamental principles of mechanics listed in Sec. 1.2 to write equations expressing the conditions of rest or motion of the bodies considered. Each equation should be clearly related to one of the free-body diagrams and should be numbered. If you do not have enough equations to solve for the unknowns, try selecting another system, or reexamine your strategy to see if you can apply other principles to the problem. Once you have obtained enough equations, you can find a numerical solution by following the usual rules of algebra, neatly recording each step and the intermediate results. Alternatively, you can solve the resulting equations with your calculator or a computer. (For multipart problems, it is sometimes convenient to present the Modeling and Analysis steps together, but they are both essential parts of the overall process.)

4. **Reflect and Think.** After you have obtained the answer, check it carefully. Does it make sense in the context of the original problem? For instance, the problem may ask for the force at a given point of a structure. If your answer is negative, what does that mean for the force at the point?

You can often detect mistakes in *reasoning* by checking the units. For example, to determine the moment of a force of 50 N about a point 0.60 m from its line of action, we write (Sec. 3.3A)

$$M = Fd = (30 \text{ N})(0.60 \text{ m}) = 30 \text{ N·m}$$

The unit N·m obtained by multiplying newtons by meters is the correct unit for the moment of a force; if you had obtained another unit, you would know that some mistake had been made.

You can often detect errors in *computation* by substituting the numerical answer into an equation that was not used in the solution and verifying that the equation is satisfied. The importance of correct computations in engineering cannot be overemphasized.

1.6   Numerical Accuracy

The accuracy of the solution to a problem depends upon two items: (1) the accuracy of the given data and (2) the accuracy of the computations performed. The solution cannot be more accurate than the less accurate of these two items.

For example, suppose the loading of a bridge is known to be 75 000 lb with a possible error of 100 lb either way. The relative error that measures the degree of accuracy of the data is

$$\frac{100 \text{ lb}}{75\,000 \text{ lb}} = 0.0013 = 0.13\%$$

In computing the reaction at one of the bridge supports, it would be meaningless to record it as 14 322 lb. The accuracy of the solution cannot be greater than 0.13%, no matter how precise the computations are, and the possible error in the answer may be as large as $(0.13/100)(14\,322 \text{ lb}) \approx 20 \text{ lb}$. The answer should be properly recorded as $14\,320 \pm 20$ lb.

In engineering problems, the data are seldom known with an accuracy greater than 0.2%. It is therefore seldom justified to write answers with an accuracy greater than 0.2%. A practical rule is to use four figures to record numbers beginning with a "1" and three figures in all other cases. Unless otherwise indicated, you should assume the data given in a problem are known with a comparable degree of accuracy. A force of 40 lb, for example, should be read as 40.0 lb, and a force of 15 lb should be read as 15.00 lb.

Electronic calculators are widely used by practicing engineers and engineering students. The speed and accuracy of these calculators facilitate the numerical computations in the solution of many problems. However, you should not record more significant figures than can be justified merely because you can obtain them easily. As noted previously, an accuracy greater than 0.2% is seldom necessary or meaningful in the solution of practical engineering problems.

2

Statics of Particles

Many engineering problems can be solved by considering the equilibrium of a "particle." In the case of this beam that is being hoisted into position, a relation between the tensions in the various cables involved can be obtained by considering the equilibrium of the hook to which the cables are attached.

Introduction

Objectives

- **Describe** force as a vector quantity.
- **Examine** vector operations useful for the analysis of forces.
- **Determine** the resultant of multiple forces acting on a particle.
- **Resolve** forces into components.
- **Add** forces that have been resolved into rectangular components.
- **Introduce** the concept of the free-body diagram.
- **Use** free-body diagrams to assist in the analysis of planar and spatial particle equilibrium problems.

Introduction

In this chapter, you will study the effect of forces acting on particles. By the word "particle" we do not mean only tiny bits of matter, like an atom or an electron. Instead, we mean that the sizes and shapes of the bodies under consideration do not significantly affect the solutions of the problems. Another way of saying this is that we assume all forces acting on a given body act at the same point. This does not mean the object must be tiny—if you were modeling the mechanics of the Milky Way galaxy, for example, you could treat the Sun and the entire Solar System as just a particle.

Our first step is to explain how to replace two or more forces acting on a given particle by a single force having the same effect as the original forces. This single equivalent force is called the *resultant* of the original forces. After this step, we will derive the relations among the various forces acting on a particle in a state of *equilibrium*. We will use these relations to determine some of the forces acting on the particle.

The first part of this chapter deals with forces contained in a single plane. Because two lines determine a plane, this situation arises any time we can reduce the problem to one of a particle subjected to two forces that support a third force, such as a crate suspended from two chains or a traffic light held in place by two cables. In the second part of this chapter, we examine the more general case of forces in three-dimensional space.

2.1 ADDITION OF PLANAR FORCES

Many important practical situations in engineering involve forces in the same plane. These include forces acting on a pulley, projectile motion, and an object in equilibrium on a flat surface. We will examine this situation first before looking at the added complications of forces acting in three-dimensional space.

2.1A Force on a Particle: Resultant of Two Forces

A force represents the action of one body on another. It is generally characterized by its **point of application**, its **magnitude**, and its **direction**. Forces acting on a given particle, however, have the same point of application. Thus, each force considered in this chapter is completely defined by its magnitude and direction.

The magnitude of a force is characterized by a certain number of units. As indicated in Chap. 1, the SI units used by engineers to measure the magnitude of a force are the newton (N) and its multiple the kilonewton (kN), which is equal to 1000 N. The U.S. customary units used for the same purpose are the pound (lb) and its multiple the kilopound (kip), which is equal to 1000 lb. We saw in Chapter 1 that a force of 445 N is equivalent to a force of 100 lb or that a force of 100 N equals a force of about 22.5 lb.

We define the direction of a force by its **line of action** and the **sense** of the force. The line of action is the infinite straight line along which the force acts; it is characterized by the angle it forms with some fixed axis (Fig. 2.1). The force itself is represented by a segment of that line; through the use of an appropriate scale, we can choose the length of this segment to represent the magnitude of the force. We indicate the sense of the force by an arrowhead. It is important in defining a force to indicate its sense. Two forces having the same magnitude and the same line of action but a different sense, such as the forces shown in Fig. 2.1*a* and *b*, have directly opposite effects on a particle.

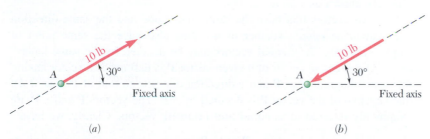

Fig. 2.1 The line of action of a force makes an angle with a given fixed axis. (*a*) The sense of the 10-lb force is away from particle *A*; (*b*) the sense of the 10-lb force is toward particle *A*.

Experimental evidence shows that two forces **P** and **Q** acting on a particle *A* (Fig. 2.2*a*) can be replaced by a single force **R** that has the same effect on the particle (Fig. 2.2*c*). This force is called the **resultant** of the forces **P** and **Q**. We can obtain **R**, as shown in Fig. 2.2*b*, by constructing a parallelogram, using **P** and **Q** as two adjacent sides. **The diagonal that passes through *A* represents the resultant**. This method for finding the resultant is known as the **parallelogram law** for the addition of two forces. This law is based on experimental evidence; it cannot be proved or derived mathematically.

2.1B Vectors

We have just seen that forces do not obey the rules of addition defined in ordinary arithmetic or algebra. For example, two forces acting at a right angle to each other, one of 4 lb and the other of 3 lb, add up to a force of

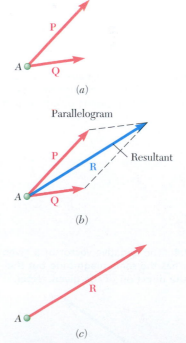

Fig. 2.2 (*a*) Two forces **P** and **Q** act on particle *A*. (*b*) Draw a parallelogram with **P** and **Q** as the adjacent sides and label the diagonal that passes through *A* as **R**. (*c*) **R** is the resultant of the two forces **P** and **Q** and is equivalent to their sum.

Photo 2.1 In its purest form, a tug-of-war pits two opposite and almost-equal forces against each other. Whichever team can generate the larger force, wins. As you can see, a competitive tug-of-war can be quite intense.

5 lb acting at an angle between them, *not* to a force of 7 lb. Forces are not the only quantities that follow the parallelogram law of addition. As you will see later, *displacements, velocities, accelerations,* and *momenta* are other physical quantities possessing magnitude and direction that add according to the parallelogram law. All of these quantities can be represented mathematically by **vectors**. Those physical quantities that have magnitude but not direction, such as *volume, mass,* or *energy*, are represented by plain numbers often called **scalars** to distinguish them from vectors.

Vectors are defined as **mathematical expressions possessing magnitude and direction, which add according to the parallelogram law**. Vectors are represented by arrows in diagrams and are distinguished from scalar quantities in this text through the use of boldface type (**P**). In longhand writing, a vector may be denoted by drawing a short arrow above the letter used to represent it ($\vec{P}$). The magnitude of a vector defines the length of the arrow used to represent it. In this text, we use italic type to denote the magnitude of a vector. Thus, the magnitude of the vector **P** is denoted by *P*.

A vector used to represent a force acting on a given particle has a well-defined point of application—namely, the particle itself. Such a vector is said to be a *fixed*, or *bound*, vector and cannot be moved without modifying the conditions of the problem. Other physical quantities, however, such as couples (see Chap. 3), are represented by vectors that may be freely moved in space; these vectors are called *free* vectors. Still other physical quantities, such as forces acting on a rigid body (see Chap. 3), are represented by vectors that can be moved along their lines of action; they are known as *sliding* vectors.

Two vectors that have the same magnitude and the same direction are said to be *equal*, whether or not they also have the same point of application (Fig. 2.3); equal vectors may be denoted by the same letter.

The *negative vector* of a given vector **P** is defined as a vector having the same magnitude as **P** and a direction opposite to that of **P** (Fig. 2.4); the negative of the vector **P** is denoted by −**P**. The vectors **P** and −**P** are commonly referred to as **equal and opposite** vectors. Clearly, we have

$$\mathbf{P} + (-\mathbf{P}) = 0$$

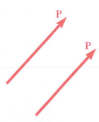

Fig. 2.3 Equal vectors have the same magnitude and the same direction, even if they have different points of application.

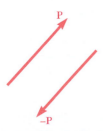

Fig. 2.4 The negative vector of a given vector has the same magnitude but the opposite direction of the given vector.

2.1C   Addition of Vectors

By definition, vectors add according to the parallelogram law. Thus, we obtain the sum of two vectors **P** and **Q** by attaching the two vectors to the same point *A* and constructing a parallelogram, using **P** and **Q** as two adjacent sides (Fig. 2.5). The diagonal that passes through *A* represents the sum of the vectors **P** and **Q**, denoted by **P** + **Q**. The fact that the sign + is used for both vector and scalar addition should not cause any confusion if vector and scalar quantities are always carefully distinguished. Note that the magnitude of the vector **P** + **Q** is *not*, in general, equal to the sum *P* + *Q* of the magnitudes of the vectors **P** and **Q**.

Since the parallelogram constructed on the vectors **P** and **Q** does not depend upon the order in which **P** and **Q** are selected, we conclude that the addition of two vectors is *commutative*, and we write

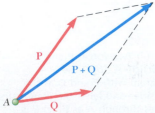

Fig. 2.5 Using the parallelogram law to add two vectors.

$$\mathbf{P} + \mathbf{Q} = \mathbf{Q} + \mathbf{P} \tag{2.1}$$

From the parallelogram law, we can derive an alternative method for determining the sum of two vectors, known as the **triangle rule**. Consider Fig. 2.5, where the sum of the vectors **P** and **Q** has been determined by the parallelogram law. Since the side of the parallelogram opposite **Q** is equal to **Q** in magnitude and direction, we could draw only half of the parallelogram (Fig. 2.6*a*). The sum of the two vectors thus can be found by **arranging P and Q in tip-to-tail fashion and then connecting the tail of P with the tip of Q**. If we draw the other half of the parallelogram, as in Fig. 2.6*b*, we obtain the same result, confirming that vector addition is commutative.

We define *subtraction* of a vector as the addition of the corresponding negative vector. Thus, we determine the vector **P** − **Q**, representing the difference between the vectors **P** and **Q**, by adding to **P** the negative vector −**Q** (Fig. 2.7). We write

$$\mathbf{P} - \mathbf{Q} = \mathbf{P} + (-\mathbf{Q}) \tag{2.2}$$

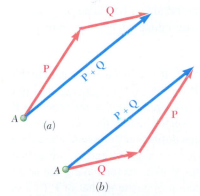

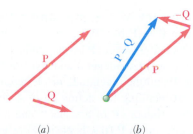

Fig. 2.6 The triangle rule of vector addition. (*a*) Adding vector **Q** to vector **P** equals (*b*) adding vector **P** to vector **Q**.

Fig. 2.7 Vector subtraction: Subtracting vector **Q** from vector **P** is the same as adding vector −**Q** to vector **P**.

Here again we should observe that, although we use the same sign to denote both vector and scalar subtraction, we avoid confusion by taking care to distinguish between vector and scalar quantities.

We now consider the *sum of three or more vectors*. The sum of three vectors **P**, **Q**, and **S** is, *by definition*, obtained by first adding the vectors **P** and **Q** and then adding the vector **S** to the vector **P** + **Q**. We write

$$\mathbf{P} + \mathbf{Q} + \mathbf{S} = (\mathbf{P} + \mathbf{Q}) + \mathbf{S} \tag{2.3}$$

Similarly, we obtain the sum of four vectors by adding the fourth vector to the sum of the first three. It follows that we can obtain the sum of any number of vectors by applying the parallelogram law repeatedly to successive pairs of vectors until all of the given vectors are replaced by a single vector.

If the given vectors are *coplanar*, i.e., if they are contained in the same plane, we can obtain their sum graphically. For this case, repeated application of the triangle rule is simpler than applying the parallelogram law. In Fig. 2.8*a*, we find the sum of three vectors **P**, **Q**, and **S** in this manner. The triangle rule is first applied to obtain the sum **P** + **Q** of the

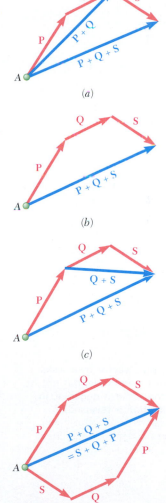

Fig. 2.8 Graphical addition of vectors. (*a*) Applying the triangle rule twice to add three vectors; (*b*) the vectors can be added in one step by the polygon rule; (*c*) vector addition is associative; (*d*) the order of addition is immaterial.

vectors **P** and **Q**; we apply it again to obtain the sum of the vectors **P** + **Q** and **S**. However, we could have omitted determining the vector **P** + **Q** and obtain the sum of the three vectors directly, as shown in Fig. 2.8*b*, by **arranging the given vectors in tip-to-tail fashion and connecting the tail of the first vector with the tip of the last one**. This is known as the **polygon rule** for the addition of vectors.

The result would be unchanged if, as shown in Fig. 2.8*c*, we had replaced the vectors **Q** and **S** by their sum **Q** + **S**. We may thus write

$$\mathbf{P} + \mathbf{Q} + \mathbf{S} = (\mathbf{P} + \mathbf{Q}) + \mathbf{S} = \mathbf{P} + (\mathbf{Q} + \mathbf{S}) \qquad (2.4)$$

which expresses the fact that vector addition is *associative*. Recalling that vector addition also has been shown to be commutative in the case of two vectors, we can write

$$\mathbf{P} + \mathbf{Q} + \mathbf{S} = (\mathbf{P} + \mathbf{Q}) + \mathbf{S} = \mathbf{S} + (\mathbf{P} + \mathbf{Q}) \qquad (2.5)$$
$$= \mathbf{S} + (\mathbf{Q} + \mathbf{P}) = \mathbf{S} + \mathbf{Q} + \mathbf{P}$$

This expression, as well as others we can obtain in the same way, shows that the order in which several vectors are added together is immaterial (Fig. 2.8*d*).

Product of a Scalar and a Vector.   It is convenient to denote the sum **P** + **P** by 2**P**, the sum **P** + **P** + **P** by 3**P**, and, in general, the sum of *n* equal vectors **P** by the product *n***P**. Therefore, we define the product *n***P** of a positive integer *n* and a vector **P** as a vector having the same direction as **P** and the magnitude *nP*. Extending this definition to include all scalars and recalling the definition of a negative vector given earlier, we define the product *k***P** of a scalar *k* and a vector **P** as a vector having the same direction as **P** (if *k* is positive) or a direction opposite to that of **P** (if *k* is negative) and a magnitude equal to the product of *P* and the absolute value of *k* (Fig. 2.9).

2.1D   Resultant of Several Concurrent Forces

Consider a particle *A* acted upon by several coplanar forces, i.e., by several forces contained in the same plane (Fig. 2.10*a*). Since the forces all pass through *A*, they are also said to be *concurrent*. We can add the vectors representing the forces acting on *A* by the polygon rule (Fig. 2.10*b*). Since the use of the polygon rule is equivalent to the repeated application of the parallelogram law, the vector **R** obtained in this way represents the resultant of the given concurrent forces. That is, the single force **R** has the same effect on the particle *A* as the given forces. As before, the order in which we add the vectors **P**, **Q**, and **S** representing the given forces is immaterial.

2.1E   Resolution of a Force into Components

We have seen that two or more forces acting on a particle may be replaced by a single force that has the same effect on the particle. Conversely, a single

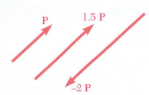

Fig. 2.9 Multiplying a vector by a scalar changes the vector's magnitude, but not its direction (unless the scalar is negative, in which case the direction is reversed).

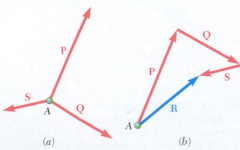

Fig. 2.10 Concurrent forces can be added by the polygon rule.

force **F** acting on a particle may be replaced by two or more forces that, together, have the same effect on the particle. These forces are called **components** of the original force **F**, and the process of substituting them for **F** is called **resolving the force F into components**.

Clearly, each force **F** can be resolved into an infinite number of possible sets of components. Sets of *two components* **P** *and* **Q** are the most important as far as practical applications are concerned. However, even then, the number of ways in which a given force **F** may be resolved into two components is unlimited (Fig. 2.11).

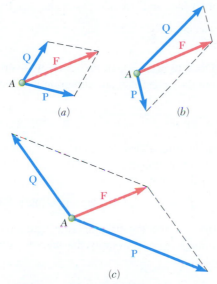

(a)

(b)

(c)

Fig. 2.11 Three possible sets of components for a given force vector **F**.

In many practical problems, we start with a given vector **F** and want to determine a useful set of components. Two cases are of particular interest:

1. **One of the Two Components, P, Is Known.** We obtain the second component, **Q**, by applying the triangle rule and joining the tip of **P** to the tip of **F** (Fig. 2.12). We can determine the magnitude and direction of **Q** graphically or by trigonometry. Once we have determined **Q**, both components **P** and **Q** should be applied at *A*.
2. **The Line of Action of Each Component Is Known.** We obtain the magnitude and sense of the components by applying the parallelogram law and drawing lines through the tip of **F** that are parallel to the given lines of action (Fig. 2.13). This process leads to two well-defined components, **P** and **Q**, which can be determined graphically or computed trigonometrically by applying the law of sines.

You will encounter many similar cases; for example, you might know the direction of one component while the magnitude of the other component is to be as small as possible (see Sample Prob. 2.2). In all cases, you need to draw the appropriate triangle or parallelogram that satisfies the given conditions.

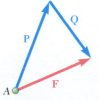

Fig. 2.12 When component **P** is known, use the triangle rule to find component **Q**.

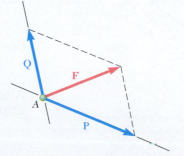

Fig. 2.13 When the lines of action are known, use the parallelogram rule to determine components **P** and **Q**.

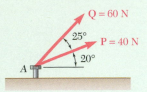

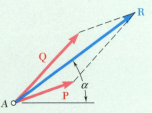

Fig. 1 Parallelogram law applied to add forces **P** and **Q**.

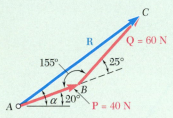

Fig. 2 Triangle rule applied to add forces **P** and **Q**.

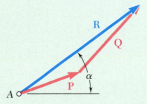

Fig. 3 Geometry of triangle rule applied to add forces **P** and **Q**.

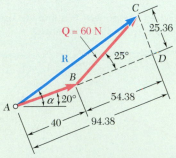

Fig. 4 Alternative geometry of triangle rule applied to add forces **P** and **Q**.

Sample Problem 2.1

Two forces **P** and **Q** act on a bolt A. Determine their resultant.

STRATEGY: Two lines determine a plane, so this is a problem of two coplanar forces. You can solve the problem graphically or by trigonometry.

MODELING: For a graphical solution, you can use the parallelogram rule or the triangle rule for addition of vectors. For a trigonometric solution, you can use the law of cosines and law of sines or use a right-triangle approach.

ANALYSIS:

Graphical Solution. Draw to scale a parallelogram with sides equal to **P** and **Q** (Fig. 1). Measure the magnitude and direction of the resultant. They are

$$R = 98 \text{ N} \qquad \alpha = 35° \qquad \mathbf{R} = 98 \text{ N} \measuredangle 35° \quad \blacktriangleleft$$

You can also use the triangle rule. Draw forces **P** and **Q** in tip-to-tail fashion (Fig. 2). Again measure the magnitude and direction of the resultant. The answers should be the same.

$$R = 98 \text{ N} \qquad \alpha = 35° \qquad \mathbf{R} = 98 \text{ N} \measuredangle 35° \quad \blacktriangleleft$$

Trigonometric Solution. Using the triangle rule again, you know two sides and the included angle (Fig. 3). Apply the law of cosines.

$$R^2 = P^2 + Q^2 - 2PQ \cos B$$
$$R^2 = (40 \text{ N})^2 + (60 \text{ N})^2 - 2(40 \text{ N})(60 \text{ N}) \cos 155°$$
$$R = 97.73 \text{ N}$$

Now apply the law of sines:

$$\frac{\sin A}{Q} = \frac{\sin B}{R} \qquad \frac{\sin A}{60 \text{ N}} = \frac{\sin 155°}{97.73 \text{ N}} \tag{1}$$

Solving Eq. (1) for sin A, you obtain

$$\sin A = \frac{(60 \text{ N}) \sin 155°}{97.73 \text{ N}}$$

Using a calculator, compute this quotient, and then obtain its arc sine:

$$A = 15.04° \qquad \alpha = 20° + A = 35.04°$$

Use three significant figures to record the answer (cf. Sec. 1.6):

$$\mathbf{R} = 97.7 \text{ N} \measuredangle 35.0° \quad \blacktriangleleft$$

Alternative Trigonometric Solution. Construct the right triangle BCD (Fig. 4) and compute

$$CD = (60 \text{ N}) \sin 25° = 25.36 \text{ N}$$
$$BD = (60 \text{ N}) \cos 25° = 54.38 \text{ N}$$

(continued)

Then, using triangle ACD, you have

$$\tan A = \frac{25.36 \text{ N}}{94.38 \text{ N}} \qquad A = 15.04°$$

$$R = \frac{25.36}{\sin A} \qquad R = 97.73 \text{ N}$$

Again,

$$\alpha = 20° + A = 35.04° \qquad \mathbf{R} = 97.7 \text{ N} \measuredangle 35.0° \blacktriangleleft$$

REFLECT and THINK: An analytical solution using trigonometry provides for greater accuracy. However, it is helpful to use a graphical solution as a check.

Sample Problem 2.2

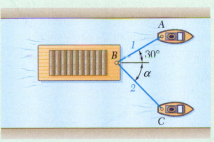

Two tugboats are pulling a barge. If the resultant of the forces exerted by the tugboats is a 5000-lb force directed along the axis of the barge, determine (a) the tension in each of the ropes, given that $\alpha = 45°$, (b) the value of α for which the tension in rope 2 is minimum.

STRATEGY: This is a problem of two coplanar forces. You can solve the first part either graphically or analytically. In the second part, a graphical approach readily shows the necessary direction for rope 2, and you can use an analytical approach to complete the solution.

MODELING: You can use the parallelogram law or the triangle rule to solve part (a). For part (b), use a variation of the triangle rule.

ANALYSIS: a. Tension for $\alpha = 45°$.

Graphical Solution. Use the parallelogram law. The resultant (the diagonal of the parallelogram) is equal to 5000 lb and is directed to the right. Draw the sides parallel to the ropes (Fig. 1). If the drawing is done to scale, you should measure

$$T_1 = 3700 \text{ lb} \qquad\qquad T_2 = 2600 \text{ lb} \blacktriangleleft$$

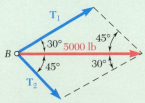

Fig. 1 Parallelogram law applied to add forces $\mathbf{T}_1$ and $\mathbf{T}_2$.

Trigonometric Solution.   Use the triangle rule. Note that the triangle in Fig. 2 represents half of the parallelogram shown in Fig. 1. Using the law of sines,

$$\frac{T_1}{\sin 45°} = \frac{T_2}{\sin 30°} = \frac{5000 \text{ lb}}{\sin 105°}$$

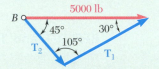

Fig. 2 Triangle rule applied to add forces $\mathbf{T}_1$ and $\mathbf{T}_2$.

With a calculator, compute and store the value of the last quotient. Multiply this value successively by sin 45° and sin 30°, obtaining

$$T_1 = 3660 \text{ lb} \qquad T_2 = 2590 \text{ lb} \quad \blacktriangleleft$$

b. Value of α for Minimum T_2.   To determine the value of α for which the tension in rope 2 is minimum, use the triangle rule again. In Fig. 3, line *1-1'* is the known direction of $\mathbf{T}_1$. Several possible directions of $\mathbf{T}_2$ are shown by the lines *2-2'*. The minimum value of T_2 occurs when $\mathbf{T}_1$ and $\mathbf{T}_2$ are perpendicular (Fig. 4). Thus, the minimum value of T_2 is

$$T_2 = (5000 \text{ lb}) \sin 30° = 2500 \text{ lb}$$

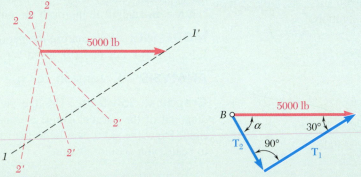

Fig. 3 Determination of direction of minimum $\mathbf{T}_2$.

Fig. 4 Triangle rule applied for minimum $\mathbf{T}_2$.

Corresponding values of T_1 and α are

$$T_1 = (5000 \text{ lb}) \cos 30° = 4330 \text{ lb}$$
$$\alpha = 90° - 30°$$
$$\alpha = 60° \quad \blacktriangleleft$$

REFLECT and THINK: Part (a) is a straightforward application of resolving a vector into components. The key to part (b) is recognizing that the minimum value of T_2 occurs when $\mathbf{T}_1$ and $\mathbf{T}_2$ are perpendicular.

SOLVING PROBLEMS
ON YOUR OWN

The preceding sections were devoted to adding vectors by using the parallelogram law, triangle rule, and polygon rule with application to forces.

We presented two sample problems. In Sample Prob. 2.1, we used the parallelogram law to determine the resultant of two forces of known magnitude and direction. In Sample Prob. 2.2, we used it to resolve a given force into two components of known direction.

You will now be asked to solve problems on your own. Some may resemble one of the sample problems; others may not. What all problems and sample problems in this section have in common is that they can be solved by direct application of the parallelogram law.

Your solution of a given problem should consist of the following steps:

1. Identify which forces are the applied forces and which is the resultant. It is often helpful to write the vector equation that shows how the forces are related. For example, in Sample Prob. 2.1 you could write

$$\mathbf{R} = \mathbf{P} + \mathbf{Q}$$

You may want to keep this relation in mind as you formulate the next part of the solution.

2. Draw a parallelogram with the applied forces as two adjacent sides and the resultant as the included diagonal (Fig. 2.2). Alternatively, you can *use the* **triangle rule** with the applied forces drawn in tip-to-tail fashion and the resultant extending from the tail of the first vector to the tip of the second (Fig. 2.6).

3. Indicate all dimensions. Using one of the triangles of the parallelogram or the triangle constructed according to the triangle rule, indicate all dimensions—whether sides or angles—and determine the unknown dimensions either graphically or by trigonometry.

4. Recall the laws of trigonometry. If you use trigonometry, remember that the law of cosines should be applied first if two sides and the included angle are known [Sample Prob. 2.1], and the law of sines should be applied first if one side and all angles are known [Sample Prob. 2.2].

If you have had prior exposure to mechanics, you might be tempted to ignore the solution techniques of this lesson in favor of resolving the forces into rectangular components. The component method is important and is considered in the next section, but use of the parallelogram law simplifies the solution of many problems and should be mastered first.

Problems

2.1 Two forces are applied as shown to a hook. Determine graphically the magnitude and direction of their resultant using (*a*) the parallelogram law, (*b*) the triangle rule.

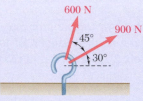

Fig. P2.1

2.2 Two forces are applied as shown to a bracket support. Determine graphically the magnitude and direction of their resultant using (*a*) the parallelogram law, (*b*) the triangle rule.

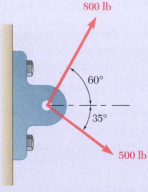

Fig. P2.2

2.3 Two structural members *B* and *C* are bolted to bracket *A*. Knowing that both members are in tension and that $P = 10$ kN and $Q = 15$ kN, determine graphically the magnitude and direction of the resultant force exerted on the bracket using (*a*) the parallelogram law, (*b*) the triangle rule.

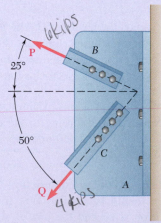

Fig. *P2.3* and P2.4

2.4 Two structural members *B* and *C* are bolted to bracket *A*. Knowing that both members are in tension and that $P = 6$ kips and $Q = 4$ kips, determine graphically the magnitude and direction of the resultant force exerted on the bracket using (*a*) the parallelogram law, (*b*) the triangle rule.

2.5 A stake is being pulled out of the ground by means of two ropes as shown. Knowing that $\alpha = 30°$, determine by trigonometry (*a*) the magnitude of the force **P** so that the resultant force exerted on the stake is vertical, (*b*) the corresponding magnitude of the resultant.

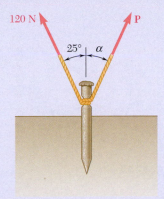

Fig. P2.5

2.6 A telephone cable is clamped at *A* to the pole *AB*. Knowing that the tension in the left-hand portion of the cable is $T_1 = 800$ lb, determine by trigonometry (*a*) the required tension T_2 in the right-hand portion if the resultant **R** of the forces exerted by the cable at *A* is to be vertical, (*b*) the corresponding magnitude of **R**.

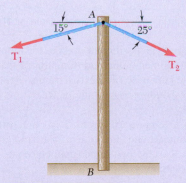

Fig. P2.6 and *P2.7*

2.7 A telephone cable is clamped at *A* to the pole *AB*. Knowing that the tension in the right-hand portion of the cable is $T_2 = 1000$ lb, determine by trigonometry (*a*) the required tension T_1 in the left-hand portion if the resultant **R** of the forces exerted by the cable at *A* is to be vertical, (*b*) the corresponding magnitude of **R**.

2.8 A disabled automobile is pulled by means of two ropes as shown. The tension in rope *AB* is 2.2 kN, and the angle α is 25°. Knowing that the resultant of the two forces applied at *A* is directed along the axis of the automobile, determine by trigonometry (*a*) the tension in rope *AC*, (*b*) the magnitude of the resultant of the two forces applied at *A*.

2.9 A disabled automobile is pulled by means of two ropes as shown. Knowing that the tension in rope *AB* is 3 kN, determine by trigonometry the tension in rope *AC* and the value of α so that the resultant force exerted at *A* is a 4.8-kN force directed along the axis of the automobile.

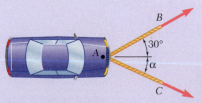

Fig. P2.8 and P2.9

2.10 Two forces are applied as shown to a hook support. Knowing that the magnitude of **P** is 35 N, determine by trigonometry (a) the required angle α if the resultant **R** of the two forces applied to the support is to be horizontal, (b) the corresponding magnitude of **R**.

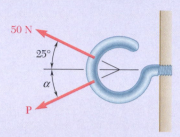

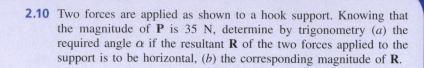

Fig. P2.10

2.11 A steel tank is to be positioned in an excavation. Knowing that $\alpha = 20°$, determine by trigonometry (a) the required magnitude of the force **P** if the resultant **R** of the two forces applied at A is to be vertical, (b) the corresponding magnitude of **R**.

2.12 A steel tank is to be positioned in an excavation. Knowing that the magnitude of **P** is 500 lb, determine by trigonometry (a) the required angle α if the resultant **R** of the two forces applied at A is to be vertical, (b) the corresponding magnitude of **R**.

2.13 A steel tank is to be positioned in an excavation. Determine by trigonometry (a) the magnitude and direction of the smallest force **P** for which the resultant **R** of the two forces applied at A is vertical, (b) the corresponding magnitude of **R**.

2.14 For the hook support of Prob. 2.10, determine by trigonometry (a) the magnitude and direction of the smallest force **P** for which the resultant **R** of the two forces applied to the support is horizontal, (b) the corresponding magnitude of **R**.

2.15 For the hook support shown, determine by trigonometry the magnitude and direction of the resultant of the two forces applied to the support.

2.16 Solve Prob. 2.1 by trigonometry.

2.17 Solve Prob. 2.4 by trigonometry.

2.18 For the stake of Prob. 2.5, knowing that the tension in one rope is 120 N, determine by trigonometry the magnitude and direction of the force **P** so that the resultant is a vertical force of 160 N.

2.19 Two forces **P** and **Q** are applied to the lid of a storage bin as shown. Knowing that $P = 48$ N and $Q = 60$ N, determine by trigonometry the magnitude and direction of the resultant of the two forces.

2.20 Two forces **P** and **Q** are applied to the lid of a storage bin as shown. Knowing that $P = 60$ N and $Q = 48$ N, determine by trigonometry the magnitude and direction of the resultant of the two forces.

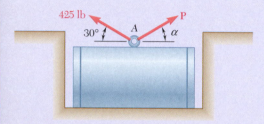
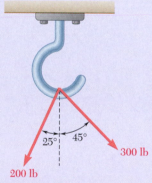

Fig. P2.11, *P2.12* and P2.13

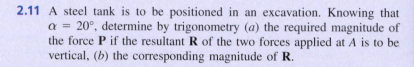
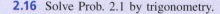

Fig. P2.15

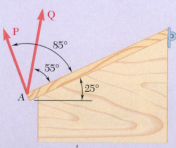

Fig. P2.19 and *P2.20*

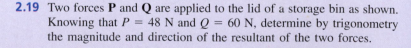

28

2.2 ADDING FORCES BY COMPONENTS

In Sec. 2.1E, we described how to resolve a force into components. Here we discuss how to add forces by using their components, especially rectangular components. This method is often the most convenient way to add forces and, in practice, is the most common approach. (Note that we can readily extend the properties of vectors established in this section to the rectangular components of any vector quantity, such as velocity or momentum.)

2.2A Rectangular Components of a Force: Unit Vectors

In many problems, it is useful to resolve a force into two components that are perpendicular to each other. Figure 2.14 shows a force **F** resolved into a component $\mathbf{F}_x$ along the x axis and a component $\mathbf{F}_y$ along the y axis. The parallelogram drawn to obtain the two components is a rectangle, and $\mathbf{F}_x$ and $\mathbf{F}_y$ are called **rectangular components**.

The x and y axes are usually chosen to be horizontal and vertical, respectively, as in Fig. 2.14; they may, however, be chosen in any two perpendicular directions, as shown in Fig. 2.15. In determining the

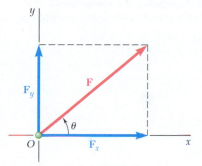

Fig. 2.14 Rectangular components of a force **F**.

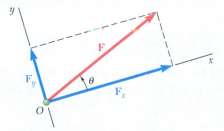

Fig. 2.15 Rectangular components of a force **F** for axes rotated away from horizontal and vertical.

rectangular components of a force, you should think of the construction lines shown in Figs. 2.14 and 2.15 as being *parallel* to the x and y axes, rather than *perpendicular* to these axes. This practice will help avoid mistakes in determining *oblique* components, as in Sec. 2.1E.

Force in Terms of Unit Vectors. To simplify working with rectangular components, we introduce two vectors of unit magnitude, directed respectively along the positive x and y axes. These vectors are called **unit vectors** and are denoted by **i** and **j**, respectively (Fig. 2.16). Recalling the

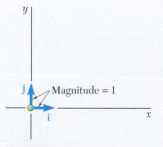

Fig. 2.16 Unit vectors along the x and y axes.

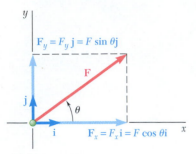

Fig. 2.17 Expressing the components of **F** in terms of unit vectors with scalar multipliers.

definition of the product of a scalar and a vector given in Sec. 2.1C, note that we can obtain the rectangular components F_x and F_y of a force **F** by multiplying respectively the unit vectors **i** and **j** by appropriate scalars (Fig. 2.17). We have

$$\mathbf{F}_x = F_x\mathbf{i} \qquad \mathbf{F}_y = F_y\mathbf{j} \qquad (2.6)$$

and

$$\mathbf{F} = F_x\mathbf{i} + F_y\mathbf{j} \qquad (2.7)$$

The scalars F_x and F_y may be positive or negative, depending upon the sense of $\mathbf{F}_x$ and of $\mathbf{F}_y$, but their absolute values are equal to the magnitudes of the component forces $\mathbf{F}_x$ and $\mathbf{F}_y$, respectively. The scalars F_x and F_y are called the **scalar components** of the force **F**, whereas the actual component forces $\mathbf{F}_x$ and $\mathbf{F}_y$ should be referred to as the **vector components** of **F**. However, when there exists no possibility of confusion, we may refer to the vector as well as the scalar components of **F** as simply the **components** of **F**. Note that the scalar component F_x is positive when the vector component $\mathbf{F}_x$ has the same sense as the unit vector **i** (i.e., the same sense as the positive x axis) and is negative when $\mathbf{F}_x$ has the opposite sense. A similar conclusion holds for the sign of the scalar component F_y.

Scalar Components. Denoting by F the magnitude of the force **F** and by θ the angle between **F** and the x axis, which is measured counterclockwise from the positive x axis (Fig. 2.17), we may express the scalar components of **F** as

$$F_x = F \cos \theta \qquad F_y = F \sin \theta \qquad (2.8)$$

These relations hold for any value of the angle θ from $0°$ to $360°$, and they define the signs as well as the absolute values of the scalar components F_x and F_y.

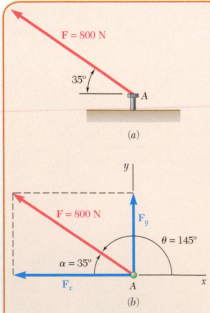

Fig. 2.18 (a) Force F exerted on a bolt; (b) rectangular components of F.

Concept Application 2.1

A force of 800 N is exerted on a bolt A as shown in Fig. 2.18a. Determine the horizontal and vertical components of the force.

Solution

In order to obtain the correct sign for the scalar components F_x and F_y, we could substitute the value $180° - 35° = 145°$ for θ in Eqs. (2.8). However, it is often more practical to determine by inspection the signs of F_x and F_y (Fig. 2.18b) and then use the trigonometric functions of the angle $\alpha = 35°$. Therefore,

$$F_x = -F \cos \alpha = -(800 \text{ N}) \cos 35° = -655 \text{ N}$$
$$F_y = +F \sin \alpha = +(800 \text{ N}) \sin 35° = +459 \text{ N}$$

The vector components of **F** are thus

$$\mathbf{F}_x = -(655 \text{ N})\mathbf{i} \qquad \mathbf{F}_y = +(459 \text{ N})\mathbf{j}$$

and we may write **F** in the form

$$\mathbf{F} = -(655 \text{ N})\mathbf{i} + (459 \text{ N})\mathbf{j} \quad \blacktriangleleft$$

Concept Application 2.2

A man pulls with a force of 300 N on a rope attached to the top of a building, as shown in Fig. 2.19a. What are the horizontal and vertical components of the force exerted by the rope at point A?

Solution

You can see from Fig. 2.19b that

$$F_x = +(300 \text{ N}) \cos \alpha \qquad F_y = -(300 \text{ N}) \sin \alpha$$

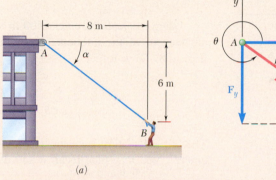

(a)   (b)

Fig. 2.19 (a) A man pulls on a rope attached to a building; (b) components of the rope's force **F**.

Observing that $AB = 10$ m, we find from Fig. 2.19a

$$\cos \alpha = \frac{8 \text{ m}}{AB} = \frac{8 \text{ m}}{10 \text{ m}} = \frac{4}{5} \qquad \sin \alpha = \frac{6 \text{ m}}{AB} = \frac{6 \text{ m}}{10 \text{ m}} = \frac{3}{5}$$

We thus obtain

$$F_x = +(300 \text{ N})\frac{4}{5} = +240 \text{ N} \qquad F_y = -(300 \text{ N})\frac{3}{5} = -180 \text{ N}$$

This gives us a total force of

$$\mathbf{F} = (240 \text{ N})\mathbf{i} - (180 \text{ N})\mathbf{j} \quad \blacktriangleleft$$

Direction of a Force.   When a force **F** is defined by its rectangular components F_x and F_y (see Fig. 2.17), we can find the angle θ defining its direction from

$$\tan \theta = \frac{F_y}{F_x} \qquad\qquad (2.9)$$

We can obtain the magnitude F of the force by applying the Pythagorean theorem,

$$F = \sqrt{F_x^2 + F_y^2} \qquad\qquad (2.10)$$

or by solving for F from one of the Eqs. (2.8).

Concept Application 2.3

A force $\mathbf{F} = (700\ \text{lb})\mathbf{i} + (1500\ \text{lb})\mathbf{j}$ is applied to a bolt A. Determine the magnitude of the force and the angle θ it forms with the horizontal.

Solution

First draw a diagram showing the two rectangular components of the force and the angle θ (Fig. 2.20). From Eq. (2.9), you obtain

$$\tan \theta = \frac{F_y}{F_x} = \frac{1500\ \text{lb}}{700\ \text{lb}}$$

Using a calculator, enter 1500 lb and divide by 700 lb; computing the arc tangent of the quotient gives you $\theta = 65.0°$. Solve the second of Eqs. (2.8) for F to get

$$F = \frac{F_y}{\sin \theta} = \frac{1500\ \text{lb}}{\sin 65.0°} = 1655\ \text{lb}$$

The last calculation is easier if you store the value of F_y when originally entered; you may then recall it and divide it by $\sin \theta$.

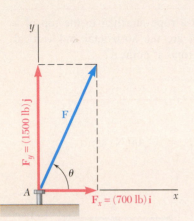

Fig. 2.20 Components of a force $\mathbf{F}$ exerted on a bolt.

2.2B Addition of Forces by Summing X and Y Components

We described in Sec. 2.1A how to add forces according to the parallelogram law. From this law, we derived two other methods that are more readily applicable to the graphical solution of problems: the triangle rule for the addition of two forces and the polygon rule for the addition of three or more forces. We also explained that the force triangle used to define the resultant of two forces could be used to obtain a trigonometric solution.

However, when we need to add three or more forces, we cannot obtain any practical trigonometric solution from the force polygon that defines the resultant of the forces. In this case, the best approach is to obtain an analytic solution of the problem by resolving each force into two rectangular components.

Consider, for instance, three forces $\mathbf{P}$, $\mathbf{Q}$, and $\mathbf{S}$ acting on a particle A (Fig. 2.21a). Their resultant $\mathbf{R}$ is defined by the relation

$$\mathbf{R} = \mathbf{P} + \mathbf{Q} + \mathbf{S} \tag{2.11}$$

Resolving each force into its rectangular components, we have

$$R_x\mathbf{i} + R_y\mathbf{j} = P_x\mathbf{i} + P_y\mathbf{j} + Q_x\mathbf{i} + Q_y\mathbf{j} + S_x\mathbf{i} + S_y\mathbf{j}$$
$$= (P_x + Q_x + S_x)\mathbf{i} + (P_y + Q_y + S_y)\mathbf{j}$$

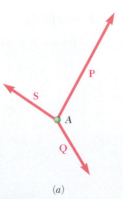

(a)

Fig. 2.21 (a) Three forces acting on a particle.

From this equation, we can see that

$$R_x = P_x + Q_x + S_x \qquad R_y = P_y + Q_y + S_y \qquad \textbf{(2.12)}$$

or for short,

$$R_x = \Sigma F_x \qquad R_y = \Sigma F_y \qquad \textbf{(2.13)}$$

We thus conclude that **when several forces are acting on a particle, we obtain the scalar components R_x and R_y of the resultant R by adding algebraically the corresponding scalar components of the given forces.** *(Clearly, this result also applies to the addition of other vector quantities, such as velocities, accelerations, or momenta.)*

In practice, determining the resultant **R** is carried out in three steps, as illustrated in Fig. 2.21.

1. Resolve the given forces (Fig. 2.21*a*) into their *x* and *y* components (Fig. 2.21*b*).

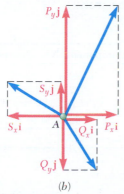

(b)

Fig. 2.21 (*b*) Rectangular components of each force.

2. Add these components to obtain the *x* and *y* components of **R** (Fig. 2.21*c*).

(c)

Fig. 2.21 (*c*) Summation of the components.

3. Apply the parallelogram law to determine the resultant $\mathbf{R} = R_x\mathbf{i} + R_y\mathbf{j}$ (Fig. 2.21*d*).

The procedure just described is most efficiently carried out if you arrange the computations in a table (see Sample Problem 2.3). Although this is the only practical analytic method for adding three or more forces, it is also often preferred to the trigonometric solution in the case of adding two forces.

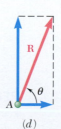

(d)

Fig. 2.21 (*d*) Determining the resultant from its components.

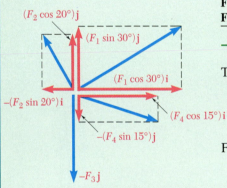

$(F_2 \cos 20°)\mathbf{j}$
$(F_1 \sin 30°)\mathbf{j}$
$(F_1 \cos 30°)\mathbf{i}$
$-(F_2 \sin 20°)\mathbf{i}$
$(F_4 \cos 15°)\mathbf{i}$
$-(F_4 \sin 15°)\mathbf{j}$
$-F_3\mathbf{j}$

Fig. 1 Rectangular components of each force.

Sample Problem 2.3

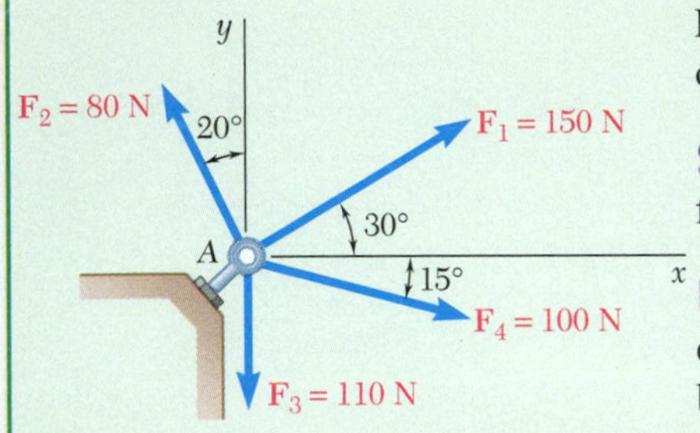

Four forces act on bolt A as shown. Determine the resultant of the forces on the bolt.

STRATEGY: The simplest way to approach a problem of adding four forces is to resolve the forces into components.

MODELING: As we mentioned, solving this kind of problem is usually easier if you arrange the components of each force in a table. In the table below, we entered the x and y components of each force as determined by trigonometry (Fig. 1). According to the convention adopted in this section, the scalar number representing a force component is positive if the force component has the same sense as the corresponding coordinate axis. Thus, x components acting to the right and y components acting upward are represented by positive numbers.

ANALYSIS:

Force	Magnitude, N	x Component, N	y Component, N
F_1	150	+129.9	+75.0
F_2	80	−27.4	+75.2
F_3	110	0	−110.0
F_4	100	+96.6	−25.9
		$R_x = +199.1$	$R_y = +14.3$

Thus, the resultant $\mathbf{R}$ of the four forces is

$$\mathbf{R} = R_x\mathbf{i} + R_y\mathbf{j} \qquad \mathbf{R} = (199.1 \text{ N})\mathbf{i} + (14.3 \text{ N})\mathbf{j} \quad \blacktriangleleft$$

You can now determine the magnitude and direction of the resultant. From the triangle shown in Fig. 2, you have

$$\tan\alpha = \frac{R_y}{R_x} = \frac{14.3 \text{ N}}{199.1 \text{ N}} \qquad \alpha = 4.1°$$

$$R = \frac{14.3 \text{ N}}{\sin\alpha} = 199.6 \text{ N} \qquad \mathbf{R} = 199.6 \text{ N} \measuredangle 4.1° \quad \blacktriangleleft$$

$R_y = (14.3 \text{ N})\mathbf{j}$
$R_x = (199.1 \text{ N})\mathbf{i}$

Fig. 2 Resultant of the given force system.

REFLECT and THINK: Arranging data in a table not only helps you keep track of the calculations, but also makes things simpler for using a calculator on similar computations.

SOLVING PROBLEMS
ON YOUR OWN

You saw in the preceding lesson that we can determine the resultant of two forces either graphically or from the trigonometry of an oblique triangle.

A. When three or more forces are involved, the best way to determine their resultant R is by first resolving each force into **rectangular components**. You may encounter either of two cases, depending upon the way in which each of the given forces is defined.

Case 1. The force F is defined by its magnitude F and the angle α it forms with the x axis. Obtain the x and y components of the force by multiplying F by $\cos \alpha$ and $\sin \alpha$, respectively [Concept Application 2.1].

Case 2. The force F is defined by its magnitude F and the coordinates of two points A and B on its line of action (Fig. 2.19). Find the angle α that **F** forms with the x axis by trigonometry, and then use the process of Case 1. However, you can also find the components of **F** directly from proportions among the various dimensions involved without actually determining α [Concept Application 2.2].

B. Rectangular components of the resultant. Obtain the components R_x and R_y of the resultant by adding the corresponding components of the given forces algebraically [Sample Prob. 2.3].

You can express the resultant in vectorial form using the unit vectors **i** and **j**, which are directed along the x and y axes, respectively:

$$\mathbf{R} = R_x\mathbf{i} + R_y\mathbf{j}$$

Alternatively, you can determine the *magnitude and direction* of the resultant by solving the right triangle of sides R_x and R_y for R and for the angle that **R** forms with the x axis.

Problems

2.21 and 2.22 Determine the x and y components of each of the forces shown.

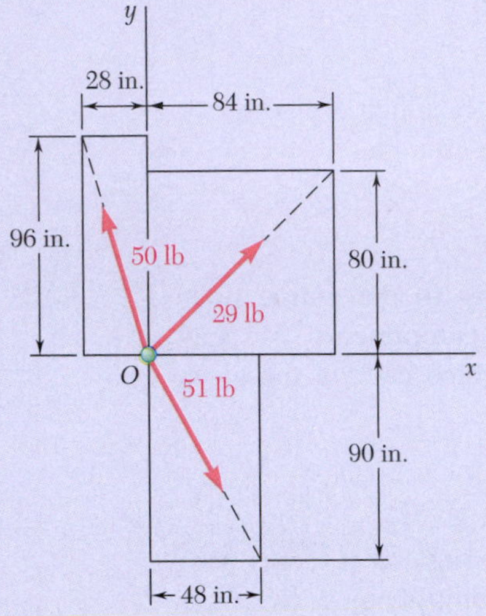

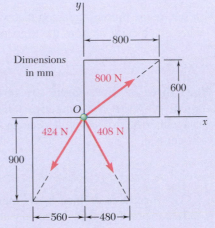

Fig. P2.21

Fig. P2.22

2.23 and 2.24 Determine the x and y components of each of the forces shown.

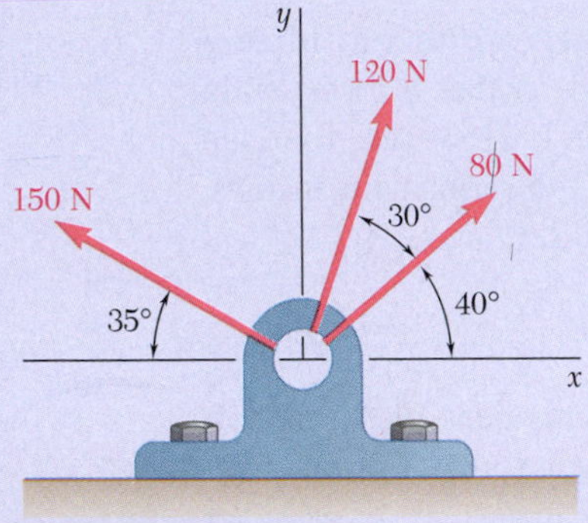

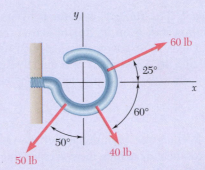

Fig. P2.23

Fig. P2.24

2.25 Member BC exerts on member AC a force $\mathbf{P}$ directed along line BC. Knowing that $\mathbf{P}$ must have a 325-N horizontal component, determine (a) the magnitude of the force $\mathbf{P}$, (b) its vertical component.

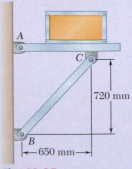

Fig. P2.25

2.26 Member *BD* exerts on member *ABC* a force **P** directed along line *BD*. Knowing that **P** must have a 300-lb horizontal component, determine (*a*) the magnitude of the force **P**, (*b*) its vertical component.

2.27 The hydraulic cylinder *BC* exerts on member *AB* a force **P** directed along line *BC*. Knowing that **P** must have a 600-N component perpendicular to member *AB*, determine (*a*) the magnitude of the force **P**, (*b*) its component along line *AB*.

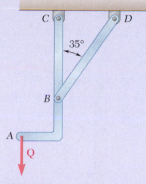

Fig. P2.26

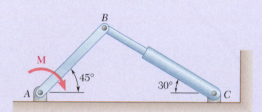

Fig. P2.27

2.28 Cable *AC* exerts on beam *AB* a force **P** directed along line *AC*. Knowing that **P** must have a 350-lb vertical component, determine (*a*) the magnitude of the force **P**, (*b*) its horizontal component.

2.29 The hydraulic cylinder *BD* exerts on member *ABC* a force **P** directed along line *BD*. Knowing that **P** must have a 750-N component perpendicular to member *ABC*, determine (*a*) the magnitude of the force **P**, (*b*) its component parallel to *ABC*.

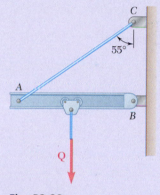

Fig. P2.28

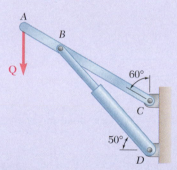

Fig. P2.29

2.30 The guy wire *BD* exerts on the telephone pole *AC* a force **P** directed along *BD*. Knowing that **P** must have a 720-N component perpendicular to the pole *AC*, determine (*a*) the magnitude of the force **P**, (*b*) its component along line *AC*.

2.31 Determine the resultant of the three forces of Prob. 2.21.

2.32 Determine the resultant of the three forces of Prob. 2.23.

2.33 Determine the resultant of the three forces of Prob. 2.24.

2.34 Determine the resultant of the three forces of Prob. 2.22.

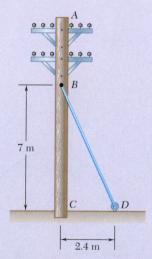

Fig. *P2.30*

2.35 Knowing that $\alpha = 35°$, determine the resultant of the three forces shown.

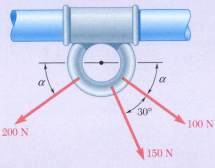

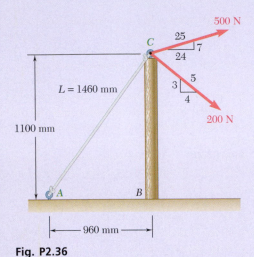

Fig. P2.36

Fig. P2.35

2.36 Knowing that the tension in rope AC is 365 N, determine the resultant of the three forces exerted at point C of post BC.

2.37 Knowing that $\alpha = 40°$, determine the resultant of the three forces shown.

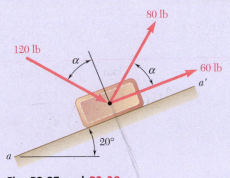
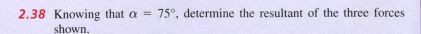

Fig. P2.37 and P2.38

2.38 Knowing that $\alpha = 75°$, determine the resultant of the three forces shown.

2.39 For the collar of Prob. 2.35, determine (a) the required value of α if the resultant of the three forces shown is to be vertical, (b) the corresponding magnitude of the resultant.

2.40 For the post of Prob. 2.36, determine (a) the required tension in rope AC if the resultant of the three forces exerted at point C is to be horizontal, (b) the corresponding magnitude of the resultant.

2.41 Determine (a) the required tension in cable AC, knowing that the resultant of the three forces exerted at point C of boom BC must be directed along BC, (b) the corresponding magnitude of the resultant.

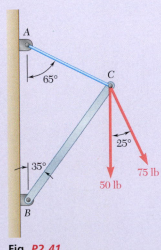

Fig. P2.41

2.42 For the block of Probs. 2.37 and 2.38, determine (a) the required value of α if the resultant of the three forces shown is to be parallel to the incline, (b) the corresponding magnitude of the resultant.

2.3 FORCES AND EQUILIBRIUM IN A PLANE

Now that we have seen how to add forces, we can proceed to one of the key concepts in this course: the equilibrium of a particle. The connection between equilibrium and the sum of forces is very direct: a particle can be in equilibrium only when the sum of the forces acting on it is zero.

2.3A   Equilibrium of a Particle

In the preceding sections, we discussed methods for determining the resultant of several forces acting on a particle. Although it has not occurred in any of the problems considered so far, it is quite possible for the resultant to be zero. In such a case, the net effect of the given forces is zero, and the particle is said to be in **equilibrium**. We thus have the definition:

> **When the resultant of all the forces acting on a particle is zero, the particle is in equilibrium**.

A particle acted upon by two forces is in equilibrium if the two forces have the same magnitude and the same line of action but opposite sense. The resultant of the two forces is then zero, as shown in Fig. 2.22.

Another case of equilibrium of a particle is represented in Fig. 2.23a, where four forces are shown acting on particle A. In Fig. 2.23b, we use the polygon rule to determine the resultant of the given forces. Starting from point O with $\mathbf{F}_1$ and arranging the forces in tip-to-tail fashion, we find that the tip of $\mathbf{F}_4$ coincides with the starting point O. Thus, the resultant $\mathbf{R}$ of the given system of forces is zero, and the particle is in equilibrium.

Photo 2.2 Forces acting on the carabiner include the weight of the girl and her harness, and the force exerted by the pulley attachment. Treating the carabiner as a particle, it is in equilibrium because the resultant of all forces acting on it is zero.

Fig. 2.22 When a particle is in equilibrium, the resultant of all forces acting on the particle is zero.

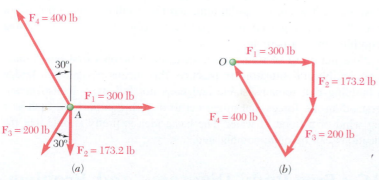

(a)   (b)

Fig. 2.23 (a) Four forces acting on particle A; (b) using the polygon law to find the resultant of the forces in (a), which is zero because the particle is in equilibrium.

The closed polygon drawn in Fig. 2.23b provides a *graphical* expression of the equilibrium of A. To express *algebraically* the conditions for the equilibrium of a particle, we write

Equilibrium of a particle $$\mathbf{R} = \Sigma \mathbf{F} = 0 \qquad\qquad (2.14)$$

Resolving each force **F** into rectangular components, we have

$$\Sigma (F_x\mathbf{i} + F_y\mathbf{j}) = 0 \qquad \text{or} \qquad (\Sigma F_x)\mathbf{i} + (\Sigma F_y)\mathbf{j} = 0$$

We conclude that the necessary and sufficient conditions for the equilibrium of a particle are

**Equilibrium of a particle
(scalar equations)**

$$\Sigma F_x = 0 \qquad \Sigma F_y = 0 \tag{2.15}$$

Returning to the particle shown in Fig. 2.23, we can check that the equilibrium conditions are satisfied. We have

$$\Sigma F_x = 300 \text{ lb} - (200 \text{ lb}) \sin 30° - (400 \text{ lb}) \sin 30°$$
$$= 300 \text{ lb} - 100 \text{ lb} - 200 \text{ lb} = 0$$
$$\Sigma F_y = -173.2 \text{ lb} - (200 \text{ lb}) \cos 30° + (400 \text{ lb}) \cos 30°$$
$$= -173.2 \text{ lb} - 173.2 \text{ lb} + 346.4 \text{ lb} = 0$$

2.3B Newton's First Law of Motion

As we discussed in Section 1.2, Sir Isaac Newton formulated three fundamental laws upon which the science of mechanics is based. The first of these laws can be stated as:

If the resultant force acting on a particle is zero, the particle will remain at rest (if originally at rest) or will move with constant speed in a straight line (if originally in motion).

From this law and from the definition of equilibrium just presented, we can see that a particle in equilibrium is either at rest or moving in a straight line with constant speed. If a particle does not behave in either of these ways, it is not in equilibrium, and the resultant force on it is not zero. In the following section, we consider various problems concerning the equilibrium of a particle.

Note that most of statics involves using Newton's first law to analyze an equilibrium situation. In practice, this means designing a bridge or a building that remains stable and does not fall over. It also means understanding the forces that might act to disturb equilibrium, such as a strong wind or a flood of water. The basic idea is pretty simple, but the applications can be quite complicated.

2.3C Free-Body Diagrams and Problem Solving

In practice, a problem in engineering mechanics is derived from an actual physical situation. A sketch showing the physical conditions of the problem is known as a **space diagram**.

The methods of analysis discussed in the preceding sections apply to a system of forces acting on a particle. A large number of problems involving actual structures, however, can be reduced to problems concerning the equilibrium of a particle. The method is to choose a significant particle and draw a separate diagram showing this particle and all the forces acting on it. Such a diagram is called a **free-body diagram**. (The name derives from the fact that when drawing the chosen body, or particle, it is "free" from all other bodies in the actual situation.)

As an example, consider the 75-kg crate shown in the space diagram of Fig. 2.24*a*. This crate was lying between two buildings, and is now being lifted onto a truck, which will remove it. The crate is supported by a vertical cable that is joined at *A* to two ropes, which pass over pulleys attached to the buildings at *B* and *C*. We want to determine the tension in each of the ropes *AB* and *AC*.

In order to solve this problem, we first draw a free-body diagram showing a particle in equilibrium. Since we are interested in the rope tensions, the free-body diagram should include at least one of these tensions or, if possible, both tensions. You can see that point *A* is a good free body for this problem. The free-body diagram of point *A* is shown in Fig. 2.24*b*. It shows point *A* and the forces exerted on *A* by the vertical cable and the two ropes. The force exerted by the cable is directed downward, and its magnitude is equal to the weight *W* of the crate. Recalling Eq. (1.4), we write

$$W = mg = (75 \text{ kg})(9.81 \text{ m/s}^2) = 736 \text{ N}$$

and indicate this value in the free-body diagram. The forces exerted by the two ropes are not known. Since they are respectively equal in magnitude to the tensions in rope *AB* and rope *AC*, we denote them by $\mathbf{T}_{AB}$ and $\mathbf{T}_{AC}$ and draw them away from *A* in the directions shown in the space diagram. No other detail is included in the free-body diagram.

Since point *A* is in equilibrium, the three forces acting on it must form a closed triangle when drawn in tip-to-tail fashion. We have drawn this **force triangle** in Fig. 2.24*c*. The values T_{AB} and T_{AC} of the tensions in the ropes may be found graphically if the triangle is drawn to scale, or they may be found by trigonometry. If we choose trigonometry, we use the law of sines:

$$\frac{T_{AB}}{\sin 60°} = \frac{T_{AC}}{\sin 40°} = \frac{736 \text{ N}}{\sin 80°}$$

$$T_{AB} = 647 \text{ N} \qquad T_{AC} = 480 \text{ N}$$

When a particle is in equilibrium under three forces, you can solve the problem by drawing a force triangle. When a particle is in equilibrium under more than three forces, you can solve the problem graphically by drawing a force polygon. If you need an analytic solution, you should solve the **equations of equilibrium** given in Sec. 2.3A:

$$\Sigma F_x = 0 \qquad \Sigma F_y = 0 \qquad \textbf{(2.15)}$$

These equations can be solved for no more than *two unknowns*. Similarly, the force triangle used in the case of equilibrium under three forces can be solved for only two unknowns.

The most common types of problems are those in which the two unknowns represent (1) the two components (or the magnitude and direction) of a single force or (2) the magnitudes of two forces, each of known direction. Problems involving the determination of the maximum or minimum value of the magnitude of a force are also encountered (see Probs. 2.57 through 2.61).

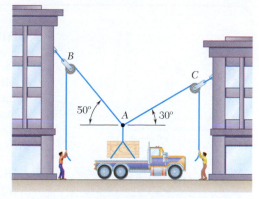

(*a*) Space diagram

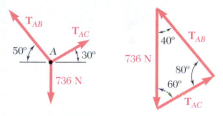

(*b*) Free-body diagram   (*c*) Force triangle

Fig. 2.24   (*a*) The space diagram shows the physical situation of the problem; (*b*) the free-body diagram shows one central particle and the forces acting on it; (c) the force triangle can be solved with the law of sines. Note that the forces form a closed triangle because the particle is in equilibrium and the resultant force is zero.

Photo 2.3   As illustrated in Fig. 2.24, it is possible to determine the tensions in the cables supporting the shaft shown by treating the hook as a particle and then applying the equations of equilibrium to the forces acting on the hook.

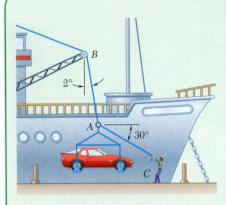

Sample Problem 2.4

In a ship-unloading operation, a 3500-lb automobile is supported by a cable. A worker ties a rope to the cable at A and pulls on it in order to center the automobile over its intended position on the dock. At the moment illustrated, the automobile is stationary, the angle between the cable and the vertical is 2°, and the angle between the rope and the horizontal is 30°. What are the tensions in the rope and cable?

STRATEGY: This is a problem of equilibrium under three coplanar forces. You can treat point A as a particle and solve the problem using a force triangle.

MODELING and ANALYSIS:

Free-Body Diagram. Choose point A as the particle and draw the complete free-body diagram (Fig. 1). T_{AB} is the tension in the cable AB, and T_{AC} is the tension in the rope.

Equilibrium Condition. Since only three forces act on point A, draw a force triangle to express that it is in equilibrium (Fig. 2). Using the law of sines,

$$\frac{T_{AB}}{\sin 120°} = \frac{T_{AC}}{\sin 2°} = \frac{3500 \text{ lb}}{\sin 58°}$$

With a calculator, compute and store the value of the last quotient. Multiplying this value successively by sin 120° and sin 2°, you obtain

$$T_{AB} = 3570 \text{ lb} \qquad T_{AC} = 144 \text{ lb} \quad \blacktriangleleft$$

REFLECT and THINK: This is a common problem of knowing one force in a three-force equilibrium problem and calculating the other forces from the given geometry. This basic type of problem will occur often as part of more complicated situations in this text.

T_{AB}

2°

A

30°

T_{AC}

3500 lb

Fig. 1 Free-body diagram of particle A.

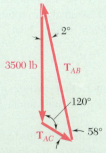

2°

3500 lb T_{AB}

120°

T_{AC} 58°

Fig. 2 Force triangle of the forces acting on particle A.

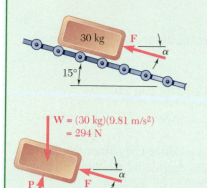

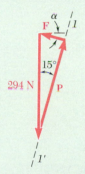

Fig. 1 Free-body diagram of package, treated as a particle.

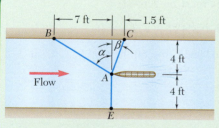

Fig. 2 Force triangle of the forces acting on package.

Sample Problem 2.5

Determine the magnitude and direction of the smallest force **F** that maintains the 30-kg package shown in equilibrium. Note that the force exerted by the rollers on the package is perpendicular to the incline.

STRATEGY: This is an equilibrium problem with three coplanar forces that you can solve with a force triangle. The new wrinkle is to determine a minimum force. You can approach this part of the solution in a way similar to Sample Problem 2.2.

MODELING and ANALYSIS:

Free-Body Diagram. Choose the package as a free body, assuming that it can be treated as a particle. Then draw the corresponding free-body diagram (Fig. 1).

Equilibrium Condition. Since only three forces act on the free body, draw a force triangle to express that it is in equilibrium (Fig. 2). Line *1-1'* represents the known direction of **P**. In order to obtain the minimum value of the force **F**, choose the direction of **F** to be perpendicular to that of **P**. From the geometry of this triangle,

$$F = (294 \text{ N}) \sin 15° = 76.1 \text{ N} \qquad \alpha = 15°$$

$$\mathbf{F} = 76.1 \text{ N} \; \angle 15° \; \blacktriangleleft$$

REFLECT and THINK: Determining maximum and minimum forces to maintain equilibrium is a common practical problem. Here the force needed is about 25% of the weight of the package, which seems reasonable for an incline of 15°.

Sample Problem 2.6

For a new sailboat, a designer wants to determine the drag force that may be expected at a given speed. To do so, she places a model of the proposed hull in a test channel and uses three cables to keep its bow on the centerline of the channel. Dynamometer readings indicate that for a given speed, the tension is 40 lb in cable *AB* and 60 lb in cable *AE*. Determine the drag force exerted on the hull and the tension in cable *AC*.

STRATEGY: The cables all connect at point *A*, so you can treat that as a particle in equilibrium. Because four forces act at *A* (tensions in three cables and the drag force), you should use the equilibrium conditions and sum forces by components to solve for the unknown forces.

(continued)

MODELING and ANALYSIS:

Determining the Angles.
First, determine the angles α and β defining the direction of cables AB and AC:

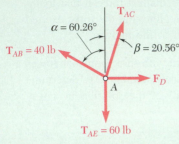

Fig. 1   Free-body diagram of particle A.

$$\tan\alpha = \frac{7 \text{ ft}}{4 \text{ ft}} = 1.75 \qquad \tan\beta = \frac{1.5 \text{ ft}}{4 \text{ ft}} = 0.375$$

$$\alpha = 60.26° \qquad \beta = 20.56°$$

Free-Body Diagram.
Choosing point A as a free body, draw the free-body diagram (Fig. 1). It includes the forces exerted by the three cables on the hull, as well as the drag force $\mathbf{F}_D$ exerted by the flow.

Equilibrium Condition.
Because point A is in equilibrium, the resultant of all forces is zero:

$$\mathbf{R} = \mathbf{T}_{AB} + \mathbf{T}_{AC} + \mathbf{T}_{AE} + \mathbf{F}_D = 0 \tag{1}$$

Because more than three forces are involved, resolve the forces into x and y components (Fig. 2):

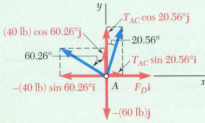

Fig. 2   Rectangular components of forces acting on particle A.

$$\begin{aligned}
\mathbf{T}_{AB} &= -(40 \text{ lb}) \sin 60.26°\mathbf{i} + (40 \text{ lb}) \cos 60.26°\mathbf{j} \\
&= -(34.73 \text{ lb})\mathbf{i} + (19.84 \text{ lb})\mathbf{j} \\
\mathbf{T}_{AC} &= T_{AC} \sin 20.56°\mathbf{i} + T_{AC} \cos 20.56°\mathbf{j} \\
&= 0.3512 T_{AC}\mathbf{i} + 0.9363 T_{AC}\mathbf{j} \\
\mathbf{T}_{AE} &= -(60 \text{ lb})\mathbf{j} \\
\mathbf{F}_D &= F_D\mathbf{i}
\end{aligned}$$

Substituting these expressions into Eq. (1) and factoring the unit vectors $\mathbf{i}$ and $\mathbf{j}$, you have

$$(-34.73 \text{ lb} + 0.3512 T_{AC} + F_D)\mathbf{i} + (19.84 \text{ lb} + 0.9363 T_{AC} - 60 \text{ lb})\mathbf{j} = 0$$

This equation is satisfied if, and only if, the coefficients of $\mathbf{i}$ and $\mathbf{j}$ are each equal to zero. You obtain the following two equilibrium equations, which express, respectively, that the sum of the x components and the sum of the y components of the given forces must be zero.

$$(\Sigma F_x = 0\text{:}) \qquad -34.73 \text{ lb} + 0.3512 T_{AC} + F_D = 0 \tag{2}$$

$$(\Sigma F_y = 0\text{:}) \qquad 19.84 \text{ lb} + 0.9363 T_{AC} - 60 \text{ lb} = 0 \tag{3}$$

From Eq. (3), you find

$$T_{AC} = +42.9 \text{ lb} \blacktriangleleft$$

Substituting this value into Eq. (2) yields

$$F_D = +19.66 \text{ lb} \blacktriangleleft$$

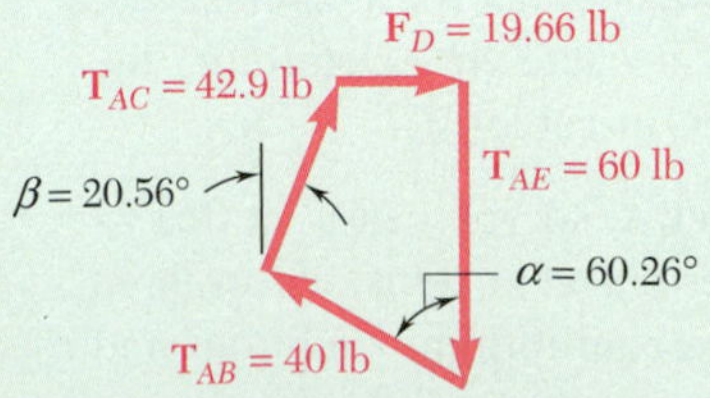

REFLECT and THINK:
In drawing the free-body diagram, you assumed a sense for each unknown force. A positive sign in the answer indicates that the assumed sense is correct. You can draw the complete force polygon (Fig. 3) to check the results.

Fig. 3   Force polygon of forces acting on particle A.

SOLVING PROBLEMS
ON YOUR OWN

When a particle is in **equilibrium**, the resultant of the forces acting on the particle must be zero. Expressing this fact in the case of a particle under *coplanar forces* provides you with two relations among these forces. As in the preceding sample problems, you can use these relations to determine two unknowns—such as the magnitude and direction of one force or the magnitudes of two forces.

Drawing a clear and accurate free-body diagram is a must in the solution of any equilibrium problem. This diagram shows the particle and all of the forces acting on it. Indicate in your free-body diagram the magnitudes of known forces, as well as any angle or dimensions that define the direction of a force. Any unknown magnitude or angle should be denoted by an appropriate symbol. Nothing else should be included in the free-body diagram. Skipping this step might save you pencil and paper, but it is very likely to lead you to a wrong solution.

Case 1. If the free-body diagram involves only **three forces**, the rest of the solution is best carried out by drawing these forces in tip-to-tail fashion to form a **force triangle**. You can solve this triangle graphically or by trigonometry for no more than two unknowns [Sample Probs. 2.4 and 2.5].

Case 2. If the free-body diagram indicates **more than three forces**, it is most practical to use an *analytic solution*. Select x and y axes and resolve each of the forces into x and y components. Setting the sum of the x components and the sum of the y components of all the forces to zero, you obtain two equations that you can solve for no more than two unknowns [Sample Prob. 2.6].

We strongly recommend that, when using an analytic solution, you write the equations of equilibrium in the same form as Eqs. (2) and (3) of Sample Prob. 2.6. The practice adopted by some students of initially placing the unknowns on the left side of the equation and the known quantities on the right side may lead to confusion in assigning the appropriate sign to each term.

Regardless of the method used to solve a two-dimensional equilibrium problem, you can determine at most two unknowns. If a two-dimensional problem involves more than two unknowns, you must obtain one or more additional relations from the information contained in the problem statement.

Problems

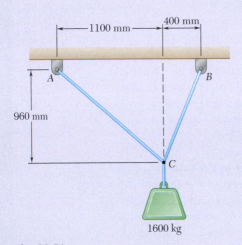

Fig. P2.F1

2.F1 Two cables are tied together at C and loaded as shown. Draw the free-body diagram needed to determine the tension in AC and BC.

2.F2 Two forces of magnitude $T_A = 8$ kips and $T_B = 15$ kips are applied as shown to a welded connection. Knowing that the connection is in equilibrium, draw the free-body diagram needed to determine the magnitudes of the forces T_C and T_D.

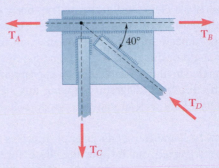

Fig. P2.F2

2.F3 The 60-lb collar A can slide on a frictionless vertical rod and is connected as shown to a 65-lb counterweight C. Draw the free-body diagram needed to determine the value of h for which the system is in equilibrium.

2.F4 A chairlift has been stopped in the position shown. Knowing that each chair weighs 250 N and that the skier in chair E weighs 765 N, draw the free-body diagrams needed to determine the weight of the skier in chair F.

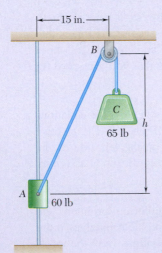

Fig. P2.F3

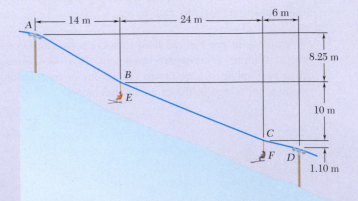

Fig. P2.F4

2.43 Two cables are tied together at C and are loaded as shown. Determine the tension (a) in cable AC, (b) in cable BC.

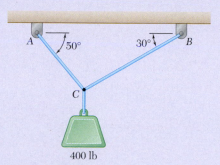

Fig. P2.43

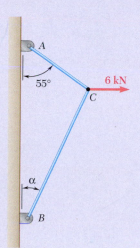

Fig. P2.44

2.44 Two cables are tied together at C and are loaded as shown. Knowing that $\alpha = 30°$, determine the tension (a) in cable AC, (b) in cable BC.

2.45 Two cables are tied together at C and loaded as shown. Determine the tension (a) in cable AC, (b) in cable BC.

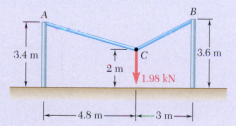

Fig. *P2.45*

2.46 Two cables are tied together at C and are loaded as shown. Knowing that $P = 500$ N and $\alpha = 60°$, determine the tension in (a) in cable AC, (b) in cable BC.

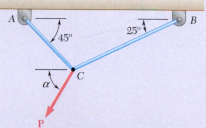

Fig. P2.46

2.47 Two cables are tied together at C and are loaded as shown. Determine the tension (a) in cable AC, (b) in cable BC.

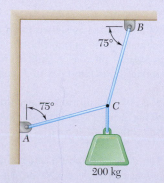

Fig. *P2.47*

47

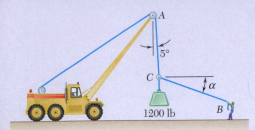

Fig. P2.48

2.48 Knowing that $\alpha = 20°$, determine the tension (*a*) in cable *AC*, (*b*) in rope *BC*.

2.49 Two cables are tied together at *C* and are loaded as shown. Knowing that $P = 300$ N, determine the tension in cables *AC* and *BC*.

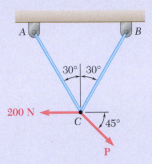

Fig. P2.49 and P2.50

2.50 Two cables are tied together at *C* and are loaded as shown. Determine the range of values of **P** for which both cables remain taut.

2.51 Two forces **P** and **Q** are applied as shown to an aircraft connection. Knowing that the connection is in equilibrium and that $P = 500$ lb and $Q = 650$ lb, determine the magnitudes of the forces exerted on rods *A* and *B*.

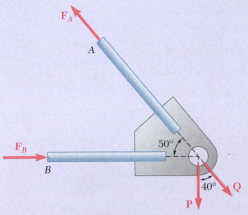

Fig. P2.51 and *P2.52*

2.52 Two forces **P** and **Q** are applied as shown to an aircraft connection. Knowing that the connection is in equilibrium and that the magnitudes of the forces exerted on rods *A* and *B* are $F_A = 750$ lb and $F_B = 400$ lb, determine the magnitudes of **P** and **Q**.

2.53 A welded connection is in equilibrium under the action of the four forces shown. Knowing that $F_A = 8$ kN and $F_B = 16$ kN, determine the magnitudes of the other two forces.

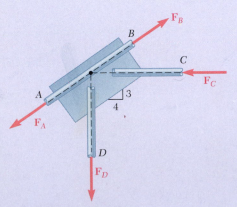

Fig. P2.53 and P2.54

2.54 A welded connection is in equilibrium under the action of the four forces shown. Knowing that $F_A = 5$ kN and $F_D = 6$ kN, determine the magnitudes of the other two forces.

2.55 A sailor is being rescued using a boatswain's chair that is suspended from a pulley that can roll freely on the support cable ACB and is pulled at a constant speed by cable CD. Knowing that $\alpha = 30°$ and $\beta = 10°$ and that the combined weight of the boatswain's chair and the sailor is 200 lb, determine the tension (a) in the support cable ACB, (b) in the traction cable CD.

2.56 A sailor is being rescued using a boatswain's chair that is suspended from a pulley that can roll freely on the support cable ACB and is pulled at a constant speed by cable CD. Knowing that $\alpha = 25°$ and $\beta = 15°$ and that the tension in cable CD is 20 lb, determine (a) the combined weight of the boatswain's chair and the sailor, (b) the tension in the support cable ACB.

2.57 For the cables of Prob. 2.44, find the value of α for which the tension is as small as possible (a) in cable BC, (b) in both cables simultaneously. In each case determine the tension in each cable.

2.58 For the cables of Prob. 2.46, it is known that the maximum allowable tension is 600 N in cable AC and 750 N in cable BC. Determine (a) the maximum force **P** that can be applied at C, (b) the corresponding value of α.

2.59 For the situation described in Fig. P2.48, determine (a) the value of α for which the tension in rope BC is as small as possible, (b) the corresponding value of the tension.

2.60 Two cables tied together at C are loaded as shown. Determine the range of values of Q for which the tension will not exceed 60 lb in either cable.

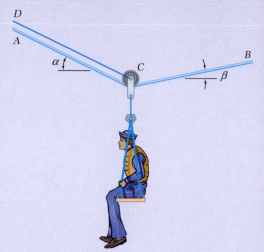

Fig. P2.55 and P2.56

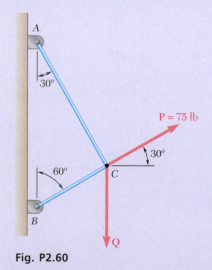

Fig. P2.60

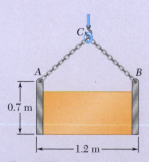

2.61 A movable bin and its contents have a combined weight of 2.8 kN. Determine the shortest chain sling ACB that can be used to lift the loaded bin if the tension in the chain is not to exceed 5 kN.

Fig. P2.61

2.62 For $W = 800$ N, $P = 200$ N, and $d = 600$ mm, determine the value of h consistent with equilibrium.

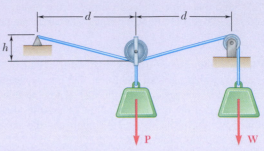

Fig. P2.62

2.63 Collar A is connected as shown to a 50-lb load and can slide on a frictionless horizontal rod. Determine the magnitude of the force **P** required to maintain the equilibrium of the collar when (a) $x = 4.5$ in., (b) $x = 15$ in.

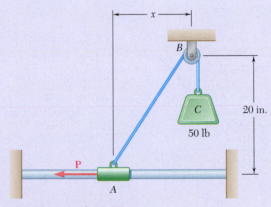

Fig. P2.63 and P2.64

2.64 Collar A is connected as shown to a 50-lb load and can slide on a frictionless horizontal rod. Determine the distance x for which the collar is in equilibrium when $P = 48$ lb.

2.65 Three forces are applied to a bracket as shown. The directions of the two 150-N forces may vary, but the angle between these forces is always 50°. Determine the range of values of α for which the magnitude of the resultant of the forces acting at A is less than 600 N.

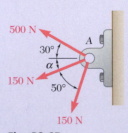

Fig. P2.65

2.66 A 200-kg crate is to be supported by the rope-and-pulley arrangement shown. Determine the magnitude and direction of the force **P** that must be exerted on the free end of the rope to maintain equilibrium. (*Hint:* The tension in the rope is the same on each side of a simple pulley. This can be proved by the methods of Chap. 4.)

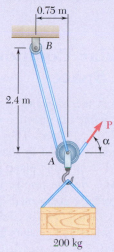

Fig. **P2.66**

2.67 A 600-lb crate is supported by several rope-and-pulley arrangements as shown. Determine for each arrangement the tension in the rope. (See the hint for Prob. 2.66.)

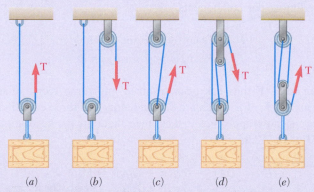

(a)   *(b)*   *(c)*   *(d)*   *(e)*

Fig. **P2.67**

2.68 Solve parts *b* and *d* of Prob. 2.67, assuming that the free end of the rope is attached to the crate.

2.69 A load **Q** is applied to pulley *C*, which can roll on the cable *ACB*. The pulley is held in the position shown by a second cable *CAD*, which passes over the pulley *A* and supports a load **P**. Knowing that *P* = 750 N, determine (*a*) the tension in cable *ACB*, (*b*) the magnitude of load **Q**.

2.70 An 1800-N load **Q** is applied to pulley *C*, which can roll on the cable *ACB*. The pulley is held in the position shown by a second cable *CAD*, which passes over the pulley *A* and supports a load **P**. Determine (*a*) the tension in cable *ACB*, (*b*) the magnitude of load **P**.

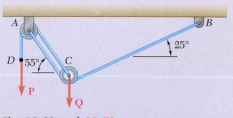

Fig. **P2.69** and **P2.70**

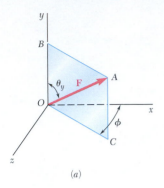

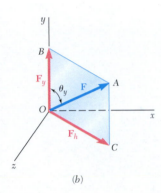

(b)

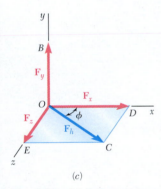

(c)

Fig. 2.25 (*a*) A force **F** in an *xyz* coordinate system; (*b*) components of **F** along the *y* axis and in the *xz* plane; (*c*) components of **F** along the three rectangular axes.

2.4 ADDING FORCES IN SPACE

The problems considered in the first part of this chapter involved only two dimensions; they were formulated and solved in a single plane. In the last part of this chapter, we discuss problems involving the three dimensions of space.

2.4A   Rectangular Components of a Force in Space

Consider a force **F** acting at the origin *O* of the system of rectangular coordinates *x*, *y*, and *z*. To define the direction of **F**, we draw the vertical plane *OBAC* containing **F** (Fig. 2.25*a*). This plane passes through the vertical *y* axis; its orientation is defined by the angle ϕ it forms with the *xy* plane. The direction of **F** within the plane is defined by the angle θ_y that **F** forms with the *y* axis. We can resolve the force **F** into a vertical component $\mathbf{F}_y$ and a horizontal component $\mathbf{F}_h$; this operation, shown in Fig. 2.25*b*, is carried out in plane *OBAC* according to the rules developed earlier. The corresponding scalar components are

$$F_y = F \cos \theta_y \qquad F_h = F \sin \theta_y \qquad \textbf{(2.16)}$$

However, we can also resolve $\mathbf{F}_h$ into two rectangular components $\mathbf{F}_x$ and $\mathbf{F}_z$ along the *x* and *z* axes, respectively. This operation, shown in Fig. 2.25*c*, is carried out in the *xz* plane. We obtain the following expressions for the corresponding scalar components:

$$F_x = F_h \cos \phi = F \sin \theta_y \cos \phi$$
$$F_z = F_h \sin \phi = F \sin \theta_y \sin \phi \qquad \textbf{(2.17)}$$

The given force **F** thus has been resolved into three rectangular vector components $\mathbf{F}_x$, $\mathbf{F}_y$, $\mathbf{F}_z$, which are directed along the three coordinate axes.

We can now apply the Pythagorean theorem to the triangles *OAB* and *OCD* of Fig. 2.25:

$$F^2 = (OA)^2 = (OB)^2 + (BA)^2 = F_y^2 + F_h^2$$
$$F_h^2 = (OC)^2 = (OD)^2 + (DC)^2 = F_x^2 + F_z^2$$

Eliminating F_h^2 from these two equations and solving for *F*, we obtain the following relation between the magnitude of **F** and its rectangular scalar components:

Magnitude of a force in space

$$F = \sqrt{F_x^2 + F_y^2 + F_z^2} \qquad \textbf{(2.18)}$$

The relationship between the force **F** and its three components $\mathbf{F}_x$, $\mathbf{F}_y$, and $\mathbf{F}_z$ is more easily visualized if we draw a "box" having $\mathbf{F}_x$, $\mathbf{F}_y$, and $\mathbf{F}_z$ for edges, as shown in Fig. 2.26. The force **F** is then represented by the main diagonal *OA* of this box. Figure 2.26*b* shows the right triangle

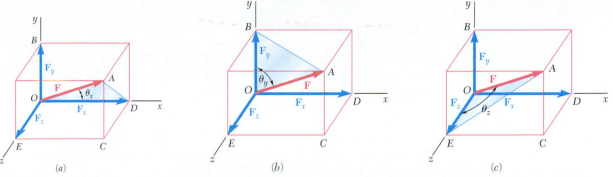

Fig. 2.26 (a) Force **F** in a three-dimensional box, showing its angle with the *x* axis; (b) force **F** and its angle with the *y* axis; (c) force **F** and its angle with the *z* axis.

OAB used to derive the first of the formulas (2.16): $F_y = F \cos \theta_y$. In Fig. 2.26*a* and *c*, two other right triangles have also been drawn: *OAD* and *OAE*. These triangles occupy positions in the box comparable with that of triangle *OAB*. Denoting by θ_x and θ_z, respectively, the angles that **F** forms with the *x* and *z* axes, we can derive two formulas similar to $F_y = F \cos \theta_y$. We thus write

**Scalar components
of a force F**

$$F_x = F \cos \theta_x \qquad F_y = F \cos \theta_y \qquad F_z = F \cos \theta_z \qquad \text{(2.19)}$$

The three angles θ_x, θ_y, and θ_z define the direction of the force **F**; they are more commonly used for this purpose than the angles θ_y and ϕ introduced at the beginning of this section. The cosines of θ_x, θ_y, and θ_z are known as the **direction cosines** of the force **F**.

Introducing the unit vectors **i**, **j**, and **k**, which are directed respectively along the *x*, *y*, and *z* axes (Fig. 2.27), we can express **F** in the form

**Vector expression
of a force F**
$$\mathbf{F} = F_x \mathbf{i} + F_y \mathbf{j} + F_z \mathbf{k} \qquad \text{(2.20)}$$

where the scalar components F_x, F_y, and F_z are defined by the relations in Eq. (2.19).

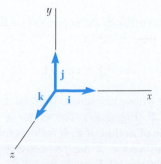

Fig. 2.27 The three unit vectors **i**, **j**, **k** lie along the three coordinate axes *x*, *y*, *z*, respectively.

Concept Application 2.4

A force of 500 N forms angles of 60°, 45°, and 120°, respectively, with the x, y, and z axes. Find the components F_x, F_y, and F_z of the force and express the force in terms of unit vectors.

Solution

Substitute $F = 500$ N, $\theta_x = 60°$, $\theta_y = 45°$, and $\theta_z = 120°$ into formulas (2.19). The scalar components of **F** are then

$$F_x = (500 \text{ N}) \cos 60° = +250 \text{ N}$$
$$F_y = (500 \text{ N}) \cos 45° = +354 \text{ N}$$
$$F_z = (500 \text{ N}) \cos 120° = -250 \text{ N}$$

Carrying these values into Eq. (2.20), you have

$$\mathbf{F} = (250 \text{ N})\mathbf{i} + (354 \text{ N})\mathbf{j} - (250 \text{ N})\mathbf{k}$$

As in the case of two-dimensional problems, a plus sign indicates that the component has the same sense as the corresponding axis, and a minus sign indicates that it has the opposite sense.

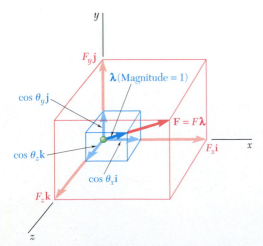

Fig. 2.28 Force **F** can be expressed as the product of its magnitude F and a unit vector **λ** in the direction of **F**. Also shown are the components of **F** and its unit vector.

The angle a force **F** forms with an axis should be measured from the positive side of the axis and is always between 0 and 180°. An angle θ_x smaller than 90° (acute) indicates that **F** (assumed attached to O) is on the same side of the yz plane as the positive x axis; $\cos \theta_x$ and F_x are then positive. An angle θ_x larger than 90° (obtuse) indicates that **F** is on the other side of the yz plane; $\cos \theta_x$ and F_x are then negative. In Concept Application 2.4, the angles θ_x and θ_y are acute and θ_z is obtuse; consequently, F_x and F_y are positive and F_z is negative.

Substituting into Eq. (2.20) the expressions obtained for F_x, F_y, and F_z in Eq. (2.19), we have

$$\mathbf{F} = F(\cos \theta_x \mathbf{i} + \cos \theta_y \mathbf{j} + \cos \theta_z \mathbf{k}) \tag{2.21}$$

This equation shows that the force **F** can be expressed as the product of the scalar F and the vector

$$\boldsymbol{\lambda} = \cos \theta_x \mathbf{i} + \cos \theta_y \mathbf{j} + \cos \theta_z \mathbf{k} \tag{2.22}$$

Clearly, the vector **λ** is a vector whose magnitude is equal to 1 and whose direction is the same as that of **F** (Fig. 2.33). The vector **λ** is referred to as the **unit vector along the line of action** of **F**. It follows from Eq. (2.22) that the components of the unit vector **λ** are respectively equal to the direction cosines of the line of action of **F**:

$$\lambda_x = \cos \theta_x \qquad \lambda_y = \cos \theta_y \qquad \lambda_z = \cos \theta_z \tag{2.23}$$

Note that the values of the three angles θ_x, θ_y, and θ_z are not independent. Recalling that the sum of the squares of the components of a vector is equal to the square of its magnitude, we can write

$$\lambda_x^2 + \lambda_y^2 + \lambda_z^2 = 1$$

Substituting for λ_x, λ_y, and λ_z from Eq. (2.23), we obtain

Relationship among direction cosines
$$\cos^2\theta_x + \cos^2\theta_y + \cos^2\theta_z = 1 \qquad \textbf{(2.24)}$$

In Concept Application 2.4, for instance, once the values $\theta_x = 60°$ and $\theta_y = 45°$ have been selected, the value of θ_z *must* be equal to 60° or 120° in order to satisfy the identity in Eq. (2.24).

When the components F_x, F_y, and F_z of a force $\mathbf{F}$ are given, we can obtain the magnitude F of the force from Eq. (2.18). We can then solve relations in Eq. (2.19) for the direction cosines as

$$\cos\theta_x = \frac{F_x}{F} \qquad \cos\theta_y = \frac{F_y}{F} \qquad \cos\theta_z = \frac{F_z}{F} \qquad \textbf{(2.25)}$$

From the direction cosines, we can find the angles θ_x, θ_y, and θ_z characterizing the direction of $\mathbf{F}$.

Concept Application 2.5

A force $\mathbf{F}$ has the components $F_x = 20$ lb, $F_y = -30$ lb, and $F_z = 60$ lb. Determine its magnitude F and the angles θ_x, θ_y, and θ_z it forms with the coordinate axes.

Solution

You can obtain the magnitude of $\mathbf{F}$ from formula (2.18):

$$F = \sqrt{F_x^2 + F_y^2 + F_z^2}$$
$$= \sqrt{(20\ \text{lb})^2 + (-30\ \text{lb})^2 + (60\ \text{lb})^2}$$
$$= \sqrt{4900}\ \text{lb} = 70\ \text{lb}$$

Substituting the values of the components and magnitude of $\mathbf{F}$ into Eqs. (2.25), the direction cosines are

$$\cos\theta_x = \frac{F_x}{F} = \frac{20\ \text{lb}}{70\ \text{lb}} \qquad \cos\theta_y = \frac{F_y}{F} = \frac{-30\ \text{lb}}{70\ \text{lb}} \qquad \cos\theta_z = \frac{F_z}{F} = \frac{60\ \text{lb}}{70\ \text{lb}}$$

Calculating each quotient and its arc cosine gives you

$$\theta_x = 73.4° \qquad \theta_y = 115.4° \qquad \theta_z = 31.0°$$

These computations can be carried out easily with a calculator.

2.4B Force Defined by its Magnitude and Two Points on its Line of Action

In many applications, the direction of a force $\mathbf{F}$ is defined by the coordinates of two points, $M(x_1, y_1, z_1)$ and $N(x_2, y_2, z_2)$, located on its line of action (Fig. 2.29). Consider the vector $\overrightarrow{MN}$ joining M and N and of the same

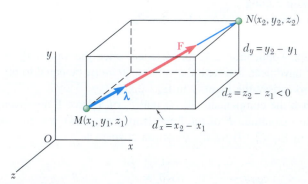

Fig. 2.29 A case where the line of action of force $\mathbf{F}$ is determined by the two points M and N. We can calculate the components of $\mathbf{F}$ and its direction cosines from the vector $\overrightarrow{MN}$.

sense as a force $\mathbf{F}$. Denoting its scalar components by d_x, d_y, and d_z, respectively, we write

$$\overrightarrow{MN} = d_x\mathbf{i} + d_y\mathbf{j} + d_z\mathbf{k} \tag{2.26}$$

We can obtain a unit vector $\boldsymbol{\lambda}$ along the line of action of $\mathbf{F}$ (i.e., along the line MN) by dividing the vector $\overrightarrow{MN}$ by its magnitude MN. Substituting for $\overrightarrow{MN}$ from Eq. (2.26) and observing that MN is equal to the distance d from M to N, we have

$$\boldsymbol{\lambda} = \frac{\overrightarrow{MN}}{MN} = \frac{1}{d}(d_x\mathbf{i} + d_y\mathbf{j} + d_z\mathbf{k}) \tag{2.27}$$

Recalling that $\mathbf{F}$ is equal to the product of F and $\boldsymbol{\lambda}$, we have

$$\mathbf{F} = F\boldsymbol{\lambda} = \frac{F}{d}(d_x\mathbf{i} + d_y\mathbf{j} + d_z\mathbf{k}) \tag{2.28}$$

It follows that the scalar components of $\mathbf{F}$ are, respectively,

Scalar components of force F

$$F_x = \frac{Fd_x}{d} \qquad F_y = \frac{Fd_y}{d} \qquad F_z = \frac{Fd_z}{d} \tag{2.29}$$

The relations in Eq. (2.29) considerably simplify the determination of the components of a force $\mathbf{F}$ of given magnitude F when the line of action of $\mathbf{F}$ is defined by two points M and N. The calculation consists of first subtracting the coordinates of M from those of N, then determining the components of the vector $\overrightarrow{MN}$ and the distance d from M to N. Thus,

$$d_x = x_2 - x_1 \qquad d_y = y_2 - y_1 \qquad d_z = z_2 - z_1$$

$$d = \sqrt{d_x^2 + d_y^2 + d_z^2}$$

Substituting for F and for d_x, d_y, d_z, and d into the relations in Eq. (2.29), we obtain the components F_x, F_y, and F_z of the force.

We can then obtain the angles θ_x, θ_y, and θ_z that $\mathbf{F}$ forms with the coordinate axes from Eqs. (2.25). Comparing Eqs. (2.22) and (2.27), we can write

**Direction cosines
of force F**

$$\cos\theta_x = \frac{d_x}{d} \qquad \cos\theta_y = \frac{d_y}{d} \qquad \cos\theta_z = \frac{d_z}{d} \qquad \text{(2.30)}$$

In other words, we can determine the angles θ_x, θ_y, and θ_z directly from the components and the magnitude of the vector $\overrightarrow{MN}$.

2.4C Addition of Concurrent Forces in Space

We can determine the resultant $\mathbf{R}$ of two or more forces in space by summing their rectangular components. Graphical or trigonometric methods are generally not practical in the case of forces in space.

The method followed here is similar to that used in Sec. 2.2B with coplanar forces. Setting

$$\mathbf{R} = \Sigma\mathbf{F}$$

we resolve each force into its rectangular components:

$$R_x\mathbf{i} + R_y\mathbf{j} + R_z\mathbf{k} = \Sigma \, (F_x\mathbf{i} + F_y\mathbf{j} + F_z\mathbf{k})$$
$$= (\Sigma F_x)\mathbf{i} + (\Sigma F_y)\mathbf{j} + (\Sigma F_z)\mathbf{k}$$

From this equation, it follows that

**Rectangular components
of the resultant**

$$R_x = \Sigma F_x \qquad R_y = \Sigma F_y \qquad R_z = \Sigma F_z \qquad \text{(2.31)}$$

The magnitude of the resultant and the angles θ_x, θ_y, and θ_z that the resultant forms with the coordinate axes are obtained using the method discussed earlier in this section. We end up with

**Resultant of concurrent
forces in space**

$$R = \sqrt{R_x^2 + R_y^2 + R_z^2} \qquad \text{(2.32)}$$

$$\cos\theta_x = \frac{R_x}{R} \qquad \cos\theta_y = \frac{R_y}{R} \qquad \cos\theta_z = \frac{R_z}{R} \qquad \text{(2.33)}$$

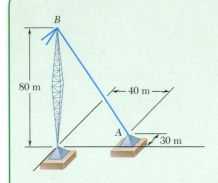

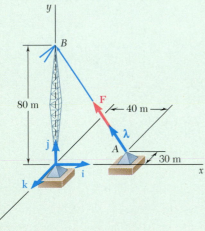

Fig. 1 Cable force acting on bolt at A, and its unit vector.

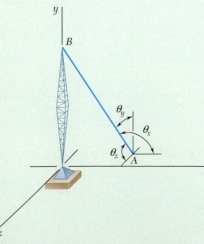

Fig. 2 Direction angles for cable AB.

Sample Problem 2.7

A tower guy wire is anchored by means of a bolt at A. The tension in the wire is 2500 N. Determine (a) the components F_x, F_y, and F_z of the force acting on the bolt and (b) the angles θ_x, θ_y, and θ_z defining the direction of the force.

STRATEGY: From the given distances, we can determine the length of the wire and the direction of a unit vector along it. From that, we can find the components of the tension and the angles defining its direction.

MODELING and ANALYSIS:

a. Components of the Force.   The line of action of the force acting on the bolt passes through points A and B, and the force is directed from A to B. The components of the vector $\overrightarrow{AB}$, which has the same direction as the force, are

$$d_x = -40 \text{ m} \qquad d_y = +80 \text{ m} \qquad d_z = +30 \text{ m}$$

The total distance from A to B is

$$AB = d = \sqrt{d_x^2 + d_y^2 + d_z^2} = 94.3 \text{ m}$$

Denoting the unit vectors along the coordinate axes by $\mathbf{i}$, $\mathbf{j}$, and $\mathbf{k}$, you have

$$\overrightarrow{AB} = -(40 \text{ m})\mathbf{i} + (80 \text{ m})\mathbf{j} + (30 \text{ m})\mathbf{k}$$

Introducing the unit vector $\boldsymbol{\lambda} = \overrightarrow{AB}/AB$ (Fig. 1), you can express $\mathbf{F}$ in terms of $\overrightarrow{AB}$ as

$$\mathbf{F} = F\boldsymbol{\lambda} = F\frac{\overrightarrow{AB}}{AB} = \frac{2500 \text{ N}}{94.3 \text{ m}}\overrightarrow{AB}$$

Substituting the expression for $\overrightarrow{AB}$ gives you

$$\mathbf{F} = \frac{2500 \text{ N}}{94.3 \text{ m}}\left[-(40 \text{ m})\mathbf{i} + (80 \text{ m})\mathbf{j} + (30 \text{ m})\mathbf{k}\right]$$
$$= -(1060 \text{ N})\mathbf{i} + (2120 \text{ N})\mathbf{j} + (795 \text{ N})\mathbf{k}$$

The components of $\mathbf{F}$, therefore, are

$$F_x = -1060 \text{ N} \qquad F_y = +2120 \text{ N} \qquad F_z = +795 \text{ N} \quad \blacktriangleleft$$

b. Direction of the Force.   Using Eqs. (2.25), you can write the direction cosines directly (Fig. 2):

$$\cos\theta_x = \frac{F_x}{F} = \frac{-1060 \text{ N}}{2500 \text{ N}} \qquad \cos\theta_y = \frac{F_y}{F} = \frac{+2120 \text{ N}}{2500 \text{ N}}$$

$$\cos\theta_z = \frac{F_z}{F} = \frac{+795 \text{ N}}{2500 \text{ N}}$$

Calculating each quotient and its arc cosine, you obtain

$$\theta_x = 115.1° \qquad \theta_y = 32.0° \qquad \theta_z = 71.5° \quad \blacktriangleleft$$

(*Note.* You could have obtained this same result by using the components and magnitude of the vector $\overrightarrow{AB}$ rather than those of the force **F**.)

REFLECT and THINK: It makes sense that, for a given geometry, only a certain set of components and angles characterize a given resultant force. The methods in this section allow you to translate back and forth between forces and geometry.

Sample Problem 2.8

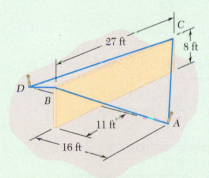

A wall section of precast concrete is temporarily held in place by the cables shown. If the tension is 840 lb in cable *AB* and 1200 lb in cable *AC*, determine the magnitude and direction of the resultant of the forces exerted by cables *AB* and *AC* on stake *A*.

STRATEGY: This is a problem in adding concurrent forces in space. The simplest approach is to first resolve the forces into components and to then sum the components and find the resultant.

MODELING and ANALYSIS:

Components of the Forces. First resolve the force exerted by each cable on stake *A* into *x*, *y*, and *z* components. To do this, determine the components and magnitude of the vectors $\overrightarrow{AB}$ and $\overrightarrow{AC}$, measuring them from *A* toward the wall section (Fig. 1). Denoting the unit vectors along the coordinate axes by **i**, **j**, **k**, these vectors are

$$\overrightarrow{AB} = -(16 \text{ ft})\mathbf{i} + (8 \text{ ft})\mathbf{j} + (11 \text{ ft})\mathbf{k} \qquad AB = 21 \text{ ft}$$
$$\overrightarrow{AC} = -(16 \text{ ft})\mathbf{i} + (8 \text{ ft})\mathbf{j} - (16 \text{ ft})\mathbf{k} \qquad AC = 24 \text{ ft}$$

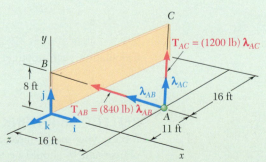

Fig. 1 Cable forces acting on stake at *A*, and their unit vectors.

Denoting by $\boldsymbol{\lambda}_{AB}$ the unit vector along AB, the tension in AB is

$$\mathbf{T}_{AB} = T_{AB}\boldsymbol{\lambda}_{AB} = T_{AB}\frac{\overrightarrow{AB}}{AB} = \frac{840 \text{ lb}}{21 \text{ ft}}\overrightarrow{AB}$$

Substituting the expression found for $\overrightarrow{AB}$, the tension becomes

$$\mathbf{T}_{AB} = \frac{840 \text{ lb}}{21 \text{ ft}}\left[-(16 \text{ ft})\mathbf{i} + (8 \text{ ft})\mathbf{j} + (11 \text{ ft})\mathbf{k}\right]$$

$$\mathbf{T}_{AB} = -(640 \text{ lb})\mathbf{i} + (320 \text{ lb})\mathbf{j} + (440 \text{ lb})\mathbf{k}$$

Similarly, denoting by $\boldsymbol{\lambda}_{AC}$ the unit vector along AC, the tension in AC is

$$\mathbf{T}_{AC} = T_{AC}\boldsymbol{\lambda}_{AC} = T_{AC}\frac{\overrightarrow{AC}}{AC} = \frac{1200 \text{ lb}}{24 \text{ ft}}\overrightarrow{AC}$$

$$\mathbf{T}_{AC} = -(800 \text{ lb})\mathbf{i} + (400 \text{ lb})\mathbf{j} - (800 \text{ lb})\mathbf{k}$$

Resultant of the Forces.   The resultant $\mathbf{R}$ of the forces exerted by the two cables is

$$\mathbf{R} = \mathbf{T}_{AB} + \mathbf{T}_{AC} = -(1440 \text{ lb})\mathbf{i} + (720 \text{ lb})\mathbf{j} - (360 \text{ lb})\mathbf{k}$$

You can now determine the magnitude and direction of the resultant as

$$R = \sqrt{R_x^2 + R_y^2 + R_z^2} = \sqrt{(-1440)^2 + (720)^2 + (-300)^2}$$

$$R = 1650 \text{ lb} \quad \blacktriangleleft$$

The direction cosines come from Eqs. (2.33):

$$\cos\theta_x = \frac{R_x}{R} = \frac{-1440 \text{ lb}}{1650 \text{ lb}} \qquad \cos\theta_y = \frac{R_y}{R} = \frac{+720 \text{ lb}}{1650 \text{ lb}}$$

$$\cos\theta_z = \frac{R_z}{R} = \frac{-360 \text{ lb}}{1650 \text{ lb}}$$

Calculating each quotient and its arc cosine, the angles are

$$\theta_x = 150.8° \qquad \theta_y = 64.1° \qquad \theta_z = 102.6° \quad \blacktriangleleft$$

REFLECT and THINK: Based on visual examination of the cable forces, you might have anticipated that θ_x for the resultant should be obtuse and θ_y should be acute. The outcome of θ_z was not as apparent.

SOLVING PROBLEMS ON YOUR OWN

In this section, we saw that we can define a **force in space** by its magnitude and direction or by the three rectangular components F_x, F_y, and F_z.

A. When a force is defined by its magnitude and direction, you can find its rectangular components F_x, F_y, and F_z as follows.

Case 1. If the direction of the force **F** is defined by the angles θ_y and ϕ shown in Fig. 2.25, projections of **F** through these angles or their complements will yield the components of **F** [Eqs. (2.17)]. Note that to find the x and z components of **F**, first project **F** onto the horizontal plane; the projection $\mathbf{F}_h$ obtained in this way is then resolved into the components $\mathbf{F}_x$ and $\mathbf{F}_z$ (Fig. 2.25c).

Case 2. If the direction of the force **F** is defined by the angles θ_x, θ_y, and θ_z that **F** forms with the coordinate axes, you can obtain each component by multiplying the magnitude F of the force by the cosine of the corresponding angle [Concept Application 2.4]:

$$F_x = F \cos \theta_x \qquad F_y = F \cos \theta_y \qquad F_z = F \cos \theta_z$$

Case 3. If the direction of the force **F** is defined by two points M and N located on its line of action (Fig. 2.29), first express the vector $\overrightarrow{MN}$ drawn from M to N in terms of its components d_x, d_y, and d_z and the unit vectors **i**, **j**, and **k**:

$$\overrightarrow{MN} = d_x\mathbf{i} + d_y\mathbf{j} + d_z\mathbf{k}$$

Then determine the unit vector $\boldsymbol{\lambda}$ along the line of action of **F** by dividing the vector $\overrightarrow{MN}$ by its magnitude MN. Multiplying $\boldsymbol{\lambda}$ by the magnitude of **F** gives you the desired expression for **F** in terms of its rectangular components [Sample Prob. 2.7]:

$$\mathbf{F} = F\boldsymbol{\lambda} = \frac{F}{d}(d_x\mathbf{i} + d_y\mathbf{j} + d_z\mathbf{k})$$

It is helpful to use a consistent and meaningful system of notation when determining the rectangular components of a force. The method used in this text is illustrated in Sample Prob. 2.8, where the force $\mathbf{T}_{AB}$ acts from stake A toward point B. Note that the subscripts have been ordered to agree with the direction of the force. We recommend that you adopt the same notation, as it will help you identify point 1 (the first subscript) and point 2 (the second subscript).

When calculating the vector defining the line of action of a force, you might think of its scalar components as the number of steps you must take in each coordinate direction to go from point 1 to point 2. It is essential that you always remember to assign the correct sign to each of the components.

(continued)

B. When a force is defined by its rectangular components F_x, F_y, and F_z, you can obtain its magnitude F from

$$F = \sqrt{F_x^2 + F_y^2 + F_z^2}$$

You can determine the direction cosines of the line of action of **F** by dividing the components of the force by F:

$$\cos \theta_x = \frac{F_x}{F} \quad \cos \theta_y = \frac{F_y}{F} \quad \cos \theta_z = \frac{F_z}{F}$$

From the direction cosines, you can obtain the angles θ_x, θ_y, and θ_z that **F** forms with the coordinate axes [Concept Application 2.5].

C. To determine the resultant R of two or more forces in three-dimensional space, first determine the rectangular components of each force by one of the procedures described previously. Adding these components will yield the components R_x, R_y, and R_z of the resultant. You can then obtain the magnitude and direction of the resultant as indicated previously for a force **F** [Sample Prob. 2.8].

Problems

END-OF-SECTION PROBLEMS

2.71 Determine (a) the x, y, and z components of the 600-N force, (b) the angles θ_x, θ_y, and θ_z that the force forms with the coordinate axes.

2.72 Determine (a) the x, y, and z components of the 450-N force, (b) the angles θ_x, θ_y, and θ_z that the force forms with the coordinate axes.

2.73 A gun is aimed at a point A located 35° east of north. Knowing that the barrel of the gun forms an angle of 40° with the horizontal and that the maximum recoil force is 400 N, determine (a) the x, y, and z components of that force, (b) the values of the angles θ_x, θ_y, and θ_z defining the direction of the recoil force. (Assume that the x, y, and z axes are directed, respectively, east, up, and south.)

2.74 Solve Prob. 2.73 assuming that point A is located 15° north of west and that the barrel of the gun forms an angle of 25° with the horizontal.

2.75 The angle between spring AB and the post DA is 30°. Knowing that the tension in the spring is 50 lb, determine (a) the x, y, and z components of the force exerted on the circular plate at B, (b) the angles θ_x, θ_y, and θ_z defining the direction of the force at B.

2.76 The angle between spring AC and the post DA is 30°. Knowing that the tension in the spring is 40 lb, determine (a) the x, y, and z components of the force exerted on the circular plate at C, (b) the angles θ_x, θ_y, and θ_z defining the direction of the force at C.

2.77 Cable AB is 65 ft long, and the tension in that cable is 3900 lb. Determine (a) the x, y, and z components of the force exerted by the cable on the anchor B, (b) the angles θ_x, θ_y, and θ_z defining the direction of that force.

2.78 Cable AC is 70 ft long, and the tension in that cable is 5250 lb. Determine (a) the x, y, and z components of the force exerted by the cable on the anchor C, (b) the angles θ_x, θ_y, and θ_z defining the direction of that force.

2.79 Determine the magnitude and direction of the force $\mathbf{F} = (240 \text{ N})\mathbf{i} - (270 \text{ N})\mathbf{j} + (680 \text{ N})\mathbf{k}$.

2.80 Determine the magnitude and direction of the force $\mathbf{F} = (320 \text{ N})\mathbf{i} + (400 \text{ N})\mathbf{j} - (250 \text{ N})\mathbf{k}$.

2.81 A force acts at the origin of a coordinate system in a direction defined by the angles $\theta_x = 69.3°$ and $\theta_z = 57.9°$. Knowing that the y component of the force is −174.0 lb, determine (a) the angle θ_y, (b) the other components and the magnitude of the force.

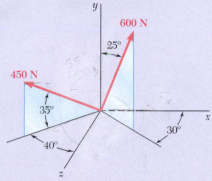

Fig. P2.71 and P2.72

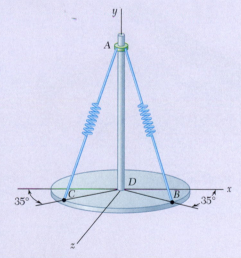

Fig. P2.75 and P2.76

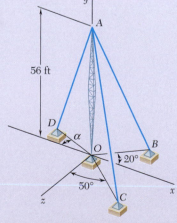

Fig. P2.77 and P2.78

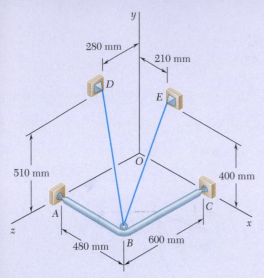

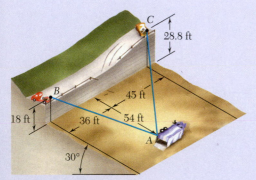

Fig. P2.85

Fig. P2.87 and P2.88

2.82 A force acts at the origin of a coordinate system in a direction defined by the angles $\theta_x = 70.9°$ and $\theta_y = 144.9°$. Knowing that the z component of the force is -52.0 lb, determine (*a*) the angle θ_z, (*b*) the other components and the magnitude of the force.

2.83 A force **F** of magnitude 210 N acts at the origin of a coordinate system. Knowing that $F_x = 80$ N, $\theta_z = 151.2°$, and $F_y < 0$, determine (*a*) the components F_y and F_z, (*b*) the angles θ_x and θ_y.

2.84 A force **F** of magnitude 1200 N acts at the origin of a coordinate system. Knowing that $\theta_x = 65°$, $\theta_y = 40°$, and $F_z > 0$, determine (*a*) the components of the force, (*b*) the angle θ_z.

2.85 A frame *ABC* is supported in part by cable *DBE* that passes through a frictionless ring at *B*. Knowing that the tension in the cable is 385 N, determine the components of the force exerted by the cable on the support at *D*.

2.86 For the frame and cable of Prob. 2.85, determine the components of the force exerted by the cable on the support at *E*.

2.87 In order to move a wrecked truck, two cables are attached at *A* and pulled by winches *B* and *C* as shown. Knowing that the tension in cable *AB* is 2 kips, determine the components of the force exerted at *A* by the cable.

2.88 In order to move a wrecked truck, two cables are attached at *A* and pulled by winches *B* and *C* as shown. Knowing that the tension in cable *AC* is 1.5 kips, determine the components of the force exerted at *A* by the cable.

2.89 A rectangular plate is supported by three cables as shown. Knowing that the tension in cable *AB* is 408 N, determine the components of the force exerted on the plate at *B*.

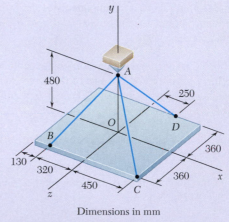

Fig. P2.89 and *P2.90*

2.90 A rectangular plate is supported by three cables as shown. Knowing that the tension in cable *AD* is 429 N, determine the components of the force exerted on the plate at *D*.

2.91 Find the magnitude and direction of the resultant of the two forces shown knowing that $P = 300$ N and $Q = 400$ N.

2.92 Find the magnitude and direction of the resultant of the two forces shown knowing that $P = 400$ N and $Q = 300$ N.

2.93 Knowing that the tension is 425 lb in cable AB and 510 lb in cable AC, determine the magnitude and direction of the resultant of the forces exerted at A by the two cables.

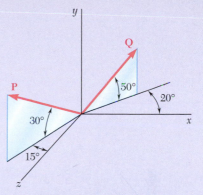

Fig. P2.91 and P2.92

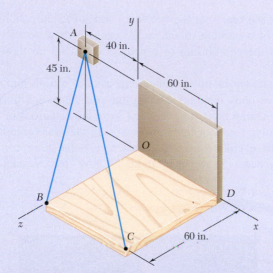

Fig. *P2.93* and P2.94

2.94 Knowing that the tension is 510 lb in cable AB and 425 lb in cable AC, determine the magnitude and direction of the resultant of the forces exerted at A by the two cables.

2.95 For the frame of Prob. 2.85, determine the magnitude and direction of the resultant of the forces exerted by the cable at B knowing that the tension in the cable is 385 N.

2.96 For the plate of Prob. 2.89, determine the tensions in cables AB and AD knowing that the tension in cable AC is 54 N and that the resultant of the forces exerted by the three cables at A must be vertical.

2.97 The boom OA carries a load P and is supported by two cables as shown. Knowing that the tension in cable AB is 183 lb and that the resultant of the load P and of the forces exerted at A by the two cables must be directed along OA, determine the tension in cable AC.

2.98 For the boom and loading of Prob. 2.97, determine the magnitude of the load P.

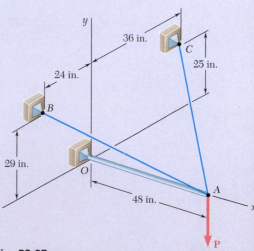

Fig. P2.97

2.5   FORCES AND EQUILIBRIUM IN SPACE

According to the definition given in Sec. 2.3, a particle A is in equilibrium if the resultant of all the forces acting on A is zero. The components R_x, R_y, and R_z of the resultant of forces in space are given by equations (2.31); when the components of the resultant are zero, we have

$$\Sigma F_x = 0 \qquad \Sigma F_y = 0 \qquad \Sigma F_z = 0 \qquad \textbf{(2.34)}$$

Equations (2.34) represent the necessary and sufficient conditions for the equilibrium of a particle in space. We can use them to solve problems dealing with the equilibrium of a particle involving no more than three unknowns.

The first step in solving three-dimensional equilibrium problems is to draw a free-body diagram showing the particle in equilibrium and *all* of the forces acting on it. You can then write the equations of equilibrium (2.34) and solve them for three unknowns. In the more common types of problems, these unknowns will represent (1) the three components of a single force or (2) the magnitude of three forces, each of known direction.

Photo 2.4   Although we cannot determine the tension in the four cables supporting the car by using the three equations (2.34), we can obtain a relation among the tensions by analyzing the equilibrium of the hook.

Sample Problem 2.9

A 200-kg cylinder is hung by means of two cables AB and AC that are attached to the top of a vertical wall. A horizontal force $\mathbf{P}$ perpendicular to the wall holds the cylinder in the position shown. Determine the magnitude of $\mathbf{P}$ and the tension in each cable.

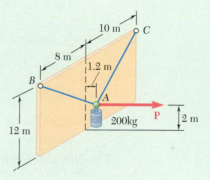

STRATEGY: Connection point A is acted upon by four forces, including the weight of the cylinder. You can use the given geometry to express the force components of the cables and then apply equilibrium conditions to calculate the tensions.

MODELING and ANALYSIS:

Free-Body Diagram.   Choose point A as a free body; this point is subjected to four forces, three of which are of unknown magnitude. Introducing the unit vectors $\mathbf{i}$, $\mathbf{j}$, and $\mathbf{k}$, resolve each force into rectangular components (Fig. 1):

$$\mathbf{P} = P\mathbf{i}$$

$$\mathbf{W} = -mg\mathbf{j} = -(200 \text{ kg})(9.81 \text{ m/s}^2)\mathbf{j} = -(1962 \text{ N})\mathbf{j} \qquad \textbf{(1)}$$

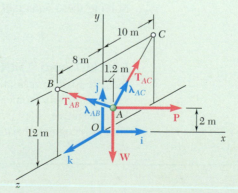

Fig. 1 Free-body diagram of particle A.

For $\mathbf{T}_{AB}$ and $\mathbf{T}_{AC}$, it is first necessary to determine the components and magnitudes of the vectors $\overrightarrow{AB}$ and $\overrightarrow{AC}$. Denoting the unit vector along AB by $\boldsymbol{\lambda}_{AB}$, you can write $\mathbf{T}_{AB}$ as

$$\overrightarrow{AB} = -(1.2 \text{ m})\mathbf{i} + (10 \text{ m})\mathbf{j} + (8 \text{ m})\mathbf{k} \qquad AB = 12.862 \text{ m}$$

$$\boldsymbol{\lambda}_{AB} = \frac{\overrightarrow{AB}}{12.862 \text{ m}} = -0.09330\mathbf{i} + 0.7775\mathbf{j} + 0.6220\mathbf{k}$$

$$\mathbf{T}_{AB} = T_{AB}\boldsymbol{\lambda}_{AB} = -0.09330T_{AB}\mathbf{i} + 0.7775T_{AB}\mathbf{j} + 0.6220T_{AB}\mathbf{k} \qquad \textbf{(2)}$$

Similarly, denoting the unit vector along AC by $\boldsymbol{\lambda}_{AC}$, you have for $\mathbf{T}_{AC}$

$$\overrightarrow{AC} = -(1.2 \text{ m})\mathbf{i} + (10 \text{ m})\mathbf{j} - (10 \text{ m})\mathbf{k} \qquad AC = 14.193 \text{ m}$$

$$\boldsymbol{\lambda}_{AC} = \frac{\overrightarrow{AC}}{14.193 \text{ m}} = -0.08455\mathbf{i} + 0.7046\mathbf{j} - 0.7046\mathbf{k}$$

$$\Gamma_{AC} = T_{AC}\boldsymbol{\lambda}_{AC} = -0.08455T_{AC}\mathbf{i} + 0.7046T_{AC}\mathbf{j} - 0.7046T_{AC}\mathbf{k} \qquad \textbf{(3)}$$

Equilibrium Condition. Since A is in equilibrium, you must have

$$\Sigma\mathbf{F} = 0: \qquad \mathbf{T}_{AB} + \mathbf{T}_{AC} + \mathbf{P} + \mathbf{W} = 0$$

or substituting from Eqs. (1), (2), and (3) for the forces and factoring $\mathbf{i}$, $\mathbf{j}$, and $\mathbf{k}$, you have

$$(-0.09330T_{AB} - 0.08455T_{AC} + P)\mathbf{i}$$
$$+ (0.7775T_{AB} + 0.7046T_{AC} - 1962 \text{ N})\mathbf{j}$$
$$+ (0.6220T_{AB} - 0.7046T_{AC})\mathbf{k} = 0$$

Setting the coefficients of $\mathbf{i}$, $\mathbf{j}$, and $\mathbf{k}$ equal to zero, you can write three scalar equations, which express that the sums of the x, y, and z components of the forces are respectively equal to zero.

$$(\Sigma F_x = 0:) \qquad -0.09330T_{AB} - 0.08455T_{AC} + P = 0$$
$$(\Sigma F_y = 0:) \qquad +0.7775T_{AB} + 0.7046T_{AC} - 1962 \text{ N} = 0$$
$$(\Sigma F_z = 0:) \qquad +0.6220T_{AB} - 0.7046T_{AC} = 0$$

Solving these equations, you obtain

$$P = 235 \text{ N} \qquad T_{AB} = 1402 \text{ N} \qquad T_{AC} = 1238 \text{ N} \quad \blacktriangleleft$$

REFLECT and THINK: The solution of the three unknown forces yielded positive results, which is completely consistent with the physical situation of this problem. Conversely, if one of the cable force results had been negative, thereby reflecting compression instead of tension, you should recognize that the solution is in error.

SOLVING PROBLEMS ON YOUR OWN

We saw earlier that when a particle is in **equilibrium**, the resultant of the forces acting on the particle must be zero. In the case of the equilibrium of a particle in three-dimensional space, this equilibrium condition provides you with three relations among the forces acting on the particle. These relations may be used to determine three unknowns—usually the magnitudes of three forces.

The solution usually consists of the following steps:

1. Draw a free-body diagram of the particle. This diagram shows the particle and all the forces acting on it. Indicate on the diagram the magnitudes of known forces, as well as any angles or dimensions that define the direction of a force. Any unknown magnitude or angle should be denoted by an appropriate symbol. Nothing else should be included in the free-body diagram.

2. Resolve each force into rectangular components. Following the method used earlier, determine for each force $\mathbf{F}$ the unit vector $\boldsymbol{\lambda}$ defining the direction of that force, and express $\mathbf{F}$ as the product of its magnitude F and $\boldsymbol{\lambda}$. You will obtain an expression of the form

$$\mathbf{F} = F\boldsymbol{\lambda} = \frac{F}{d}(d_x\mathbf{i} + d_y\mathbf{j} + d_z\mathbf{k})$$

where d, d_x, d_y, and d_z are dimensions obtained from the free-body diagram of the particle. If you know the magnitude as well as the direction of the force, then F is known and the expression obtained for $\mathbf{F}$ is well defined; otherwise F is one of the three unknowns that should be determined.

3. Set the resultant, or sum, of the forces exerted on the particle equal to zero. You will obtain a vector equation consisting of terms containing the unit vectors $\mathbf{i}$, $\mathbf{j}$, or $\mathbf{k}$. Group the terms containing the same unit vector and factor that vector. For the vector equation to be satisfied, you must set the coefficient of each of the unit vectors equal to zero. This yields three scalar equations that you can solve for no more than three unknowns [Sample Prob. 2.9].

Problems

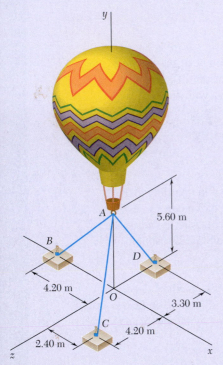

Fig. P2.F5

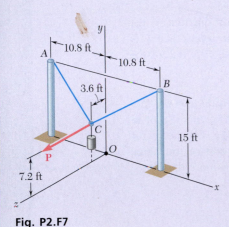

Fig. P2.F7

2.F5 Three cables are used to tether a balloon as shown. Knowing that the tension in cable AC is 444 N, draw the free-body diagram needed to determine the vertical force **P** exerted by the balloon at A.

2.F6 A container of mass $m = 120$ kg is supported by three cables as shown. Draw the free-body diagram needed to determine the tension in each cable.

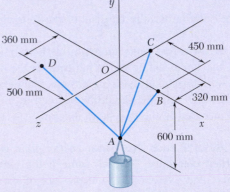

Fig. P2.F6

2.F7 A 150-lb cylinder is supported by two cables AC and BC that are attached to the top of vertical posts. A horizontal force **P**, which is perpendicular to the plane containing the posts, holds the cylinder in the position shown. Draw the free-body diagram needed to determine the magnitude of **P** and the force in each cable.

2.F8 A transmission tower is held by three guy wires attached to a pin at A and anchored by bolts at B, C, and D. Knowing that the tension in wire AB is 630 lb, draw the free-body diagram needed to determine the vertical force **P** exerted by the tower on the pin at A.

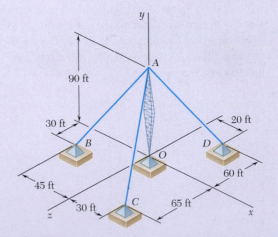

Fig. P2.F8

2.99 A container is supported by three cables that are attached to a ceiling as shown. Determine the weight W of the container knowing that the tension in cable AB is 6 kN.

2.100 A container is supported by three cables that are attached to a ceiling as shown. Determine the weight W of the container, knowing that the tension in cable AD is 4.3 kN.

2.101 Three cables are used to tether a balloon as shown. Determine the vertical force **P** exerted by the balloon at A knowing that the tension in cable AD is 481 N.

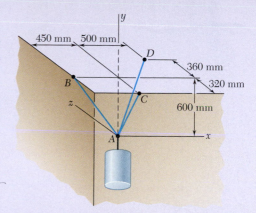

Fig. P2.99 and *P2.100*

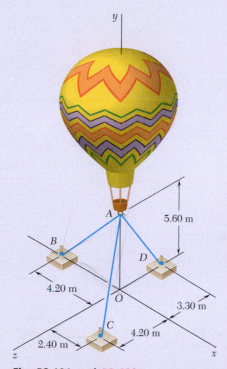

Fig. P2.101 and *P2.102*

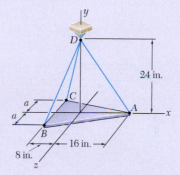

Fig. P2.103

2.102 Three cables are used to tether a balloon as shown. Knowing that the balloon exerts an 800-N vertical force at A, determine the tension in each cable.

2.103 A 36-lb triangular plate is supported by three wires as shown. Determine the tension in each wire, knowing that $a = 6$ in.

2.104 Solve Prob. 2.103, assuming that $a = 8$ in.

2.105 A crate is supported by three cables as shown. Determine the weight of the crate knowing that the tension in cable AC is 544 lb.

2.106 A 1600-lb crate is supported by three cables as shown. Determine the tension in each cable.

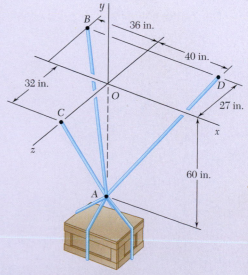

Fig. *P2.105* and P2.106

2.107 Three cables are connected at A, where the forces **P** and **Q** are applied as shown. Knowing that $Q = 0$, find the value of P for which the tension in cable AD is 305 N.

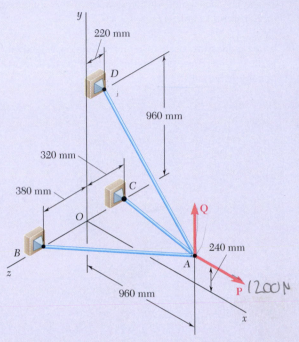

Fig. P2.107 and P2.108

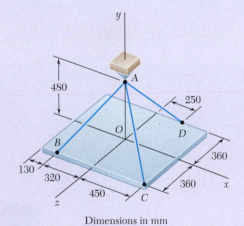

Dimensions in mm

Fig. P2.109 and P2.110

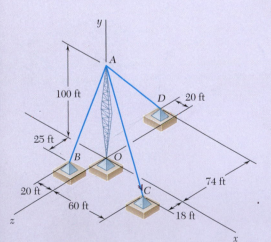

Fig. P2.111 and P2.112

2.108 Three cables are connected at A, where the forces **P** and **Q** are applied as shown. Knowing that $P = 1200$ N, determine the values of Q for which cable AD is taut.

2.109 A rectangular plate is supported by three cables as shown. Knowing that the tension in cable AC is 60 N, determine the weight of the plate.

2.110 A rectangular plate is supported by three cables as shown. Knowing that the tension in cable AD is 520 N, determine the weight of the plate.

2.111 A transmission tower is held by three guy wires attached to a pin at A and anchored by bolts at B, C, and D. If the tension in wire AB is 840 lb, determine the vertical force **P** exerted by the tower on the pin at A.

2.112 A transmission tower is held by three guy wires attached to a pin at A and anchored by bolts at B, C, and D. If the tension in wire AC is 590 lb, determine the vertical force **P** exerted by the tower on the pin at A.

2.113 In trying to move across a slippery icy surface, a 175-lb man uses two ropes AB and AC. Knowing that the force exerted on the man by the icy surface is perpendicular to that surface, determine the tension in each rope.

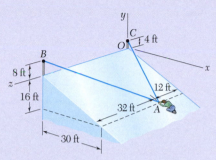

Fig. P2.113

2.114 Solve Prob. 2.113 assuming that a friend is helping the man at A by pulling on him with a force $\mathbf{P} = -(45 \text{ lb})\mathbf{k}$.

2.115 For the rectangular plate of Probs. 2.109 and 2.110, determine the tension in each of the three cables knowing that the weight of the plate is 792 N.

2.116 For the cable system of Probs. 2.107 and 2.108, determine the tension in each cable knowing that $P = 2880$ N and $Q = 0$.

2.117 For the cable system of Probs. 2.107 and 2.108, determine the tension in each cable knowing that $P = 2880$ N and $Q = 576$ N.

2.118 For the cable system of Probs. 2.107 and 2.108, determine the tension in each cable knowing that $P = 2880$ N and $Q = -576$ N ($\mathbf{Q}$ is directed downward).

2.119 For the transmission tower of Probs. 2.111 and 2.112, determine the tension in each guy wire knowing that the tower exerts on the pin at A an upward vertical force of 1800 lb.

2.120 Three wires are connected at point D, which is located 18 in. below the T-shaped pipe support ABC. Determine the tension in each wire when a 180-lb cylinder is suspended from point D as shown.

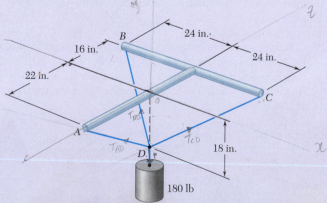

Fig. P2.120

2.121 A container of weight W is suspended from ring A, to which cables AC and AE are attached. A force $\mathbf{P}$ is applied to the end F of a third cable that passes over a pulley at B and through ring A and that is attached to a support at D. Knowing that $W = 1000$ N, determine the magnitude of $\mathbf{P}$. (*Hint:* The tension is the same in all portions of cable *FBAD*.)

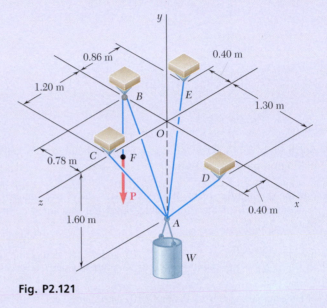

Fig. P2.121

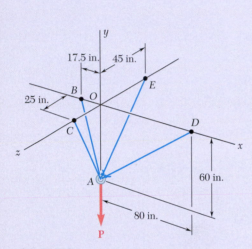

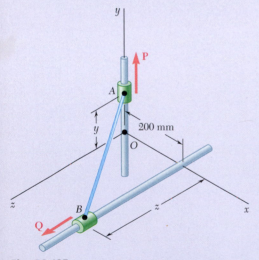

Fig. P2.123

Fig. P2.125

2.122 Knowing that the tension in cable AC of the system described in Prob. 2.121 is 150 N, determine (*a*) the magnitude of the force $\mathbf{P}$, (*b*) the weight W of the container.

2.123 Cable BAC passes through a frictionless ring A and is attached to fixed supports at B and C, while cables AD and AE are both tied to the ring and are attached, respectively, to supports at D and E. Knowing that a 200-lb vertical load $\mathbf{P}$ is applied to ring A, determine the tension in each of the three cables.

2.124 Knowing that the tension in cable AE of Prob. 2.123 is 75 lb, determine (*a*) the magnitude of the load $\mathbf{P}$, (*b*) the tension in cables BAC and AD.

2.125 Collars A and B are connected by a 525-mm-long wire and can slide freely on frictionless rods. If a force $\mathbf{P} = (341$ N$)\mathbf{j}$ is applied to collar A, determine (*a*) the tension in the wire when $y = 155$ mm, (*b*) the magnitude of the force $\mathbf{Q}$ required to maintain the equilibrium of the system.

2.126 Solve Prob. 2.125 assuming that $y = 275$ mm.

Review and Summary

In this chapter, we have studied the effect of forces on particles, i.e., on bodies of such shape and size that we may assume all forces acting on them apply at the same point.

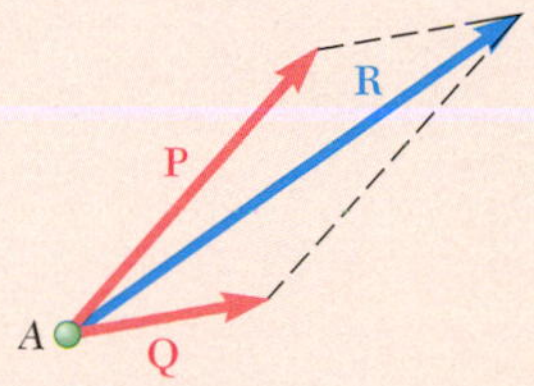

Resultant of Two Forces

Forces are *vector quantities;* they are characterized by a point of application, a magnitude, and a direction, and they add according to the parallelogram law (Fig. 2.30). We can determine the magnitude and direction of the resultant **R** of two forces **P** and **Q** either graphically or by trigonometry using the law of cosines and the law of sines [Sample Prob. 2.1].

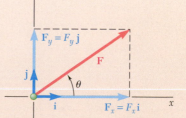

Fig. 2.30

Components of a Force

Any given force acting on a particle can be resolved into two or more components, i.e., it can be replaced by two or more forces that have the same effect on the particle. A force **F** can be resolved into two components **P** and **Q** by drawing a parallelogram with **F** for its diagonal; the components **P** and **Q** are then represented by the two adjacent sides of the parallelogram (Fig. 2.31). Again, we can determine the components either graphically or by trigonometry [Sec. 2.1E].

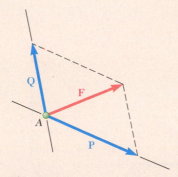

Fig. 2.31

Rectangular Components; Unit Vectors

A force **F** is resolved into two rectangular components if its components $\mathbf{F}_x$ and $\mathbf{F}_y$ are perpendicular to each other and are directed along the coordinate axes (Fig. 2.32). Introducing the unit vectors **i** and **j** along the x and y axes, respectively, we can write the components and the vector as [Sec. 2.2A]

$$\mathbf{F}_x = F_x\mathbf{i} \qquad \mathbf{F}_y = F_y\mathbf{j} \tag{2.6}$$

and

$$\mathbf{F} = F_x\mathbf{i} + F_y\mathbf{j} \tag{2.7}$$

where F_x and F_y are the *scalar components* of **F**. These components, which can be positive or negative, are defined by the relations

$$F_x = F\cos\theta \qquad F_y = F\sin\theta \tag{2.8}$$

Fig. 2.32

When the rectangular components F_x and F_y of a force $\mathbf{F}$ are given, we can obtain the angle θ defining the direction of the force from

$$\tan \theta = \frac{F_y}{F_x} \qquad (2.9)$$

We can obtain the magnitude F of the force by solving one of the equations (2.8) for F or by applying the Pythagorean theorem:

$$F = \sqrt{F_x^2 + F_y^2} \qquad (2.10)$$

Resultant of Several Coplanar Forces

When three or more coplanar forces act on a particle, we can obtain the rectangular components of their resultant $\mathbf{R}$ by adding the corresponding components of the given forces algebraically [Sec. 2.2B]:

$$R_x = \Sigma F_x \qquad R_y = \Sigma F_y \qquad (2.13)$$

The magnitude and direction of $\mathbf{R}$ then can be determined from relations similar to Eqs. (2.9) and (2.10) [Sample Prob. 2.3].

Forces in Space

A force $\mathbf{F}$ in three-dimensional space can be resolved into rectangular components $\mathbf{F}_x$, $\mathbf{F}_y$, and $\mathbf{F}_z$ [Sec. 2.4A]. Denoting by θ_x, θ_y, and θ_z, respectively, the angles that $\mathbf{F}$ forms with the x, y, and z axes (Fig. 2.33), we have

$$F_x = F \cos \theta_x \qquad F_y = F \cos \theta_y \qquad F_z = F \cos \theta_z \qquad (2.19)$$

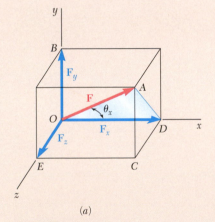 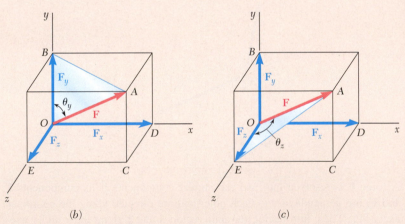

(a)   (b)   (c)

Fig. 2.33

Direction Cosines

The cosines of θ_x, θ_y, and θ_z are known as the *direction cosines* of the force $\mathbf{F}$. Introducing the unit vectors $\mathbf{i}$, $\mathbf{j}$, and $\mathbf{k}$ along the coordinate axes, we can write $\mathbf{F}$ as

$$\mathbf{F} = F_x \mathbf{i} + F_y \mathbf{j} + F_z \mathbf{k} \qquad (2.20)$$

or

$$\mathbf{F} = F(\cos \theta_x \mathbf{i} + \cos \theta_y \mathbf{j} + \cos \theta_z \mathbf{k}) \qquad (2.21)$$

This last equation shows (Fig. 2.34) that **F** is the product of its magnitude F and the unit vector expressed by

$$\boldsymbol{\lambda} = \cos\theta_x\mathbf{i} + \cos\theta_y\mathbf{j} + \cos\theta_z\mathbf{k}$$

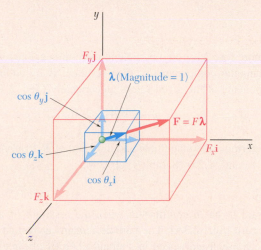

Fig. 2.34

Since the magnitude of **λ** is equal to unity, we must have

$$\cos^2\theta_x + \cos^2\theta_y + \cos^2\theta_z = 1 \qquad (2.24)$$

When we are given the rectangular components F_x, F_y, and F_z of a force **F**, we can find the magnitude F of the force by

$$F = \sqrt{F_x^2 + F_y^2 + F_z^2} \qquad (2.18)$$

and the direction cosines of **F** are obtained from Eqs. (2.19). We have

$$\cos\theta_x = \frac{F_x}{F} \qquad \cos\theta_y = \frac{F_y}{F} \qquad \cos\theta_z = \frac{F_z}{F} \qquad (2.25)$$

When a force **F** is defined in three-dimensional space by its magnitude F and two points M and N on its line of action [Sec. 2.4B], we can obtain its rectangular components by first expressing the vector $\overrightarrow{MN}$ joining points M and N in terms of its components d_x, d_y, and d_z (Fig. 2.35):

$$\overrightarrow{MN} = d_x\mathbf{i} + d_y\mathbf{j} + d_z\mathbf{k} \qquad (2.26)$$

We next determine the unit vector **λ** along the line of action of **F** by dividing $\overrightarrow{MN}$ by its magnitude $MN = d$:

$$\boldsymbol{\lambda} = \frac{\overrightarrow{MN}}{MN} = \frac{1}{d}(d_x\mathbf{i} + d_y\mathbf{j} + d_z\mathbf{k}) \qquad (2.27)$$

Recalling that **F** is equal to the product of F and **λ**, we have

$$\mathbf{F} = F\boldsymbol{\lambda} = \frac{F}{d}(d_x\mathbf{i} + d_y\mathbf{j} + d_z\mathbf{k}) \qquad (2.28)$$

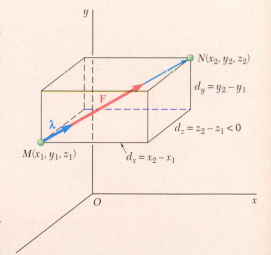

Fig. 2.35

From this equation it follows [Sample Probs. 2.7 and 2.8] that the scalar components of $\mathbf{F}$ are, respectively,

$$F_x = \frac{Fd_x}{d} \qquad F_y = \frac{Fd_y}{d} \qquad F_z = \frac{Fd_z}{d} \qquad \textbf{(2.29)}$$

Resultant of Forces in Space

When two or more forces act on a particle in three-dimensional space, we can obtain the rectangular components of their resultant $\mathbf{R}$ by adding the corresponding components of the given forces algebraically [Sec. 2.4C]. We have

$$R_x = \Sigma F_x \qquad R_y = \Sigma F_y \qquad R_z = \Sigma F_z \qquad \textbf{(2.31)}$$

We can then determine the magnitude and direction of $\mathbf{R}$ from relations similar to Eqs. (2.18) and (2.25) [Sample Prob. 2.8].

Equilibrium of a Particle

A particle is said to be in equilibrium when the resultant of all the forces acting on it is zero [Sec. 2.3A]. The particle remains at rest (if originally at rest) or moves with constant speed in a straight line (if originally in motion) [Sec. 2.3B].

Free-Body Diagram

To solve a problem involving a particle in equilibrium, first draw a free-body diagram of the particle showing all of the forces acting on it [Sec. 2.3C]. If only three coplanar forces act on the particle, you can draw a force triangle to express that the particle is in equilibrium. Using graphical methods of trigonometry, you can solve this triangle for no more than two unknowns [Sample Prob. 2.4]. If more than three coplanar forces are involved, you should use the equations of equilibrium:

$$\Sigma F_x = 0 \qquad \Sigma F_y = 0 \qquad \textbf{(2.15)}$$

These equations can be solved for no more than two unknowns [Sample Prob. 2.6].

Equilibrium in Space

When a particle is in equilibrium in three-dimensional space [Sec. 2.5], use the three equations of equilibrium:

$$\Sigma F_x = 0 \qquad \Sigma F_y = 0 \qquad \Sigma F_z = 0 \qquad \textbf{(2.34)}$$

These equations can be solved for no more than three unknowns [Sample Prob. 2.9].

Review Problems

2.127 Two structural members *A* and *B* are bolted to a bracket as shown. Knowing that both members are in compression and that the force is 15 kN in member *A* and 10 kN in member *B*, determine by trigonometry the magnitude and direction of the resultant of the forces applied to the bracket by members *A* and *B*.

2.128 Determine the *x* and *y* components of each of the forces shown.

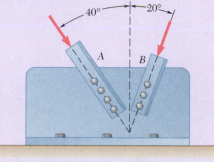

Fig. P2.127

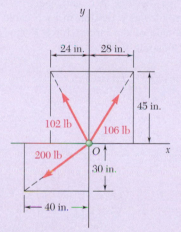

Fig. P2.128

2.129 A hoist trolley is subjected to the three forces shown. Knowing that $\alpha = 40°$, determine (*a*) the required magnitude of the force **P** if the resultant of the three forces is to be vertical, (*b*) the corresponding magnitude of the resultant.

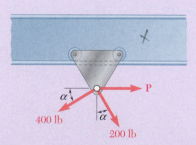

Fig. P2.129

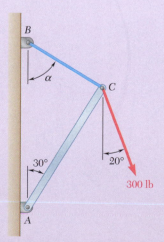

Fig. P2.130

2.130 Knowing that $\alpha = 55°$ and that boom *AC* exerts on pin *C* a force directed along line *AC*, determine (*a*) the magnitude of that force, (*b*) the tension in cable *BC*.

2.131 Two cables are tied together at C and loaded as shown. Knowing that $P = 360$ N, determine the tension (*a*) in cable AC, (*b*) in cable BC.

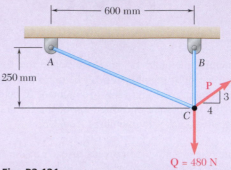

Fig. P2.131

2.132 Two cables tied together at C are loaded as shown. Knowing that the maximum allowable tension in each cable is 800 N, determine (*a*) the magnitude of the largest force **P** that can be applied at C, (*b*) the corresponding value of α.

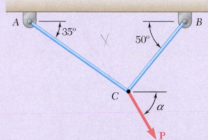

Fig. *P2.132*

2.133 The end of the coaxial cable AE is attached to the pole AB, which is strengthened by the guy wires AC and AD. Knowing that the tension in wire AC is 120 lb, determine (*a*) the components of the force exerted by this wire on the pole, (*b*) the angles θ_x, θ_y, and θ_z that the force forms with the coordinate axes.

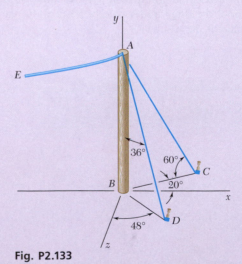

Fig. P2.133

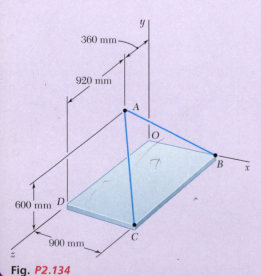

Fig. *P2.134*

2.134 Knowing that the tension in cable AC is 2130 N, determine the components of the force exerted on the plate at C.

2.135 Find the magnitude and direction of the resultant of the two forces shown knowing that $P = 600$ N and $Q = 450$ N.

2.136 A container of weight W is suspended from ring A. Cable BAC passes through the ring and is attached to fixed supports at B and C. Two forces $\mathbf{P} = P\mathbf{i}$ and $\mathbf{Q} = Q\mathbf{k}$ are applied to the ring to maintain the container in the position shown. Knowing that $W = 376$ N, determine P and Q. (*Hint*: The tension is the same in both portions of cable BAC.)

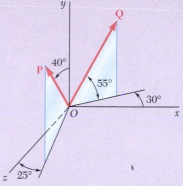

Fig. P2.135

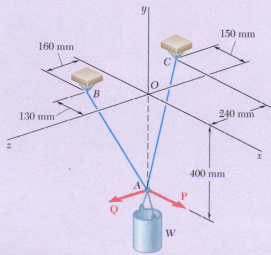

Fig. P2.136

2.137 Collars A and B are connected by a 25-in.-long wire and can slide freely on frictionless rods. If a 60-lb force $\mathbf{Q}$ is applied to collar B as shown, determine (*a*) the tension in the wire when $x = 9$ in., (*b*) the corresponding magnitude of the force $\mathbf{P}$ required to maintain the equilibrium of the system.

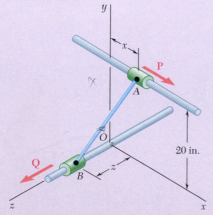

Fig. P2.137 and *P2.138*

2.138 Collars A and B are connected by a 25-in.-long wire and can slide freely on frictionless rods. Determine the distances x and z for which the equilibrium of the system is maintained when $P = 120$ lb and $Q = 60$ lb.

3

Rigid Bodies: Equivalent Systems of Forces

Four tugboats work together to free the oil tanker Coastal Eagle Point that ran aground while attempting to navigate a channel in Tampa Bay. It will be shown in this chapter that the forces exerted on the ship by the tugboats could be replaced by an equivalent force exerted by a single, more powerful, tugboat.

Objectives

- **Discuss** the principle of transmissibility that enables a force to be treated as a sliding vector.
- **Define** the moment of a force about a point.
- **Examine** vector and scalar products, useful in analysis involving moments.
- **Apply** Varignon's Theorem to simplify certain moment analyses.
- **Define** the mixed triple product and use it to determine the moment of a force about an axis.
- **Define** the moment of a couple, and consider the particular properties of couples.
- **Resolve** a given force into an equivalent force-couple system at another point.
- **Reduce** a system of forces into an equivalent force-couple system.
- **Examine** circumstances where a system of forces can be reduced to a single force.
- **Define** a wrench and consider how any general system of forces can be reduced to a wrench.

Introduction

In Chapter 2, we assumed that each of the bodies considered could be treated as a single particle. Such a view, however, is not always possible. In general, a body should be treated as a combination of a large number of particles. In this case, we need to consider the size of the body as well as the fact that forces act on different parts of the body and thus have different points of application.

Most of the bodies considered in elementary mechanics are assumed to be rigid. We define a **rigid body** as one that does not deform. Actual structures and machines are never absolutely rigid and deform under the loads to which they are subjected. However, these deformations are usually small and do not appreciably affect the conditions of equilibrium or the motion of the structure under consideration. They are important, though, as far as the resistance of the structure to failure is concerned and are considered in the study of mechanics of materials.

In this chapter, you will study the effect of forces exerted on a rigid body, and you will learn how to replace a given system of forces by a simpler equivalent system. This analysis rests on the fundamental assumption that the effect of a given force on a rigid body remains unchanged if that force is moved along its line of action (*principle of transmissibility*). It follows that forces acting on a rigid body can be represented by *sliding vectors,* as indicated earlier in Sec. 2.1B.

Two important concepts associated with the effect of a force on a rigid body are the *moment of a force about a point* (Sec. 3.1E) and the *moment of a force about an axis* (Sec. 3.2C). The determination of these quantities involves computing vector products and scalar products of two vectors, so in this chapter, we introduce the fundamentals of vector algebra and apply them to the solution of problems involving forces acting on rigid bodies.

Another concept introduced in this chapter is that of a *couple,* i.e., the combination of two forces that have the same magnitude, parallel lines of action, and opposite sense (Sec. 3.3A). As you will see, we can replace any system of forces acting on a rigid body by an equivalent system consisting of one force acting at a given point and one couple. This basic combination is called a *force-couple system.* In the case of concurrent, coplanar, or parallel forces, we can further reduce the equivalent force-couple system to a single force, called the *resultant* of the system, or to a single couple, called the *resultant couple* of the system.

3.1 FORCES AND MOMENTS

The basic definition of a force does not change if the force acts on a point or on a rigid body. However, the effects of the force can be very different, depending on factors such as the point of application or line of action of that force. As a result, calculations involving forces acting on a rigid body are generally more complicated than situations involving forces acting on a point. We begin by examining some general classifications of forces acting on rigid bodies.

3.1A External and Internal Forces

Forces acting on rigid bodies can be separated into two groups: (1) *external forces* and (2) *internal forces.*

1. **External forces** are exerted by other bodies on the rigid body under consideration. They are entirely responsible for the external behavior of the rigid body, either causing it to move or ensuring that it remains at rest. We shall be concerned only with external forces in this chapter and in Chaps. 4 and 5.
2. **Internal forces** hold together the particles forming the rigid body. If the rigid body is structurally composed of several parts, the forces holding the component parts together are also defined as internal forces. We will consider internal forces in Chaps. 6 and 7.

As an example of external forces, consider the forces acting on a disabled truck that three people are pulling forward by means of a rope attached to the front bumper (Fig. 3.1a). The external forces acting on the truck are shown in a **free-body diagram** (Fig. 3.1b). Note that this free-body diagram shows the entire object, not just a particle representing the object. Let us first consider the **weight** of the truck. Although it embodies the effect of the earth's pull on each of the particles forming the truck, the weight can be represented by the single force **W**. The **point of application** of this force—that is, the point at which the force acts—is defined as the **center of gravity** of the truck. (In Chap. 5, we will show how to determine the location of centers of gravity.) The weight **W** tends to make the truck move vertically downward. In fact, it would actually cause the truck to

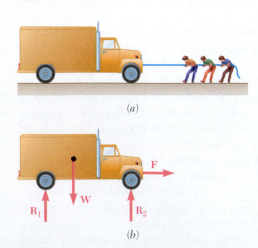

Fig. 3.1 (a) Three people pulling on a truck with a rope; (b) free-body diagram of the truck, shown as a rigid body instead of a particle.

move downward, i.e., to fall, if it were not for the presence of the ground. The ground opposes the downward motion of the truck by means of the reactions $\mathbf{R}_1$ and $\mathbf{R}_2$. These forces are exerted *by* the ground *on* the truck and must therefore be included among the external forces acting on the truck.

The people pulling on the rope exert the force $\mathbf{F}$. The point of application of $\mathbf{F}$ is on the front bumper. The force $\mathbf{F}$ tends to make the truck move forward in a straight line and does actually make it move, since no external force opposes this motion. (We are ignoring rolling resistance here for simplicity.) This forward motion of the truck, during which each straight line keeps its original orientation (the floor of the truck remains horizontal, and the walls remain vertical), is known as a **translation**. Other forces might cause the truck to move differently. For example, the force exerted by a jack placed under the front axle would cause the truck to pivot about its rear axle. Such a motion is a **rotation**. We conclude, therefore, that each *external force* acting on a *rigid body* can, if unopposed, impart to the rigid body a motion of translation or rotation, or both.

3.1B Principle of Transmissibility: Equivalent Forces

The **principle of transmissibility** states that the conditions of equilibrium or motion of a rigid body remain unchanged if a force $\mathbf{F}$ acting at a given point of the rigid body is replaced by a force $\mathbf{F}'$ of the same magnitude and same direction, but acting at a different point, *provided that the two forces have the same line of action* (Fig. 3.2). The two forces $\mathbf{F}$ and $\mathbf{F}'$ have the same effect on the rigid body and are said to be **equivalent forces**. This principle, which states that the action of a force may be *transmitted* along its line of action, is based on experimental evidence. It *cannot* be derived from the properties established so far in this text and therefore must be accepted as an experimental law. (You will see in Sec. 16.1D that we *can* derive the principle of transmissibility from the study of the dynamics of rigid bodies, but this study requires the use of Newton's second and third laws and of several other concepts as well.) Therefore, our study of the statics of rigid bodies is based on the three principles introduced so far: the parallelogram law of vector addition, Newton's first law, and the principle of transmissibility.

We indicated in Chap. 2 that we could represent the forces acting on a particle by vectors. These vectors had a well-defined point of application—namely, the particle itself—and were therefore fixed, or bound, vectors. In the case of forces acting on a rigid body, however, the point of application of the force does not matter, as long as the line of action remains unchanged. Thus, forces acting on a rigid body must be represented by a different kind of vector, known as a **sliding vector**, since forces are allowed to slide along their lines of action. Note that all of the properties we derive in the following sections for the forces acting on a rigid body are valid more generally for any system of sliding vectors. In order to keep our presentation more intuitive, however, we will carry it out in terms of physical forces rather than in terms of mathematical sliding vectors.

Returning to the example of the truck, we first observe that the line of action of the force $\mathbf{F}$ is a horizontal line passing through both the front

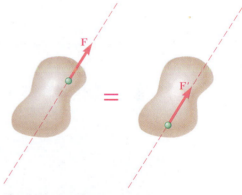

Fig. 3.2 Two forces $\mathbf{F}$ and $\mathbf{F}'$ are equivalent if they have the same magnitude and direction and the same line of action, even if they act at different points.

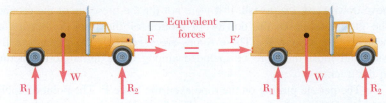

Fig. 3.3 Force **F'** is equivalent to force **F**, so the motion of the truck is the same whether you pull it or push it.

and rear bumpers of the truck (Fig. 3.3). Using the principle of transmissibility, we can therefore replace **F** by an *equivalent force* **F'** acting on the rear bumper. In other words, the conditions of motion are unaffected, and all of the other external forces acting on the truck (**W**, **R**$_1$, **R**$_2$) remain unchanged if the people push on the rear bumper instead of pulling on the front bumper.

The principle of transmissibility and the concept of equivalent forces have limitations. Consider, for example, a short bar *AB* acted upon by equal and opposite axial forces **P**$_1$ and **P**$_2$, as shown in Fig. 3.4*a*. According

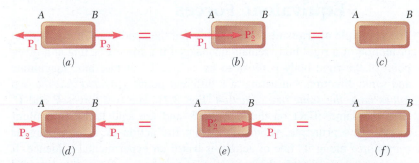

Fig. 3.4 (*a–c*) A set of equivalent forces acting on bar *AB*; (*d–f*) another set of equivalent forces acting on bar *AB*. Both sets produce the same external effect (equilibrium in this case) but different internal forces and deformations.

to the principle of transmissibility, we can replace force **P**$_2$ by a force **P**$_2'$ having the same magnitude, the same direction, and the same line of action but acting at *A* instead of *B* (Fig. 3.4*b*). The forces **P**$_1$ and **P**$_2'$ acting on the same particle can be added according to the rules of Chap. 2, and since these forces are equal and opposite, their sum is equal to zero. Thus, in terms of the external behavior of the bar, the original system of forces shown in Fig. 3.4*a* is equivalent to no force at all (Fig. 3.4*c*).

Consider now the two equal and opposite forces **P**$_1$ and **P**$_2$ acting on the bar *AB* as shown in Fig. 3.4*d*. We can replace the force **P**$_2$ by a force **P**$_2'$ having the same magnitude, the same direction, and the same line of action but acting at *B* instead of at *A* (Fig. 3.4*e*). We can add forces **P**$_1$ and **P**$_2'$, and their sum is again zero (Fig. 3.4*f*). From the point of view of the mechanics of rigid bodies, the systems shown in Fig. 3.4*a* and *d* are thus equivalent. However, the *internal forces* and *deformations* produced by the two systems are clearly different. The bar of Fig. 3.4*a* is in *tension* and, if not absolutely rigid, increases in length slightly; the bar of Fig. 3.4*d* is in *compression* and, if not absolutely rigid, decreases in length slightly. Thus, although we can use the principle of transmissibility to determine the conditions of motion or equilibrium of rigid bodies and to compute the external forces acting on these bodies, it should be avoided, or at least used with care, in determining internal forces and deformations.

3.1C Vector Products

In order to gain a better understanding of the effect of a force on a rigid body, we need to introduce a new concept, the *moment of a force about a point.* However, this concept is more clearly understood and is applied more effectively if we first add to the mathematical tools at our disposal the vector product of two vectors.

The **vector product** of two vectors **P** and **Q** is defined as the vector **V** that satisfies the following conditions.

1. The line of action of **V** is perpendicular to the plane containing **P** and **Q** (Fig. 3.5*a*).
2. The magnitude of **V** is the product of the magnitudes of **P** and **Q** and of the sine of the angle θ formed by **P** and **Q** (the measure of which is always 180° or less). We thus have

Magnitude of a vector product

$$V = PQ \sin \theta \tag{3.1}$$

3. The direction of **V** is obtained from the **right-hand rule**. Close your right hand and hold it so that your fingers are curled in the same sense as the rotation through θ that brings the vector **P** in line with the vector **Q**. Your thumb then indicates the direction of the vector **V** (Fig. 3.5*b*). Note that if **P** and **Q** do not have a common point of application, you should first redraw them from the same point. The three vectors **P, Q,** and **V**—taken in that order—are said to form a *right-handed triad.*[†]

As stated previously, the vector **V** satisfying these three conditions (which define it uniquely) is referred to as the *vector product* of **P** and **Q**. It is represented by the mathematical expression

Vector product

$$\mathbf{V} = \mathbf{P} \times \mathbf{Q} \tag{3.2}$$

Because of this notation, the vector product of two vectors **P** and **Q** is also referred to as the *cross product* of **P** and **Q**.

It follows from Eq. (3.1) that if the vectors **P** and **Q** have either the same direction or opposite directions, their vector product is zero. In the general case when the angle θ formed by the two vectors is neither 0° nor 180°, Eq. (3.1) has a simple geometric interpretation: The magnitude V of the vector product of **P** and **Q** is equal to the area of the parallelogram that has **P** and **Q** for sides (Fig. 3.6). The vector product **P** × **Q** is

[†]Note that the *x, y,* and *z* axes used in Chap. 2 form a right-handed system of orthogonal axes and that the unit vectors **i, j,** and **k** defined in Sec. 2.4A form a right-handed orthogonal triad.

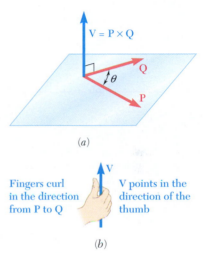

(a)

(b)

Fig. 3.5 (a) The vector product **V** has the magnitude $PQ \sin \theta$ and is perpendicular to the plane of **P** and **Q**; (b) you can determine the direction of **V** by using the right-hand rule.

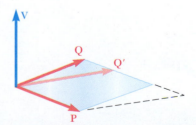

Fig. 3.6 The magnitude of the vector product **V** equals the area of the parallelogram formed by **P** and **Q**. If you change **Q** to **Q′** in such a way that the parallelogram changes shape but **P** and the area are still the same, then the magnitude of **V** remains the same.

therefore unchanged if we replace **Q** by a vector **Q′** that is coplanar with **P** and **Q** such that the line joining the tips of **Q** and **Q′** is parallel to **P**:

$$\mathbf{V} = \mathbf{P} \times \mathbf{Q} = \mathbf{P} \times \mathbf{Q}' \tag{3.3}$$

From the third condition used to define the vector product **V** of **P** and **Q**—namely, that **P**, **Q**, and **V** must form a right-handed triad—it follows that vector products *are not commutative*; i.e., **Q** × **P** is not equal to **P** × **Q**. Indeed, we can easily check that **Q** × **P** is represented by the vector −**V**, which is equal and opposite to **V**:

$$\mathbf{Q} \times \mathbf{P} = -(\mathbf{P} \times \mathbf{Q}) \tag{3.4}$$

Concept Application 3.1

Let us compute the vector product $\mathbf{V} = \mathbf{P} \times \mathbf{Q}$, where the vector **P** is of magnitude 6 and lies in the zx plane at an angle of 30° with the x axis, and where the vector **Q** is of magnitude 4 and lies along the x axis (Fig. 3.7).

Solution

It follows immediately from the definition of the vector product that the vector **V** must lie along the y axis, directed upward, with the magnitude

$$V = PQ \sin \theta = (6)(4) \sin 30° = 12 \quad \blacksquare$$

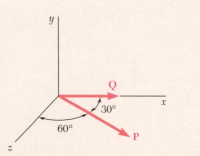

Fig. 3.7 Two vectors P and Q with angle between them.

We saw that the commutative property does not apply to vector products. However, it can be demonstrated that the *distributive* property

$$\mathbf{P} \times (\mathbf{Q}_1 + \mathbf{Q}_2) = \mathbf{P} \times \mathbf{Q}_1 + \mathbf{P} \times \mathbf{Q}_2 \tag{3.5}$$

does hold.

A third property, the associative property, does not apply to vector products; we have in general

$$(\mathbf{P} \times \mathbf{Q}) \times \mathbf{S} \neq \mathbf{P} \times (\mathbf{Q} \times \mathbf{S}) \tag{3.6}$$

3.1D   Rectangular Components of Vector Products

Before we turn back to forces acting on rigid bodes, let's look at a more convenient way to express vector products using rectangular components. To do this, we use the unit vectors **i**, **j**, and **k** that were defined in Chap. 2.

Consider first the vector product **i** × **j** (Fig. 3.8*a*). Since both vectors have a magnitude equal to 1 and since they are at a right angle to each other, their vector product is also a unit vector. This unit vector must be **k**, since the vectors **i**, **j**, and **k** are mutually perpendicular and form a

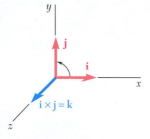

(a)

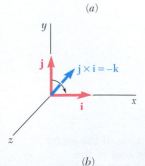

(b)

Fig. 3.8 (*a*) The vector product of the **i** and **j** unit vectors is the **k** unit vector; (*b*) the vector product of the **j** and **i** unit vectors is the −**k** unit vector.

right-handed triad. Similarly, it follows from the right-hand rule given in Sec. 3.1C that the product $\mathbf{j} \times \mathbf{i}$ is equal to $-\mathbf{k}$ (Fig. 3.8b). Finally, note that the vector product of a unit vector with itself, such as $\mathbf{i} \times \mathbf{i}$, is equal to zero, since both vectors have the same direction. Thus, we can list the vector products of all the various possible pairs of unit vectors:

$$
\begin{array}{lll}
\mathbf{i} \times \mathbf{i} = 0 & \mathbf{j} \times \mathbf{i} = -\mathbf{k} & \mathbf{k} \times \mathbf{i} = \mathbf{j} \\
\mathbf{i} \times \mathbf{j} = \mathbf{k} & \mathbf{j} \times \mathbf{j} = 0 & \mathbf{k} \times \mathbf{j} = -\mathbf{i} \quad (3.7) \\
\mathbf{i} \times \mathbf{k} = -\mathbf{j} & \mathbf{j} \times \mathbf{k} = \mathbf{i} & \mathbf{k} \times \mathbf{k} = 0
\end{array}
$$

We can determine the sign of the vector product of two unit vectors simply by arranging them in a circle and reading them in the order of the multiplication (Fig. 3.9). The product is positive if they follow each other in counterclockwise order and is negative if they follow each other in clockwise order.

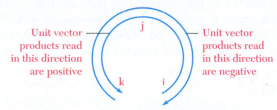

Unit vector products read in this direction are positive

Unit vector products read in this direction are negative

Fig. 3.9 Arrange the three letters **i**, **j**, **k** in a counterclockwise circle. You can use the order of letters for the three unit vectors in a vector product to determine its sign.

We can now easily express the vector product $\mathbf{V}$ of two given vectors $\mathbf{P}$ and $\mathbf{Q}$ in terms of the rectangular components of these vectors. Resolving $\mathbf{P}$ and $\mathbf{Q}$ into components, we first write

$$
\mathbf{V} = \mathbf{P} \times \mathbf{Q} = (P_x\mathbf{i} + P_y\mathbf{j} + P_z\mathbf{k}) \times (Q_x\mathbf{i} + Q_y\mathbf{j} + Q_z\mathbf{k})
$$

Making use of the distributive property, we express $\mathbf{V}$ as the sum of vector products, such as $P_x\mathbf{i} \times Q_y\mathbf{j}$. We find that each of the expressions obtained is equal to the vector product of two unit vectors, such as $\mathbf{i} \times \mathbf{j}$, multiplied by the product of two scalars, such as P_xQ_y. Recalling the identities of Eq. (3.7) and factoring out $\mathbf{i}$, $\mathbf{j}$, and $\mathbf{k}$, we obtain

$$
\mathbf{V} = (P_yQ_z - P_zQ_y)\mathbf{i} + (P_zQ_x - P_xQ_z)\mathbf{j} + (P_xQ_y - P_yQ_x)\mathbf{k} \quad (3.8)
$$

Thus, the rectangular components of the vector product $\mathbf{V}$ are

Rectangular components of a vector product

$$
\begin{aligned}
V_x &= P_yQ_z - P_zQ_y \\
V_y &= P_zQ_x - P_xQ_z \quad (3.9) \\
V_z &= P_xQ_y - P_yQ_x
\end{aligned}
$$

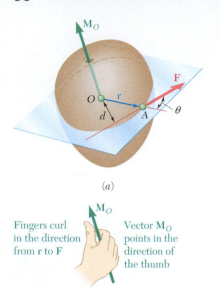

(a)

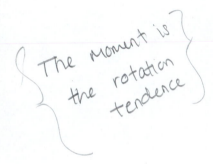

Fingers curl in the direction from **r** to **F**

Vector $\mathbf{M}_O$ points in the direction of the thumb

(b)

Fig. 3.10 Moment of a force about a point. (a) The moment $\mathbf{M}_O$ is the vector product of the position vector **r** and the force **F**; (b) a right-hand rule indicates the sense of $\mathbf{M}_O$.

Returning to Eq. (3.8), notice that the right-hand side represents the expansion of a determinant. Thus, we can express the vector product **V** in the following form, which is more easily memorized:[†]

Rectangular components of a vector product (determinant form)

$$\mathbf{V} = \begin{vmatrix} \mathbf{i} & \mathbf{j} & \mathbf{k} \\ P_x & P_y & P_z \\ Q_x & Q_y & Q_z \end{vmatrix} \qquad (3.10)$$

3.1E   Moment of a Force about a Point

We are now ready to consider a force **F** acting on a rigid body (Fig. 3.10a). As we know, the force **F** is represented by a vector that defines its magnitude and direction. However, the effect of the force on the rigid body depends also upon its point of application A. The position of A can be conveniently defined by the vector **r** that joins the fixed reference point O with A; this vector is known as the *position vector* of A. The position vector **r** and the force **F** define the plane shown in Fig. 3.10a.

We define the **moment of F about O** as the vector product of **r** and **F**:

Moment of a force about a point O

$$\mathbf{M}_O = \mathbf{r} \times \mathbf{F} \qquad (3.11)$$

According to the definition of the vector product given in Sec. 3.1C, the moment $\mathbf{M}_O$ must be perpendicular to the plane containing O and force **F**. The sense of $\mathbf{M}_O$ is defined by the sense of the rotation that will bring vector **r** in line with vector **F**; this rotation is observed as *counterclockwise* by an observer located at the tip of $\mathbf{M}_O$. Another way of defining the sense of $\mathbf{M}_O$ is furnished by a variation of the right-hand rule: Close your right hand and hold it so that your fingers curl in the sense of the rotation that **F** would impart to the rigid body about a fixed axis directed along the line of action of $\mathbf{M}_O$. Then your thumb indicates the sense of the moment $\mathbf{M}_O$ (Fig. 3.10b).

Finally, denoting by θ the angle between the lines of action of the position vector **r** and the force **F**, we find that the magnitude of the moment of **F** about O is

Magnitude of the moment of a force

$$M_O = rF \sin \theta = Fd \qquad (3.12)$$

[†]Any determinant consisting of three rows and three columns can be evaluated by repeating the first and second columns and forming products along each diagonal line. The sum of the products obtained along the red lines is then subtracted from the sum of the products obtained along the black lines.

where d represents the perpendicular distance from O to the line of action of $\mathbf{F}$ (see Fig. 3.10). Experimentally, the tendency of a force $\mathbf{F}$ to make a rigid body rotate about a fixed axis perpendicular to the force depends upon the distance of $\mathbf{F}$ from that axis, as well as upon the magnitude of $\mathbf{F}$. For example, a child's breath can exert enough force to make a toy propeller spin (Fig. 3.11*a*), but a wind turbine requires the force of a substantial wind to rotate the blades and generate electrical power (Fig. 3.11*b*). However, the perpendicular distance between the rotation point and the line of action of the force (often called the *moment arm*) is just as important. If you want to apply a small moment to turn a nut on a pipe without breaking it, you might use a small pipe wrench that gives you a small moment

(*a*) Small force

(*b*) Large force

(*c*) Small moment arm

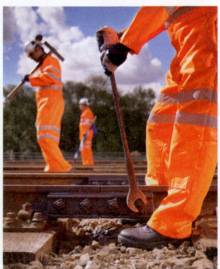

(*d*) Large moment arm

Fig. 3.11 (*a, b*) The moment of a force depends on the magnitude of the force; (*c, d*) it also depends on the length of the moment arm.

arm (Fig. 3.11c). But if you need a larger moment, you could use a large wrench with a long moment arm (Fig. 3.11d). Therefore,

> *The magnitude of M_O measures the tendency of the force F to make the rigid body rotate about a fixed axis directed along M_O.*

In the SI system of units, where a force is expressed in newtons (N) and a distance in meters (m), the moment of a force is expressed in newton-meters (N·m). In the U.S. customary system of units, where a force is expressed in pounds and a distance in feet or inches, the moment of a force is expressed in lb·ft or lb·in.

Note that although the moment M_O of a force about a point depends upon the magnitude, the line of action, and the sense of the force, it does *not* depend upon the actual position of the point of application of the force along its line of action. Conversely, the moment M_O of a force F does not characterize the position of the point of application of F.

However, as we will see shortly, the moment M_O of a force F of a given magnitude and direction *completely defines the line of action of* F. Indeed, the line of action of F must lie in a plane through O perpendicular to the moment M_O; its distance d from O must be equal to the quotient M_O/F of the magnitudes of M_O and F; and the sense of M_O determines whether the line of action of F occurs on one side or the other of the point O.

Recall from Sec. 3.1B that the principle of transmissibility states that two forces F and F' are equivalent (i.e., have the same effect on a rigid body) if they have the same magnitude, same direction, and same line of action. We can now restate this principle:

> *Two forces F and F' are equivalent if, and only if, they are equal (i.e., have the same magnitude and same direction) and have equal moments about a given point O.*

The necessary and sufficient conditions for two forces F and F' to be equivalent are thus

$$F = F' \qquad \text{and} \qquad M_O = M'_O \tag{3.13}$$

We should observe that if the relations of Eqs. (3.13) hold for a given point O, they hold for any other point.

Two-Dimensional Problems. Many applications in statics deal with two-dimensional structures. Such structures have length and breadth but only negligible depth. Often, they are subjected to forces contained in the plane of the structure. We can easily represent two-dimensional structures and the forces acting on them on a sheet of paper or on a blackboard. Their analysis is therefore considerably simpler than that of three-dimensional structures and forces.

Consider, for example, a rigid slab acted upon by a force F in the plane of the slab (Fig. 3.12). The moment of F about a point O, which is chosen in the plane of the figure, is represented by a vector M_O perpendicular to that plane and of magnitude Fd. In the case of Fig. 3.12a, the vector M_O points *out of* the page, whereas in the case of Fig. 3.12b, it points *into* the page. As we look at the figure, we observe in the first case

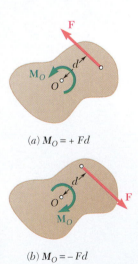

(a) $M_O = + Fd$

(b) $M_O = -Fd$

Fig. 3.12 (a) A moment that tends to produce a counterclockwise rotation is positive; (b) a moment that tends to produce a clockwise rotation is negative.

that **F** tends to rotate the slab counterclockwise and in the second case that it tends to rotate the slab clockwise. Therefore, it is natural to refer to the sense of the moment of **F** about O in Fig. 3.12a as counterclockwise ↰, and in Fig. 3.12b as clockwise ↲.

Since the moment of a force **F** acting in the plane of the figure must be perpendicular to that plane, we need only specify the *magnitude* and the *sense* of the moment of **F** about O. We do this by assigning to the magnitude M_O of the moment a positive or negative sign according to whether the vector **M_O** points out of or into the page.

3.1F Rectangular Components of the Moment of a Force

We can use the distributive property of vector products to determine the moment of the resultant of several *concurrent forces*. If several forces **F_1**, **F_2**, . . . are applied at the same point A (Fig. 3.13) and if we denote by **r** the position vector of A, it follows immediately from Eq. (3.5) that

$$\mathbf{r} \times (\mathbf{F}_1 + \mathbf{F}_2 + \cdots) = \mathbf{r} \times \mathbf{F}_1 + \mathbf{r} \times \mathbf{F}_2 + \cdots \qquad (3.14)$$

In words,

> *The moment about a given point O of the resultant of several concurrent forces is equal to the sum of the moments of the various forces about the same point O.*

This property, which was originally established by the French mathematician Pierre Varignon (1654–1722) long before the introduction of vector algebra, is known as **Varignon's theorem**.

The relation in Eq. (3.14) makes it possible to replace the direct determination of the moment of a force **F** by determining the moments of two or more component forces. As you will see shortly, **F** is generally resolved into components parallel to the coordinate axes. However, it may be more expeditious in some instances to resolve **F** into components that are not parallel to the coordinate axes (see Sample Prob. 3.3).

In general, determining the moment of a force in space is considerably simplified if the force and the position vector of its point of application are resolved into rectangular x, y, and z components. Consider, for example, the moment **M_O** about O of a force **F** whose components are F_x, F_y, and F_z and that is applied at a point A with coordinates x, y, and z (Fig. 3.14). Since the components of the position vector **r** are respectively equal to the coordinates x, y, and z of the point A, we can write **r** and **F** as

$$\mathbf{r} = x\mathbf{i} + y\mathbf{j} + z\mathbf{k} \qquad (3.15)$$
$$\mathbf{F} = F_x\mathbf{i} + F_y\mathbf{j} + F_z\mathbf{k} \qquad (3.16)$$

Substituting for **r** and **F** from Eqs. (3.15) and (3.16) into

$$\mathbf{M}_O = \mathbf{r} \times \mathbf{F} \qquad (3.11)$$

and recalling Eqs. (3.8) and (3.9), we can write the moment **M_O** of **F** about O in the form

$$\mathbf{M}_O = M_x\mathbf{i} + M_y\mathbf{j} + M_z\mathbf{k} \qquad (3.17)$$

where the components M_x, M_y, and M_z are defined by the relations

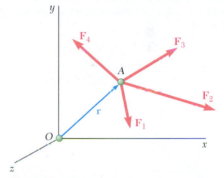

Fig. 3.13 Varignon's theorem says that the moment about point O of the resultant of these four forces equals the sum of the moments about point O of the individual forces.

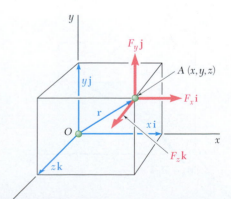

Fig. 3.14 The moment **M_O** about point O of a force **F** applied at point A is the vector product of the position vector **r** and the force **F**, which can both be expressed in rectangular components.

you can add moments

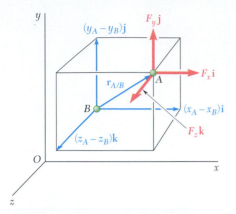

Fig. 3.15 The moment $\mathbf{M}_B$ about the point B of a force $\mathbf{F}$ applied at point A is the vector product of the position vector $\mathbf{r}_{A/B}$ and force $\mathbf{F}$.

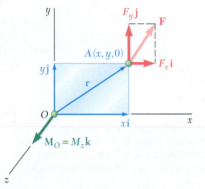

Fig. 3.16 In a two-dimensional problem, the moment $\mathbf{M}_O$ of a force $\mathbf{F}$ applied at A in the xy plane reduces to the z component of the vector product of $\mathbf{r}$ with $\mathbf{F}$.

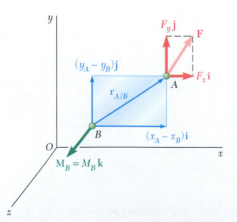

Fig. 3.17 In a two-dimensional problem, the moment $\mathbf{M}_B$ about a point B of a force $\mathbf{F}$ applied at A in the xy plane reduces to the z component of the vector product of $\mathbf{r}_{A/B}$ with $\mathbf{F}$.

Rectangular components of a moment

$$M_x = yF_z - zF_y$$
$$M_y = zF_x - xF_z \qquad \text{(3.18)}$$
$$M_z = xF_y - yF_x$$

As you will see in Sec. 3.2C, the scalar components M_x, M_y, and M_z of the moment $\mathbf{M}_O$ measure the tendency of the force $\mathbf{F}$ to impart to a rigid body a rotation about the x, y, and z axes, respectively. Substituting from Eq. (3.18) into Eq. (3.17), we can also write $\mathbf{M}_O$ in the form of the determinant, as

$$\mathbf{M}_O = \begin{vmatrix} \mathbf{i} & \mathbf{j} & \mathbf{k} \\ x & y & z \\ F_x & F_y & F_z \end{vmatrix} \qquad \text{(3.19)}$$

To compute the moment $\mathbf{M}_B$ about an arbitrary point B of a force $\mathbf{F}$ applied at A (Fig. 3.15), we must replace the position vector $\mathbf{r}$ in Eq. (3.11) by a vector drawn from B to A. This vector is the *position vector of A relative to B*, denoted by $\mathbf{r}_{A/B}$. Observing that $\mathbf{r}_{A/B}$ can be obtained by subtracting $\mathbf{r}_B$ from $\mathbf{r}_A$, we write

to move the point →

$$\mathbf{M}_B = \mathbf{r}_{A/B} \times \mathbf{F} = (\mathbf{r}_A - \mathbf{r}_B) \times \mathbf{F} \qquad \text{(3.20)}$$

or using the determinant form,

$$\mathbf{M}_B = \begin{vmatrix} \mathbf{i} & \mathbf{j} & \mathbf{k} \\ x_{A/B} & y_{A/B} & z_{A/B} \\ F_x & F_y & F_z \end{vmatrix} \qquad \text{(3.21)}$$

where $x_{A/B}$, $y_{A/B}$, and $z_{A/B}$ denote the components of the vector $\mathbf{r}_{A/B}$:

$$x_{A/B} = x_A - x_B \qquad y_{A/B} = y_A - y_B \qquad z_{A/B} = z_A - z_B$$

In the case of two-dimensional problems, we can assume without loss of generality that the force $\mathbf{F}$ lies in the xy plane (Fig. 3.16). Setting $z = 0$ and $F_z = 0$ in Eq. (3.19), we obtain

$$\mathbf{M}_O = (xF_y - yF_x)\mathbf{k}$$

We can verify that the moment of $\mathbf{F}$ about O is perpendicular to the plane of the figure and that it is completely defined by the scalar

$$M_O = M_z = xF_y - yF_x \qquad \text{(3.22)}$$

As noted earlier, a positive value for M_O indicates that the vector $\mathbf{M}_O$ points out of the paper (the force $\mathbf{F}$ tends to rotate the body counterclockwise about O), and a negative value indicates that the vector $\mathbf{M}_O$ points into the paper (the force $\mathbf{F}$ tends to rotate the body clockwise about O).

To compute the moment about $B(x_B, y_B)$ of a force lying in the xy plane and applied at $A(x_A, y_A)$ (Fig. 3.17), we set $z_{A/B} = 0$ and $F_z = 0$ in Eq. (3.21) and note that the vector $\mathbf{M}_B$ is perpendicular to the xy plane and is defined in magnitude and sense by the scalar

$$M_B = (x_A - x_B)F_y - (y_A - y_B)F_x \qquad \text{(3.23)}$$

Sample Problem 3.1

A 100-lb vertical force is applied to the end of a lever, which is attached to a shaft at O. Determine (a) the moment of the 100-lb force about O; (b) the horizontal force applied at A that creates the same moment about O; (c) the smallest force applied at A that creates the same moment about O; (d) how far from the shaft a 240-lb vertical force must act to create the same moment about O; (e) whether any one of the forces obtained in parts b, c, or d is equivalent to the original force.

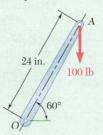

STRATEGY: The calculations asked for all involve variations on the basic defining equation of a moment, $M_O = Fd$.

MODELING and ANALYSIS:

a. Moment about O. The perpendicular distance from O to the line of action of the 100-lb force (Fig. 1) is

$$d = (24 \text{ in.}) \cos 60° = 12 \text{ in.}$$

The magnitude of the moment about O of the 100-lb force is

$$M_O = Fd = (100 \text{ lb})(12 \text{ in.}) = 1200 \text{ lb·in.}$$

Since the force tends to rotate the lever clockwise about O, represent the moment by a vector $\mathbf{M}_O$ perpendicular to the plane of the figure and pointing *into* the paper. You can express this fact with the notation

$$\mathbf{M}_O = 1200 \text{ lb·in. } \circlearrowright \quad \blacktriangleleft$$

b. Horizontal Force. In this case, you have (Fig. 2)

$$d = (24 \text{ in.}) \sin 60° = 20.8 \text{ in.}$$

Since the moment about O must be 1200 lb·in., you obtain

$$M_O = Fd$$
$$1200 \text{ lb·in.} = F(20.8 \text{ in.})$$
$$F = 57.7 \text{ lb} \qquad \mathbf{F} = 57.7 \text{ lb} \rightarrow \quad \blacktriangleleft$$

c. Smallest Force. Since $M_O = Fd$, the smallest value of F occurs when d is maximum. Choose the force perpendicular to OA and note that $d = 24$ in. (Fig. 3); thus

$$M_O = Fd$$
$$1200 \text{ lb·in.} = F(24 \text{ in.})$$
$$F = 50 \text{ lb} \qquad \mathbf{F} = 50 \text{ lb } \diagdown 30° \quad \blacktriangleleft$$

(continued)

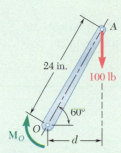

Fig. 1 Determination of the moment of the 100-lb force about O using perpendicular distance d.

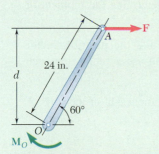

Fig. 2 Determination of horizontal force at A that creates same moment about O.

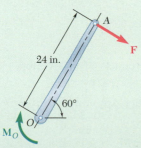

Fig. 3 Determination of smallest force at A that creates same moment about O.

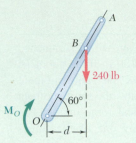

Fig. 4 Position of vertical 240-lb force that creates same moment about O.

d. 240-lb Vertical Force. In this case (Fig. 4), $M_O = Fd$ yields

$$1200 \text{ lb·in.} = (240 \text{ lb})d \qquad d = 5 \text{ in.}$$

but

$$OB \cos 60° = d$$

so $\qquad\qquad\qquad\qquad\qquad\qquad\qquad\qquad\qquad\qquad OB = 10 \text{ in.} \blacktriangleleft$

e. None of the forces considered in parts *b*, *c*, or *d* is equivalent to the original 100-lb force. Although they have the same moment about *O*, they have different *x* and *y* components. In other words, although each force tends to rotate the shaft in the same direction, each causes the lever to pull on the shaft in a different way.

REFLECT and THINK: Various combinations of force and lever arm can produce equivalent moments, but the system of force and moment produces a different overall effect in each case.

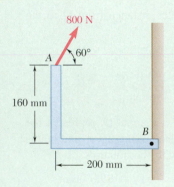

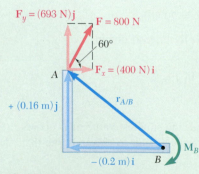

Fig. 1 The moment M_B is determined from the vector product of position vector $r_{A/B}$ and force vector **F**.

Sample Problem 3.2

A force of 800 N acts on a bracket as shown. Determine the moment of the force about *B*.

STRATEGY: You can resolve both the force and the position vector from *B* to *A* into rectangular components and then use a vector approach to complete the solution.

MODELING and ANALYSIS: Obtain the moment M_B of the force **F** about *B* by forming the vector product

$$M_B = r_{A/B} \times F$$

where $r_{A/B}$ is the vector drawn from *B* to *A* (Fig. 1). Resolving $r_{A/B}$ and **F** into rectangular components, you have

$$r_{A/B} = -(0.2 \text{ m})i + (0.16 \text{ m})j$$
$$F = (800 \text{ N}) \cos 60°i + (800 \text{ N}) \sin 60°j$$
$$= (400 \text{ N})i + (693 \text{ N})j$$

Recalling the relations in Eq. (3.7) for the cross products of unit vectors (Sec. 3.5), you obtain

$$M_B = r_{A/B} \times F = [-(0.2 \text{ m})i + (0.16 \text{ m})j] \times [(400 \text{ N})i + (693 \text{ N})j]$$
$$= -(138.6 \text{ N·m})k - (64.0 \text{ N·m})k$$
$$= -(202.6 \text{ N·m})k \qquad\qquad M_B = 203 \text{ N·m} \downarrow \blacktriangleleft$$

The moment M_B is a vector perpendicular to the plane of the figure and pointing *into* the page.

(continued)

REFLECT and THINK: We can also use a scalar approach to solve this problem using the components for the force $\mathbf{F}$ and the position vector $\mathbf{r}_{A/B}$. Following the right-hand rule for assigning signs, we have

$$+\text{\textcurrency}M_B = \Sigma M_B = \Sigma Fd = -(400 \text{ N})(0.16 \text{ m}) - (693 \text{ N})(0.2 \text{ m}) = -202.6 \text{ N·m}$$

$$\mathbf{M}_B = 203 \text{ N·m} \downarrow \quad \blacktriangleleft$$

Sample Problem 3.3

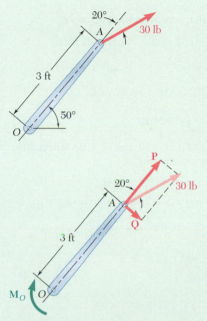

A 30-lb force acts on the end of the 3-ft lever as shown. Determine the moment of the force about O.

STRATEGY: Resolving the force into components that are perpendicular and parallel to the axis of the lever greatly simplifies the moment calculation.

MODELING and ANALYSIS: Replace the force by two components: one component $\mathbf{P}$ in the direction of OA and one component $\mathbf{Q}$ perpendicular to OA (Fig. 1). Since O is on the line of action of $\mathbf{P}$, the moment of $\mathbf{P}$ about O is zero. Thus, the moment of the 30-lb force reduces to the moment of $\mathbf{Q}$, which is clockwise and can be represented by a negative scalar.

$$Q = (30 \text{ lb}) \sin 20° = 10.26 \text{ lb}$$
$$M_O = -Q(3 \text{ ft}) = -(10.26 \text{ lb})(3 \text{ ft}) = -30.8 \text{ lb·ft}$$

Since the value obtained for the scalar M_O is negative, the moment $\mathbf{M}_O$ points *into* the page. You can write it as

$$M_O = 30.8 \text{ lb·ft} \downarrow \quad \blacktriangleleft$$

Fig. 1 30-lb force at A resolved into components $\mathbf{P}$ and $\mathbf{Q}$ to simplify the determination of the moment $\mathbf{M}_O$.

REFLECT and THINK: Always be alert for simplifications that can reduce the amount of computation.

Sample Problem 3.4

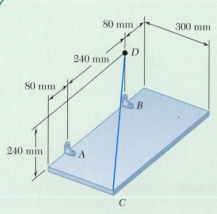

A rectangular plate is supported by brackets at A and B and by a wire CD. If the tension in the wire is 200 N, determine the moment about A of the force exerted by the wire on point C.

STRATEGY: The solution requires resolving the tension in the wire and the position vector from A to C into rectangular components. You will need a unit vector approach to determine the force components.

MODELING and ANALYSIS: Obtain the moment $\mathbf{M}_A$ about A of the force $\mathbf{F}$ exerted by the wire on point C by forming the vector product

$$\mathbf{M}_A = \mathbf{r}_{C/A} \times \mathbf{F} \quad (1)$$

(continued)

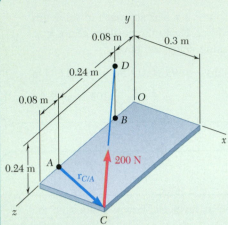

Fig. 1 The moment $\mathbf{M}_A$ is determined from position vector $\mathbf{r}_{C/A}$ and force vector $\mathbf{F}$.

where $\mathbf{r}_{C/A}$ is the vector from A to C

$$\mathbf{r}_{C/A} = \overrightarrow{AC} = (0.3 \text{ m})\mathbf{i} + (0.08 \text{ m})\mathbf{k} \qquad (2)$$

and $\mathbf{F}$ is the 200-N force directed along CD (Fig. 1). Introducing the unit vector

$$\boldsymbol{\lambda} = \overrightarrow{CD}/CD,$$

you can express $\mathbf{F}$ as

$$\mathbf{F} = F\boldsymbol{\lambda} = (200 \text{N}) \frac{\overrightarrow{CD}}{CD} \qquad (3)$$

Resolving the vector $\overrightarrow{CD}$ into rectangular components, you have

$$\overrightarrow{CD} = -(0.3 \text{ m})\mathbf{i} + (0.24 \text{ m})\mathbf{j} - (0.32 \text{ m})\mathbf{k} \qquad CD = 0.50 \text{ m}$$

Substituting into (3) gives you

$$\mathbf{F} = \frac{200 \text{N}}{0.50 \text{m}} [-(0.3 \text{ m})\mathbf{i} + (0.24 \text{ m})\mathbf{j} - (0.32 \text{ m})\mathbf{k}]$$

$$= -(120 \text{ N})\mathbf{i} + (96 \text{ N})\mathbf{j} - (128 \text{ N})\mathbf{k} \qquad (4)$$

Substituting for $\mathbf{r}_{C/A}$ and $\mathbf{F}$ from (2) and (4) into (1) and recalling the relations in Eq. (3.7) of Sec. 3.1D, you obtain (Fig. 2)

$$\mathbf{M}_A = \mathbf{r}_{C/A} \times \mathbf{F} = (0.3\mathbf{i} + 0.08\mathbf{k}) \times (-120\mathbf{i} + 96\mathbf{j} - 128\mathbf{k})$$

$$= (0.3)(96)\mathbf{k} + (0.3)(-128)(-\mathbf{j}) + (0.08)(-120)\mathbf{j} + (0.08)(96)(-\mathbf{i})$$

$$\mathbf{M}_A = -(7.68 \text{ N·m})\mathbf{i} + (28.8 \text{ N·m})\mathbf{j} + (28.8 \text{ N·m})\mathbf{k} \quad ◄$$

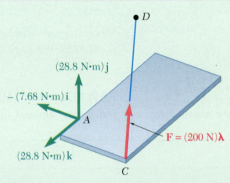

Fig. 2 Components of moment $\mathbf{M}_A$ applied at A.

Alternative Solution. As indicated in Sec. 3.1F, you can also express the moment $\mathbf{M}_A$ in the form of a determinant:

$$\mathbf{M}_A = \begin{vmatrix} \mathbf{i} & \mathbf{j} & \mathbf{k} \\ x_C - x_A & y_C - y_A & z_C - z_A \\ F_x & F_y & F_z \end{vmatrix} = \begin{vmatrix} \mathbf{i} & \mathbf{j} & \mathbf{k} \\ 0.3 & 0 & 0.08 \\ -120 & 96 & -128 \end{vmatrix}$$

$$\mathbf{M}_A = -(7.68 \text{N·m})\mathbf{i} + (28.8 \text{ N·m})\mathbf{j} + (28.8 \text{ N·m})\mathbf{k} \quad ◄$$

REFLECT and THINK: Two-dimensional problems often are solved easily using a scalar approach, but the versatility of a vector analysis is quite apparent in a three-dimensional problem such as this.

SOLVING PROBLEMS
ON YOUR OWN

In this section, we introduced the *vector product* or *cross product* of two vectors. In the following problems, you will use the vector product to compute the *moment of a force about a point* and also to determine the *perpendicular distance* from a point to a line.

We defined the moment of the force $\mathbf{F}$ about the point O of a rigid body as

$$\mathbf{M}_O = \mathbf{r} \times \mathbf{F} \tag{3.11}$$

where $\mathbf{r}$ is the position vector *from O to any point* on the line of action of $\mathbf{F}$. Since the vector product is not commutative, it is absolutely necessary when computing such a product that you place the vectors in the proper order and that each vector have the correct sense. The moment $\mathbf{M}_O$ is important because its magnitude is a measure of the tendency of the force $\mathbf{F}$ to cause the rigid body to rotate about an axis directed along $\mathbf{M}_O$.

1. Computing the moment $\mathbf{M}_O$ of a force in two dimensions. You can use one of the following procedures:

    a. Use Eq. (3.12), $M_O = Fd$, which expresses the magnitude of the moment as the product of the magnitude of $\mathbf{F}$ and the *perpendicular distance d* from O to the line of action of $\mathbf{F}$ [Sample Prob. 3.1].

    b. Express $\mathbf{r}$ and $\mathbf{F}$ in component form and formally evaluate the vector product $\mathbf{M}_O = \mathbf{r} \times \mathbf{F}$ [Sample Prob. 3.2].

    c. Resolve $\mathbf{F}$ into components respectively parallel and perpendicular to the position vector $\mathbf{r}$. Only the perpendicular component contributes to the moment of $\mathbf{F}$ [Sample Prob. 3.3].

    d. Use Eq. (3.22), $M_O = M_z = xF_y - yF_x$. When applying this method, the simplest approach is to treat the scalar components of $\mathbf{r}$ and $\mathbf{F}$ as positive and then to assign, by observation, the proper sign to the moment produced by each force component [Sample Prob. 3.2].

2. Computing the moment $\mathbf{M}_O$ of a force $\mathbf{F}$ in three dimensions. Following the method of Sample Prob. 3.4, the first step in the calculation is to select the most convenient (simplest) position vector $\mathbf{r}$. You should next express $\mathbf{F}$ in terms of its rectangular components. The final step is to evaluate the vector product $\mathbf{r} \times \mathbf{F}$ to determine the moment. In most three-dimensional problems you will find it easiest to calculate the vector product using a determinant.

3. Determining the perpendicular distance d from a point A to a given line. First assume that a force $\mathbf{F}$ of known magnitude F lies along the given line. Next determine its moment about A by forming the vector product $\mathbf{M}_A = \mathbf{r} \times \mathbf{F}$, and calculate this product as indicated above. Then compute its magnitude M_A. Finally, substitute the values of F and M_A into the equation $M_A = Fd$ and solve for d.

Problems

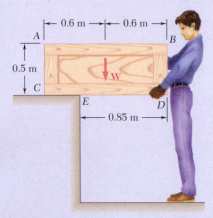

Fig. P3.1 and P3.2

3.1 A crate of mass 80 kg is held in the position shown. Determine (*a*) the moment produced by the weight **W** of the crate about *E*, (*b*) the smallest force applied at *B* that creates a moment of equal magnitude and opposite sense about *E*.

3.2 A crate of mass 80 kg is held in the position shown. Determine (*a*) the moment produced by the weight **W** of the crate about *E*, (*b*) the smallest force applied at *A* that creates a moment of equal magnitude and opposite sense about *E*, (*c*) the magnitude, sense, and point of application on the bottom of the crate of the smallest vertical force that creates a moment of equal magnitude and opposite sense about *E*.

3.3 It is known that a vertical force of 200 lb is required to remove the nail at *C* from the board. As the nail first starts moving, determine (*a*) the moment about *B* of the force exerted on the nail, (*b*) the magnitude of the force **P** that creates the same moment about *B* if $\alpha = 10°$, and (*c*) the smallest force **P** that creates the same moment about *B*.

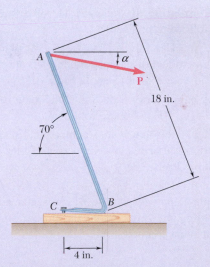

Fig. *P3.3*

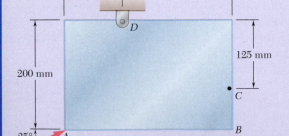

Fig. P3.4 and P3.5

3.4 A 300-N force is applied at *A* as shown. Determine (*a*) the moment of the 300-N force about *D*, (*b*) the smallest force applied at *B* that creates the same moment about *D*.

3.5 A 300-N force is applied at *A* as shown. Determine (*a*) the moment of the 300-N force about *D*, (*b*) the magnitude and sense of the horizontal force applied at *C* that creates the same moment about *D*, (*c*) the smallest force applied at *C* that creates the same moment about *D*.

3.6 A 20-lb force is applied to the control rod AB as shown. Knowing that the length of the rod is 9 in. and that $\alpha = 25°$, determine the moment of the force about point B by resolving the force into horizontal and vertical components.

3.7 A 20-lb force is applied to the control rod AB as shown. Knowing that the length of the rod is 9 in. and that $\alpha = 25°$, determine the moment of the force about point B by resolving the force into components along AB and in a direction perpendicular to AB.

3.8 A 20-lb force is applied to the control rod AB as shown. Knowing that the length of the rod is 9 in. and that the moment of the force about B is 120 lb·in. clockwise, determine the value of α.

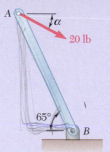

Fig. P3.6 through *P3.8*

3.9 Rod AB is held in place by the cord AC. Knowing that the tension in the cord is 1350 N and that $c = 360$ mm, determine the moment about B of the force exerted by the cord at point A by resolving that force into horizontal and vertical components applied (*a*) at point A, (*b*) at point C.

3.10 Rod AB is held in place by the cord AC. Knowing that $c = 840$ mm and that the moment about B of the force exerted by the cord at point A is 756 N·m, determine the tension in the cord.

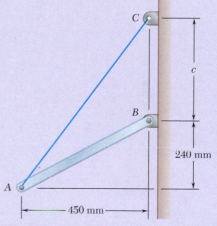

Fig. P3.9 and *P3.10*

3.11 and 3.12 The tailgate of a car is supported by the hydraulic lift BC. If the lift exerts a 125-lb force directed along its centerline on the ball and socket at B, determine the moment of the force about A.

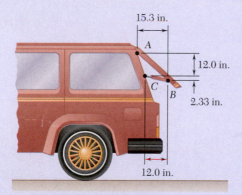

Fig. P3.11

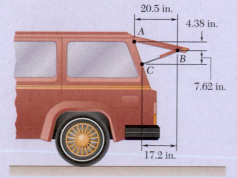

Fig. P3.12

3.13 and 3.14 It is known that the connecting rod AB exerts on the crank BC a 2.5-kN force directed down and to the left along the centerline of AB. Determine the moment of the force about C.

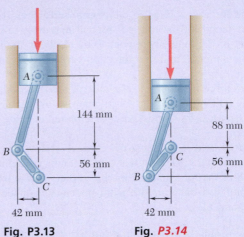

Fig. P3.13   Fig. *P3.14*

144 mm
56 mm
42 mm

88 mm
56 mm
42 mm

3.15 Form the vector products $\mathbf{B} \times \mathbf{C}$ and $\mathbf{B}' \times \mathbf{C}$, where $B = B'$, and use the results obtained to prove the identity

$$\sin \alpha \cos \beta = \tfrac{1}{2}\sin (\alpha + \beta) + \tfrac{1}{2}\sin (\alpha - \beta).$$

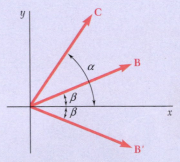

Fig. P3.15

3.16 The vectors $\mathbf{P}$ and $\mathbf{Q}$ are two adjacent sides of a parallelogram. Determine the area of the parallelogram when (*a*) $\mathbf{P} = -8\mathbf{i} + 4\mathbf{j} - 4\mathbf{k}$ and $\mathbf{Q} = 3\mathbf{i} + 3\mathbf{j} + 6\mathbf{k}$, (*b*) $\mathbf{P} = 7\mathbf{i} - 6\mathbf{j} - 3\mathbf{k}$ and $\mathbf{Q} = -3\mathbf{i} + 6\mathbf{j} - 2\mathbf{k}$.

3.17 A plane contains the vectors $\mathbf{A}$ and $\mathbf{B}$. Determine the unit vector normal to the plane when $\mathbf{A}$ and $\mathbf{B}$ are equal to, respectively, (*a*) $2\mathbf{i} + 3\mathbf{j} - 6\mathbf{k}$ and $5\mathbf{i} - 8\mathbf{j} - 6\mathbf{k}$, (*b*) $4\mathbf{i} - 4\mathbf{j} + 3\mathbf{k}$ and $-3\mathbf{i} + 7\mathbf{j} - 5\mathbf{k}$.

3.18 A line passes through the points (12 m, 8 m) and (−3 m, −5 m). Determine the perpendicular distance d from the line to the origin O of the system of coordinates.

3.19 Determine the moment about the origin O of the force $\mathbf{F} = 4\mathbf{i} - 3\mathbf{j} + 5\mathbf{k}$ that acts at a point A. Assume that the position vector of A is (*a*) $\mathbf{r} = 2\mathbf{i} + 3\mathbf{j} - 4\mathbf{k}$, (*b*) $\mathbf{r} = -8\mathbf{i} + 6\mathbf{j} - 10\mathbf{k}$, (*c*) $\mathbf{r} = 8\mathbf{i} - 6\mathbf{j} + 5\mathbf{k}$.

3.20 Determine the moment about the origin O of the force $\mathbf{F} = 2\mathbf{i} + 3\mathbf{j} - 4\mathbf{k}$ that acts at a point A. Assume that the position vector of A is (*a*) $\mathbf{r} = 3\mathbf{i} - 6\mathbf{j} + 5\mathbf{k}$, (*b*) $\mathbf{r} = \mathbf{i} - 4\mathbf{j} - 2\mathbf{k}$, (*c*) $\mathbf{r} = 4\mathbf{i} + 6\mathbf{j} - 8\mathbf{k}$.

3.21 Before the trunk of a large tree is felled, cables AB and BC are attached as shown. Knowing that the tensions in cables AB and BC are 555 N and 660 N, respectively, determine the moment about O of the resultant force exerted on the tree by the cables at B.

3.22 The 12-ft boom AB has a fixed end A. A steel cable is stretched from the free end B of the boom to a point C located on the vertical wall. If the tension in the cable is 380 lb, determine the moment about A of the force exerted by the cable at B.

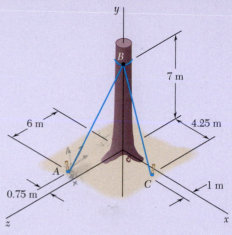

Fig. P3.21

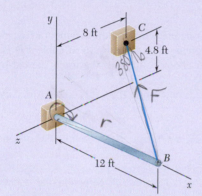

Fig. P3.22

3.23 A 200-N force is applied as shown to the bracket ABC. Determine the moment of the force about A.

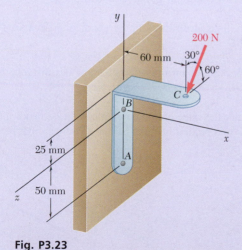

Fig. P3.23

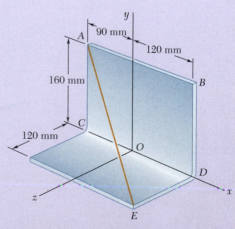

Fig. P3.24

3.24 The wire AE is stretched between the corners A and E of a bent plate. Knowing that the tension in the wire is 435 N, determine the moment about O of the force exerted by the wire (*a*) on corner A, (*b*) on corner E.

3.25 A 6-ft-long fishing rod *AB* is securely anchored in the sand of a beach. After a fish takes the bait, the resulting force in the line is 6 lb. Determine the moment about *A* of the force exerted by the line at *B*.

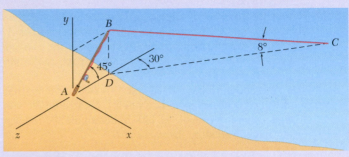

Fig. P3.25

3.26 A precast concrete wall section is temporarily held by two cables as shown. Knowing that the tension in cable *BD* is 900 N, determine the moment about point *O* of the force exerted by the cable at *B*.

3.27 In Prob. 3.22, determine the perpendicular distance from point *A* to cable *BC*.

3.28 In Prob. 3.24, determine the perpendicular distance from point *O* to wire *AE*.

3.29 In Prob. 3.24, determine the perpendicular distance from point *B* to wire *AE*.

3.30 In Prob. 3.25, determine the perpendicular distance from point *A* to a line drawn through points *B* and *C*.

3.31 In Prob. 3.25, determine the perpendicular distance from point *D* to a line drawn through points *B* and *C*.

3.32 In Prob. 3.26, determine the perpendicular distance from point *O* to cable *BD*.

3.33 In Prob. 3.26, determine the perpendicular distance from point *C* to cable *BD*.

3.34 Determine the value of *a* that minimizes the perpendicular distance from point *C* to a section of pipeline that passes through points *A* and *B*.

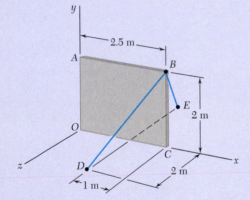

Fig. P3.26

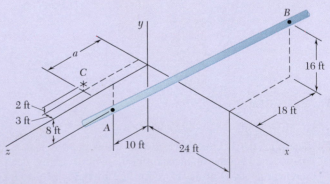

Fig. P3.34

3.2   MOMENT OF A FORCE ABOUT AN AXIS

We want to extend the idea of the moment about a point to the often useful concept of the moment about an axis. However, first we need to introduce another tool of vector mathematics. We have seen that the vector product multiplies two vectors together and produces a new vector. Here we examine the scalar product, which multiplies two vectors together and produces a scalar quantity.

3.2A   Scalar Products

The **scalar product** of two vectors **P** and **Q** is defined as the product of the magnitudes of **P** and **Q** and of the cosine of the angle θ formed between them (Fig. 3.18). The scalar product of **P** and **Q** is denoted by **P · Q**.

Fig. 3.18   Two vectors **P** and **Q** and the angle θ between them.

Scalar product
$$\mathbf{P} \cdot \mathbf{Q} = PQ\cos\theta \qquad (3.24)$$

Note that this expression is not a vector but a *scalar*, which explains the name *scalar product*. Because of the notation used, **P · Q** is also referred to as the *dot product* of the vectors **P** and **Q**.

It follows from its very definition that the scalar product of two vectors is *commutative*, i.e., that

$$\mathbf{P} \cdot \mathbf{Q} = \mathbf{Q} \cdot \mathbf{P} \qquad (3.25)$$

It can also be proven that the scalar product is *distributive*, as shown by

$$\mathbf{P} \cdot (\mathbf{Q}_1 + \mathbf{Q}_2) = \mathbf{P} \cdot \mathbf{Q}_1 + \mathbf{P} \cdot \mathbf{Q}_2 \qquad (3.26)$$

As far as the associative property is concerned, this property cannot apply to scalar products. Indeed, $(\mathbf{P} \cdot \mathbf{Q}) \cdot \mathbf{S}$ has no meaning, because $\mathbf{P} \cdot \mathbf{Q}$ is not a vector but a scalar.

We can also express the scalar product of two vectors **P** and **Q** in terms of their rectangular components. Resolving **P** and **Q** into components, we first write

$$\mathbf{P} \cdot \mathbf{Q} = (P_x\mathbf{i} + P_y\mathbf{j} + P_z\mathbf{k}) \cdot (Q_x\mathbf{i} + Q_y\mathbf{j} + Q_z\mathbf{k})$$

Making use of the distributive property, we express **P · Q** as the sum of scalar products, such as $P_x\mathbf{i} \cdot Q_x\mathbf{i}$ and $P_x\mathbf{i} \cdot Q_y\mathbf{j}$. However, from the definition of the scalar product, it follows that the scalar products of the unit vectors are either zero or one.

$$\begin{array}{lll} \mathbf{i} \cdot \mathbf{i} = 1 & \mathbf{j} \cdot \mathbf{j} = 1 & \mathbf{k} \cdot \mathbf{k} = 1 \\ \mathbf{i} \cdot \mathbf{j} = 0 & \mathbf{j} \cdot \mathbf{k} = 0 & \mathbf{k} \cdot \mathbf{i} = 0 \end{array} \qquad (3.27)$$

Thus, the expression for **P · Q** reduces to

Scalar product

$$\mathbf{P} \cdot \mathbf{Q} = P_xQ_x + P_yQ_y + P_zQ_z \qquad (3.28)$$

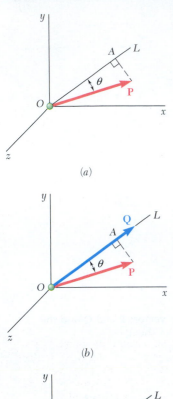

(a)

(b)

(c)

Fig. 3.19 *(a)* The projection of vector **P** at an angle θ to a line *OL*; *(b)* the projection of **P** and a vector **Q** along *OL*; *(c)* the projection of **P**, a unit vector $\boldsymbol{\lambda}$ along *OL*, and the angles of *OL* with the coordinate axes.

In the particular case when **P** and **Q** are equal, we note that

$$\mathbf{P} \cdot \mathbf{P} = P_x^2 + P_y^2 + P_z^2 = P^2 \tag{3.29}$$

Applications of the Scalar Product

1. **Angle formed by two given vectors.** Let two vectors be given in terms of their components:

$$\mathbf{P} = P_x\mathbf{i} + P_y\mathbf{j} + P_z\mathbf{k}$$
$$\mathbf{Q} = Q_x\mathbf{i} + Q_y\mathbf{j} + Q_z\mathbf{k}$$

To determine the angle formed by the two vectors, we equate the expressions obtained in Eqs. (3.24) and (3.28) for their scalar product,

$$PQ \cos \theta = P_xQ_x + P_yQ_y + P_zQ_z$$

Solving for $\cos \theta$, we have

$$\cos \theta = \frac{P_xQ_x + P_yQ_y + P_zQ_z}{PQ} \tag{3.30}$$

2. **Projection of a vector on a given axis.** Consider a vector **P** forming an angle θ with an axis, or directed line, *OL* (Fig. 3.19*a*). We define the *projection of* **P** *on the axis OL* as the scalar

$$P_{OL} = P \cos \theta \tag{3.31}$$

The projection P_{OL} is equal in absolute value to the length of the segment *OA*. It is positive if *OA* has the same sense as the axis *OL*—that is, if θ is acute—and negative otherwise. If **P** and *OL* are at a right angle, the projection of **P** on *OL* is zero.

Now consider a vector **Q** directed along *OL* and of the same sense as *OL* (Fig. 3.19*b*). We can express the scalar product of **P** and **Q** as

$$\mathbf{P} \cdot \mathbf{Q} = PQ \cos \theta = P_{OL}Q \tag{3.32}$$

from which it follows that

$$P_{OL} = \frac{\mathbf{P} \cdot \mathbf{Q}}{Q} = \frac{P_xQ_x + P_yQ_y + P_zQ_z}{Q} \tag{3.33}$$

In the particular case when the vector selected along *OL* is the unit vector $\boldsymbol{\lambda}$ (Fig. 3.19*c*), we have

$$P_{OL} = \mathbf{P} \cdot \boldsymbol{\lambda} \tag{3.34}$$

Recall from Sec. 2.4A that the components of $\boldsymbol{\lambda}$ along the coordinate axes are respectively equal to the direction cosines of *OL*. Resolving **P** and $\boldsymbol{\lambda}$ into rectangular components, we can express the projection of **P** on *OL* as

$$P_{OL} = P_x \cos \theta_x + P_y \cos \theta_y + P_z \cos \theta_z \tag{3.35}$$

where θ_x, θ_y, and θ_z denote the angles that the axis *OL* forms with the coordinate axes.

3.2B Mixed Triple Products

We have now seen both forms of multiplying two vectors together: the vector product and the scalar product. Here we define the **mixed triple product** of the three vectors **S**, **P**, and **Q** as the scalar expression

Mixed triple product

$$\mathbf{S} \cdot (\mathbf{P} \times \mathbf{Q}) \qquad (3.36)$$

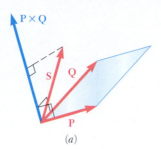

(a)

This is obtained by forming the scalar product of **S** with the vector product of **P** and **Q**. [In Chapter 15, we will introduce another kind of triple product, called the vector triple product, $\mathbf{S} \times (\mathbf{P} \times \mathbf{Q})$.]

The mixed triple product of **S**, **P**, and **Q** has a simple geometrical interpretation (Fig. 3.20a). Recall from Sec. 3.4 that the vector $\mathbf{P} \times \mathbf{Q}$ is perpendicular to the plane containing **P** and **Q** and that its magnitude is equal to the area of the parallelogram that has **P** and **Q** for sides. Also, Eq. (3.32) indicates that we can obtain the scalar product of **S** and $\mathbf{P} \times \mathbf{Q}$ by multiplying the magnitude of $\mathbf{P} \times \mathbf{Q}$ (i.e., the area of the parallelogram defined by **P** and **Q**) by the projection of **S** on the vector $\mathbf{P} \times \mathbf{Q}$ (i.e., by the projection of **S** on the normal to the plane containing the parallelogram). The mixed triple product is thus equal, in absolute value, to the volume of the parallelepiped having the vectors **S**, **P**, and **Q** for sides (Fig. 3.20b). The sign of the mixed triple product is positive if **S**, **P**, and **Q** form a right-handed triad and negative if they form a left-handed triad. [That is, $\mathbf{S} \cdot (\mathbf{P} \times \mathbf{Q})$ is negative if the rotation that brings **P** into line with **Q** is observed as clockwise from the tip of **S**]. The mixed triple product is zero if **S**, **P**, and **Q** are coplanar.

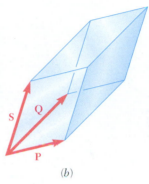

(b)

Fig. 3.20 (a) The mixed triple product is equal to the magnitude of the cross product of two vectors multiplied by the projection of the third vector onto that cross product; (b) the result equals the volume of the parallelepiped formed by the three vectors.

Since the parallelepiped defined in this way is independent of the order in which the three vectors are taken, the six mixed triple products that can be formed with **S**, **P**, and **Q** all have the same absolute value, although not the same sign. It is easily shown that

$$\mathbf{S} \cdot (\mathbf{P} \times \mathbf{Q}) = \mathbf{P} \cdot (\mathbf{Q} \times \mathbf{S}) = \mathbf{Q} \cdot (\mathbf{S} \times \mathbf{P})$$
$$= -\mathbf{S} \cdot (\mathbf{Q} \times \mathbf{P}) = -\mathbf{P} \cdot (\mathbf{S} \times \mathbf{Q}) = -\mathbf{Q} \cdot (\mathbf{P} \times \mathbf{S}) \qquad (3.37)$$

Arranging the letters representing the three vectors counterclockwise in a circle (Fig. 3.21), we observe that the sign of the mixed triple product remains unchanged if the vectors are permuted in such a way that they still read in counterclockwise order. Such a permutation is said to be a *circular permutation*. It also follows from Eq. (3.37) and from the commutative property of scalar products that the mixed triple product of **S**, **P**, and **Q** can be defined equally well as $\mathbf{S} \cdot (\mathbf{P} \times \mathbf{Q})$ or $(\mathbf{S} \times \mathbf{P}) \cdot \mathbf{Q}$.

Fig. 3.21 Counterclockwise arrangement for determining the sign of the mixed triple product of three vectors **P**, **Q**, and **S**.

We can also express the mixed triple product of the vectors **S**, **P**, and **Q** in terms of the rectangular components of these vectors. Denoting $\mathbf{P} \times \mathbf{Q}$ by **V** and using formula (3.28) to express the scalar product of **S** and **V**, we have

$$\mathbf{S} \cdot (\mathbf{P} \times \mathbf{Q}) = \mathbf{S} \cdot \mathbf{V} = S_x V_x + S_y V_y + S_z V_z$$

Substituting from the relations in Eq. (3.9) for the components of **V**, we obtain

$$\mathbf{S} \cdot (\mathbf{P} \times \mathbf{Q}) = S_x(P_y Q_z - P_z Q_y) + S_y(P_z Q_x - P_x Q_z)$$
$$+ S_z(P_x Q_y - P_y Q_x) \qquad (3.38)$$

We can write this expression in a more compact form if we observe that it represents the expansion of a determinant:

Mixed triple product, determinant form

$$\mathbf{S} \cdot (\mathbf{P} \times \mathbf{Q}) = \begin{vmatrix} S_x & S_y & S_z \\ P_x & P_y & P_z \\ Q_x & Q_y & Q_z \end{vmatrix} \tag{3.39}$$

By applying the rules governing the permutation of rows in a determinant, we could easily verify the relations in Eq. (3.37), which we derived earlier from geometrical considerations.

3.2C Moment of a Force about a Given Axis

Now that we have the necessary mathematical tools, we can introduce the concept of moment of a force about an axis. Consider again a force $\mathbf{F}$ acting on a rigid body and the moment $\mathbf{M}_O$ of that force about O (Fig. 3.22). Let OL be an axis through O.

We define the moment M_{OL} of $\mathbf{F}$ about OL as the projection OC of the moment M_O onto the axis OL.

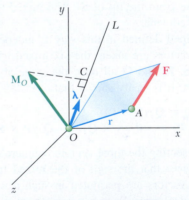

Fig. 3.22 The moment $\mathbf{M}_{OL}$ of a force $\mathbf{F}$ about the axis OL is the projection on OL of the moment $\mathbf{M}_O$. The calculation involves the unit vector $\boldsymbol{\lambda}$ along OL and the position vector $\mathbf{r}$ from O to A, the point upon which the force $\mathbf{F}$ acts.

Suppose we denote the unit vector along OL by $\boldsymbol{\lambda}$ and recall the expressions (3.34) and (3.11) for the projection of a vector on a given axis and for the moment $\mathbf{M}_O$ of a force $\mathbf{F}$. Then we can express M_{OL} as

Moment about an axis through the origin

$$M_{OL} = \boldsymbol{\lambda} \cdot \mathbf{M}_O = \boldsymbol{\lambda} \cdot (\mathbf{r} \times \mathbf{F}) \tag{3.40}$$

This shows that the moment M_{OL} of **F** about the axis OL is the scalar obtained by forming the mixed triple product of $\boldsymbol{\lambda}$, **r**, and **F**. We can also express M_{OL} in the form of a determinant,

$$M_{OL} = \begin{vmatrix} \lambda_x & \lambda_y & \lambda_z \\ x & y & z \\ F_x & F_y & F_z \end{vmatrix} \tag{3.41}$$

where $\lambda_x, \lambda_y, \lambda_z$ = direction cosines of axis OL

$\quad\quad x, y, z$ = coordinates of point of application of **F**

$\quad F_x, F_y, F_z$ = components of force **F**

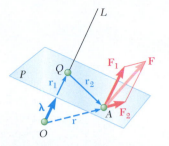

Fig. 3.23 By resolving the force **F** into components parallel to the axis OL and in a plane perpendicular to the axis, we can show that the moment $\mathbf{M}_{OL}$ of **F** about OL measures the tendency of **F** to rotate the rigid body about the axis.

The physical significance of the moment M_{OL} of a force **F** about a fixed axis OL becomes more apparent if we resolve **F** into two rectangular components $\mathbf{F}_1$ and $\mathbf{F}_2$, with $\mathbf{F}_1$ parallel to OL and $\mathbf{F}_2$ lying in a plane P perpendicular to OL (Fig. 3.23). Resolving **r** similarly into two components $\mathbf{r}_1$ and $\mathbf{r}_2$ and substituting for **F** and **r** into Eq. (3.40), we get

$$M_{OL} = \boldsymbol{\lambda} \cdot [(\mathbf{r}_1 + \mathbf{r}_2) \times (\mathbf{F}_1 + \mathbf{F}_2)]$$
$$= \boldsymbol{\lambda} \cdot (\mathbf{r}_1 \times \mathbf{F}_1) + \boldsymbol{\lambda} \cdot (\mathbf{r}_1 \times \mathbf{F}_2) + \boldsymbol{\lambda} \cdot (\mathbf{r}_2 \times \mathbf{F}_1) + \boldsymbol{\lambda} \cdot (\mathbf{r}_2 \times \mathbf{F}_2)$$

Note that all of the mixed triple products except the last one are equal to zero because they involve vectors that are coplanar when drawn from a common origin (Sec. 3.2B). Therefore, this expression reduces to

$$M_{OL} = \boldsymbol{\lambda} \cdot (\mathbf{r}_2 \times \mathbf{F}_2) \tag{3.42}$$

The vector product $\mathbf{r}_2 \times \mathbf{F}_2$ is perpendicular to the plane P and represents the moment of the component $\mathbf{F}_2$ of **F** about the point Q where OL intersects P. Therefore, the scalar M_{OL}, which is positive if $\mathbf{r}_2 \times \mathbf{F}_2$ and OL have the same sense and is negative otherwise, measures the tendency of $\mathbf{F}_2$ to make the rigid body rotate about the fixed axis OL. The other component $\mathbf{F}_1$ of **F** does not tend to make the body rotate about OL, because $\mathbf{F}_1$ and OL are parallel. Therefore, we conclude that

*The moment M_{OL} of **F** about OL measures the tendency of the force **F** to impart to the rigid body a rotation about the fixed axis OL.*

From the definition of the moment of a force about an axis, it follows that the moment of **F** about a coordinate axis is equal to the component of $\mathbf{M}_O$ along that axis. If we substitute each of the unit vectors **i**, **j**, and **k** for $\boldsymbol{\lambda}$ in Eq. (3.40), we obtain expressions for the *moments of **F** about the coordinate axes*. These expressions are respectively equal to those obtained earlier for the components of the moment $\mathbf{M}_O$ of **F** about O:

$$\begin{aligned} M_x &= yF_z - zF_y \\ M_y &= zF_x - xF_z \\ M_z &= xF_y - yF_x \end{aligned} \tag{3.18}$$

Just as the components F_x, F_y, and F_z of a force **F** acting on a rigid body measure, respectively, the tendency of **F** to move the rigid body in the x, y, and z directions, the moments M_x, M_y, and M_z of **F** about the coordinate axes measure the tendency of **F** to impart to the rigid body a rotation about the x, y, and z axes, respectively.

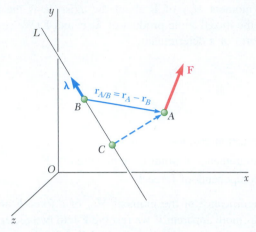

Fig. 3.24 The moment of a force about an axis or line L can be found by evaluating the mixed triple product at a point B on the line. The choice of B is arbitrary, since using any other point on the line, such as C, yields the same result.

More generally, we can obtain the moment of a force $\mathbf{F}$ applied at A about an axis that does not pass through the origin by choosing an arbitrary point B on the axis (Fig. 3.24) and determining the projection on the axis BL of the moment $\mathbf{M}_B$ of $\mathbf{F}$ about B. The equation for this projection is given here.

Moment about an arbitrary axis

$$M_{BL} = \boldsymbol{\lambda} \cdot \mathbf{M}_B = \boldsymbol{\lambda} \cdot (\mathbf{r}_{A/B} \times \mathbf{F}) \tag{3.43}$$

where $\mathbf{r}_{A/B} = \mathbf{r}_A - \mathbf{r}_B$ represents the vector drawn from B to A. Expressing M_{BL} in the form of a determinant, we have

$$M_{BL} = \begin{vmatrix} \lambda_x & \lambda_y & \lambda_z \\ x_{A/B} & y_{A/B} & z_{A/B} \\ F_x & F_y & F_z \end{vmatrix} \tag{3.44}$$

where λ_x, λ_y, λ_z = direction cosines of axis BL

$$x_{A/B} = x_A - x_B \qquad y_{A/B} = y_A - y_B \qquad z_{A/B} = z_A - z_B$$

F_x, F_y, F_z = components of force $\mathbf{F}$

Note that this result is independent of the choice of the point B on the given axis. Indeed, denoting by M_{CL} the moment obtained with a different point C, we have

$$M_{CL} = \boldsymbol{\lambda} \cdot [(\mathbf{r}_A - \mathbf{r}_C) \times \mathbf{F}]$$
$$= \boldsymbol{\lambda} \cdot [(\mathbf{r}_A - \mathbf{r}_B) \times \mathbf{F}] + \boldsymbol{\lambda} \cdot [(\mathbf{r}_B - \mathbf{r}_C) \times \mathbf{F}]$$

However, since the vectors $\boldsymbol{\lambda}$ and $\mathbf{r}_B - \mathbf{r}_C$ lie along the same line, the volume of the parallelepiped having the vectors $\boldsymbol{\lambda}$, $\mathbf{r}_B - \mathbf{r}_C$, and $\mathbf{F}$ for sides is zero, as is the mixed triple product of these three vectors (Sec. 3.2B). The expression obtained for M_{CL} thus reduces to its first term, which is the expression used earlier to define M_{BL}. In addition, it follows from Sec. 3.1E that, when computing the moment of $\mathbf{F}$ about the given axis, A can be any point on the line of action of $\mathbf{F}$.

Sample Problem 3.5

A cube of side a is acted upon by a force $\mathbf{P}$ along the diagonal of a face, as shown. Determine the moment of $\mathbf{P}$ (*a*) about A, (*b*) about the edge AB, (*c*) about the diagonal AG of the cube. (*d*) Using the result of part *c*, determine the perpendicular distance between AG and FC.

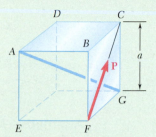

STRATEGY: Use the equations presented in this section to compute the moments asked for. You can find the distance between AG and FC from the expression for the moment M_{AG}.

MODELING and ANALYSIS:

a. Moment about *A*. Choosing x, y, and z axes as shown (Fig. 1), resolve into rectangular components the force $\mathbf{P}$ and the vector $\mathbf{r}_{F/A} = \overrightarrow{AF}$ drawn from A to the point of application F of $\mathbf{P}$.

$$\mathbf{r}_{F/A} = a\mathbf{i} - a\mathbf{j} = a(\mathbf{i} - \mathbf{j})$$
$$\mathbf{P} = (P/\sqrt{2})\mathbf{j} - (P/\sqrt{2})\mathbf{k} = (P/\sqrt{2})(\mathbf{j} - \mathbf{k})$$

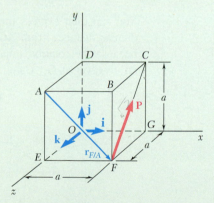

Fig. 1 Position vector $\mathbf{r}_{F/A}$ and force vector $\mathbf{P}$ relative to chosen coordinate system.

The moment of $\mathbf{P}$ about A is the vector product of these two vectors:

$$\mathbf{M}_A = \mathbf{r}_{F/A} \times \mathbf{P} = a(\mathbf{i} - \mathbf{j}) \times (P/\sqrt{2})(\mathbf{j} - \mathbf{k})$$
$$\mathbf{M}_A = (aP/\sqrt{2})(\mathbf{i} + \mathbf{j} + \mathbf{k}) \quad \blacktriangleleft$$

b. Moment about *AB*. You want the projection of $\mathbf{M}_A$ on AB:

$$M_{AB} = \mathbf{i} \cdot \mathbf{M}_A = \mathbf{i} \cdot (aP/\sqrt{2})(\mathbf{i} + \mathbf{j} + \mathbf{k})$$
$$M_{AB} = aP/\sqrt{2} \quad \blacktriangleleft$$

(continued)

You can verify that since AB is parallel to the x axis, M_{AB} is also the x component of the moment $\mathbf{M}_A$.

c. Moment about diagonal AG.
You obtain the moment of $\mathbf{P}$ about AG by projecting $\mathbf{M}_A$ on AG. If you denote the unit vector along AG by $\boldsymbol{\lambda}$ (Fig. 2), the calculation looks like this:

$$\boldsymbol{\lambda} = \frac{\overrightarrow{AG}}{AG} = \frac{a\mathbf{i} - a\mathbf{j} - a\mathbf{k}}{a\sqrt{3}} = (1/\sqrt{3})(\mathbf{i} - \mathbf{j} - \mathbf{k})$$

$$M_{AG} = \boldsymbol{\lambda} \cdot \mathbf{M}_A = (1/\sqrt{3})(\mathbf{i} - \mathbf{j} - \mathbf{k}) \cdot (aP/\sqrt{2})(\mathbf{i} + \mathbf{j} + \mathbf{k})$$

$$M_{AG} = (aP/\sqrt{6})(1 - 1 - 1) \quad M_{AG} = -aP/\sqrt{6} \quad \blacktriangleleft$$

Alternative Method.
You can also calculate the moment of $\mathbf{P}$ about AG from the determinant form:

$$M_{AG} = \begin{vmatrix} \lambda_x & \lambda_y & \lambda_z \\ x_{F/A} & y_{F/A} & z_{F/A} \\ F_x & F_y & F_z \end{vmatrix} = \begin{vmatrix} 1/\sqrt{3} & -1/\sqrt{3} & -1/\sqrt{3} \\ a & -a & 0 \\ 0 & P/\sqrt{2} & -P/\sqrt{2} \end{vmatrix} = -aP/\sqrt{6}$$

d. Perpendicular Distance between AG and FC.
First note that $\mathbf{P}$ is perpendicular to the diagonal AG. You can check this by forming the scalar product $\mathbf{P} \cdot \boldsymbol{\lambda}$ and verifying that it is zero:

$$\mathbf{P} \cdot \boldsymbol{\lambda} = (P/\sqrt{2})(\mathbf{j} - \mathbf{k}) \cdot (1/\sqrt{3})(\mathbf{i} - \mathbf{j} - \mathbf{k}) = (P\sqrt{6})(0 - 1 + 1) = 0$$

You can then express the moment M_{AG} as $-Pd$, where d is the perpendicular distance from AG to FC (Fig. 3). (The negative sign is needed because the rotation imparted to the cube by $\mathbf{P}$ appears as clockwise to an observer at G.) Using the value found for M_{AG} in part c,

$$M_{AG} = -Pd = -aP/\sqrt{6} \qquad d = a/\sqrt{6} \quad \blacktriangleleft$$

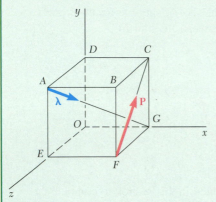

Fig. 2 Unit vector $\boldsymbol{\lambda}$ used to determine moment of **P** about AG.

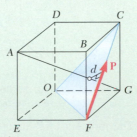

Fig. 3 Perpendicular distance d from AG to FC.

REFLECT and THINK: In a problem like this, it is important to visualize the forces and moments in three dimensions so you can choose the appropriate equations for finding them and also recognize the geometric relationships between them.

SOLVING PROBLEMS ON YOUR OWN

In the problems for this section, you will apply the *scalar product* (or *dot product*) of two vectors to determine the *angle formed by two given vectors* and the *projection of a force on a given axis*. You will also use the *mixed triple product* of three vectors to find the *moment of a force about a given axis* and the *perpendicular distance between two lines*.

1. Calculating the angle formed by two given vectors. First express the vectors in terms of their components and determine the magnitudes of the two vectors. Then find the cosine of the desired angle by dividing the scalar product of the two vectors by the product of their magnitudes [Eq. (3.30)].

2. Computing the projection of a vector P on a given axis *OL*. In general, begin by expressing $\mathbf{P}$ and the unit vector $\boldsymbol{\lambda}$, which defines the direction of the axis, in component form. Take care that $\boldsymbol{\lambda}$ has the correct sense (that is, $\boldsymbol{\lambda}$ is directed from O to L). The required projection is then equal to the scalar product $\mathbf{P} \cdot \boldsymbol{\lambda}$. However, if you know the angle θ formed by $\mathbf{P}$ and $\boldsymbol{\lambda}$, the projection is also given by $P \cos \theta$.

3. Determining the moment M_{OL} of a force about a given axis *OL*. We defined M_{OL} as

$$M_{OL} = \boldsymbol{\lambda} \cdot \mathbf{M}_O = \boldsymbol{\lambda} \cdot (\mathbf{r} \times \mathbf{F}) \qquad (3.40)$$

where $\boldsymbol{\lambda}$ is the unit vector along *OL* and $\mathbf{r}$ is a position vector *from any point* on the line *OL to any point* on the line of action of $\mathbf{F}$. As was the case for the moment of a force about a point, choosing the most convenient position vector will simplify your calculations. Also, recall the warning of the preceding section: The vectors $\mathbf{r}$ and $\mathbf{F}$ must have the correct sense, and they must be placed in the proper order. The procedure you should follow when computing the moment of a force about an axis is illustrated in part c of Sample Prob. 3.5. The two essential steps in this procedure are (1) express $\boldsymbol{\lambda}$, $\mathbf{r}$, and $\mathbf{F}$ in terms of their rectangular components and (2) evaluate the mixed triple product $\boldsymbol{\lambda} \cdot (\mathbf{r} \times \mathbf{F})$ to determine the moment about the axis. In most three-dimensional problems, the most convenient way to compute the mixed triple product is by using a determinant.

As noted in the text, when $\boldsymbol{\lambda}$ is directed along one of the coordinate axes, M_{OL} is equal to the scalar component of $\mathbf{M}_O$ along that axis.

4. Determining the perpendicular distance between two lines. Remember that it is the perpendicular component $\mathbf{F}_2$ of the force $\mathbf{F}$ that tends to make a body rotate about a given axis *OL* (Fig. 3.23). It then follows that

$$M_{OL} = F_2 \, d$$

(continued)

where M_{OL} is the moment of $\mathbf{F}$ about axis OL and d is the perpendicular distance between OL and the line of action of $\mathbf{F}$. This last equation provides a simple technique for determining d. First assume that a force $\mathbf{F}$ of known magnitude F lies along one of the given lines and that the unit vector $\boldsymbol{\lambda}$ lies along the other line. Next compute the moment M_{OL} of the force $\mathbf{F}$ about the second line using the method discussed above. The magnitude of the parallel component, F_1, of $\mathbf{F}$ is obtained using the scalar product:

$$F_1 = \mathbf{F} \cdot \boldsymbol{\lambda}$$

The value of F_2 is then determined from

$$F_2 = \sqrt{F^2 - F_1^2}$$

Finally, substitute the values of M_{OL} and F_2 into the equation $M_{OL} = F_2\, d$ and solve for d.

You should now realize that the calculation of the perpendicular distance in part d of Sample Prob. 3.5 was simplified by $\mathbf{P}$ being perpendicular to the diagonal AG. In general, the two given lines will not be perpendicular, so you will have to use the technique just outlined when determining the perpendicular distance between them.

Problems

3.35 Given the vectors $\mathbf{P} = 2\mathbf{i} + 3\mathbf{j} - \mathbf{k}$, $\mathbf{Q} = 5\mathbf{i} - 4\mathbf{j} + 3\mathbf{k}$, and $\mathbf{S} = -3\mathbf{i} + 2\mathbf{j} - 5\mathbf{k}$, compute the scalar products $\mathbf{P} \cdot \mathbf{Q}$, $\mathbf{P} \cdot \mathbf{S}$, and $\mathbf{Q} \cdot \mathbf{S}$.

3.36 Form the scalar product $\mathbf{B} \cdot \mathbf{C}$ and use the result obtained to prove the identity

$$\cos(\alpha - \beta) = \cos\alpha\cos\beta + \sin\alpha\sin\beta$$

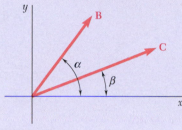

Fig. P3.36

3.37 Three cables are used to support a container as shown. Determine the angle formed by cables AB and AD.

3.38 Three cables are used to support a container as shown. Determine the angle formed by cables AC and AD.

3.39 Knowing that the tension in cable AC is 280 lb, determine (a) the angle between cable AC and the boom AB, (b) the projection on AB of the force exerted by cable AC at point A.

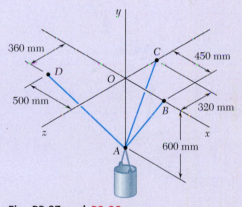

Fig. P3.37 and *P3.38*

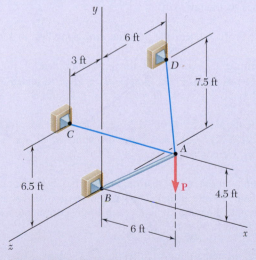

Fig. P3.39 and P3.40

3.40 Knowing that the tension in cable AD is 180 lb, determine (a) the angle between cable AD and the boom AB, (b) the projection on AB of the force exerted by cable AD at point A.

3.41 Ropes *AB* and *BC* are two of the ropes used to support a tent. The two ropes are attached to a stake at *B*. If the tension in rope *AB* is 540 N, determine (*a*) the angle between rope *AB* and the stake, (*b*) the projection on the stake of the force exerted by rope *AB* at point *B*.

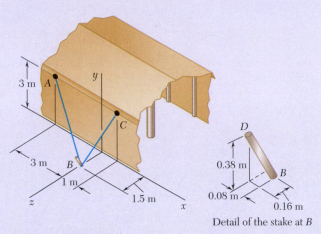

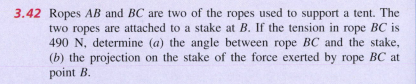

Fig. P3.41 and *P3.42*

3.42 Ropes *AB* and *BC* are two of the ropes used to support a tent. The two ropes are attached to a stake at *B*. If the tension in rope *BC* is 490 N, determine (*a*) the angle between rope *BC* and the stake, (*b*) the projection on the stake of the force exerted by rope *BC* at point *B*.

3.43 The 20-in. tube *AB* can slide along a horizontal rod. The ends *A* and *B* of the tube are connected by elastic cords to the fixed point *C*. For the position corresponding to *x* = 11 in., determine the angle formed by the two cords, (*a*) using Eq. (3.30), (*b*) applying the law of cosines to triangle *ABC*.

3.44 Solve Prob. 3.43 for the position corresponding to *x* = 4 in.

3.45 Determine the volume of the parallelepiped of Fig. 3.20*b* when (*a*) **P** = 4**i** − 3**j** + 2**k**, **Q** = −2**i** − 5**j** + **k**, and **S** = 7**i** + **j** − **k**, (*b*) **P** = 5**i** − **j** + 6**k**, **Q** = 2**i** + 3**j** + **k**, and **S** = −3**i** − 2**j** + 4**k**.

3.46 Given the vectors **P** = 3**i** − **j** + **k**, **Q** = 4**i** + Q_y**j** − 2**k**, and **S** = 2**i** − 2**j** + 2**k**, determine the value of Q_y for which the three vectors are coplanar.

3.47 A crane is oriented so that the end of the 25-m boom *AO* lies in the *yz* plane. At the instant shown, the tension in cable *AB* is 4 kN. Determine the moment about each of the coordinate axes of the force exerted on *A* by cable *AB*.

3.48 The 25-m crane boom *AO* lies in the *yz* plane. Determine the maximum permissible tension in cable *AB* if the absolute value of moments about the coordinate axes of the force exerted on *A* by cable *AB* must be

$$|M_x| \leq 60 \text{ kN·m}, \quad |M_y| \leq 12 \text{ kN·m}, \quad |M_z| \leq 8 \text{ kN·m}$$

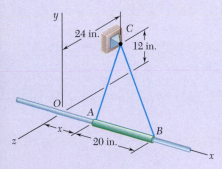

Fig. P3.43

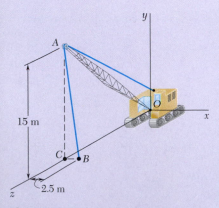

Fig. P3.47 and P3.48

3.49 To loosen a frozen valve, a force **F** with a magnitude of 70 lb is applied to the handle of the valve. Knowing that $\theta = 25°$, $M_x = -61$ lb·ft, and $M_z = -43$ lb·ft, determine ϕ and d.

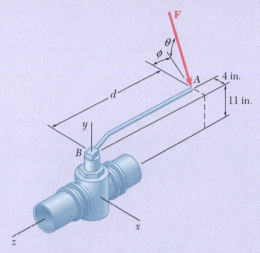

Fig. P3.49 and P3.50

3.50 When a force **F** is applied to the handle of the valve shown, its moments about the x and z axes are $M_x = -77$ lb·ft and $M_z = -81$ lb·ft, respectively. For $d = 27$ in., determine the moment M_y of **F** about the y axis.

3.51 To lift a heavy crate, a man uses a block and tackle attached to the bottom of an I-beam at hook B. Knowing that the moments about the y and the z axes of the force exerted at B by portion AB of the rope are, respectively, 120 N·m and -460 N·m, determine the distance a.

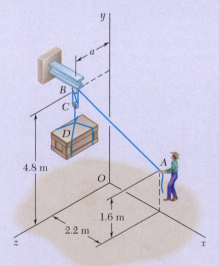

Fig. P3.51 and P3.52

3.52 To lift a heavy crate, a man uses a block and tackle attached to the bottom of an I-beam at hook B. Knowing that the man applies a 195-N force to end A of the rope and that the moment of that force about the y axis is 132 N·m, determine the distance a.

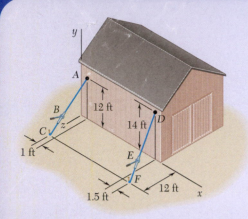

Fig. P3.53

3.53 A farmer uses cables and winch pullers B and E to plumb one side of a small barn. If it is known that the sum of the moments about the x axis of the forces exerted by the cables on the barn at points A and D is equal to 4728 lb·ft, determine the magnitude of $\mathbf{T}_{DE}$ when $T_{AB} = 255$ lb.

3.54 Solve Prob. 3.53 when the tension in cable AB is 306 lb.

3.55 The 23-in. vertical rod CD is welded to the midpoint C of the 50-in. rod AB. Determine the moment about AB of the 235-lb force $\mathbf{P}$.

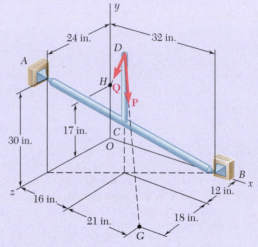

Fig. P3.55 and *P3.56*

3.56 The 23-in. vertical rod CD is welded to the midpoint C of the 50-in. rod AB. Determine the moment about AB of the 174-lb force $\mathbf{Q}$.

3.57 The frame ACD is hinged at A and D and is supported by a cable that passes through a ring at B and is attached to hooks at G and H. Knowing that the tension in the cable is 450 N, determine the moment about the diagonal AD of the force exerted on the frame by portion BH of the cable.

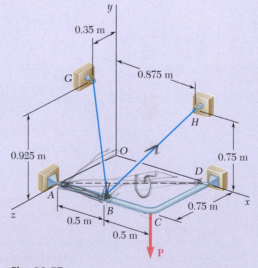

Fig. P3.57

3.58 In Prob. 3.57, determine the moment about the diagonal *AD* of the force exerted on the frame by portion *BG* of the cable.

3.59 The triangular plate *ABC* is supported by ball-and-socket joints at *B* and *D* and is held in the position shown by cables *AE* and *CF*. If the force exerted by cable *AE* at *A* is 55 N, determine the moment of that force about the line joining points *D* and *B*.

3.60 The triangular plate *ABC* is supported by ball-and-socket joints at *B* and *D* and is held in the position shown by cables *AE* and *CF*. If the force exerted by cable *CF* at *C* is 33 N, determine the moment of that force about the line joining points *D* and *B*.

3.61 A regular tetrahedron has six edges of length *a*. A force **P** is directed as shown along edge *BC*. Determine the moment of **P** about edge *OA*.

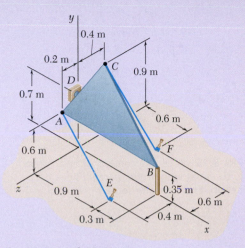

Fig. P3.59 and P3.60

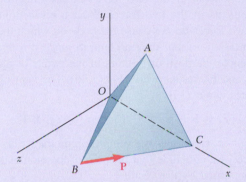

Fig. P3.61 and *P3.62*

3.62 A regular tetrahedron has six edges of length *a*. (*a*) Show that two opposite edges, such as *OA* and *BC*, are perpendicular to each other. (*b*) Use this property and the result obtained in Prob. 3.61 to determine the perpendicular distance between edges *OA* and *BC*.

3.63 Two forces **F₁** and **F₂** in space have the same magnitude *F*. Prove that the moment of **F₁** about the line of action of **F₂** is equal to the moment of **F₂** about the line of action of **F₁**.

*3.64 In Prob. 3.55, determine the perpendicular distance between rod *AB* and the line of action of **P**.

*3.65 In Prob. 3.56, determine the perpendicular distance between rod *AB* and the line of action of **Q**.

*3.66 In Prob. 3.57, determine the perpendicular distance between portion *BH* of the cable and the diagonal *AD*.

*3.67 In Prob. 3.58, determine the perpendicular distance between portion *BG* of the cable and the diagonal *AD*.

*3.68 In Prob. 3.59, determine the perpendicular distance between cable *AE* and the line joining points *D* and *B*.

*3.69 In Prob. 3.60, determine the perpendicular distance between cable *CF* and the line joining points *D* and *B*.

Fig. 3.25 A couple consists of two forces with equal magnitude, parallel lines of action, and opposite sense.

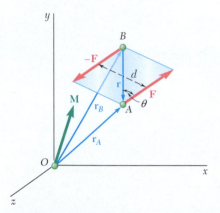

Fig. 3.26 The moment **M** of the couple about *O* is the sum of the moments of **F** and of −**F** about *O*.

Photo 3.1 The parallel upward and downward forces of equal magnitude exerted on the arms of the lug nut wrench are an example of a couple.

3.3 COUPLES AND FORCE-COUPLE SYSTEMS

Now that we have studied the effects of forces and moments on a rigid body, we can ask if it is possible to simplify a system of forces and moments without changing these effects. It turns out that we *can* replace a system of forces and moments with a simpler and equivalent system. One of the key ideas used in such a transformation is called a couple.

3.3A Moment of a Couple

Two forces **F** *and* −**F**, *having the same magnitude, parallel lines of action, and opposite sense, are said to form a* **couple** (Fig. 3.25). Clearly, the sum of the components of the two forces in any direction is zero. The sum of the moments of the two forces about a given point, however, is not zero. The two forces do not cause the body on which they act to move along a line (translation), but they do tend to make it rotate.

Let us denote the position vectors of the points of application of **F** and −**F** by $\mathbf{r}_A$ and $\mathbf{r}_B$, respectively (Fig. 3.26). The sum of the moments of the two forces about *O* is

$$\mathbf{r}_A \times \mathbf{F} + \mathbf{r}_B \times (-\mathbf{F}) = (\mathbf{r}_A - \mathbf{r}_B) \times \mathbf{F}$$

Setting $\mathbf{r}_A - \mathbf{r}_B = \mathbf{r}$, where **r** is the vector joining the points of application of the two forces, we conclude that the sum of the moments of **F** and −**F** about *O* is represented by the vector

$$\mathbf{M} = \mathbf{r} \times \mathbf{F} \qquad (3.45)$$

The vector **M** is called the *moment of the couple*. It is perpendicular to the plane containing the two forces, and its magnitude is

$$M = rF \sin \theta = Fd \qquad (3.46)$$

where *d* is the perpendicular distance between the lines of action of **F** and −**F** and θ is the angle between **F** (or −**F**) and **r**. The sense of **M** is defined by the right-hand rule.

Note that the vector **r** in Eq. (3.45) is independent of the choice of the origin *O* of the coordinate axes. Therefore, we would obtain the same result if the moments of **F** and −**F** had been computed about a different point *O′*. Thus, the moment **M** of a couple is a *free vector* (Sec. 2.1B), which can be applied at any point (Fig. 3.27).

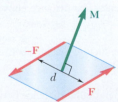

Fig. 3.27 The moment **M** of a couple equals the product of **F** and *d*, is perpendicular to the plane of the couple, and may be applied at any point of that plane.

From the definition of the moment of a couple, it also follows that two couples—one consisting of the forces $\mathbf{F}_1$ and $-\mathbf{F}_1$, the other of the forces $\mathbf{F}_2$ and $-\mathbf{F}_2$ (Fig. 3.28)—have equal moments if

$$F_1 d_1 = F_2 d_2 \qquad (3.47)$$

provided that the two couples lie in parallel planes (or in the same plane) and have the same sense (i.e., clockwise or counterclockwise).

3.3B Equivalent Couples

Imagine that three couples act successively on the same rectangular box (Fig. 3.29). As we have just seen, the only motion a couple can impart to a rigid body is a rotation. Since each of the three couples shown has the same moment $\mathbf{M}$ (same direction and same magnitude $M = 120$ lb·in.), we can expect each couple to have the same effect on the box.

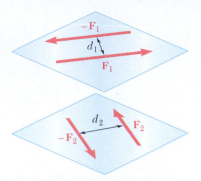

Fig. 3.28 Two couples have the same moment if they lie in parallel planes, have the same sense, and if $F_1 d_1 = F_2 d_2$.

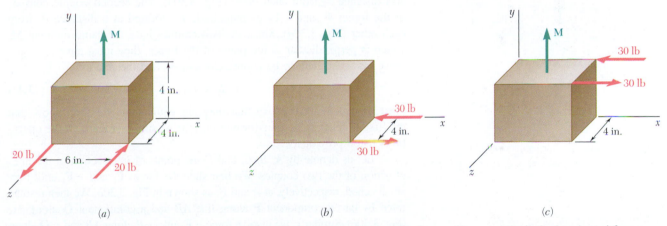

(a) (b) (c)

Fig. 3.29 Three equivalent couples. (a) A couple acting on the bottom of the box, acting counterclockwise viewed from above; (b) a couple in the same plane and with the same sense but larger forces than in (a); (c) a couple acting in a different plane but same sense.

As reasonable as this conclusion appears, we should not accept it hastily. Although intuition is of great help in the study of mechanics, it should not be accepted as a substitute for logical reasoning. Before stating that two systems (or groups) of forces have the same effect on a rigid body, we should prove that fact on the basis of the experimental evidence introduced so far. This evidence consists of the parallelogram law for the addition of two forces (Sec. 2.1A) and the principle of transmissibility (Sec. 3.1B). Therefore, we state that **two systems of forces are equivalent** (i.e., they have the same effect on a rigid body) **if we can transform one of them into the other by means of one or several of the following operations**: (1) replacing two forces acting on the same particle by their resultant; (2) resolving a force into two components; (3) canceling two equal and opposite forces acting on the same particle; (4) attaching to the same particle two equal and opposite forces; and (5) moving a force along its line of action. Each of these operations is easily justified on the basis of the parallelogram law or the principle of transmissibility.

Let us now prove that **two couples having the same moment M are equivalent**. First consider two couples contained in the same plane, and

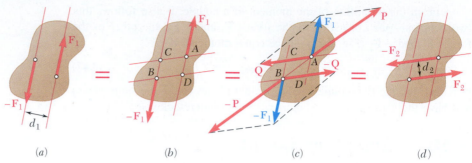

Fig. 3.30   Four steps in transforming one couple to another couple in the same plane by using simple operations. (a) Starting couple; (b) label points of intersection of lines of action of the two couples; (c) resolve forces from first couple into components; (d) final couple.

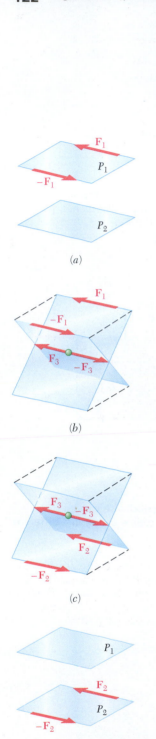

Fig. 3.31   Four steps in transforming one couple to another couple in a parallel plane by using simple operations. (a) Initial couple; (b) add a force pair along the line of intersection of two diagonal planes; (c) replace two couples with equivalent couples in the same planes; (d) final couple.

assume that this plane coincides with the plane of the figure (Fig. 3.30). The first couple consists of the forces $\mathbf{F}_1$ and $-\mathbf{F}_1$ of magnitude F_1, located at a distance d_1 from each other (Fig. 3.30a). The second couple consists of the forces $\mathbf{F}_2$ and $-\mathbf{F}_2$ of magnitude F_2, located at a distance d_2 from each other (Fig. 3.30d). Since the two couples have the same moment $\mathbf{M}$, which is perpendicular to the plane of the figure, they must have the same sense (assumed here to be counterclockwise), and the relation

$$F_1 d_1 = F_2 d_2 \qquad\qquad (3.47)$$

must be satisfied. To prove that they are equivalent, we shall show that the first couple can be transformed into the second by means of the operations listed previously.

Let us denote by *A, B, C,* and *D* the points of intersection of the lines of action of the two couples. We first slide the forces $\mathbf{F}_1$ and $-\mathbf{F}_1$ until they are attached, respectively, at *A* and *B*, as shown in Fig. 3.30b. We then resolve force $\mathbf{F}_1$ into a component $\mathbf{P}$ along line *AB* and a component $\mathbf{Q}$ along *AC* (Fig. 3.30c). Similarly, we resolve force $-\mathbf{F}_1$ into $-\mathbf{P}$ along *AB* and $-\mathbf{Q}$ along *BD*. The forces $\mathbf{P}$ and $-\mathbf{P}$ have the same magnitude, the same line of action, and opposite sense; we can move them along their common line of action until they are applied at the same point and may then be canceled. Thus, the couple formed by $\mathbf{F}_1$ and $-\mathbf{F}_1$ reduces to a couple consisting of $\mathbf{Q}$ and $-\mathbf{Q}$.

We now show that the forces $\mathbf{Q}$ and $-\mathbf{Q}$ are respectively equal to the forces $-\mathbf{F}_2$ and $\mathbf{F}_2$. We obtain the moment of the couple formed by $\mathbf{Q}$ and $-\mathbf{Q}$ by computing the moment of $\mathbf{Q}$ about *B*. Similarly, the moment of the couple formed by $\mathbf{F}_1$ and $-\mathbf{F}_1$ is the moment of $\mathbf{F}_1$ about *B*. However, by Varignon's theorem, the moment of $\mathbf{F}_1$ is equal to the sum of the moments of its components $\mathbf{P}$ and $\mathbf{Q}$. Since the moment of $\mathbf{P}$ about *B* is zero, the moment of the couple formed by $\mathbf{Q}$ and $-\mathbf{Q}$ must be equal to the moment of the couple formed by $\mathbf{F}_1$ and $-\mathbf{F}_1$. Recalling Eq. (3.47), we have

$$Qd_2 = F_1 d_1 = F_2 d_2 \qquad \text{and} \qquad Q = F_2$$

Thus, the forces $\mathbf{Q}$ and $-\mathbf{Q}$ are respectively equal to the forces $-\mathbf{F}_2$ and $\mathbf{F}_2$, and the couple of Fig. 3.30a is equivalent to the couple of Fig. 3.30d.

Now consider two couples contained in parallel planes P_1 and P_2. We prove that they are equivalent if they have the same moment. In view of the preceding discussion, we can assume that the couples consist of forces of the same magnitude F acting along parallel lines (Fig. 3.31a and d). We propose to show that the couple contained in plane P_1 can be transformed into the couple contained in plane P_2 by means of the standard operations listed previously.

Let us consider the two diagonal planes defined respectively by the lines of action of $\mathbf{F}_1$ and $-\mathbf{F}_2$ and by those of $-\mathbf{F}_1$ and $\mathbf{F}_2$ (Fig. 3.31*b*). At a point on their line of intersection, we attach two forces $\mathbf{F}_3$ and $-\mathbf{F}_3$, which are respectively equal to $\mathbf{F}_1$ and $-\mathbf{F}_1$. The couple formed by $\mathbf{F}_1$ and $-\mathbf{F}_3$ can be replaced by a couple consisting of $\mathbf{F}_3$ and $-\mathbf{F}_2$ (Fig. 3.31*c*), because both couples clearly have the same moment and are contained in the same diagonal plane. Similarly, the couple formed by $-\mathbf{F}_1$ and $\mathbf{F}_3$ can be replaced by a couple consisting of $-\mathbf{F}_3$ and $\mathbf{F}_2$. Canceling the two equal and opposite forces $\mathbf{F}_3$ and $-\mathbf{F}_3$, we obtain the desired couple in plane P_2 (Fig. 3.31*d*). Thus, we conclude that two couples having the same moment $\mathbf{M}$ are equivalent, whether they are contained in the same plane or in parallel planes.

The property we have just established is very important for the correct understanding of the mechanics of rigid bodies. It indicates that when a couple acts on a rigid body, it does not matter where the two forces forming the couple act or what magnitude and direction they have. The only thing that counts is the *moment* of the couple (magnitude and direction). Couples with the same moment have the same effect on the rigid body.

3.3C   Addition of Couples

Consider two intersecting planes P_1 and P_2 and two couples acting respectively in P_1 and P_2. Recall that each couple is a free vector in its respective plane and can be represented within this plane by any combination of equal, opposite, and parallel forces and of perpendicular distance of separation that provides the same sense and magnitude for this couple. Thus, we can assume, without any loss of generality, that the couple in P_1 consists of two forces $\mathbf{F}_1$ and $-\mathbf{F}_1$ perpendicular to the line of intersection of the two planes and acting respectively at A and B (Fig. 3.32*a*). Similarly, we can assume that the couple in P_2 consists of two forces $\mathbf{F}_2$ and $-\mathbf{F}_2$ perpendicular to AB and acting respectively at A and B. It is clear that the resultant $\mathbf{R}$ of $\mathbf{F}_1$ and $\mathbf{F}_2$ and the resultant $-\mathbf{R}$ of $-\mathbf{F}_1$ and $-\mathbf{F}_2$ form a couple. Denoting the vector joining B to A by $\mathbf{r}$ and recalling the definition of the moment of a couple (Sec. 3.3A), we express the moment $\mathbf{M}$ of the resulting couple as

$$\mathbf{M} = \mathbf{r} \times \mathbf{R} = \mathbf{r} \times (\mathbf{F}_1 + \mathbf{F}_2)$$

By Varignon's theorem, we can expand this expression as

$$\mathbf{M} = \mathbf{r} \times \mathbf{F}_1 + \mathbf{r} \times \mathbf{F}_2$$

The first term in this expression represents the moment $\mathbf{M}_1$ of the couple in P_1, and the second term represents the moment $\mathbf{M}_2$ of the couple in P_2. Therefore, we have

$$\mathbf{M} = \mathbf{M}_1 + \mathbf{M}_2 \tag{3.48}$$

We conclude that the sum of two couples of moments $\mathbf{M}_1$ and $\mathbf{M}_2$ is a couple of moment $\mathbf{M}$ equal to the vector sum of $\mathbf{M}_1$ and $\mathbf{M}_2$ (Fig. 3.32*b*). We can extend this conclusion to state that any number of couples can be added to produce one resultant couple, as

$$\mathbf{M} = \Sigma \mathbf{M} = \Sigma (\mathbf{r} \times \mathbf{F})$$

3.3D   Couple Vectors

We have seen that couples with the same moment, whether they act in the same plane or in parallel planes, are equivalent. Therefore, we have no need to draw the actual forces forming a given couple in order to define its effect

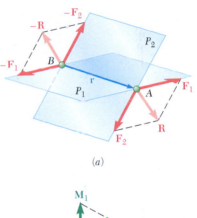

(a)

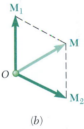

(b)

Fig. 3.32 (*a*) We can add two couples, each acting in one of two intersecting planes, to form a new couple. (*b*) The moment of the resultant couple is the vector sum of the moments of the component couples.

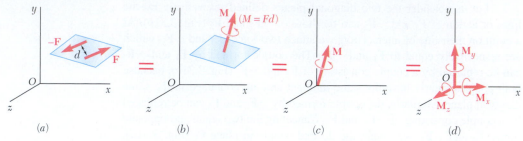

Fig. 3.33 (a) A couple formed by two forces can be represented by (b) a couple vector, oriented perpendicular to the plane of the couple. (c) The couple vector is a free vector and can be moved to other points of application, such as the origin. (d) A couple vector can be resolved into components along the coordinate axes.

on a rigid body (Fig. 3.33a). It is sufficient to draw an arrow equal in magnitude and direction to the moment **M** of the couple (Fig. 3.33b). We have also seen that the sum of two couples is itself a couple and that we can obtain the moment **M** of the resultant couple by forming the vector sum of the moments **M**₁ and **M**₂ of the given couples. Thus, couples obey the law of addition of vectors, so the arrow used in Fig. 3.33b to represent the couple defined in Fig. 3.33a truly can be considered a vector.

The vector representing a couple is called a **couple vector**. Note that, in Fig. 3.33, we use a red arrow to distinguish the couple vector, *which represents the couple itself*, from the *moment* of the couple, which was represented by a green arrow in earlier figures. Also note that we added the symbol ↱ to this red arrow to avoid any confusion with vectors representing forces. A couple vector, like the moment of a couple, is a free vector. Therefore, we can choose its point of application at the origin of the system of coordinates, if so desired (Fig. 3.33c). Furthermore, we can resolve the couple vector **M** into component vectors **M**ₓ, **M**_y, and **M**_z that are directed along the coordinate axes (Fig. 3.33d). These component vectors represent couples acting, respectively, in the *yz, zx,* and *xy* planes.

3.3E Resolution of a Given Force into a Force at *O* and a Couple

Consider a force **F** acting on a rigid body at a point *A* defined by the position vector **r** (Fig. 3.34a). Suppose that for some reason it would simplify the analysis to have the force act at point *O* instead. Although we can move **F** along its line of action (principle of transmissibility), we cannot move it to a point *O* that does not lie on the original line of action without modifying the action of **F** on the rigid body.

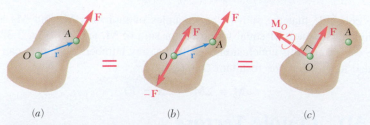

Fig. 3.34 Replacing a force with a force and a couple. (a) Initial force **F** acting at point *A*; (b) attaching equal and opposite forces at *O*; (c) force **F** acting at point *O* and a couple.

We can, however, attach two forces at point O, one equal to $\mathbf{F}$ and the other equal to $-\mathbf{F}$, without modifying the action of the original force on the rigid body (Fig. 3.34b). As a result of this transformation, we now have a force $\mathbf{F}$ applied at O; the other two forces form a couple of moment $\mathbf{M}_O = \mathbf{r} \times \mathbf{F}$. Thus,

> **Any force F acting on a rigid body can be moved to an arbitrary point O provided that we add a couple whose moment is equal to the moment of F about O.**

The couple tends to impart to the rigid body the same rotational motion about O that force $\mathbf{F}$ tended to produce before it was transferred to O. We represent the couple by a couple vector $\mathbf{M}_O$ that is perpendicular to the plane containing $\mathbf{r}$ and $\mathbf{F}$. Since $\mathbf{M}_O$ is a free vector, it may be applied anywhere; for convenience, however, the couple vector is usually attached at O together with $\mathbf{F}$. This combination is referred to as a **force-couple system** (Fig. 3.34c).

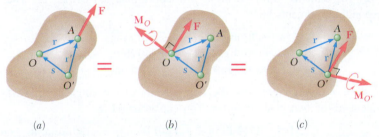

(a)   (b)   (c)

Fig. 3.35 Moving a force to different points. (a) Initial force F acting at A; (b) force F acting at O and a couple; (c) force F acting at O' and a different couple.

If we move force $\mathbf{F}$ from A to a different point O' (Fig. 3.35a and c), we have to compute the moment $\mathbf{M}_{O'} = \mathbf{r}' \times \mathbf{F}$ of $\mathbf{F}$ about O' and add a new force-couple system consisting of $\mathbf{F}$ and the couple vector $\mathbf{M}_{O'}$ at O'. We can obtain the relation between the moments of $\mathbf{F}$ about O and O' as

$$\mathbf{M}_{O'} = \mathbf{r}' \times \mathbf{F} = (\mathbf{r} + \mathbf{s}) \times \mathbf{F} = \mathbf{r} \times \mathbf{F} + \mathbf{s} \times \mathbf{F}$$

$$\mathbf{M}_{O'} = \mathbf{M}_O + \mathbf{s} \times \mathbf{F} \tag{3.49}$$

where $\mathbf{s}$ is the vector joining O' to O. Thus, we obtain the moment $\mathbf{M}_{O'}$ of $\mathbf{F}$ about O' by adding to the moment $\mathbf{M}_O$ of $\mathbf{F}$ about O the vector product $\mathbf{s} \times \mathbf{F}$, representing the moment about O' of the force $\mathbf{F}$ applied at O.

We also could have established this result by observing that, in order to transfer to O' the force-couple system attached at O (Fig. 3.35b and c), we could freely move the couple vector $\mathbf{M}_O$ to O'. However, to move force $\mathbf{F}$ from O to O', we need to add to $\mathbf{F}$ a couple vector whose moment is equal to the moment about O' of force $\mathbf{F}$ applied at O. Thus, the couple vector $\mathbf{M}_{O'}$ must be the sum of $\mathbf{M}_O$ and the vector $\mathbf{s} \times \mathbf{F}$.

As noted here, the force-couple system obtained by transferring a force $\mathbf{F}$ from a point A to a point O consists of $\mathbf{F}$ and a couple vector $\mathbf{M}_O$ perpendicular to $\mathbf{F}$. Conversely, any force-couple system consisting of a force $\mathbf{F}$ and a couple vector $\mathbf{M}_O$ that are *mutually perpendicular* can be replaced by a single equivalent force. This is done by moving force $\mathbf{F}$ in the plane perpendicular to $\mathbf{M}_O$ until its moment about O is equal to the moment of the couple being replaced.

Photo 3.2 The force exerted by each hand on the wrench could be replaced with an equivalent force-couple system acting on the nut.

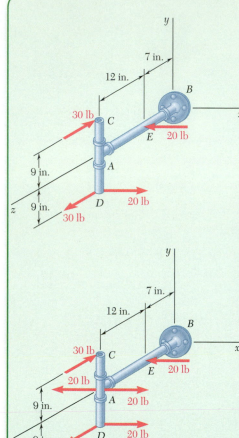

Fig. 1 Placing two equal and opposite 20-lb forces at A to simplify calculations.

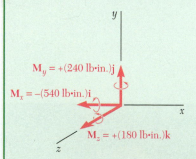

Fig. 2 The three couples represented as couple vectors.

Sample Problem 3.6

Determine the components of the single couple equivalent to the two couples shown.

STRATEGY: Look for ways to add equal and opposite forces to the diagram that, along with already known perpendicular distances, will produce new couples with moments along the coordinate axes. These can be combined into a single equivalent couple.

MODELING: You can simplify the computations by attaching two equal and opposite 20-lb forces at A (Fig. 1). This enables you to replace the original 20-lb-force couple by two new 20-lb-force couples: one lying in the zx plane and the other in a plane parallel to the xy plane.

ANALYSIS: You can represent these three couples by three couple vectors $\mathbf{M}_x$, $\mathbf{M}_y$, and $\mathbf{M}_z$ directed along the coordinate axes (Fig. 2). The corresponding moments are

$$M_x = -(30 \text{ lb})(18 \text{ in.}) = -540 \text{ lb·in.}$$
$$M_y = +(20 \text{ lb})(12 \text{ in.}) = +240 \text{ lb·in.}$$
$$M_z = +(20 \text{ lb})(9 \text{ in.}) = +180 \text{ lb·in.}$$

These three moments represent the components of the single couple $\mathbf{M}$ equivalent to the two given couples. You can write $\mathbf{M}$ as

$$\mathbf{M} = -(540 \text{ lb·in.})\mathbf{i} + (240 \text{ lb·in.})\mathbf{j} + (180 \text{ lb·in.})\mathbf{k} \quad \blacktriangleleft$$

REFLECT and THINK: You can also obtain the components of the equivalent single couple $\mathbf{M}$ by computing the sum of the moments of the four given forces about an arbitrary point. Selecting point D, the moment is (Fig. 3)

$$\mathbf{M} = \mathbf{M}_D = (18 \text{ in.})\mathbf{j} \times (-30 \text{ lb})\mathbf{k} + [(9 \text{ in.})\mathbf{j} - (12 \text{ in.})\mathbf{k}] \times (-20 \text{ lb})\mathbf{i}$$

After computing the various cross products, you get the same result, as

$$\mathbf{M} = -(540 \text{ lb·in.})\mathbf{i} + (240 \text{ lb·in.})\mathbf{j} + (180 \text{ lb·in.})\mathbf{k} \quad \blacktriangleleft$$

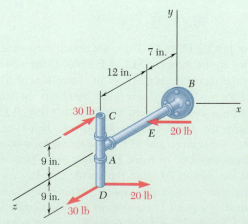

Fig. 3 Using the given force system, the equivalent single couple can also be determined from the sum of moments of the forces about any point, such as point D.

Sample Problem 3.7

Replace the couple and force shown by an equivalent single force applied to the lever. Determine the distance from the shaft to the point of application of this equivalent force.

STRATEGY: First replace the given force and couple by an equivalent force-couple system at O. By moving the force of this force-couple system a distance that creates the same moment as the couple, you can then replace the system with one equivalent force.

MODELING and ANALYSIS: To replace the given force and couple, move the force $\mathbf{F} = -(400\ \text{N})\mathbf{j}$ to O, and at the same time, add a couple of moment $\mathbf{M}_O$ that is equal to the moment about O of the force in its original position (Fig. 1). Thus,

$$\mathbf{M}_O = \overrightarrow{OB} \times \mathbf{F} = [(0.150\ \text{m})\mathbf{i} + (0.260\ \text{m})\mathbf{j}] \times (-400\ \text{N})\mathbf{j}$$
$$= -(60\ \text{N·m})\mathbf{k}$$

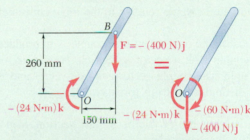

Fig. 1 Replacing given force and couple with an equivalent force-couple at O.

When you add this new couple to the couple of moment $-(24\ \text{N·m})\mathbf{k}$ formed by the two 200-N forces, you obtain a couple of moment $-(84\ \text{N·m})\mathbf{k}$ (Fig. 2). You can replace this last couple by applying $\mathbf{F}$ at a point C chosen in such a way that

$$-(84\ \text{N·m})\mathbf{k} = \overrightarrow{OC} \times \mathbf{F}$$
$$= [(OC)\cos 60°\mathbf{i} + (OC)\sin 60°\mathbf{j}] \times (-400\ \text{N})\mathbf{j}$$
$$= -(OC)\cos 60°(400\ \text{N})\mathbf{k}$$

The result is

$$(OC)\cos 60° = 0.210\ \text{m} = 210\ \text{mm} \qquad OC = 420\ \text{mm} \blacktriangleleft$$

REFLECT and THINK: Since the effect of a couple does not depend on its location, you can move the couple of moment $-(24\ \text{N·m})\mathbf{k}$ to B, obtaining a force-couple system at B (Fig. 3). Now you can eliminate this couple by applying $\mathbf{F}$ at a point C chosen in such a way that

$$-(24\ \text{N·m})\mathbf{k} = \overrightarrow{BC} \times \mathbf{F}$$
$$= -(BC)\cos 60°(400\ \text{N})\mathbf{k}$$

The conclusion is

$$(BC)\cos 60° = 0.060\ \text{m} = 60\ \text{mm} \qquad BC = 120\ \text{mm}$$
$$OC = OB + BC = 300\ \text{mm} + 120\ \text{mm} \qquad OC = 420\ \text{mm} \blacktriangleleft$$

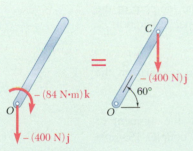

Fig. 2 Resultant couple eliminated by moving force **F**.

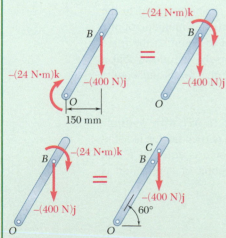

Fig. 3 Couple can be moved to B with no change in effect. This couple can then be eliminated by moving force **F**.

SOLVING PROBLEMS ON YOUR OWN

In this section, we discussed the properties of *couples*. To solve the following problems, remember that the net effect of a couple is to produce a moment **M**. Since this moment is independent of the point about which it is computed, **M** is a *free vector* and remains unchanged if you move it from point to point. Also, two couples are *equivalent* (that is, they have the same effect on a given rigid body) if they produce the same moment.

When determining the moment of a couple, all previous techniques for computing moments apply. Also, since the moment of a couple is a free vector, you should compute its value relative to the most convenient point.

Because the only effect of a couple is to produce a moment, it is possible to represent a couple with a vector, called the *couple vector*, that is equal to the moment of the couple. The couple vector is a free vector and is represented by a special symbol, ⤸, to distinguish it from force vectors.

In solving the problems in this section, you will be called upon to perform the following operations:

1. Adding two or more couples. This results in a new couple, the moment of which is obtained by adding vectorially the moments of the given couples [Sample Prob. 3.6].

2. Replacing a force with an equivalent force-couple system at a specified point. As explained in Sec. 3.3E, the force of a force-couple system is equal to the original force, whereas the required couple vector is equal to the moment of the original force about the given point. In addition, it is important to note that the force and the couple vector are perpendicular to each other. Conversely, it follows that a force-couple system can be reduced to a single force only if the force and couple vector are mutually perpendicular (see the next paragraph).

3. Replacing a force-couple system (with F perpendicular to M) with a single equivalent force. The requirement that **F** and **M** be mutually perpendicular is satisfied in all two-dimensional problems. The single equivalent force is equal to **F** and is applied in such a way that its moment about the original point of application is equal to **M** [Sample Prob. 3.7].

Problems

3.70 Two 80-N forces are applied as shown to the corners B and D of a rectangular plate. (*a*) Determine the moment of the couple formed by the two forces by resolving each force into horizontal and vertical components and adding the moments of the two resulting couples. (*b*) Use the result obtained to determine the perpendicular distance between lines BE and DF.

3.71 Two parallel 40-N forces are applied to a lever as shown. Determine the moment of the couple formed by the two forces (*a*) by resolving each force into horizontal and vertical components and adding the moments of the two resulting couples, (*b*) by using the perpendicular distance between the two forces, (*c*) by summing the moments of the two forces about point A.

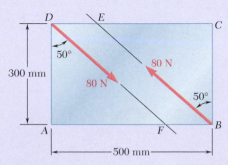

Fig. P3.70

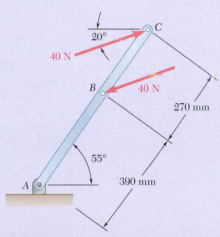

Fig. P3.71

3.72 Four $1\frac{1}{2}$-in.-diameter pegs are attached to a board as shown. Two strings are passed around the pegs and pulled with the forces indicated. (*a*) Determine the resultant couple acting on the board. (*b*) If only one string is used, around which pegs should it pass and in what directions should it be pulled to create the same couple with the minimum tension in the string? (*c*) What is the value of that minimum tension?

3.73 Four pegs of the same diameter are attached to a board as shown. Two strings are passed around the pegs and pulled with the forces indicated. Determine the diameter of the pegs knowing that the resultant couple applied to the board is 1132.5 lb·in. counterclockwise.

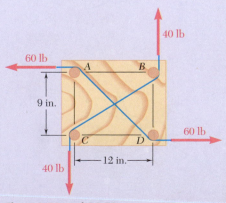

Fig. *P3.72* and P3.73

3.74 A piece of plywood in which several holes are being drilled successively has been secured to a workbench by means of two nails. Knowing that the drill exerts a 12-N·m couple on the piece of plywood, determine the magnitude of the resulting forces applied to the nails if they are located (*a*) at *A* and *B*, (*b*) at *B* and *C*, (*c*) at *A* and *C*.

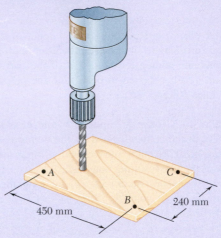

Fig. P3.74

3.75 The two shafts of a speed-reducer unit are subjected to couples of magnitude $M_1 = 15$ lb·ft and $M_2 = 3$ lb·ft, respectively. Replace the two couples with a single equivalent couple, specifying its magnitude and the direction of its axis.

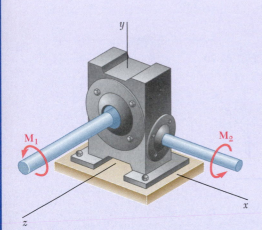

Fig. P3.75

3.76 If $P = 0$ in the figure, replace the two remaining couples with a single equivalent couple, specifying its magnitude and the direction of its axis.

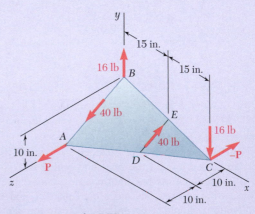

Fig. P3.76 and P3.77

3.77 If $P = 20$ lb in the figure, replace the three couples with a single equivalent couple, specifying its magnitude and the direction of its axis.

3.78 Replace the two couples shown with a single equivalent couple, specifying its magnitude and the direction of its axis.

3.79 Solve Prob. 3.78, assuming that two 10-N vertical forces have been added, one acting upward at *C* and the other downward at *B*.

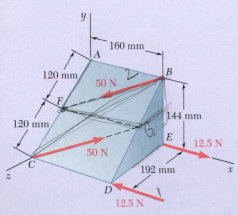

Fig. P3.78

3.80 Shafts A and B connect the gear box to the wheel assemblies of a tractor, and shaft C connects it to the engine. Shafts A and B lie in the vertical yz plane, while shaft C is directed along the x axis. Replace the couples applied to the shafts by a single equivalent couple, specifying its magnitude and the direction of its axis.

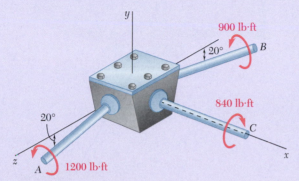

Fig. P3.80

3.81 A 500-N force is applied to a bent plate as shown. Determine (a) an equivalent force-couple system at B, (b) an equivalent system formed by a vertical force at A and a force at B.

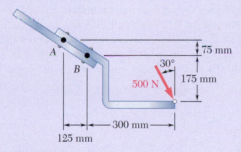

Fig. P3.81

3.82 The tension in the cable attached to the end C of an adjustable boom ABC is 560 lb. Replace the force exerted by the cable at C with an equivalent force-couple system (a) at A, (b) at B.

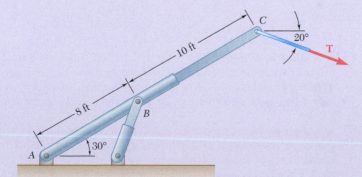

Fig. P3.82

3.83 A dirigible is tethered by a cable attached to its cabin at *B*. If the tension in the cable is 1040 N, replace the force exerted by the cable at *B* with an equivalent system formed by two parallel forces applied at *A* and *C*.

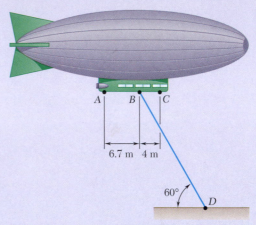

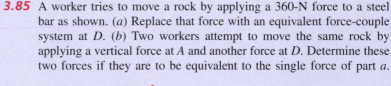

Fig. P3.83

3.84 A 30-lb vertical force **P** is applied at *A* to the bracket shown, which is held by screws at *B* and *C*. (*a*) Replace **P** with an equivalent force-couple system at *B*. (*b*) Find the two horizontal forces at *B* and *C* that are equivalent to the couple obtained in part *a*.

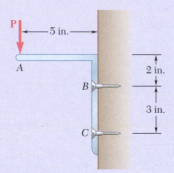

Fig. P3.84

3.85 A worker tries to move a rock by applying a 360-N force to a steel bar as shown. (*a*) Replace that force with an equivalent force-couple system at *D*. (*b*) Two workers attempt to move the same rock by applying a vertical force at *A* and another force at *D*. Determine these two forces if they are to be equivalent to the single force of part *a*.

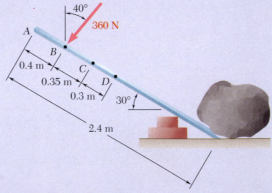

Fig. *P3.85* and P3.86

3.86 A worker tries to move a rock by applying a 360-N force to a steel bar as shown. If two workers attempt to move the same rock by applying a force at *A* and a parallel force at *C*, determine these two forces so that they will be equivalent to the single 360-N force shown in the figure.

3.87 The shearing forces exerted on the cross section of a steel channel can be represented by a 900-N vertical force and two 250-N horizontal forces as shown. Replace this force and couple with a single force **F** applied at point *C*, and determine the distance *x* from *C* to line *BD*. (Point *C* is defined as the *shear center* of the section.)

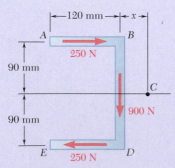

Fig. P3.87

3.88 A force and a couple are applied as shown to the end of a cantilever beam. (*a*) Replace this system with a single force **F** applied at point *C*, and determine the distance *d* from *C* to a line drawn through points *D* and *E*. (*b*) Solve part *a* if the directions of the two 360-N forces are reversed.

3.89 Three control rods attached to a lever *ABC* exert on it the forces shown. (*a*) Replace the three forces with an equivalent force-couple system at *B*. (*b*) Determine the single force that is equivalent to the force-couple system obtained in part *a*, and specify its point of application on the lever.

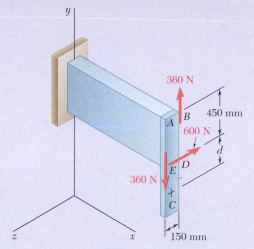

Fig. P3.88

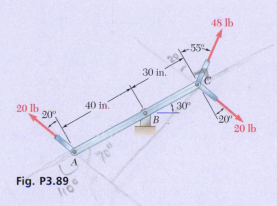

Fig. P3.89

3.90 A rectangular plate is acted upon by the force and couple shown. This system is to be replaced with a single equivalent force. (*a*) For $\alpha = 40°$, specify the magnitude and line of action of the equivalent force. (*b*) Specify the value of α if the line of action of the equivalent force is to intersect line *CD* 300 mm to the right of *D*.

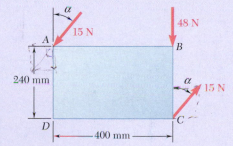

Fig. P3.90

3.91 While tapping a hole, a machinist applies the horizontal forces shown to the handle of the tap wrench. Show that these forces are equivalent to a single force, and specify, if possible, the point of application of the single force on the handle.

3.92 A hexagonal plate is acted upon by the force **P** and the couple shown. Determine the magnitude and the direction of the smallest force **P** for which this system can be replaced with a single force at *E*.

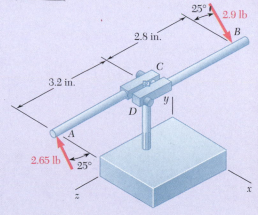

Fig. P3.91

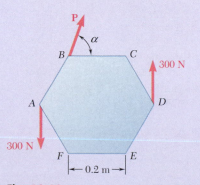

Fig. P3.92

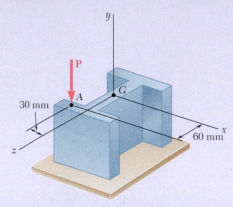

Fig. P3.93

3.93 Replace the 250-kN force **P** with an equivalent force-couple system at *G*.

3.94 A 2.6-kip force is applied at point *D* of the cast-iron post shown. Replace that force with an equivalent force-couple system at the center *A* of the base section.

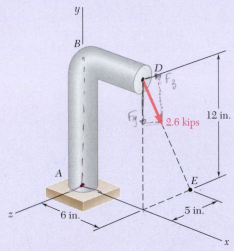

Fig. *P3.94*

3.95 Replace the 150-N force with an equivalent force-couple system at *A*.

3.96 To keep a door closed, a wooden stick is wedged between the floor and the doorknob. The stick exerts at *B* a 175-N force directed along line *AB*. Replace that force with an equivalent force-couple system at *C*.

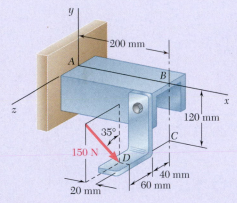

Fig. P3.95

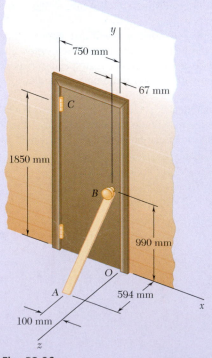

Fig. P3.96

3.97 A 46-lb force **F** and a 2120-lb·in. couple **M** are applied to corner *A* of the block shown. Replace the given force-couple system with an equivalent force-couple system at corner *H*.

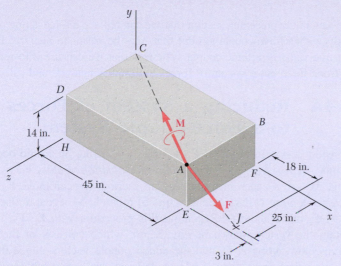

Fig. P3.97

3.98 A 110-N force acting in a vertical plane parallel to the *yz* plane is applied to the 220-mm-long horizontal handle *AB* of a socket wrench. Replace the force with an equivalent force-couple system at the origin *O* of the coordinate system.

3.99 An antenna is guyed by three cables as shown. Knowing that the tension in cable *AB* is 288 lb, replace the force exerted at *A* by cable *AB* with an equivalent force-couple system at the center *O* of the base of the antenna.

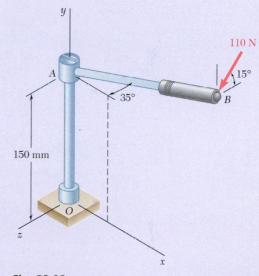

Fig. P3.98

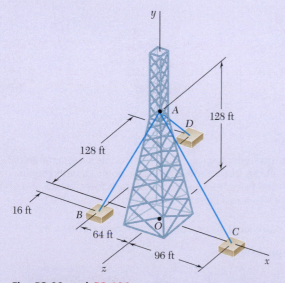

Fig. P3.99 and *P3.100*

3.100 An antenna is guyed by three cables as shown. Knowing that the tension in cable *AD* is 270 lb, replace the force exerted at *A* by cable *AD* with an equivalent force-couple system at the center *O* of the base of the antenna.

3.4 SIMPLIFYING SYSTEMS OF FORCES

We saw in the preceding section that we can replace a force acting on a rigid body with a force-couple system that may be easier to analyze. However, the true value of a force-couple system is that we can use it to replace not just one force but a system of forces to simplify analysis and calculations.

3.4A Reducing a System of Forces to a Force-Couple System

Consider a system of forces $\mathbf{F}_1$, $\mathbf{F}_2$, $\mathbf{F}_3$, . . . , acting on a rigid body at the points A_1, A_2, A_3, . . . , *defined by the position vectors r_1, r_2, r_3, etc.* (Fig. 3.36*a*). As seen in the preceding section, we can move $\mathbf{F}_1$ from A_1 to a given point O if we add a couple of moment $\mathbf{M}_1$ equal to the moment $\mathbf{r}_1 \times \mathbf{F}_1$ of $\mathbf{F}_1$ about O. Repeating this procedure with $\mathbf{F}_2$, $\mathbf{F}_3$, . . . , we obtain the system shown in Fig. 3.36*b*, which consists of the original forces, now acting at O, and the added couple vectors. Since the forces are now concurrent, they can be added vectorially and replaced by their resultant $\mathbf{R}$. Similarly, the couple vectors $\mathbf{M}_1$, $\mathbf{M}_2$, $\mathbf{M}_3$, . . . , can be added vectorially and replaced by a single couple vector $\mathbf{M}_O^R$. Thus,

We can reduce any system of forces, however complex, to an equivalent force-couple system acting at a given point O.

Note that, although each of the couple vectors $\mathbf{M}_1$, $\mathbf{M}_2$, $\mathbf{M}_3$, . . . in Fig. 3.36*b* is perpendicular to its corresponding force, the resultant force $\mathbf{R}$ and the resultant couple vector $\mathbf{M}_O^R$ shown in Fig. 3.36*c* are not, in general, perpendicular to each other.

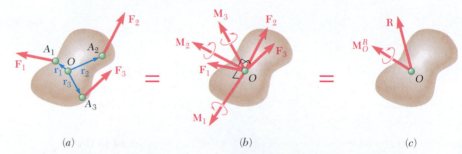

(*a*)   (*b*)   (*c*)

Fig. 3.36 Reducing a system of forces to a force-couple system. (*a*) Initial system of forces; (*b*) all the forces moved to act at point *O*, with couple vectors added; (*c*) all the forces reduced to a resultant force vector and all the couple vectors reduced to a resultant couple vector.

The equivalent force-couple system is defined by

Force-couple system

$$\mathbf{R} = \Sigma\mathbf{F} \qquad \mathbf{M}_O^R = \Sigma\mathbf{M}_O = \Sigma(\mathbf{r} \times \mathbf{F}) \qquad (3.50)$$

These equations state that we obtain force $\mathbf{R}$ by adding all of the forces of the system, whereas we obtain the moment of the resultant couple vector $\mathbf{M}_O^R$, called the **moment resultant** of the system, by adding the moments about O of all the forces of the system.

Once we have reduced a given system of forces to a force and a couple at a point O, we can replace it with a force and a couple at another point O'. The resultant force $\mathbf{R}$ will remain unchanged, whereas the new moment resultant $\mathbf{M}_{O'}^R$ will be equal to the sum of $\mathbf{M}_O^R$ and the moment about O' of force $\mathbf{R}$ attached at O (Fig. 3.37). We have

$$\mathbf{M}_{O'}^R = \mathbf{M}_O^R + \mathbf{s} \times \mathbf{R} \qquad (3.51)$$

In practice, the reduction of a given system of forces to a single force $\mathbf{R}$ at O and a couple vector $\mathbf{M}_O^R$ is carried out in terms of components. Resolving each position vector $\mathbf{r}$ and each force $\mathbf{F}$ of the system into rectangular components, we have

$$\mathbf{r} = x\mathbf{i} + y\mathbf{j} + z\mathbf{k} \qquad (3.52)$$
$$\mathbf{F} = F_x\mathbf{i} + F_y\mathbf{j} + F_z\mathbf{k} \qquad (3.53)$$

Substituting for $\mathbf{r}$ and $\mathbf{F}$ in Eq. (3.50) and factoring out the unit vectors $\mathbf{i}$, $\mathbf{j}$, and $\mathbf{k}$, we obtain $\mathbf{R}$ and $\mathbf{M}_O^R$ in the form

$$\mathbf{R} = R_x\mathbf{i} + R_y\mathbf{j} + R_z\mathbf{k} \qquad \mathbf{M}_O^R = M_x^R\mathbf{i} + M_y^R\mathbf{j} + M_z^R\mathbf{k} \qquad (3.54)$$

The components R_x, R_y, and R_z represent, respectively, the sums of the x, y, and z components of the given forces and measure the tendency of the system to impart to the rigid body a translation in the x, y, or z direction. Similarly, the components M_x^R, M_y^R, and M_z^R represent, respectively, the sum of the moments of the given forces about the x, y, and z axes and measure the tendency of the system to impart to the rigid body a rotation about the x, y, or z axis.

If we need to know the magnitude and direction of force $\mathbf{R}$, we can obtain them from the components R_x, R_y, and R_z by means of the relations in Eqs. (2.18) and (2.19) of Sec. 2.4A. Similar computations yield the magnitude and direction of the couple vector $\mathbf{M}_O^R$.

3.4B Equivalent and Equipollent Systems of Forces

We have just seen that any system of forces acting on a rigid body can be reduced to a force-couple system at a given point O. This equivalent force-couple system characterizes completely the effect of the given force system on the rigid body.

Two systems of forces are equivalent if they can be reduced to the same force-couple system at a given point O.

Recall that the force-couple system at O is defined by the relations in Eq. (3.50). Therefore, we can state that

Two systems of forces, $F_1, F_2, F_3, \ldots$, and $F_1', F_2', F_3', \ldots$, that act on the same rigid body are equivalent if, and only if, the sums of the forces and the sums of the moments about a given point O of the forces of the two systems are, respectively, equal.

Mathematically, the necessary and sufficient conditions for the two systems of forces to be equivalent are

Conditions for equivalent systems of forces

$$\Sigma\mathbf{F} = \Sigma\mathbf{F}' \qquad \text{and} \qquad \Sigma\mathbf{M}_O = \Sigma\mathbf{M}_O' \qquad (3.55)$$

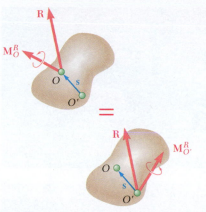

Fig. 3.37 Once a system of forces has been reduced to a force-couple system at one point, we can replace it with an equivalent force-couple system at another point. The force resultant stays the same, but we have to add the moment of the resultant force about the new point to the resultant couple vector.

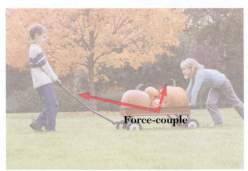

Photo 3.3 The forces exerted by the children upon the wagon can be replaced with an equivalent force-couple system when analyzing the motion of the wagon.

Note that to prove that two systems of forces are equivalent, we must establish the second of the relations in Eq. (3.55) with respect to *only one point O*. It will hold, however, with respect to *any point* if the two systems are equivalent.

Resolving the forces and moments in Eqs. (3.55) into their rectangular components, we can express the necessary and sufficient conditions for the equivalence of two systems of forces acting on a rigid body as

$$\Sigma F_x = \Sigma F'_x \qquad \Sigma F_y = \Sigma F'_y \qquad \Sigma F_z = \Sigma F'_z$$
$$\Sigma M_x = \Sigma M'_x \qquad \Sigma M_y = \Sigma M'_y \qquad \Sigma M_z = \Sigma M'_z \qquad \textbf{(3.56)}$$

These equations have a simple physical significance. They express that

Two systems of forces are equivalent if they tend to impart to the rigid body (1) the same translation in the x, y, and z directions, respectively, and (2) the same rotation about the x, y, and z axes, respectively.

In general, when two systems of vectors satisfy Eqs. (3.55) or (3.56), i.e., when their resultants and their moment resultants about an arbitrary point O are respectively equal, the two systems are said to be **equipollent**. The result just established can thus be restated as

If two systems of forces acting on a rigid body are equipollent, they are also equivalent.

It is important to note that this statement does not apply to *any* system of vectors. Consider, for example, a system of forces acting on a set of independent particles that do *not* form a rigid body. A different system of forces acting on the same particles may happen to be equipollent to the first one; i.e., it may have the same resultant and the same moment resultant. Yet, since different forces now act on the various particles, their effects on these particles are different; the two systems of forces, while equipollent, are *not equivalent*.

3.4C   Further Reduction of a System of Forces

We have now seen that any given system of forces acting on a rigid body can be reduced to an equivalent force-couple system at O, consisting of a force **R** equal to the sum of the forces of the system, and a couple vector $\mathbf{M}^R_O$ of moment equal to the moment resultant of the system.

When $\mathbf{R} = 0$, the force-couple system reduces to the couple vector $\mathbf{M}^R_O$. The given system of forces then can be reduced to a single couple called the **resultant couple** of the system.

What are the conditions under which a given system of forces can be reduced to a single force? It follows from the preceding section that we can replace the force-couple system at O by a single force **R** acting along a new line of action if **R** and $\mathbf{M}^R_O$ are mutually perpendicular. The systems of forces that can be reduced to a single force, or *resultant*, are therefore the systems for which force **R** and the couple vector $\mathbf{M}^R_O$ are mutually perpendicular. This condition *is generally not satisfied* by systems of forces in space, but it *is satisfied* by systems consisting of (1) concurrent forces, (2) coplanar forces, or (3) parallel forces. Let's look at each case separately.

1. **Concurrent forces** act at the same point; therefore, we can add them directly to obtain their resultant **R**. Thus, they always reduce to a single force. Concurrent forces were discussed in detail in Chap. 2.

2. **Coplanar forces** act in the same plane, which we assume to be the plane of the figure (Fig. 3.38*a*). The sum **R** of the forces of the system also lies in the plane of the figure, whereas the moment of each force about O and thus the moment resultant $\mathbf{M}_O^R$ are perpendicular to that plane. The force-couple system at O consists, therefore, of a force **R** and a couple vector $\mathbf{M}_O^R$ that are mutually perpendicular (Fig. 3.38*b*).[†] We can reduce them to a single force **R** by moving **R** in the plane of the figure until its moment about O becomes equal to $\mathbf{M}_O^R$. The distance from O to the line of action of **R** is $d = M_O^R/R$ (Fig. 3.38*c*).

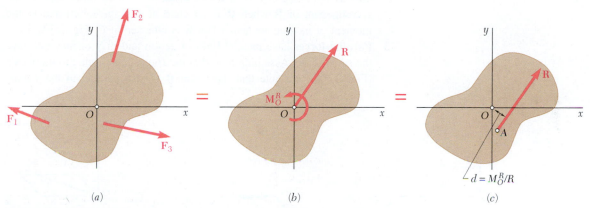

Fig. 3.38 Reducing a system of coplanar forces. (*a*) Initial system of forces; (*b*) equivalent force-couple system at O; (*c*) moving the resultant force to a point A such that the moment of **R** about O equals the couple vector.

As noted earlier, the reduction of a system of forces is considerably simplified if we resolve the forces into rectangular components. The force-couple system at O is then characterized by the components (Fig. 3.39*a*)

$$R_x = \Sigma F_x \qquad R_y = \Sigma F_y \qquad M_z^R = M_O^R = \Sigma M_O \qquad \textbf{(3.57)}$$

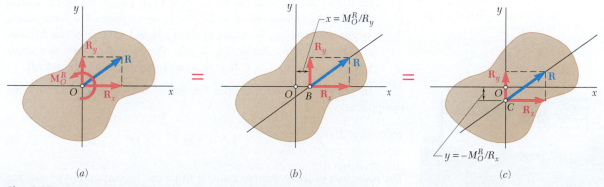

Fig. 3.39 Reducing a system of coplanar forces by using rectangular components. (*a*) From Fig. 3.38(*b*), resolve the resultant into components along the *x* and *y* axes; (*b*) determining the *x* intercept of the final line of action of the resultant; (*c*) determining the *y* intercept of the final line of action of the resultant.

[†]Because the couple vector $\mathbf{M}_O^R$ is perpendicular to the plane of the figure, we represent it by the symbol ↰. A counterclockwise couple ↰ represents a vector pointing out of the page and a clockwise couple ↲ represents a vector pointing into the page.

To reduce the system to a single force $\mathbf{R}$, the moment of $\mathbf{R}$ about O must be equal to $\mathbf{M}_O^R$. If we denote the coordinates of the point of application of the resultant by x and y and apply equation (3.22) of Sec. 3.1F, we have

$$xR_y - yR_x = M_O^R$$

This represents the equation of the line of action of $\mathbf{R}$. We can also determine the x and y intercepts of the line of action of the resultant directly by noting that $\mathbf{M}_O^R$ must be equal to the moment about O of the y component of $\mathbf{R}$ when $\mathbf{R}$ is attached at B (Fig. 3.39b) and to the moment of its x component when $\mathbf{R}$ is attached at C (Fig. 3.39c).

3. **Parallel forces** have parallel lines of action and may or may not have the same sense. Assuming here that the forces are parallel to the y axis (Fig. 3.40a), we note that their sum $\mathbf{R}$ is also parallel to the y axis.

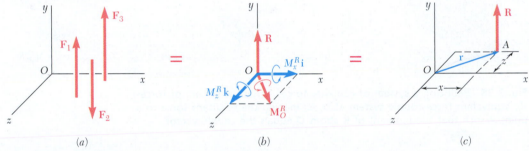

(a)   (b)   (c)

Fig. 3.40 Reducing a system of parallel forces. (a) Initial system of forces; (b) equivalent force-couple system at O, resolved into components; (c) moving $\mathbf{R}$ to point A, chosen so that the moment of $\mathbf{R}$ about O equals the resultant moment about O.

Photo 3.4 The parallel wind forces acting on the highway signs can be reduced to a single equivalent force. Determining this force can simplify the calculation of the forces acting on the supports of the frame to which the signs are attached.

On the other hand, since the moment of a given force must be perpendicular to that force, the moment about O of each force of the system and thus the moment resultant $\mathbf{M}_O^R$ lie in the zx plane. The force-couple system at O consists, therefore, of a force $\mathbf{R}$ and a couple vector $\mathbf{M}_O^R$ that are mutually perpendicular (Fig. 3.40b). We can reduce them to a single force $\mathbf{R}$ (Fig. 3.40c) or, if $\mathbf{R} = 0$, to a single couple of moment $\mathbf{M}_O^R$.

In practice, the force-couple system at O is characterized by the components

$$R_y = \Sigma F_y \qquad M_x^R = \Sigma M_x \qquad M_z^R = \Sigma M_z \qquad \textbf{(3.58)}$$

The reduction of the system to a single force can be carried out by moving $\mathbf{R}$ to a new point of application $A(x, 0, z)$, which is chosen so that the moment of $\mathbf{R}$ about O is equal to $\mathbf{M}_O^R$.

$$\mathbf{r} \times \mathbf{R} = \mathbf{M}_O^R$$

$$(x\mathbf{i} + z\mathbf{k}) \times R_y\mathbf{j} = M_x^R\mathbf{i} + M_z^R\mathbf{k}$$

By computing the vector products and equating the coefficients of the corresponding unit vectors in both sides of the equation, we obtain two scalar equations that define the coordinates of A:

$$-zR_y = M_x^R \quad \text{and} \quad xR_y = M_z^R$$

These equations express the fact that the moments of $\mathbf{R}$ about the x and z axes must be equal, respectively, to M_x^R and M_z^R.

*3.4D   Reduction of a System of Forces to a Wrench

In the general case of a system of forces in space, the equivalent force-couple system at O consists of a force $\mathbf{R}$ and a couple vector $\mathbf{M}_O^R$ that are not perpendicular and where neither is zero (Fig. 3.41a). This system of forces *cannot* be reduced to a single force or to a single couple. However, we still have a way of simplifying this system further.

The simplification method consists of first replacing the couple vector by two other couple vectors that are obtained by resolving $\mathbf{M}_O^R$ into a component $\mathbf{M}_1$ along $\mathbf{R}$ and a component $\mathbf{M}_2$ in a plane perpendicular to $\mathbf{R}$ (Fig. 3.41b). Then we can replace the couple vector $\mathbf{M}_2$ and force $\mathbf{R}$ by a single force $\mathbf{R}$ acting along a new line of action. The original system of forces thus reduces to $\mathbf{R}$ and to the couple vector $\mathbf{M}_1$ (Fig. 3.41c), i.e., to $\mathbf{R}$ and a couple acting in the plane perpendicular to $\mathbf{R}$.

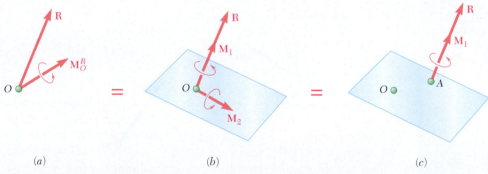

(a)   (b)   (c)

Fig. 3.41 Reducing a system of forces to a wrench. (a) General force system reduced to a single force and a couple vector, not perpendicular to each other; (b) resolving the couple vector into components along the line of action of the force and perpendicular to it; (c) moving the force and collinear couple vector (the wrench) to eliminate the couple vector perpendicular to the force.

This particular force-couple system is called a **wrench** because the resulting combination of push and twist is the same as that caused by an actual wrench. The line of action of $\mathbf{R}$ is known as the *axis of the wrench,* and the ratio $p = M_1/R$ is called the *pitch of the wrench.* A wrench therefore consists of two collinear vectors: a force $\mathbf{R}$ and a couple vector

$$\mathbf{M}_1 = p\mathbf{R} \qquad\qquad (3.59)$$

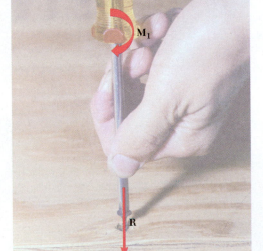

Photo 3.5 The pushing-turning action associated with the tightening of a screw illustrates the collinear lines of action of the force and couple vector that constitute a wrench.

Recall the expression in Eq. (3.33) for the projection of a vector on the line of action of another vector. Using this equation, we note that the projection of $\mathbf{M}_O^R$ on the line of action of $\mathbf{R}$ is

$$M_1 = \frac{\mathbf{R} \cdot \mathbf{M}_O^R}{R}$$

Thus, we can express the pitch of the wrench as[†]

$$p = \frac{M_1}{R} = \frac{\mathbf{R} \cdot \mathbf{M}_O^R}{R^2} \tag{3.60}$$

To define the axis of the wrench, we can write a relation involving the position vector $\mathbf{r}$ of an arbitrary point P located on that axis. We first attach the resultant force $\mathbf{R}$ and couple vector $\mathbf{M}_1$ at P (Fig. 3.42). Then, since the moment about O of this force-couple system must be equal to the moment resultant $\mathbf{M}_O^R$ of the original force system, we have

$$\mathbf{M}_1 + \mathbf{r} \times \mathbf{R} = \mathbf{M}_O^R \tag{3.61}$$

Alternatively, using Eq. (3.59), we have

$$p\mathbf{R} + \mathbf{r} \times \mathbf{R} = \mathbf{M}_O^R \tag{3.62}$$

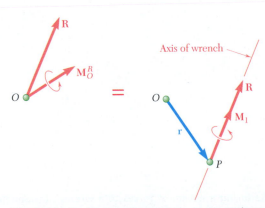

Fig. 3.42 By finding the position vector $\mathbf{r}$ that locates any arbitrary point on the axis of the wrench, you can define the axis.

[†]The expressions obtained for the projection of the couple vector on the line of action of $\mathbf{R}$ and for the pitch of the wrench are independent of the choice of point O. Using the relation (3.51) of Sec. 3.4A, we note that if a different point O' had been used, the numerator in (3.60) would have been

$$\mathbf{R} \cdot \mathbf{M}_{O'}^R = \mathbf{R} \cdot (\mathbf{M}_O^R + \mathbf{s} \times \mathbf{R}) = \mathbf{R} \cdot \mathbf{M}_O^R + \mathbf{R} \cdot (\mathbf{s} \times \mathbf{R})$$

Since the mixed triple product $\mathbf{R} \cdot (\mathbf{s} \times \mathbf{R})$ is identically equal to zero, we have

$$\mathbf{R} \cdot \mathbf{M}_{O'}^R = \mathbf{R} \cdot \mathbf{M}_O^R$$

Thus, the scalar product $\mathbf{R} \cdot \mathbf{M}_O^R$ is independent of the choice of point O.

Sample Problem 3.8

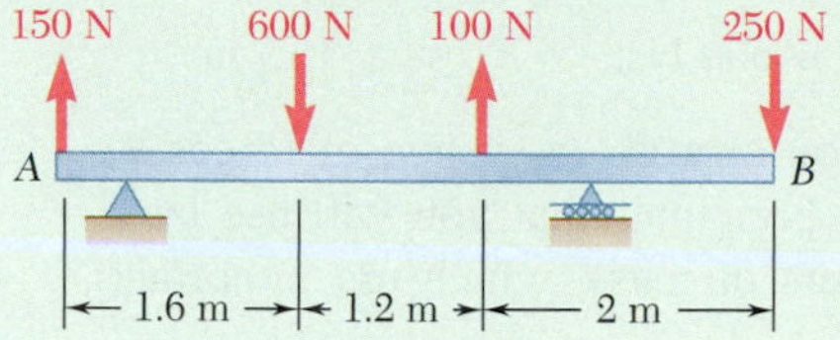

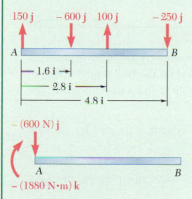

A 4.80-m-long beam is subjected to the forces shown. Reduce the given system of forces to *(a)* an equivalent force-couple system at *A, (b)* an equivalent force-couple system at *B, (c)* a single force or resultant. *Note:* Since the reactions at the supports are not included in the given system of forces, the given system will not maintain the beam in equilibrium.

STRATEGY: The *force* part of an equivalent force-couple system is simply the sum of the forces involved. The *couple* part is the sum of the moments caused by each force relative to the point of interest. Once you find the equivalent force-couple at one point, you can transfer it to any other point by a moment calculation.

MODELING and ANALYSIS:

a. Force-Couple System at *A*. The force-couple system at *A* equivalent to the given system of forces consists of a force $\mathbf{R}$ and a couple $\mathbf{M}_A^R$ defined as (Fig. 1):

$$\mathbf{R} = \Sigma\mathbf{F}$$
$$= (150\ \text{N})\mathbf{j} - (600\ \text{N})\mathbf{j} + (100\ \text{N})\mathbf{j} - (250\ \text{N})\mathbf{j} = -(600\ \text{N})\mathbf{j}$$
$$\mathbf{M}_A^R = \Sigma(\mathbf{r} \times \mathbf{F})$$
$$= (1.6\mathbf{i}) \times (-600\mathbf{j}) + (2.8\mathbf{i}) \times (100\mathbf{j}) + (4.8\mathbf{i}) \times (-250\mathbf{j})$$
$$= -(1880\ \text{N·m})\mathbf{k}$$

The equivalent force-couple system at *A* is thus

$$\mathbf{R} = 600\ \text{N} \downarrow \qquad \mathbf{M}_A^R = 1880\ \text{N·m} \downarrow \quad \blacktriangleleft$$

Fig. 1 Force-couple system at *A* that is equivalent to given system of forces.

b. Force-Couple System at *B*. You want to find a force-couple system at *B* equivalent to the force-couple system at *A* determined in part *a*. The force $\mathbf{R}$ is unchanged, but you must determine a new couple $\mathbf{M}_B^R$, the moment of which is equal to the moment about *B* of the force-couple system determined in part *a* (Fig. 2). You have

$$\mathbf{M}_B^R = \mathbf{M}_A^R + \overrightarrow{BA} \times \mathbf{R}$$
$$= -(1880\ \text{N·m})\mathbf{k} + (-4.8\ \text{m})\mathbf{i} \times (-600\ \text{N})\mathbf{j}$$
$$= -(1880\ \text{N·m})\mathbf{k} + (2880\ \text{N·m})\mathbf{k} = +(1000\ \text{N·m})\mathbf{k}$$

The equivalent force-couple system at *B* is thus

$$\mathbf{R} = 600\ \text{N} \downarrow \qquad \mathbf{M}_B^R = 1000\ \text{N·m} \ \curvearrowright \quad \blacktriangleleft$$

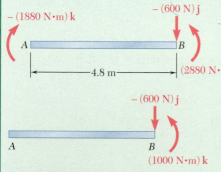

Fig. 2 Finding force-couple system at *B* equivalent to that determined in part *a*.

c. Single Force or Resultant. The resultant of the given system of forces is equal to $\mathbf{R}$, and its point of application must be such that the moment of $\mathbf{R}$ about *A* is equal to $\mathbf{M}_A^R$ (Fig. 3). This equality of moments leads to

$$\mathbf{r} \times \mathbf{R} = \mathbf{M}_A^R$$
$$x\mathbf{i} \times (-600\ \text{N})\mathbf{j} = -(1880\ \text{N·m})\mathbf{k}$$
$$-x(600\ \text{N})\mathbf{k} = -(1880\ \text{N·m})\mathbf{k}$$

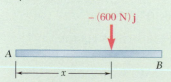

Fig. 3 Single force that is equivalent to given system of forces.

(continued)

Solving for x, you get $x = 3.13$ m. Thus, the single force equivalent to the given system is defined as

$$\mathbf{R} = 600 \text{ N} \downarrow \qquad x = 3.13 \text{ m} \blacktriangleleft$$

REFLECT and THINK: This reduction of a given system of forces to a single equivalent force uses the same principles that you will use later for finding centers of gravity and centers of mass, which are important parameters in engineering mechanics.

Sample Problem 3.9

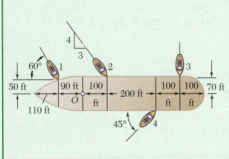

Four tugboats are bringing an ocean liner to its pier. Each tugboat exerts a 5000-lb force in the direction shown. Determine (*a*) the equivalent force-couple system at the foremast *O*, (*b*) the point on the hull where a single, more powerful tugboat should push to produce the same effect as the original four tugboats.

STRATEGY: The equivalent force-couple system is defined by the sum of the given forces and the sum of the moments of those forces at a particular point. A single tugboat could produce this system by exerting the resultant force at a point of application that produces an equivalent moment.

MODELING and ANALYSIS:

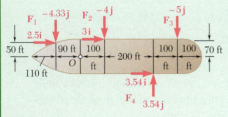

Fig. 1 Given forces resolved into components.

a. Force-Couple System at *O*. Resolve each of the given forces into components, as in Fig. 1 (kip units are used). The force-couple system at *O* equivalent to the given system of forces consists of a force **R** and a couple $\mathbf{M}_O^R$ defined as

$$\mathbf{R} = \Sigma\mathbf{F}$$
$$= (2.50\mathbf{i} - 4.33\mathbf{j}) + (3.00\mathbf{i} - 4.00\mathbf{j}) + (-5.00\mathbf{j}) + (3.54\mathbf{i} + 3.54\mathbf{j})$$
$$= 9.04\mathbf{i} - 9.79\mathbf{j}$$

$$\mathbf{M}_O^R = \Sigma(\mathbf{r} \times \mathbf{F})$$
$$= (-90\mathbf{i} + 50\mathbf{j}) \times (2.50\mathbf{i} - 4.33\mathbf{j})$$
$$\quad + (100\mathbf{i} + 70\mathbf{j}) \times (3.00\mathbf{i} - 4.00\mathbf{j})$$
$$\quad + (400\mathbf{i} + 70\mathbf{j}) \times (-5.00\mathbf{j})$$
$$\quad + (300\mathbf{i} - 70\mathbf{j}) \times (3.54\mathbf{i} + 3.54\mathbf{j})$$
$$= (390 - 125 - 400 - 210 - 2000 + 1062 + 248)\mathbf{k}$$
$$= -1035\mathbf{k}$$

The equivalent force-couple system at *O* is thus (Fig. 2)

$$\mathbf{R} = (9.04 \text{ kips})\mathbf{i} - (9.79 \text{ kips})\mathbf{j} \qquad \mathbf{M}_O^R = -(1035 \text{ kip·ft})\mathbf{k}$$

or

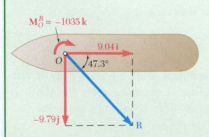

Fig. 2 Equivalent force-couple system at *O*.

$$\mathbf{R} = 13.33 \text{ kips} \ \diagdown 47.3° \qquad \mathbf{M}_O^R = 1035 \text{ kip·ft} \ \downcurvearrow \blacktriangleleft$$

(continued)

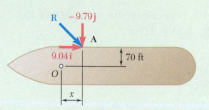

Fig. 3 Point of application of single tugboat to create same effect as given force system.

Remark: Since all the forces are contained in the plane of the figure, you would expect the sum of their moments to be perpendicular to that plane. Note that you could obtain the moment of each force component directly from the diagram by first forming the product of its magnitude and perpendicular distance to O and then assigning to this product a positive or a negative sign, depending upon the sense of the moment.

b. Single Tugboat.   The force exerted by a single tugboat must be equal to $\mathbf{R}$, and its point of application A must be such that the moment of $\mathbf{R}$ about O is equal to $\mathbf{M}_O^R$ (Fig. 3). Observing that the position vector of A is

$$\mathbf{r} = x\mathbf{i} + 70\mathbf{j}$$

you have

$$\mathbf{r} \times \mathbf{R} = \mathbf{M}_O^R$$
$$(x\mathbf{i} + 70\mathbf{j}) \times (9.04\mathbf{i} - 9.79\mathbf{j}) = -1035\mathbf{k}$$
$$-x(9.79)\mathbf{k} - 633\mathbf{k} = -1035\mathbf{k} \quad x = 41.1 \text{ ft} \quad \blacktriangleleft$$

REFLECT and THINK: Reducing the given situation to that of a single force makes it easier to visualize the overall effect of the tugboats in maneuvering the ocean liner. But in practical terms, having four boats applying force allows for greater control in slowing and turning a large ship in a crowded harbor.

Sample Problem 3.10

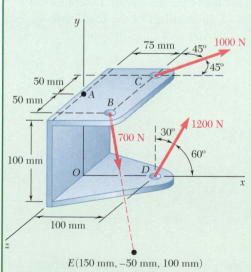

Three cables are attached to a bracket as shown. Replace the forces exerted by the cables with an equivalent force-couple system at A.

STRATEGY: First determine the relative position vectors drawn from point A to the points of application of the various forces and resolve the forces into rectangular components. Then sum the forces and moments.

MODELING and ANALYSIS: Note that $\mathbf{F}_B = (700 \text{ N})\boldsymbol{\lambda}_{BE}$ where

$$\boldsymbol{\lambda}_{BE} = \frac{\overrightarrow{BE}}{BE} = \frac{75\mathbf{i} - 150\mathbf{j} + 50\mathbf{k}}{175}$$

Using meters and newtons, the position and force vectors are

$$\mathbf{r}_{B/A} = \overrightarrow{AB} = 0.075\mathbf{i} + 0.050\mathbf{k} \qquad \mathbf{F}_B = 300\mathbf{i} - 600\mathbf{j} + 200\mathbf{k}$$
$$\mathbf{r}_{C/A} = \overrightarrow{AC} = 0.075\mathbf{i} - 0.050\mathbf{k} \qquad \mathbf{F}_C = 707\mathbf{i} \qquad\quad - 707\mathbf{k}$$
$$\mathbf{r}_{D/A} = \overrightarrow{AD} = 0.100\mathbf{i} - 0.100\mathbf{j} \qquad \mathbf{F}_D = 600\mathbf{i} + 1039\mathbf{j}$$

The force-couple system at A equivalent to the given forces consists of a force $\mathbf{R} = \Sigma\mathbf{F}$ and a couple $\mathbf{M}_A^R = \Sigma(\mathbf{r} \times \mathbf{F})$. Obtain the force $\mathbf{R}$ by adding respectively the x, y, and z components of the forces:

$$\mathbf{R} = \Sigma\mathbf{F} = (1607 \text{ N})\mathbf{i} + (439 \text{ N})\mathbf{j} - (507 \text{ N})\mathbf{k} \quad \blacktriangleleft$$

(continued)

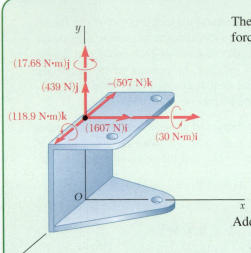

Fig. 1 Rectangular components of equivalent force-couple system at A.

The computation of $\mathbf{M}_A^R$ is facilitated by expressing the moments of the forces in the form of determinants (Sec. 3.1F). Thus,

$$\mathbf{r}_{B/A} \times \mathbf{F}_B = \begin{vmatrix} \mathbf{i} & \mathbf{j} & \mathbf{k} \\ 0.075 & 0 & 0.050 \\ 300 & -600 & 200 \end{vmatrix} = 30\mathbf{i} \qquad -45\mathbf{k}$$

$$\mathbf{r}_{C/A} \times \mathbf{F}_C = \begin{vmatrix} \mathbf{i} & \mathbf{j} & \mathbf{k} \\ 0.075 & 0 & -0.050 \\ 707 & 0 & -707 \end{vmatrix} = \qquad 17.68\mathbf{j}$$

$$\mathbf{r}_{D/A} \times \mathbf{F}_D = \begin{vmatrix} \mathbf{i} & \mathbf{j} & \mathbf{k} \\ 0.100 & -0.100 & 0 \\ 600 & 1039 & 0 \end{vmatrix} = \qquad 163.9\mathbf{k}$$

Adding these expressions, you have

$$\mathbf{M}_A^R = \Sigma(\mathbf{r} \times \mathbf{F}) = (30 \text{ N·m})\mathbf{i} + (17.68 \text{ N·m})\mathbf{j} + (118.9 \text{ N·m})\mathbf{k} \quad \blacktriangleleft$$

Figure 1 shows the rectangular components of the force $\mathbf{R}$ and the couple $\mathbf{M}_A^R$.

REFLECT and THINK: The determinant approach to calculating moments shows its advantages in a general three-dimensional problem such as this.

Sample Problem 3.11

A square foundation mat supports the four columns shown. Determine the magnitude and point of application of the resultant of the four loads.

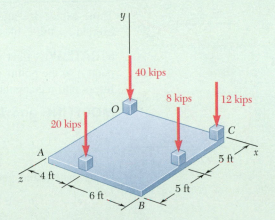

STRATEGY: Start by reducing the given system of forces to a force-couple system at the origin O of the coordinate system. Then reduce the system further to a single force applied at a point with coordinates x, z.

MODELING: The force-couple system consists of a force $\mathbf{R}$ and a couple vector $\mathbf{M}_O^R$ defined as

$$\mathbf{R} = \Sigma\mathbf{F} \qquad \mathbf{M}_O^R = \Sigma(\mathbf{r} \times \mathbf{F})$$

(continued)

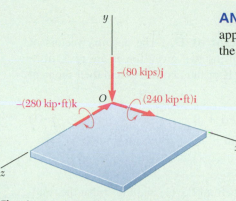

Fig. 1 Force-couple system at O that is equivalent to given force system.

ANALYSIS: After determining the position vectors of the points of application of the various forces, you may find it convenient to arrange the computations in tabular form. The results are shown in Fig. 1.

r, ft	F, kips	r × F, kip·ft
0	$-40\mathbf{j}$	0
$10\mathbf{i}$	$-12\mathbf{j}$	$-120\mathbf{k}$
$10\mathbf{i} + 5\mathbf{k}$	$-8\mathbf{j}$	$40\mathbf{i} - 80\mathbf{k}$
$4\mathbf{i} + 10\mathbf{k}$	$-20\mathbf{j}$	$200\mathbf{i} - 80\mathbf{k}$
	$\mathbf{R} = -80\mathbf{j}$	$\mathbf{M}_O^R = 240\mathbf{i} - 280\mathbf{k}$

The force $\mathbf{R}$ and the couple vector $\mathbf{M}_O^R$ are mutually perpendicular, so you can reduce the force-couple system further to a single force $\mathbf{R}$. Select the new point of application of $\mathbf{R}$ in the plane of the mat and in such a way that the moment of $\mathbf{R}$ about O is equal to $\mathbf{M}_O^R$. Denote the position vector of the desired point of application by $\mathbf{r}$ and its coordinates by x and z (Fig. 2). Then

$$\mathbf{r} \times \mathbf{R} = \mathbf{M}_O^R$$
$$(x\mathbf{i} + z\mathbf{k}) \times (-80\mathbf{j}) = 240\mathbf{i} - 280\mathbf{k}$$
$$-80x\mathbf{k} + 80z\mathbf{i} = 240\mathbf{i} - 280\mathbf{k}$$

It follows that

$$-80x = -280 \qquad 80z = 240$$
$$x = 3.50 \text{ ft} \qquad z = 3.00 \text{ ft}$$

The resultant of the given system of forces is

$$\mathbf{R} = 80 \text{ kips}\downarrow \qquad \text{at } x = 3.50 \text{ ft}, z = 3.00 \text{ ft} \blacktriangleleft$$

Fig. 2 Single force that is equivalent to given force system.

REFLECT and THINK: The fact that the given forces are all parallel simplifies the calculations, so the final step becomes just a two-dimensional analysis.

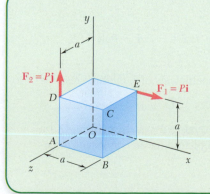

Sample Problem 3.12

Two forces of the same magnitude P act on a cube of side a as shown. Replace the two forces by an equivalent wrench, and determine (a) the magnitude and direction of the resultant force $\mathbf{R}$, (b) the pitch of the wrench, (c) the point where the axis of the wrench intersects the yz plane.

STRATEGY: The first step is to determine the equivalent force-couple system at the origin O. Then you can reduce this system to a wrench and determine its properties.

(continued)

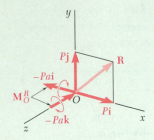

Fig. 1 Force-couple system at O that is equivalent to the given force system.

MODELING and ANALYSIS:

Equivalent Force-Couple System at O.
The position vectors of the points of application E and D of the two given forces are $\mathbf{r}_E = a\mathbf{i} + a\mathbf{j}$ and $\mathbf{r}_D = a\mathbf{j} + a\mathbf{k}$. The resultant $\mathbf{R}$ of the two forces and their moment resultant $\mathbf{M}_O^R$ about O are (Fig. 1)

$$\mathbf{R} = \mathbf{F}_1 + \mathbf{F}_2 = P\mathbf{i} + P\mathbf{j} = P(\mathbf{i} + \mathbf{j}) \tag{1}$$

$$\mathbf{M}_O^R = \mathbf{r}_E \times \mathbf{F}_1 + \mathbf{r}_D \times \mathbf{F}_2 = (a\mathbf{i} + a\mathbf{j}) \times P\mathbf{i} + (a\mathbf{j} + a\mathbf{k}) \times P\mathbf{j}$$

$$= -Pa\mathbf{k} - Pa\mathbf{i} = -Pa(\mathbf{i} + \mathbf{k}) \tag{2}$$

a. Resultant Force R. It follows from Eq. (1) and Fig. 1 that the resultant force $\mathbf{R}$ has a magnitude of $R = P\sqrt{2}$, lies in the xy plane, and forms angles of $45°$ with the x and y axes. Thus

$$R = P\sqrt{2} \qquad \theta_x = \theta_y = 45° \qquad \theta_z = 90° \quad \blacktriangleleft$$

b. Pitch of the Wrench. Using equation (3.60) of Sec. 3.4D and Eqs. (1) and (2) above, the pitch p of the wrench is

$$p = \frac{\mathbf{R} \cdot \mathbf{M}_O^R}{R^2} = \frac{P(\mathbf{i} + \mathbf{j}) \cdot (-Pa)(\mathbf{i} + \mathbf{k})}{(P\sqrt{2})^2} = \frac{-P^2a(1 + 0 + 0)}{2P^2} \qquad p = -\frac{a}{2} \quad \blacktriangleleft$$

c. Axis of the Wrench. From the pitch and from Eq. (3.59), the wrench consists of the force $\mathbf{R}$ found in Eq. (1) and the couple vector

$$\mathbf{M}_1 = p\mathbf{R} = -\frac{a}{2}P(\mathbf{i} + \mathbf{j}) = -\frac{Pa}{2}(\mathbf{i} + \mathbf{j}) \tag{3}$$

To find the point where the axis of the wrench intersects the yz plane, set the moment of the wrench about O equal to the moment resultant $\mathbf{M}_O^R$ of the original system:

$$\mathbf{M}_1 + \mathbf{r} \times \mathbf{R} = \mathbf{M}_O^R$$

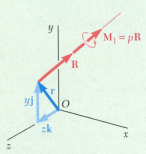

Fig. 2 Wrench that is equivalent to the given force system.

Alternatively, noting that $\mathbf{r} = y\mathbf{j} + z\mathbf{k}$ (Fig. 2) and substituting for $\mathbf{R}$, $\mathbf{M}_O^R$, and $\mathbf{M}_1$ from Eqs. (1), (2), and (3), we have

$$-\frac{Pa}{2}(\mathbf{i} + \mathbf{j}) + (y\mathbf{j} + z\mathbf{k}) \times P(\mathbf{i} + \mathbf{j}) = -Pa(\mathbf{i} + \mathbf{k})$$

$$-\frac{Pa}{2}\mathbf{i} - \frac{Pa}{2}\mathbf{j} - Py\mathbf{k} + Pz\mathbf{j} - Pz\mathbf{i} = -Pa\mathbf{i} - Pa\mathbf{k}$$

Equating the coefficients of $\mathbf{k}$ and then the coefficients of $\mathbf{j}$, the final result is

$$y = a \qquad z = a/2 \quad \blacktriangleleft$$

REFLECT and THINK: Conceptually, reducing a system of forces to a wrench is simply an additional application of finding an equivalent force-couple system.

SOLVING PROBLEMS
ON YOUR OWN

In this section you studied the reduction and simplification of force systems. In solving the problems that follow, you will be asked to perform the following operations.

1. Reducing a force system to a force and a couple at a given point A. The force is the *resultant* $\mathbf{R}$ of the system that is obtained by adding the various forces. The moment of the couple is the *moment resultant* of the system that is obtained by adding the moments about A of the various forces. We have

$$\mathbf{R} = \Sigma \mathbf{F} \qquad \mathbf{M}_A^R = \Sigma(\mathbf{r} \times \mathbf{F})$$

where the position vector $\mathbf{r}$ is drawn from A to *any point* on the line of action of $\mathbf{F}$.

2. Moving a force-couple system from point A to point B. If you wish to reduce a given force system to a force-couple system at point B, you need not recompute the moments of the forces about B after you have reduced it to a force-couple system at point A. The resultant $\mathbf{R}$ remains unchanged, and you can obtain the new moment resultant $\mathbf{M}_B^R$ by adding the moment about B of the force $\mathbf{R}$ applied at A to $\mathbf{M}_A^R$ [Sample Prob. 3.8]. Denoting the vector drawn from B to A as $\mathbf{s}$, you have

$$\mathbf{M}_B^R = \mathbf{M}_A^R + \mathbf{s} \times \mathbf{R}$$

3. Checking whether two force systems are equivalent. First reduce each force system to a force-couple system *at the same, but arbitrary, point A* (as explained in the first operation). The two force systems are equivalent (that is, they have the same effect on the given rigid body) if the two reduced force-couple systems are identical; that is, if

$$\Sigma \mathbf{F} = \Sigma \mathbf{F}' \qquad \text{and} \qquad \Sigma \mathbf{M}_A = \Sigma \mathbf{M}'_A$$

You should recognize that if the first of these equations is not satisfied—that is, if the two systems do not have the same resultant $\mathbf{R}$—the two systems cannot be equivalent, and there is no need to check whether or not the second equation is satisfied.

4. Reducing a given force system to a single force. First reduce the given system to a force-couple system consisting of the resultant $\mathbf{R}$ and the couple vector $\mathbf{M}_A^R$ at

(continued)

some convenient point A (as explained in the first operation). Recall from Section 3.4 that further reduction to a single force is possible *only if the force* $\mathbf{R}$ *and the couple vector* $\mathbf{M}_A^R$ *are mutually perpendicular*. This will certainly be the case for systems of forces that are either *concurrent, coplanar,* or *parallel*. You can then obtain the required single force by moving $\mathbf{R}$ until its moment about A is equal to $\mathbf{M}_A^R$, as you did in several problems in Section 3.4. More formally, the position vector $\mathbf{r}$ drawn from A to any point on the line of action of the single force $\mathbf{R}$ must satisfy the equation

$$\mathbf{r} \times \mathbf{R} = \mathbf{M}_A^R$$

This procedure was illustrated in Sample Probs. 3.8, 3.9, and 3.11.

5. Reducing a given force system to a wrench. If the given system includes forces that are not concurrent, coplanar, or parallel, the equivalent force-couple system at a point A will consist of a force $\mathbf{R}$ and a couple vector $\mathbf{M}_A^R$ that, in general, *are not mutually perpendicular*. (To check whether $\mathbf{R}$ and $\mathbf{M}_A^R$ are mutually perpendicular, form their scalar product. If this product is zero, they are mutually perpendicular; otherwise, they are not.) If $\mathbf{R}$ and $\mathbf{M}_A^R$ are not mutually perpendicular, the force-couple system (and thus the given system of forces) *cannot be reduced to a single force*. However, the system can be reduced to a *wrench*—the combination of a force $\mathbf{R}$ and a couple vector $\mathbf{M}_1$ directed along a common line of action called the *axis of the wrench* (Fig. 3.42). The ratio $p = M_1/R$ is called the *pitch* of the wrench.

To reduce a given force system to a wrench, you should follow these steps,

    a. Reduce the given system to an equivalent force-couple system $(\mathbf{R}, \mathbf{M}_O^R)$, typically located at the origin O.

    b. Determine the pitch p from Eq. (3.60),

$$p = \frac{M_1}{R} = \frac{\mathbf{R} \cdot \mathbf{M}_O^R}{R^2}$$

    and the couple vector from $\mathbf{M}_1 = p\mathbf{R}$.

    c. Set the moment about O of the wrench equal to the moment resultant $\mathbf{M}_O^R$ of the force-couple system at O:

$$\mathbf{M}_1 + \mathbf{r} \times \mathbf{R} = \mathbf{M}_O^R \tag{3.61}$$

This equation allows you to determine the point where the line of action of the wrench intersects a specified plane, since the position vector $\mathbf{r}$ is directed from O to that point. These steps are illustrated in Sample Prob. 3.12. Although determining a wrench and the point where its axis intersects a plane may appear difficult, the process is simply the application of several of the ideas and techniques developed in this chapter. Once you have mastered the wrench, you can feel confident that you understand much of Chap. 3.

Problems

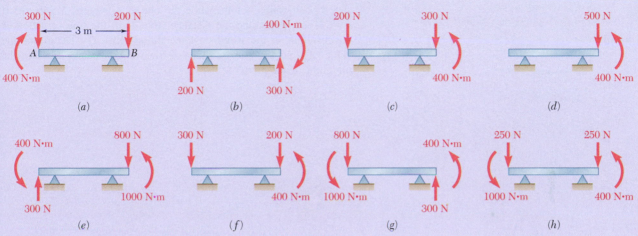

Fig. P3.101

3.101 A 3-m-long beam is subjected to a variety of loadings. (*a*) Replace each loading with an equivalent force-couple system at end *A* of the beam. (*b*) Which of the loadings are equivalent?

3.102 A 3-m-long beam is loaded as shown. Determine the loading of Prob. 3.101 that is equivalent to this loading.

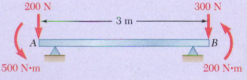

Fig. P3.102

3.103 Determine the single equivalent force and the distance from point *A* to its line of action for the beam and loading of (*a*) Prob. 3.101*a*, (*b*) Prob. 3.101*b*, (*c*) Prob. 3.102.

3.104 Five separate force-couple systems act at the corners of a piece of sheet metal that has been bent into the shape shown. Determine which of these systems is equivalent to a force **F** = (10 lb)**i** and a couple of moment **M** = (15 lb·ft)**j** + (15 lb·ft)**k** located at the origin.

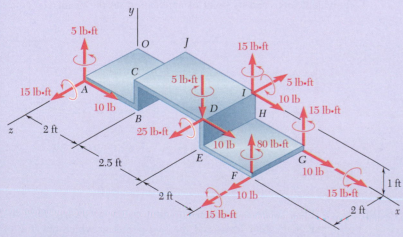

Fig. P3.104

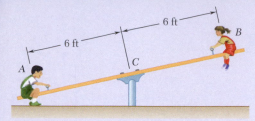

Fig. P3.105

3.105 The weights of two children sitting at ends *A* and *B* of a seesaw are 84 lb and 64 lb, respectively. Where should a third child sit so that the resultant of the weights of the three children will pass through *C* if she weighs (*a*) 60 lb, (*b*) 52 lb?

3.106 Three stage lights are mounted on a pipe as shown. The lights at *A* and *B* each weigh 4.1 lb, while the one at *C* weighs 3.5 lb. (*a*) If *d* = 25 in., determine the distance from *D* to the line of action of the resultant of the weights of the three lights. (*b*) Determine the value of *d* so that the resultant of the weights passes through the midpoint of the pipe.

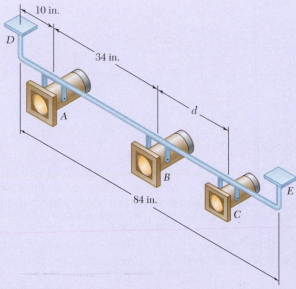

Fig. P3.106

3.107 A beam supports three loads of given magnitude and a fourth load whose magnitude is a function of position. If *b* = 1.5 m and the loads are to be replaced with a single equivalent force, determine (*a*) the value of *a* so that the distance from support *A* to the line of action of the equivalent force is maximum, (*b*) the magnitude of the equivalent force and its point of application on the beam.

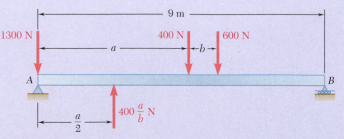

Fig. *P3.107*

3.108 A 6 × 12-in. plate is subjected to four loads as shown. Find the resultant of the four loads and the two points at which the line of action of the resultant intersects the edge of the plate.

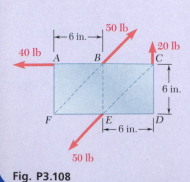

Fig. P3.108

3.109 A 32-lb motor is mounted on the floor. Find the resultant of the weight and the forces exerted on the belt, and determine where the line of action of the resultant intersects the floor.

3.110 To test the strength of a 625 × 500-mm suitcase, forces are applied as shown. If $P = 88$ N, (a) determine the resultant of the applied forces, (b) locate the two points where the line of action of the resultant intersects the edge of the suitcase.

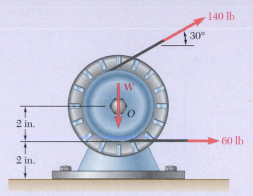

Fig. P3.109

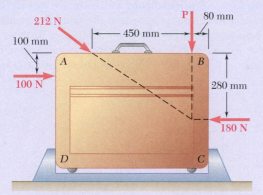

Fig. P3.110

3.111 Solve Prob. 3.110, assuming that $P = 138$ N.

3.112 Pulleys A and B are mounted on bracket $CDEF$. The tension on each side of the two belts is as shown. Replace the four forces with a single equivalent force, and determine where its line of action intersects the bottom edge of the bracket.

3.113 A truss supports the loading shown. Determine the equivalent force acting on the truss and the point of intersection of its line of action with a line drawn through points A and G.

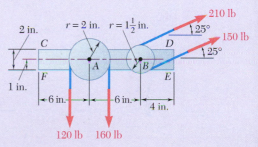

Fig. P3.112

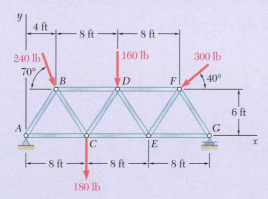

Fig. P3.113

3.114 A couple of magnitude $M = 80$ lb·in. and the three forces shown are applied to an angle bracket. (a) Find the resultant of this system of forces. (b) Locate the points where the line of action of the resultant intersects line AB and line BC.

3.115 A couple $\mathbf{M}$ and the three forces shown are applied to an angle bracket. Find the moment of the couple if the line of action of the resultant of the force system is to pass through (a) point A, (b) point B, (c) point C.

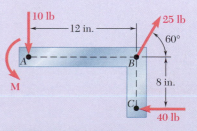

Fig. P3.114 and P3.115

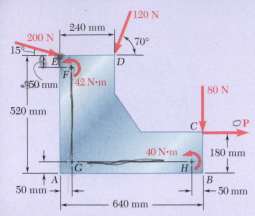

Fig. P3.116

3.116 A machine component is subjected to the forces and couples shown. The component is to be held in place by a single rivet that can resist a force but not a couple. For $P = 0$, determine the location of the rivet hole if it is to be located (*a*) on line *FG*, (*b*) on line *GH*.

3.117 Solve Prob. 3.116, assuming that $P = 60$ N.

3.118 As follower *AB* rolls along the surface of member *C*, it exerts a constant force **F** perpendicular to the surface. (*a*) Replace **F** with an equivalent force-couple system at the point *D* obtained by drawing the perpendicular from the point of contact to the *x* axis. (*b*) For $a = 1$ m and $b = 2$ m, determine the value of *x* for which the moment of the equivalent force-couple system at *D* is maximum.

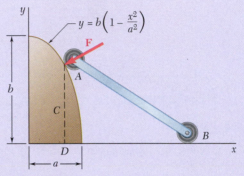

Fig. P3.118

3.119 A machine component is subjected to the forces shown, each of which is parallel to one of the coordinate axes. Replace these forces with an equivalent force-couple system at *A*.

3.120 Two 150-mm-diameter pulleys are mounted on line shaft *AD*. The belts at *B* and *C* lie in vertical planes parallel to the *yz* plane. Replace the belt forces shown with an equivalent force-couple system at *A*.

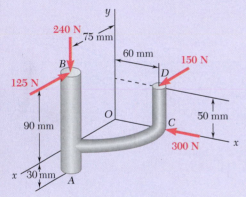

Fig. P3.119

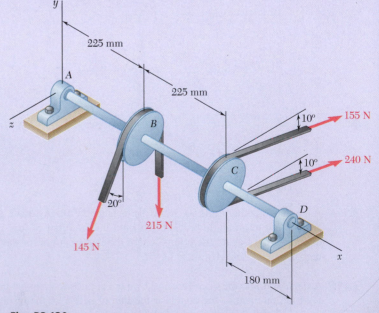

Fig. P3.120

3.121 As an adjustable brace *BC* is used to bring a wall into plumb, the force-couple system shown is exerted on the wall. Replace this force-couple system with an equivalent force-couple system at *A* if *R* = 21.2 lb and *M* = 13.25 lb·ft.

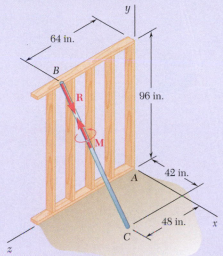

Fig. *P3.121*

3.122 In order to unscrew the tapped faucet *A*, a plumber uses two pipe wrenches as shown. By exerting a 40-lb force on each wrench at a distance of 10 in. from the axis of the pipe and in a direction perpendicular to the pipe and to the wrench, he prevents the pipe from rotating, and thus he avoids loosening or further tightening the joint between the pipe and the tapped elbow *C*. Determine (*a*) the angle *θ* that the wrench at *A* should form with the vertical if elbow *C* is not to rotate about the vertical, (*b*) the force-couple system at *C* equivalent to the two 40-lb forces when this condition is satisfied.

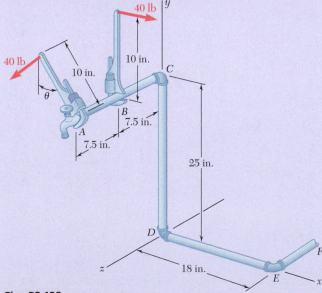

Fig. P3.122

3.123 Assuming *θ* = 60° in Prob. 3.122, replace the two 40-lb forces with an equivalent force-couple system at *D* and determine whether the plumber's action tends to tighten or loosen the joint between (*a*) pipe *CD* and elbow *D*, (*b*) elbow *D* and pipe *DE*. Assume all threads to be right-handed.

3.124 Four forces are applied to the machine component *ABDE* as shown. Replace these forces with an equivalent force-couple system at *A*.

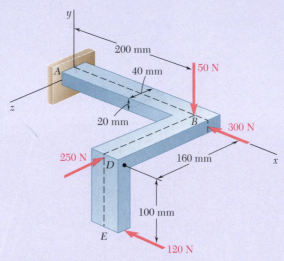

Fig. P3.124

3.125 A blade held in a brace is used to tighten a screw at *A*. (*a*) Determine the forces exerted at *B* and *C*, knowing that these forces are equivalent to a force-couple system at *A* consisting of $\mathbf{R} = -(25 \text{ N})\mathbf{i} + R_y\mathbf{j} + R_z\mathbf{k}$ and $\mathbf{M}_A^R = -(13.5 \text{ N·m})\mathbf{i}$. (*b*) Find the corresponding values of R_y and R_z. (*c*) What is the orientation of the slot in the head of the screw for which the blade is least likely to slip when the brace is in the position shown?

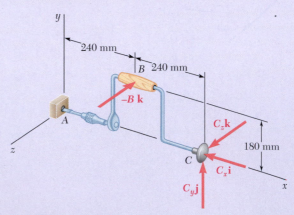

Fig. P3.125

3.126 A mechanic uses a crowfoot wrench to loosen a bolt at *C*. The mechanic holds the socket wrench handle at points *A* and *B* and applies forces at these points. Knowing that these forces are equivalent to a force-couple system at *C* consisting of the force $\mathbf{C} = -(8 \text{ lb})\mathbf{i} + (4 \text{ lb})\mathbf{k}$ and the couple $\mathbf{M}_C = (360 \text{ lb·in.})\mathbf{i}$, determine the forces applied at *A* and at *B* when $A_z = 2$ lb.

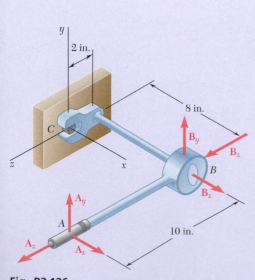

Fig. P3.126

3.127 Three children are standing on a 5 × 5-m raft. If the weights of the children at points *A*, *B*, and *C* are 375 N, 260 N, and 400 N, respectively, determine the magnitude and the point of application of the resultant of the three weights.

3.128 Three children are standing on a 5 × 5-m raft. The weights of the children at points *A*, *B*, and *C* are 375 N, 260 N, and 400 N, respectively. If a fourth child weighing 425 N climbs onto the raft, determine where she should stand if the other children remain in the positions shown and if the line of action of the resultant of the four weights is to pass through the center of the raft.

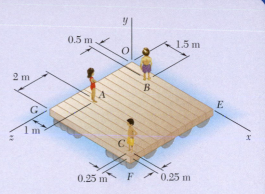

Fig. P3.127 and P3.128

3.129 Four signs are mounted on a frame spanning a highway, and the magnitudes of the horizontal wind forces acting on the signs are as shown. Determine the magnitude and the point of application of the resultant of the four wind forces when *a* = 1 ft and *b* = 12 ft.

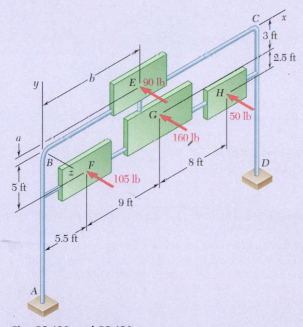

Fig. P3.129 and P3.130

3.130 Four signs are mounted on a frame spanning a highway, and the magnitudes of the horizontal wind forces acting on the signs are as shown. Determine *a* and *b* so that the point of application of the resultant of the four forces is at *G*.

3.131 A concrete foundation mat of 5-m radius supports four equally spaced columns, each of which is located 4 m from the center of the mat. Determine the magnitude and the point of application of the resultant of the four loads.

3.132 Determine the magnitude and the point of application of the smallest additional load that must be applied to the foundation mat of Prob. 3.131 if the resultant of the five loads is to pass through the center of the mat.

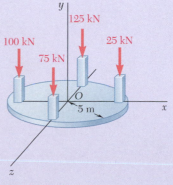

Fig. P3.131

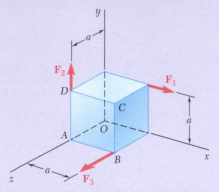

Fig. P3.133

*3.133 Three forces of the same magnitude P act on a cube of side a as shown. Replace the three forces with an equivalent wrench and determine (a) the magnitude and direction of the resultant force $\mathbf{R}$, (b) the pitch of the wrench, (c) the axis of the wrench.

*3.134 A piece of sheet metal is bent into the shape shown and is acted upon by three forces. If the forces have the same magnitude P, replace them with an equivalent wrench and determine (a) the magnitude and the direction of the resultant force $\mathbf{R}$, (b) the pitch of the wrench, (c) the axis of the wrench.

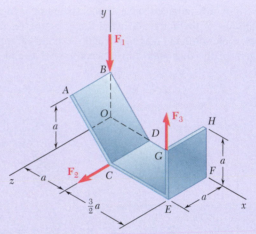

Fig. P3.134

*3.135 and *3.136 The forces and couples shown are applied to two screws as a piece of sheet metal is fastened to a block of wood. Reduce the forces and the couples to an equivalent wrench and determine (a) the resultant force $\mathbf{R}$, (b) the pitch of the wrench, (c) the point where the axis of the wrench intersects the xz plane.

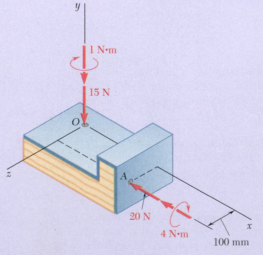

Fig. *P3.135*

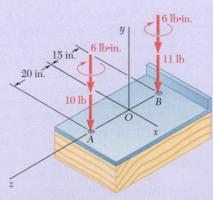

Fig. P3.136

158

*3.137 and *3.138 Two bolts at *A* and *B* are tightened by applying the forces and couples shown. Replace the two wrenches with a single equivalent wrench and determine (*a*) the resultant **R**, (*b*) the pitch of the single equivalent wrench, (*c*) the point where the axis of the wrench intersects the *xz* plane.

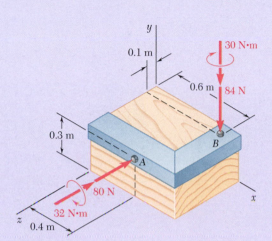

Fig. P3.137

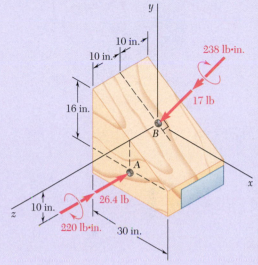

Fig. P3.138

*3.139 Two ropes attached at *A* and *B* are used to move the trunk of a fallen tree. Replace the forces exerted by the ropes with an equivalent wrench and determine (*a*) the resultant force **R**, (*b*) the pitch of the wrench, (*c*) the point where the axis of the wrench intersects the *yz* plane.

*3.140 A flagpole is guyed by three cables. If the tensions in the cables have the same magnitude *P*, replace the forces exerted on the pole with an equivalent wrench and determine (*a*) the resultant force **R**, (*b*) the pitch of the wrench, (*c*) the point where the axis of the wrench intersects the *xz* plane.

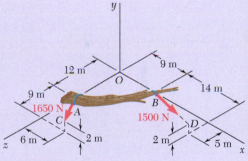

Fig. P3.139

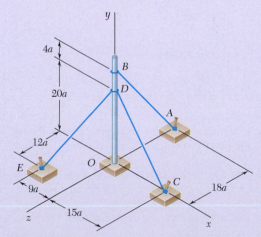

Fig. P3.140

159

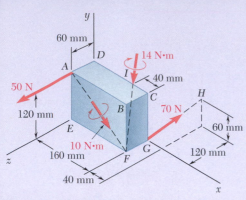

Fig. P3.141

*3.141 and *3.142 Determine whether the force-and-couple system shown can be reduced to a single equivalent force **R**. If it can, determine **R** and the point where the line of action of **R** intersects the *yz* plane. If it cannot be reduced, replace the given system with an equivalent wrench and determine its resultant, its pitch, and the point where its axis intersects the *yz* plane.

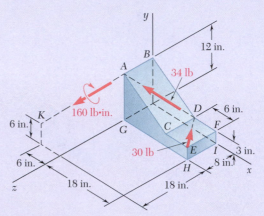

Fig. *P3.142*

*3.143 Replace the wrench shown with an equivalent system consisting of two forces perpendicular to the *y* axis and applied respectively at *A* and *B*.

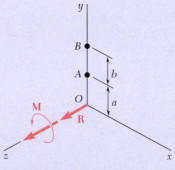

Fig. P3.143

*3.144 Show that, in general, a wrench can be replaced with two forces chosen in such a way that one force passes through a given point while the other force lies in a given plane.

*3.145 Show that a wrench can be replaced with two perpendicular forces, one of which is applied at a given point.

*3.146 Show that a wrench can be replaced with two forces, one of which has a prescribed line of action.

Review and Summary

Principle of Transmissibility

In this chapter, we presented the effects of forces exerted on a rigid body. We began by distinguishing between **external** and **internal** forces [Sec. 3.1A]. We then explained that, according to the **principle of transmissibility**, the effect of an external force on a rigid body remains unchanged if we move that force along its line of action [Sec. 3.1B]. In other words, two forces **F** and **F'** acting on a rigid body at two different points have the same effect on that body if they have the same magnitude, same direction, and same line of action (Fig. 3.43). Two such forces are said to be **equivalent.**

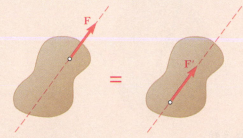

Fig. 3.43

Vector Product

Before proceeding with the discussion of **equivalent systems of forces**, we introduced the concept of the **vector product of two vectors** [Sec. 3.1C]. We defined the vector product

$$\mathbf{V} = \mathbf{P} \times \mathbf{Q}$$

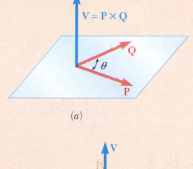

of the vectors **P** and **Q** as a vector perpendicular to the plane containing **P** and **Q** (Fig. 3.44) with a magnitude of

$$V = PQ \sin \theta \qquad (3.1)$$

and directed in such a way that a person located at the tip of **V** will observe the rotation to be counterclockwise through θ, bringing the vector **P** in line with the vector **Q**. The three vectors **P**, **Q**, and **V**—taken in that order—are said to form a *right-handed triad*. It follows that the vector products **Q** × **P** and **P** × **Q** are represented by equal and opposite vectors:

$$\mathbf{Q} \times \mathbf{P} = -(\mathbf{P} \times \mathbf{Q}) \qquad (3.4)$$

Fig. 3.44

It also follows from the definition of the vector product of two vectors that the vector products of the unit vectors **i**, **j**, and **k** are

$$\mathbf{i} \times \mathbf{i} = 0 \qquad \mathbf{i} \times \mathbf{j} = \mathbf{k} \qquad \mathbf{j} \times \mathbf{i} = -\mathbf{k}$$

and so on. You can determine the sign of the vector product of two unit vectors by arranging in a circle and in counterclockwise order the three letters representing the unit vectors (Fig. 3.45): The vector product of two unit vectors is positive if they follow each other in counterclockwise order and negative if they follow each other in clockwise order.

Fig. 3.45

Rectangular Components of Vector Product

The **rectangular components of the vector product V** of two vectors **P** and **Q** are expressed [Sec. 3.1D] as

$$\begin{aligned}
V_x &= P_y Q_z - P_z Q_y \\
V_y &= P_z Q_x - P_x Q_z \\
V_z &= P_x Q_y - P_y Q_x
\end{aligned} \qquad (3.9)$$

We can also express the components of a vector product as a determinant:

$$\mathbf{V} = \begin{vmatrix} \mathbf{i} & \mathbf{j} & \mathbf{k} \\ P_x & P_y & P_z \\ Q_x & Q_y & Q_z \end{vmatrix} \quad (3.10)$$

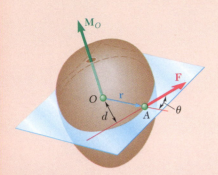

Fig. 3.46

Moment of a Force about a Point

We defined the **moment of a force F about a point** O [Sec. 3.1E] as the vector product

$$\mathbf{M}_O = \mathbf{r} \times \mathbf{F} \quad (3.11)$$

where $\mathbf{r}$ is the *position vector* drawn from O to the point of application A of the force $\mathbf{F}$ (Fig. 3.46). Denoting the angle between the lines of action of $\mathbf{r}$ and $\mathbf{F}$ as θ, we found that the magnitude of the moment of $\mathbf{F}$ about O is

$$M_O = rF \sin \theta = Fd \quad (3.12)$$

where d represents the perpendicular distance from O to the line of action of $\mathbf{F}$.

Rectangular Components of Moment

The **rectangular components of the moment $\mathbf{M}_O$ of a force F** [Sec. 3.1F] are

$$\begin{aligned} M_x &= yF_z - zF_y \\ M_y &= zF_x - xF_z \\ M_z &= xF_y - yF_x \end{aligned} \quad (3.18)$$

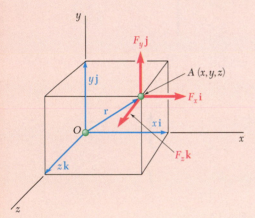

Fig. 3.47

where x, y, and z are the components of the position vector $\mathbf{r}$ (Fig. 3.47). Using a determinant form, we also wrote

$$\mathbf{M}_O = \begin{vmatrix} \mathbf{i} & \mathbf{j} & \mathbf{k} \\ x & y & z \\ F_x & F_y & F_z \end{vmatrix} \quad (3.19)$$

In the more general case of the moment about an arbitrary point B of a force $\mathbf{F}$ applied at A, we had

$$\mathbf{M}_B = \begin{vmatrix} \mathbf{i} & \mathbf{j} & \mathbf{k} \\ x_{A/B} & y_{A/B} & z_{A/B} \\ F_x & F_y & F_z \end{vmatrix} \quad (3.21)$$

where $x_{A/B}$, $y_{A/B}$, and $z_{A/B}$ denote the components of the vector $\mathbf{r}_{A/B}$:

$$x_{A/B} = x_A - x_B \qquad y_{A/B} = y_A - y_B \qquad z_{A/B} = z_A - z_B$$

In the case of *problems involving only two dimensions*, we can assume the force $\mathbf{F}$ lies in the xy plane. Its moment $\mathbf{M}_B$ about a point B in the same plane is perpendicular to that plane (Fig. 3.48) and is completely defined by the scalar

$$M_B = (x_A - x_B)F_y - (y_A - y_B)F_x \quad (3.23)$$

Various methods for computing the moment of a force about a point were illustrated in Sample Probs. 3.1 through 3.4.

Scalar Product of Two Vectors

The **scalar product** of two vectors $\mathbf{P}$ and $\mathbf{Q}$ [Sec. 3.2A], denoted by $\mathbf{P} \cdot \mathbf{Q}$, is defined as the scalar quantity

$$\mathbf{P} \cdot \mathbf{Q} = PQ \cos \theta \quad (3.24)$$

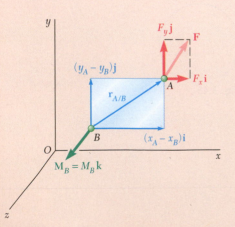

Fig. 3.48

where θ is the angle between $\mathbf{P}$ and $\mathbf{Q}$ (Fig. 3.49). By expressing the scalar product of $\mathbf{P}$ and $\mathbf{Q}$ in terms of the rectangular components of the two vectors, we determined that

$$\mathbf{P} \cdot \mathbf{Q} = P_x Q_x + P_y Q_y + P_z Q_z \quad (3.28)$$

Fig. 3.49

Projection of a Vector on an Axis

We obtain the **projection of a vector P on an axis** OL (Fig. 3.50) by forming the scalar product of $\mathbf{P}$ and the unit vector $\boldsymbol{\lambda}$ along OL. We have

$$P_{OL} = \mathbf{P} \cdot \boldsymbol{\lambda} \quad (3.34)$$

Using rectangular components, this becomes

$$P_{OL} = P_x \cos \theta_x + P_y \cos \theta_y + P_z \cos \theta_z \quad (3.35)$$

where θ_x, θ_y, and θ_z denote the angles that the axis OL forms with the coordinate axes.

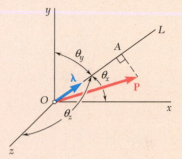

Fig. 3.50

Mixed Triple Product of Three Vectors

We defined the **mixed triple product** of the three vectors $\mathbf{S}$, $\mathbf{P}$, and $\mathbf{Q}$ as the scalar expression

$$\mathbf{S} \cdot (\mathbf{P} \times \mathbf{Q}) \quad (3.36)$$

obtained by forming the scalar product of $\mathbf{S}$ with the vector product of $\mathbf{P}$ and $\mathbf{Q}$ [Sec. 3.2B]. We showed that

$$\mathbf{S} \cdot (\mathbf{P} \times \mathbf{Q}) = \begin{vmatrix} S_x & S_y & S_z \\ P_x & P_y & P_z \\ Q_x & Q_y & Q_z \end{vmatrix} \quad (3.39)$$

where the elements of the determinant are the rectangular components of the three vectors.

Moment of a Force about an Axis

We defined the **moment of a force F about an axis** OL [Sec. 3.2C] as the projection OC on OL of the moment $\mathbf{M}_O$ of the force $\mathbf{F}$ (Fig. 3.51), i.e., as the mixed triple product of the unit vector $\boldsymbol{\lambda}$, the position vector $\mathbf{r}$, and the force $\mathbf{F}$:

$$M_{OL} = \boldsymbol{\lambda} \cdot \mathbf{M}_O = \boldsymbol{\lambda} \cdot (\mathbf{r} \times \mathbf{F}) \quad (3.40)$$

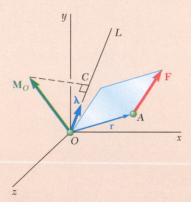

Fig. 3.51

The determinant form for the mixed triple product is

$$M_{OL} = \begin{vmatrix} \lambda_x & \lambda_y & \lambda_z \\ x & y & z \\ F_x & F_y & F_z \end{vmatrix} \tag{3.41}$$

where

$\lambda_x, \lambda_y, \lambda_z$ = direction cosines of axis OL

x, y, z = components of $\mathbf{r}$

F_x, F_y, F_z = components of $\mathbf{F}$

An example of determining the moment of a force about a skew axis appears in Sample Prob. 3.5.

Couples

Two forces $\mathbf{F}$ *and* $-\mathbf{F}$ *having the same magnitude, parallel lines of action, and opposite sense are said to form a* **couple** [Sec. 3.3A]. The moment of a couple is independent of the point about which it is computed; it is a vector $\mathbf{M}$ perpendicular to the plane of the couple and equal in magnitude to the product of the common magnitude F of the forces and the perpendicular distance d between their lines of action (Fig. 3.52).

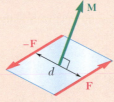

Fig. 3.52

Two couples having the same moment $\mathbf{M}$ are *equivalent,* i.e., they have the same effect on a given rigid body [Sec. 3.3B]. The sum of two couples is itself a couple [Sec. 3.3C], and we can obtain the moment $\mathbf{M}$ of the resultant couple by adding vectorially the moments $\mathbf{M}_1$ and $\mathbf{M}_2$ of the original couples [Sample Prob. 3.6]. It follows that we can represent a couple by a vector, called a **couple vector**, equal in magnitude and direction to the moment $\mathbf{M}$ of the couple [Sec. 3.3D]. A couple vector is a *free vector* that can be attached to the origin O if so desired and resolved into components (Fig. 3.53).

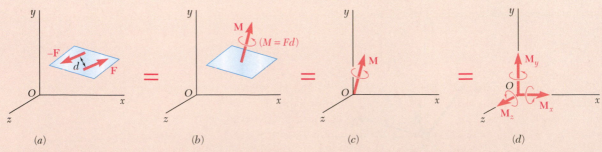

(a) (b) (c) (d)

Fig. 3.53

Force-Couple System

Any force $\mathbf{F}$ acting at a point A of a rigid body can be replaced by a **force-couple system** at an arbitrary point O consisting of the force $\mathbf{F}$ applied at O

and a couple of moment $\mathbf{M}_O$, which is equal to the moment about O of the force $\mathbf{F}$ in its original position [Sec. 3.3E]. Note that the force $\mathbf{F}$ and the couple vector $\mathbf{M}_O$ are always perpendicular to each other (Fig. 3.54).

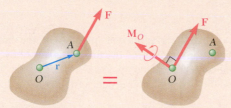

Fig. 3.54

Reduction of a System of Forces to a Force-Couple System

It follows [Sec. 3.4A] that *any system of forces can be reduced to a force-couple system at a given point O* by first replacing each of the forces of the system by an equivalent force-couple system at O (Fig. 3.55) and then adding all of the forces and all of the couples to obtain a resultant force $\mathbf{R}$ and a resultant couple vector $\mathbf{M}_O^R$ [Sample Probs. 3.8 through 3.11]. In general, the resultant $\mathbf{R}$ and the couple vector $\mathbf{M}_O^R$ will not be perpendicular to each other.

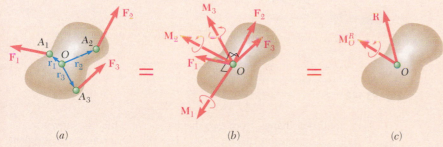

(*a*) (*b*) (*c*)

Fig. 3.55

Equivalent Systems of Forces

We concluded [Sec. 3.4B] that, as far as rigid bodies are concerned, *two systems of forces, $\mathbf{F}_1$, $\mathbf{F}_2$, $\mathbf{F}_3$, . . . and $\mathbf{F}_1'$, $\mathbf{F}_2'$, $\mathbf{F}_3'$, . . . , are equivalent if, and only if,*

$$\Sigma\mathbf{F} = \Sigma\mathbf{F}' \qquad \text{and} \qquad \Sigma\mathbf{M}_O = \Sigma\mathbf{M}_O' \qquad (3.55)$$

Further Reduction of a System of Forces

If the resultant force $\mathbf{R}$ and the resultant couple vector $\mathbf{M}_O^R$ are perpendicular to each other, we can further reduce the force-couple system at O to a single resultant force [Sec. 3.4C]. This is the case for systems consisting of (*a*) concurrent forces (cf. Chap. 2), (*b*) coplanar forces [Sample Probs. 3.8 and 3.9], or (*c*) parallel forces [Sample Prob. 3.11]. If the resultant $\mathbf{R}$ and the couple vector $\mathbf{M}_O^R$ are *not* perpendicular to each other, the system *cannot* be reduced to a single force. We can, however, reduce it to a special type of force-couple system called a *wrench,* consisting of the resultant $\mathbf{R}$ and a couple vector $\mathbf{M}_1$ directed along $\mathbf{R}$ [Sec. 3.4D and Sample Prob. 3.12].

Review Problems

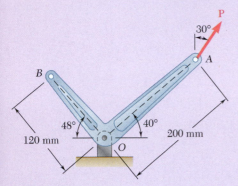

Fig. P3.147

3.147 A 300-N force **P** is applied at point *A* of the bell crank shown. (*a*) Compute the moment of the force **P** about *O* by resolving it into horizontal and vertical components. (*b*) Using the result of part *a*, determine the perpendicular distance from *O* to the line of action of **P**.

3.148 A winch puller *AB* is used to straighten a fence post. Knowing that the tension in cable *BC* is 1040 N and length *d* is 1.90 m, determine the moment about *D* of the force exerted by the cable at *C* by resolving that force into horizontal and vertical components applied (*a*) at point *C*, (*b*) at point *E*.

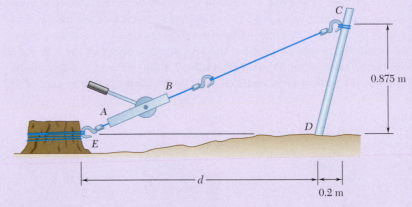

Fig. P3.148

3.149 A small boat hangs from two davits, one of which is shown in the figure. The tension in line *ABAD* is 82 lb. Determine the moment about *C* of the resultant force **R**$_A$ exerted on the davit at *A*.

3.150 Consider the volleyball net shown. Determine the angle formed by guy wires *AB* and *AC*.

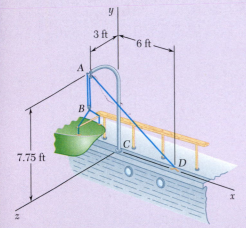

Fig. *P3.149*

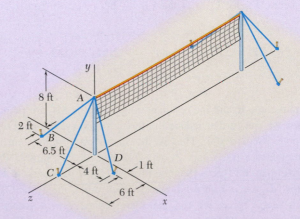

Fig. P3.150

3.151 A single force **P** acts at C in a direction perpendicular to the handle BC of the crank shown. Determine the moment M_x of **P** about the x axis when $\theta = 65°$, knowing that $M_y = -15$ N·m and $M_z = -36$ N·m.

3.152 A small boat hangs from two davits, one of which is shown in the figure. It is known that the moment about the z axis of the resultant force $\mathbf{R}_A$ exerted on the davit at A must not exceed 279 lb·ft in absolute value. Determine the largest allowable tension in line $ABAD$ when $x = 6$ ft.

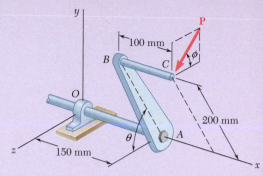

Fig. P3.151

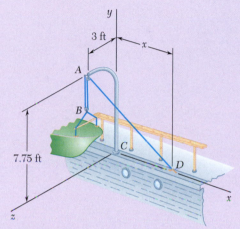

Fig. P3.152

3.153 In a manufacturing operation, three holes are drilled simultaneously in a workpiece. If the holes are perpendicular to the surfaces of the workpiece, replace the couples applied to the drills with a single equivalent couple, specifying its magnitude and the direction of its axis.

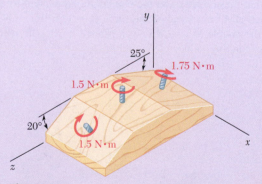

Fig. P3.153

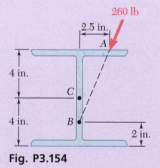

Fig. P3.154

3.154 A 260-lb force is applied at A to the rolled-steel section shown. Replace that force with an equivalent force-couple system at the center C of the section.

3.155 The force and couple shown are to be replaced by an equivalent single force. Knowing that $P = 2Q$, determine the required value of α if the line of action of the single equivalent force is to pass through (a) point A, (b) point C.

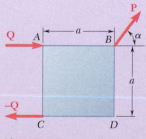

Fig. P3.155

3.156 A 77-N force $\mathbf{F}_1$ and a 31-N·m couple $\mathbf{M}_1$ are applied to corner E of the bent plate shown. If $\mathbf{F}_1$ and $\mathbf{M}_1$ are to be replaced with an equivalent force-couple system ($\mathbf{F}_2$, $\mathbf{M}_2$) at corner B and if $(M_2)_z = 0$, determine (a) the distance d, (b) $\mathbf{F}_2$ and $\mathbf{M}_2$.

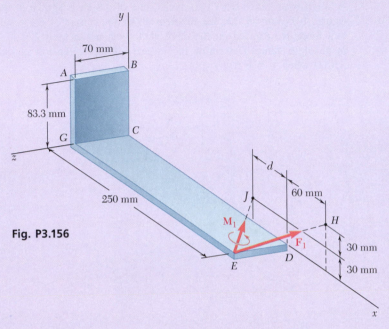

Fig. P3.156

3.157 Three horizontal forces are applied as shown to a vertical cast-iron arm. Determine the resultant of the forces and the distance from the ground to its line of action when (a) $P = 200$ N, (b) $P = 2400$ N, (c) $P = 1000$ N.

3.158 While using a pencil sharpener, a student applies the forces and couple shown. (a) Determine the forces exerted at B and C knowing that these forces and the couple are equivalent to a force-couple system at A consisting of the force $\mathbf{R} = (2.6\text{ lb})\mathbf{i} + R_y\mathbf{j} - (0.7\text{ lb})\mathbf{k}$ and the couple $\mathbf{M}_A^R = M_x\mathbf{i} + (1.0\text{ lb·ft})\mathbf{j} - (0.72\text{ lb·ft})\mathbf{k}$. (b) Find the corresponding values of R_y and M_x.

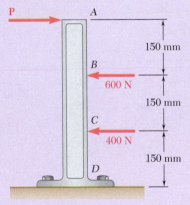

Fig. P3.157

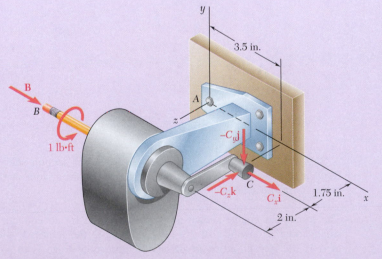

Fig. P3.158

4

Equilibrium of Rigid Bodies

The Tianjin Eye is a Ferris wheel that straddles a bridge over the Hai River in China. The structure is designed so that the support reactions at the wheel bearings as well as those at the base of the frame maintain equilibrium under the effects of vertical gravity and horizontal wind forces.

Objectives

- **Analyze** the static equilibrium of rigid bodies in two and three dimensions.
- **Consider** the attributes of a properly drawn free-body diagram, an essential tool for the equilibrium analysis of rigid bodies.
- **Examine** rigid bodies supported by statically indeterminate reactions and partial constraints.
- **Study** two cases of particular interest: the equilibrium of two-force and three-force bodies.

Introduction

We saw in Chapter 3 how to reduce the external forces acting on a rigid body to a force-couple system at some arbitrary point O. When the force and the couple are both equal to zero, the external forces form a system equivalent to zero, and the rigid body is said to be in **equilibrium**.

We can obtain the necessary and sufficient conditions for the equilibrium of a rigid body by setting $\mathbf{R}$ and $\mathbf{M}_O^R$ equal to zero in the relations of Eq. (3.50) of Sec. 3.4A:

$$\Sigma \mathbf{F} = 0 \qquad \Sigma \mathbf{M}_O = \Sigma\,(\mathbf{r} \times \mathbf{F}) = 0 \tag{4.1}$$

Resolving each force and each moment into its rectangular components, we can replace these vector equations for the equilibrium of a rigid body with the following six scalar equations:

$$\Sigma F_x = 0 \qquad \Sigma F_y = 0 \qquad \Sigma F_z = 0 \tag{4.2}$$
$$\Sigma M_x = 0 \qquad \Sigma M_y = 0 \qquad \Sigma M_z = 0 \tag{4.3}$$

We can use these equations to determine unknown forces applied to the rigid body or unknown reactions exerted on it by its supports. Note that Eqs. (4.2) express the fact that the components of the external forces in the x, y, and z directions are balanced; Eqs. (4.3) express the fact that the moments of the external forces about the x, y, and z axes are balanced. Therefore, for a rigid body in equilibrium, the system of external forces imparts no translational or rotational motion to the body.

In order to write the equations of equilibrium for a rigid body, we must first identify all of the forces acting on that body and then draw the corresponding **free-body diagram**. In this chapter, we first consider the equilibrium of *two-dimensional structures* subjected to forces contained in their planes and study how to draw their free-body diagrams. In addition to the forces *applied* to a structure, we must also consider the *reactions* exerted on the structure by its supports. A specific reaction is associated with each type of support. You will see how to determine whether the structure is properly supported, so that you can know in advance whether you can solve the equations of equilibrium for the unknown forces and reactions.

Later in this chapter, we consider the equilibrium of three-dimensional structures, and we provide the same kind of analysis to these structures and their supports.

Free-Body Diagrams

In solving a problem concerning a rigid body in equilibrium, it is essential to consider *all* of the forces acting on the body. It is equally important to exclude any force that is *not* directly applied to the body. Omitting a force or adding an extraneous one would destroy the conditions of equilibrium. Therefore, the first step in solving the problem is to draw a **free-body diagram** of the rigid body under consideration.

We have already used free-body diagrams on many occasions in Chap. 2. However, in view of their importance to the solution of equilibrium problems, we summarize here the steps you must follow in drawing a correct free-body diagram.

1. Start with a clear decision regarding the choice of the free body to be analyzed. Mentally, you need to detach this body from the ground and separate it from all other bodies. Then you can sketch the contour of this isolated body.
2. Indicate all external forces on the free-body diagram. These forces represent the actions exerted *on* the free body *by* the ground and *by* the bodies that have been detached. In the diagram, apply these forces at the various points where the free body was supported by the ground or

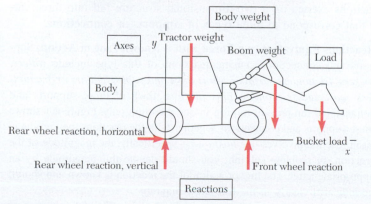

Photo 4.1 A tractor supporting a bucket load. As shown, its free-body diagram should include all external forces acting on the tractor.

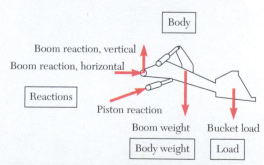

Photo 4.2 Tractor bucket and boom. In Chap. 6, we will see how to determine the internal forces associated with interconnected members such as these using free-body diagrams like the one shown.

was connected to the other bodies. Generally, you should include the *weight* of the free body among the external forces, since it represents the attraction exerted by the earth on the various particles forming the free body. You will see in Chapter 5 that you should draw the weight so it acts at the center of gravity of the body. If the free body is made of several parts, do *not* include the forces the various parts exert on each other among the external forces. These forces are internal forces as far as the free body is concerned.

3. Clearly mark the magnitudes and directions of the *known external forces* on the free-body diagram. Recall that when indicating the directions of these forces, the forces are those exerted *on,* and not *by,* the free body. Known external forces generally include the *weight* of the free body and *forces applied* for a given purpose.

4. *Unknown external forces* usually consist of the **reactions** through which the ground and other bodies oppose a possible motion of the free body. The reactions constrain the free body to remain in the same position; for that reason, they are sometimes called *constraining forces*. Reactions are exerted at the points where the free body is *supported by* or con- *nected to* other bodies; you should clearly indicate these points. Reac- tions are discussed in detail in Secs. 4.1 and 4.3.

5. The free-body diagram should also include dimensions, since these may be needed for computing moments of forces. Any other detail, however, should be omitted.

4.1   EQUILIBRIUM IN TWO DIMENSIONS

In the first part of this chapter, we consider the equilibrium of two-dimensional structures; i.e., we assume that the structure being analyzed and the forces applied to it are contained in the same plane. Clearly, the reactions needed to maintain the structure in the same position are also contained in this plane.

4.1A   Reactions for a Two-Dimensional Structure

The reactions exerted on a two-dimensional structure fall into three cat- egories that correspond to three types of **supports** or **connections**.

1. **Reactions Equivalent to a Force with a Known Line of Action.** Sup- ports and connections causing reactions of this type include *rollers, rockers, frictionless surfaces, short links and cables, collars on friction- less rods,* and *frictionless pins in slots.* Each of these supports and connections can prevent motion in one direction only. Figure 4.1 shows these supports and connections together with the reactions they produce. Each reaction involves *one unknown*—specifically, the magnitude of the reaction. In problem solving, you should denote this magnitude by an appropriate letter. The line of action of the reaction is known and should be indicated clearly in the free-body diagram.

    The sense of the reaction must be as shown in Fig. 4.1 for cases of a frictionless surface (toward the free body) or a cable (away from the free body). The reaction can be directed either way in the cases of double-track rollers, links, collars on rods, or pins in slots. Generally, we

Support or Connection				Reaction	Number of Unknowns
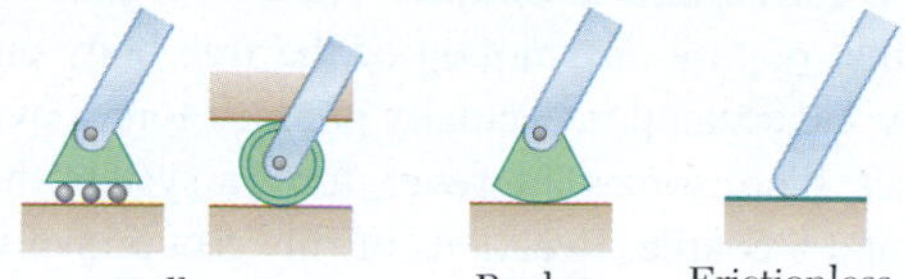 Rollers	Rocker	Frictionless surface		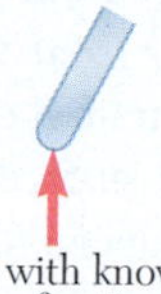 Force with known line of action perpendicular to surface	1
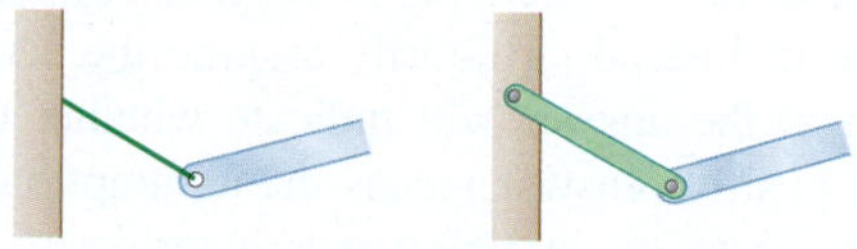 Short cable		Short link		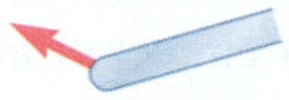 Force with known line of action along cable or link	1
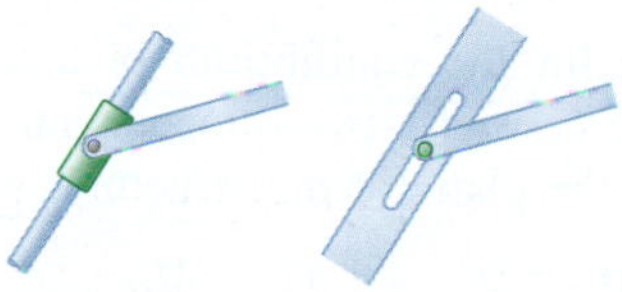 Collar on frictionless rod		Frictionless pin in slot		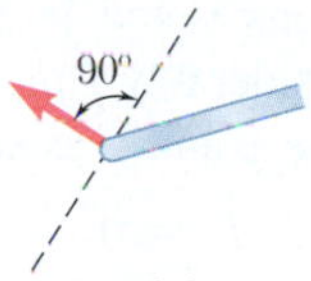 Force with known line of action perpendicular to rod or slot	1
Frictionless pin or hinge		Rough surface		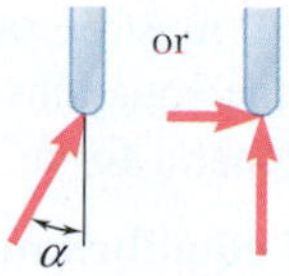 or — Force of unknown direction	2
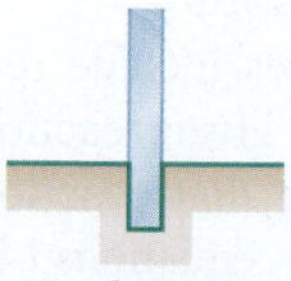 Fixed support				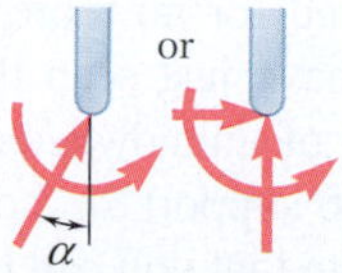 or — Force and couple	3

Fig. 4.1 Reactions of supports and connections in two dimensions.

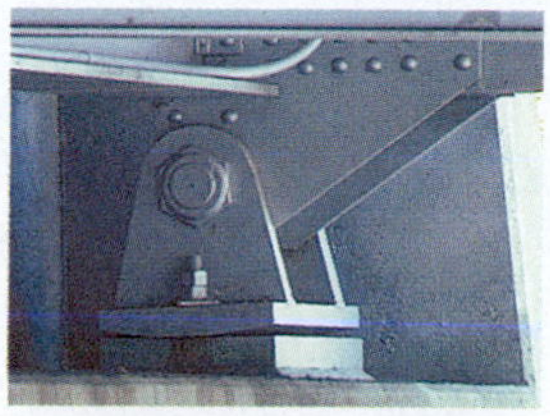

This rocker bearing supports the weight of a bridge. The convex surface of the rocker allows the bridge to move slightly horizontally.

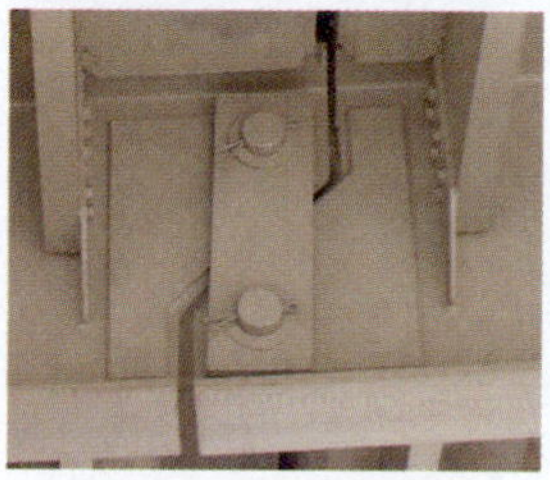

Links are often used to support suspended spans of highway bridges.

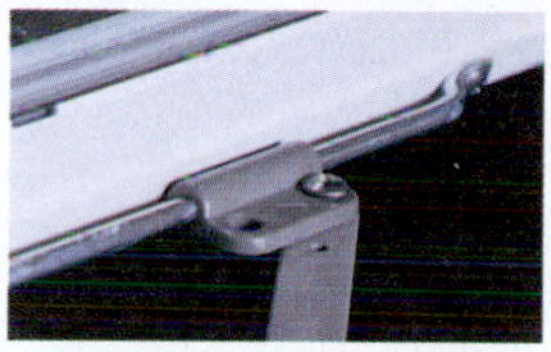

Force applied to the slider exerts a normal force on the rod, causing the window to open.

Pin supports are common on bridges and overpasses.

This cantilever support is fixed at one end and extends out into space at the other end.

assume that single-track rollers and rockers are reversible, so the corresponding reactions can be directed either way.

2. **Reactions Equivalent to a Force of Unknown Direction and Magnitude.** Supports and connections causing reactions of this type include *frictionless pins in fitted holes, hinges,* and *rough surfaces.* They can prevent translation of the free body in all directions, but they cannot prevent the body from rotating about the connection. Reactions of this group involve *two unknowns* and are usually represented by their *x* and

y components. In the case of a rough surface, the component normal to the surface must be directed away from the surface.

3. **Reactions Equivalent to a Force and a Couple.** These reactions are caused by *fixed supports* that oppose any motion of the free body and thus constrain it completely. Fixed supports actually produce forces over the entire surface of contact; these forces, however, form a system that can be reduced to a force and a couple. Reactions of this group involve *three unknowns* usually consisting of the two components of the force and the moment of the couple.

When the sense of an unknown force or couple is not readily apparent, do not attempt to determine it. Instead, arbitrarily assume the sense of the force or couple; the sign of the answer will indicate whether the assumption is correct or not. (A positive answer means the assumption is correct, while a negative answer means the assumption is incorrect.)

4.1B   Rigid-Body Equilibrium in Two Dimensions

The conditions stated in Sec. 4.1A for the equilibrium of a rigid body become considerably simpler for the case of a two-dimensional structure. Choosing the *x* and *y* axes to be in the plane of the structure, we have

$$F_z = 0 \qquad M_x = M_y = 0 \qquad M_z = M_O$$

for each of the forces applied to the structure. Thus, the six equations of equilibrium stated in Sec. 4.1 reduce to three equations:

$$\Sigma F_x = 0 \qquad \Sigma F_y = 0 \qquad \Sigma M_O = 0 \qquad \textbf{(4.4)}$$

Since $\Sigma M_O = 0$ must be satisfied regardless of the choice of the origin O, we can write the equations of equilibrium for a two-dimensional structure in the more general form

Equations of equilibrium in two dimensions

$$\Sigma F_x = 0 \qquad \Sigma F_y = 0 \qquad \Sigma M_A = 0 \qquad \textbf{(4.5)}$$

where A is any point in the plane of the structure. These three equations can be solved for no more than *three unknowns*.

You have just seen that unknown forces include reactions and that the number of unknowns corresponding to a given reaction depends upon the type of support or connection causing that reaction. Referring to Fig. 4.1, note that you can use the equilibrium equations (4.5) to determine the reactions associated with two rollers and one cable, or one fixed support, or one roller and one pin in a fitted hole, etc.

For example, consider Fig. 4.2*a*, in which the truss shown is in equilibrium and is subjected to the given forces **P**, **Q**, and **S**. The truss is held in place by a pin at *A* and a roller at *B*. The pin prevents point *A* from moving by exerting a force on the truss that can be resolved into the components $\mathbf{A}_x$ and $\mathbf{A}_y$. The roller keeps the truss from rotating about *A* by exerting the vertical force **B**. The free-body diagram of the truss is shown in Fig. 4.2*b*; it includes the reactions $\mathbf{A}_x$, $\mathbf{A}_y$, and **B** as well as the applied forces **P**, **Q**, and **S** (in *x* and *y* component form) and the weight **W** of the truss.

Since the truss is in equilibrium, the sum of the moments about A of all of the forces shown in Fig. 4.2*b* is zero, or $\Sigma M_A = 0$. We can use

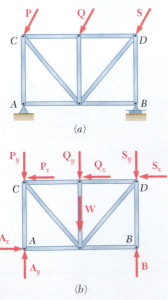

Fig. 4.2   *(a)* A truss supported by a pin and a roller; *(b)* free-body diagram of the truss.

this equation to determine the magnitude B because the equation does not contain A_x or A_y. Then, since the sum of the x components and the sum of the y components of the forces are zero, we write the equations $\Sigma F_x = 0$ and $\Sigma F_y = 0$. From these equations, we can obtain the components A_x and A_y, respectively.

We could obtain an additional equation by noting that the sum of the moments of the external forces about a point other than A is zero. We could write, for instance, $\Sigma M_B = 0$. This equation, however, does not contain any new information, because we have already established that the system of forces shown in Fig. 4.2b is equivalent to zero. The additional equation *is not independent* and cannot be used to determine a fourth unknown. It can be useful, however, for checking the solution obtained from the original three equations of equilibrium.

Although the three equations of equilibrium cannot be *augmented* by additional equations, any of them can be *replaced* by another equation. Properly chosen, the new system of equations still describes the equilibrium conditions but may be easier to work with. For example, an alternative system of equations for equilibrium is

$$\Sigma F_x = 0 \qquad \Sigma M_A = 0 \qquad \Sigma M_B = 0 \qquad (4.6)$$

Here the second point about which the moments are summed (in this case, point B) cannot lie on the line parallel to the y axis that passes through point A (Fig. 4.2b). These equations are sufficient conditions for the equilibrium of the truss. The first two equations indicate that the external forces must reduce to a single vertical force at A. Since the third equation requires that the moment of this force be zero about a point B that is not on its line of action, the force must be zero, and the rigid body is in equilibrium.

A third possible set of equilibrium equations is

$$\Sigma M_A = 0 \qquad \Sigma M_B = 0 \qquad \Sigma M_C = 0 \qquad (4.7)$$

where the points A, B, and C do not lie in a straight line (Fig. 4.2b). The first equation requires that the external forces reduce to a single force at A; the second equation requires that this force pass through B; and the third equation requires that it pass through C. Since the points A, B, C do not lie in a straight line, the force must be zero, and the rigid body is in equilibrium.

Notice that the equation $\Sigma M_A = 0$, stating that the sum of the moments of the forces about pin A is zero, possesses a more definite physical meaning than either of the other two equations (4.7). These two equations express a similar idea of balance but with respect to points about which the rigid body is not actually hinged. They are, however, as useful as the first equation. The choice of equilibrium equations should not be unduly influenced by their physical meaning. Indeed, in practice, it is desirable to choose equations of equilibrium containing only one unknown, since this eliminates the necessity of solving simultaneous equations. You can obtain equations containing only one unknown by summing moments about the point of intersection of the lines of action of two unknown forces or, if these forces are parallel, by summing force components in a direction perpendicular to their common direction.

For example, in Fig. 4.3, in which the truss shown is held by rollers at A and B and a short link at D, we can eliminate the reactions at A and B by summing x components. We can eliminate the reactions at A and D

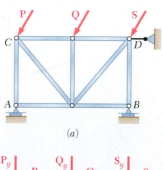
(a)

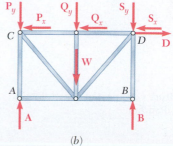
(b)

Fig. 4.3 (a) A truss supported by two rollers and a short link; (b) free-body diagram of the truss.

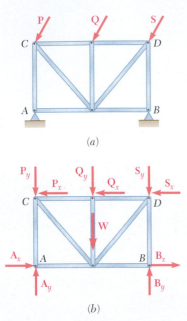

(a)

(b)

Fig. 4.4 (a) Truss with statically indeterminate reactions; (b) free-body diagram.

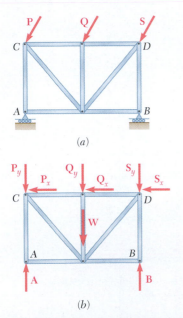

(a)

(b)

Fig. 4.5 (a) Truss with partial constraints; (b) free-body diagram.

by summing moments about C and the reactions at B and D by summing moments about D. The resulting equations are

$$\Sigma F_x = 0 \qquad \Sigma M_C = 0 \qquad \Sigma M_D = 0$$

Each of these equations contains only one unknown.

4.1C Statically Indeterminate Reactions and Partial Constraints

In the two examples considered in Figs. 4.2 and 4.3, the types of supports used were such that the rigid body could not possibly move under the given loads or under any other loading conditions. In such cases, the rigid body is said to be **completely constrained**. Recall that the reactions corresponding to these supports involved *three unknowns* and could be determined by solving the three equations of equilibrium. When such a situation exists, the reactions are said to be **statically determinate**.

Consider Fig. 4.4a, in which the truss shown is held by pins at A and B. These supports provide more constraints than are necessary to keep the truss from moving under the given loads or under any other loading conditions. Note from the free-body diagram of Fig. 4.4b that the corresponding reactions involve *four unknowns*. We pointed out in Sec. 4.1D that only three independent equilibrium equations are available; therefore, in this case, we have *more unknowns than equations*. As a result, we cannot determine all of the unknowns. The equations $\Sigma M_A = 0$ and $\Sigma M_B = 0$ yield the vertical components B_y and A_y, respectively, but the equation $\Sigma F_x = 0$ gives only the sum $A_x + B_x$ of the horizontal components of the reactions at A and B. The components A_x and B_x are **statically indeterminate**. We could determine their magnitudes by considering the deformations produced in the truss by the given loading, but this method is beyond the scope of statics and belongs to the study of mechanics of materials.

Let's consider the opposite situation. The supports holding the truss shown in Fig. 4.5a consist of rollers at A and B. Clearly, the constraints provided by these supports are not sufficient to keep the truss from moving. Although they prevent any vertical motion, the truss is free to move horizontally. The truss is said to be **partially constrained**.[†] From the free-body diagram in Fig. 4.5b, note that the reactions at A and B involve only *two unknowns*. Since three equations of equilibrium must still be satisfied, we have *fewer unknowns than equations*. In such a case, one of the equilibrium equations will not be satisfied in general. The equations $\Sigma M_A = 0$ and $\Sigma M_B = 0$ can be satisfied by a proper choice of reactions at A and B, but the equation $\Sigma F_x = 0$ is not satisfied unless the sum of the horizontal components of the applied forces happens to be zero. We thus observe that the equilibrium of the truss of Fig. 4.5 cannot be maintained under general loading conditions.

From these examples, it would appear that, if a rigid body is to be completely constrained and if the reactions at its supports are to be statically determinate, **there must be as many unknowns as there are equations of equilibrium**. When this condition is *not* satisfied, we can be certain that either the rigid body is not completely constrained or that the reactions at its supports

[†]Partially constrained bodies are often referred to as *unstable*. However, to avoid confusion between this type of instability, due to insufficient constraints, and the type of instability considered in Chap. 10, which relates to the behavior of a rigid body when its equilibrium is disturbed, we shall restrict the use of the words *stable* and *unstable* to the latter case.

are not statically determinate. It is also possible that the rigid body is not completely constrained *and* that the reactions are statically indeterminate.

You should note, however, that, although this condition is *necessary*, it is *not sufficient*. In other words, the fact that the number of unknowns is equal to the number of equations is no guarantee that a body is completely constrained or that the reactions at its supports are statically determinate. Consider Fig. 4.6*a*, which shows a truss held by rollers at *A*, *B*, and *E*. We have three unknown reactions of **A**, **B**, and **E** (Fig. 4.6*b*), but the equation $\Sigma F_x = 0$ is not satisfied unless the sum of the horizontal components of the applied forces happens to be zero. Although there are a sufficient number of constraints, these constraints are not properly arranged, so the truss is free to move horizontally. We say that the truss is **improperly constrained**. Since only two equilibrium equations are left for determining three unknowns, the reactions are statically indeterminate. Thus, improper constraints also produce static indeterminacy.

The truss shown in Fig. 4.7 is another example of improper constraints— and of static indeterminacy. This truss is held by a pin at *A* and by rollers at *B* and *C*, which altogether involve four unknowns. Since only three independent equilibrium equations are available, the reactions at the supports are statically indeterminate. On the other hand, we note that the equation $\Sigma M_A = 0$ cannot be satisfied under general loading conditions, since the lines of action of the reactions **B** and **C** pass through *A*. We conclude that the truss can rotate about *A* and that it is improperly constrained.[†]

The examples of Figs. 4.6 and 4.7 lead us to conclude that

A rigid body is improperly constrained whenever the supports (even though they may provide a sufficient number of reactions) are arranged in such a way that the reactions must be either concurrent or parallel.[‡]

In summary, to be sure that a two-dimensional rigid body is completely constrained and that the reactions at its supports are statically determinate, you should verify that the reactions involve three—and only three—unknowns and that the supports are arranged in such a way that they do not require the reactions to be either concurrent or parallel.

Supports involving statically indeterminate reactions should be used with care in the design of structures and only with a full knowledge of the problems they may cause. On the other hand, the analysis of structures possessing statically indeterminate reactions often can be partially carried out by the methods of statics. In the case of the truss of Fig. 4.4, for example, we can determine the vertical components of the reactions at *A* and *B* from the equilibrium equations.

For obvious reasons, supports producing partial or improper constraints should be avoided in the design of stationary structures. However, a partially or improperly constrained structure will not necessarily collapse; under particular loading conditions, equilibrium can be maintained. For example, the trusses of Figs. 4.5 and 4.6 will be in equilibrium if the applied forces **P**, **Q**, and **S** are vertical. Besides, structures designed to move *should* be only partially constrained. A railroad car, for instance, would be of little use if it were completely constrained by having its brakes applied permanently.

[†]Rotation of the truss about *A* requires some "play" in the supports at *B* and *C*. In practice such play will always exist. In addition, we note that if the play is kept small, the displacements of the rollers *B* and *C* and, thus, the distances from *A* to the lines of action of the reactions **B** and **C** will also be small. The equation $\Sigma M_A = 0$ then requires that **B** and **C** be very large, a situation which can result in the failure of the supports at *B* and *C*.

[‡]Because this situation arises from an inadequate arrangement or *geometry* of the supports, it is often referred to as *geometric instability*.

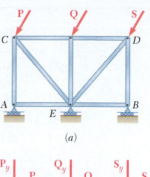

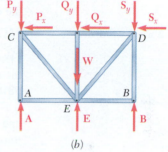

Fig. 4.6 *(a)* Truss with improper constraints; *(b)* free-body diagram.

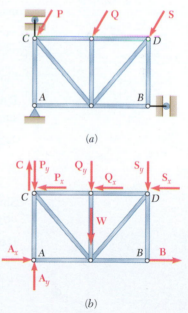

Fig. 4.7 *(a)* Truss with improper constraints; *(b)* free-body diagram.

Sample Problem 4.1

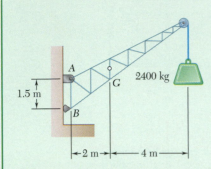

A fixed crane has a mass of 1000 kg and is used to lift a 2400-kg crate. It is held in place by a pin at A and a rocker at B. The center of gravity of the crane is located at G. Determine the components of the reactions at A and B.

STRATEGY: Draw a free-body diagram to show all of the forces acting on the crane, then use the equilibrium equations to calculate the values of the unknown forces.

MODELING:

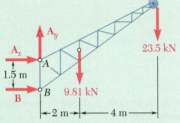

Fig. 1 Free-body diagram of crane.

Free-Body Diagram. By multiplying the masses of the crane and of the crate by $g = 9.81$ m/s^2, you obtain the corresponding weights—that is, 9810 N or 9.81 kN, and 23 500 N or 23.5 kN (Fig. 1). The reaction at pin A is a force of unknown direction; you can represent it by components A_x and A_y. The reaction at the rocker B is perpendicular to the rocker surface; thus, it is horizontal. Assume that A_x, A_y, and B act in the directions shown.

ANALYSIS:

Determination of B. The sum of the moments of all external forces about point A is zero. The equation for this sum contains neither A_x nor A_y, since the moments of A_x and A_y about A are zero. Multiplying the magnitude of each force by its perpendicular distance from A, you have

$$+\circlearrowleft \Sigma M_A = 0: \qquad +B(1.5 \text{ m}) - (9.81 \text{ kN})(2 \text{ m}) - (23.5 \text{ kN})(6 \text{ m}) = 0$$
$$B = +107.1 \text{ kN} \qquad \mathbf{B} = 107.1 \text{ kN} \rightarrow ◄$$

Since the result is positive, the reaction is directed as assumed.

Determination of A_x. Determine the magnitude of A_x by setting the sum of the horizontal components of all external forces to zero.

$$\xrightarrow{+} \Sigma F_x = 0: \qquad A_x + B = 0$$
$$A_x + 107.1 \text{ kN} = 0$$
$$A_x = -107.1 \text{ kN} \qquad \mathbf{A}_x = 107.1 \text{ kN} \leftarrow ◄$$

Since the result is negative, the sense of A_x is opposite to that assumed originally.

Determination of A_y. The sum of the vertical components must also equal zero. Therefore,

$$+\uparrow \Sigma F_y = 0: \qquad A_y - 9.81 \text{ kN} - 23.5 \text{ kN} = 0$$
$$A_y = +33.3 \text{ kN} \qquad \mathbf{A}_y = 33.3 \text{ kN} \uparrow ◄$$

Adding the components A_x and A_y vectorially, you can find that the reaction at A is 112.2 kN ⬊17.3°.

REFLECT and THINK: You can check the values obtained for the reactions by recalling that the sum of the moments of all the external forces about any point must be zero. For example, considering point B (Fig. 2), you can show

$$+\circlearrowleft \Sigma M_B = -(9.81 \text{ kN})(2 \text{ m}) - (23.5 \text{ kN})(6 \text{ m}) + (107.1 \text{ kN})(1.5 \text{ m}) = 0$$

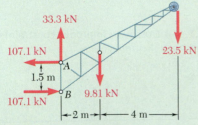

Fig. 2 Free-body diagram of crane with solved reactions.

Sample Problem 4.2

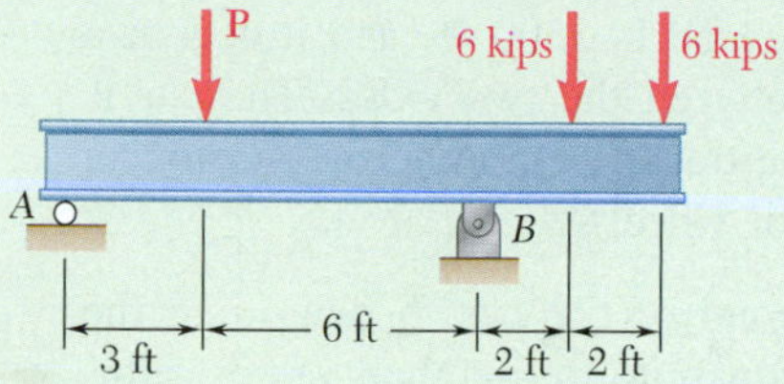

Three loads are applied to a beam as shown. The beam is supported by a roller at A and by a pin at B. Neglecting the weight of the beam, determine the reactions at A and B when $P = 15$ kips.

STRATEGY: Draw a free-body diagram of the beam, then write the equilibrium equations, first summing forces in the x direction and then summing moments at A and at B.

MODELING:

Free-Body Diagram. The reaction at A is vertical and is denoted by **A** (Fig. 1). Represent the reaction at B by components $\mathbf{B}_x$ and $\mathbf{B}_y$. Assume that each component acts in the direction shown.

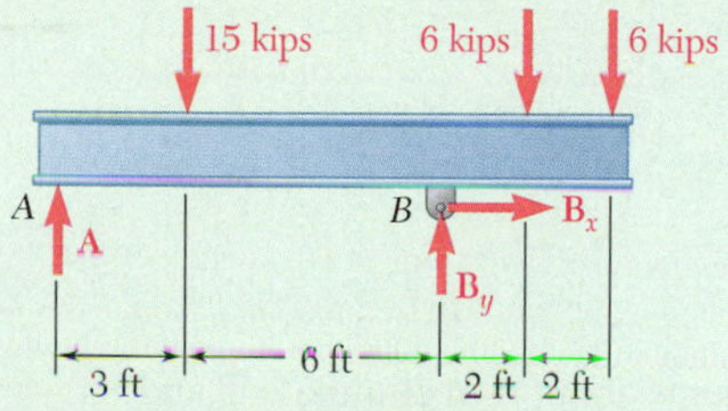

ANALYSIS:

Equilibrium Equations. Write the three equilibrium equations and solve for the reactions indicated:

$$\xrightarrow{+}\Sigma F_x = 0: \qquad\qquad\qquad B_x = 0 \qquad\qquad \mathbf{B}_x = 0 \blacktriangleleft$$

$$+\curvearrowleft\Sigma M_A = 0:$$
$$-(15 \text{ kips})(3 \text{ ft}) + B_y(9 \text{ ft}) - (6 \text{ kips})(11 \text{ ft}) - (6 \text{ kips})(13 \text{ ft}) = 0$$
$$B_y = +21.0 \text{ kips} \qquad \mathbf{B}_y = 21.0 \text{ kips} \uparrow \blacktriangleleft$$

$$+\curvearrowleft\Sigma M_B = 0:$$
$$-A(9 \text{ ft}) + (15 \text{ kips})(6 \text{ ft}) - (6 \text{ kips})(2 \text{ ft}) - (6 \text{ kips})(4 \text{ ft}) = 0$$
$$A = +6.00 \text{ kips} \qquad \mathbf{A} = 6.00 \text{ kips} \uparrow \blacktriangleleft$$

REFLECT and THINK: Check the results by adding the vertical components of all of the external forces:

$$+\uparrow\Sigma F_y = +6.00 \text{ kips} - 15 \text{ kips} + 21.0 \text{ kips} - 6 \text{ kips} - 6 \text{ kips} = 0$$

Remark. In this problem, the reactions at both A and B are vertical; however, these reactions are vertical for different reasons. At A, the beam is supported by a roller; hence, the reaction cannot have any horizontal component. At B, the horizontal component of the reaction is zero because it must satisfy the equilibrium equation $\Sigma F_x = 0$ and none of the other forces acting on the beam has a horizontal component.

You might have noticed at first glance that the reaction at B was vertical and dispensed with the horizontal component $\mathbf{B}_x$. This, however, is bad practice. In following it, you run the risk of forgetting the component $\mathbf{B}_x$ when the loading conditions require such a component (i.e., when a horizontal load is included). Also, you found the component $\mathbf{B}_x$ to be zero by using and solving an equilibrium equation, $\Sigma F_x = 0$. By setting $\mathbf{B}_x$ equal to zero immediately, you might not realize that you actually made use of this equation. Thus, you might lose track of the number of equations available for solving the problem.

Fig.1 Free-body diagram of beam.

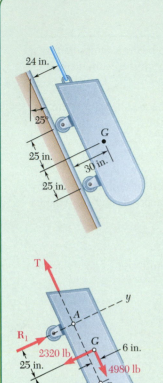

Sample Problem 4.3

A loading car is at rest on a track forming an angle of 25° with the vertical. The gross weight of the car and its load is 5500 lb, and it acts at a point 30 in. from the track, halfway between the two axles. The car is held by a cable attached 24 in. from the track. Determine the tension in the cable and the reaction at each pair of wheels.

STRATEGY: Draw a free-body diagram of the car to determine the unknown forces, and write equilibrium equations to find their values, summing moments at A and B and then summing forces.

MODELING:

Free-Body Diagram. The reaction at each wheel is perpendicular to the track, and the tension force T is parallel to the track. Therefore, for convenience, choose the x axis parallel to the track and the y axis perpendicular to the track (Fig. 1). Then resolve the 5500-lb weight into x and y components.

$$W_x = +(5500 \text{ lb}) \cos 25° = +4980 \text{ lb}$$
$$W_y = -(5500 \text{ lb}) \sin 25° = -2320 \text{ lb}$$

ANALYSIS:

Equilibrium Equations. Take moments about A to eliminate T and R_1 from the computation.

$+\curvearrowleft \Sigma M_A = 0$: $-(2320 \text{ lb})(25 \text{ in.}) - (4980 \text{ lb})(6 \text{ in.}) + R_2(50 \text{ in.}) = 0$

$R_2 = +1758 \text{ lb}$ $\mathbf{R_2} = 1758 \text{ lb} \nearrow$ ◄

Then take moments about B to eliminate T and R_2 from the computation.

$+\curvearrowleft \Sigma M_B = 0$: $(2320 \text{ lb})(25 \text{ in.}) - (4980 \text{ lb})(6 \text{ in.}) - R_1(50 \text{ in.}) = 0$

$R_1 = +562 \text{ lb}$ $\mathbf{R_1} = +562 \text{ lb} \nearrow$ ◄

Determine the value of T by summing forces in the x direction.

$\searrow + \Sigma F_x = 0$: $+4980 \text{ lb} - T = 0$

$T = +4980 \text{ lb}$ $\mathbf{T} = 4980 \text{ lb} \nwarrow$ ◄

Figure 2 shows the computed values of the reactions.

REFLECT and THINK: You can verify the computations by summing forces in the y direction.

$$\nearrow + \Sigma F_y = +562 \text{ lb} + 1758 \text{ lb} - 2320 \text{ lb} = 0$$

You could also check the solution by computing moments about any point other than A or B.

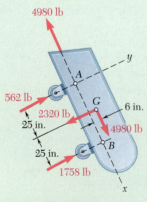

Fig. 1 Free-body diagram of car.

Fig. 2 Free-body diagram of car with solved reactions.

Sample Problem 4.4

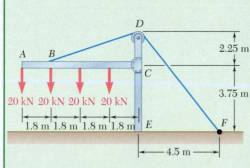

The frame shown supports part of the roof of a small building. Knowing that the tension in the cable is 150 kN, determine the reaction at the fixed end E.

STRATEGY: Draw a free-body diagram of the frame and of the cable BDF. The support at E is fixed, so the reactions here include a moment; to determine its value, sum moments about point E.

MODELING:

Free-Body Diagram. Represent the reaction at the fixed end E by the force components $\mathbf{E}_x$ and $\mathbf{E}_y$ and the couple $\mathbf{M}_E$ (Fig. 1). The other forces acting on the free body are the four 20-kN loads and the 150-kN force exerted at end F of the cable.

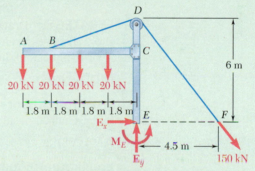

Fig. 1 Free-body diagram of frame.

ANALYSIS:

Equilibrium Equations. First note that

$$DF = \sqrt{(4.5 \text{ m})^2 + (6 \text{ m})^2} = 7.5 \text{ m}$$

Then you can write the three equilibrium equations and solve for the reactions at E.

$\xrightarrow{+}\Sigma F_x = 0:$ $\qquad E_x + \dfrac{4.5}{7.5}(150 \text{ kN}) = 0$

$\qquad\qquad\qquad\qquad E_x = -90.0 \text{ kN} \qquad \mathbf{E}_x = 90.0 \text{ kN} \leftarrow$ ◄

$+\uparrow\Sigma F_y = 0:$ $\qquad E_y - 4(20 \text{ kN}) - \dfrac{6}{7.5}(150 \text{ kN}) = 0$

$\qquad\qquad\qquad\qquad E_y = +200 \text{ kN} \qquad \mathbf{E}_y = 200 \text{ kN} \uparrow$ ◄

$+\curvearrowleft\Sigma M_E = 0:$ $\quad (20 \text{ kN})(7.2 \text{ m}) + (20 \text{ kN})(5.4 \text{ m}) + (20 \text{ kN})(3.6 \text{ m})$

$\qquad\qquad\qquad + (20 \text{ kN})(1.8 \text{ m}) - \dfrac{6}{7.5}(150 \text{ kN})(4.5 \text{ m}) + M_E = 0$

$\qquad\qquad\qquad M_E = +180.0 \text{ kN·m} \quad \mathbf{M}_E = 180.0 \text{ kN·m} \curvearrowleft$ ◄

REFLECT and THINK: The cable provides a fourth constraint, making this situation statically indeterminate. This problem therefore gave us the value of the cable tension, which would have been determined by means other than statics. We could then use the three available independent static equilibrium equations to solve for the remaining three reactions.

Sample Problem 4.5

A 400-lb weight is attached at A to the lever shown. The constant of the spring BC is $k = 250$ lb/in., and the spring is unstretched when $\theta = 0$. Determine the position of equilibrium.

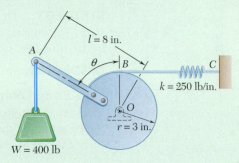

STRATEGY: Draw a free-body diagram of the lever and cylinder to show all forces acting on the body (Fig. 1), then sum moments about O. Your final answer should be the angle θ.

MODELING:

Free-Body Diagram. Denote by s the deflection of the spring from its unstretched position and note that $s = r\theta$. Then $F = ks = kr\theta$.

ANALYSIS:

Equilibrium Equation. Sum the moments of $\mathbf{W}$ and $\mathbf{F}$ about O to eliminate the reactions supporting the cylinder. The result is

$$+\curvearrowleft\Sigma M_O = 0: \qquad Wl \sin \theta - r(kr\theta) = 0 \qquad \sin \theta = \frac{kr^2}{Wl}\theta$$

Substituting the given data yields

$$\sin \theta = \frac{(250 \text{ lb/in.})(3 \text{ in.})^2}{(400 \text{ lb})(8 \text{ in.})}\theta \quad \sin \theta = 0.703\,\theta$$

Solving by trial and error, the angle is $\qquad \theta = 0 \qquad \theta = 80.3° \blacktriangleleft$

REFLECT and THINK: The weight could represent any vertical force acting on the lever. The key to the problem is to express the spring force as a function of the angle θ.

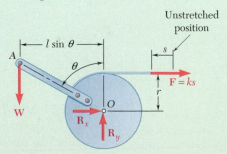

Fig. 1 Free-body diagram of the lever and cylinder.

SOLVING PROBLEMS ON YOUR OWN

You saw that, for a rigid body in equilibrium, the system of external forces is equivalent to zero. To solve an equilibrium problem, your first task is to draw a neat, reasonably large **free-body diagram** on which you show all external forces. You should include both known and unknown forces.

For a two-dimensional rigid body, the reactions at the supports can involve one, two, or three unknowns, depending on the type of support (Fig. 4.1). A correct free-body diagram is essential for the successful solution of a problem. Never proceed with the solution of a problem until you are sure that your free-body diagram includes all loads, all reactions, and the weight of the body (if appropriate).

1. You can write three equilibrium equations and solve them for *three unknowns.* The three equations might be

$$\Sigma F_x = 0 \qquad \Sigma F_y = 0 \qquad \Sigma M_O = 0$$

However, usually several alternative sets of equations are possible, such as

$$\Sigma F_x = 0 \qquad \Sigma M_A = 0 \qquad \Sigma M_B = 0$$

where point B is chosen in such a way that the line AB is not parallel to the y axis, or

$$\Sigma M_A = 0 \qquad \Sigma M_B = 0 \qquad \Sigma M_C = 0$$

where the points A, B, and C do not lie along a straight line.

2. To simplify your solution, it may be helpful to use one of the following solution techniques.

   a. By summing moments about the point of intersection of the lines of action of two unknown forces, you obtain an equation in a single unknown.

   b. By summing components in a direction perpendicular to two unknown parallel forces, you also obtain an equation in a single unknown.

3. After drawing your free-body diagram, you may find that one of the following special situations arises.

   a. The reactions involve fewer than three unknowns. The body is said to be **partially constrained** and motion of the body is possible.

   b. The reactions involve more than three unknowns. The reactions are said to be **statically indeterminate.** Although you may be able to calculate one or two reactions, you cannot determine all of them.

   c. The reactions pass through a single point or are parallel. The body is said to be **improperly constrained** and motion can occur under a general loading condition.

Problems

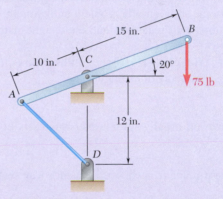

FREE-BODY PRACTICE PROBLEMS

4.F1 Two crates, each of mass 350 kg, are placed as shown in the bed of a 1400-kg pick-up truck. Draw the free-body diagram needed to determine the reactions at each of the two rear wheels A and front wheels B.

4.F2 A lever AB is hinged at C and attached to a control cable at A. If the lever is subjected to a 75-lb vertical force at B, draw the free-body diagram needed to determine the tension in the cable and the reaction at C.

Fig. P4.F1

Fig. P4.F2

4.F3 A light rod AD is supported by frictionless pegs at B and C and rests against a frictionless wall at A. A vertical 120-lb force is applied at D. Draw the free-body diagram needed to determine the reactions at A, B, and C.

4.F4 A tension of 20 N is maintained in a tape as it passes through the support system shown. Knowing that the radius of each pulley is 10 mm, draw the free-body diagram needed to determine the reaction at C.

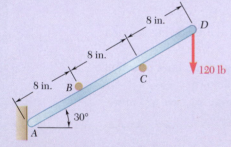

Fig. P4.F3

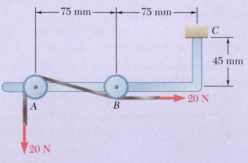

Fig. P4.F4

4.1 A gardener uses a 60-N wheelbarrow to transport a 250-N bag of fertilizer. What force must she exert on each handle?

4.2 The gardener of Prob. 4.1 wishes to transport a second 250-N bag of fertilizer at the same time as the first one. Determine the maximum allowable horizontal distance from the axle A of the wheelbarrow to the center of gravity of the second bag if she can hold only 75 N with each arm.

4.3 A 2100-lb tractor is used to lift 900 lb of gravel. Determine the reaction at each of the two (a) rear wheels A, (b) front wheels B.

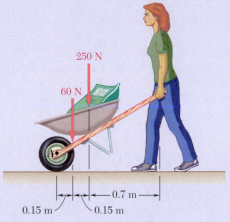
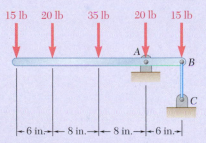

Fig. P4.1

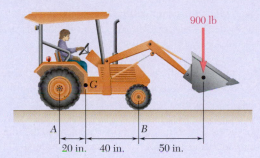

Fig. P4.3

4.4 For the beam and loading shown, determine (a) the reaction at A, (b) the tension in cable BC.

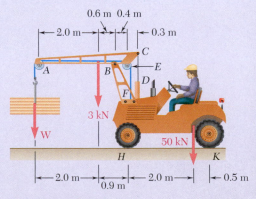

Fig. P4.4

4.5 A load of lumber of weight $W = 25$ kN is being raised by a mobile crane. The weight of boom ABC and the combined weight of the truck and driver are as shown. Determine the reaction at each of the two (a) front wheels H, (b) rear wheels K.

4.6 A load of lumber of weight $W = 25$ kN is being raised by a mobile crane. Knowing that the tension is 25 kN in all portions of cable AEF and that the weight of boom ABC is 3 kN, determine (a) the tension in rod CD, (b) the reaction at pin B.

4.7 A T-shaped bracket supports the four loads shown. Determine the reactions at A and B (a) if $a = 10$ in., (b) if $a = 7$ in.

Fig. P4.5 and P4.6

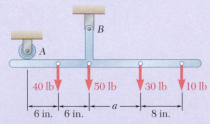

Fig. P4.7

4.8 For the bracket and loading of Prob. 4.7, determine the smallest distance a if the bracket is not to move.

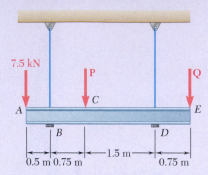

Fig. P4.9 and P4.10

4.9 Three loads are applied as shown to a light beam supported by cables attached at B and D. Neglecting the weight of the beam, determine the range of values of Q for which neither cable becomes slack when $P = 0$.

4.10 Three loads are applied as shown to a light beam supported by cables attached at B and D. Knowing that the maximum allowable tension in each cable is 12 kN and neglecting the weight of the beam, determine the range of values of Q for which the loading is safe when $P = 0$.

4.11 For the beam of Prob. 4.10, determine the range of values of Q for which the loading is safe when $P = 5$ kN.

4.12 For the beam of Sample Prob. 4.2, determine the range of values of P for which the beam will be safe, knowing that the maximum allowable value of each of the reactions is 25 kips and that the reaction at A must be directed upward.

4.13 The maximum allowable value of each of the reactions is 180 N. Neglecting the weight of the beam, determine the range of the distance d for which the beam is safe.

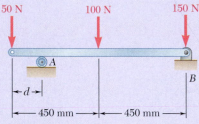

Fig. P4.13

4.14 For the beam and loading shown, determine the range of the distance a for which the reaction at B does not exceed 100 lb downward or 200 lb upward.

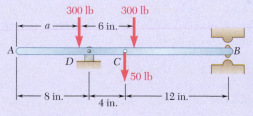

Fig. P4.14

4.15 Two links AB and DE are connected by a bell crank as shown. Knowing that the tension in link AB is 720 N, determine (a) the tension in link DE, (b) the reaction at C.

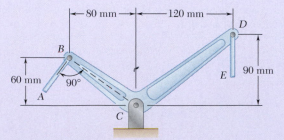

Fig. P4.15 and P4.16

4.16 Two links AB and DE are connected by a bell crank as shown. Determine the maximum force that can be safely exerted by link AB on the bell crank if the maximum allowable value for the reaction at C is 1600 N.

4.17 The required tension in cable AB is 200 lb. Determine (*a*) the vertical force **P** that must be applied to the pedal, (*b*) the corresponding reaction at C.

4.18 Determine the maximum tension that can be developed in cable AB if the maximum allowable value of the reaction at C is 250 lb.

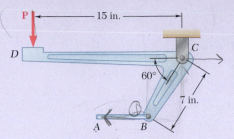

Fig. P4.17 and P4.18

4.19 The bracket BCD is hinged at C and attached to a control cable at B. For the loading shown, determine (*a*) the tension in the cable, (*b*) the reaction at C.

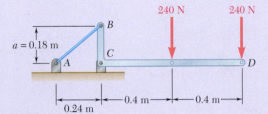

Fig. P4.19

4.20 Solve Prob. 4.19, assuming that $a = 0.32$ m.

4.21 The 40-ft boom AB weighs 2 kips; the distance from the axle A to the center of gravity G of the boom is 20 ft. For the position shown, determine (*a*) the tension T in the cable, (*b*) the reaction at A.

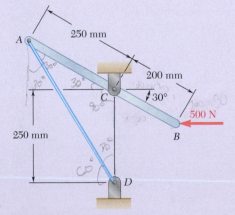

Fig. P4.21

4.22 A lever AB is hinged at C and attached to a control cable at A. If the lever is subjected to a 500-N horizontal force at B, determine (*a*) the tension in the cable, (*b*) the reaction at C.

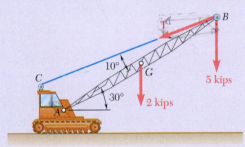

Fig. P4.22

187

4.23 and 4.24 For each of the plates and loadings shown, determine the reactions at A and B.

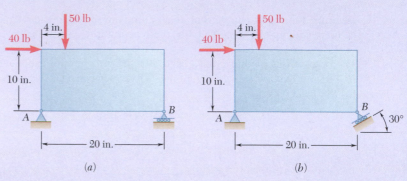

(a)   (b)

Fig. P4.23

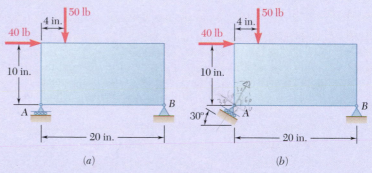

(a)   (b)

Fig. P4.24

4.25 A rod AB, hinged at A and attached at B to cable BD, supports the loads shown. Knowing that $d = 200$ mm, determine (a) the tension in cable BD, (b) the reaction at A.

4.26 A rod AB, hinged at A and attached at B to cable BD, supports the loads shown. Knowing that $d = 150$ mm, determine (a) the tension in cable BD, (b) the reaction at A.

4.27 Determine the reactions at A and B when (a) $\alpha = 0$, (b) $\alpha = 90°$, (c) $\alpha = 30°$.

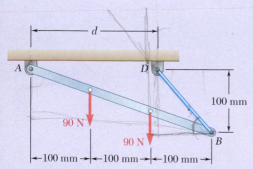

Fig. P4.25 and P4.26

$$\tan(\theta) = \frac{100}{100} = 1$$

$$\theta = 45°$$

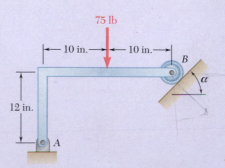

Fig. P4.27

4.28 Determine the reactions at A and C when (a) $\alpha = 0$, (b) $\alpha = 30°$.

4.29 Rod ABC is bent in the shape of an arc of circle of radius R. Knowing that $\theta = 30°$, determine the reaction (a) at B, (b) at C.

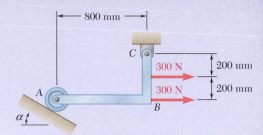

Fig. P4.28

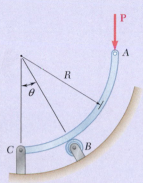

Fig. P4.29 and P4.30

4.30 Rod ABC is bent in the shape of an arc of circle of radius R. Knowing that $\theta = 60°$, determine the reaction (a) at B, (b) at C.

4.31 Neglecting friction, determine the tension in cable ABD and the reaction at C when $\theta = 60°$.

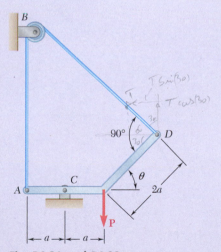

Fig. P4.31 and P4.32

4.32 Neglecting friction, determine the tension in cable ABD and the reaction at C when $\theta = 45°$.

4.33 A force P of magnitude 90 lb is applied to member $ACDE$ that is supported by a frictionless pin at D and by the cable ABE. Since the cable passes over a small pulley at B, the tension may be assumed to be the same in portions AB and BE of the cable. For the case when $a = 3$ in., determine (a) the tension in the cable, (b) the reaction at D.

4.34 Solve Prob. 4.33 for $a = 6$ in.

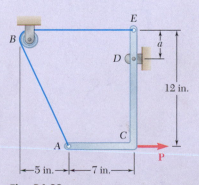

Fig. P4.33

4.35 Bar AC supports two 400-N loads as shown. Rollers at A and C rest against frictionless surfaces and a cable BD is attached at B. Determine (a) the tension in cable BD, (b) the reaction at A, (c) the reaction at C.

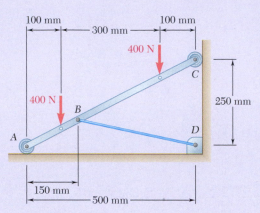

Fig. P4.35

4.36 A light bar AD is suspended from a cable BE and supports a 20-kg block at C. The ends A and D of the bar are in contact with frictionless vertical walls. Determine the tension in cable BE and the reactions at A and D.

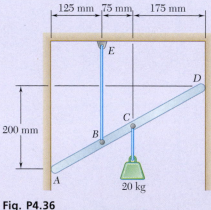

Fig. P4.36

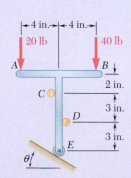

Fig. P4.37 and P4.38

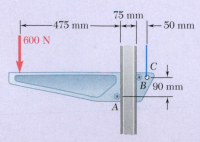

Fig. P4.39

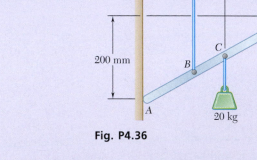

4.37 The T-shaped bracket shown is supported by a small wheel at E and pegs at C and D. Neglecting the effect of friction, determine the reactions at C, D, and E when $\theta = 30°$.

4.38 The T-shaped bracket shown is supported by a small wheel at E and pegs at C and D. Neglecting the effect of friction, determine (a) the smallest value of θ for which the equilibrium of the bracket is maintained, (b) the corresponding reactions at C, D, and E.

4.39 A movable bracket is held at rest by a cable attached at C and by frictionless rollers at A and B. For the loading shown, determine (a) the tension in the cable, (b) the reactions at A and B.

4.40 A light bar *AB* supports a 15-kg block at its midpoint *C*. Rollers at *A* and *B* rest against frictionless surfaces, and a horizontal cable *AD* is attached at *A*. Determine (*a*) the tension in cable *AD*, (*b*) the reactions at *A* and *B*.

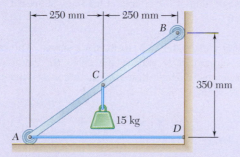

Fig. P4.40

4.41 Two slots have been cut in plate *DEF*, and the plate has been placed so that the slots fit two fixed, frictionless pins *A* and *B*. Knowing that *P* = 15 lb, determine (*a*) the force each pin exerts on the plate, (*b*) the reaction at *F*.

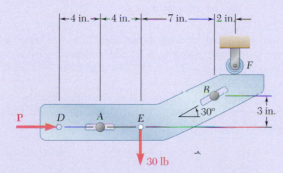

Fig. P4.41

4.42 For the plate of Prob. 4.41, the reaction at *F* must be directed downward, and its maximum value is 20 lb. Neglecting friction at the pins, determine the required range of values of *P*.

4.43 The rig shown consists of a 1200-lb horizontal member *ABC* and a vertical member *DBE* welded together at *B*. The rig is being used to raise a 3600-lb crate at a distance $x = 12$ ft from the vertical member *DBE*. If the tension in the cable is 4 kips, determine the reaction at *E*, assuming that the cable is (*a*) anchored at *F* as shown in the figure, (*b*) attached to the vertical member at a point located 1 ft above *E*.

4.44 For the rig and crate of Prob. 4.43 and assuming that cable is anchored at *F* as shown, determine (*a*) the required tension in cable *ADCF* if the maximum value of the couple at *E* as *x* varies from 1.5 to 17.5 ft is to be as small as possible, (*b*) the corresponding maximum value of the couple.

4.45 A 175-kg utility pole is used to support at *C* the end of an electric wire. The tension in the wire is 600 N, and the wire forms an angle of 15° with the horizontal at *C*. Determine the largest and smallest allowable tensions in the guy cable *BD* if the magnitude of the couple at *A* may not exceed 500 N·m.

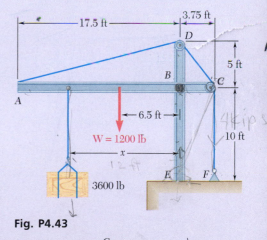

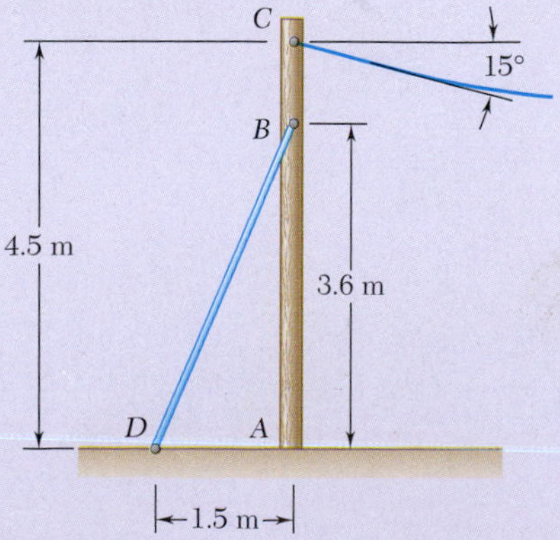

Fig. P4.43

Fig. P4.45

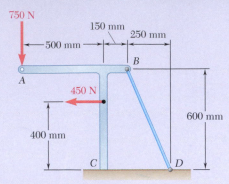

750 N
150 mm
250 mm
500 mm
B
A
450 N
600 mm
400 mm
C   D

Fig. P4.46 and P4.47

4.46 Knowing that the tension in wire *BD* is 1300 N, determine the reaction at the fixed support *C* of the frame shown.

4.47 Determine the range of allowable values of the tension in wire *BD* if the magnitude of the couple at the fixed support *C* is not to exceed 100 N·m.

4.48 Beam *AD* carries the two 40-lb loads shown. The beam is held by a fixed support at *D* and by the cable *BE* that is attached to the counterweight *W*. Determine the reaction at *D* when (*a*) *W* = 100 lb, (*b*) *W* = 90 lb.

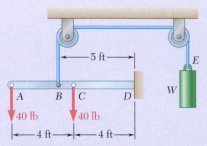

Fig. P4.48 and *P4.49*

4.49 For the beam and loading shown, determine the range of values of *W* for which the magnitude of the couple at *D* does not exceed 40 lb·ft.

4.50 An 8-kg mass can be supported in the three different ways shown. Knowing that the pulleys have a 100-mm radius, determine the reaction at *A* in each case.

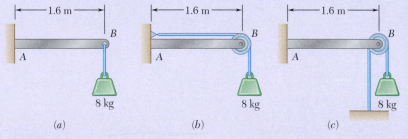

Fig. P4.50

4.51 A uniform rod *AB* with a length of *l* and weight of *W* is suspended from two cords *AC* and *BC* of equal length. Determine the angle θ corresponding to the equilibrium position when a couple **M** is applied to the rod.

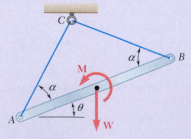

Fig. P4.51

4.52 Rod AD is acted upon by a vertical force **P** at end A and by two equal and opposite horizontal forces of magnitude Q at points B and C. Neglecting the weight of the rod, express the angle θ corresponding to the equilibrium position in terms of P and Q.

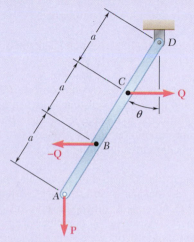

Fig. P4.52

4.53 A slender rod AB with a weight of W is attached to blocks A and B that move freely in the guides shown. The blocks are connected by an elastic cord that passes over a pulley at C. (*a*) Express the tension in the cord in terms of W and θ. (*b*) Determine the value of θ for which the tension in the cord is equal to $3W$.

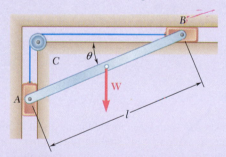

Fig. P4.53

4.54 and 4.55 A vertical load **P** is applied at end B of rod BC. (*a*) Neglecting the weight of the rod, express the angle θ corresponding to the equilibrium position in terms of P, l, and the counterweight W. (*b*) Determine the value of θ corresponding to equilibrium if $P = 2W$.

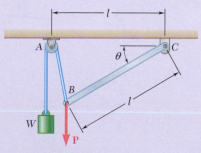

Fig. P4.54

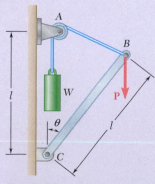

Fig. P4.55

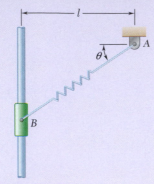

Fig. P4.56

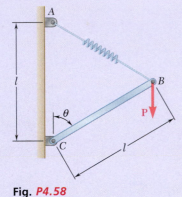

Fig. P4.58

4.56 A collar *B* with a weight of *W* can move freely along the vertical rod shown. The constant of the spring is *k*, and the spring is unstretched when $\theta = 0$. (*a*) Derive an equation in θ, *W*, *k*, and *l* that must be satisfied when the collar is in equilibrium. (*b*) Knowing that *W* = 300 N, *l* = 500 mm, and *k* = 800 N/m, determine the value of θ corresponding to equilibrium.

4.57 Solve Sample Prob. 4.5, assuming that the spring is unstretched when $\theta = 90°$.

4.58 A vertical load **P** is applied at end *B* of rod *BC*. The constant of the spring is *k*, and the spring is unstretched when $\theta = 60°$. (*a*) Neglecting the weight of the rod, express the angle θ corresponding to the equilibrium position in terms of *P*, *k*, and *l*. (*b*) Determine the value of θ corresponding to equilibrium if $P = \frac{1}{4}kl$.

4.59 Eight identical 500 × 750-mm rectangular plates, each of mass *m* = 40 kg, are held in a vertical plane as shown. All connections consist of frictionless pins, rollers, or short links. In each case, determine whether (*a*) the plate is completely, partially, or improperly constrained, (*b*) the reactions are statically determinate or indeterminate, (*c*) the equilibrium of the plate is maintained in the position shown. Also, wherever possible, compute the reactions.

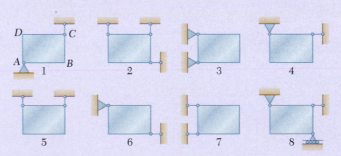

Fig. P4.59

4.60 The bracket *ABC* can be supported in the eight different ways shown. All connections consist of smooth pins, rollers, or short links. For each case, answer the questions listed in Prob. 4.59, and, wherever possible, compute the reactions, assuming that the magnitude of the force **P** is 100 lb.

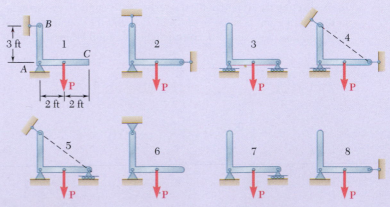

Fig. P4.60

4.2 TWO SPECIAL CASES

In practice, some simple cases of equilibrium occur quite often, either as part of a more complicated analysis or as the complete models of a situation. By understanding the characteristics of these cases, you can often simplify the overall analysis.

4.2A Equilibrium of a Two-Force Body

A particular case of equilibrium of considerable interest in practical applications is that of a rigid body subjected to two forces. Such a body is commonly called a **two-force body**. We show here that, **if a two-force body is in equilibrium, the two forces must have the same magnitude, the same line of action, and opposite sense**.

Consider a corner plate subjected to two forces $\mathbf{F}_1$ and $\mathbf{F}_2$ acting at A and B, respectively (Fig. 4.8a). If the plate is in equilibrium, the sum of the moments of $\mathbf{F}_1$ and $\mathbf{F}_2$ about any axis must be zero. First, we sum moments about A. Since the moment of $\mathbf{F}_1$ is obviously zero, the moment of $\mathbf{F}_2$ also must be zero and the line of action of $\mathbf{F}_2$ must pass through A (Fig. 4.8b). Similarly, summing moments about B, we can show that the line of action of $\mathbf{F}_1$ must pass through B (Fig. 4.8c). Therefore, both forces have the same line of action (line AB). You can see from either of the equations $\Sigma F_x = 0$ and $\Sigma F_y = 0$ that they must also have the same magnitude but opposite sense.

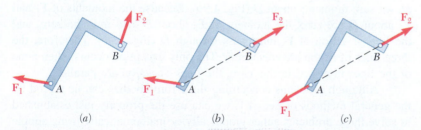

(a) (b) (c)

Fig. 4.8 A two-force body in equilibrium. *(a)* Forces act at two points of the body; *(b)* summing moments about point *A* shows that the line of action of **F**₂ must pass through *A*; *(c)* summing moments about point *B* shows that the line of action of **F**₁ must pass through *B*.

If several forces act at two points A and B, the forces acting at A can be replaced by their resultant $\mathbf{F}_1$, and those acting at B can be replaced by their resultant $\mathbf{F}_2$. Thus, a two-force body can be more generally defined as **a rigid body subjected to forces acting at only two points**. The resultants $\mathbf{F}_1$ and $\mathbf{F}_2$ then must have the same line of action, the same magnitude, and opposite sense (Fig. 4.8).

Later, in the study of structures, frames, and machines, you will see how the recognition of two-force bodies simplifies the solution of certain problems.

4.2B   Equilibrium of a Three-Force Body

Another case of equilibrium that is of great practical interest is that of a **three-force body**, i.e., a rigid body subjected to three forces or, more generally, **a rigid body subjected to forces acting at only three points**. Consider a rigid body subjected to a system of forces that can be reduced to three forces $\mathbf{F}_1$, $\mathbf{F}_2$, and $\mathbf{F}_3$ acting at A, B, and C, respectively (Fig. 4.9a). We show that if the body is in equilibrium, **the lines of action of the three forces must be either concurrent or parallel**.

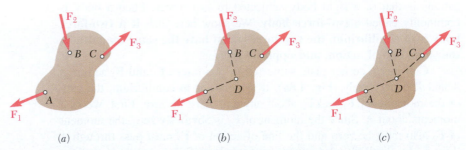

(a) (b) (c)

Fig. 4.9   A three-force body in equilibrium. *(a–c)* Demonstration that the lines of action of the three forces must be either concurrent or parallel.

Since the rigid body is in equilibrium, the sum of the moments of $\mathbf{F}_1$, $\mathbf{F}_2$, and $\mathbf{F}_3$ about any axis must be zero. Assuming that the lines of action of $\mathbf{F}_1$ and $\mathbf{F}_2$ intersect and denoting their point of intersection by D, we sum moments about D (Fig. 4.9b). Because the moments of $\mathbf{F}_1$ and $\mathbf{F}_2$ about D are zero, the moment of $\mathbf{F}_3$ about D also must be zero, and the line of action of $\mathbf{F}_3$ must pass through D (Fig. 4.9c). Therefore, the three lines of action are concurrent. The only exception occurs when none of the lines intersect; in this case, the lines of action are parallel.

Although problems concerning three-force bodies can be solved by the general methods of Sec. 4.1, we can use the property just established to solve these problems either graphically or mathematically using simple trigonometric or geometric relations (see Sample Problem 4.6).

Sample Problem 4.6

A man raises a 10-kg joist with a length of 4 m by pulling on a rope. Find the tension T in the rope and the reaction at A.

STRATEGY: The joist is acted upon by three forces: its weight $\mathbf{W}$, the force $\mathbf{T}$ exerted by the rope, and the reaction $\mathbf{R}$ of the ground at A. Therefore, it is a three-force body, and you can compute the forces by using a force triangle.

MODELING: First note that

$$W = mg = (10 \text{ kg})(9.81 \text{ m/s}^2) = 98.1 \text{ N}$$

Since the joist is a three-force body, the forces acting on it must be concurrent. The reaction $\mathbf{R}$ therefore must pass through the point of intersection C of the lines of action of the weight $\mathbf{W}$ and the tension force $\mathbf{T}$, as shown in the free-body diagram (Fig. 1). You can use this fact to determine the angle α that $\mathbf{R}$ forms with the horizontal.

ANALYSIS: Draw the vertical line BF through B and the horizontal line CD through C (Fig. 2). Then

$$AF = BF = (AB) \cos 45° = (4 \text{ m}) \cos 45° = 2.828 \text{ m}$$
$$CD = EF = AE = \tfrac{1}{2}(AF) = 1.414 \text{ m}$$
$$BD = (CD) \cot (45° + 25°) = (1.414 \text{ m}) \tan 20° = 0.515 \text{ m}$$
$$CE = DF = BF - BD = 2.828 \text{ m} - 0.515 \text{ m} = 2.313 \text{ m}$$

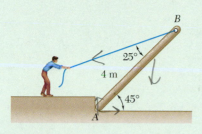

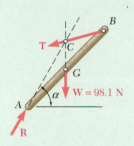

Fig. 1 Free-body diagram of joist.

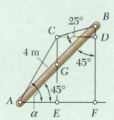

Fig. 2 Geometry analysis of the lines of action for the three forces acting on joist, concurrent at point C.

From these calculations, you can determine the angle α as

$$\tan \alpha = \frac{CE}{AE} = \frac{2.313 \text{ m}}{1.414 \text{ m}} = 1.636$$

$$\alpha = 58.6° \blacktriangleleft$$

You now know the directions of all the forces acting on the joist.

Force Triangle. Draw a force triangle as shown (Fig. 3) with its interior angles computed from the known directions of the forces. You can then use the law of sines to find the unknown forces.

$$\frac{T}{\sin 31.4°} = \frac{R}{\sin 110°} = \frac{98.1 \text{ N}}{\sin 38.6°}$$

$$T = 81.9 \text{ N} \blacktriangleleft$$
$$R = 147.8 \text{ N} \measuredangle 58.6° \blacktriangleleft$$

Fig. 3 Force triangle.

REFLECT and THINK: In practice, three-force members occur often, so learning this method of analysis is useful in many situations.

SOLVING PROBLEMS ON YOUR OWN

This section covered two particular cases of equilibrium of a rigid body.

1. A two-force body is subjected to forces at only two points. The resultants of the forces acting at each of these points must have the *same magnitude, the same line of action, and opposite sense.* This property allows you to simplify the solutions of some problems by replacing the two unknown components of a reaction by a single force of unknown magnitude but of *known direction.*

2. A three-force body is subjected to forces at only three points. The resultants of the forces acting at each of these points must be *concurrent or parallel.* To solve a problem involving a three-force body with concurrent forces, draw the free-body diagram showing that these three forces pass through the same point. You may be able to complete the solution by using simple geometry, such as a force triangle and the law of sines [see Sample Prob. 4.6].

This method for solving problems involving three-force bodies is not difficult to understand, but in practice, it can be difficult to sketch the necessary geometric constructions. If you encounter difficulty, first draw a reasonably large free-body diagram and then seek a relation between known or easily calculated lengths and a dimension that involves an unknown. Sample Prob. 4.6 illustrates this technique, where we used the easily calculated dimensions AE and CE to determine the angle α.

Problems

4.61 A 500-lb cylindrical tank, 8 ft in diameter, is to be raised over a 2-ft obstruction. A cable is wrapped around the tank and pulled horizontally as shown. Knowing that the corner of the obstruction at A is rough, find the required tension in the cable and the reaction at A.

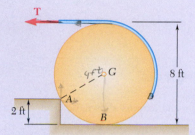

Fig. P4.61

4.62 Determine the reactions at A and B when $a = 180$ mm.

4.63 For the bracket and loading shown, determine the range of values of the distance a for which the magnitude of the reaction at B does not exceed 600 N.

4.64 The spanner shown is used to rotate a shaft. A pin fits in a hole at A, while a flat, frictionless surface rests against the shaft at B. If a 60-lb force **P** is exerted on the spanner at D, find the reactions at A and B.

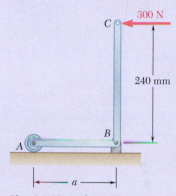

Fig. P4.62 and P4.63

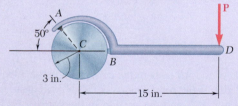

Fig. P4.64

4.65 Determine the reactions at B and C when $a = 30$ mm.

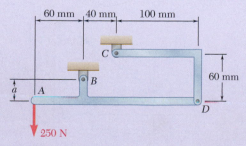

Fig. P4.65

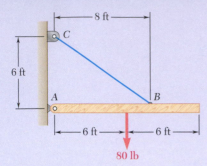

Fig. P4.66

4.66 A 12-ft wooden beam weighing 80 lb is supported by a pin and bracket at *A* and by cable *BC*. Find the reaction at *A* and the tension in the cable.

4.67 Determine the reactions at *B* and *D* when *b* = 60 mm.

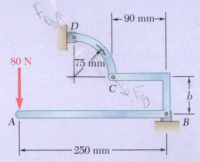

Fig. P4.67

4.68 For the frame and loading shown, determine the reactions at *C* and *D*.

4.69 A 50-kg crate is attached to the trolley-beam system shown. Knowing that *a* = 1.5 m, determine (*a*) the tension in cable *CD*, (*b*) the reaction at *B*.

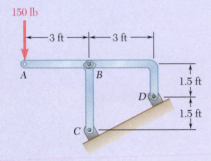

Fig. *P4.68*

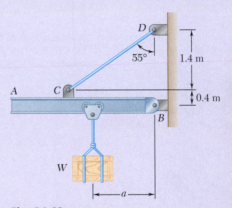

Fig. P4.69

4.70 One end of rod *AB* rests in the corner *A* and the other end is attached to cord *BD*. If the rod supports a 150-N load at its midpoint *C*, find the reaction at *A* and the tension in the cord.

4.71 For the boom and loading shown, determine (*a*) the tension in cord *BD*, (*b*) the reaction at *C*.

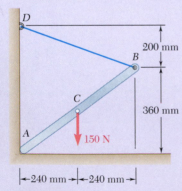

Fig. *P4.70*

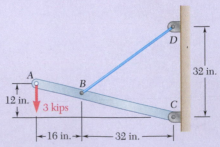

Fig. P4.71

4.72 A 40-lb roller of 8-in. diameter, which is to be used on a tile floor, is resting directly on the subflooring as shown. Knowing that the thickness of each tile is 0.3 in., determine the force **P** required to move the roller onto the tiles if the roller is (*a*) pushed to the left, (*b*) pulled to the right.

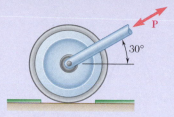

Fig. P4.72

4.73 A T-shaped bracket supports a 300-N load as shown. Determine the reactions at *A* and *C* when $\alpha = 45°$.

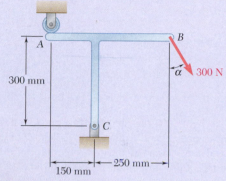

Fig. P4.73 and *P4.74*

4.74 A T-shaped bracket supports a 300-N load as shown. Determine the reactions at *A* and *C* when $\alpha = 60°$.

4.75 Rod *AB* is supported by a pin and bracket at *A* and rests against a frictionless peg at *C*. Determine the reactions at *A* and *C* when a 170-N vertical force is applied at *B*.

4.76 Solve Prob. 4.75, assuming that the 170-N force applied at *B* is horizontal and directed to the left.

4.77 Member *ABC* is supported by a pin and bracket at *B* and by an inextensible cord attached at *A* and *C* and passing over a frictionless pulley at *D*. The tension may be assumed to be the same in portions *AD* and *CD* of the cord. For the loading shown and neglecting the size of the pulley, determine the tension in the cord and the reaction at *B*.

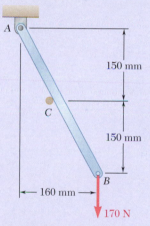

Fig. P4.75

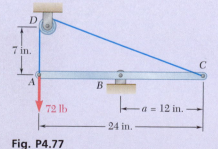

Fig. P4.77

4.78 Using the method of Sec. 4.2B, solve Prob. 4.22.

4.79 Knowing that $\theta = 30°$, determine the reaction (*a*) at *B*, (*b*) at *C*.

4.80 Knowing that $\theta = 60°$, determine the reaction (*a*) at *B*, (*b*) at *C*.

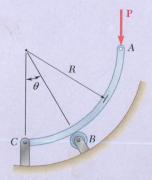

Fig. P4.79 and P4.80

4.81 Determine the reactions at A and B when $\beta = 50°$.

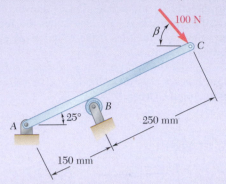

Fig. P4.81 and P4.82

4.82 Determine the reactions at A and B when $\beta = 80°$.

4.83 Rod AB is bent into the shape of an arc of circle and is lodged between two pegs D and E. It supports a load $\mathbf{P}$ at end B. Neglecting friction and the weight of the rod, determine the distance c corresponding to equilibrium when $a = 20$ mm and $R = 100$ mm.

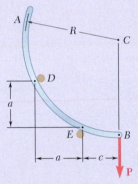

Fig. P4.83

4.84 A slender rod of length L is attached to collars that can slide freely along the guides shown. Knowing that the rod is in equilibrium, derive an expression for angle θ in terms of angle β.

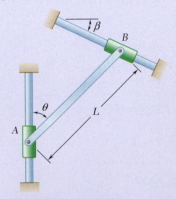

Fig. P4.84 and P4.85

4.85 An 8-kg slender rod of length L is attached to collars that can slide freely along the guides shown. Knowing that the rod is in equilibrium and that $\beta = 30°$, determine (a) the angle θ that the rod forms with the vertical, (b) the reactions at A and B.

4.86 A uniform rod *AB* of length 2*R* rests inside a hemispherical bowl of radius *R* as shown. Neglecting friction, determine the angle θ corresponding to equilibrium.

4.87 A slender rod *BC* with a length of *L* and weight *W* is held by two cables as shown. Knowing that cable *AB* is horizontal and that the rod forms an angle of 40° with the horizontal, determine (*a*) the angle θ that cable *CD* forms with the horizontal, (*b*) the tension in each cable.

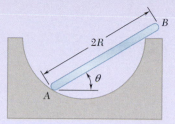

Fig. P4.86

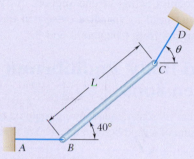

Fig. P4.87

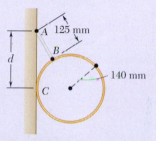

Fig. P4.88

4.88 A thin ring with a mass of 2 kg and radius *r* = 140 mm is held against a frictionless wall by a 125-mm string *AB*. Determine (*a*) the distance *d*, (*b*) the tension in the string, (*c*) the reaction at *C*.

4.89 A slender rod with a length of *L* and weight *W* is attached to a collar at *A* and is fitted with a small wheel at *B*. Knowing that the wheel rolls freely along a cylindrical surface of radius *R*, and neglecting friction, derive an equation in θ, *L*, and *R* that must be satisfied when the rod is in equilibrium.

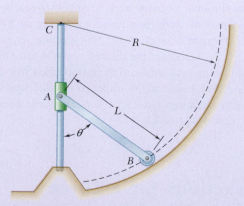

Fig. P4.89

4.90 Knowing that for the rod of Prob. 4.89, *L* = 15 in., *R* = 20 in., and *W* = 10 lb, determine (*a*) the angle θ corresponding to equilibrium, (*b*) the reactions at *A* and *B*.

4.3 EQUILIBRIUM IN THREE DIMENSIONS

The most general situation of rigid-body equilibrium occurs in three dimensions. The approach to modeling and analyzing these situations is the same as in two dimensions: Draw a free-body diagram and then write and solve the equilibrium equations. However, you now have more equations and more variables to deal with. In addition, reactions at supports and connections can be more varied, having as many as three force components and three couples acting at one support. As you will see in the Sample Problems, you need to visualize clearly in three dimensions and recall the vector analysis from Chapters 2 and 3.

4.3A Rigid-Body Equilibrium in Three Dimensions

We saw in Sec. 4.1 that six scalar equations are required to express the conditions for the equilibrium of a rigid body in the general three-dimensional case:

$$\Sigma F_x = 0 \qquad \Sigma F_y = 0 \qquad \Sigma F_z = 0 \qquad \text{(4.2)}$$
$$\Sigma M_x = 0 \qquad \Sigma M_y = 0 \qquad \Sigma M_z = 0 \qquad \text{(4.3)}$$

We can solve these equations for no more than *six unknowns,* which generally represent reactions at supports or connections.

In most problems, we can obtain the scalar equations (4.2) and (4.3) more conveniently if we first write the conditions for the equilibrium of the rigid body considered in vector form:

$$\Sigma \mathbf{F} = 0 \qquad \Sigma \mathbf{M}_O = \Sigma(\mathbf{r} \times \mathbf{F}) = 0 \qquad \text{(4.1)}$$

Then we can express the forces $\mathbf{F}$ and position vectors $\mathbf{r}$ in terms of scalar components and unit vectors. This enables us to compute all vector products either by direct calculation or by means of determinants (see Sec. 3.1F). Note that we can eliminate as many as three unknown reaction components from these computations through a judicious choice of the point O. By equating to zero the coefficients of the unit vectors in each of the two relations in Eq. (4.1), we obtain the desired scalar equations.[†]

Some equilibrium problems and their associated free-body diagrams might involve individual couples $\mathbf{M}_i$ either as applied loads or as support reactions. In such situations, you can accommodate these couples by expressing the second part of Eq. (4.1) as

$$\Sigma \mathbf{M}_O = \Sigma(\mathbf{r} \times \mathbf{F}) + \Sigma \mathbf{M}_i = 0 \qquad \text{(4.1')}$$

4.3B Reactions for a Three-Dimensional Structure

The reactions on a three-dimensional structure range from a single force of known direction exerted by a frictionless surface to a force-couple system

---

[†]In some problems, it may be convenient to eliminate from the solution the reactions at two points A and B by writing the equilibrium equation $\Sigma M_{AB} = 0$. This involves determining the moments of the forces about the axis AB joining points A and B (see Sample Prob. 4.10).

exerted by a fixed support. Consequently, in problems involving the equilibrium of a three-dimensional structure, between one and six unknowns may be associated with the reaction at each support or connection.

Figure 4.10 shows various types of supports and connections with their corresponding reactions. A simple way of determining the type of reaction corresponding to a given support or connection and the number of unknowns involved is to find which of the six fundamental motions (translation in the *x, y,* and *z* directions and rotation about the *x, y,* and *z* axes) are allowed and which motions are prevented. The number of motions prevented equals the number of reactions.

Ball supports, frictionless surfaces, and cables, for example, prevent translation in one direction only and thus exert a single force whose line of action is known. Therefore, each of these supports involves one unknown—namely, the magnitude of the reaction. Rollers on rough surfaces and wheels on rails prevent translation in two directions; the corresponding reactions consist of two unknown force components. Rough surfaces in direct contact and ball-and-socket supports prevent translation in three directions while still allowing rotation; these supports involve three unknown force components.

Some supports and connections can prevent rotation as well as translation; the corresponding reactions include couples as well as forces. For example, the reaction at a fixed support, which prevents any motion (rotation as well as translation) consists of three unknown forces and three unknown couples. A universal joint, which is designed to allow rotation about two axes, exerts a reaction consisting of three unknown force components and one unknown couple.

Other supports and connections are primarily intended to prevent translation; their design, however, is such that they also prevent some rotations. The corresponding reactions consist essentially of force components but *may* also include couples. One group of supports of this type includes hinges and bearings designed to support radial loads only (for example, journal bearings or roller bearings). The corresponding reactions consist of two force components but may also include two couples. Another group includes pin-and-bracket supports, hinges, and bearings designed to support an axial thrust as well as a radial load (for example, ball bearings). The corresponding reactions consist of three force components but may include two couples. However, these supports do not exert any appreciable couples under normal conditions of use. Therefore, *only* force components should be included in their analysis *unless* it is clear that couples are necessary to maintain the equilibrium of the rigid body or unless the support is known to have been specifically designed to exert a couple (see Probs. 4.119 through 4.122).

If the reactions involve more than six unknowns, you have more unknowns than equations, and some of the reactions are **statically indeterminate**. If the reactions involve fewer than six unknowns, you have more equations than unknowns, and some of the equations of equilibrium cannot be satisfied under general loading conditions. In this case, the rigid body is only **partially constrained**. Under the particular loading conditions corresponding to a given problem, however, the extra equations often reduce to trivial identities, such as 0 = 0, and can be disregarded; although only partially constrained, the rigid body remains in equilibrium (see Sample Probs. 4.7 and 4.8). Even with six or more unknowns, it is possible that some equations of equilibrium are not satisfied. This can occur when the reactions associated with the given supports either are parallel or intersect the same line; the rigid body is then **improperly constrained**.

Photo 4.3 Universal joints, seen on the drive shafts of rear-wheel-drive cars and trucks, allow rotational motion to be transferred between two noncollinear shafts.

Photo 4.4 This pillow block bearing supports the shaft of a fan used in an industrial facility.

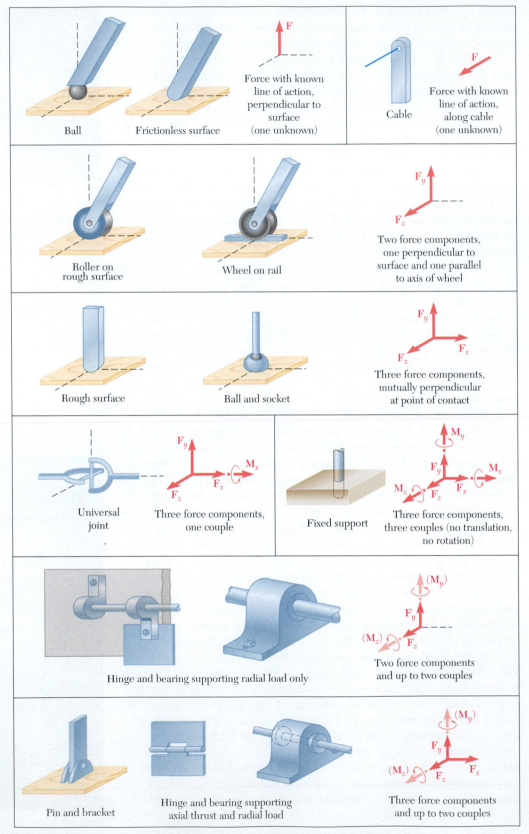

Fig. 4.10 Reactions at supports and connections in three dimensions.

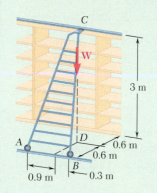

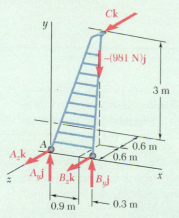

Fig. 1 Free-body diagram of ladder.

Sample Problem 4.7

A 20-kg ladder used to reach high shelves in a storeroom is supported by two flanged wheels A and B mounted on a rail and by a flangeless wheel C resting against a rail fixed to the wall. An 80-kg man stands on the ladder and leans to the right. The line of action of the combined weight $\mathbf{W}$ of the man and ladder intersects the floor at point D. Determine the reactions at A, B, and C.

STRATEGY: Draw a free-body diagram of the ladder, then write and solve the equilibrium equations in three dimensions.

MODELING:

Free-Body Diagram. The combined weight of the man and ladder is

$$\mathbf{W} = -mg\mathbf{j} = -(80 \text{ kg} + 20 \text{ kg})(9.81 \text{ m/s}^2)\mathbf{j} = -(981 \text{ N})\mathbf{j}$$

You have five unknown reaction components: two at each flanged wheel and one at the flangeless wheel (Fig. 1). The ladder is thus only partially constrained; it is free to roll along the rails. It is, however, in equilibrium under the given load because the equation $\Sigma F_x = 0$ is satisfied.

ANALYSIS:

Equilibrium Equations. The forces acting on the ladder form a system equivalent to zero:

$$\Sigma \mathbf{F} = 0: \quad A_y\mathbf{j} + A_z\mathbf{k} + B_y\mathbf{j} + B_z\mathbf{k} - (981 \text{ N})\mathbf{j} + C\mathbf{k} = 0$$
$$(A_y + B_y - 981 \text{ N})\mathbf{j} + (A_z + B_z + C)\mathbf{k} = 0 \qquad \textbf{(1)}$$

$$\Sigma \mathbf{M}_A = \Sigma(\mathbf{r} \times \mathbf{F}) = 0: \quad 1.2\mathbf{i} \times (B_y\mathbf{j} + B_z\mathbf{k}) + (0.9\mathbf{i} - 0.6\mathbf{k}) \times (-981\mathbf{j})$$
$$+ (0.6\mathbf{i} + 3\mathbf{j} - 1.2\mathbf{k}) \times C\mathbf{k} = 0$$

Computing the vector products gives you[†]

$$1.2B_y\mathbf{k} - 1.2B_z\mathbf{j} - 882.9\mathbf{k} - 588.6\mathbf{i} - 0.6C\mathbf{j} + 3C\mathbf{i} = 0$$
$$(3C - 588.6)\mathbf{i} - (1.2B_z + 0.6C)\mathbf{j} + (1.2B_y - 882.9)\mathbf{k} = 0 \qquad \textbf{(2)}$$

Setting the coefficients of $\mathbf{i}$, $\mathbf{j}$, and $\mathbf{k}$ equal to zero in Eq. (2) produces the following three scalar equations, which state that the sum of the moments about each coordinate axis must be zero:

$$3C - 588.6 = 0 \qquad C = +196.2 \text{ N}$$
$$1.2B_z + 0.6C = 0 \qquad B_z = -98.1 \text{ N}$$
$$1.2B_y - 882.9 = 0 \qquad B_y = +736 \text{ N}$$

The reactions at B and C are therefore

$$\mathbf{B} = +(736 \text{ N})\mathbf{j} - (98.1 \text{ N})\mathbf{k} \qquad \mathbf{C} = +(196.2 \text{ N})\mathbf{k} \quad \blacktriangleleft$$

[†]The moments in this sample problem, as well as in Sample Probs. 4.8 and 4.9, also can be expressed as determinants (see Sample Prob. 3.10).

Setting the coefficients of **j** and **k** equal to zero in Eq. (1), you obtain two scalar equations stating that the sums of the components in the y and z directions are zero. Substitute the values above for B_y, B_z, and C to get

$$A_y + B_y - 981 = 0 \qquad A_y + 736 - 981 = 0 \qquad A_y = +245 \text{ N}$$
$$A_z + B_z + C = 0 \qquad A_z - 98.1 + 196.2 = 0 \qquad A_z = -98.1 \text{ N}$$

Therefore, the reaction at A is

$$\mathbf{A} = +(245 \text{ N})\mathbf{j} - (98.1 \text{ N})\mathbf{k} \qquad \blacktriangleleft$$

REFLECT and THINK: You summed moments about A as part of the analysis. As a check, you could now use these results and demonstrate that the sum of moments about any other point, such as point B, is also zero.

Sample Problem 4.8

A 5×8-ft sign of uniform density weighs 270 lb and is supported by a ball-and-socket joint at A and by two cables. Determine the tension in each cable and the reaction at A.

STRATEGY: Draw a free-body diagram of the sign, and express the unknown cable tensions as Cartesian vectors. Then determine the cable tensions and the reaction at A by writing and solving the equilibrium equations.

MODELING:

Free-Body Diagram. The forces acting on the sign are its weight $\mathbf{W} = -(270 \text{ lb})\mathbf{j}$ and the reactions at A, B, and E (Fig. 1). The reaction at A is a force of unknown direction represented by three unknown components. Since the directions of the forces exerted by the cables are known, these forces involve only one unknown each: specifically, the magnitudes T_{BD} and T_{EC}. The total of five unknowns means that the sign is partially constrained. It can rotate freely about the x axis; it is, however, in equilibrium under the given loading, since the equation $\Sigma M_x = 0$ is satisfied.

ANALYSIS: You can express the components of the forces $\mathbf{T}_{BD}$ and $\mathbf{T}_{EC}$ in terms of the unknown magnitudes T_{BD} and T_{EC} as follows:

$$\overrightarrow{BD} = -(8 \text{ ft})\mathbf{i} + (4 \text{ ft})\mathbf{j} - (8 \text{ ft})\mathbf{k} \qquad BD = 12 \text{ ft}$$
$$\overrightarrow{EC} = -(6 \text{ ft})\mathbf{i} + (3 \text{ ft})\mathbf{j} + (2 \text{ ft})\mathbf{k} \qquad EC = 7 \text{ ft}$$

$$\mathbf{T}_{BD} = T_{BD}\left(\frac{\overrightarrow{BD}}{BD}\right) = T_{BD}(-\tfrac{2}{3}\mathbf{i} + \tfrac{1}{3}\mathbf{j} - \tfrac{2}{3}\mathbf{k})$$

$$\mathbf{T}_{EC} = T_{EC}\left(\frac{\overrightarrow{EC}}{EC}\right) = T_{EC}(-\tfrac{6}{7}\mathbf{i} + \tfrac{3}{7}\mathbf{j} - \tfrac{2}{7}\mathbf{k})$$

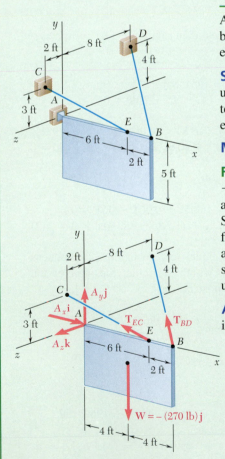

Fig. 1 Free-body diagram of sign.

Equilibrium Equations. The forces acting on the sign form a system equivalent to zero:

$\Sigma \mathbf{F} = 0$: $A_x\mathbf{i} + A_y\mathbf{j} + A_z\mathbf{k} + \mathbf{T}_{BD} + \mathbf{T}_{EC} - (270 \text{ lb})\mathbf{j} = 0$

$(A_x - \frac{2}{3}T_{BD} - \frac{6}{7}T_{EC})\mathbf{i} + (A_y + \frac{1}{3}T_{BD} + \frac{3}{7}T_{EC} - 270 \text{ lb})\mathbf{j}$
$$+ (A_z - \frac{2}{3}T_{BD} + \frac{2}{7}T_{EC})\mathbf{k} = 0 \quad \textbf{(1)}$$

$\Sigma \mathbf{M}_A = \Sigma(\mathbf{r} \times \mathbf{F}) = 0$:

$(8 \text{ ft})\mathbf{i} \times T_{BD}(-\frac{2}{3}\mathbf{i} + \frac{1}{3}\mathbf{j} - \frac{2}{3}\mathbf{k}) + (6 \text{ ft})\mathbf{i} \times T_{EC}(-\frac{6}{7}\mathbf{i} + \frac{3}{7}\mathbf{j} + \frac{2}{7}\mathbf{k})$
$$+ (4 \text{ ft})\mathbf{i} \times (-270 \text{ lb})\mathbf{j} = 0$$

$$(2.667T_{BD} + 2.571T_{EC} - 1080 \text{ lb})\mathbf{k} + (5.333T_{BD} - 1.714T_{EC})\mathbf{j} = 0 \quad \textbf{(2)}$$

Setting the coefficients of $\mathbf{j}$ and $\mathbf{k}$ equal to zero in Eq. (2) yields two scalar equations that can be solved for T_{BD} and T_{EC}:

$$T_{BD} = 101.3 \text{ lb} \qquad T_{EC} = 315 \text{ lb} \quad \blacktriangleleft$$

Setting the coefficients of $\mathbf{i}$, $\mathbf{j}$, and $\mathbf{k}$ equal to zero in Eq. (1) produces three more equations, which yield the components of $\mathbf{A}$.

$$\mathbf{A} = +(338 \text{ lb})\mathbf{i} + (101.2 \text{ lb})\mathbf{j} - (22.5 \text{ lb})\mathbf{k} \quad \blacktriangleleft$$

REFLECT and THINK: Cables can only act in tension, and the free-body diagram and Cartesian vector expressions for the cables were consistent with this. The solution yielded positive results for the cable forces, which confirms that they are in tension and validates the analysis.

Sample Problem 4.9

A uniform pipe cover of radius $r = 240$ mm and mass 30 kg is held in a horizontal position by the cable CD. Assuming that the bearing at B does not exert any axial thrust, determine the tension in the cable and the reactions at A and B.

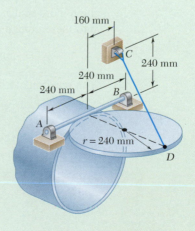

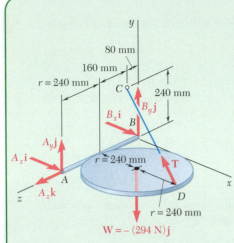

Fig. 1 Free-body diagram of pipe cover.

STRATEGY: Draw a free-body diagram with the coordinate axes shown (Fig. 1) and express the unknown cable tension as a Cartesian vector. Then apply the equilibrium equations to determine this tension and the support reactions.

MODELING:

Free-Body Diagram. The forces acting on the free body include its weight, which is

$$\mathbf{W} = -mg\mathbf{j} = -(30\ \text{kg})(9.81\ \text{m/s}^2)\mathbf{j} = -(294\ \text{N})\mathbf{j}$$

The reactions involve six unknowns: the magnitude of the force **T** exerted by the cable, three force components at hinge A, and two at hinge B. Express the components of **T** in terms of the unknown magnitude T by resolving the vector $\overrightarrow{DC}$ into rectangular components:

$$\overrightarrow{DC} = -(480\ \text{mm})\mathbf{i} + (240\ \text{mm})\mathbf{j} - (160\ \text{mm})\mathbf{k} \qquad DC = 560\ \text{mm}$$

$$\mathbf{T} = T\frac{\overrightarrow{DC}}{DC} = -\tfrac{6}{7}T\mathbf{i} + \tfrac{3}{7}T\mathbf{j} - \tfrac{2}{7}T\mathbf{k}$$

ANALYSIS:

Equilibrium Equations. The forces acting on the pipe cover form a system equivalent to zero. Thus,

$$\Sigma\mathbf{F} = 0: \qquad A_x\mathbf{i} + A_y\mathbf{j} + A_z\mathbf{k} + B_x\mathbf{i} + B_y\mathbf{j} + \mathbf{T} - (294\ \text{N})\mathbf{j} = 0$$

$$(A_x + B_x - \tfrac{6}{7}T)\mathbf{i} + (A_y + B_y + \tfrac{3}{7}T - 294\ \text{N})\mathbf{j} + (A_z - \tfrac{2}{7}T)\mathbf{k} = 0 \quad \textbf{(1)}$$

$$\Sigma\mathbf{M}_B = \Sigma(\mathbf{r} \times \mathbf{F}) = 0:$$
$$2r\mathbf{k} \times (A_x\mathbf{i} + A_y\mathbf{j} + A_z\mathbf{k})$$
$$+ (2r\mathbf{i} + r\mathbf{k}) \times (-\tfrac{6}{7}T\mathbf{i} + \tfrac{3}{7}T\mathbf{j} - \tfrac{2}{7}T\mathbf{k})$$
$$+ (r\mathbf{i} + r\mathbf{k}) \times (-294\ \text{N})\mathbf{j} = 0$$
$$(-2A_y - \tfrac{3}{7}T + 294\ \text{N})r\mathbf{i} + (2A_x - \tfrac{2}{7}T)r\mathbf{j} + (\tfrac{6}{7}T - 294\ \text{N})r\mathbf{k} = 0 \quad \textbf{(2)}$$

Setting the coefficients of the unit vectors equal to zero in Eq. (2) gives three scalar equations, which yield

$$A_x = +49.0\ \text{N} \qquad A_y = +73.5\ \text{N} \qquad T = 343\ \text{N} \ \blacktriangleleft$$

Setting the coefficients of the unit vectors equal to zero in Eq. (1) produces three more scalar equations. After substituting the values of T, A_x, and A_y into these equations, you obtain

$$A_z = +98.0\ \text{N} \qquad B_x = +245\ \text{N} \qquad B_y = +73.5\ \text{N}$$

The reactions at A and B are therefore

$$\mathbf{A} = +(49.0\ \text{N})\mathbf{i} + (73.5\ \text{N})\mathbf{j} + (98.0\ \text{N})\mathbf{k} \ \blacktriangleleft$$

$$\mathbf{B} = +(245\ \text{N})\mathbf{i} + (73.5\ \text{N})\mathbf{j} \ \blacktriangleleft$$

REFLECT and THINK: As a check, you can determine the tension in the cable using a scalar analysis. Assigning signs by the right-hand rule (rhr), we have

$$(+\text{rhr}) \quad \Sigma M_z = 0: \quad \tfrac{3}{7}T(0.48\ \text{m}) - (294\ \text{N})(0.24\ \text{m}) = 0 \quad T = 343\ \text{N} \ \blacktriangleleft$$

Sample Problem 4.10

A 450-lb load hangs from the corner C of a rigid piece of pipe $ABCD$ that has been bent as shown. The pipe is supported by ball-and-socket joints A and D, which are fastened, respectively, to the floor and to a vertical wall, and by a cable attached at the midpoint E of the portion BC of the pipe and at a point G on the wall. Determine (a) where G should be located if the tension in the cable is to be minimum, (b) the corresponding minimum value of the tension.

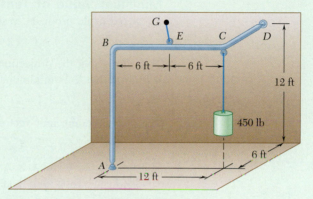

STRATEGY: Draw the free-body diagram of the pipe showing the reactions at A and D. Isolate the unknown tension $\mathbf{T}$ and the known weight $\mathbf{W}$ by summing moments about the diagonal line AD, and compute values from the equilibrium equations.

MODELING and ANALYSIS:

Free-Body Diagram. The free-body diagram of the pipe includes the load $\mathbf{W} = (-450 \text{ lb})\mathbf{j}$, the reactions at A and D, and the force $\mathbf{T}$ exerted by the cable (Fig. 1). To eliminate the reactions at A and D from the computations, take the sum of the moments of the forces about the line AD and set it equal to zero. Denote the unit vector along AD by $\boldsymbol{\lambda}$, which enables you to write

$$\Sigma M_{AD} = 0: \quad \boldsymbol{\lambda} \cdot (\overrightarrow{AE} \times \mathbf{T}) + \boldsymbol{\lambda} \cdot (\overrightarrow{AC} \times \mathbf{W}) = 0 \quad \textbf{(1)}$$

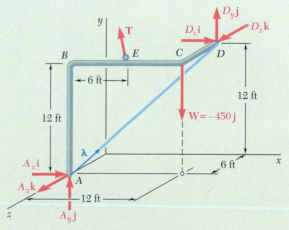

Fig. 1 Free-body diagram of pipe.

You can compute the second term in Eq. (1) as follows:

$$\vec{AC} \times \mathbf{W} = (12\mathbf{i} + 12\mathbf{j}) \times (-450\mathbf{j}) = -5400\mathbf{k}$$

$$\boldsymbol{\lambda} = \frac{\vec{AD}}{AD} = \frac{12\mathbf{i} + 12\mathbf{j} - 6\mathbf{k}}{18} = \tfrac{2}{3}\mathbf{i} + \tfrac{2}{3}\mathbf{j} - \tfrac{1}{3}\mathbf{k}$$

$$\boldsymbol{\lambda} \cdot (\vec{AC} \times \mathbf{W}) = (\tfrac{2}{3}\mathbf{i} + \tfrac{2}{3}\mathbf{j} - \tfrac{1}{3}\mathbf{k}) \cdot (-5400\mathbf{k}) = +1800$$

Substituting this value into Eq. (1) gives

$$\boldsymbol{\lambda} \cdot (\vec{AE} \times \mathbf{T}) = -1800 \text{ lb·ft} \qquad (2)$$

Minimum Value of Tension.   Recalling the commutative property for mixed triple products, you can rewrite Eq. (2) in the form

$$\mathbf{T} \cdot (\boldsymbol{\lambda} \times \vec{AE}) = -1800 \text{ lb·ft} \qquad (3)$$

This shows that the projection of **T** on the vector $\boldsymbol{\lambda} \times \vec{AE}$ is a constant. It follows that **T** is minimum when it is parallel to the vector

$$\boldsymbol{\lambda} \times \vec{AE} = (\tfrac{2}{3}\mathbf{i} + \tfrac{2}{3}\mathbf{j} - \tfrac{1}{3}\mathbf{k}) \times (6\mathbf{i} + 12\mathbf{j}) = 4\mathbf{i} - 2\mathbf{j} + 4\mathbf{k}$$

The corresponding unit vector is $\tfrac{2}{3}\mathbf{i} - \tfrac{1}{3}\mathbf{j} + \tfrac{2}{3}\mathbf{k}$, which gives

$$\mathbf{T}_{min} = T(\tfrac{2}{3}\mathbf{i} - \tfrac{1}{3}\mathbf{j} + \tfrac{2}{3}\mathbf{k}) \qquad (4)$$

Substituting for **T** and $\boldsymbol{\lambda} \times \vec{AE}$ in Eq. (3) and computing the dot products yields $6T = -1800$ and, thus, $T = -300$. Carrying this value into Eq. (4) gives you

$$\mathbf{T}_{min} = -200\mathbf{i} + 100\mathbf{j} - 200\mathbf{k} \qquad T_{min} = 300 \text{ lb} \quad \blacktriangleleft$$

Location of G.   Since the vector $\vec{EG}$ and the force $\mathbf{T}_{min}$ have the same direction, their components must be proportional. Denoting the coordinates of G by x, y, and 0 (Fig. 2), you get

$$\frac{x - 6}{-200} = \frac{y - 12}{+100} = \frac{0 - 6}{-200} \qquad x = 0 \qquad y = 15 \text{ ft} \quad \blacktriangleleft$$

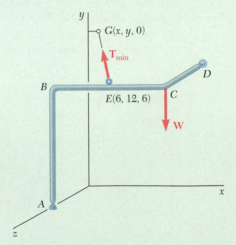

Fig. 2 Location of point G for minimum tension in cable.

REFLECT and THINK: Sometimes you have to rely on the vector analysis presented in Chapters 2 and 3 as much as on the conditions for equilibrium described in this chapter.

SOLVING PROBLEMS
ON YOUR OWN

In this section, you considered the equilibrium of a *three-dimensional body*. It is again most important that you draw a complete *free-body diagram* as the first step of your solution.

1. Pay particular attention to the reactions at the supports as you draw the free-body diagram. The number of unknowns at a support can range from one to six (Fig. 4.10). To decide whether an unknown reaction or reaction component exists at a support, ask yourself whether the support prevents motion of the body in a certain direction or about a certain axis.

    a. If motion is prevented in a certain direction, include in your free-body diagram an unknown *reaction* or *reaction component* that acts in the *same direction*.

    b. If a support prevents rotation about a certain axis, include in your free-body diagram a *couple* of unknown magnitude that acts about the *same axis*.

2. The external forces acting on a three-dimensional body form a system equivalent to zero. Writing $\Sigma \mathbf{F} = 0$ and $\Sigma \mathbf{M}_A = 0$ about an appropriate point A and setting the coefficients of $\mathbf{i}$, $\mathbf{j}$, $\mathbf{k}$ in both equations equal to zero provides you with six scalar equations. In general, these equations contain six unknowns and may be solved for these unknowns.

3. After completing your free-body diagram, you may want to seek equations involving as few unknowns as possible. The following strategies may help you.

    a. By summing moments about a ball-and-socket support or a hinge, you obtain equations from which three unknown reaction components have been eliminated [Sample Probs. 4.8 and 4.9].

    b. If you can draw an axis through the points of application of all but one of the unknown reactions, summing moments about that axis will yield an equation in a single unknown [Sample Prob. 4.10].

4. After drawing your free-body diagram, you may find that one of the following situations exists.

    a. The reactions involve fewer than six unknowns. The body is partially constrained and motion of the body is possible. However, you may be able to determine the reactions for a given loading condition [Sample Prob. 4.7].

    b. The reactions involve more than six unknowns. The reactions are statically indeterminate. Although you may be able to calculate one or two reactions, you cannot determine all of them [Sample Prob. 4.10].

    c. The reactions are parallel or intersect the same line. The body is improperly constrained, and motion can occur under a general loading condition.

Problems

FREE-BODY PRACTICE PROBLEMS

4.F5 Two tape spools are attached to an axle supported by bearings at A and D. The radius of spool B is 1.5 in. and the radius of spool C is 2 in. Knowing that $T_B = 20$ lb and that the system rotates at a constant rate, draw the free-body diagram needed to determine the reactions at A and D. Assume that the bearing at A does not exert any axial thrust and neglect the weights of the spools and axle.

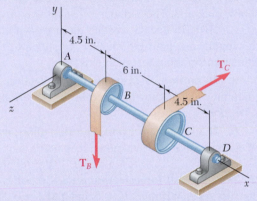

Fig. P4.F5

4.F6 A 12-m pole supports a horizontal cable CD and is held by a ball and socket at A and two cables BE and BF. Knowing that the tension in cable CD is 14 kN and assuming that CD is parallel to the x axis ($\phi = 0$), draw the free-body diagram needed to determine the tension in cables BE and BF and the reaction at A.

4.F7 A 20-kg cover for a roof opening is hinged at corners A and B. The roof forms an angle of 30° with the horizontal, and the cover is maintained in a horizontal position by the brace CE. Draw the free-body diagram needed to determine the magnitude of the force exerted by the brace and the reactions at the hinges. Assume that the hinge at A does not exert any axial thrust.

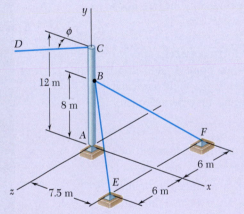

Fig. P4.F6

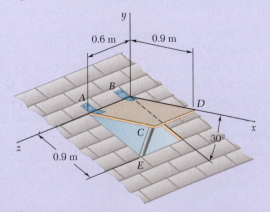

Fig. P4.F7

4.91 Two transmission belts pass over a double-sheaved pulley that is attached to an axle supported by bearings at A and D. The radius of the inner sheave is 125 mm and the radius of the outer sheave is 250 mm. Knowing that when the system is at rest, the tension is 90 N in both portions of belt B and 150 N in both portions of belt C, determine the reactions at A and D. Assume that the bearing at D does not exert any axial thrust.

4.92 Solve Prob. 4.91, assuming that the pulley rotates at a constant rate and that $T_B = 104$ N, $T'_B = 84$ N, and $T_C = 175$ N.

4.93 A small winch is used to raise a 120-lb load. Find (a) the magnitude of the vertical force **P** that should be applied at C to maintain equilibrium in the position shown, (b) the reactions at A and B, assuming that the bearing at B does not exert any axial thrust.

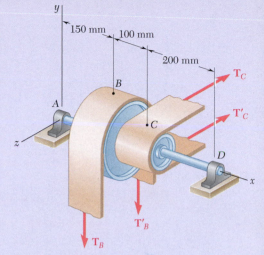

Fig. P4.91

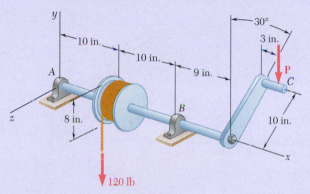

Fig. P4.93

4.94 A 4×8-ft sheet of plywood weighing 34 lb has been temporarily placed among three pipe supports. The lower edge of the sheet rests on small collars at A and B and its upper edge leans against pipe C. Neglecting friction at all surfaces, determine the reactions at A, B, and C.

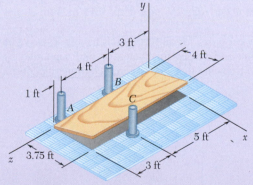

Fig. P4.94

4.95 A 250×400-mm plate of mass 12 kg and a 300-mm-diameter pulley are welded to axle AC that is supported by bearings at A and B. For $\beta = 30°$, determine (a) the tension in the cable, (b) the reactions at A and B. Assume that the bearing at B does not exert any axial thrust.

4.96 Solve Prob. 4.95 for $\beta = 60°$.

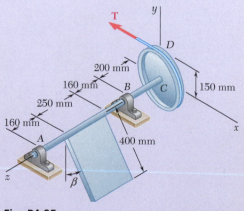

Fig. P4.95

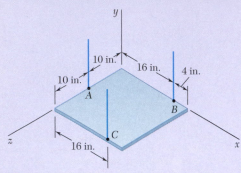

Fig. P4.97 and *P4.98*

4.97 The 20 × 20-in. square plate shown weighs 56 lb and is supported by three vertical wires. Determine the tension in each wire.

4.98 The 20 × 20-in. square plate shown weighs 56 lb and is supported by three vertical wires. Determine the weight and location of the lightest block that should be placed on the plate if the tensions in the three wires are to be equal.

4.99 An opening in a floor is covered by a 1 × 1.2-m sheet of plywood with a mass of 18 kg. The sheet is hinged at A and B and is maintained in a position slightly above the floor by a small block C. Determine the vertical component of the reaction (*a*) at A, (*b*) at B, (*c*) at C.

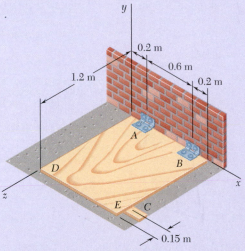

Fig. P4.99

4.100 Solve Prob. 4.99, assuming that the small block C is moved and placed under edge DE at a point 0.15 m from corner E.

4.101 Two steel pipes AB and BC, each having a mass per unit length of 8 kg/m, are welded together at B and supported by three vertical wires. Knowing that *a* = 0.4 m, determine the tension in each wire.

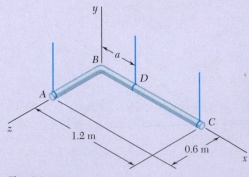

Fig. P4.101

4.102 For the pipe assembly of Prob. 4.101, determine (a) the largest permissible value of a if the assembly is not to tip, (b) the corresponding tension in each wire.

4.103 The 24-lb square plate shown is supported by three vertical wires. Determine (a) the tension in each wire when $a = 10$ in., (b) the value of a for which the tension in each wire is 8 lb.

4.104 The table shown weighs 30 lb and has a diameter of 4 ft. It is supported by three legs equally spaced around the edge. A vertical load **P** with a magnitude of 100 lb is applied to the top of the table at D. Determine the maximum value of a if the table is not to tip over. Show, on a sketch, the area of the table over which **P** can act without tipping the table.

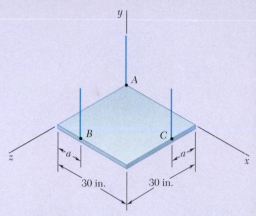

Fig. P4.103

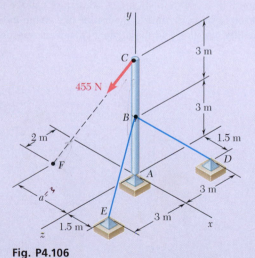

Fig. _P4.104_

4.105 A 10-ft boom is acted upon by the 840-lb force shown. Determine the tension in each cable and the reaction at the ball-and-socket joint at A.

4.106 The 6-m pole ABC is acted upon by a 455-N force as shown. The pole is held by a ball-and-socket joint at A and by two cables BD and BE. For $a = 3$ m, determine the tension in each cable and the reaction at A.

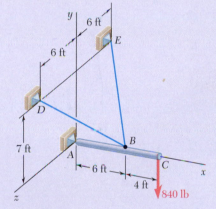

Fig. P4.105

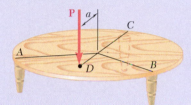

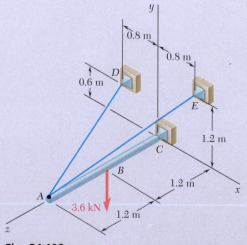

Fig. P4.106

4.107 Solve Prob. 4.106 for $a = 1.5$ m.

4.108 A 2.4-m boom is held by a ball-and-socket joint at C and by two cables AD and AE. Determine the tension in each cable and the reaction at C.

4.109 Solve Prob. 4.108, assuming that the 3.6-kN load is applied at point A.

Fig. P4.108

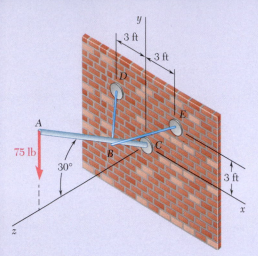

Fig. P4.110

4.110 The 10-ft flagpole *AC* forms an angle of 30° with the *z* axis. It is held by a ball-and-socket joint at *C* and by two thin braces *BD* and *BE*. Knowing that the distance *BC* is 3 ft, determine the tension in each brace and the reaction at *C*.

4.111 A 48-in. boom is held by a ball-and-socket joint at *C* and by two cables *BF* and *DAE*; cable *DAE* passes around a frictionless pulley at *A*. For the loading shown, determine the tension in each cable and the reaction at *C*.

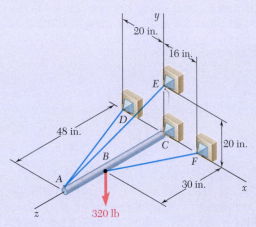

Fig. *P4.111*

4.112 Solve Prob. 4.111, assuming that the 320-lb load is applied at *A*.

4.113 A 10-kg storm window measuring 900 × 1500 mm is held by hinges at *A* and *B*. In the position shown, it is held away from the side of the house by a 600-mm stick *CD*. Assuming that the hinge at *A* does not exert any axial thrust, determine the magnitude of the force exerted by the stick and the components of the reactions at *A* and *B*.

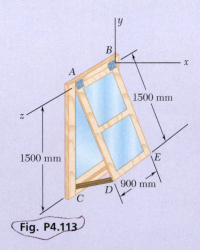

Fig. P4.113

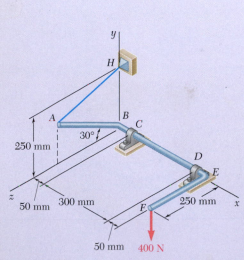

Fig. *P4.114*

4.114 The bent rod *ABEF* is supported by bearings at *C* and *D* and by wire *AH*. Knowing that portion *AB* of the rod is 250 mm long, determine (*a*) the tension in wire *AH*, (*b*) the reactions at *C* and *D*. Assume that the bearing at *D* does not exert any axial thrust.

4.115 The horizontal platform $ABCD$ weighs 60 lb and supports a 240-lb load at its center. The platform is normally held in position by hinges at A and B and by braces CE and DE. If brace DE is removed, determine the reactions at the hinges and the force exerted by the remaining brace CE. The hinge at A does not exert any axial thrust.

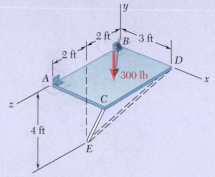

Fig. P4.115

4.116 The lid of a roof scuttle weighs 75 lb. It is hinged at corners A and B and maintained in the desired position by a rod CD pivoted at C. A pin at end D of the rod fits into one of several holes drilled in the edge of the lid. For $\alpha = 50°$, determine (a) the magnitude of the force exerted by rod CD, (b) the reactions at the hinges. Assume that the hinge at B does not exert any axial thrust.

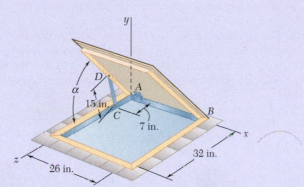

Fig. P4.116

4.117 A 100-kg uniform rectangular plate is supported in the position shown by hinges A and B and by cable DCE that passes over a frictionless hook at C. Assuming that the tension is the same in both parts of the cable, determine (a) the tension in the cable, (b) the reactions at A and B. Assume that the hinge at B does not exert any axial thrust.

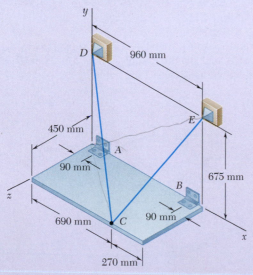

Fig. P4.117

4.118 Solve Prob. 4.117, assuming that cable DCE is replaced by a cable attached to point E and hook C.

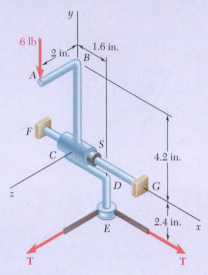

6 lb
2 in.
1.6 in.
B
A
F
S
C
4.2 in.
D G
z
E
2.4 in. x
T   T

Fig. P4.121

4.119 Solve Prob. 4.113, assuming that the hinge at *A* has been removed and that the hinge at *B* can exert couples about axes parallel to the *x* and *y* axes.

4.120 Solve Prob. 4.115, assuming that the hinge at *B* has been removed and that the hinge at *A* can exert an axial thrust, as well as couples about axes parallel to the *x* and *y* axes.

4.121 The assembly shown is used to control the tension *T* in a tape that passes around a frictionless spool at *E*. Collar *C* is welded to rods *ABC* and *CDE*. It can rotate about shaft *FG* but its motion along the shaft is prevented by a washer *S*. For the loading shown, determine (*a*) the tension *T* in the tape, (*b*) the reaction at *C*.

4.122 The assembly shown is welded to collar *A* that fits on the vertical pin shown. The pin can exert couples about the *x* and *z* axes but does not prevent motion about or along the *y* axis. For the loading shown, determine the tension in each cable and the reaction at *A*.

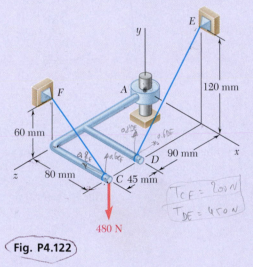

$T_{CF} = 200 N$
$T_{DE} = 450 N$

480 N

Fig. P4.122

4.123 The rigid L-shaped member *ABC* is supported by a ball-and-socket joint at *A* and by three cables. If a 1.8-kN load is applied at *F*, determine the tension in each cable.

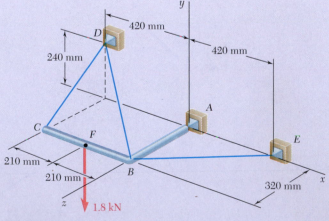

Fig. P4.123

4.124 Solve Prob. 4.123, assuming that the 1.8-kN load is applied at C.

4.125 The rigid L-shaped member ABF is supported by a ball-and-socket joint at A and by three cables. For the loading shown, determine the tension in each cable and the reaction at A.

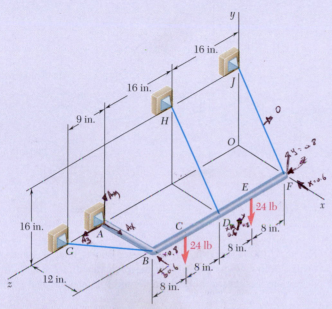

Fig. P4.125

4.126 Solve Prob. 4.125, assuming that the load at C has been removed.

4.127 Three rods are welded together to form a "corner" that is supported by three eyebolts. Neglecting friction, determine the reactions at A, B, and C when $P = 240$ lb, $a = 12$ in., $b = 8$ in., and $c = 10$ in.

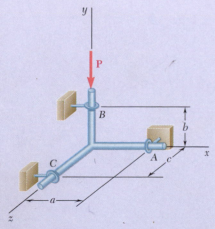

Fig. P4.127

4.128 Solve Prob. 4.127, assuming that the force **P** is removed and is replaced by a couple $\mathbf{M} = +\,(600\ \text{lb}\cdot\text{in.})\mathbf{j}$ acting at B.

4.129 Frame *ABCD* is supported by a ball-and-socket joint at *A* and by three cables. For *a* = 150 mm, determine the tension in each cable and the reaction at *A*.

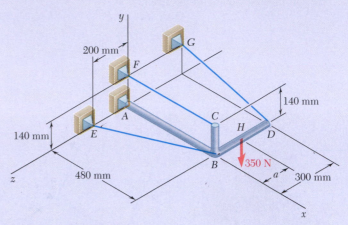

Fig. P4.129 and *P4.130*

4.130 Frame *ABCD* is supported by a ball-and-socket joint at *A* and by three cables. Knowing that the 350-N load is applied at *D* (*a* = 300 mm), determine the tension in each cable and the reaction at *A*.

4.131 The assembly shown consists of an 80-mm rod *AF* that is welded to a cross frame consisting of four 200-mm arms. The assembly is supported by a ball-and-socket joint at *F* and by three short links, each of which forms an angle of 45° with the vertical. For the loading shown, determine (*a*) the tension in each link, (*b*) the reaction at *F*.

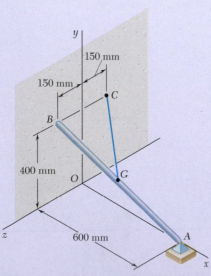

Fig. P4.132

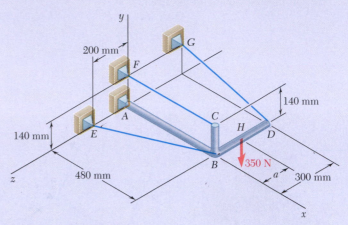

Fig. P4.131

4.132 The uniform 10-kg rod *AB* is supported by a ball-and-socket joint at *A* and by the cord *CG* that is attached to the midpoint *G* of the rod. Knowing that the rod leans against a frictionless vertical wall at *B*, determine (*a*) the tension in the cord, (*b*) the reactions at *A* and *B*.

4.133 The frame *ACD* is supported by ball-and-socket joints at *A* and *D* and by a cable that passes through a ring at *B* and is attached to hooks at *G* and *H*. Knowing that the frame supports at point *C* a load of magnitude *P* = 268 N, determine the tension in the cable.

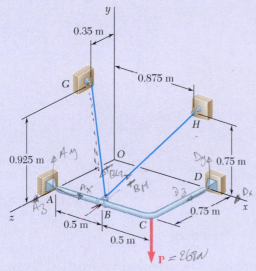

Fig. P4.133

4.134 Solve Prob. 4.133, assuming that cable *GBH* is replaced by a cable *GB* attached at *G* and *B*.

4.135 The bent rod *ABDE* is supported by ball-and-socket joints at *A* and *E* and by the cable *DF*. If a 60-lb load is applied at *C* as shown, determine the tension in the cable.

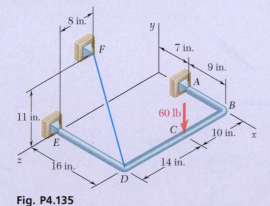

Fig. P4.135

4.136 Solve Prob. 4.135, assuming that cable *DF* is replaced by a cable connecting *B* and *F*.

4.137 Two rectangular plates are welded together to form the assembly shown. The assembly is supported by ball-and-socket joints at *B* and *D* and by a ball on a horizontal surface at *C*. For the loading shown, determine the reaction at *C*.

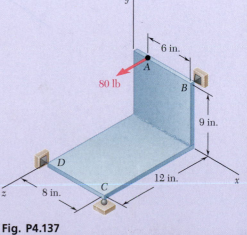

Fig. P4.137

4.138 The pipe *ACDE* is supported by ball-and-socket joints at *A* and *E* and by the wire *DF*. Determine the tension in the wire when a 640-N load is applied at *B* as shown.

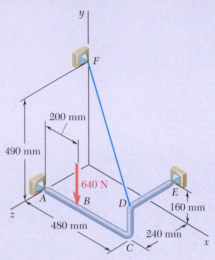

Fig. P4.138

4.139 Solve Prob. 4.138, assuming that wire *DF* is replaced by a wire connecting *C* and *F*.

4.140 Two 2 × 4-ft plywood panels, each with a weight of 12 lb, are nailed together as shown. The panels are supported by ball-and-socket joints at *A* and *F* and by the wire *BH*. Determine (*a*) the location of *H* in the *xy* plane if the tension in the wire is to be minimum, (*b*) the corresponding minimum tension.

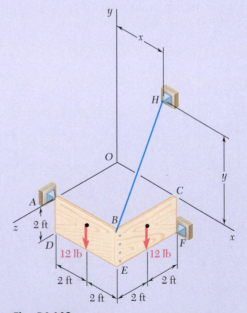

Fig. P4.140

4.141 Solve Prob. 4.140, subject to the restriction that *H* must lie on the *y* axis.

Review and Summary

Equilibrium Equations

This chapter was devoted to the study of the **equilibrium of rigid bodies**, i.e., to the situation when the external forces acting on a rigid body *form a system equivalent to zero* [Introduction]. We then have

$$\Sigma \mathbf{F} = 0 \qquad \Sigma \mathbf{M}_O = \Sigma(\mathbf{r} \times \mathbf{F}) = 0 \qquad \textbf{(4.1)}$$

Resolving each force and each moment into its rectangular components, we can express the necessary and sufficient conditions for the equilibrium of a rigid body with the following six scalar equations:

$$\Sigma F_x = 0 \qquad \Sigma F_y = 0 \qquad \Sigma F_z = 0 \qquad \textbf{(4.2)}$$
$$\Sigma M_x = 0 \qquad \Sigma M_y = 0 \qquad \Sigma M_z = 0 \qquad \textbf{(4.3)}$$

We can use these equations to determine unknown forces applied to the rigid body or unknown reactions exerted by its supports.

Free-Body Diagram

When solving a problem involving the equilibrium of a rigid body, it is essential to consider *all* of the forces acting on the body. Therefore, the first step in the solution of the problem should be to draw a **free-body diagram** showing the body under consideration and all of the unknown as well as known forces acting on it.

Equilibrium of a Two-Dimensional Structure

In the first part of this chapter, we considered the **equilibrium of a two-dimensional structure**; i.e., we assumed that the structure considered and the forces applied to it were contained in the same plane. We saw that each of the reactions exerted on the structure by its supports could involve one, two, or three unknowns, depending upon the type of support [Sec. 4.1A].

In the case of a two-dimensional structure, the equations given previously reduce to *three equilibrium equations*:

$$\Sigma F_x = 0 \qquad \Sigma F_y = 0 \qquad \Sigma M_A = 0 \qquad \textbf{(4.5)}$$

where A is an arbitrary point in the plane of the structure [Sec. 4.1B]. We can use these equations to solve for three unknowns. Although the three equilibrium equations (4.5) cannot be *augmented* with additional equations, any of them can be *replaced* by another equation. Therefore, we can write alternative sets of equilibrium equations, such as

$$\Sigma F_x = 0 \qquad \Sigma M_A = 0 \qquad \Sigma M_B = 0 \qquad \textbf{(4.6)}$$

where point B is chosen in such a way that the line AB is not parallel to the y axis, or

$$\Sigma M_A = 0 \qquad \Sigma M_B = 0 \qquad \Sigma M_C = 0 \qquad \textbf{(4.7)}$$

where the points A, B, and C do not lie in a straight line.

Static Indeterminacy, Partial Constraints, Improper Constraints

Since any set of equilibrium equations can be solved for only three unknowns, the reactions at the supports of a rigid two-dimensional structure cannot be completely determined if they involve *more than three unknowns;* they are said to be *statically indeterminate* [Sec. 4.1C]. On the other hand, if the reactions involve *fewer than three unknowns,* equilibrium is not maintained under general loading conditions; the structure is said to be *partially constrained.* The fact that the reactions involve exactly three unknowns is no guarantee that you can solve the equilibrium equations for all three unknowns. If the supports are arranged in such a way that the reactions are *either concurrent or parallel,* the reactions are statically indeterminate, and the structure is said to be *improperly constrained.*

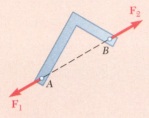

Fig. 4.11

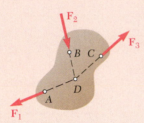

Fig. 4.12

Two-Force Body, Three-Force Body

We gave special attention in Sec. 4.2 to two particular cases of equilibrium of a rigid body. We defined a **two-force body** as a rigid body subjected to forces at only two points, and we showed that the resultants $\mathbf{F}_1$ and $\mathbf{F}_2$ of these forces must have the *same magnitude, the same line of action, and opposite sense* (Fig. 4.11), which is a property that simplifies the solution of certain problems in later chapters. We defined a **three-force body** as a rigid body subjected to forces at only three points, and we demonstrated that the resultants $\mathbf{F}_1$, $\mathbf{F}_2$, and $\mathbf{F}_3$ of these forces must be *either concurrent* (Fig. 4.12) *or parallel.* This property provides us with an alternative approach to the solution of problems involving a three-force body [Sample Prob. 4.6].

Equilibrium of a Three-Dimensional Body

In the second part of this chapter, we considered the *equilibrium of a three-dimensional body.* We saw that each of the reactions exerted on the body by its supports could involve between one and six unknowns, depending upon the type of support [Sec. 4.3A].

In the general case of the equilibrium of a three-dimensional body, all six of the scalar equilibrium equations (4.2) and (4.3) should be used and solved for *six unknowns* [Sec. 4.3B]. In most problems, however, we can obtain these equations more conveniently if we start from

$$\Sigma\mathbf{F} = 0 \qquad \Sigma\mathbf{M}_O = \Sigma\,(\mathbf{r} \times \mathbf{F}) = 0 \qquad \textbf{(4.1)}$$

and then express the forces $\mathbf{F}$ and position vectors $\mathbf{r}$ in terms of scalar components and unit vectors. We can compute the vector products either directly or by means of determinants, and obtain the desired scalar equations by equating to zero the coefficients of the unit vectors [Sample Probs. 4.7 through 4.9].

We noted that we may eliminate as many as three unknown reaction components from the computation of $\Sigma\mathbf{M}_O$ in the second of the relations (4.1) through a judicious choice of point O. Also, we can eliminate the reactions at two points A and B from the solution of some problems by writing the equation $\Sigma M_{AB} = 0$, which involves the computation of the moments of the forces about an axis AB joining points A and B [Sample Prob. 4.10].

We observed that when a body is subjected to individual couples $\mathbf{M}_i$, either as applied loads or as support reactions, we can include these couples by expressing the second part of Eq. (4.1) as

$$\Sigma\mathbf{M}_O = \Sigma(\mathbf{r} \times \mathbf{F}) + \Sigma\mathbf{M}_i = 0 \qquad \textbf{(4.1$'$)}$$

If the reactions involve more than six unknowns, some of the reactions are *statically indeterminate;* if they involve fewer than six unknowns, the rigid body is only *partially constrained.* Even with six or more unknowns, the rigid body is *improperly constrained* if the reactions associated with the given supports are either parallel or intersect the same line.

Review Problems

4.142 A 3200-lb forklift truck is used to lift a 1700-lb crate. Determine the reaction at each of the two (a) front wheels A, (b) rear wheels B.

4.143 The lever BCD is hinged at C and attached to a control rod at B. If P = 100 lb, determine (a) the tension in rod AB, (b) the reaction at C.

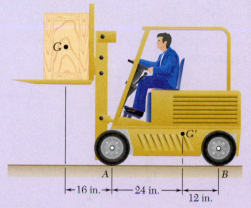

Fig. P4.142

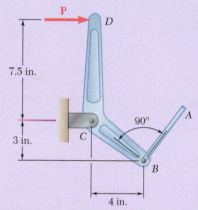

Fig. P4.143

4.144 Determine the reactions at A and B when (a) h = 0, (b) h = 200 mm.

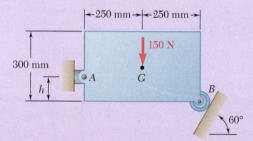

Fig. P4.144

4.145 Neglecting friction and the radius of the pulley, determine (a) the tension in cable ADB, (b) the reaction at C.

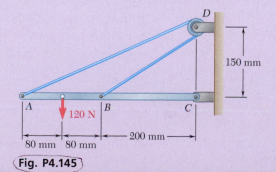

Fig. P4.145

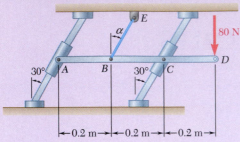

Fig. P4.146

4.146 Bar *AD* is attached at *A* and *C* to collars that can move freely on the rods shown. If the cord *BE* is vertical ($\alpha = 0$), determine the tension in the cord and the reactions at *A* and *C*.

4.147 A slender rod *AB*, with a weight of *W*, is attached to blocks *A* and *B* that move freely in the guides shown. The constant of the spring is *k*, and the spring is unstretched when $\theta = 0$. (*a*) Neglecting the weight of the blocks, derive an equation in *W*, *k*, *l*, and θ that must be satisfied when the rod is in equilibrium. (*b*) Determine the value of θ when *W* = 75 lb, *l* = 30 in., and *k* = 3 lb/in.

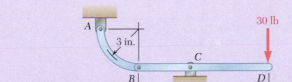

Fig. *P4.147*

4.148 Determine the reactions at *A* and *B* when *a* = 150 mm.

4.149 For the frame and loading shown, determine the reactions at *A* and *C*.

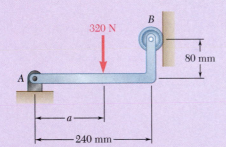

Fig. P4.148

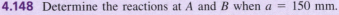

Fig. P4.149

4.150 A 200-mm lever and a 240-mm-diameter pulley are welded to the axle *BE* that is supported by bearings at *C* and *D*. If a 720-N vertical load is applied at *A* when the lever is horizontal, determine (*a*) the tension in the cord, (*b*) the reactions at *C* and *D*. Assume that the bearing at *D* does not exert any axial thrust.

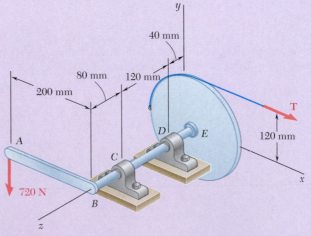

Fig. *P4.150*

4.151 The 45-lb square plate shown is supported by three vertical wires. Determine the tension in each wire.

4.152 The rectangular plate shown weighs 75 lb and is held in the position shown by hinges at A and B and by cable EF. Assuming that the hinge at B does not exert any axial thrust, determine (a) the tension in the cable, (b) the reactions at A and B.

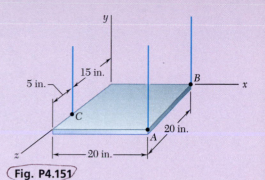

Fig. P4.151

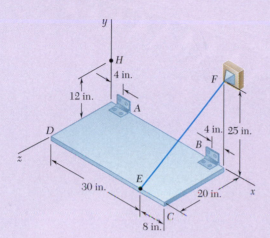

Fig. P4.152

4.153 A force $\mathbf{P}$ is applied to a bent rod ABC, which may be supported in four different ways as shown. In each case, if possible, determine the reactions at the supports.

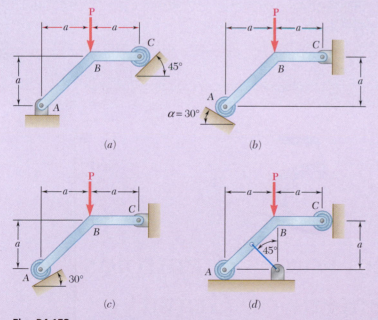

(a)

(b)

(c)

(d)

Fig. P4.153

5

Distributed Forces: Centroids and Centers of Gravity

Loads on dams include three types of distributed forces: the weights of its constituent elements, the pressure forces exerted by the water on its submerged face, and the pressure forces exerted by the ground on its base.

Introduction

Photo 5.1   The precise balancing of the components of a mobile requires an understanding of centers of gravity and centroids, the main topics of this chapter.

Objectives

- **Describe** the centers of gravity of two and three-dimensional bodies.
- **Define** the centroids of lines, areas, and volumes.
- **Consider** the first moments of lines and areas, and examine their properties.
- **Determine** centroids of composite lines, areas, and volumes by summation methods.
- **Determine** centroids of composite lines, areas, and volumes by integration.
- **Apply** the theorems of Pappus-Guldinus to analyze surfaces and bodies of revolution.
- **Analyze** distributed loads on beams and forces on submerged surfaces.

Introduction

We have assumed so far that we could represent the attraction exerted by the earth on a rigid body by a single force **W**. This force, called the force due to gravity or the weight of the body, is applied at the **center of gravity** of the body (Sec. 3.1A). Actually, the earth exerts a force on each of the particles forming the body, so we should represent the attraction of the earth on a rigid body by a large number of small forces distributed over the entire body. You will see in this chapter, however, that all of these small forces can be replaced by a single equivalent force **W**. You will also see how to determine the center of gravity—i.e., the point of application of the resultant **W**—for bodies of various shapes.

In the first part of this chapter, we study two-dimensional bodies, such as flat plates and wires contained in a given plane. We introduce two concepts closely associated with determining the center of gravity of a plate or a wire: the **centroid** of an area or a line and the **first moment** of an area or a line with respect to a given axis. Computing the area of a surface of revolution or the volume of a body of revolution is directly related to determining the centroid of the line or area used to generate that surface or body of revolution (theorems of Pappus-Guldinus). Also, as we show in Sec. 5.3, the determination of the centroid of an area simplifies the analysis of beams subjected to distributed loads and the computation of the forces exerted on submerged rectangular surfaces, such as hydraulic gates and portions of dams.

In the last part of this chapter, you will see how to determine the center of gravity of a three-dimensional body as well as how to calculate the centroid of a volume and the first moments of that volume with respect to the coordinate planes.

5.1 PLANAR CENTERS OF GRAVITY AND CENTROIDS

In Chapter 4, we showed how the locations of the lines of action of forces affects the replacement of a system of forces with an equivalent system of forces and couples. In this section, we extend this idea to show how a distributed system of forces (in particular, the elements of an object's weight) can be replaced by a single resultant force acting at a specific point on an object. The specific point is called the object's center of gravity.

5.1A Center of Gravity of a Two-Dimensional Body

Let us first consider a flat horizontal plate (Fig. 5.1). We can divide the plate into n small elements. We denote the coordinates of the first element by x_1 and y_1, those of the second element by x_2 and y_2, etc. The forces exerted by the earth on the elements of the plate are denoted, respectively, by $\Delta \mathbf{W}_1, \Delta \mathbf{W}_2, \ldots, \Delta \mathbf{W}_n$. These forces or weights are directed toward the center of the earth; however, for all practical purposes, we can assume them to be parallel. Their resultant is therefore a single force in the same direction. The magnitude W of this force is obtained by adding the magnitudes of the elemental weights.

$$\Sigma F_z: \qquad W = \Delta W_1 + \Delta W_2 + \cdots + \Delta W_n$$

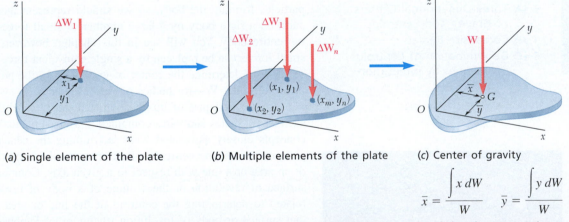

(a) Single element of the plate

(b) Multiple elements of the plate

(c) Center of gravity

$$\bar{x} = \frac{\int x \, dW}{W} \qquad \bar{y} = \frac{\int y \, dW}{W}$$

Fig. 5.1 The center of gravity of a plate is the point where the resultant weight of the plate acts. It is the weighted average of all the elements of weight that make up the plate.

To obtain the coordinates $\bar{x}$ and $\bar{y}$ of point G where the resultant $\mathbf{W}$ should be applied, we note that the moments of $\mathbf{W}$ about the y and x axes are equal to the sum of the corresponding moments of the elemental weights:

$$\Sigma M_y: \qquad \bar{x}W = x_1\Delta W_1 + x_2\Delta W_2 + \cdots + x_n\Delta W_n$$
$$\Sigma M_x: \qquad \bar{y}W = y_1\Delta W_1 + y_2\Delta W_2 + \cdots + y_n\Delta W_n$$

(5.1)

Solving these equations for $\bar{x}$ and $\bar{y}$ gives us

$$\bar{x} = \frac{x_1\Delta W_1 + x_2\Delta W_2 + \cdots + x_n\Delta W_n}{W}$$

$$\bar{y} = \frac{y_1\Delta W_1 + y_2\Delta W_2 + \cdots + y_n\Delta W_n}{W}$$

We could use these equations in this form to find the center of gravity of a collection of n objects, each with a weight of W_i.

If we now increase the number of elements into which we divide the plate and simultaneously decrease the size of each element, in the limit of infinitely many elements of infinitesimal size, we obtain the expressions

Weight, center of gravity of a flat plate

$$W = \int dW \qquad \bar{x}W = \int x\, dW \qquad \bar{y}W = \int y\, dW \qquad \textbf{(5.2)}$$

Or, solving for $\bar{x}$ and $\bar{y}$, we have

$$W = \int dW \qquad \bar{x} = \frac{\int x\, dW}{W} \qquad \bar{y} = \frac{\int y\, dW}{W} \qquad \textbf{(5.2′)}$$

These equations define the weight **W** and the coordinates $\bar{x}$ and $\bar{y}$ of the **center of gravity** G of a flat plate. The same equations can be derived for a wire lying in the xy plane (Fig. 5.2). Note that the center of gravity G of a wire is usually not located on the wire.

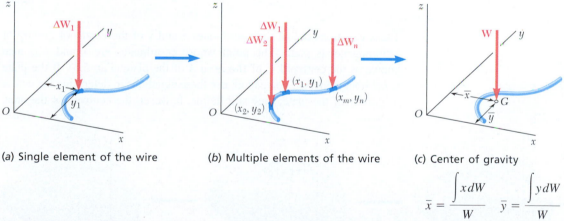

(a) Single element of the wire   (b) Multiple elements of the wire   (c) Center of gravity

$$\bar{x} = \frac{\int x\, dW}{W} \qquad \bar{y} = \frac{\int y\, dW}{W}$$

Fig. 5.2 The center of gravity of a wire is the point where the resultant weight of the wire acts. The center of gravity may not actually be located on the wire.

Photo 5.2 The center of gravity of a boomerang is not located on the object itself.

5.1B   Centroids of Areas and Lines

In the case of a flat homogeneous plate of uniform thickness, we can express the magnitude ΔW of the weight of an element of the plate as

$$\Delta W = \gamma t\, \Delta A$$

where γ = specific weight (weight per unit volume) of the material
$\quad t$ = thickness of the plate
$\quad \Delta A$ = area of the element

Similarly, we can express the magnitude W of the weight of the entire plate as

$$W = \gamma t A$$

where A is the total area of the plate.

If U.S. customary units are used, the specific weight γ should be expressed in lb/ft^3, the thickness t in feet, and the areas ΔA and A in square feet. Then ΔW and W are expressed in pounds. If SI units are used, γ should be expressed in N/m^3, t in meters, and the areas ΔA and A in square meters; the weights ΔW and W are then expressed in newtons.[†]

Substituting for ΔW and W in the moment equations (5.1) and dividing throughout by γt, we obtain

$$\Sigma M_y: \quad \bar{x}A = x_1\,\Delta A_1 + x_2\,\Delta A_2 + \cdots + x_n\,\Delta A_n$$
$$\Sigma M_x: \quad \bar{y}A = y_1\,\Delta A_1 + y_2\,\Delta A_2 + \cdots + y_n\,\Delta A_n$$

If we increase the number of elements into which the area A is divided and simultaneously decrease the size of each element, in the limit we obtain

Centroid of an area A

$$\bar{x}A = \int x\,dA \qquad \bar{y}A = \int y\,dA \tag{5.3}$$

Or, solving for $\bar{x}$ and $\bar{y}$, we obtain

$$\bar{x} = \frac{\int x\,dA}{A} \qquad \bar{y} = \frac{\int y\,dA}{A} \tag{5.3'}$$

These equations define the coordinates $\bar{x}$ and $\bar{y}$ of the center of gravity of a homogeneous plate. The point whose coordinates are $\bar{x}$ and $\bar{y}$ is also known as the **centroid C of the area A** of the plate (Fig. 5.3). If the plate is not homogeneous, you cannot use these equations to determine the center of gravity of the plate; they still define, however, the centroid of the area.

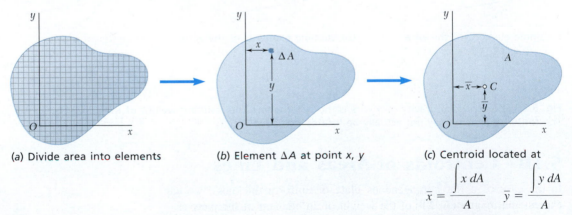

(a) Divide area into elements   (b) Element ΔA at point x, y (c) Centroid located at

$$\bar{x} = \frac{\int x\,dA}{A} \qquad \bar{y} = \frac{\int y\,dA}{A}$$

Fig. 5.3   The centroid of an area is the point where a homogeneous plate of uniform thickness would balance.

[†]We should note that in the SI system of units, a given material is generally characterized by its density ρ (mass per unit volume) rather than by its specific weight γ. You can obtain the specific weight of the material from the relation

$$\gamma = \rho g$$

where $g = 9.81$ m/s^2. Note that since ρ is expressed in kg/m^3, the units of γ are (kg/m^3)(m/s^2), or N/m^3.

In the case of a homogeneous wire of uniform cross section, we can express the magnitude ΔW of the weight of an element of wire as

$$\Delta W = \gamma a\, \Delta L$$

where γ = specific weight of the material
a = cross-sectional area of the wire
ΔL = length of the element

The center of gravity of the wire then coincides with the **centroid C of the line** L defining the shape of the wire (Fig. 5.4). We can obtain the coordinates $\bar{x}$ and $\bar{y}$ of the centroid of line L from the equations

Centroid of a line L

$$\bar{x}L = \int x\, dL \qquad \bar{y}L = \int y\, dL \qquad (5.4)$$

Solving for $\bar{x}$ and $\bar{y}$ gives us

$$\bar{x} = \frac{\int x\, dL}{L} \qquad \bar{y} = \frac{\int y\, dL}{L} \qquad (5.4')$$

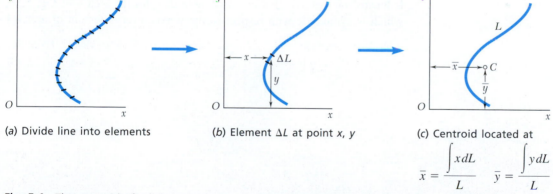

(a) Divide line into elements (b) Element ΔL at point x, y (c) Centroid located at

$$\bar{x} = \frac{\int x\, dL}{L} \qquad \bar{y} = \frac{\int y\, dL}{L}$$

Fig. 5.4 The centroid of a line is the point where a homogeneous wire of uniform cross section would balance.

5.1C First Moments of Areas and Lines

The integral $\int x\, dA$ in Eqs. (5.3) is known as the **first moment of the area A with respect to the y axis** and is denoted by Q_y. Similarly, the integral $\int y\, dA$ defines the **first moment of A with respect to the x axis** and is denoted by Q_x. That is,

First moments of area A

$$Q_y = \int x\, dA \qquad Q_x = \int y\, dA \qquad (5.5)$$

Comparing Eqs. (5.3) with Eqs. (5.5), we note that we can express the first moments of the area A as the products of the area and the coordinates of its centroid:

$$Q_y = \bar{x}A \qquad Q_x = \bar{y}A \qquad\qquad \textbf{(5.6)}$$

It follows from Eqs. (5.6) that we can obtain the coordinates of the centroid of an area by dividing the first moments of that area by the area itself. The first moments of the area are also useful in mechanics of materials for determining the shearing stresses in beams under transverse loadings. Finally, we observe from Eqs. (5.6) that, if the centroid of an area is located on a coordinate axis, the first moment of the area with respect to that axis is zero. Conversely, if the first moment of an area with respect to a coordinate axis is zero, the centroid of the area is located on that axis.

We can use equations similar to Eqs. (5.5) and (5.6) to define the first moments of a line with respect to the coordinate axes and to express these moments as the products of the length L of the line and the coordinates $\bar{x}$ and $\bar{y}$ of its centroid.

An area A is said to be **symmetric with respect to an axis** BB' if for every point P of the area there exists a point P' of the same area such that the line PP' is perpendicular to BB' and is divided into two equal parts by that axis (Fig. 5.5a). The axis BB' is called an **axis of symmetry**. A line L is said to be symmetric with respect to an axis BB' if it satisfies similar conditions. When an area A or a line L possesses an axis of symmetry BB', its first moment with respect to BB' is zero, and its centroid is located on that axis. For example, note that, for the area A of Fig. 5.5b, which is symmetric with respect to the y axis, every element of area dA

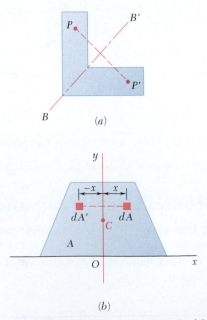

Fig. 5.5 Symmetry about an axis. (a) The area is symmetric about the axis BB'. (b) The centroid of the area is located on the axis of symmetry.

with abscissa x corresponds to an element dA' of equal area and with abscissa $-x$. It follows that the integral in the first of Eqs. (5.5) is zero and, thus, that $Q_y = 0$. It also follows from the first of the relations in Eq. (5.3) that $\bar{x} = 0$. Thus, if an area A or a line L possesses an axis of symmetry, its centroid C is located on that axis.

We further note that if an area or line possesses two axes of symmetry, its centroid C must be located at the intersection of the two axes (Fig. 5.6). This property enables us to determine immediately the centroids of areas such as circles, ellipses, squares, rectangles, equilateral triangles, or other symmetric figures, as well as the centroids of lines in the shape of the circumference of a circle, the perimeter of a square, etc.

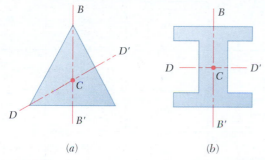

(a)   (b)

Fig. 5.6 If an area has two axes of symmetry, the centroid is located at their intersection. (a) An area with two axes of symmetry but no center of symmetry; (b) an area with two axes of symmetry and a center of symmetry.

We say that an area A is **symmetric with respect to a center** O if, for every element of area dA of coordinates x and y, there exists an element dA' of equal area with coordinates $-x$ and $-y$ (Fig. 5.7). It then follows that the integrals in Eqs. (5.5) are both zero and that $Q_x = Q_y = 0$. It also follows from Eqs. (5.3) that $\bar{x} = \bar{y} = 0$; that is, that the centroid of the area coincides with its center of symmetry O. Similarly, if a line possesses a center of symmetry O, the centroid of the line coincides with the center O.

Note that a figure possessing a center of symmetry does not necessarily possess an axis of symmetry (Fig. 5.7), whereas a figure possessing two axes of symmetry does not necessarily possess a center of symmetry (Fig. 5.6a). However, if a figure possesses two axes of symmetry at right angles to each other, the point of intersection of these axes is a center of symmetry (Fig. 5.6b).

Determining the centroids of unsymmetrical areas and lines and of areas and lines possessing only one axis of symmetry will be discussed in the next section. Centroids of common shapes of areas and lines are shown in Fig. 5.8A and B.

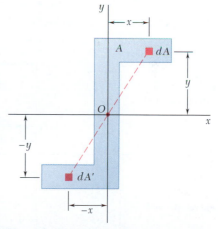

Fig. 5.7 An area may have a center of symmetry but no axis of symmetry.

5.1D Composite Plates and Wires

In many instances, we can divide a flat plate into rectangles, triangles, or the other common shapes shown in Fig. 5.8A. We can determine the abscissa $\overline{X}$ of the plate's center of gravity G from the abscissas $\bar{x}_1, \bar{x}_2, \ldots, \bar{x}_n$ of the centers of gravity of the various parts. To do this, we equate the moment of the weight of the whole plate about the y axis to the sum of

Shape		$\overline{x}$	$\overline{y}$	Area
Triangular area	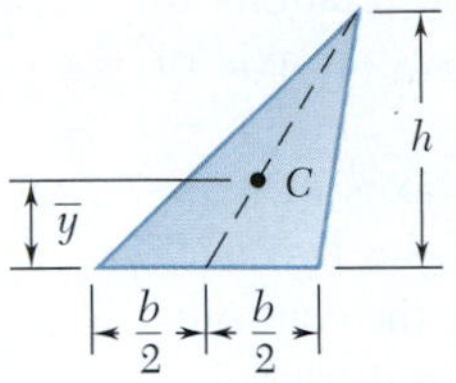		$\dfrac{h}{3}$	$\dfrac{bh}{2}$
Quarter-circular area	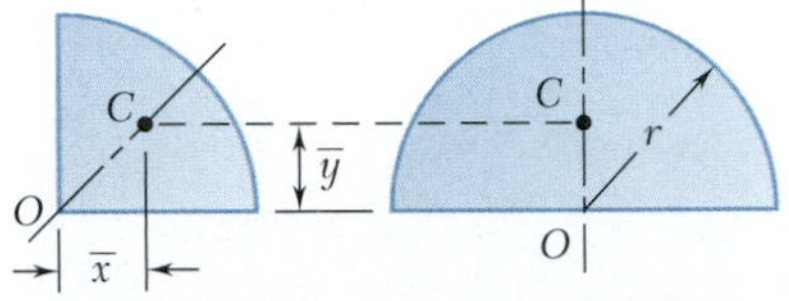	$\dfrac{4r}{3\pi}$	$\dfrac{4r}{3\pi}$	$\dfrac{\pi r^2}{4}$
Semicircular area		0	$\dfrac{4r}{3\pi}$	$\dfrac{\pi r^2}{2}$
Quarter-elliptical area	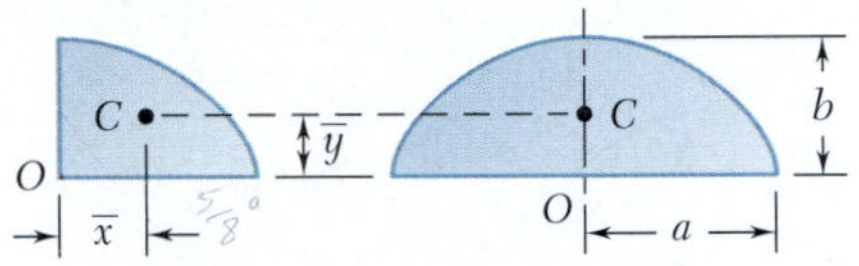	$\dfrac{4a}{3\pi}$	$\dfrac{4b}{3\pi}$	$\dfrac{\pi ab}{4}$
Semielliptical area		0	$\dfrac{4b}{3\pi}$	$\dfrac{\pi ab}{2}$
Semiparabolic area	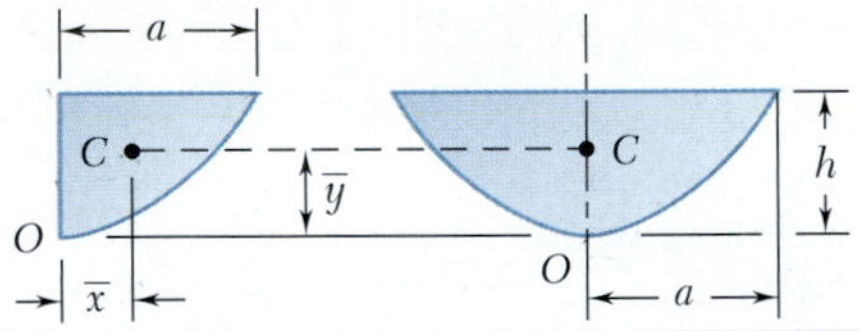	$\dfrac{3a}{8}$	$\dfrac{3h}{5}$	$\dfrac{2ah}{3}$
Parabolic area		0	$\dfrac{3h}{5}$	$\dfrac{4ah}{3}$
Parabolic spandrel	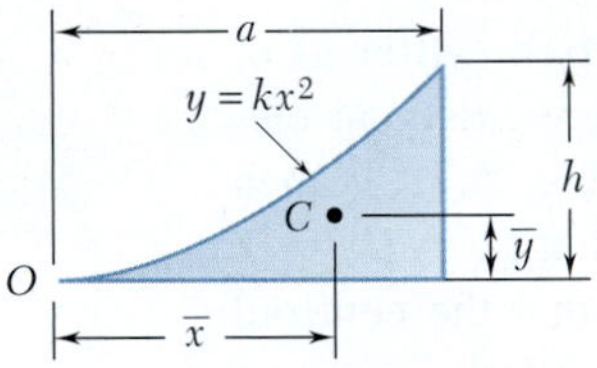	$\dfrac{3a}{4}$	$\dfrac{3h}{10}$	$\dfrac{ah}{3}$
General spandrel	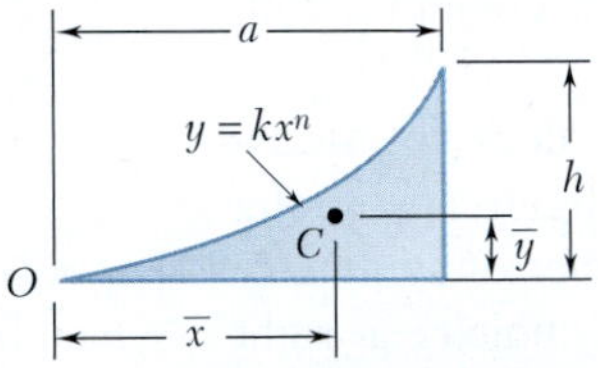	$\dfrac{n+1}{n+2}a$	$\dfrac{n+1}{4n+2}h$	$\dfrac{ah}{n+1}$
Circular sector	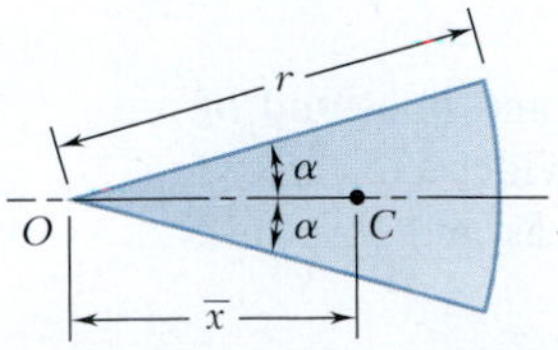	$\dfrac{2r\sin\alpha}{3\alpha}$	0	αr^2

Fig. 5.8A Centroids of common shapes of areas.

Shape		$\bar{x}$	$\bar{y}$	Length
Quarter-circular arc	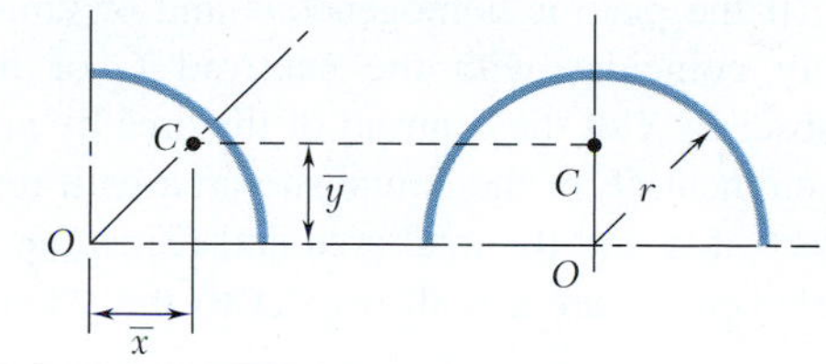 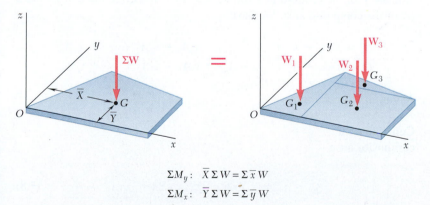	$\dfrac{2r}{\pi}$	$\dfrac{2r}{\pi}$	$\dfrac{\pi r}{2}$
Semicircular arc		0	$\dfrac{2r}{\pi}$	πr
Arc of circle	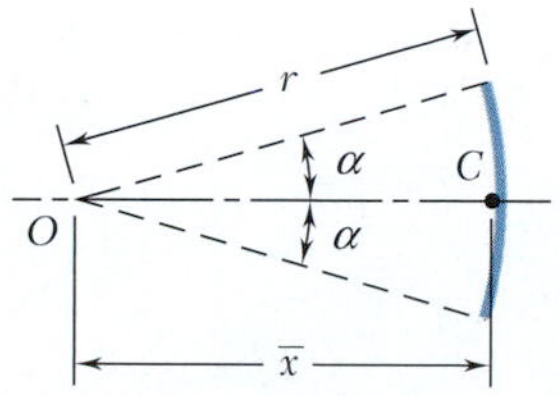	$\dfrac{r\sin\alpha}{\alpha}$	0	$2\alpha r$

Fig. 5.8B Centroids of common shapes of lines.

the moments of the weights of the various parts about the same axis (Fig. 5.9). We can obtain the ordinate $\bar{Y}$ of the center of gravity of the plate in a similar way by equating moments about the x axis. Mathematically, we have

$$\Sigma M_y: \quad \bar{X}(W_1 + W_2 + \cdots + W_n) = \bar{x}_1 W_1 + \bar{x}_2 W_2 + \cdots + \bar{x}_n W_n$$
$$\Sigma M_x: \quad \bar{Y}(W_1 + W_2 + \cdots + W_n) = \bar{y}_1 W_1 + \bar{y}_2 W_2 + \cdots + \bar{y}_n W_n$$

$$\Sigma M_y: \quad \bar{X}\Sigma W = \Sigma \bar{x}\, W$$
$$\Sigma M_x: \quad \bar{Y}\Sigma W = \Sigma \bar{y}\, W$$

Fig. 5.9. We can determine the location of the center of gravity G of a composite plate from the centers of gravity G_1, G_2, ... of the component plates.

In more condensed notation, this is

**Center of gravity
of a composite plate**

$$\bar{X} = \frac{\Sigma \bar{x} W}{W} \qquad \bar{Y} = \frac{\Sigma \bar{y} W}{W} \tag{5.7}$$

We can use these equations to find the coordinates $\overline{X}$ and $\overline{Y}$ of the center of gravity of the plate from the centers of gravity of its component parts.

If the plate is homogeneous and of uniform thickness, the center of gravity coincides with the centroid C of its area. We can determine the abscissa $\overline{X}$ of the centroid of the area by noting that we can express the first moment Q_y of the composite area with respect to the y axis as (1) the product of $\overline{X}$ and the total area and (2) as the sum of the first moments of the elementary areas with respect to the y axis (Fig. 5.10). We obtain the

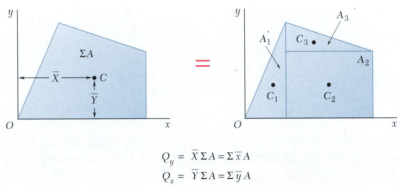

$$Q_y = \overline{X}\Sigma A = \Sigma\overline{x}A$$
$$Q_x = \overline{Y}\Sigma A = \Sigma\overline{y}A$$

Fig. 5.10 We can find the location of the centroid of a composite area from the centroids of the component areas.

ordinate $\overline{Y}$ of the centroid in a similar way by considering the first moment Q_x of the composite area. We have

$$Q_y = \overline{X}(A_1 + A_2 + \cdots + A_n) = \overline{x}_1A_1 + \overline{x}_2A_2 + \cdots + \overline{x}_nA_n$$
$$Q_x = \overline{Y}(A_1 + A_2 + \cdots + A_n) = \overline{y}_1A_1 + \overline{y}_2A_2 + \cdots + \overline{y}_nA_n$$

Again, in shorter form,

Centroid of a composite area

$$Q_y = \overline{X}\Sigma A = \Sigma\overline{x}A \qquad Q_x = \overline{Y}\Sigma A = \Sigma\overline{y}A \qquad (5.8)$$

These equations yield the first moments of the composite area, or we can use them to obtain the coordinates $\overline{X}$ and $\overline{Y}$ of its centroid.

First moments of areas, like moments of forces, can be positive or negative. Thus, you need to take care to assign the appropriate sign to the moment of each area. For example, an area whose centroid is located to the left of the y axis has a negative first moment with respect to that axis. Also, the area of a hole should be assigned a negative sign (Fig. 5.11).

Similarly, it is possible in many cases to determine the center of gravity of a composite wire or the centroid of a composite line by dividing the wire or line into simpler elements (see Sample Prob. 5.2).

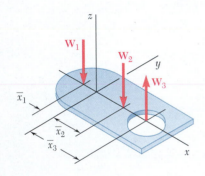

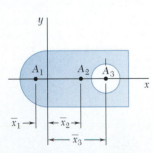

	$\overline{x}$	A	$\overline{x}A$
A_1 Semicircle	−	+	−
A_2 Full rectangle	+	+	+
A_3 Circular hole	+	−	−

Fig. 5.11 When calculating the centroid of a composite area, note that if the centroid of a component area has a negative coordinate distance relative to the origin, or if the area represents a hole, then the first moment is negative.

Sample Problem 5.1

For the plane area shown, determine (*a*) the first moments with respect to the *x* and *y* axes; (*b*) the location of the centroid.

STRATEGY: Break up the given area into simple components, find the centroid of each component, and then find the overall first moments and centroid.

MODELING: As shown in Fig. 1, you obtain the given area by adding a rectangle, a triangle, and a semicircle and then subtracting a circle. Using the coordinate axes shown, find the area and the coordinates of the centroid of each of the component areas. To keep track of the data, enter them in a table. The area of the circle is indicated as negative because it is subtracted from the other areas. The coordinate $\bar{y}$ of the centroid of the triangle is negative for the axes shown. Compute the first moments of the component areas with respect to the coordinate axes and enter them in your table.

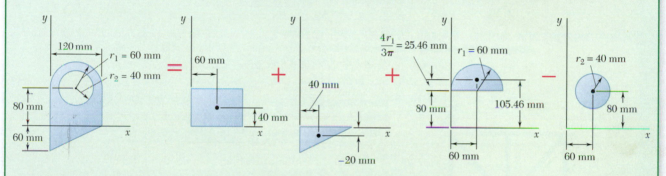

Component	A, mm^2	$\bar{x}$, mm	$\bar{y}$, mm	$\bar{x}A$, mm^3	$\bar{y}A$, mm^3
Rectangle	$(120)(80) = 9.6 \times 10^3$	60	40	$+576 \times 10^3$	$+384 \times 10^3$
Triangle	$\frac{1}{2}(120)(60) = 3.6 \times 10^3$	40	-20	$+144 \times 10^3$	-72×10^3
Semicircle	$\frac{1}{2}\pi(60)^2 = 5.655 \times 10^3$	60	105.46	$+339.3 \times 10^3$	$+596.4 \times 10^3$
Circle	$-\pi(40)^2 = -5.027 \times 10^3$	60	80	-301.6×10^3	-402.2×10^3
	$\Sigma A = 13.828 \times 10^3$			$\Sigma\bar{x}A = +757.7 \times 10^3$	$\Sigma\bar{y}A = +506.2 \times 10^3$

Fig. 1 Given area modeled as the combination of simple geometric shapes.

ANALYSIS:

a. First Moments of the Area. Using Eqs. (5.8), you obtain

$$Q_x = \Sigma\bar{y}A = 506.2 \times 10^3 \text{ mm}^3 \qquad Q_x = 506 \times 10^3 \text{ mm}^3 \blacktriangleleft$$
$$Q_y = \Sigma\bar{x}A = 757.7 \times 10^3 \text{ mm}^3 \qquad Q_y = 758 \times 10^3 \text{ mm}^3 \blacktriangleleft$$

b. Location of Centroid. Substituting the values given in the table into the equations defining the centroid of a composite area yields (Fig. 2)

$$\bar{X}\Sigma A = \Sigma\bar{x}A: \quad \bar{X}(13.828 \times 10^3 \text{ mm}^2) = 757.7 \times 10^3 \text{ mm}^3$$
$$\bar{X} = 54.8 \text{ mm} \blacktriangleleft$$

$$\bar{Y}\Sigma A = \Sigma\bar{y}A: \quad \bar{Y}(13.828 \times 10^3 \text{ mm}^2) = 506.2 \times 10^3 \text{ mm}^3$$
$$\bar{Y} = 36.6 \text{ mm} \blacktriangleleft$$

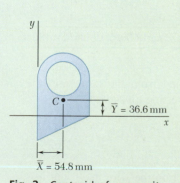

Fig. 2 Centroid of composite area.

REFLECT and THINK: Given that the lower portion of the shape has more area to the left and that the upper portion has a hole, the location of the centroid seems reasonable upon visual inspection.

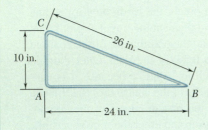

Sample Problem 5.2

The figure shown is made from a piece of thin, homogeneous wire. Determine the location of its center of gravity.

STRATEGY: Since the figure is formed of homogeneous wire, its center of gravity coincides with the centroid of the corresponding line. Therefore, you can simply determine that centroid.

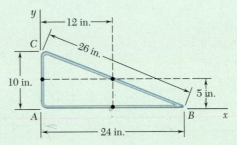

Fig. 1 Location of each line segment's centroid.

MODELING: Choosing the coordinate axes shown in Fig. 1 with the origin at *A*, determine the coordinates of the centroid of each line segment and compute the first moments with respect to the coordinate axes. You may find it convenient to list the data in a table.

Segment	L, in.	$\bar{x}$, in.	$\bar{y}$, in.	$\bar{x}L$, in^2	$\bar{y}L$, in^2
AB	24	12	0	288	0
BC	26	12	5	312	130
CA	10	0	5	0	50
	$\Sigma L = 60$			$\Sigma \bar{x}L = 600$	$\Sigma \bar{y}L = 180$

ANALYSIS: Substituting the values obtained from the table into the equations defining the centroid of a composite line gives

$\bar{X}\Sigma L = \Sigma \bar{x}L$: $\bar{X}(60 \text{ in.}) = 600 \text{ in}^2$ $\bar{X} = 10$ in. ◄

$\bar{Y}\Sigma L = \Sigma \bar{y}L$: $\bar{Y}(60 \text{ in.}) = 180 \text{ in}^2$ $\bar{Y} = 3$ in. ◄

REFLECT and THINK: The centroid is not on the wire itself, but it is within the area enclosed by the wire.

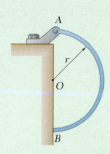

Sample Problem 5.3

A uniform semicircular rod of weight W and radius r is attached to a pin at A and rests against a frictionless surface at B. Determine the reactions at A and B.

STRATEGY: The key to solving the problem is finding where the weight W of the rod acts. Since the rod is a simple geometrical shape, you can look in Fig. 5.8 for the location of the wire's centroid.

MODELING: Draw a free-body diagram of the rod (Fig. 1). The forces acting on the rod are its weight $\mathbf{W}$, which is applied at the center of gravity G (whose position is obtained from Fig. 5.8B); a reaction at A, represented by its components $\mathbf{A}_x$ and $\mathbf{A}_y$; and a horizontal reaction at B.

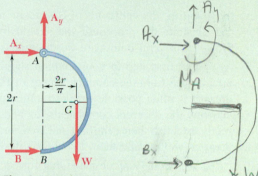

Fig. 1   Free-body diagram of rod.

ANALYSIS:

$+\gamma \; \Sigma M_A = 0$: $\qquad B(2r) - W\left(\dfrac{2r}{\pi}\right) = 0$

$$B = +\frac{W}{\pi} \qquad\qquad \boxed{\mathbf{B} = \frac{W}{\pi} \rightarrow}$$

$\xrightarrow{+} \Sigma F_x = 0$: $\qquad A_x + B = 0$

$$A_x = -B = -\frac{W}{\pi} \qquad \mathbf{A}_x = \frac{W}{\pi} \leftarrow$$

$+\uparrow \Sigma F_y = 0$: $\qquad A_y - W = 0 \qquad \boxed{\mathbf{A}_y = W \uparrow}$

Adding the two components of the reaction at A (Fig. 2), we have

$$A = \left[W^2 + \left(\frac{W}{\pi}\right)^2 \right]^{1/2} \qquad\qquad \boxed{A = W\left(1 + \frac{1}{\pi^2}\right)^{1/2}}$$

$$\tan \alpha = \frac{W}{W/\pi} = \pi \qquad\qquad \boxed{\alpha = \tan^{-1}\pi}$$

The answers can also be expressed as

$$\boxed{\mathbf{A} = 1.049W \;\angle\; 72.3° \qquad \mathbf{B} = 0.318W \rightarrow}$$

Fig. 2   Reaction at A.

REFLECT and THINK: Once you know the location of the rod's center of gravity, the problem is a straightforward application of the concepts in Chapter 4.

SOLVING PROBLEMS
ON YOUR OWN

In this section, we developed the general equations for locating the centers of gravity of two-dimensional bodies and wires [Eqs. (5.2)] and the centroids of plane areas [Eqs. (5.3)] and lines [Eqs. (5.4)]. In the following problems, you will have to locate the centroids of composite areas and lines or determine the first moments of the area for composite plates [Eqs. (5.8)].

1. Locating the centroids of composite areas and lines. Sample Problems 5.1 and 5.2 illustrate the procedure you should follow when solving problems of this type. However, several points are worth emphasizing.

   a. The first step in your solution should be to decide how to construct the given area or line from the common shapes of Fig. 5.8. You should recognize that for plane areas it is often possible to construct a particular shape in more than one way. Also, showing the different components (as is done in Sample Prob. 5.1) can help you correctly establish their centroids and areas or lengths. Do not forget that you can subtract areas as well as add them to obtain a desired shape.

   b. We strongly recommend that for each problem you construct a table listing the areas or lengths and the respective coordinates of the centroids. Remember, any areas that are "removed" (such as holes) are treated as negative. Also, the sign of negative coordinates must be included. Therefore, you should always carefully note the location of the origin of the coordinate axes.

   c. When possible, use symmetry [Sec. 5.1C] to help you determine the location of a centroid.

   d. In the formulas for the circular sector and for the arc of a circle in Fig. 5.8, the angle α must always be expressed in radians.

2. Calculating the first moments of an area. The procedures for locating the centroid of an area and for determining the first moments of an area are similar; however, it is not necessary to compute the total area for finding first moments. Also, as noted in Sec. 5.1C, you should recognize that the first moment of an area relative to a centroidal axis is zero.

3. Solving problems involving the center of gravity. The bodies considered in the following problems are homogeneous; thus, their centers of gravity and centroids coincide. In addition, when a body that is suspended from a single pin is in equilibrium, the pin and the body's center of gravity must lie on the same vertical line.

It may appear that many of the problems in this section have little to do with the study of mechanics. However, being able to locate the centroid of composite shapes will be essential in several topics that you will study later in this course.

Problems

5.1 through 5.9 Locate the centroid of the plane area shown.

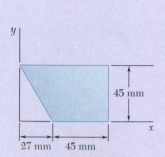

Fig. P5.1

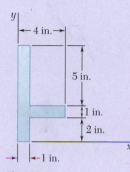

Fig. P5.2

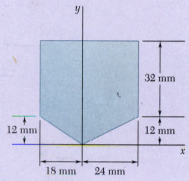

Fig. P5.3

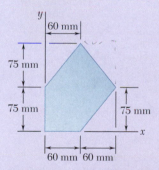

Fig. P5.4

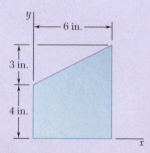

Fig. P5.5

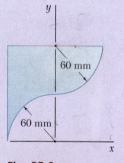

Fig. P5.6

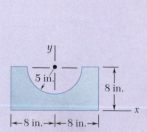

Fig. P5.7

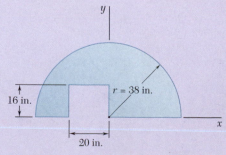

Fig. P5.8

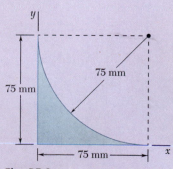

Fig. P5.9

5.10 through 5.15 Locate the centroid of the plane area shown.

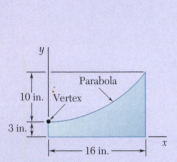

Fig. P5.10

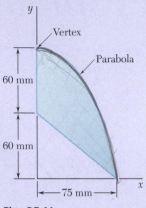

Fig. P5.11

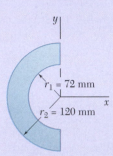

Fig. *P5.12*

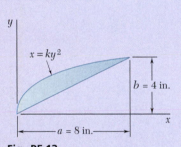

Fig. P5.13

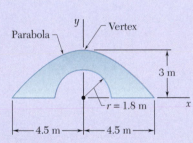

Fig. P5.14

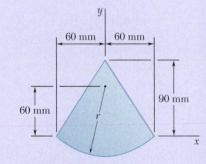

Fig. *P5.15*

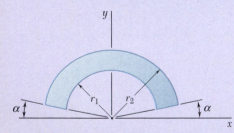

Fig. P5.16 and Fig. P5.17

5.16 Determine the y coordinate of the centroid of the shaded area in terms of r_1, r_2, and α.

5.17 Show that as r_1 approaches r_2, the location of the centroid approaches that for an arc of circle of radius $(r_1 + r_2)/2$.

5.18 Determine the x coordinate of the centroid of the trapezoid shown in terms of h_1, h_2, and a.

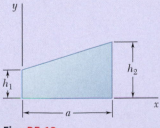

Fig. *P5.18*

5.19 For the semiannular area of Prob. 5.12, determine the ratio r_1 to r_2 so that the centroid of the area is located at $x = -\frac{1}{2}r_2$ and $y = 0$.

5.20 A composite beam is constructed by bolting four plates to four 60 × 60 × 12-mm angles as shown. The bolts are equally spaced along the beam, and the beam supports a vertical load. As proved in mechanics of materials, the shearing forces exerted on the bolts at A and B are proportional to the first moments with respect to the centroidal x axis of the red shaded areas shown, respectively, in parts a and b of the figure. Knowing that the force exerted on the bolt at A is 280 N, determine the force exerted on the bolt at B.

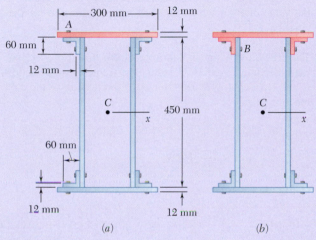

Fig. P5.20

5.21 and 5.22 The horizontal x axis is drawn through the centroid C of the area shown, and it divides the area into two component areas A_1 and A_2. Determine the first moment of each component area with respect to the x axis, and explain the results obtained.

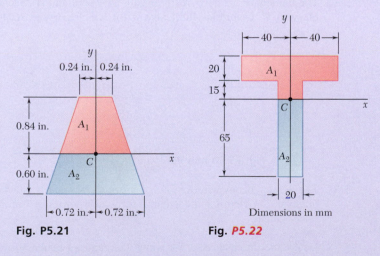

Fig. P5.21   **Fig. P5.22**

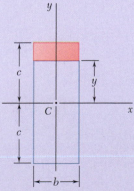

5.23 The first moment of the shaded area with respect to the x axis is denoted by Q_x. (a) Express Q_x in terms of b, c, and the distance y from the base of the shaded area to the x axis. (b) For what value of y is Q_x maximum, and what is that maximum value?

Fig. P5.23

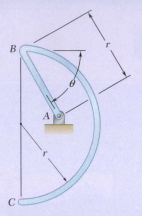

Fig. P5.28

5.24 through 5.27 A thin, homogeneous wire is bent to form the perimeter of the figure indicated. Locate the center of gravity of the wire figure thus formed.
5.24 Fig. P5.1.
5.25 Fig. P5.3.
5.26 Fig. P5.5.
5.27 Fig. P5.8.

5.28 The homogeneous wire *ABC* is bent into a semicircular arc and a straight section as shown and is attached to a hinge at *A*. Determine the value of θ for which the wire is in equilibrium for the indicated position.

5.29 The frame for a sign is fabricated from thin, flat steel bar stock of mass per unit length 4.73 kg/m. The frame is supported by a pin at *C* and by a cable *AB*. Determine (*a*) the tension in the cable, (*b*) the reaction at *C*.

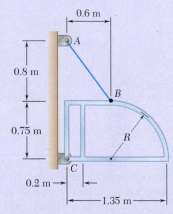

Fig. P5.29

5.30 The homogeneous wire *ABCD* is bent as shown and is attached to a hinge at *C*. Determine the length *L* for which portion *BCD* of the wire is horizontal.

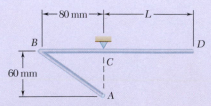

Fig. P5.30 and P5.31

5.31 The homogeneous wire *ABCD* is bent as shown and is attached to a hinge at *C*. Determine the length *L* for which portion *AB* of the wire is horizontal.

5.32 Determine the distance *h* for which the centroid of the shaded area is as far above line *BB'* as possible when (*a*) $k = 0.10$, (*b*) $k = 0.80$.

5.33 Knowing that the distance *h* has been selected to maximize the distance $\bar{y}$ from line *BB'* to the centroid of the shaded area, show that $\bar{y} = 2h/3$.

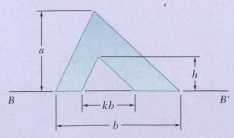

Fig. P5.32 and P5.33

5.2 FURTHER CONSIDERATIONS OF CENTROIDS

The objects we analyzed in Sec. 5.1 were composites of basic geometric shapes like rectangles, triangles, and circles. The same idea of locating a center of gravity or centroid applies for an object with a more complicated shape, but the mathematical techniques for finding the location are a little more difficult.

5.2A Determination of Centroids by Integration

For an area bounded by analytical curves (i.e., curves defined by algebraic equations), we usually determine the centroid by evaluating the integrals in Eqs. (5.3′):

$$\bar{x} = \frac{\displaystyle\int x\,dA}{A} \qquad \bar{y} = \frac{\displaystyle\int y\,dA}{A} \qquad\qquad (5.3')$$

If the element of area dA is a small rectangle of sides dx and dy, evaluating each of these integrals requires a *double integration* with respect to x and y. A double integration is also necessary if we use polar coordinates for which dA is a small element with sides dr and $r\,d\theta$.

In most cases, however, it is possible to determine the coordinates of the centroid of an area by performing a single integration. We can achieve this by choosing dA to be a thin rectangle or strip, or it can be a thin sector or pie-shaped element (Fig. 5.12). The centroid of the thin rectangle is located at its center, and the centroid of the thin sector is located at a distance $(2/3)r$ from its vertex (as it is for a triangle). Then we obtain the coordinates of the centroid of the area under consideration

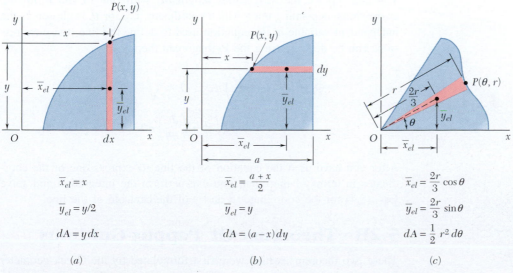

$$\bar{x}_{el} = x$$
$$\bar{y}_{el} = y/2$$
$$dA = y\,dx$$

(a)

$$\bar{x}_{el} = \frac{a+x}{2}$$
$$\bar{y}_{el} = y$$
$$dA = (a-x)\,dy$$

(b)

$$\bar{x}_{el} = \frac{2r}{3}\cos\theta$$
$$\bar{y}_{el} = \frac{2r}{3}\sin\theta$$
$$dA = \frac{1}{2}r^2\,d\theta$$

(c)

Fig. 5.12 Centroids and areas of differential elements. (a) Vertical rectangular strip; (b) horizontal rectangular strip; (c) triangular sector.

by setting the first moment of the entire area with respect to each of the coordinate axes equal to the sum (or integral) of the corresponding moments of the elements of the area. Denoting the coordinates of the centroid of the element dA by $\bar{x}_{el}$ and $\bar{y}_{el}$, we have

First moments of area

$$Q_y = \bar{x}A = \int \bar{x}_{el}\, dA$$

$$Q_x = \bar{y}A = \int \bar{y}_{el}\, dA$$

(5.9)

If we do not already know the area A, we can also compute it from these elements.

In order to carry out the integration, we need to express the coordinates $\bar{x}_{el}$ and $\bar{y}_{el}$ of the centroid of the element of area dA in terms of the coordinates of a point located on the curve bounding the area under consideration. Also, we should express the area of the element dA in terms of the coordinates of that point and the appropriate differentials. This has been done in Fig. 5.12 for three common types of elements; the pie-shaped element of part (c) should be used when the equation of the curve bounding the area is given in polar coordinates. You can substitute the appropriate expressions into formulas (5.9), and then use the equation of the bounding curve to express one of the coordinates in terms of the other. This process reduces the double integration to a single integration. Once you have determined the area and evaluated the integrals in Eqs. (5.9), you can solve these equations for the coordinates $\bar{x}$ and $\bar{y}$ of the centroid of the area.

When a line is defined by an algebraic equation, you can determine its centroid by evaluating the integrals in Eqs. (5.4′):

$$\bar{x} = \frac{\int x\, dL}{L} \qquad \bar{y} = \frac{\int y\, dL}{L}$$

(5.4′)

You can replace the differential length dL with one of the following expressions, depending upon which coordinate, x, y, or θ, is chosen as the independent variable in the equation used to define the line (these expressions can be derived using the Pythagorean theorem):

$$dL = \sqrt{1 + \left(\frac{dy}{dx}\right)^2}\, dx \qquad dL = \sqrt{1 + \left(\frac{dx}{dy}\right)^2}\, dy$$

$$dL = \sqrt{r^2 + \left(\frac{dr}{d\theta}\right)^2}\, d\theta$$

After you have used the equation of the line to express one of the coordinates in terms of the other, you can perform the integration and solve Eqs. (5.4) for the coordinates $\bar{x}$ and $\bar{y}$ of the centroid of the line.

5.2B Theorems Of Pappus-Guldinus

These two theorems, which were first formulated by the Greek geometer Pappus during the third century C.E. and later restated by the Swiss mathematician Guldinus or Guldin (1577–1643), deal with surfaces and bodies

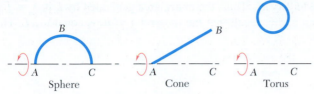

Fig. 5.13. Rotating plane curves about an axis generates surfaces of revolution.

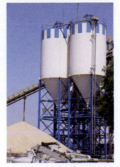

Photo 5.3 The storage tanks shown are bodies of revolution. Thus, their surface areas and volumes can be determined using the theorems of Pappus-Guldinus.

of revolution. A **surface of revolution** is a surface that can be generated by rotating a plane curve about a fixed axis. For example, we can obtain the surface of a sphere by rotating a semicircular arc *ABC* about the diameter *AC* (Fig. 5.13). Similarly, rotating a straight line *AB* about an axis *AC* produces the surface of a cone, and rotating the circumference of a circle about a nonintersecting axis generates the surface of a torus or ring. A **body of revolution** is a body that can be generated by rotating a plane area about a fixed axis. As shown in Fig. 5.14, we can generate a sphere, a cone, and a torus by rotating the appropriate shape about the indicated axis.

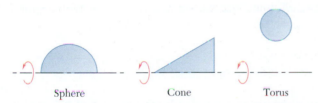

Fig. 5.14. Rotating plane areas about an axis generates volumes of revolution.

Theorem I. *The area of a surface of revolution is equal to the length of the generating curve times the distance traveled by the centroid of the curve while the surface is being generated.*

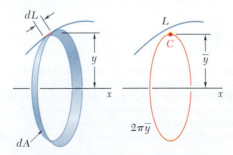

Fig. 5.15 An element of length *dL* rotated about the *x* axis generates a circular strip of area *dA*. The area of the entire surface of revolution equals the length of the line *L* multiplied by the distance traveled by the centroid *C* of the line during one revolution.

Proof. Consider an element *dL* of the line *L* (Fig. 5.15) that is revolved about the *x* axis. The circular strip generated by the element *dL* has an area

dA equal to $2\pi y\, dL$. Thus, the entire area generated by L is $A = \int 2\pi y\, dL$. Recall our earlier result that the integral $\int y\, dL$ is equal to $\bar{y}L$. Therefore, we have

$$A = 2\pi\bar{y}L \qquad\qquad (5.10)$$

Here $2\pi\bar{y}$ is the distance traveled by the centroid C of L (Fig. 5.15). $\square$

Note that the generating curve must not cross the axis about which it is rotated; if it did, the two sections on either side of the axis would generate areas having opposite signs, and the theorem would not apply.

Theorem II.   *The volume of a body of revolution is equal to the generating area times the distance traveled by the centroid of the area while the body is being generated.*

Proof.   Consider an element dA of the area A that is revolved about the x axis (Fig. 5.16). The circular ring generated by the element dA has a volume dV equal to $2\pi y\, dA$. Thus, the entire volume generated by A is $V = \int 2\pi y\, dA$, and since we showed earlier that the integral $\int y\, dA$ is equal to $\bar{y}A$, we have

$$V = 2\pi\bar{y}A \qquad\qquad (5.11)$$

Here $2\pi\bar{y}$ is the distance traveled by the centroid of A. $\square$

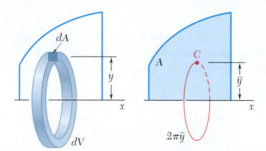

Fig. 5.16   An element of area dA rotated about the x axis generates a circular ring of volume dV. The volume of the entire body of revolution equals the area of the region A multiplied by the distance traveled by the centroid C of the region during one revolution.

Again, note that the theorem does not apply if the axis of rotation intersects the generating area.

The theorems of Pappus-Guldinus offer a simple way to compute the areas of surfaces of revolution and the volumes of bodies of revolution. Conversely, they also can be used to determine the centroid of a plane curve if you know the area of the surface generated by the curve or to determine the centroid of a plane area if you know the volume of the body generated by the area (see Sample Prob. 5.8).

Sample Problem 5.4

Determine the location of the centroid of a parabolic spandrel by direct integration.

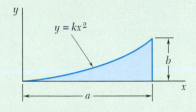

STRATEGY: First express the parabolic curve using the parameters a and b. Then choose a differential element of area and express its area in terms of a, b, x, and y. We illustrate the solution first with a vertical element and then a horizontal element.

MODELING:

Determination of the Constant k. Determine the value of k by substituting $x = a$ and $y = b$ into the given equation. We have $b = ka^2$ or $k = b/a^2$. The equation of the curve is thus

$$y = \frac{b}{a^2}x^2 \qquad \text{or} \qquad x = \frac{a}{b^{1/2}}y^{1/2}$$

ANALYSIS:

Vertical Differential Element. Choosing the differential element shown in Fig. 1, the total area of the region is

$$A = \int dA = \int y\, dx = \int_0^a \frac{b}{a^2}x^2\, dx = \left[\frac{b}{a^2}\frac{x^3}{3}\right]_0^a = \frac{ab}{3}$$

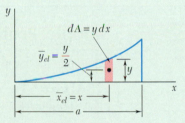

Fig. 1 Vertical differential element used to determine centroid.

The first moment of the differential element with respect to the y axis is $\bar{x}_{el}\, dA$; hence, the first moment of the entire area with respect to this axis is

$$Q_y = \int \bar{x}_{el}\, dA = \int xy\, dx = \int_0^a x\left(\frac{b}{a^2}x^2\right)dx = \left[\frac{b}{a^2}\frac{x^4}{4}\right]_0^a = \frac{a^2 b}{4}$$

Since $Q_y = \bar{x}A$, you have

$$\bar{x}A = \int \bar{x}_{el}\, dA \qquad \bar{x}\frac{ab}{3} = \frac{a^2 b}{4} \qquad \bar{x} = \tfrac{3}{4}a \quad \blacktriangleleft$$

Likewise, the first moment of the differential element with respect to the x axis is $\bar{y}_{el}\, dA$, so the first moment of the entire area about the x axis is

$$Q_x = \int \bar{y}_{el}\, dA = \int \frac{y}{2}y\, dx = \int_0^a \frac{1}{2}\left(\frac{b}{a^2}x^2\right)^2 dx = \left[\frac{b^2}{2a^4}\frac{x^5}{5}\right]_0^a = \frac{ab^2}{10}$$

Since $Q_x = \bar{y}A$, you get

$$\bar{y}A = \int \bar{y}_{el}\, dA \qquad \bar{y}\frac{ab}{3} = \frac{ab^2}{10} \qquad \bar{y} = \tfrac{3}{10}b \quad \blacktriangleleft$$

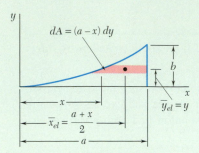

Fig. 2 Horizontal differential element used to determine centroid.

Horizontal Differential Element. You obtain the same results by considering a horizontal element (Fig. 2). The first moments of the area are

$$Q_y = \int \bar{x}_{el} \, dA = \int \frac{a+x}{2}(a-x) \, dy = \int_0^b \frac{a^2-x^2}{2} \, dy$$

$$= \frac{1}{2}\int_0^b \left(a^2 - \frac{a^2}{b}y\right) dy = \frac{a^2 b}{4}$$

$$Q_x = \int \bar{y}_{el} \, dA = \int y(a-x) \, dy = \int y\left(a - \frac{a}{b^{1/2}}y^{1/2}\right) dy$$

$$= \int_0^b \left(ay - \frac{a}{b^{1/2}}y^{3/2}\right) dy = \frac{ab^2}{10}$$

To determine $\bar{x}$ and $\bar{y}$, again substitute these expressions into the equations defining the centroid of the area.

REFLECT and THINK: You obtain the same results whether you choose a vertical or a horizontal element of area, as you should. You can use both methods as a check against making a mistake in your calculations.

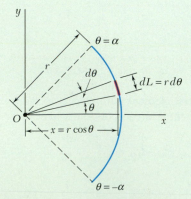

Fig. 1 Differential element used to determine centroid.

Sample Problem 5.5

Determine the location of the centroid of the circular arc shown.

STRATEGY: For a simple figure with circular geometry, you should use polar coordinates.

MODELING: The arc is symmetrical with respect to the x axis, so $\bar{y} = 0$. Choose a differential element, as shown in Fig. 1.

ANALYSIS: Determine the length of the arc by integration.

$$L = \int dL = \int_{-\alpha}^{\alpha} r \, d\theta = r \int_{-\alpha}^{\alpha} d\theta = 2r\alpha$$

The first moment of the arc with respect to the y axis is

$$Q_y = \int x \, dL = \int_{-\alpha}^{\alpha} (r\cos\theta)(r\,d\theta) = r^2 \int_{-\alpha}^{\alpha} \cos\theta \, d\theta$$

$$= r^2 [\sin\theta]_{-\alpha}^{\alpha} = 2r^2 \sin\alpha$$

Since $Q_y = \bar{x}L$, you obtain

$$\bar{x}(2r\alpha) = 2r^2\sin\alpha \qquad \bar{x} = \frac{r\sin\alpha}{\alpha} \quad \blacktriangleleft$$

REFLECT and THINK: Observe that this result matches that given for this case in Fig. 5.8B.

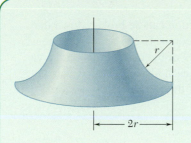

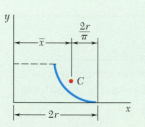

Fig. 1 Centroid location of arc.

Sample Problem 5.6

Determine the area of the surface of revolution shown that is obtained by rotating a quarter-circular arc about a vertical axis.

STRATEGY: According to the first Pappus-Guldinus theorem, the area of the surface of revolution is equal to the product of the length of the arc and the distance traveled by its centroid.

MODELING and ANALYSIS: Referring to Fig. 5.8B and Fig. 1, you have

$$\bar{x} = 2r - \frac{2r}{\pi} = 2r\left(1 - \frac{1}{\pi}\right)$$

$$A = 2\pi\bar{x}L = 2\pi\left[2r\left(1 - \frac{1}{\pi}\right)\right]\left(\frac{\pi r}{2}\right)$$

$$A = 2\pi r^2(\pi - 1) \blacktriangleleft$$

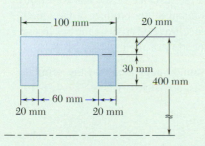

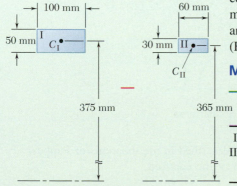

Fig. 1 Modeling the given area by subtracting area II from area I.

Sample Problem 5.7

The outside diameter of a pulley is 0.8 m, and the cross section of its rim is as shown. Knowing that the pulley is made of steel and that the density of steel is $\rho = 7.85 \times 10^3$ kg/m³, determine the mass and weight of the rim.

STRATEGY: You can determine the volume of the rim by applying the second Pappus-Guldinus theorem, which states that the volume equals the product of the given cross-sectional area and the distance traveled by its centroid in one complete revolution. However, you can find the volume more easily by observing that the cross section can be formed from rectangle I with a positive area and from rectangle II with a negative area (Fig. 1).

MODELING: Use a table to keep track of the data, as you did in Sec. 5.1.

	Area, mm²	$\bar{y}$, mm	Distance Traveled by C, mm	Volume, mm³
I	+5000	375	$2\pi(375) = 2356$	$(5000)(2356) = 11.78 \times 10^6$
II	−1800	365	$2\pi(365) = 2293$	$(-1800)(2293) = -4.13 \times 10^6$
				Volume of rim $= 7.65 \times 10^6$

ANALYSIS: Since 1 mm $= 10^{-3}$ m, you have 1 mm$^{-3} = (10^{-3}$ m$)^3 = 10^{-9}$ m³. Thus you obtain $V = 7.65 \times 10^6$ mm³ $= (7.65 \times 10^6)(10^{-9}$ m³$) = 7.65 \times 10^{-3}$ m³.

$$m = \rho V = (7.85 \times 10^3 \text{ kg/m}^3)(7.65 \times 10^{-3} \text{ m}^3) \qquad m = 60.0 \text{ kg} \blacktriangleleft$$

$$W = mg = (60.0 \text{ kg})(9.81 \text{ m/s}^2) = 589 \text{ kg·m/s}^2 \qquad W = 589 \text{ N} \blacktriangleleft$$

REFLECT and THINK: When a cross section can be broken down into multiple common shapes, you can apply Theorem II of Pappus–Guldinus in a manner that involves finding the products of the centroid ($\bar{y}$) and area (A), or the first moments of area ($\bar{y}A$), for each shape. Thus, it was not necessary to find the centroid or the area of the overall cross section.

Sample Problem 5.8

Using the theorems of Pappus-Guldinus, determine (*a*) the centroid of a semicircular area and (*b*) the centroid of a semicircular arc. Recall that the volume and the surface area of a sphere are $\frac{4}{3}\pi r^3$ and $4\pi r^2$, respectively.

STRATEGY: The volume of a sphere is equal to the product of the area of a semicircle and the distance traveled by the centroid of the semicircle in one revolution about the *x* axis. Given the volume, you can determine the distance traveled by the centroid and thus the distance of the centroid from the axis. Similarly, the area of a sphere is equal to the product of the length of the generating semicircle and the distance traveled by its centroid in one revolution. You can use this to find the location of the centroid of the arc.

MODELING: Draw diagrams of the semicircular area and the semicircular arc (Fig. 1) and label the important geometries.

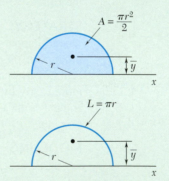

Fig. 1 Semicircular area and semicircular arc.

ANALYSIS: Set up the equalities described in the theorems of Pappus-Guldinus and solve for the location of the centroid.

$$V = 2\pi\bar{y}A \qquad \frac{4}{3}\pi r^3 = 2\pi\bar{y}\left(\frac{1}{2}\pi r^2\right) \qquad \bar{y} = \frac{4r}{3\pi} \ \blacktriangleleft$$

$$A = 2\pi\bar{y}L \qquad 4\pi r^2 = 2\pi\bar{y}(\pi r) \qquad \bar{y} = \frac{2r}{\pi} \ \blacktriangleleft$$

REFLECT and THINK: Observe that this result matches those given for these cases in Fig. 5.8.

SOLVING PROBLEMS
ON YOUR OWN

In the problems for this section, you will use the equations

$$\bar{x} = \frac{\int x\,dA}{A} \qquad \bar{y} = \frac{\int y\,dA}{A} \tag{5.3'}$$

$$\bar{x} = \frac{\int x\,dL}{L} \qquad \bar{y} = \frac{\int y\,dL}{L} \tag{5.4'}$$

to locate the centroids of plane areas and lines, respectively. You will also apply the theorems of Pappus-Guldinus to determine the areas of surfaces of revolution and the volumes of bodies of revolution.

1. Determining the centroids of areas and lines by direct integration. When solving problems of this type, you should follow the method of solution shown in Sample Probs. 5.4 and 5.5. To compute A or L, determine the first moments of the area or the line, and solve Eqs. (5.3) or (5.4) for the coordinates of the centroid. In addition, you should pay particular attention to the following points.

 a. Begin your solution by carefully defining or determining each term in the applicable integral formulas. We strongly encourage you to show on your sketch of the given area or line your choice for dA or dL and the distances to its centroid.

 b. As explained in Sec. 5.2A, x and y in Eqs. (5.3) and (5.4) represent the *coordinates of the centroid* of the differential elements dA and dL. It is important to recognize that the coordinates of the centroid of dA are not equal to the coordinates of a point located on the curve bounding the area under consideration. You should carefully study Fig. 5.12 until you fully understand this important point.

 c. To possibly simplify or minimize your computations, always examine the shape of the given area or line before defining the differential element that you will use. For example, sometimes it may be preferable to use horizontal rectangular elements instead of vertical ones. Also, it is usually advantageous to use polar coordinates when a line or an area has circular symmetry.

 d. Although most of the integrations in this section are straightforward, at times it may be necessary to use more advanced techniques, such as trigonometric substitution or integration by parts. Using a table of integrals is often the fastest method to evaluate difficult integrals.

2. Applying the theorems of Pappus-Guldinus. As shown in Sample Probs. 5.6 through 5.8, these simple, yet very useful theorems allow you to apply your knowledge of centroids to the computation of areas and volumes. Although the theorems refer to the distance traveled by the centroid and to the length of the generating curve or to the generating area, the resulting equations [Eqs. (5.10) and (5.11)] contain the products of these quantities, which are simply the first moments of a line ($\bar{y}L$) and an area ($\bar{y}A$), respectively. Thus, for those problems for which the generating line or area consists of more than one common shape, you need only determine $\bar{y}L$ or $\bar{y}A$; you do not have to calculate the length of the generating curve or the generating area.

Problems

5.34 through *5.36* Determine by direct integration the centroid of the area shown. Express your answer in terms of a and h.

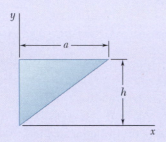

Fig. P5.34

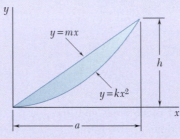

Fig. P5.35

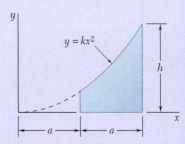

Fig. *P5.36*

5.37 through 5.39 Determine by direct integration the centroid of the area shown.

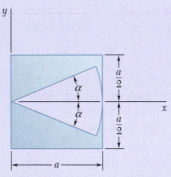

Fig. P5.37

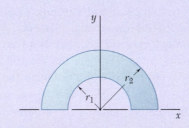

Fig. *P5.38*

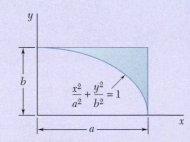

Fig. P5.39

5.40 and 5.41 Determine by direct integration the centroid of the area shown. Express your answer in terms of a and b.

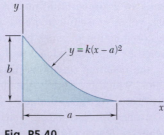

Fig. P5.40

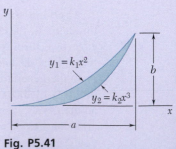

Fig. P5.41

5.42 Determine by direct integration the centroid of the area shown.

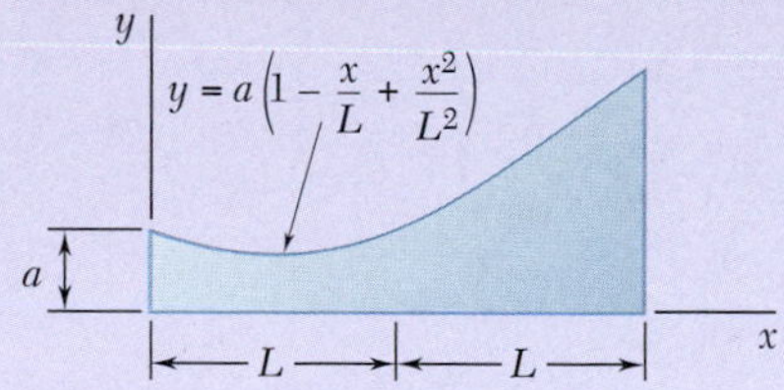

5.43 and 5.44 Determine by direct integration the centroid of the area shown. Express your answer in terms of a and b.

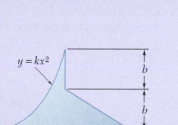

Fig. P5.42

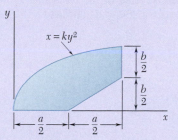

Fig. P5.43

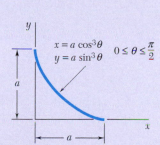

Fig. P5.44

5.45 and 5.46 A homogeneous wire is bent into the shape shown. Determine by direct integration the x coordinate of its centroid.

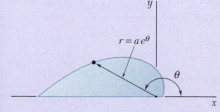

Fig. P5.45

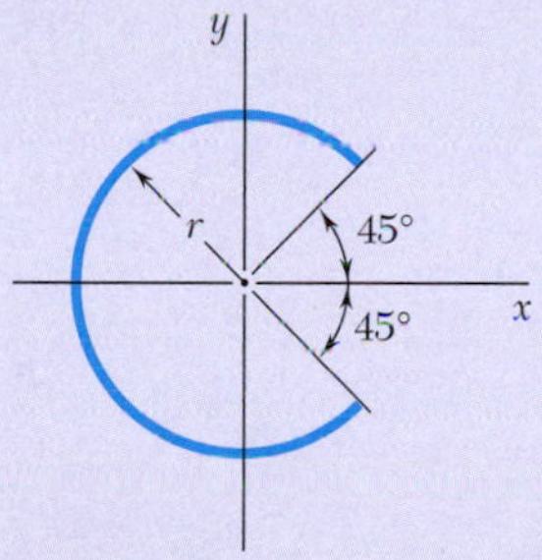

Fig. P5.46

*5.47 A homogeneous wire is bent into the shape shown. Determine by direct integration the x coordinate of its centroid. Express your answer in terms of a.

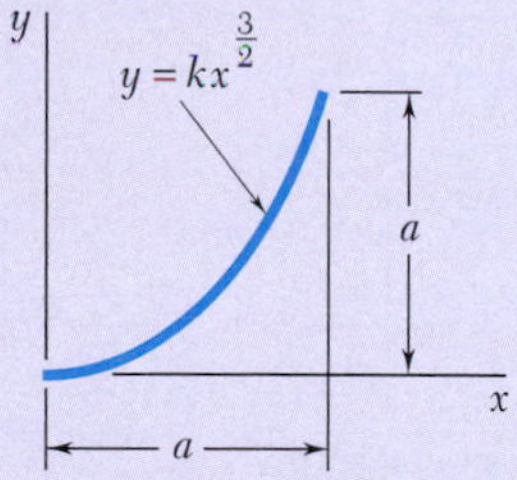

*5.48 and *5.49 Determine by direct integration the centroid of the area shown.

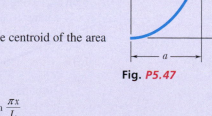

Fig. P5.47

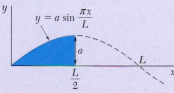

Fig. P5.48

Fig. P5.49

5.50 Determine the centroid of the area shown in terms of a.

5.51 Determine the centroid of the area shown when $a = 4$ in.

5.52 Determine the volume and the surface area of the solid obtained by rotating the area of Prob. 5.1 about (a) the x axis, (b) the line $x = 72$ mm.

5.53 Determine the volume and the surface area of the solid obtained by rotating the area of Prob. 5.2 about (a) the x axis, (b) the y axis.

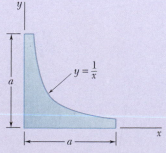

Fig. P5.50 and P5.51

5.54 Determine the volume and the surface area of the solid obtained by rotating the area of Prob. 5.6 about (*a*) the line *x* = –60 mm, (*b*) the line *y* = 120 mm.

5.55 Determine the volume and the surface area of the chain link shown, which is made from a 6-mm-diameter bar, if $R = 10$ mm and $L = 30$ mm.

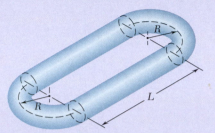
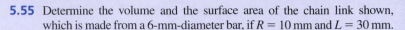

Fig. P5.55

5.56 Determine the volume of the solid generated by rotating the parabolic area shown about (*a*) the *x* axis, (*b*) the axis *AA′*.

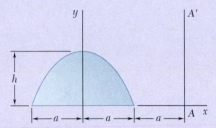

Fig. *P5.56*

5.57 Verify that the expressions for the volumes of the first four shapes in Fig. 5.21 are correct.

5.58 Knowing that two equal caps have been removed from a 10-in.-diameter wooden sphere, determine the total surface area of the remaining portion.

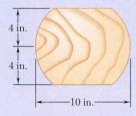

Fig. P5.58

5.59 Three different drive belt profiles are to be studied. If at any given time each belt makes contact with one-half of the circumference of its pulley, determine the *contact area* between the belt and the pulley for each design.

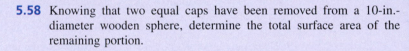
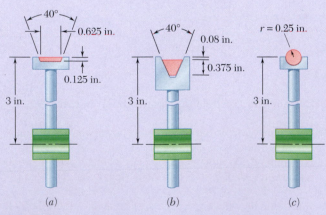

(*a*)   (*b*)   (*c*)

Fig. *P5.59*

5.60 Determine the capacity, in liters, of the punch bowl shown if $R = 250$ mm.

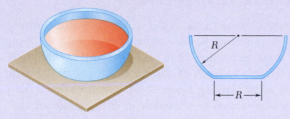

Fig. P5.60

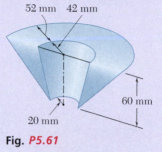

Fig. P5.61

5.61 Determine the volume and total surface area of the bushing shown.

5.62 Determine the volume and weight of the solid brass knob shown, knowing that the specific weight of brass is 0.306 lb/in^3.

5.63 Determine the total surface area of the solid brass knob shown.

5.64 The aluminum shade for the small high-intensity lamp shown has a uniform thickness of 1 mm. Knowing that the density of aluminum is 2800 kg/m^3, determine the mass of the shade.

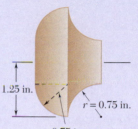

Fig. P5.62 and P5.63

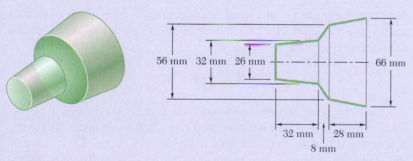

Fig. P5.64

***5.65** The shade for a wall-mounted light is formed from a thin sheet of translucent plastic. Determine the surface area of the outside of the shade, knowing that it has the parabolic cross section shown.

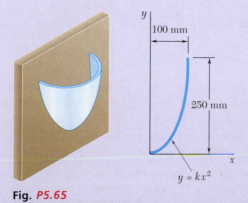

Fig. P5.65

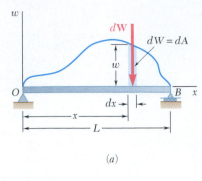

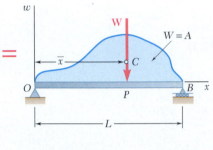

Fig. 5.17 (a) A load curve representing the distribution of load forces along a horizontal beam, with an element of length dx; (b) the resultant load W has magnitude equal to the area A under the load curve and acts through the centroid of the area.

Photo 5.4 The roof of a building shown must be able to support not only the total weight of the snow but also the nonsymmetric distributed loads resulting from drifting of the snow.

5.3 ADDITIONAL APPLICATIONS OF CENTROIDS

We can use the concept of the center of gravity or the centroid of an area to solve other problems besides those dealing with the weights of flat plates. The same techniques allow us to deal with other kinds of distributed loads on objects, such as the forces on a straight beam (a bridge girder or the main carrying beam of a house floor) or a flat plate under water (the side of a dam or a window in an aquarium tank).

5.3A Distributed Loads on Beams

Consider a beam supporting a **distributed load**; this load may consist of the weight of materials supported directly or indirectly by the beam, or it may be caused by wind or hydrostatic pressure. We can represent the distributed load by plotting the load w supported per unit length (Fig. 5.17); this load is expressed in N/m or in lb/ft. The magnitude of the force exerted on an element of the beam with length dx is $dW = w\, dx$, and the total load supported by the beam is

$$W = \int_0^L w\, dx$$

Note that the product $w\, dx$ is equal in magnitude to the element of area dA shown in Fig. 5.17a. The load W is thus equal in magnitude to the total area A under the load curve, as

$$W = \int dA = A$$

We now want to determine where a *single concentrated load* $\mathbf{W}$, of the same magnitude W as the total distributed load, should be applied on the beam if it is to produce the same reactions at the supports (Fig. 5.17b). However, this concentrated load $\mathbf{W}$, which represents the resultant of the given distributed loading, is equivalent to the loading only when considering the free-body diagram of the entire beam. We obtain the point of application P of the equivalent concentrated load $\mathbf{W}$ by setting the moment of $\mathbf{W}$ about point O equal to the sum of the moments of the elemental loads $d\mathbf{W}$ about O. Thus,

$$(OP)W = \int x\, dW$$

Then, since $dW = w\, dx = dA$ and $W = A$, we have

$$(OP)A = \int_0^L x\, dA \qquad\qquad \textbf{(5.12)}$$

Since this integral represents the first moment with respect to the w axis of the area under the load curve, we can replace it with the product $\bar{x}A$. We therefore have $OP = \bar{x}$, where $\bar{x}$ is the distance from the w axis to the centroid C of the area A (this is *not* the centroid of the beam).

We can summarize this result:

We can replace a distributed load on a beam by a concentrated load; the magnitude of this single load is equal to the area under the load curve, and its line of action passes through the centroid of that area.

Note, however, that the concentrated load is equivalent to the given loading only so far as external forces are concerned. It can be used to determine reactions, but should not be used to compute internal forces and deflections.

*5.3B   Forces on Submerged Surfaces

The approach used for distributed loads on beams works in other applications as well. Here, we use it to determine the resultant of the hydrostatic pressure forces exerted on a *rectangular surface* submerged in a liquid. We can use these methods to determine the resultant of the hydrostatic forces exerted on the surfaces of dams, rectangular gates, and vanes. In Chap. 9, we discuss the resultants of forces on submerged surfaces of variable width.

Consider a rectangular plate with a length of L and width of b, where b is measured perpendicular to the plane of the figure (Fig. 5.18). As for the case of distributed loads on a beam, the load exerted on an element of the plate with a length of dx is $w\,dx$, where w is the load per unit length and x is the distance along the length. However, this load also can be expressed as $p\,dA = pb\,dx$, where p is the gage pressure in the liquid[†] and b is the width of the plate; thus, $w = bp$. Since the gage pressure in a liquid is $p = \gamma h$, where γ is the specific weight of the liquid and h is the vertical distance from the free surface, it follows that

$$w = bp = b\gamma h \tag{5.13}$$

This equation shows that the load per unit length w is proportional to h and, thus, varies linearly with x.

From the results of Sec. 5.3A, the resultant **R** of the hydrostatic forces exerted on one side of the plate is equal in magnitude to the trapezoidal area under the load curve, and its line of action passes through the centroid C of that area. The point P of the plate where **R** is applied is known as the *center of pressure*.[‡]

Now consider the forces exerted by a liquid on a curved surface of constant width (Fig. 5.19a). Since determining the resultant **R** of these forces by direct integration would not be easy, we consider the free body obtained by detaching the volume of liquid ABD bounded by the curved surface AB and by the two plane surfaces AD and DB shown in Fig. 5.19b. The forces acting on the free body ABD are the weight **W** of the detached volume of liquid, the resultant $\mathbf{R}_1$ of the forces exerted on AD, the resultant $\mathbf{R}_2$ of the forces exerted on BD, and the resultant $-\mathbf{R}$ of the forces exerted *by the curved surface on the liquid*. The resultant $-\mathbf{R}$ is both equal and opposite to and has the same line of action as the resultant **R** of the forces exerted *by the liquid on the curved surface*. We can determine the forces **W**, $\mathbf{R}_1$, and $\mathbf{R}_2$ by standard methods. After their values have been found, we obtain the force $-\mathbf{R}$ by solving the equations of equilibrium for the free body of Fig. 5.19b. The resultant **R** of the hydrostatic forces exerted on the curved surface is just the reverse of $-\mathbf{R}$.

[†]The pressure p, which represents a load per unit area, is measured in N/m^2 or in lb/ft^2. The derived SI unit N/m^2 is called a pascal (Pa).

[‡]The area under the load curve is equal to $w_E L$, where w_E is the load per unit length at the center E of the plate. Then from Eq. (5.13), we have

$$R = w_E L = (bp_E)L = p_E(bL) = p_E A$$

where A denotes the area of the plate. Thus, we can obtain the magnitude of **R** by multiplying the area of the plate by the pressure at its center E. Note, however, that the resultant **R** *should be applied at P, not at E.*

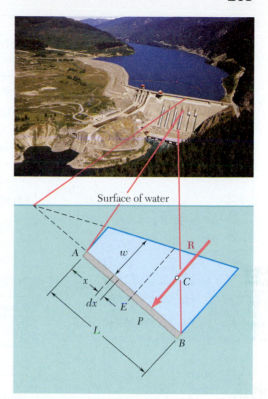

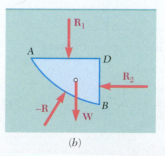

Fig. 5.18 The waterside face of a hydroelectric dam can be modeled as a rectangular plate submerged under water. Shown is a side view of the plate.

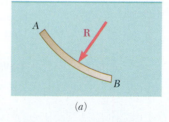

(a)

(b)

Fig. 5.19 (a) Force R exerted by a liquid on a submerged curved surface of constant width; (b) free-body diagram of the volume of liquid ABD.

Sample Problem 5.9

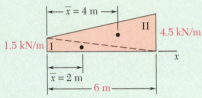

A beam supports a distributed load as shown. (*a*) Determine the equivalent concentrated load. (*b*) Determine the reactions at the supports.

STRATEGY: The magnitude of the resultant of the load is equal to the area under the load curve, and the line of action of the resultant passes through the centroid of the same area. Break down the area into pieces for easier calculation, and determine the resultant load. Then, use the calculated forces or their resultant to determine the reactions.

MODELING and ANALYSIS:

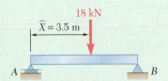

Fig. 1 The load modeled as two triangular areas.

a. *Equivalent Concentrated Load.* Divide the area under the load curve into two triangles (Fig. 1), and construct the table below. To simplify the computations and tabulation, the given loads per unit length have been converted into kN/m.

Component	A, kN	$\bar{x}$, m	$\bar{x}A$, kN·m
Triangle I	4.5	2	9
Triangle II	13.5	4	54
	$\Sigma A = 18.0$		$\Sigma \bar{x}A = 63$

Thus, $\bar{X}\Sigma A = \Sigma\bar{x}A$: $\bar{X}(18 \text{ kN}) = 63 \text{ kN·m}$ $\bar{X} = 3.5 \text{ m}$

The equivalent concentrated load (Fig. 2) is

$$\mathbf{W} = 18 \text{ kN} \downarrow \quad \blacktriangleleft$$

Its line of action is located at a distance

$$\bar{X} = 3.5 \text{ m to the right of } A \quad \blacktriangleleft$$

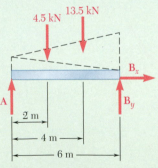

Fig. 2 Equivalent concentrated load.

b. *Reactions.* The reaction at *A* is vertical and is denoted by **A**. Represent the reaction at *B* by its components $\mathbf{B}_x$ and $\mathbf{B}_y$. Consider the given load to be the sum of two triangular loads (see the free-body diagram, Fig. 3). The resultant of each triangular load is equal to the area of the triangle and acts at its centroid.

Write the following equilibrium equations from the free-body diagram:

$\xrightarrow{+} \Sigma F_x = 0$: $\mathbf{B}_x = 0 \quad \blacktriangleleft$

$+\circlearrowleft \Sigma M_A = 0$: $-(4.5 \text{ kN})(2 \text{ m}) - (13.5 \text{ kN})(4 \text{ m}) + B_y(6 \text{ m}) = 0$

$$\mathbf{B}_y = 10.5 \text{ kN} \uparrow \quad \blacktriangleleft$$

$+\circlearrowleft \Sigma M_B = 0$: $+(4.5 \text{ kN})(4 \text{ m}) + (13.5 \text{ kN})(2 \text{ m}) - A(6 \text{ m}) = 0$

$$\mathbf{A} = 7.5 \text{ kN} \uparrow \quad \blacktriangleleft$$

Fig. 3 Free-body diagram of beam.

REFLECT and THINK: You can replace the given distributed load by its resultant, which you found in part *a*. Then you can determine the reactions from the equilibrium equations $\Sigma F_x = 0$, $\Sigma M_A = 0$, and $\Sigma M_B = 0$. Again the results are

$$\mathbf{B}_x = 0 \qquad \mathbf{B}_y = 10.5 \text{ kN} \uparrow \qquad \mathbf{A} = 7.5 \text{ kN} \uparrow \quad \blacktriangleleft$$

Sample Problem 5.10

The cross section of a concrete dam is shown. Consider a 1-ft-thick section of the dam, and determine (a) the resultant of the reaction forces exerted by the ground on the base AB of the dam, (b) the resultant of the pressure forces exerted by the water on the face BC of the dam. The specific weights of concrete and water are 150 lb/ft^3 and 62.4 lb/ft^3, respectively.

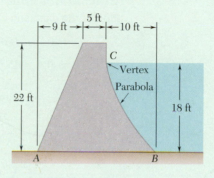

STRATEGY: Draw a free-body diagram of the section of the dam, breaking it into parts to simplify the calculations. Model the resultant of the reactions as a force-couple system at A. Use the method described in Sec. 5.3B to find the force exerted by the dam on the water and reverse it to find the force exerted by the water on face BC.

MODELING and ANALYSIS:

a. Ground Reaction. Choose as a free body the 1-ft-thick section $AEFCDB$ of the dam and water (Fig. 1). The reaction forces exerted by the ground on the base AB are represented by an equivalent force-couple system at A. Other forces acting on the free body are the weight of the dam represented by the weights of its components $\mathbf{W}_1$, $\mathbf{W}_2$, and $\mathbf{W}_3$; the weight of the water $\mathbf{W}_4$; and the resultant $\mathbf{P}$ of the pressure forces exerted on section BD by the water to the right of section BD.

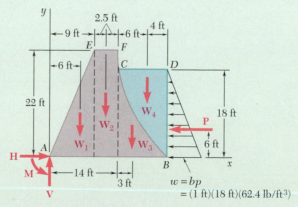

Fig. 1 Free-body diagram of dam and water.

Calculate each of the forces that appear in the free-body diagram, Fig. 1:

$$W_1 = \tfrac{1}{2}(9 \text{ ft})(22 \text{ ft})(1 \text{ ft})(150 \text{ lb/ft}^3) = 14,850 \text{ lb}$$
$$W_2 = (5 \text{ ft})(22 \text{ ft})(1 \text{ ft})(150 \text{ lb/ft}^3) = 16,500 \text{ lb}$$
$$W_3 = \tfrac{1}{3}(10 \text{ ft})(18 \text{ ft})(1 \text{ ft})(150 \text{ lb/ft}^3) = 9000 \text{ lb}$$
$$W_4 = \tfrac{2}{3}(10 \text{ ft})(18 \text{ ft})(1 \text{ ft})(62.4 \text{ lb/ft}^3) = 7488 \text{ lb}$$
$$P = \tfrac{1}{2}(18 \text{ ft})(1 \text{ ft})(18 \text{ ft})(62.4 \text{ lb/ft}^3) = 10,109 \text{ lb}$$

Equilibrium Equations. Write the equilibrium equations for the section of the dam, and calculate the forces and moment labeled at A in Fig. 1.

$\xrightarrow{+} \Sigma F_x = 0$: $H - 10,109 \text{ lb} = 0$ $\mathbf{H} = 10,110 \text{ lb} \rightarrow$ ◀

$+\uparrow \Sigma F_y = 0$: $V - 14,850 \text{ lb} - 16,500 \text{ lb} - 9000 \text{ lb} - 7488 \text{ lb} = 0$

$$\mathbf{V} = 47,840 \text{ lb} \uparrow ◀$$

$+\curvearrowleft \Sigma M_A = 0$: $-(14,850 \text{ lb})(6 \text{ ft}) - (16,500 \text{ lb})(11.5 \text{ ft})$
$- (9000 \text{ lb})(17 \text{ ft}) - (7488 \text{ lb})(20 \text{ ft}) + (10,109 \text{ lb})(6 \text{ ft}) + M = 0$

$$\mathbf{M} = 520,960 \text{ lb·ft} \curvearrowleft ◀$$

You can replace the force-couple system by a single force acting at a distance d to the right of A, where

$$d = \frac{520,960 \text{ lb·ft}}{47,840 \text{ lb}} = 10.89 \text{ ft}$$

b. Resultant R of Water Forces. Draw a free-body diagram for the parabolic section of water *BCD* (Fig. 2). The forces involved are the resultant $-\mathbf{R}$ of the forces exerted by the dam on the water, the weight $\mathbf{W}_4$, and the force $\mathbf{P}$. Since these forces must be concurrent, $-\mathbf{R}$ passes through the point of intersection G of $\mathbf{W}_4$ and $\mathbf{P}$. Draw a force triangle to determine the magnitude and direction of $-\mathbf{R}$. The resultant $\mathbf{R}$ of the forces exerted by the water on the face *BC* is equal and opposite. Hence,

$$\mathbf{R} = 12,580 \text{ lb} \nearrow 36.5° ◀$$

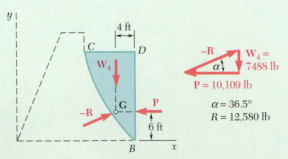

Fig. 2 Free-body diagram of parabolic section of water *BCD*.

REFLECT and THINK: Note that if you found the distance d to be negative—that is, if the moment reaction at A had been acting in the opposite direction—this would have indicated an instability condition of the dam. In this situation, the effects of the water pressure would overcome the weight of the dam, causing it to tip about A.

SOLVING PROBLEMS
ON YOUR OWN

The problems in this section involve two common and very important types of loading: distributed loads on beams and forces on submerged surfaces of constant width. As we discussed in Sec. 5.3 and illustrated in Sample Probs. 5.9 and 5.10, determining the single equivalent force for each of these loadings requires a knowledge of centroids.

1. Analyzing beams subjected to distributed loads. In Sec. 5.3A, we showed that a distributed load on a beam can be replaced by a single equivalent force. The magnitude of this force is equal to the area under the distributed load curve, and its line of action passes through the centroid of that area. Thus, you should begin solving this kind of problem by replacing the various distributed loads on a given beam by their respective single equivalent forces. You can then determine the reactions at the supports of the beam by using the methods of Chap. 4.

When possible, divide complex distributed loads into the common-shape areas shown in Fig. 5.8A (Sample Prob. 5.9). You can replace each of these areas under the loading curve by a single equivalent force. If required, you can further reduce the system of equivalent forces to a single equivalent force. As you study Sample Prob. 5.9, note how we used the analogy between force and area under the loading curve and applied the techniques for locating the centroid of a composite area to analyze a beam subjected to a distributed load.

2. Solving problems involving forces on submerged bodies. Remember the following points and techniques when solving problems of this type.

   a. The pressure p at a depth h below the free surface of a liquid is equal to γh or $\rho g h$, where γ and ρ are the specific weight and the density of the liquid, respectively. The load per unit length w acting on a submerged surface of constant width b is then

$$w = bp = b\gamma h = b\rho gh$$

   b. The line of action of the resultant force **R** acting on a submerged plane surface is perpendicular to the surface.

   c. For a vertical or inclined plane rectangular surface with a width of b, you can represent the loading on the surface using a linearly distributed load that is trapezoidal in shape (Fig. 5.18). The magnitude of the resultant **R** is given by

$$R = \gamma h_E A$$

where h_E is the vertical distance to the center of the surface and A is the area of the surface.

d. The load curve is triangular (rather than trapezoidal) when the top edge of a plane rectangular surface coincides with the free surface of the liquid, since the pressure of the liquid at the free surface is zero. For this case, it is straightforward to determine the line of action of **R**, because it passes through the centroid of a *triangular* distributed load.

e. For the general case, rather than analyzing a trapezoid, we suggest that you use the method indicated in part *b* of Sample Prob. 5.9. First divide the trapezoidal distributed load into two triangles, and then compute the magnitude of the resultant of each triangular load. (The magnitude is equal to the area of the triangle times the width of the plate.) Note that the line of action of each resultant force passes through the centroid of the corresponding triangle and that the sum of these forces is equivalent to **R**. Thus, rather than using **R**, you can use the two equivalent resultant forces whose points of application are easily calculated. You should use the equation given for *R* here in paragraph **c** when you need only the magnitude of **R**.

f. When the submerged surface of a constant width is curved, you can obtain the resultant force acting on the surface by considering the equilibrium of the volume of liquid bounded by the curved surface and by using horizontal and vertical planes (Fig. 5.19). Observe that the force $\mathbf{R}_1$ of Fig. 5.19 is equal to the weight of the liquid lying above the plane *AD*. The method of solution for problems involving curved surfaces is shown in part *b* of Sample Prob. 5.10.

In subsequent mechanics courses (in particular, mechanics of materials and fluid mechanics), you will have ample opportunity to use the ideas introduced in this section.

Problems

5.66 and 5.67 For the beam and loading shown, determine (a) the magnitude and location of the resultant of the distributed load, (b) the reactions at the beam supports.

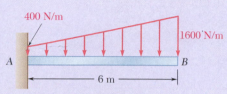

400 N/m
1600 N/m
A
B
6 m

Fig. P5.66

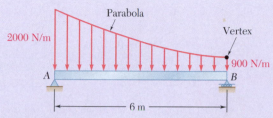

Parabola
Vertex
2000 N/m
900 N/m
A
B
6 m

Fig. P5.67

5.68 through 5.73 Determine the reactions at the beam supports for the given loading.

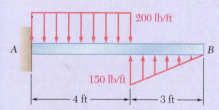

200 lb/ft
A
B
150 lb/ft
4 ft
3 ft

Fig. *P5.68*

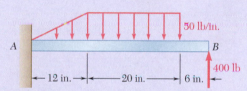

50 lb/in.
A
B
400 lb
12 in.
20 in.
6 in.

Fig. P5.69

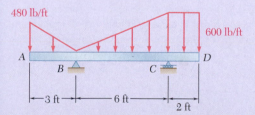

480 lb/ft
600 lb/ft
A
B
C
D
3 ft
6 ft
2 ft

Fig. P5.70

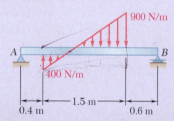

900 N/m
A
B
400 N/m
0.4 m
1.5 m
0.6 m

Fig. P5.71

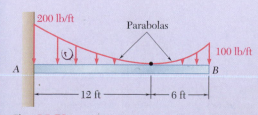

200 lb/ft
Parabolas
100 lb/ft
A
B
12 ft
6 ft

Fig. *P5.72*

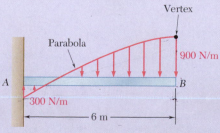

Vertex
Parabola
900 N/m
A
B
300 N/m
6 m

Fig. P5.73

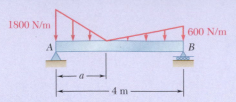

1800 N/m

600 N/m

A   B

|← a →|

|←——— 4 m ———→|

Fig. P5.74 and _P5.75_

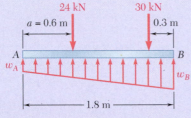

24 kN   30 kN

a = 0.6 m   0.3 m

A   B

w_A w_B

|←———— 1.8 m ————→|

Fig. P5.78

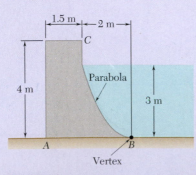

|1.5 m|← 2 m →|

C

Parabola

4 m

3 m

A   B

Vertex

Fig. P5.81

5.74 Determine (_a_) the distance _a_ so that the vertical reactions at supports _A_ and _B_ are equal, (_b_) the corresponding reactions at the supports.

5.75 Determine (_a_) the distance _a_ so that the reaction at support _B_ is minimum, (_b_) the corresponding reactions at the supports.

5.76 Determine the reactions at the beam supports for the given loading when $w_0 = 400$ lb/ft.

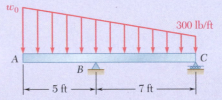

w_0

300 lb/ft

A

B   C

|←—— 5 ft ——→|←—— 7 ft ——→|

Fig. P5.76 and P5.77

5.77 Determine (_a_) the distributed load w_0 at the end _A_ of the beam _ABC_ for which the reaction at _C_ is zero, (_b_) the corresponding reaction at _B_.

5.78 The beam _AB_ supports two concentrated loads and rests on soil that exerts a linearly distributed upward load as shown. Determine the values of w_A and w_B corresponding to equilibrium.

5.79 For the beam and loading of Prob. 5.78, determine (_a_) the distance _a_ for which $w_A = 20$ kN/m, (_b_) the corresponding value of w_B.

In the following problems, use $\gamma = 62.4$ lb/ft^3 for the specific weight of fresh water and $\gamma_c = 150$ lb/ft^3 for the specific weight of concrete if U.S. customary units are used. With SI units, use $\rho = 10^3$ kg/m^3 for the density of fresh water and $\rho_c = 2.40 \times 10^3$ kg/m^3 for the density of concrete. (See the footnote on page 234 for how to determine the specific weight of a material given its density.)

5.80 The cross section of a concrete dam is as shown. For a 1-ft-wide dam section determine (_a_) the resultant of the reaction forces exerted by the ground on the base _AB_ of the dam, (_b_) the point of application of the resultant of part _a_, (_c_) the resultant of the pressure forces exerted by the water on the face _BC_ of the dam.

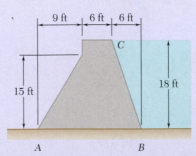

|← 9 ft →|← 6 ft →|← 6 ft →|

C

15 ft   18 ft

A   B

Fig. P5.80

5.81 The cross section of a concrete dam is as shown. For a 1-m-wide dam section determine (_a_) the resultant of the reaction forces exerted by the ground on the base _AB_ of the dam, (_b_) the point of application of the resultant of part _a_, (_c_) the resultant of the pressure forces exerted by the water on the face _BC_ of the dam.

5.82 The dam for a lake is designed to withstand the additional force caused by silt that has settled on the lake bottom. Assuming that silt is equivalent to a liquid of density $\rho_s = 1.76 \times 10^3$ kg/m³ and considering a 1-m-wide section of dam, determine the percentage increase in the force acting on the dam face for a silt accumulation of depth 2 m.

5.83 The base of a dam for a lake is designed to resist up to 120 percent of the horizontal force of the water. After construction, it is found that silt (that is equivalent to a liquid of density $\rho_s = 1.76 \times 10^3$ kg/m³) is settling on the lake bottom at the rate of 12 mm/year. Considering a 1-m-wide section of dam, determine the number of years of service until the dam becomes unsafe.

Fig. P5.82 and *P5.83*

5.84 An automatic valve consists of a 9 × 9-in. square plate that is pivoted about a horizontal axis through A located at a distance $h = 3.6$ in. above the lower edge. Determine the depth of water d for which the valve will open.

5.85 An automatic valve consists of a 9 × 9-in. square plate that is pivoted about a horizontal axis through A. If the valve is to open when the depth of water is $d = 18$ in., determine the distance h from the bottom of the valve to the pivot A.

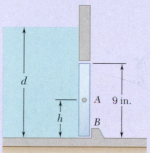

Fig. P5.84 and P5.85

5.86 The 3 × 4-m side AB of a tank is hinged at its bottom A and is held in place by a thin rod BC. The maximum tensile force the rod can withstand without breaking is 200 kN, and the design specifications require the force in the rod not to exceed 20 percent of this value. If the tank is slowly filled with water, determine the maximum allowable depth of water d in the tank.

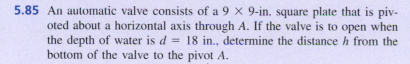

Fig. *P5.86* and P5.87

5.87 The 3 × 4-m side of an open tank is hinged at its bottom A and is held in place by a thin rod BC. The tank is to be filled with glycerine with a density of 1263 kg/m³. Determine the force $\mathbf{T}$ in the rod and the reaction at the hinge after the tank is filled to a depth of 2.9 m.

5.88 A 0.5 × 0.8-m gate AB is located at the bottom of a tank filled with water. The gate is hinged along its top edge A and rests on a frictionless stop at B. Determine the reactions at A and B when cable BCD is slack.

5.89 A 0.5 × 0.8-m gate AB is located at the bottom of a tank filled with water. The gate is hinged along its top edge A and rests on a frictionless stop at B. Determine the minimum tension required in cable BCD to open the gate.

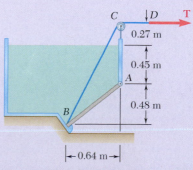

Fig. P5.88 and P5.89

271

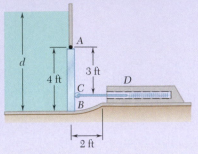

Fig. P5.90

5.90 A 4 × 2-ft gate is hinged at A and is held in position by rod CD. End D rests against a spring whose constant is 828 lb/ft. The spring is undeformed when the gate is vertical. Assuming that the force exerted by rod CD on the gate remains horizontal, determine the minimum depth of water d for which the bottom B of the gate will move to the end of the cylindrical portion of the floor.

5.91 Solve Prob. 5.90 if the gate weighs 1000 lb.

5.92 A prismatically shaped gate placed at the end of a freshwater channel is supported by a pin and bracket at A and rests on a frictionless support at B. The pin is located at a distance $h = 0.10$ m below the center of gravity C of the gate. Determine the depth of water d for which the gate will open.

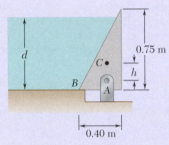

Fig. P5.92 and P5.93

5.93 A prismatically shaped gate placed at the end of a freshwater channel is supported by a pin and bracket at A and rests on a frictionless support at B. The pin is located at a distance h below the center of gravity C of the gate. Determine the distance h if the gate is to open when $d = 0.75$ m.

5.94 A long trough is supported by a continuous hinge along its lower edge and by a series of horizontal cables attached to its upper edge. Determine the tension in each of the cables at a time when the trough is completely full of water.

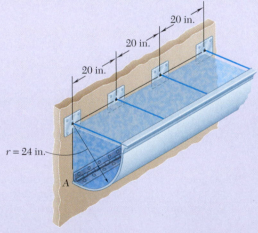

Fig. P5.94

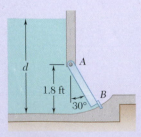

Fig. P5.95

5.95 The square gate AB is held in the position shown by hinges along its top edge A and by a shear pin at B. For a depth of water $d = 3.5$ ft, determine the force exerted on the gate by the shear pin.

5.4 CENTERS OF GRAVITY AND CENTROIDS OF VOLUMES

So far in this chapter, we have dealt with finding centers of gravity and centroids of two-dimensional areas and objects such as flat plates and plane surfaces. However, the same ideas apply to three-dimensional objects as well. The most general situations require the use of multiple integration for analysis, but we can often use symmetry considerations to simplify the calculations. In this section, we show how to do this.

5.4A Three-Dimensional Centers of Gravity and Centroids

Photo 5.5 To predict the flight characteristics of the modified Boeing 747 when used to transport a space shuttle, engineers had to determine the center of gravity of each craft.

For a three-dimensional body, we obtain the center of gravity G by dividing the body into small elements. The weight $\mathbf{W}$ of the body acting at G is equivalent to the system of distributed forces $\Delta\mathbf{W}$ representing the weights of the small elements. Choosing the y axis to be vertical with positive sense upward (Fig. 5.20) and denoting the position vector of G to be $\bar{\mathbf{r}}$, we set $\mathbf{W}$ equal to the sum of the elemental weights $\Delta\mathbf{W}$ and set its moment about O equal to the sum of the moments about O of the elemental weights. Thus,

$$\Sigma\mathbf{F}: \qquad -W\mathbf{j} = \Sigma(-\Delta W\mathbf{j})$$
$$\Sigma\mathbf{M}_O: \qquad \bar{\mathbf{r}} \times (-W\mathbf{j}) = \Sigma[\mathbf{r} \times (-\Delta W\mathbf{j})] \tag{5.14}$$

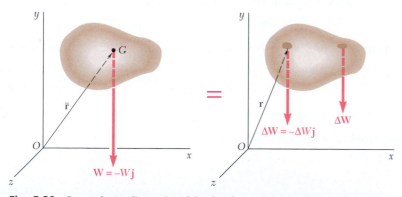

Fig. 5.20 For a three-dimensional body, the weight $\mathbf{W}$ acting through the center of gravity G and its moment about O is equivalent to the system of distributed weights acting on all the elements of the body and the sum of their moments about O.

We can rewrite the last equation in the form

$$\bar{\mathbf{r}}W \times (-\mathbf{j}) = (\Sigma\mathbf{r}\,\Delta W) \times (-\mathbf{j}) \tag{5.15}$$

From these equations, we can see that the weight $\mathbf{W}$ of the body is equivalent to the system of the elemental weights $\Delta\mathbf{W}$ if the following conditions are satisfied:

$$W = \Sigma\,\Delta W \qquad \bar{\mathbf{r}}W = \Sigma\mathbf{r}\,\Delta W$$

Increasing the number of elements and simultaneously decreasing the size of each element, we obtain in the limit as

**Weight, center of gravity
of a three-dimensional body**

$$W = \int dW \qquad \mathbf{\bar{r}} W = \int \mathbf{r}\, dW \tag{5.16}$$

Note that these relations are independent of the orientation of the body. For example, if the body and the coordinate axes were rotated so that the z axis pointed upward, the unit vector $-\mathbf{j}$ would be replaced by $-\mathbf{k}$ in Eqs. (5.14) and (5.15), but the relations in Eqs. (5.16) would remain unchanged.

Resolving the vectors $\mathbf{\bar{r}}$ and $\mathbf{r}$ into rectangular components, we note that the second of the relations in Eqs. (5.16) is equivalent to the three scalar equations

$$\bar{x} W = \int x\, dW \qquad \bar{y} W = \int y\, dW \qquad \bar{z} W = \int z\, dW \tag{5.17}$$

or

$$\bar{x} = \frac{\int x\, dW}{W} \qquad \bar{y} = \frac{\int y\, dW}{W} \qquad \bar{z} = \frac{\int z\, dW}{W} \tag{5.17'}$$

If the body is made of a homogeneous material of specific weight γ, we can express the magnitude dW of the weight of an infinitesimal element in terms of the volume dV of the element and express the magnitude W of the total weight in terms of the total volume V. We obtain

$$dW = \gamma\, dV \qquad W = \gamma V$$

Substituting for dW and W in the second of the relations in Eqs. (5.16), we have

$$\mathbf{\bar{r}} V = \int \mathbf{r}\, dV \tag{5.18}$$

In scalar form, this becomes

**Centroid of a
volume V**

$$\bar{x} V = \int x\, dV \qquad \bar{y} V = \int y\, dV \qquad \bar{z} V = \int z\, dV \tag{5.19}$$

or

$$\bar{x} = \frac{\int x\, dV}{V} \qquad \bar{y} = \frac{\int y\, dV}{V} \qquad \bar{z} = \frac{\int z\, dV}{V} \tag{5.19'}$$

The center of gravity of a homogeneous body whose coordinates are $\bar{x}, \bar{y}, \bar{z}$ is also known as the **centroid C of the volume** V of the body. If the body is not homogeneous, we cannot use Eqs. (5.19) to determine the center of gravity of the body; however, Eqs. (5.19) still define the centroid of the volume.

The integral $\int x\, dV$ is known as the **first moment of the volume with respect to the yz plane**. Similarly, the integrals $\int y\, dV$ and $\int z\, dV$ define the first moments of the volume with respect to the zx plane and the xy plane, respectively. You can see from Eqs. (5.19) that if the centroid of a volume is located in a coordinate plane, the first moment of the volume with respect to that plane is zero.

A volume is said to be symmetrical with respect to a given plane if, for every point P of the volume, there exists a point P' of the same volume such that the line PP' is perpendicular to the given plane and is bisected by that plane. We say the plane is a **plane of symmetry** for the given volume. When a volume V possesses a plane of symmetry, the first moment of V with respect to that plane is zero, and the centroid of the volume is located in the plane of symmetry. If a volume possesses two planes of symmetry, the centroid of the volume is located on the line of intersection of the two planes. Finally, if a volume possesses three planes of symmetry that intersect at a well-defined point (i.e., not along a common line), the point of intersection of the three planes coincides with the centroid of the volume. This property enables us to determine immediately the locations of the centroids of spheres, ellipsoids, cubes, rectangular parallelepipeds, etc.

For unsymmetrical volumes or volumes possessing only one or two planes of symmetry, we can determine the location of the centroid by integration (Sec. 5.4C). The centroids of several common volumes are shown in Fig. 5.21. Note that, in general, the centroid of a volume of revolution *does not coincide* with the centroid of its cross section. Thus, the centroid of a hemisphere is different from that of a semicircular area, and the centroid of a cone is different from that of a triangle.

5.4B Composite Bodies

If a body can be divided into several of the common shapes shown in Fig. 5.21, we can determine its center of gravity G by setting the moment about O of its total weight equal to the sum of the moments about O of the weights of the various component parts. Proceeding in this way, we obtain the following equations defining the coordinates $\overline{X}, \overline{Y}, \overline{Z}$ of the center of gravity G as

Center of gravity of a body with weight W

$$\overline{X}\Sigma W = \Sigma \overline{x}\, W \quad \overline{Y}\Sigma W = \Sigma \overline{y}\, W \quad \overline{Z}\Sigma W = \Sigma \overline{z}\, W \qquad \textbf{(5.20)}$$

or

$$\overline{X} = \frac{\Sigma \overline{x}W}{\Sigma W} \quad \overline{Y} = \frac{\Sigma \overline{y}W}{\Sigma W} \quad \overline{Z} = \frac{\Sigma \overline{z}W}{\Sigma W} \qquad \textbf{(5.20')}$$

If the body is made of a homogeneous material, its center of gravity coincides with the centroid of its volume, and we obtain

Centroid of a volume V

$$\overline{X}\Sigma V = \Sigma \overline{x}\, V \quad \overline{Y}\Sigma V = \Sigma \overline{y}\, V \quad \overline{Z}\Sigma V = \Sigma \overline{z}\, V \qquad \textbf{(5.21)}$$

or

$$\overline{X} = \frac{\Sigma \overline{x}V}{\Sigma V} \quad \overline{Y} = \frac{\Sigma \overline{y}V}{\Sigma V} \quad \overline{Z} = \frac{\Sigma \overline{z}V}{\Sigma V} \qquad \textbf{(5.21')}$$

Shape		$\bar{x}$	Volume
Hemisphere	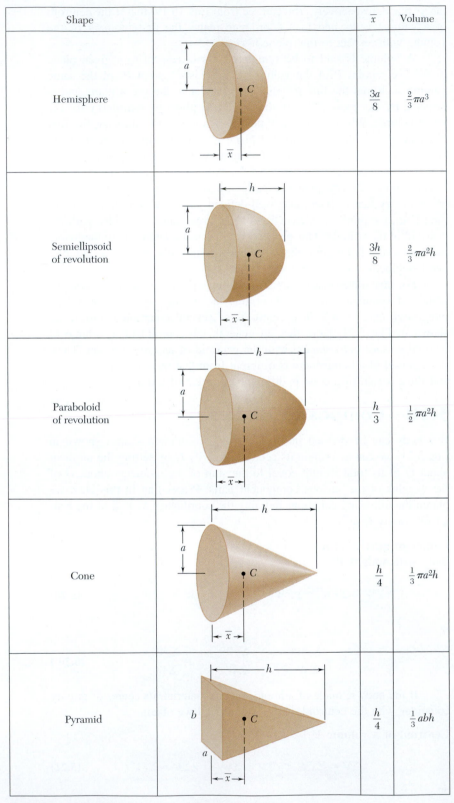	$\dfrac{3a}{8}$	$\dfrac{2}{3}\pi a^3$
Semiellipsoid of revolution		$\dfrac{3h}{8}$	$\dfrac{2}{3}\pi a^2 h$
Paraboloid of revolution		$\dfrac{h}{3}$	$\dfrac{1}{2}\pi a^2 h$
Cone		$\dfrac{h}{4}$	$\dfrac{1}{3}\pi a^2 h$
Pyramid		$\dfrac{h}{4}$	$\dfrac{1}{3}abh$

Fig. 5.21 Centroids of common shapes and volumes.

5.4C Determination of Centroids of Volumes by Integration

We can determine the centroid of a volume bounded by analytical surfaces by evaluating the integrals given earlier in this section:

$$\bar{x}V = \int x\,dV \qquad \bar{y}V = \int y\,dV \qquad \bar{z}V = \int z\,dV \qquad \textbf{(5.22)}$$

If we choose the element of volume dV to be equal to a small cube with sides dx, dy, and dz, the evaluation of each of these integrals requires a *triple integration*. However, it is possible to determine the coordinates of the centroid of most volumes by *double integration* if we choose dV to be equal to the volume of a thin filament (Fig. 5.22). We then obtain the coordinates of the centroid of the volume by rewriting Eqs. (5.22) as

$$\bar{x}V = \int \bar{x}_{el}\,dV \qquad \bar{y}V = \int \bar{y}_{el}\,dV \qquad \bar{z}V = \int \bar{z}_{el}\,dV \qquad \textbf{(5.23)}$$

Then we substitute the expressions given in Fig. 5.22 for the volume dV and the coordinates $\bar{x}_{el}$, $\bar{y}_{el}$, $\bar{z}_{el}$. By using the equation of the surface to express z in terms of x and y, we reduce the integration to a double integration in x and y.

If the volume under consideration possesses *two planes of symmetry*, its centroid must be located on the line of intersection of the two planes. Choosing the x axis to lie along this line, we have

$$\bar{y} = \bar{z} = 0$$

and the only coordinate to determine is $\bar{x}$. This can be done with a *single integration* by dividing the given volume into thin slabs parallel to the yz plane and expressing dV in terms of x and dx in the equation

$$\bar{x}V = \int \bar{x}_{el}\,dV \qquad \textbf{(5.24)}$$

For a body of revolution, the slabs are circular, and their volume is given in Fig. 5.23.

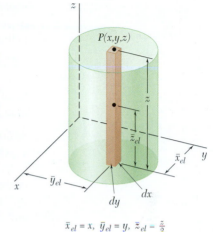

$\bar{x}_{el} = x,\ \bar{y}_{el} = y,\ \bar{z}_{el} = \frac{z}{2}$
$dV = z\,dx\,dy$

Fig. 5.22 Determining the centroid of a volume by double integration.

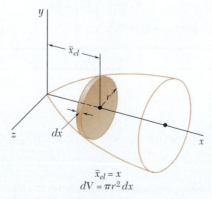

$\bar{x}_{el} = x$
$dV = \pi r^2\,dx$

Fig. 5.23 Determining the centroid of a body of revolution.

Sample Problem 5.11

Determine the location of the center of gravity of the homogeneous body of revolution shown that was obtained by joining a hemisphere and a cylinder and carving out a cone.

STRATEGY: The body is homogeneous, so the center of gravity coincides with the centroid. Since the body was formed from a composite of three simple shapes, you can find the centroid of each shape and combine them using Eq. (5.21).

MODELING: Because of symmetry, the center of gravity lies on the x axis. As shown in Fig. 1, the body is formed by adding a hemisphere to a cylinder and then subtracting a cone. Find the volume and the abscissa of the centroid of each of these components from Fig. 5.21 and enter them in a table (below). Then you can determine the total volume of the body and the first moment of its volume with respect to the yz plane.

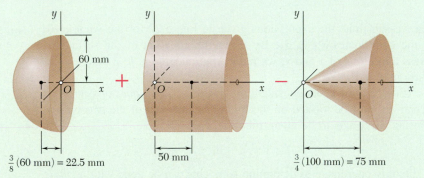

Fig. 1   The given body modeled as the combination of simple geometric shapes.

ANALYSIS: Note that the location of the centroid of the hemisphere is negative because it lies to the left of the origin.

Component	Volume, mm^3		$\bar{x}$, mm	$\bar{x}V$, mm^4
Hemisphere	$\dfrac{1}{2}\dfrac{4\pi}{3}(60)^3 =$	0.4524×10^6	-22.5	-10.18×10^6
Cylinder	$\pi(60)^2(100) =$	1.1310×10^6	$+50$	$+56.55 \times 10^6$
Cone	$-\dfrac{\pi}{3}(60)^2(100) =$	-0.3770×10^6	$+75$	-28.28×10^6
	$\Sigma V =$	1.206×10^6		$\Sigma \bar{x}V = +18.09 \times 10^6$

Thus,

$$\bar{X}\Sigma V = \Sigma \bar{x}V: \qquad \bar{X}(1.206 \times 10^6 \text{ mm}^3) = 18.09 \times 10^6 \text{ mm}^4$$

$$\bar{X} = 15 \text{ mm} \blacktriangleleft$$

REFLECT and THINK: Adding the hemisphere and subtracting the cone have the effect of shifting the centroid of the composite shape to the left of that for the cylinder (50 mm). However, because the first moment of volume for the cylinder is larger than for the hemisphere and cone combined, you should expect the centroid for the composite to still be in the positive x domain. Thus, as a rough visual check, the result of $+15$ mm is reasonable.

Sample Problem 5.12

Locate the center of gravity of the steel machine part shown. The diameter of each hole is 1 in.

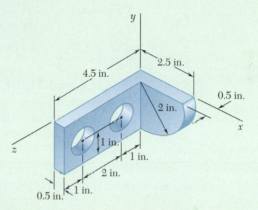

STRATEGY: This part can be broken down into the sum of two volumes minus two smaller volumes (holes). Find the volume and centroid of each volume and combine them using Eq. (5.21) to find the overall centroid.

MODELING: As shown in Fig. 1, the machine part can be obtained by adding a rectangular parallelepiped (I) to a quarter cylinder (II) and then subtracting two 1-in.-diameter cylinders (III and IV). Determine the volume and the coordinates of the centroid of each component and enter them in a table (below). Using the data in the table, determine the total volume and the moments of the volume with respect to each of the coordinate planes.

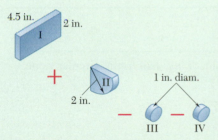

Fig. 1 The given body modeled as the combination of simple geometric shapes.

ANALYSIS: You can treat each component volume as a planar shape using Fig. 5.8A to find the volumes and centroids, but the right-angle joining of components I and II requires calculations in three dimensions. You may find it helpful to draw more detailed sketches of components with the centroids carefully labeled (Fig. 2).

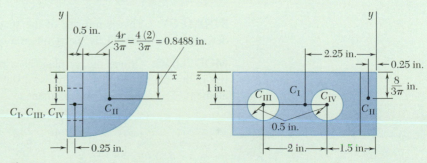

Fig. 2 Centroids of components.

	V, in^3	$\bar{x}$, in.	$\bar{y}$, in.	$\bar{z}$, in.	$\bar{x}V$, in^4	$\bar{y}V$, in^4	$\bar{z}V$, in^4
I	$(4.5)(2)(0.5) = 4.5$	0.25	-1	2.25	1.125	-4.5	10.125
II	$\frac{1}{4}\pi(2)^2(0.5) = 1.571$	1.3488	-0.8488	0.25	2.119	-1.333	0.393
III	$-\pi(0.5)^2(0.5) = -0.3927$	0.25	-1	3.5	-0.098	0.393	-1.374
IV	$-\pi(0.5)^2(0.5) = -0.3927$	0.25	-1	1.5	-0.098	0.393	-0.589
	$\Sigma V = 5.286$				$\Sigma\bar{x}V = 3.048$	$\Sigma\bar{y}V = -5.047$	$\Sigma\bar{z}V = 8.555$

Thus,

$$\bar{X}\Sigma V = \Sigma\bar{x}V: \qquad \bar{X}(5.286 \text{ in}^3) = 3.048 \text{ in}^4 \qquad \bar{X} = 0.577 \text{ in.} \blacktriangleleft$$

$$\bar{Y}\Sigma V = \Sigma\bar{y}V: \qquad \bar{Y}(5.286 \text{ in}^3) = -5.047 \text{ in}^4 \qquad \bar{Y} = -0.955 \text{ in.} \blacktriangleleft$$

$$\bar{Z}\Sigma V = \Sigma\bar{z}V: \qquad \bar{Z}(5.286 \text{ in}^3) = 8.555 \text{ in}^4 \qquad \bar{Z} = 1.618 \text{ in.} \blacktriangleleft$$

REFLECT and THINK: By inspection, you should expect $\bar{X}$ and $\bar{Z}$ to be considerably less than $(1/2)(2.5 \text{ in.})$ and $(1/2)(4.5 \text{ in.})$, respectively, and $\bar{Y}$ to be slightly less in magnitude than $(1/2)(2 \text{ in.})$. Thus, as a rough visual check, the results obtained are as expected.

Sample Problem 5.13

Determine the location of the centroid of the half right circular cone shown.

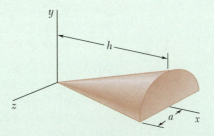

STRATEGY: This is not one of the shapes in Fig. 5.21, so you have to determine the centroid by using integration.

MODELING: Since the xy plane is a plane of symmetry, the centroid lies in this plane, and $\bar{z} = 0$. Choose a slab of thickness dx as a differential element. The volume of this element is

$$dV = \frac{1}{2}\pi r^2 \, dx$$

Obtain the coordinates $\bar{x}_{el}$ and $\bar{y}_{el}$ of the centroid of the element from Fig. 5.8 (semicircular area):

$$\bar{x}_{el} = x \qquad \bar{y}_{el} = \frac{4r}{3\pi}$$

Noting that r is proportional to x, use similar triangles (Fig. 1) to write

$$\frac{r}{x} = \frac{a}{h} \qquad r = \frac{a}{h}x$$

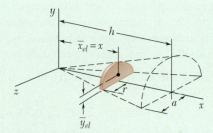

Fig. 1 Geometry of the differential element.

ANALYSIS: The volume of the body is

$$V = \int dV = \int_0^h \tfrac{1}{2}\pi r^2\, dx = \int_0^h \tfrac{1}{2}\pi\left(\frac{a}{h}x\right)^2 dx = \frac{\pi a^2 h}{6}$$

The moment of the differential element with respect to the yz plane is $\bar{x}_{el}\, dV$; the total moment of the body with respect to this plane is

$$\int \bar{x}_{el}\, dV = \int_0^h x(\tfrac{1}{2}\pi r^2)\, dx = \int_0^h x(\tfrac{1}{2}\pi)\left(\frac{a}{h}\right)^2 x^2\, dx = \frac{\pi a^2 h^2}{8}$$

Thus,

$$\bar{x}V = \int \bar{x}_{el}\, dV \qquad \bar{x}\,\frac{\pi a^2 h}{6} = \frac{\pi a^2 h^2}{8} \qquad \bar{x} = \tfrac{3}{4}h \quad \blacktriangleleft$$

Similarly, the moment of the differential element with respect to the zx plane is $\bar{y}_{el}\, dV$; the total moment is

$$\int \bar{y}_{el}\, dV = \int_0^h \frac{4r}{3\pi}(\tfrac{1}{2}\pi r^2)\, dx = \frac{2}{3}\int_0^h \left(\frac{a}{h}x\right)^3 dx = \frac{a^3 h}{6}$$

Thus,

$$\bar{y}V = \int \bar{y}_{el}\, dV \qquad \bar{y}\,\frac{\pi a^2 h}{6} = \frac{a^3 h}{6} \qquad \bar{y} = \frac{a}{\pi} \quad \blacktriangleleft$$

REFLECT and THINK: Since a full right circular cone is a body of revolution, its $\bar{x}$ is unchanged for any portion of the cone bounded by planes intersecting along the x axis. The same centroid location in the x direction was therefore obtained for the half cone that Fig. 5.21 shows for the full cone. Similarly, the same $\bar{x}$ result would be obtained for a quarter cone.

SOLVING PROBLEMS
ON YOUR OWN

In the problems for this section, you will be asked to locate the centers of gravity of three-dimensional bodies or the centroids of their volumes. All of the techniques we previously discussed for two-dimensional bodies—using symmetry, dividing the body into common shapes, choosing the most efficient differential element, etc.— also may be applied to the general three-dimensional case.

1. Locating the centers of gravity of composite bodies. In general, you must use Eqs. (5.20):

$$\overline{X}\Sigma W = \Sigma \overline{x}W \qquad \overline{Y}\Sigma W = \Sigma \overline{y}W \qquad \overline{Z}\Sigma W = \Sigma \overline{z}W \qquad \textbf{(5.20)}$$

However, for the case of a *homogeneous body*, the center of gravity of the body coincides with the *centroid of its volume*. Therefore, for this special case, you can also use Eqs. (5.21) to locate the center of gravity of the body:

$$\overline{X}\Sigma V = \Sigma \overline{x}V \qquad \overline{Y}\Sigma V = \Sigma \overline{y}V \qquad \overline{Z}\Sigma V = \Sigma \overline{z}V \qquad \textbf{(5.21)}$$

You should realize that these equations are simply an extension of the equations used for the two-dimensional problems considered earlier in the chapter. As the solutions of Sample Probs. 5.11 and 5.12 illustrate, the methods of solution for two- and three-dimensional problems are identical. Thus, we once again strongly encourage you to construct appropriate diagrams and tables when analyzing composite bodies. Also, as you study Sample Prob. 5.12, observe how we obtained the x and y coordinates of the centroid of the quarter cylinder using the equations for the centroid of a quarter circle.

Two special cases of interest occur when the given body consists of either uniform wires or uniform plates made of the same material.

    a. For a body made of *several wire elements* of the *same uniform cross section*, the cross-sectional area A of the wire elements factors out of Eqs. (5.21) when V is replaced with the product AL, where L is the length of a given element. Equations (5.21) thus reduce in this case to

$$\overline{X}\Sigma L = \Sigma \overline{x}L \qquad \overline{Y}\Sigma L = \Sigma \overline{y}L \qquad \overline{Z}\Sigma L = \Sigma \overline{z}L$$

    b. For a body made of *several plates* of the *same uniform thickness*, the thickness t of the plates factors out of Eqs. (5.21) when V is replaced with the product tA, where A is the area of a given plate. Equations (5.21) thus reduce in this case to

$$\overline{X}\Sigma A = \Sigma \overline{x}A \qquad \overline{Y}\Sigma A = \Sigma \overline{y}A \qquad \overline{Z}\Sigma A = \Sigma \overline{z}A$$

2. Locating the centroids of volumes by direct integration. As explained in Sec. 5.4C, you can simplify evaluating the integrals of Eqs. (5.22) by choosing either a thin filament (Fig. 5.22) or a thin slab (Fig. 5.23) for the element of volume dV. Thus, you should begin your solution by identifying, if possible, the dV that produces the single or double integrals that are easiest to compute. For bodies of revolution, this may be a thin slab (as in Sample Prob. 5.13) or a thin cylindrical shell. However, it is important to remember that the relationship that you establish among the variables (like the relationship between r and x in Sample Prob. 5.13) directly affects the complexity of the integrals you have to compute. Finally, we again remind you that $\overline{x}_{el}$, $\overline{y}_{el}$, and $\overline{z}_{el}$ in Eqs. (5.23) are the coordinates of the centroid of dV.

Problems

5.96 Consider the composite body shown. Determine (*a*) the value of $\bar{x}$ when $h = L/2$, (*b*) the ratio h/L for which $\bar{x} = L$.

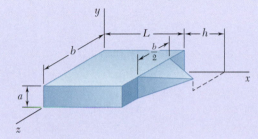

Fig. P5.96

5.97 Determine the location of the centroid of the composite body shown when (*a*) $h = 2b$, (*b*) $h = 2.5b$.

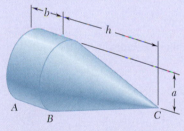

Fig. P5.97

5.98 The composite body shown is formed by removing a semiellipsoid of revolution of semimajor axis h and semiminor axis $a/2$ from a hemisphere of radius a. Determine (*a*) the y coordinate of the centroid when $h = a/2$, (*b*) the ratio h/a for which $\bar{y} = -0.4a$.

5.99 Locate the centroid of the frustum of a right circular cone when $r_1 = 40$ mm, $r_2 = 50$ mm, and $h = 60$ mm.

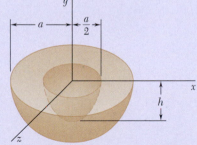

Fig. P5.98

Fig. P5.99

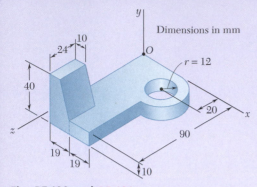

Dimensions in mm

Fig. P5.100 and *P5.101*

5.100 For the machine element shown, locate the *x* coordinate of the center of gravity.

5.101 For the machine element shown, locate the *z* coordinate of the center of gravity.

5.102 For the machine element shown, locate the *y* coordinate of the center of gravity.

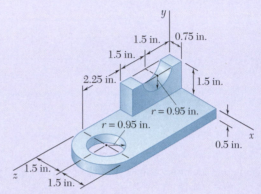

Fig. P5.102 and P5.103

5.103 For the machine element shown, locate the *z* coordinate of the center of gravity.

5.104 For the machine element shown, locate the *y* coordinate of the center of gravity.

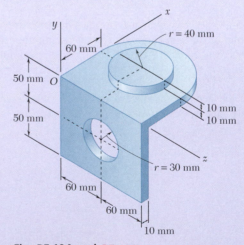

Fig. P5.104 and *P5.105*

5.105 For the machine element shown, locate the *x* coordinate of the center of gravity.

5.106 and 5.107 Locate the center of gravity of the sheet-metal form shown.

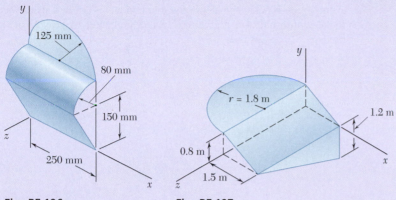

Fig. P5.106   Fig. P5.107

5.108 A corner reflector for tracking by radar has two sides in the shape of a quarter circle with a radius of 15 in. and one side in the shape of a triangle. Locate the center of gravity of the reflector, knowing that it is made of sheet metal with a uniform thickness.

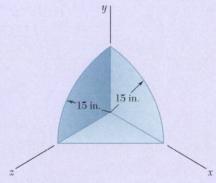

Fig. *P5.108*

5.109 A wastebasket, designed to fit in the corner of a room, is 16 in. high and has a base in the shape of a quarter circle with a radius of 10 in. Locate the center of gravity of the wastebasket, knowing that it is made of sheet metal with a uniform thickness.

5.110 An elbow for the duct of a ventilating system is made of sheet metal with a uniform thickness. Locate the center of gravity of the elbow.

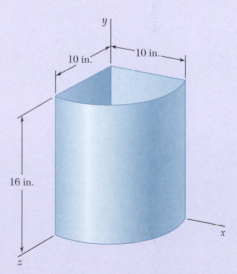

Fig. P5.109

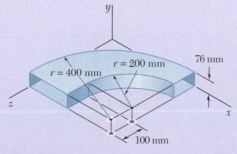

Fig. P5.110

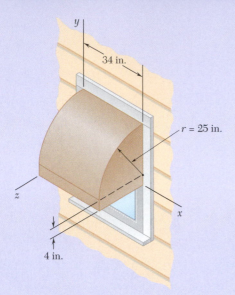

Fig. P5.111

5.111 A window awning is fabricated from sheet metal with a uniform thickness. Locate the center of gravity of the awning.

5.112 A mounting bracket for electronic components is formed from sheet metal with a uniform thickness. Locate the center of gravity of the bracket.

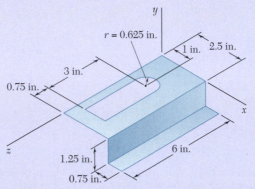

Fig. *P5.112*

5.113 A thin sheet of plastic with a uniform thickness is bent to form a desk organizer. Locate the center of gravity of the organizer.

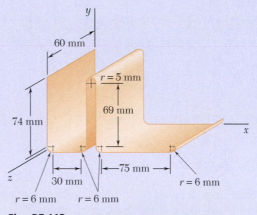

Fig. P5.113

5.114 A thin steel wire with a uniform cross section is bent into the shape shown. Locate its center of gravity.

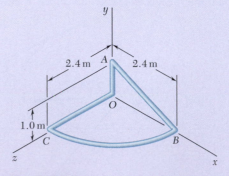

Fig. P5.114

5.115 The frame of a greenhouse is constructed from uniform aluminum channels. Locate the center of gravity of the portion of the frame shown.

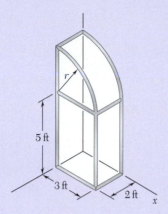

Fig. P5.115

5.116 and 5.117 Locate the center of gravity of the figure shown, knowing that it is made of thin brass rods with a uniform diameter.

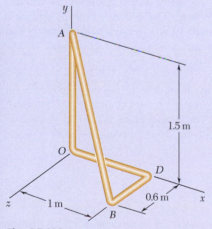

Fig. P5.116

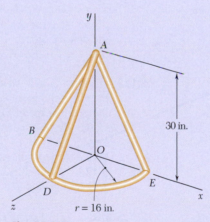

Fig. P5.117

5.118 A scratch awl has a plastic handle and a steel blade and shank. Knowing that the density of plastic is 1030 kg/m³ and of steel is 7860 kg/m³, locate the center of gravity of the awl.

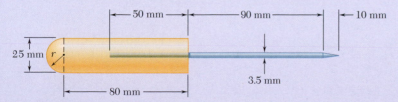

Fig. P5.118

5.119 A bronze bushing is mounted inside a steel sleeve. Knowing that the specific weight of bronze is 0.318 lb/in.³ and of steel is 0.284 lb/in.³, determine the location of the center of gravity of the assembly.

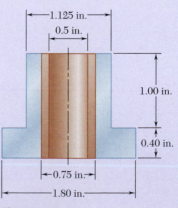

Fig. P5.119

287

5.120 A brass collar with a length of 2.5 in. is mounted on an aluminum rod with a length of 4 in. Locate the center of gravity of the composite body. (Specific weights: brass = 0.306 lb/in³, aluminum = 0.101 lb/in³.)

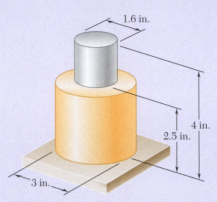

Fig. P5.120

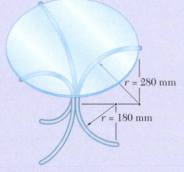

Fig. P5.121

5.121 The three legs of a small glass-topped table are equally spaced and are made of steel tubing that has an outside diameter of 24 mm and a cross-sectional area of 150 mm². The diameter and the thickness of the table top are 600 mm and 10 mm, respectively. Knowing that the density of steel is 7860 kg/m³ and of glass is 2190 kg/m³, locate the center of gravity of the table.

5.122 through 5.124 Determine by direct integration the values of $\bar{x}$ for the two volumes obtained by passing a vertical cutting plane through the given shape of Fig. 5.21. The cutting plane is parallel to the base of the given shape and divides the shape into two volumes of equal height.

5.122 A hemisphere
5.123 A semiellipsoid of revolution
5.124 A paraboloid of revolution.

5.125 and 5.126 Locate the centroid of the volume obtained by rotating the shaded area about the x axis.

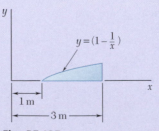

Fig. P5.125

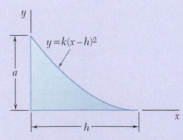

Fig. P5.126

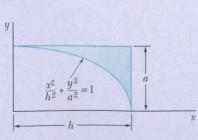

Fig. P5.127

5.127 Locate the centroid of the volume obtained by rotating the shaded area about the line $x = h$.

*5.128 Locate the centroid of the volume generated by revolving the portion of the sine curve shown about the x axis.

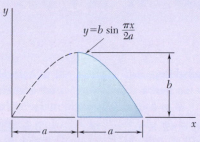

$$y = b \sin \frac{\pi x}{2a}$$

Fig. P5.128 and P5.129

*5.129 Locate the centroid of the volume generated by revolving the portion of the sine curve shown about the y axis. (*Hint:* Use a thin cylindrical shell of radius r and thickness dr as the element of volume.)

*5.130 Show that for a regular pyramid of height h and n sides ($n = 3, 4, \ldots$) the centroid of the volume of the pyramid is located at a distance $h/4$ above the base.

5.131 Determine by direct integration the location of the centroid of one-half of a thin, uniform hemispherical shell of radius R.

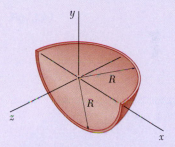

Fig. P5.131

5.132 The sides and the base of a punch bowl are of uniform thickness t. If $t \ll R$ and $R = 250$ mm, determine the location of the center of gravity of (*a*) the bowl, (*b*) the punch.

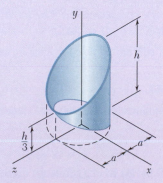

Fig. P5.132

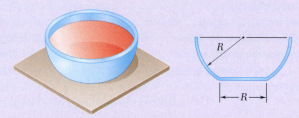

5.133 Locate the centroid of the section shown, which was cut from a thin circular pipe by two oblique planes.

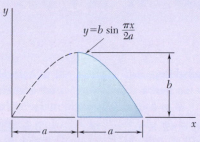

Fig. P5.133

*5.134 Locate the centroid of the section shown, which was cut from an elliptical cylinder by an oblique plane.

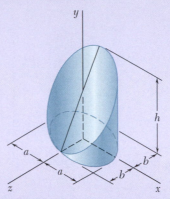

Fig. P5.134

5.135 Determine by direct integration the location of the centroid of the volume between the xz plane and the portion shown of the surface $y = 16h(ax - x^2)(bz - z^2)/a^2b^2$.

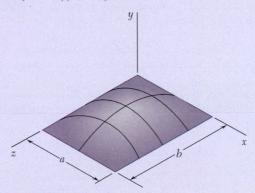

Fig. P5.135

5.136 After grading a lot, a builder places four stakes to designate the corners of the slab for a house. To provide a firm, level base for the slab, the builder places a minimum of 3 in. of gravel beneath the slab. Determine the volume of gravel needed and the x coordinate of the centroid of the volume of the gravel. (*Hint:* The bottom surface of the gravel is an oblique plane, which can be represented by the equation $y = a + bx + cz$.)

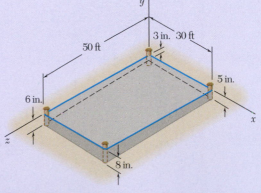

Fig. P5.136

Review and Summary

This chapter was devoted chiefly to determining the **center of gravity** of a rigid body, i.e., to determining the point G where we can apply a single force **W**—the *weight* of the body—to represent the effect of Earth's attraction on the body.

Center of Gravity of a Two-Dimensional Body

In the first part of this chapter, we considered *two-dimensional bodies*, such as flat plates and wires contained in the xy plane. By adding force components in the vertical z direction and moments about the horizontal y and x axes [Sec. 5.1A], we derived the relations

$$W = \int dW \qquad \bar{x}W = \int x\, dW \qquad \bar{y}W = \int y\, dW \qquad \textbf{(5.2)}$$

These equations define the weight of the body and the coordinates $\bar{x}$ and $\bar{y}$ of its center of gravity.

Centroid of an Area or Line

In the case of a *homogeneous flat* plate of uniform thickness [Sec. 5.1B], the center of gravity G of the plate coincides with the **centroid C of the area A** of the plate. The coordinates are defined by the relations

$$\bar{x}A = \int x\, dA \qquad \bar{y}A = \int y\, dA \qquad \textbf{(5.3)}$$

Similarly, determining the center of gravity of a *homogeneous wire of uniform cross section* contained in a plane reduces to determining the **centroid C of the line L** representing the wire; we have

$$\bar{x}L = \int x\, dL \qquad \bar{y}L = \int y\, dL \qquad \textbf{(5.4)}$$

First Moments

The integrals in Eqs. (5.3) are referred to as the **first moments** of the area A with respect to the y and x axes and are denoted by Q_y and Q_x, respectively [Sec. 5.1C]. We have

$$Q_y = \bar{x}A \qquad Q_x = \bar{y}A \qquad \textbf{(5.6)}$$

The first moments of a line can be defined in a similar way.

Properties of Symmetry

Determining the centroid C of an area or line is simplified when the area or line possesses certain properties of symmetry. If the area or line is symmetric with respect to an axis, its centroid C lies on that axis; if it is symmetric with respect to two axes, C is located at the intersection of the two axes; if it is symmetric with respect to a center O, C coincides with O.

Center of Gravity of a Composite Body

The areas and the centroids of various common shapes are tabulated in Fig. 5.8. When a flat plate can be divided into several of these shapes, the coordinates $\overline{X}$ and $\overline{Y}$ of its center of gravity G can be determined from the coordinates $\overline{x}_1, \overline{x}_2, \ldots$ and $\overline{y}_1, \overline{y}_2, \ldots$ of the centers of gravity $G_1, G_2, \ldots$ of the various parts [Sec. 5.1D]. Equating moments about the y and x axes, respectively (Fig. 5.24), we have

$$\overline{X}\Sigma W = \Sigma \overline{x}W \qquad \overline{Y}\Sigma W = \Sigma \overline{y}W \tag{5.7}$$

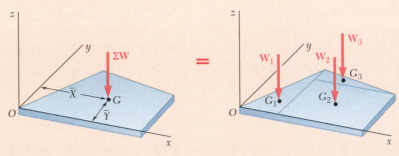

Fig. 5.24

If the plate is homogeneous and of uniform thickness, its center of gravity coincides with the centroid C of the area of the plate, and Eqs. (5.7) reduce to

$$Q_y = \overline{X}\Sigma A = \Sigma \overline{x}A \qquad Q_x = \overline{Y}\Sigma A = \Sigma \overline{y}A \tag{5.8}$$

These equations yield the first moments of the composite area, or they can be solved for the coordinates $\overline{X}$ and $\overline{Y}$ of its centroid [Sample Prob. 5.1]. Determining the center of gravity of a composite wire is carried out in a similar fashion [Sample Prob. 5.2].

Determining a Centroid by Integration

When an area is bounded by analytical curves, you can determine the coordinates of its centroid by *integration* [Sec. 5.2A]. This can be done by evaluating either the double integrals in Eqs. (5.3) or a single integral that uses one of the thin rectangular or pie-shaped elements of area shown in Fig. 5.12. Denoting by $\overline{x}_{el}$ and $\overline{y}_{el}$ the coordinates of the centroid of the element dA, we have

$$Q_y = \overline{x}A = \int \overline{x}_{el}\, dA \qquad Q_x = \overline{y}A = \int \overline{y}_{el}\, dA \tag{5.9}$$

It is advantageous to use the same element of area to compute both of the first moments Q_y and Q_x; we can also use the same element to determine the area A [Sample Prob. 5.4].

Theorems of Pappus–Guldinus

The **theorems of Pappus-Guldinus** relate the area of a surface of revolution or the volume of a body of revolution to the centroid of the generating curve or area [Sec. 5.2B]. The area A of the surface generated by rotating a curve of length L about a fixed axis (Fig. 5.25a) is

$$A = 2\pi \overline{y}L \tag{5.10}$$

where $\overline{y}$ represents the distance from the centroid C of the curve to the fixed axis. Similarly, the volume V of the body generated by rotating an area A

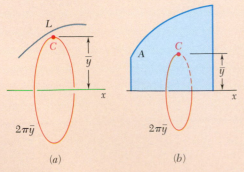

Fig. 5.25

about a fixed axis (Fig. 5.25*b*) is

$$V = 2\pi \bar{y}A \tag{5.11}$$

where $\bar{y}$ represents the distance from the centroid C of the area to the fixed axis.

Distributed Loads

The concept of centroid of an area also can be used to solve problems other than those dealing with the weight of flat plates. For example, to determine the reactions at the supports of a beam [Sec. 5.3A], we can replace a **distributed load** w by a concentrated load $\mathbf{W}$ equal in magnitude to the area A under the load curve and passing through the centroid C of that area (Fig. 5.26). We can use this same approach to determine the resultant of the hydrostatic forces exerted on a **rectangular plate submerged in a liquid** [Sec. 5.3B].

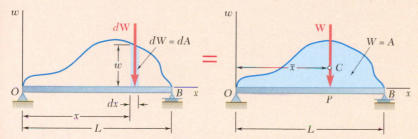

Fig. 5.26

Center of Gravity of a Three-Dimensional Body

The last part of this chapter was devoted to determining the center of gravity G of a three-dimensional body. We defined the coordinates $\bar{x}, \bar{y}, \bar{z}$ of G by the relations

$$\bar{x}W = \int x\, dW \qquad \bar{y}W = \int y\, dW \qquad \bar{z}W = \int z\, dW \tag{5.17}$$

Centroid of a Volume

In the case of a homogeneous body, the center of gravity G coincides with the centroid C of the volume V of the body. The coordinates of C are defined by the relations

$$\bar{x}V = \int x\, dV \qquad \bar{y}V = \int y\, dV \qquad \bar{z}V = \int z\, dV \tag{5.19}$$

If the volume possesses a *plane of symmetry*, its centroid C lies in that plane; if it possesses two planes of symmetry, C is located on the line of intersection of the two planes; if it possesses three planes of symmetry that intersect at only one point, C coincides with that point [Sec. 5.4A].

Center of Gravity of a Composite Body

The volumes and centroids of various common three-dimensional shapes are tabulated in Fig. 5.21. When a body can be divided into several of these shapes, we can determine the coordinates $\overline{X}, \overline{Y}, \overline{Z}$ of its center of gravity G from the corresponding coordinates of the centers of gravity of its various parts [Sec. 5.4B]. We have

$$\overline{X}\Sigma W = \Sigma \bar{x}W \qquad \overline{Y}\Sigma W = \Sigma \bar{y}W \qquad \overline{Z}\Sigma W = \Sigma \bar{z}W \tag{5.20}$$

If the body is made of a homogeneous material, its center of gravity coincides with the centroid C of its volume, and we have [Sample Probs. 5.11 and 5.12]

$$\overline{X}\Sigma V = \Sigma \overline{x}V \qquad \overline{Y}\Sigma V = \Sigma \overline{y}V \qquad \overline{Z}\Sigma V = \Sigma \overline{z}V \qquad \textbf{(5.21)}$$

Determining a Centroid by Integration

When a volume is bounded by analytical surfaces, we can find the coordinates of its centroid by *integration* [Sec. 5.4C]. To avoid the computation of triple integrals in Eqs. (5.19), we can use elements of volume in the shape of thin filaments, as shown in Fig. 5.27. Denoting the coordinates of the centroid of the element dV as $\overline{x}_{el}, \overline{y}_{el}, \overline{z}_{el}$, we rewrite Eqs. (5.19) as

$$\overline{x}V = \int \overline{x}_{el}\, dV \qquad \overline{y}V = \int \overline{y}_{el}\, dV \qquad \overline{z}V = \int \overline{z}_{el}\, dV \qquad \textbf{(5.23)}$$

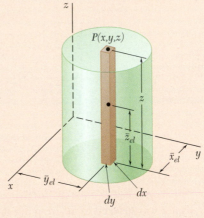

$$\overline{x}_{el} = x, \ \overline{y}_{el} = y, \ \overline{z}_{el} = \tfrac{z}{2}$$
$$dV = z\, dx\, dy$$

Fig. 5.27

that involve only double integrals. If the volume possesses *two planes of symmetry,* its centroid C is located on their line of intersection. Choosing the x axis to lie along that line and dividing the volume into thin slabs parallel to the yz plane, we can determine C from the relation

$$\overline{x}V = \int \overline{x}_{el}\, dV \qquad \textbf{(5.24)}$$

with a *single integration* [Sample Prob. 5.13]. For a body of revolution, these slabs are circular and their volume is given in Fig. 5.28.

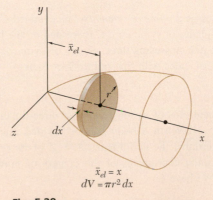

$$\overline{x}_{el} = x$$
$$dV = \pi r^2\, dx$$

Fig. 5.28

Review Problems

5.137 and 5.138 Locate the centroid of the plane area shown.

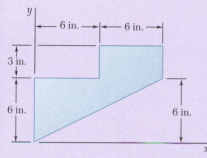

Fig. P5.137

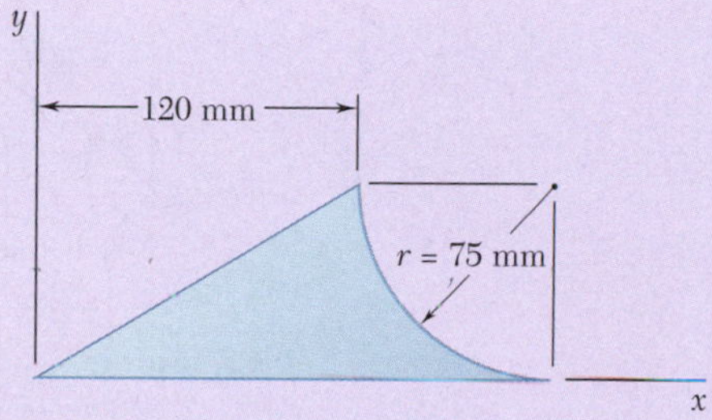

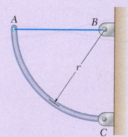

Fig. P5.138

5.139 A uniform circular rod with a weight of 8 lb and radius of 10 in. is attached to a pin at C and to the cable AB. Determine (a) the tension in the cable, (b) the reaction at C.

5.140 Determine by direct integration the centroid of the area shown. Express your answer in terms of a and h.

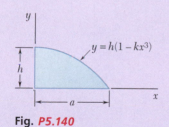

Fig. P5.140

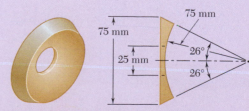

Fig. P5.139

5.141 Determine by direct integration the centroid of the area shown.

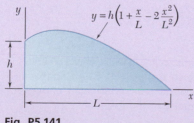

Fig. P5.141

5.142 The escutcheon (a decorative plate placed on a pipe where the pipe exits from a wall) shown is cast from brass. Knowing that the density of brass is 8470 kg/m³, determine the mass of the escutcheon.

Fig. P5.142

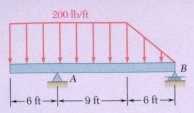

200 lb/ft

6 ft — 9 ft — 6 ft

Fig. P5.143

5.143 Determine the reactions at the beam supports for the given loading.

5.144 A beam is subjected to a linearly distributed downward load and rests on two wide supports BC and DE that exert uniformly distributed upward loads as shown. Determine the values of w_{BC} and w_{DE} corresponding to equilibrium when $w_A = 600$ N/m.

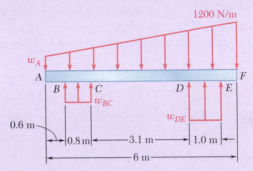

1200 N/m

w_A
A
B
C
w_{BC}
D
E
w_{DE}
F

0.6 m
0.8 m
3.1 m
1.0 m
6 m

Fig. P5.144

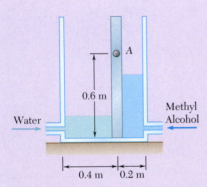

A

0.6 m

Water

Methyl Alcohol

0.4 m   0.2 m

Fig. P5.145

5.145 A tank is divided into two sections by a 1×1-m square gate that is hinged at A. A couple with a magnitude of 490 N·m is required for the gate to rotate. If one side of the tank is filled with water at the rate of 0.1 m³/min and the other side is filled simultaneously with methyl alcohol (density $\rho_{ma} = 789$ kg/m³) at the rate of 0.2 m³/min, determine at what time and in which direction the gate will rotate.

5.146 Determine the y coordinate of the centroid of the body shown.

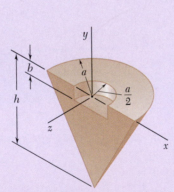

y
b
a
$\frac{a}{2}$
h
z
x

Fig. P5.146

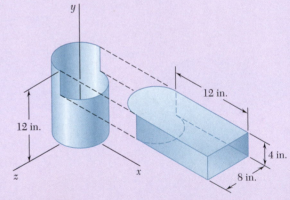

y
12 in.
12 in.
z
x
8 in.
4 in.

Fig. P5.147

5.147 An 8-in.-diameter cylindrical duct and a 4×8-in. rectangular duct are to be joined as indicated. Knowing that the ducts were fabricated from the same sheet metal, which is of uniform thickness, locate the center of gravity of the assembly.

5.148 Three brass plates are brazed to a steel pipe to form the flagpole base shown. Knowing that the pipe has a wall thickness of 8 mm and that each plate is 6 mm thick, determine the location of the center of gravity of the base. (Densities: brass = 8470 kg/m³; steel = 7860 kg/m³.)

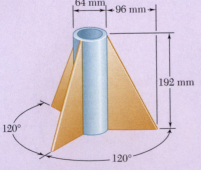

64 mm   96 mm

192 mm

120°

120°

Fig. P5.148

6

Analysis of Structures

Trusses, such as this cantilever arch bridge over Deception Pass in Washington State, provide both a practical and an economical solution to many engineering problems.

Objectives

- **Define** an ideal truss, and consider the attributes of simple trusses.
- **Analyze** plane and space trusses by the method of joints.
- **Simplify** certain truss analyses by recognizing special loading and geometry conditions.
- **Analyze** trusses by the method of sections.
- **Consider** the characteristics of compound trusses.
- **Analyze** structures containing multiforce members, such as frames and machines.

Introduction

In the preceding chapters, we studied the equilibrium of a single rigid body, where all forces involved were external to the rigid body. We now consider the equilibrium of structures made of several connected parts. This situation calls for determining not only the external forces acting on the structure, but also the forces that hold together the various parts of the structure. From the point of view of the structure as a whole, these forces are **internal forces**.

Consider, for example, the crane shown in Fig. 6.1*a* that supports a load W. The crane consists of three beams *AD, CF,* and *BE* connected by frictionless pins; it is supported by a pin at A and by a cable *DG*. The free-body diagram of the crane is drawn in Fig. 6.1*b*. The external forces shown in the diagram include the weight $\mathbf{W}$, the two components $\mathbf{A}_x$ and $\mathbf{A}_y$ of the reaction at A, and the force $\mathbf{T}$ exerted by the cable at D. The internal forces holding the various parts of the crane together do not appear in the free-body diagram. If, however, we dismember the crane and draw a free-body diagram for each of its component parts, we can see the forces holding the three beams together, since these forces are external forces from the point of view of each component part (Fig. 6.1*c*).

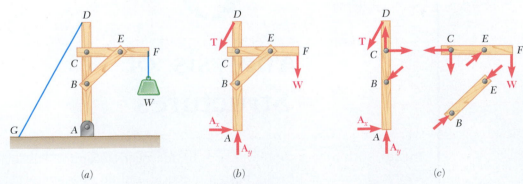

(a) (b) (c)

Fig. 6.1 A structure in equilibrium. (a) Diagram of a crane supporting a load; (b) free-body diagram of the crane; (c) free-body diagrams of the components of the crane.

Note that we represent the force exerted at B by member BE on member AD as equal and opposite to the force exerted at the same point by member AD on member BE. Similarly, the force exerted at E by BE on CF is shown equal and opposite to the force exerted by CF on BE, and the components of the force exerted at C by CF on AD are shown equal and opposite to the components of the force exerted by AD on CF. These representations agree with Newton's third law, which states that

The forces of action and reaction between two bodies in contact have the same magnitude, same line of action, and opposite sense.

We pointed out in Chap. 1 that this law, which is based on experimental evidence, is one of the six fundamental principles of elementary mechanics. Its application is essential for solving problems involving connected bodies.

In this chapter, we consider three broad categories of engineering structures:

1. **Trusses**, which are designed to support loads and are usually stationary, fully constrained structures. Trusses consist exclusively of straight members connected at joints located at the ends of each member. Members of a truss, therefore, are **two-force members**, i.e., members acted upon by two equal and opposite forces directed along the member.
2. **Frames**, which are also designed to support loads and are also usually stationary, fully constrained structures. However, like the crane of Fig. 6.1, frames always contain at least one **multi-force member**, i.e., a member acted upon by three or more forces that, in general, are not directed along the member.
3. **Machines**, which are designed to transmit and modify forces and are structures containing moving parts. Machines, like frames, always contain at least one multi-force member.

Two-force member

Multi-force member

Multi-force member

(a) A truss bridge

(b) A bicycle frame

(c) A hydraulic machine arm

Photo 6.1   The structures you see around you to support loads or transmit forces are generally trusses, frames, or machines.

6.1   ANALYSIS OF TRUSSES

The truss is one of the major types of engineering structures. It provides a practical and economical solution to many engineering situations, especially in the design of bridges and buildings. In this section, we describe the basic elements of a truss and study a common method for analyzing the forces acting in a truss.

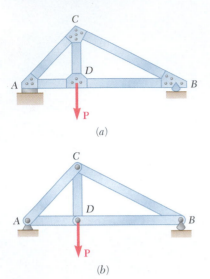

Fig. 6.2 (*a*) A typical truss consists of straight members connected at joints; (*b*) we can model a truss as two-force members connected by pins.

Photo 6.2 Shown is a pin-jointed connection on the approach span to the San Francisco–Oakland Bay Bridge.

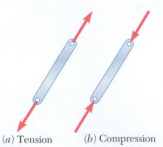

(*a*) Tension (*b*) Compression

Fig. 6.4 A two-force member of a truss can be in tension or compression.

6.1A Simple Trusses

A truss consists of straight members connected at joints, as shown in Fig. 6.2*a*. Truss members are connected at their extremities only; no member is continuous through a joint. In Fig. 6.2*a*, for example, there is no member *AB*; instead we have two distinct members *AD* and *DB*. Most actual structures are made of several trusses joined together to form a space framework. Each truss is designed to carry those loads that act in its plane and thus may be treated as a two-dimensional structure.

In general, the members of a truss are slender and can support little lateral load; all loads, therefore, must be applied at the various joints and not to the members themselves. When a concentrated load is to be applied between two joints or when the truss must support a distributed load, as in the case of a bridge truss, a floor system must be provided. The floor transmits the load to the joints through the use of stringers and floor beams (Fig. 6.3).

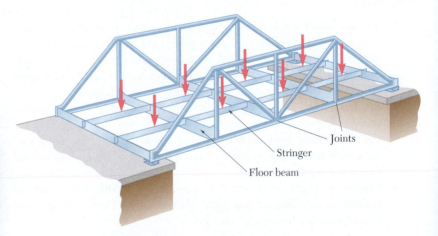

Fig. 6.3 A floor system of a truss uses stringers and floor beams to transmit an applied load to the joints of the truss.

We assume that the weights of the truss members can be applied to the joints, with half of the weight of each member applied to each of the two joints the member connects. Although the members are actually joined together by means of welded, bolted, or riveted connections, it is customary to assume that the members are pinned together; therefore, the forces acting at each end of a member reduce to a single force and no couple. This enables us to model the forces applied to a truss member as a single force at each end of the member. We can then treat each member as a two-force member, and we can consider the entire truss as a group of pins and two-force members (Fig. 6.2*b*). An individual member can be acted upon as shown in either of the two sketches of Fig. 6.4. In Fig. 6.4*a*, the forces tend to pull the member apart, and the member is in tension; in Fig. 6.4*b*, the forces tend to push the member together, and the member is in compression. Some typical trusses are shown in Fig. 6.5.

Consider the truss of Fig. 6.6*a*, which is made of four members connected by pins at *A, B, C,* and *D.* If we apply a load at *B,* the truss will greatly deform, completely losing its original shape. In contrast, the truss of Fig. 6.6*b*, which is made of three members connected by pins at *A, B,* and *C,* will deform only slightly under a load applied at *B.* The only possible

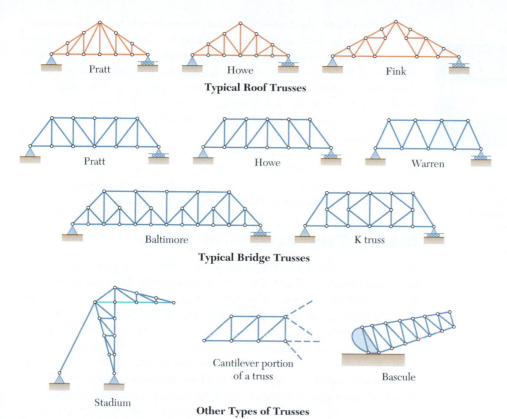

Typical Roof Trusses

Pratt   Howe   Fink

Typical Bridge Trusses

Pratt   Howe   Warren

Baltimore   K truss

Stadium   **Other Types of Trusses**

Cantilever portion
of a truss   Bascule

Fig. 6.5 You can often see trusses in the design of a building roof, a bridge, or other other larger structures.

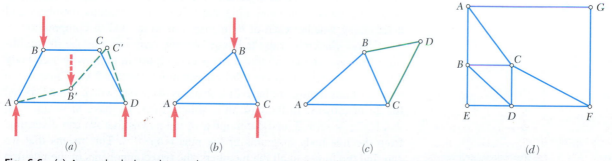

(a)   (b)   (c)   (d)

Fig. 6.6 (a) A poorly designed truss that cannot support a load; (b) the most elementary rigid truss consists of a simple triangle; (c) a larger rigid truss built up from the triangle in (b); (d) a rigid truss not made up of triangles alone.

deformation for this truss is one involving small changes in the length of its members. The truss of Fig. 6.6b is said to be a **rigid truss**, the term 'rigid' being used here to indicate that the truss *will not collapse.*

As shown in Fig. 6.6c, we can obtain a larger rigid truss by adding two members BD and CD to the basic triangular truss of Fig. 6.6b. We can repeat this procedure as many times as we like, and the resulting truss will be rigid if each time we add two new members they are attached to two existing joints and connected at a new joint. (The three joints must not be in a straight line.) A truss that can be constructed in this manner is called a **simple truss**.

Note that a simple truss is not necessarily made only of triangles. The truss of Fig. 6.6d, for example, is a simple truss that we constructed from triangle ABC by adding successively the joints D, E, F, and G.

Photo 6.3 Two K trusses were used as the main components of the movable bridge shown, which moved above a large stockpile of ore. The bucket below the trusses picked up ore and redeposited it until the ore was thoroughly mixed. The ore was then sent to the mill for processing into steel.

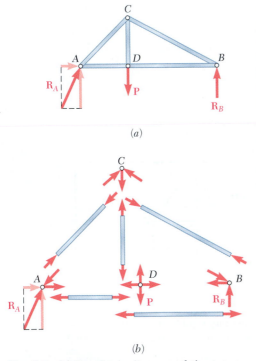

Fig. 6.7 (a) Free-body diagram of the truss as a rigid body; (b) free-body diagrams of the five members and four pins that make up the truss.

On the other hand, rigid trusses are not always simple trusses, even when they appear to be made of triangles. The Fink and Baltimore trusses shown in Fig. 6.5, for instance, are not simple trusses, because they cannot be constructed from a single triangle in the manner just described. All of the other trusses shown in Fig. 6.5 are simple trusses, as you may easily check. (For the K truss, start with one of the central triangles.)

Also note that the basic triangular truss of Fig. 6.6b has three members and three joints. The truss of Fig. 6.6c has two more members and one more joint; i.e., five members and four joints altogether. Observing that every time we add two new members, we increase the number of joints by one, we find that in a simple truss the total number of members is $m = 2n - 3$, where n is the total number of joints.

6.1B The Method of Joints

We have just seen that a truss can be considered as a group of pins and two-force members. Therefore, we can dismember the truss of Fig. 6.2, whose free-body diagram is shown in Fig. 6.7a, and draw a free-body diagram for each pin and each member (Fig. 6.7b). Each member is acted upon by two forces, one at each end; these forces have the same magnitude, same line of action, and opposite sense (Sec. 4.2A). Furthermore, Newton's third law states that the forces of action and reaction between a member and a pin are equal and opposite. Therefore, the forces exerted by a member on the two pins it connects must be directed along that member and be equal and opposite. The common magnitude of the forces exerted by a member on the two pins it connects is commonly referred to as the *force in the member,* even though this quantity is actually a scalar. Since we know the lines of action of all the internal forces in a truss, the analysis of a truss reduces to computing the forces in its various members and determining whether each of its members is in tension or compression.

Since the entire truss is in equilibrium, each pin must be in equilibrium. We can use the fact that a pin is in equilibrium to draw its free-body diagram and write two equilibrium equations (Sec. 2.3A). Thus, if the truss contains n pins, we have $2n$ equations available, which can be solved for $2n$ unknowns. In the case of a simple truss, we have $m = 2n - 3$; that is, $2n = m + 3$, and the number of unknowns that we can determine from the free-body diagrams of the pins is $m + 3$. This means that we can find the forces in all the members, the two components of the reaction $\mathbf{R}_A$, and the reaction $\mathbf{R}_B$ by considering the free-body diagrams of the pins.

We can also use the fact that the entire truss is a rigid body in equilibrium to write three more equations involving the forces shown in the free-body diagram of Fig. 6.7a. Since these equations do not contain any new information, they are not independent of the equations associated with the free-body diagrams of the pins. Nevertheless, we can use them to determine the components of the reactions at the supports. The arrangement of pins and members in a simple truss is such that it is always possible to find a joint involving only two unknown forces. We can determine these forces by using the methods of Sec. 2.3C and then transferring their values to the adjacent joints, treating them as known quantities at these joints. We repeat this procedure until we have determined all unknown forces.

As an example, let's analyze the truss of Fig. 6.7 by considering the equilibrium of each pin successively, starting with a joint at which only two forces are unknown. In this truss, all pins are subjected to at least three unknown forces. Therefore, we must first determine the reactions at the supports by considering the entire truss as a free body and using the equations of equilibrium of a rigid body. In this way we find that $\mathbf{R}_A$ is vertical, and we determine the magnitudes of $\mathbf{R}_A$ and $\mathbf{R}_B$.

This reduces the number of unknown forces at joint A to two, and we can determine these forces by considering the equilibrium of pin A. The reaction $\mathbf{R}_A$ and the forces $\mathbf{F}_{AC}$ and $\mathbf{F}_{AD}$ exerted on pin A by members AC and AD, respectively, must form a force triangle. First we draw $\mathbf{R}_A$ (Fig. 6.8); noting that $\mathbf{F}_{AC}$ and $\mathbf{F}_{AD}$ are directed along AC and AD, respectively, we complete the triangle and determine the magnitude and sense of $\mathbf{F}_{AC}$ and $\mathbf{F}_{AD}$. The magnitudes F_{AC} and F_{AD} represent the forces in members AC and AD. Since $\mathbf{F}_{AC}$ is directed down and to the left—that is, *toward* joint A—member AC pushes on pin A and is in compression. (From Newton's third law, pin A pushes *on* member AC.) Since $\mathbf{F}_{AD}$ is directed *away* from joint A, member AD pulls on pin A and is in tension. (From Newton's third law, pin A pulls *away* from member AD.)

Photo 6.4 Because roof trusses, such as those shown, require support only at their ends, it is possible to construct buildings with large, unobstructed interiors.

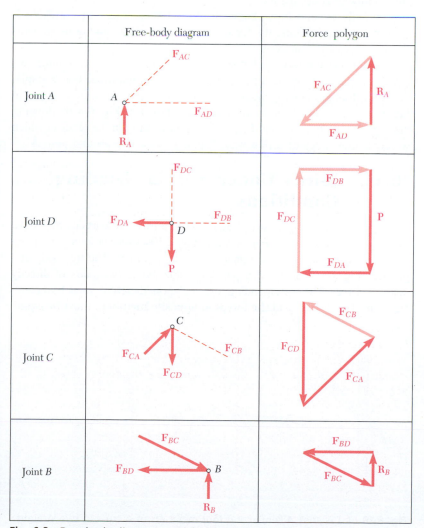

Fig. 6.8 Free-body diagrams and force polygons used to determine the forces on the pins and in the members of the truss in Fig. 6.7.

We can now proceed to joint D, where only two forces, $\mathbf{F}_{DC}$ and $\mathbf{F}_{DB}$, are still unknown. The other forces are the load $\mathbf{P}$, which is given, and the force $\mathbf{F}_{DA}$ exerted on the pin by member AD. As indicated previously, this force is equal and opposite to the force $\mathbf{F}_{AD}$ exerted by the same member on pin A. We can draw the force polygon corresponding to joint D, as shown in Fig. 6.8, and determine the forces $\mathbf{F}_{DC}$ and $\mathbf{F}_{DB}$ from that polygon. However, when more than three forces are involved, it is usually more convenient to solve the equations of equilibrium $\Sigma F_x = 0$ and $\Sigma F_y = 0$ for the two unknown forces. Since both of these forces are directed away from joint D, members DC and DB pull on the pin and are in tension.

Next, we consider joint C; its free-body diagram is shown in Fig. 6.8. Both $\mathbf{F}_{CD}$ and $\mathbf{F}_{CA}$ are known from the analysis of the preceding joints, so only $\mathbf{F}_{CB}$ is unknown. Since the equilibrium of each pin provides sufficient information to determine two unknowns, we can check our analysis at this joint. We draw the force triangle and determine the magnitude and sense of $\mathbf{F}_{CB}$. Since $\mathbf{F}_{CB}$ is directed toward joint C, member CB pushes on pin C and is in compression. The check is obtained by verifying that the force $\mathbf{F}_{CB}$ and member CB are parallel.

Finally, at joint B, we know all of the forces. Since the corresponding pin is in equilibrium, the force triangle must close, giving us an additional check of the analysis.

Note that the force polygons shown in Fig. 6.8 are not unique; we could replace each of them by an alternative configuration. For example, the force triangle corresponding to joint A could be drawn as shown in Fig. 6.9. We obtained the triangle actually shown in Fig. 6.8 by drawing the three forces $\mathbf{R}_A$, $\mathbf{F}_{AC}$, and $\mathbf{F}_{AD}$ in tip-to-tail fashion in the order in which we cross their lines of action when moving clockwise around joint A.

Fig. 6.9 Alternative force polygon for joint A in Fig. 6.8.

*6.1C Joints Under Special Loading Conditions

Some geometric arrangements of members in a truss are particularly simple to analyze by observation. For example, Fig. 6.10a shows a joint connecting four members lying along two intersecting straight lines. The free-body diagram of Fig. 6.10b shows that pin A is subjected to two pairs of directly opposite forces. The corresponding force polygon, therefore, must be a parallelogram (Fig. 6.10c), and **the forces in opposite members must be equal**.

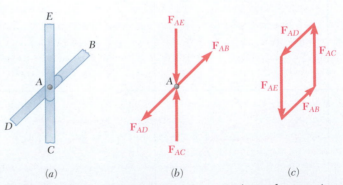

Fig. 6.10 (a) A joint A connecting four members of a truss in two straight lines; (b) free-body diagram of pin A; (c) force polygon (parallelogram) for pin A. Forces in opposite members are equal.

Consider next Fig. 6.11*a*, in which a joint connects three members and supports a load **P**. Two members lie along the same line, and load **P** acts along the third member. The free-body diagram of pin *A* and the corresponding force polygon are the same as in Fig. 6.10*b* and *c*, with F_{AE} replaced by load **P**. Thus, **the forces in the two opposite members must be equal, and the force in the other member must equal** *P*. Figure 6.11*b* shows a particular case of special interest. Since, in this case, no external load is applied to the joint, we have *P* = 0, and the force in member *AC* is zero. Member *AC* is said to be a **zero-force member**.

Now consider a joint connecting two members only. From Sec. 2.3A, we know that a particle acted upon by two forces is in equilibrium if the two forces have the same magnitude, same line of action, and opposite sense. In the case of the joint of Fig. 6.12*a*, which connects two members *AB* and *AD* lying along the same line, the forces in the two members must be equal for pin *A* to be in equilibrium. In the case of the joint of Fig. 6.12*b*, pin *A* cannot be in equilibrium unless the forces in both members are zero. Members connected as shown in Fig. 6.12*b*, therefore, must be **zero-force members**.

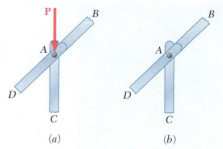

Fig. 6.11 (*a*) Joint *A* in a truss connects three members, two in a straight line and the third along the line of a load. Force in the third member equals the load. (*b*) If the load is zero, the third member is a zero-force member.

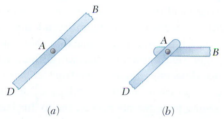

Fig. 6.12 (*a*) A joint in a truss connecting two members in a straight line. Forces in the members are equal. (*b*) If the two members are not in a straight line, they must be zero-force members.

Spotting joints that are under the special loading conditions just described will expedite the analysis of a truss. Consider, for example, a Howe truss loaded as shown in Fig. 6.13. We can recognize all of the members represented by green lines as zero-force members. Joint *C* connects three members, two of which lie in the same line, and is not subjected to any external load; member *BC* is thus a zero-force member. Applying the same reasoning to joint *K*, we find that member *JK* is also a zero-force member. But joint *J* is now in the same situation as joints *C* and *K*, so member *IJ* also must be a zero-force member. Examining joints *C*, *J*, and *K* also shows that the forces in members *AC* and *CE* are equal, that the forces in members *HJ* and *JL* are equal, and that the forces in members *IK* and *KL* are equal. Turning our attention to joint *I*, where the 20-kN load and member *HI* are collinear, we note that the force in member *HI* is 20 kN (tension) and that the forces in members *GI* and *IK* are equal. Hence, the forces in members *GI*, *IK*, and *KL* are equal.

Note that the conditions described here do not apply to joints *B* and *D* in Fig. 6.13, so it is wrong to assume that the force in member *DE* is 25 kN or that the forces in members *AB* and *BD* are equal. To determine the forces in these members and in all remaining members, you need to

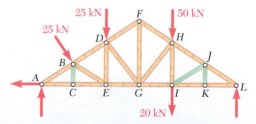

Fig. 6.13 An example of loading on a Howe truss; identifying special loading conditions.

Photo 6.5 Three-dimensional or space trusses are used for broadcast and power transmission line towers, roof framing, and spacecraft applications, such as components of the *International Space Station*.

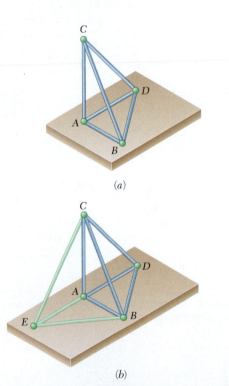

Fig. 6.14 (*a*) The most elementary space truss consists of six members joined at their ends to form a tetrahedron. (*b*) We can add three members at a time to three joints of an existing space truss, connecting the new members at a new joint, to build a larger simple space truss.

carry out the analysis of joints *A, B, D, E, F, G, H,* and *L* in the usual manner. Thus, until you have become thoroughly familiar with the conditions under which you can apply the rules described in this section, you should draw the free-body diagrams of all pins and write the corresponding equilibrium equations (or draw the corresponding force polygons) whether or not the joints being considered are under one of these special loading conditions.

A final remark concerning zero-force members: These members are not useless. For example, although the zero-force members of Fig. 6.13 do not carry any loads under the loading conditions shown, the same members would probably carry loads if the loading conditions were changed. Besides, even in the case considered, these members are needed to support the weight of the truss and to maintain the truss in the desired shape.

*6.1D Space Trusses

When several straight members of a truss are joined together at their extremities to form a three-dimensional configuration, the resulting structure is called a **space truss**. Recall from Sec. 6.1A that the most elementary two-dimensional rigid truss consists of three members joined at their extremities to form the sides of a triangle. By adding two members at a time to this basic configuration and connecting them at a new joint, we could obtain a larger rigid structure that we defined as a simple truss. Similarly, the most elementary rigid space truss consists of six members joined at their extremities to form the edges of a tetrahedron *ABCD* (Fig. 6.14*a*). By adding three members at a time to this basic configuration, such as *AE, BE,* and *CE* (Fig. 6.14*b*), attaching them to three existing joints, and connecting them at a new joint, we can obtain a larger rigid structure that we define as a **simple space truss**. (The four joints must not lie in a plane.) Note that the basic tetrahedron has six members and four joints, and every time we add three members, the number of joints increases by one. Therefore, we conclude that in a simple space truss the total number of members is $m = 3n - 6$, where n is the total number of joints.

If a space truss is to be completely constrained and if the reactions at its supports are to be statically determinate, the supports should consist of a combination of balls, rollers, and balls and sockets, providing six unknown reactions (see Sec. 4.3B). We can determine these unknown reactions by solving the six equations expressing that the three-dimensional truss is in equilibrium.

Although the members of a space truss are actually joined together by means of bolted or welded connections, we assume for analysis purposes that each joint consists of a ball-and-socket connection. Thus, no couple is applied to the members of the truss, and we can treat each member as a two-force member. The conditions of equilibrium for each joint are expressed by the three equations $\Sigma F_x = 0$, $\Sigma F_y = 0$, and $\Sigma F_z = 0$. Thus, in the case of a simple space truss containing n joints, writing the conditions of equilibrium for each joint yields $3n$ equations. Since $m = 3n - 6$, these equations suffice to determine all unknown forces (forces in m members and six reactions at the supports). However, to avoid the necessity of solving simultaneous equations, you should take care to select joints in such an order that no selected joint involves more than three unknown forces.

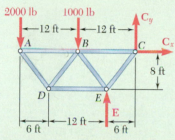

Sample Problem 6.1

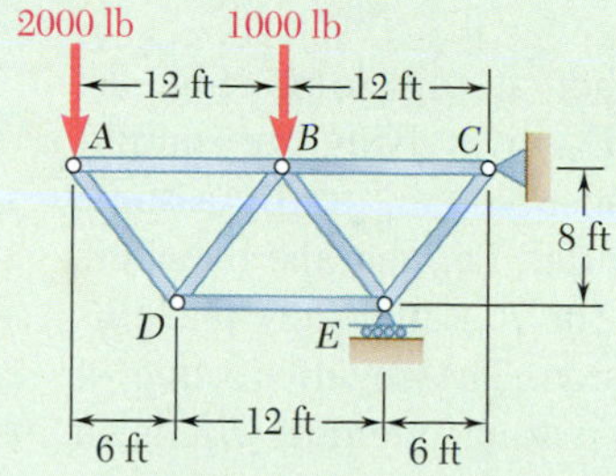

Using the method of joints, determine the force in each member of the truss shown.

STRATEGY: To use the method of joints, you start with an analysis of the free-body diagram of the entire truss. Then look for a joint connecting only two members as a starting point for the calculations. In this example, we start at joint A and proceed through joints D, B, E, and C, but you could also start at joint C and proceed through joints E, B, D, and A.

MODELING and ANALYSIS: You can combine these steps for each joint of the truss in turn. Draw a free-body diagram; draw a force polygon or write the equilibrium equations; and solve for the unknown forces.

Entire Truss. Draw a free-body diagram of the entire truss (Fig. 1); external forces acting on this free body are the applied loads and the reactions at C and E. Write the equilibrium equations, taking moments about C.

$$+\curvearrowleft \Sigma M_C = 0: \quad (2000 \text{ lb})(24 \text{ ft}) + (1000 \text{ lb})(12 \text{ ft}) - E(6 \text{ ft}) = 0$$
$$E = +10{,}000 \text{ lb} \qquad\qquad \mathbf{E} = 10{,}000 \text{ lb}\uparrow$$

$$\xrightarrow{+} \Sigma F_x = 0: \qquad\qquad\qquad\qquad\qquad \mathbf{C}_x = 0$$
$$+\uparrow \Sigma F_y = 0: \quad -2000 \text{ lb} - 1000 \text{ lb} + 10{,}000 \text{ lb} + C_y = 0$$
$$C_y = -7000 \text{ lb} \qquad\qquad \mathbf{C}_y = 7000 \text{ lb}\downarrow$$

Fig. 1 Free-body diagram of entire truss.

Joint A. This joint is subject to only two unknown forces: the forces exerted by AB and those by AD. Use a force triangle to determine $\mathbf{F}_{AB}$ and $\mathbf{F}_{AD}$ (Fig. 2). Note that member AB pulls on the joint so AB is in tension, and member AD pushes on the joint so AD is in compression. Obtain the magnitudes of the two forces from the proportion

$$\frac{2000 \text{ lb}}{4} = \frac{F_{AB}}{3} = \frac{F_{AD}}{5}$$

$$F_{AB} = 1500 \text{ lb } T \quad \blacktriangleleft$$
$$F_{AD} = 2500 \text{ lb } C \quad \blacktriangleleft$$

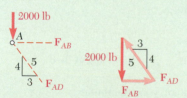

Fig. 2 Free-body diagram of joint A.

Joint D. Since you have already determined the force exerted by member AD, only two unknown forces are now involved at this joint. Again, use a force triangle to determine the unknown forces in members DB and DE (Fig. 3).

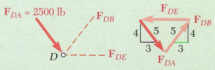

Fig. 3 Free-body diagram of joint D.

(continued)

$$F_{DB} = F_{DA} \qquad\qquad\qquad F_{DB} = 2500 \text{ lb } T \;\blacktriangleleft$$
$$F_{DE} = 2\left(\tfrac{3}{5}\right)F_{DA} \qquad\qquad F_{DE} = 3000 \text{ lb } C \;\blacktriangleleft$$

Joint *B*. Since more than three forces act at this joint (Fig. 4), determine the two unknown forces $\mathbf{F}_{BC}$ and $\mathbf{F}_{BE}$ by solving the equilibrium equations $\Sigma F_x = 0$ and $\Sigma F_y = 0$. Suppose you arbitrarily assume that both unknown forces act away from the joint; i.e., that the members are in tension. The positive value obtained for F_{BC} indicates that this assumption is correct; member *BC* is in tension. The negative value of F_{BE} indicates that the second assumption is wrong; member *BE* is in compression.

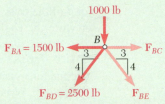

Fig. 4 Free-body diagram of joint *B*.

$$+\uparrow\Sigma F_y = 0: \quad -1000 - \tfrac{4}{5}(2500) - \tfrac{4}{5}F_{BE} = 0$$
$$F_{BE} = -3750 \text{ lb} \qquad F_{BE} = 3750 \text{ lb } C \;\blacktriangleleft$$

$$\xrightarrow{+}\Sigma F_x = 0: \quad F_{BC} - 1500 - \tfrac{3}{5}(2500) - \tfrac{3}{5}(3750) = 0$$
$$F_{BC} = +5250 \text{ lb} \qquad F_{BC} = 5250 \text{ lb } T \;\blacktriangleleft$$

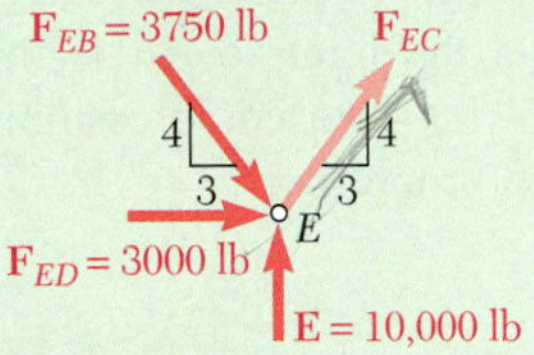

Joint *E*. Assume the unknown force $\mathbf{F}_{EC}$ acts away from the joint (Fig. 5). Summing *x* components, you obtain

$$\xrightarrow{+}\Sigma F_x = 0: \quad \tfrac{3}{5}F_{EC} + 3000 + \tfrac{3}{5}(3750) = 0$$
$$F_{EC} = -8750 \text{ lb} \qquad F_{EC} = 8750 \text{ lb } C \;\blacktriangleleft$$

Summing *y* components, you obtain a check of your computations:

$$+\uparrow\Sigma F_y = 10{,}000 - \tfrac{4}{5}(3750) - \tfrac{4}{5}(8750)$$
$$= 10{,}000 - 3000 - 7000 = 0 \qquad \text{(checks)}$$

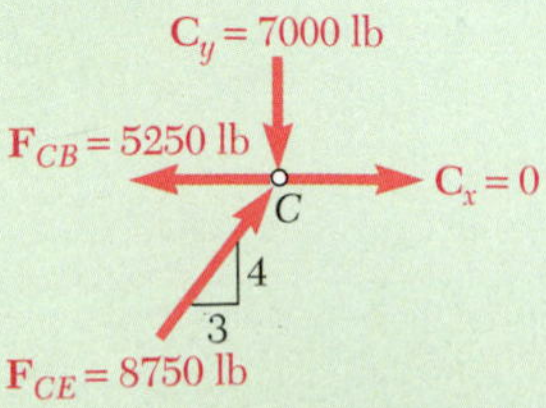

REFLECT and THINK: Using the computed values of $\mathbf{F}_{CB}$ and $\mathbf{F}_{CE}$, you can determine the reactions $\mathbf{C}_x$ and $\mathbf{C}_y$ by considering the equilibrium of Joint *C* (Fig. 6). Since these reactions have already been determined from the equilibrium of the entire truss, this provides two checks of your computations. You can also simply use the computed values of all forces acting on the joint (forces in members and reactions) and check that the joint is in equilibrium:

$$\xrightarrow{+}\Sigma F_x = -5250 + \tfrac{3}{5}(8750) = -5250 + 5250 = 0 \qquad \text{(checks)}$$
$$+\uparrow\Sigma F_y = -7000 + \tfrac{4}{5}(8750) = -7000 + 7000 = 0 \qquad \text{(checks)}$$

$F_{EB} = 3750 \text{ lb}$ F_{EC}

$F_{ED} = 3000 \text{ lb}$

$\mathbf{E} = 10{,}000 \text{ lb}$

Fig. 5 Free-body diagram of joint *E*.

$C_y = 7000 \text{ lb}$

$F_{CB} = 5250 \text{ lb}$ $C_x = 0$

$F_{CE} = 8750 \text{ lb}$

Fig. 6 Free-body diagram of joint *C*.

SOLVING PROBLEMS
ON YOUR OWN

In this section, you learned to use the **method of joints** to determine the forces in the members of a **simple truss**; that is, a truss that can be constructed from a basic triangular truss by adding to it two new members at a time and connecting them at a new joint.

The method consists of the following steps:

1. Draw a free-body diagram of the entire truss, and use this diagram to determine the reactions at the supports.

2. Locate a joint connecting only two members, and draw the free-body diagram of its pin. Use this free-body diagram to determine the unknown force in each of the two members. If only three forces are involved (the two unknown forces and a known one), you will probably find it more convenient to draw and solve the corresponding force triangle. If more than three forces are involved, you should write and solve the equilibrium equations for the pin, $\Sigma F_x = 0$ and $\Sigma F_y = 0$, assuming that the members are in tension. A positive answer means that the member is in tension, a negative answer means that the member is in compression. Once you have found the forces, enter their values on a sketch of the truss with T for tension and C for compression.

3. Next, locate a joint where the forces in only two of the connected members are still unknown. Draw the free-body diagram of the pin and use it as indicated in Step 2 to determine the two unknown forces.

4. Repeat this procedure until you have found the forces in all the members of the truss. Since you previously used the three equilibrium equations associated with the free-body diagram of the entire truss to determine the reactions at the supports, you will end up with three extra equations. These equations can be used to check your computations.

5. Note that the choice of the first joint is not unique. Once you have determined the reactions at the supports of the truss, you can choose either of two joints as a starting point for your analysis. In Sample Prob. 6.1, we started at joint A and proceeded through joints D, B, E, and C, but we could also have started at joint C and proceeded through joints E, B, D, and A. On the other hand, having selected a first joint, you may in some cases reach a point in your analysis beyond which you cannot proceed. You must then start again from another joint to complete your solution.

Keep in mind that the analysis of a simple truss always can be carried out by the method of joints. Also remember that it is helpful to outline your solution *before* starting any computations.

PROBLEMS

6.1 through 6.8 Using the method of joints, determine the force in each member of the truss shown. State whether each member is in tension or compression.

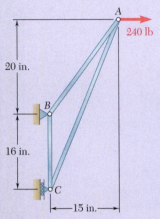

Fig. P6.1

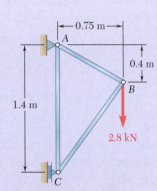

Fig. P6.2

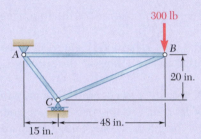

Fig. P6.3

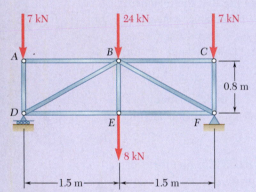

Fig. P6.4

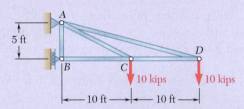

Fig. P6.5

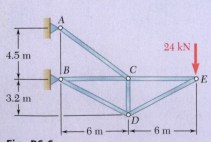

Fig. P6.6

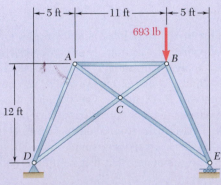

Fig. P6.7

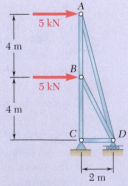

Fig. P6.8

6.9 and 6.10 Determine the force in each member of the truss shown. State whether each member is in tension or compression.

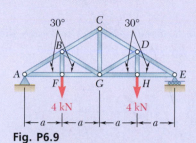

Fig. P6.9

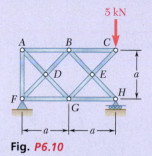

Fig. *P6.10*

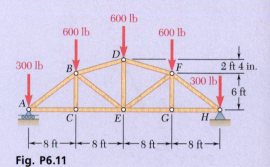

Fig. P6.11

6.11 Determine the force in each member of the Gambrel roof truss shown. State whether each member is in tension or compression.

6.12 Determine the force in each member of the Howe roof truss shown. State whether each member is in tension or compression.

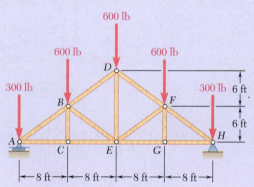

Fig. P6.12

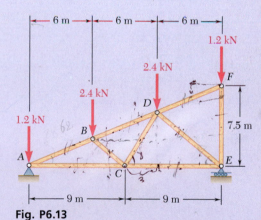

Fig. P6.13

6.13 Determine the force in each member of the roof truss shown. State whether each member is in tension or compression.

6.14 Determine the force in each member of the Fink roof truss shown. State whether each member is in tension or compression.

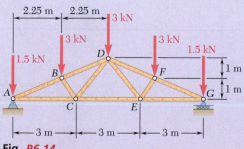

Fig. *P6.14*

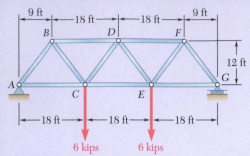

Fig. P6.15

6.15 Determine the force in each member of the Warren bridge truss shown. State whether each member is in tension or compression.

6.16 Solve Prob. 6.15 assuming that the load applied at E has been removed.

6.17 Determine the force in each member of the Pratt roof truss shown. State whether each member is in tension or compression.

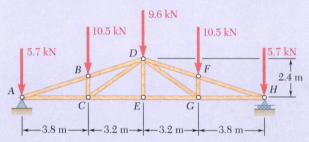

Fig. P6.17

6.18 The truss shown is one of several supporting an advertising panel. Determine the force in each member of the truss for a wind load equivalent to the two forces shown. State whether each member is in tension or compression.

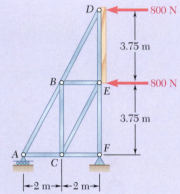

Fig. P6.18

6.19 Determine the force in each member of the Pratt bridge truss shown. State whether each member is in tension or compression.

6.20 Solve Prob. 6.19 assuming that the load applied at G has been removed.

6.21 Determine the force in each of the members located to the left of FG for the scissors roof truss shown. State whether each member is in tension or compression.

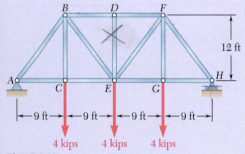

Fig. P6.19

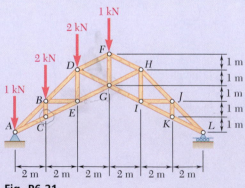

Fig. P6.21

6.22 Determine the force in member DE and in each of the members located to the left of DE for the inverted Howe roof truss shown. State whether each member is in tension or compression.

6.23 Determine the force in each of the members located to the right of DE for the inverted Howe roof truss shown. State whether each member is in tension or compression.

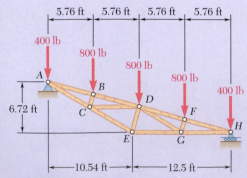

Fig. P6.22 and P6.23

6.24 The portion of truss shown represents the upper part of a power transmission line tower. For the given loading, determine the force in each of the members located above *HJ*. State whether each member is in tension or compression.

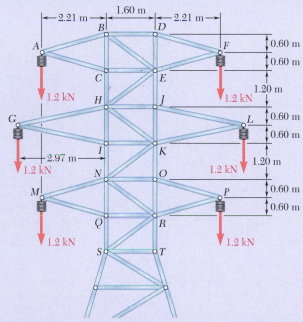

Fig. P6.24

6.25 For the tower and loading of Prob. 6.24 and knowing that $F_{CH} = F_{EJ} = 1.2$ kN *C* and $F_{EH} = 0$, determine the force in member *HJ* and in each of the members located between *HJ* and *NO*. State whether each member is in tension or compression.

6.26 Solve Prob. 6.24 assuming that the cables hanging from the right side of the tower have fallen to the ground.

6.27 and 6.28 Determine the force in each member of the truss shown. State whether each member is in tension or compression.

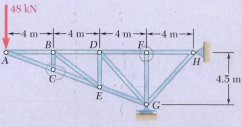

Fig. P6.28

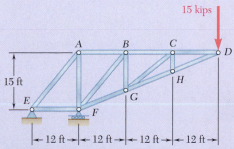

Fig. P6.27

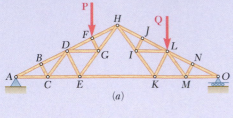

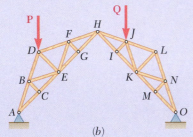

Fig. **P6.31**

6.29 Determine whether the trusses of Probs. 6.31a, 6.32a, and 6.33a are simple trusses.

6.30 Determine whether the trusses of Probs. 6.31b, 6.32b, and 6.33b are simple trusses.

6.31 For the given loading, determine the zero-force members in each of the two trusses shown.

6.32 For the given loading, determine the zero-force members in each of the two trusses shown.

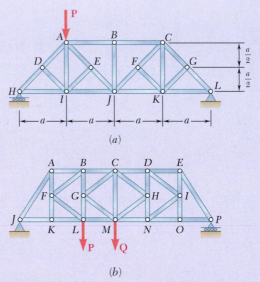

Fig. **P6.32**

6.33 For the given loading, determine the zero-force members in each of the two trusses shown.

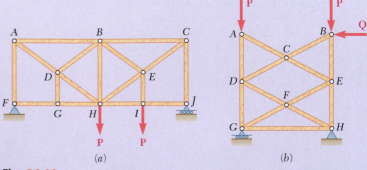

Fig. **P6.33**

6.34 Determine the zero-force members in the truss of (a) Prob. 6.21, (b) Prob. 6.27.

*6.35 The truss shown consists of six members and is supported by a short link at *A*, two short links at *B*, and a ball-and-socket at *D*. Determine the force in each of the members for the given loading.

*6.36 The truss shown consists of six members and is supported by a ball-and-socket at *B*, a short link at *C*, and two short links at *D*. Determine the force in each of the members for **P** = (−2184 N)**j** and **Q** = 0.

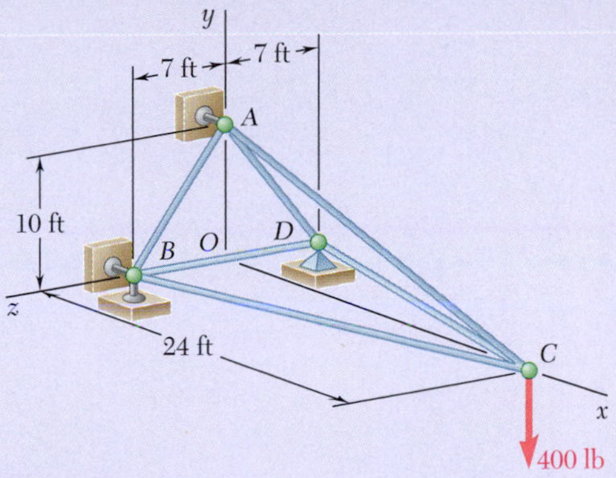

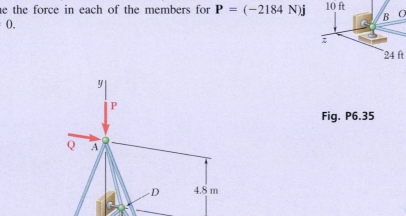

Fig. P6.35

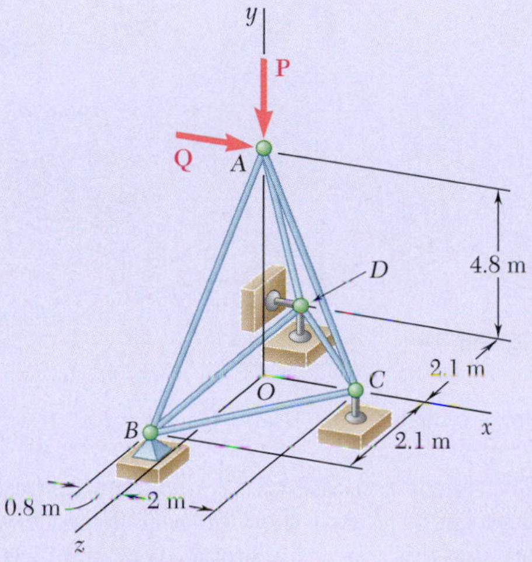

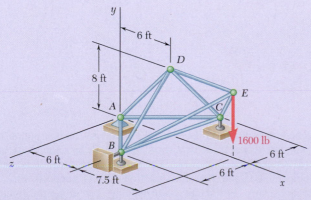

Fig. P6.36 and P6.37

*6.37 The truss shown consists of six members and is supported by a ball-and-socket at *B*, a short link at *C*, and two short links at *D*. Determine the force in each of the members for **P** = 0 and **Q** = (2968 N)**i**.

*6.38 The truss shown consists of nine members and is supported by a ball-and-socket at *A*, two short links at *B*, and a short link at *C*. Determine the force in each of the members for the given loading.

Fig. P6.38

*6.39 The truss shown consists of nine members and is supported by a ball-and-socket at *B*, a short link at *C*, and two short links at *D*. (*a*) Check that this truss is a simple truss, that it is completely constrained, and that the reactions at its supports are statically determinate. (*b*) Determine the force in each member for **P** = (−1200 N)**j** and **Q** = 0.

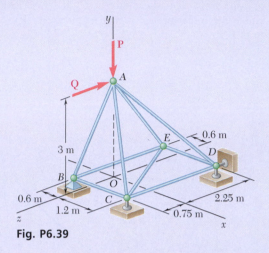

Fig. P6.39

*6.40 Solve Prob. 6.39 for **P** = 0 and **Q** = (−900 N)**k**.

*6.41 The truss shown consists of 18 members and is supported by a ball-and-socket at *A*, two short links at *B*, and one short link at *G*. (*a*) Check that this truss is a simple truss, that it is completely constrained, and that the reactions at its supports are statically determinate. (*b*) For the given loading, determine the force in each of the six members joined at *E*.

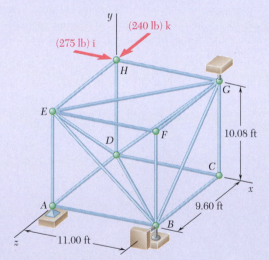

Fig. P6.41 and P6.42

*6.42 The truss shown consists of 18 members and is supported by a ball-and-socket at *A*, two short links at *B*, and one short link at *G*. (*a*) Check that this truss is a simple truss, that it is completely constrained, and that the reactions at its supports are statically determinate. (*b*) For the given loading, determine the force in each of the six members joined at *G*.

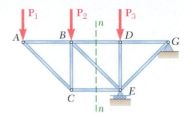

6.2   OTHER TRUSS ANALYSES

The method of joints is most effective when we want to determine the forces in all the members of a truss. If, however, we need to determine the force in only one member or in a very few members, the method of sections is more efficient.

6.2A   The Method of Sections

Assume, for example, that we want to determine the force in member BD of the truss shown in Fig. 6.15a. To do this, we must determine the force with which member BD acts on either joint B or joint D. If we were to use the method of joints, we would choose either joint B or joint D as a free body. However, we can also choose a larger portion of the truss that is composed of several joints and members, provided that the force we want to find is one of the external forces acting on that portion. If, in addition, we choose the portion of the truss as a free body where a total of only three unknown forces act upon it, we can obtain the desired force by solving the equations of equilibrium for this portion of the truss. In practice, we isolate a portion of the truss by *passing a section* through three members of the truss, one of which is the desired member. That is, we draw a line that divides the truss into two completely separate parts but does not intersect more than three members. We can then use as a free body either of the two portions of the truss obtained after the intersected members have been removed.[†]

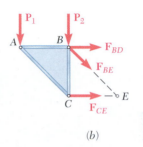

(b)

Fig. 6.15 (a) We can pass a section nn through the truss, dividing the three members BD, BE, and CE. (b) Free-body diagram of portion ABC of the truss. We assume that members BD, BE, and CE are in tension.

In Fig. 6.15a, we have passed the section nn through members BD, BE, and CE, and we have chosen the portion ABC of the truss as the free body (Fig. 6.15b). The forces acting on this free body are the loads $\mathbf{P}_1$ and $\mathbf{P}_2$ at points A and B and the three unknown forces $\mathbf{F}_{BD}$, $\mathbf{F}_{BE}$, and $\mathbf{F}_{CE}$. Since we do not know whether the members removed are in tension or compression, we have arbitrarily drawn the three forces away from the free body as if the members are in tension.

We use the fact that the rigid body ABC is in equilibrium to write three equations that we can solve for the three unknown forces. If we want to determine only force $\mathbf{F}_{BD}$, say, we need write only one equation, provided that the equation does not contain the other unknowns. Thus, the equation $\Sigma M_E = 0$ yields the value of the magnitude F_{BD} (Fig. 6.15b). A positive sign in the answer will indicate that our original assumption regarding the sense of $\mathbf{F}_{BD}$ was correct and that member BD is in tension; a negative sign will indicate that our assumption was incorrect and that BD is in compression.

On the other hand, if we want to determine only force $\mathbf{F}_{CE}$, we need to write an equation that does not involve $\mathbf{F}_{BD}$ or $\mathbf{F}_{BE}$; the appropriate equation is $\Sigma M_B = 0$. Again, a positive sign for the magnitude F_{CE} of the desired force indicates a correct assumption, that is, tension; and a negative sign indicates an incorrect assumption, that is, compression.

If we want to determine only force $\mathbf{F}_{BE}$, the appropriate equation is $\Sigma F_y = 0$. Whether the member is in tension or compression is again determined from the sign of the answer.

[†]In the analysis of some trusses, we can pass sections through more than three members, provided we can write equilibrium equations involving only one unknown that we can use to determine the forces in one, or possibly two, of the intersected members. See Probs. 6.61 through 6.64.

If we determine the force in only one member, no independent check of the computation is available. However, if we calculate all of the unknown forces acting on the free body, we can check the computations by writing an additional equation. For instance, if we determine $\mathbf{F}_{BD}$, $\mathbf{F}_{BE}$, and $\mathbf{F}_{CE}$ as indicated previously, we can check the work by verifying that $\Sigma F_x = 0$.

6.2B Trusses Made of Several Simple Trusses

Consider two simple trusses *ABC* and *DEF*. If we connect them by three bars *BD*, *BE*, and *CE* as shown in Fig. 6.16*a*, together they form a rigid truss *ABDF*. We can also combine trusses *ABC* and *DEF* into a single rigid truss by joining joints *B* and *D* at a single joint *B* and connecting joints *C* and *E* by a bar *CE* (Fig. 6.16*b*). This is known as a *Fink truss*. The trusses of Fig. 6.16*a* and *b* are *not* simple trusses; you cannot construct them from a triangular truss by adding successive pairs of members as described in Sec. 6.1A. They are rigid trusses, however, as you can check by comparing the systems of connections used to hold the simple trusses *ABC* and *DEF* together (three bars in Fig. 6.16*a*, one pin and one bar in Fig. 6.16*b*) with the systems of supports discussed in Sec. 4.1. Trusses made of several simple trusses rigidly connected are known as **compound trusses**.

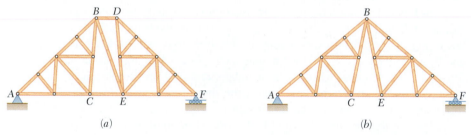

Fig. 6.16 Compound trusses. (*a*) Two simple trusses *ABC* and *DEF* connected by three bars. (*b*) Two simple trusses *ABC* and *DEF* connected by one joint and one bar (a Fink truss).

In a compound truss, the number of members *m* and the number of joints *n* are still related by the formula $m = 2n - 3$. You can verify this by observing that if a compound truss is supported by a frictionless pin and a roller (involving three unknown reactions), the total number of unknowns is $m + 3$, and this number must be equal to the number $2n$ of equations obtained by expressing that the *n* pins are in equilibrium. It follows that $m = 2n - 3$.

Compound trusses supported by a pin and a roller or by an equivalent system of supports are *statically determinate, rigid,* and *completely constrained*. This means that we can determine all of the unknown reactions and the forces in all of the members by using the methods of statics, and the truss will neither collapse nor move. However, the only way to determine all of the forces in the members using the method of joints requires solving a large number of simultaneous equations. In the case of the compound truss of Fig. 6.16*a*, for example, it is more efficient to pass a section through members *BD, BE*, and *CE* to determine the forces in these members.

Suppose, now, that the simple trusses *ABC* and *DEF* are connected by *four* bars; *BD, BE, CD*, and *CE* (Fig. 6.17). The number of members *m* is now larger than $2n - 3$. This truss is said to be **overrigid**, and one of the four members *BD, BE, CD*, or *CE* is **redundant**. If the truss is supported by a pin at *A* and a roller at *F*, the total number of unknowns is $m + 3$. Since $m > 2n - 3$, the number $m + 3$ of unknowns is now larger than the number $2n$ of available independent equations; the truss is *statically indeterminate*.

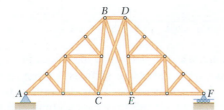

Fig. 6.17 A statically indeterminate, overrigid compound truss, due to a redundant member.

Finally, let us assume that the two simple trusses *ABC* and *DEF* are joined by a single pin, as shown in Fig. 6.18*a*. The number of members, *m*, is now smaller than $2n - 3$. If the truss is supported by a pin at *A* and a roller at *F*, the total number of unknowns is $m + 3$. Since $m < 2n - 3$, the number $m + 3$ of unknowns is now smaller than the number $2n$ of equilibrium equations that need to be satisfied. This truss is **nonrigid** and will collapse under its own weight. However, if two pins are used to support it, the truss becomes *rigid* and will not collapse (Fig. 6.18*b*). Note that the total number of unknowns is now $m + 4$ and is equal to the number $2n$ of equations.

More generally, if the reactions at the supports involve *r* unknowns, the condition for a compound truss to be statically determinate, rigid, and completely constrained is $m + r = 2n$. However, although this condition is necessary, it is not sufficient for the equilibrium of a structure that ceases to be rigid when detached from its supports (see Sec. 6.3B).

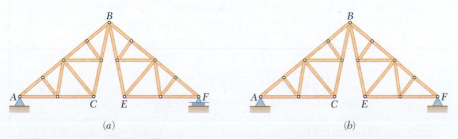

(a)   (b)

Fig. 6.18 Two simple trusses joined by a pin. (*a*) Supported by a pin and a roller, the truss will collapse under its own weight. (*b*) Supported by two pins, the truss becomes rigid and does not collapse.

Sample Problem 6.2

Determine the forces in members *EF* and *GI* of the truss shown.

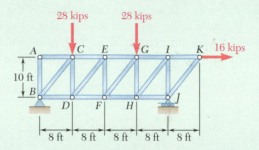

STRATEGY: You are asked to determine the forces in only two of the members in this truss, so the method of sections is more appropriate than the method of joints. You can use a free-body diagram of the entire truss to help determine the reactions, and then pass sections through the truss to isolate parts of it for calculating the desired forces.

MODELING and ANALYSIS: You can go through the steps that follow for the determination of the support reactions, and then for the analysis of portions of the truss.

Free-Body: Entire Truss. Draw a free-body diagram of the entire truss. External forces acting on this free body consist of the applied loads and the reactions at *B* and *J* (Fig. 1). Write and solve the following equilibrium equations.

$+\circlearrowleft \Sigma M_B = 0$:

$$-(28 \text{ kips})(8 \text{ ft}) - (28 \text{ kips})(24 \text{ ft}) - (16 \text{ kips})(10 \text{ ft}) + J(32 \text{ ft}) = 0$$
$$J = +33 \text{ kips} \qquad \mathbf{J} = 33 \text{ kips} \uparrow$$

$\xrightarrow{+} \Sigma F_x = 0$: $\qquad B_x + 16 \text{ kips} = 0$

$$B_x = -16 \text{ kips} \qquad \mathbf{B}_x = 16 \text{ kips} \leftarrow$$

$+\circlearrowleft \Sigma M_J = 0$:

$$(28 \text{ kips})(24 \text{ ft}) + (28 \text{ kips})(8 \text{ ft}) - (16 \text{ kips})(10 \text{ ft}) - B_y(32 \text{ ft}) = 0$$
$$B_y = +23 \text{ kips} \qquad \mathbf{B}_y = 23 \text{ kips} \uparrow$$

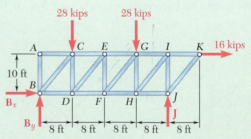

Fig. 1 Free-body diagram of entire truss.

Force in Member *EF*. Pass section *nn* through the truss diagonally so that it intersects member *EF* and only two additional members (Fig. 2).

Remove the intersected members and choose the left-hand portion of the truss as a free body (Fig. 3). Three unknowns are involved; to eliminate the two horizontal forces, we write

$$+\uparrow \Sigma F_y = 0: \qquad +23 \text{ kips} - 28 \text{ kips} - F_{EF} = 0$$
$$F_{EF} = -5 \text{ kips}$$

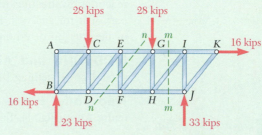

Fig. 2 Sections *nn* and *mm* that will be used to analyze members *EF* and *GI*.

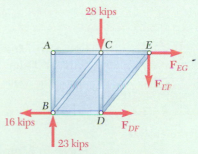

Fig. 3 Free-body diagram to analyze member *EF*.

The sense of $\mathbf{F}_{EF}$ was chosen assuming member *EF* to be in tension; the negative sign indicates that the member is in compression.

$$F_{EF} = 5 \text{ kips } C \quad \blacktriangleleft$$

Force in Member *GI*. Pass section *mm* through the truss vertically so that it intersects member *GI* and only two additional members (Fig. 2). Remove the intersected members and choose the right-hand portion of the truss as a free body (Fig. 4). Again, three unknown forces are involved; to eliminate the two forces passing through point *H*, sum the moments about that point.

$$+\curvearrowleft \Sigma M_H = 0: \qquad (33 \text{ kips})(8 \text{ ft}) - (16 \text{ kips})(10 \text{ ft}) + F_{GI}(10 \text{ ft}) = 0$$
$$F_{GI} = -10.4 \text{ kips} \qquad F_{GI} = 10.4 \text{ kips } C \quad \blacktriangleleft$$

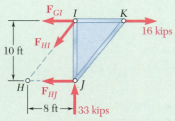

Fig. 4 Free-body diagram to analyze member *GI*.

REFLECT and THINK: Note that a section passed through a truss does not have to be vertical or horizontal; it can be diagonal as well. Choose the orientation that cuts through no more than three members of unknown force and also gives you the simplest part of the truss for which you can write equilibrium equations and determine the unknowns.

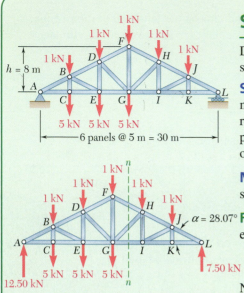

5 kN 5 kN 5 kN

6 panels @ 5 m = 30 m

Fig. 1 Free-body diagram of entire truss.

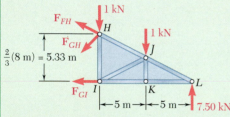

1 kN 1 kN 1 kN

5 kN 5 kN 5 kN

12.50 kN

$\alpha = 28.07°$

7.50 kN

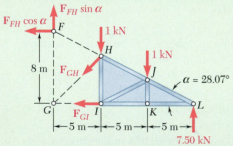

Fig. 2 Free-body diagram to analyze member *GI*.

F_{FH} F_{GH}

$\frac{2}{3}(8\text{ m}) = 5.33\text{ m}$

F_{GI} 5 m 5 m 7.50 kN

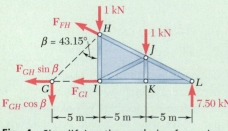

Fig. 3 Simplifying the analysis of member *FH* by first sliding its force to point *F*.

$F_{FH}\cos\alpha$ $F_{FH}\sin\alpha$

8 m F_{GH}

F_{GI} 5 m 5 m 5 m $\alpha = 28.07°$

7.50 kN

Fig. 4 Simplifying the analysis of member *GH* by first sliding its force to point *G*.

F_{FH} $F_{GH}\sin\beta$ $\beta = 43.15°$

$F_{GH}\cos\beta$ F_{GI} 5 m 5 m 5 m

7.50 kN

Sample Problem 6.3

Determine the forces in members *FH, GH,* and *GI* of the roof truss shown.

STRATEGY: You are asked to determine the forces in only three members of the truss, so use the method of sections. Determine the reactions by treating the entire truss as a free body and then isolate part of it for analysis. In this case, you can use the same smaller part of the truss to determine all three desired forces.

MODELING and ANALYSIS: Your reasoning and computation should go something like the sequence given here.

Free Body: Entire Truss. From the free-body diagram of the entire truss (Fig. 1), find the reactions at *A* and *L*:

$$\mathbf{A} = 12.50\text{ kN}\uparrow \qquad \mathbf{L} = 7.50\text{ kN}\uparrow$$

Note that

$$\tan\alpha = \frac{FG}{GL} = \frac{8\text{ m}}{15\text{ m}} = 0.5333 \qquad \alpha = 28.07°$$

Force in Member *GI*. Pass section *nn* vertically through the truss (Fig. 1). Using the portion *HLI* of the truss as a free body (Fig. 2), obtain the value of F_{GI}:

$$+\curvearrowleft\Sigma M_H = 0: \qquad (7.50\text{ kN})(10\text{ m}) - (1\text{ kN})(5\text{ m}) - F_{GI}(5.33\text{ m}) = 0$$
$$F_{GI} = +13.13\text{ kN} \qquad F_{GI} = 13.13\text{ kN } T \quad \blacktriangleleft$$

Force in Member *FH*. Determine the value of F_{FH} from the equation $\Sigma M_G = 0$. To do this, move $\mathbf{F}_{FH}$ along its line of action until it acts at point *F*, where you can resolve it into its *x* and *y* components (Fig. 3). The moment of $\mathbf{F}_{FH}$ with respect to point *G* is now $(F_{FH}\cos\alpha)(8\text{ m})$.

$$+\curvearrowleft\Sigma M_G = 0:$$
$$(7.50\text{ kN})(15\text{ m}) - (1\text{ kN})(10\text{ m}) - (1\text{ kN})(5\text{ m}) + (F_{FH}\cos\alpha)(8\text{ m}) = 0$$
$$F_{FH} = -13.81\text{ kN} \qquad F_{FH} = 13.81\text{ kN } C \quad \blacktriangleleft$$

Force in Member *GH*. First note that

$$\tan\beta = \frac{GI}{HI} = \frac{5\text{ m}}{\frac{2}{3}(8\text{ m})} = 0.9375 \qquad \beta = 43.15°$$

Then determine the value of F_{GH} by resolving the force $\mathbf{F}_{GH}$ into *x* and *y* components at point *G* (Fig. 4) and solving the equation $\Sigma M_L = 0$.

$$+\curvearrowleft\Sigma M_L = 0: \qquad (1\text{ kN})(10\text{ m}) + (1\text{ kN})(5\text{ m}) + (F_{GH}\cos\beta)(15\text{ m}) = 0$$
$$F_{GH} = -1.371\text{ kN} \qquad F_{GH} = 1.371\text{ kN } C \quad \blacktriangleleft$$

REFLECT and THINK: Sometimes you should resolve a force into components to include it in the equilibrium equations. By first sliding this force along its line of action to a more strategic point, you might eliminate one of its components from a moment equilibrium equation.

SOLVING PROBLEMS ON YOUR OWN

The **method of joints** that you studied in Sec. 6.1 is usually the best method to use when you need to find the forces *in all of the members* of a simple truss. However, the **method of sections**, which was covered in this section, is more efficient when you need to find the force *in only one member* or the forces *in a very few members* of a simple truss. The method of sections also must be used when the truss *is not a simple truss.*

A. To determine the force in a given truss member by the method of sections, follow these steps:

1. Draw a free-body diagram of the entire truss, and use this diagram to determine the reactions at the supports.

2. Pass a section through three members of the truss, one of which is the member whose force you want to find. After you cut through these members, you will have two separate portions of truss.

3. Select one of these two portions of truss and draw its free-body diagram. This diagram should include the external forces applied to the selected portion as well as the forces exerted on it by the intersected members that were removed.

4. You can now write three equilibrium equations that can be solved for the forces in the three intersected members.

5. An alternative approach is to write a single equation that can be solved for the force in the desired member. To do so, first observe whether the forces exerted by the other two members on the free body are parallel or whether their lines of action intersect.

  a. If these forces are parallel, you can eliminate them by writing an equilibrium equation involving *components in a direction perpendicular* to these two forces.

  b. If their lines of action intersect at a point *H*, you can eliminate them by writing an equilibrium equation involving *moments about H.*

6. Keep in mind that the section you use must intersect three members only. The reason is that the equilibrium equations in Step 4 can be solved for only three unknowns. However, you can pass a section through more than three members to find the force in one of those members if you can write an equilibrium equation containing only that force as an unknown. Such special situations are found in Probs. 6.61 through 6.64.

B. About completely constrained and determinate trusses:

1. Any simple truss that is simply supported is a completely constrained and determinate truss.

2. To determine whether any other truss is or is not completely constrained and determinate, count the number m of its members, the number n of its joints, and the number r of the reaction components at its supports. Compare the sum $m + r$ representing the number of unknowns and the product $2n$ representing the number of available independent equilibrium equations.

    a. If $m + r < 2n$, there are fewer unknowns than equations. Thus, some of the equations cannot be satisfied, and the truss is only *partially constrained*.

    b. If $m + r > 2n$, there are more unknowns than equations. Thus, some of the unknowns cannot be determined, and the truss is *indeterminate*.

    c. If $m + r = 2n$, there are as many unknowns as there are equations. This, however, does not mean that all of the unknowns can be determined and that all of the equations can be satisfied. To find out whether the truss is *completely* or *improperly constrained,* try to determine the reactions at its supports and the forces in its members. If you can find all of them, the truss is *completely constrained and determinate*.

Problems

6.43 A Mansard roof truss is loaded as shown. Determine the force in members *DF*, *DG*, and *EG*.

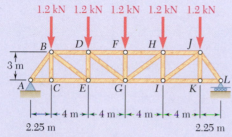

Fig. P6.43 and P6.44

6.44 A Mansard roof truss is loaded as shown. Determine the force in members *GI*, *HI*, and *HJ*.

6.45 Determine the force in members *BD* and *CD* of the truss shown.

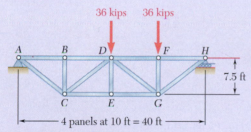

Fig. P6.45 and P6.46

6.46 Determine the force in members *DF* and *DG* of the truss shown.

6.47 Determine the force in members *CD* and *DF* of the truss shown.

6.48 Determine the force in members *FG* and *FH* of the truss shown.

6.49 Determine the force in members *CD* and *DF* of the truss shown.

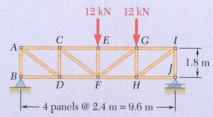

Fig. P6.47 and P6.48

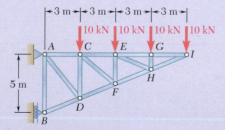

Fig. P6.49 and P6.50

6.50 Determine the force in members *CE* and *EF* of the truss shown.

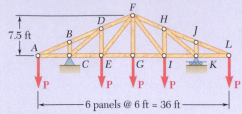

7.5 ft

6 panels @ 6 ft = 36 ft

6.51 Determine the force in members *DE* and *DF* of the truss shown when *P* = 20 kips.

6.52 Determine the force in members *EG* and *EF* of the truss shown when *P* = 20 kips.

6.53 Determine the force in members *DF* and *DE* of the truss shown.

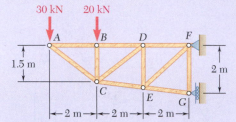

Fig. P6.53 and P6.54

30 kN   20 kN

1.5 m   2 m

2 m   2 m   2 m

6.54 Determine the force in members *CD* and *CE* of the truss shown.

6.55 A monosloped roof truss is loaded as shown. Determine the force in members *CE*, *DE*, and *DF*.

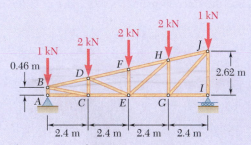

Fig. P6.55 and P6.56

1 kN   2 kN   2 kN   2 kN   1 kN

0.46 m   2.62 m

2.4 m   2.4 m   2.4 m   2.4 m

6.56 A monosloped roof truss is loaded as shown. Determine the force in members *EG*, *GH*, and *HJ*.

6.57 A Howe scissors roof truss is loaded as shown. Determine the force in members *DF*, *DG*, and *EG*.

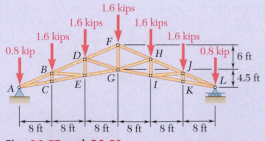

Fig. P6.57 and P6.58

1.6 kips   1.6 kips   1.6 kips   1.6 kips   1.6 kips

0.8 kip   0.8 kip   6 ft

4.5 ft

8 ft   8 ft   8 ft   8 ft   8 ft   8 ft

6.58 A Howe scissors roof truss is loaded as shown. Determine the force in members *GI*, *HI*, and *HJ*.

6.59 Determine the force in members *AD*, *CD*, and *CE* of the truss shown.

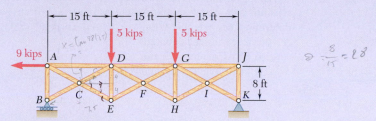

Fig. P6.59 and P6.60

6.60 Determine the force in members *DG*, *FG*, and *FH* of the truss shown.

6.61 Determine the force in members *DG* and *FI* of the truss shown. (*Hint:* Use section *aa*.)

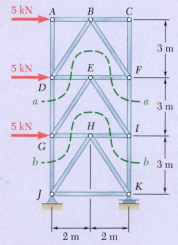

Fig. P6.61 and P6.62

6.62 Determine the force in members *GJ* and *IK* of the truss shown. (*Hint:* Use section *bb*.)

6.63 Determine the force in members *EH* and *GI* of the truss shown. (*Hint:* Use section *aa*.)

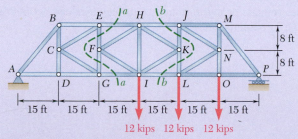

Fig. P6.63 and P6.64

6.64 Determine the force in members *HJ* and *IL* of the truss shown. (*Hint:* Use section *bb*.)

6.65 and 6.66 The diagonal members in the center panels of the power transmission line tower shown are very slender and can act only in tension; such members are known as *counters*. For the given loading, determine (*a*) which of the two counters listed below is acting, (*b*) the force in that counter.

6.65 Counters *CJ* and *HE*.

6.66 Counters *IO* and *KN*.

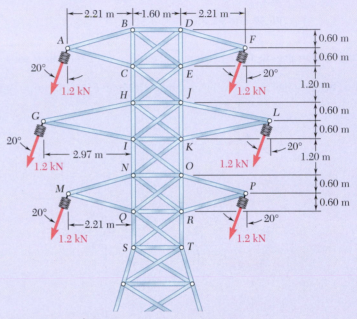

Fig. P6.65 and P6.66

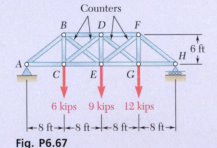

Fig. P6.67

6.67 The diagonal members in the center panels of the truss shown are very slender and can act only in tension; such members are known as *counters*. Determine the force in member *DE* and in the counters that are acting under the given loading.

6.68 Solve Prob. 6.67 assuming that the 9-kip load has been removed.

6.69 Classify each of the structures shown as completely, partially, or improperly constrained; if completely constrained, further classify as determinate or indeterminate. (All members can act both in tension and in compression.)

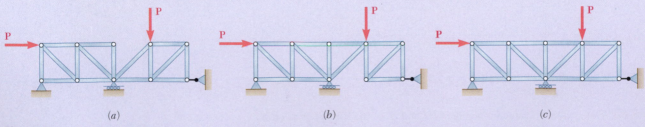

Fig. P6.69

6.70 through 6.74 Classify each of the structures shown as completely, partially, or improperly constrained; if completely constrained, further classify as determinate or indeterminate. (All members can act both in tension and in compression.)

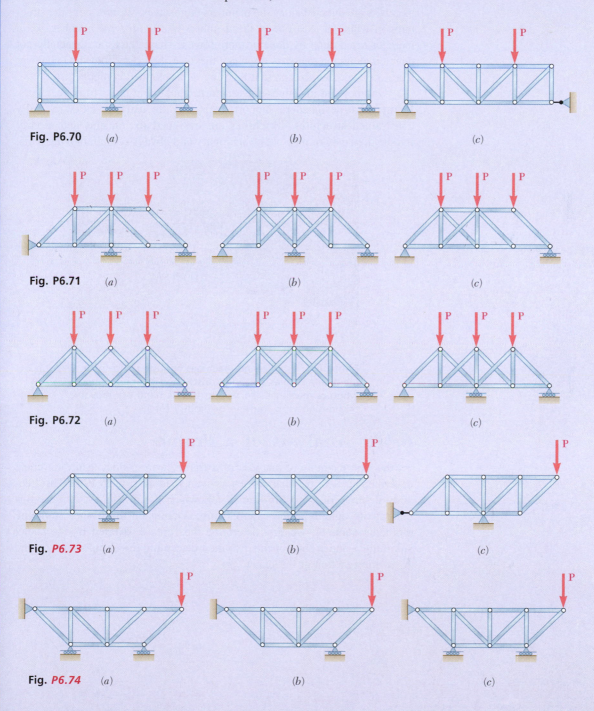

Fig. P6.70   (a)   (b)   (c)

Fig. P6.71   (a)   (b)   (c)

Fig. P6.72   (a)   (b)   (c)

Fig. P6.73   (a)   (b)   (c)

Fig. P6.74   (a)   (b)   (c)

6.3   FRAMES

When we study trusses, we are looking at structures consisting entirely of pins and straight two-force members. The forces acting on the two-force members are directed along the members themselves. We now consider structures in which at least one of the members is a *multi-force* member, i.e., a member acted upon by three or more forces. These forces are generally not directed along the members on which they act; their directions are unknown; therefore, we need to represent them by two unknown components.

Frames and machines are structures containing multi-force members. **Frames** are designed to support loads and are usually stationary, fully constrained structures. **Machines** are designed to transmit and modify forces; they may or may not be stationary and always contain moving parts.

Photo 6.6 Frames and machines contain multi-force members. Frames are fully constrained structures, whereas machines like this prosthetic hand are movable and designed to transmit or modify forces.

6.3A   Analysis of a Frame

As the first example of analysis of a frame, we consider again the crane described in Sec. 6.1 that carries a given load W (Fig. 6.19a). The free-body diagram of the entire frame is shown in Fig. 6.19b. We can use this diagram to determine the external forces acting on the frame. Summing moments about A, we first determine the force **T** exerted by the cable; summing x and y components, we then determine the components $\mathbf{A}_x$ and $\mathbf{A}_y$ of the reaction at the pin A.

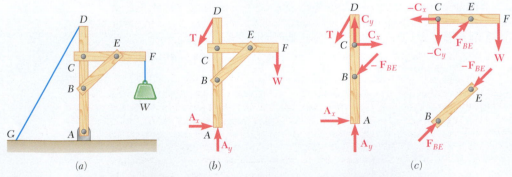

(a)   (b)   (c)

Fig. 6.19 A frame in equilibrium. (*a*) Diagram of a crane supporting a load; (*b*) free-body diagram of the crane; (*c*) free-body diagrams of the components of the crane.

In order to determine the internal forces holding the various parts of a frame together, we must dismember it and draw a free-body diagram for each of its component parts (Fig. 6.19c). First, we examine the two-force members. In this frame, member *BE* is the only two-force member. The forces acting at each end of this member must have the same magnitude, same line of action, and opposite sense (Sec. 4.2A). They are therefore directed along *BE* and are denoted, respectively, by $\mathbf{F}_{BE}$ and $-\mathbf{F}_{BE}$. We arbitrarily assume their sense as shown in Fig. 6.19c; the sign obtained for the common magnitude F_{BE} of the two forces will confirm or deny this assumption.

Next, we consider the multi-force members, i.e., the members that are acted upon by three or more forces. According to Newton's third law, the force exerted at *B* by member *BE* on member *AD* must be equal and opposite to the force $\mathbf{F}_{BE}$ exerted by *AD* on *BE*. Similarly, the force exerted at *E* by member *BE* on member *CF* must be equal and opposite to the force $-\mathbf{F}_{BE}$ exerted by *CF* on *BE*. Thus, the forces that the two-force member *BE* exerts on *AD* and *CF* are, respectively, equal to $-\mathbf{F}_{BE}$ and $\mathbf{F}_{BE}$; they have the same magnitude F_{BE}, opposite sense, and should be directed as shown in Fig. 6.19c.

Joint *C* connects two multi-force members. Since neither the direction nor the magnitude of the forces acting at *C* are known, we represent these forces by their *x* and *y* components. The components $\mathbf{C}_x$ and $\mathbf{C}_y$ of the force acting on member *AD* are arbitrarily directed to the right and upward. Since, according to Newton's third law, the forces exerted by member *CF* on *AD* and by member *AD* on *CF* are equal and opposite, the components of the force acting on member *CF must* be directed to the left and downward; we denote them, respectively, by $-\mathbf{C}_x$ and $-\mathbf{C}_y$. Whether the force $\mathbf{C}_x$ is actually directed to the right and the force $-\mathbf{C}_x$ is actually directed to the left will be determined later from the sign of their common magnitude C_x with a plus sign indicating that the assumption was correct and a minus sign that it was wrong. We complete the free-body diagrams of the multi-force members by showing the external forces acting at *A, D,* and *F*.[†]

We can now determine the internal forces by considering the free-body diagram of either of the two multi-force members. Choosing the free-body diagram of *CF*, for example, we write the equations $\Sigma M_C = 0$, $\Sigma M_E = 0$, and $\Sigma F_x = 0$, which yield the values of the magnitudes F_{BE}, C_y, and C_x, respectively. We can check these values by verifying that member *AD* is also in equilibrium.

Note that we assume the pins in Fig. 6.19 form an integral part of one of the two members they connected, so it is not necessary to show their free-body diagrams. We can always use this assumption to simplify the analysis of frames and machines. However, when a pin connects three

---

[†] It is not strictly necessary to use a minus sign to distinguish the force exerted by one member on another from the equal and opposite force exerted by the second member on the first, since the two forces belong to different free-body diagrams and thus are not easily confused. In the Sample Problems, we use the same symbol to represent equal and opposite forces that are applied to different free bodies. Note that, under these conditions, the sign obtained for a given force component does not directly relate the sense of that component to the sense of the corresponding coordinate axis. Rather, a positive sign indicates that *the sense assumed for that component in the free-body diagram is correct*, and a negative sign indicates that it is wrong.

or more members, connects a support and two or more members, or when a load is applied to a pin, we must make a clear decision in choosing the member to which we assume the pin belongs. (If multi-force members are involved, the pin should be attached to one of these members.) We then need to identify clearly the various forces exerted on the pin. This is illustrated in Sample Prob. 6.6.

6.3B Frames That Collapse Without Supports

The crane we just analyzed was constructed so that it could keep the same shape without the help of its supports; we therefore considered it to be a rigid body. Many frames, however, will collapse if detached from their supports; such frames cannot be considered rigid bodies. Consider, for example, the frame shown in Fig. 6.20a that consists of two members AC and CB carrying loads **P** and **Q** at their midpoints. The members are supported by pins at A and B and are connected by a pin at C. If we detach this frame from its supports, it will not maintain its shape. Therefore, we should consider it to be made of *two distinct rigid parts AC and CB.*

The equations $\Sigma F_x = 0$, $\Sigma F_y = 0$, and $\Sigma M = 0$ (about any given point) express the conditions for the *equilibrium of a rigid body* (Chap. 4); we should use them, therefore, in connection with the free-body diagrams of members AC and CB (Fig. 6.20b). Since these members are multi-force members and since pins are used at the supports and at the connection, we represent each of the reactions at A and B and the forces at C by two components. In accordance with Newton's third law, we represent the components of the force exerted by CB on AC and the components of the force exerted by AC on CB by vectors of the same magnitude and opposite sense. Thus, if the first pair of components consists of $\mathbf{C}_x$ and $\mathbf{C}_y$, the second pair is represented by $-\mathbf{C}_x$ and $-\mathbf{C}_y$.

Note that four unknown force components act on free body AC, whereas we need only three independent equations to express that the body is in equilibrium. Similarly, four unknowns, but only three equations, are associated with CB. However, only six different unknowns are involved in the analysis of the two members, and altogether, six equations

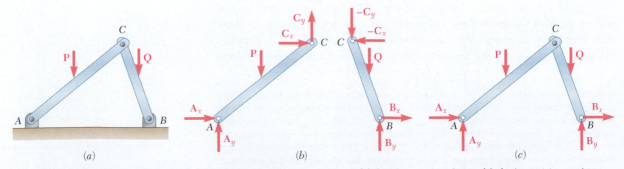

Fig. 6.20 (a) A frame of two members supported by two pins and joined together by a third pin. Without the supports, the frame would collapse and is therefore not a rigid body. (b) Free-body diagrams of the two members. (c) Free-body diagram of the whole frame.

are available to express that the members are in equilibrium. Setting $\Sigma M_A = 0$ for free body AC and $\Sigma M_B = 0$ for CB, we obtain two simultaneous equations that we can solve for the common magnitude C_x of the components $\mathbf{C}_x$ and $-\mathbf{C}_x$ and for the common magnitude C_y of the components $\mathbf{C}_y$ and $-\mathbf{C}_y$. We then have $\Sigma F_x = 0$ and $\Sigma F_y = 0$ for each of the two free bodies, successively obtaining the magnitudes A_x, A_y, B_x, and B_y.

Observe that, since the equations of equilibrium $\Sigma F_x = 0$, $\Sigma F_y = 0$, and $\Sigma M = 0$ (about any given point) are satisfied by the forces acting on free body AC and since they are also satisfied by the forces acting on free body CB, they must be satisfied when the forces acting on the two free bodies are considered simultaneously. Since the internal forces at C cancel each other, we find that the equations of equilibrium must be satisfied by the external forces shown on the free-body diagram of the frame ACB itself (Fig. 6.20c), even though the frame is not a rigid body. We can use these equations to determine some of the components of the reactions at A and B. We will find, however, that **the reactions cannot be completely determined from the free-body diagram of the whole frame**. It is thus necessary to dismember the frame and consider the free-body diagrams of its component parts (Fig. 6.20b), even when we are interested in determining external reactions only. The reason is that the equilibrium equations obtained for free body ACB *are necessary conditions* for the equilibrium of a nonrigid structure, *but these are not sufficient conditions*.

The method of solution outlined here involved simultaneous equations. We now present a more efficient method that utilizes the free body ACB as well as the free bodies AC and CB. Writing $\Sigma M_A = 0$ and $\Sigma M_B = 0$ for free body ACB, we obtain B_y and A_y. From $\Sigma M_C = 0$, $\Sigma F_x = 0$, and $\Sigma F_y = 0$ for free body AC, we successively obtain A_x, C_x, and C_y. Finally, setting $\Sigma F_x = 0$ for ACB gives us B_x.

We noted previously that the analysis of the frame in Fig. 6.20 involves six unknown force components and six independent equilibrium equations. (The equilibrium equations for the whole frame were obtained from the original six equations and, therefore, are not independent.) Moreover, we checked that all unknowns could be actually determined and that all equations could be satisfied. This frame is **statically determinate and rigid**. (We use the word "rigid" here to indicate that the frame maintains its shape as long as it remains attached to its supports.) In general, to determine whether a structure is statically determinate and rigid, you should draw a free-body diagram for each of its component parts and count the reactions and internal forces involved. You should then determine the number of independent equilibrium equations (excluding equations expressing the equilibrium of the whole structure or of groups of component parts already analyzed). If you have more unknowns than equations, the structure is *statically indeterminate*. If you have fewer unknowns than equations, the structure is *nonrigid*. If you have as many unknowns as equations *and if all unknowns can be determined and all equations satisfied* under general loading conditions, the structure is statically determinate and rigid. If, however, due to an improper arrangement of members and supports, all unknowns cannot be determined and all equations cannot be satisfied, the structure is **statically indeterminate and nonrigid**.

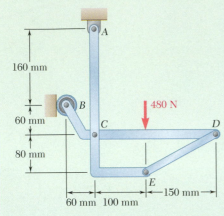

Sample Problem 6.4

In the frame shown, members *ACE* and *BCD* are connected by a pin at *C* and by the link *DE*. For the loading shown, determine the force in link *DE* and the components of the force exerted at *C* on member *BCD*.

STRATEGY: Follow the general procedure discussed in this section. First treat the entire frame as a free body, which will enable you to find the reactions at *A* and *B*. Then dismember the frame and treat each member as a free body, which will give you the equations needed to find the force at *C*.

MODELING and ANALYSIS: Since the external reactions involve only three unknowns, compute the reactions by considering the free-body diagram of the entire frame (Fig. 1).

$$+\uparrow\Sigma F_y = 0: \qquad A_y - 480\text{ N} = 0 \qquad A_y = +480\text{ N} \qquad \mathbf{A}_y = 480\text{ N}\uparrow$$
$$+\circlearrowleft\Sigma M_A = 0: \qquad -(480\text{ N})(100\text{ mm}) + B(160\text{ mm}) = 0$$
$$B = +300\text{ N} \qquad \mathbf{B} = 300\text{ N}\rightarrow$$
$$\xrightarrow{+}\Sigma F_x = 0: \qquad B + A_x = 0$$
$$300\text{ N} + A_x = 0 \qquad A_x = -300\text{ N} \qquad \mathbf{A}_x = 300\text{ N}\leftarrow$$

Now dismember the frame (Figs. 2 and 3). Since only two members are connected at *C*, the components of the unknown forces acting on *ACE* and *BCD* are, respectively, equal and opposite. Assume that link *DE* is in tension (Fig. 3) and exerts equal and opposite forces at *D* and *E*, directed as shown.

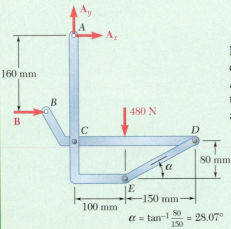

$$\alpha = \tan^{-1}\frac{80}{150} = 28.07°$$

Fig. 1 Free-body diagram of entire frame.

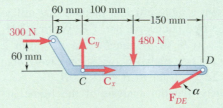

Fig. 2 Free-body diagram of member *BCD*.

Free Body: Member BCD. Using the free body *BCD* (Fig. 2), you can write and solve three equilibrium equations:

$$+\circlearrowleft\Sigma M_C = 0:$$
$$(F_{DE}\sin\alpha)(250\text{ mm}) + (300\text{ N})(80\text{ mm}) + (480\text{ N})(100\text{ mm}) = 0$$
$$F_{DE} = -561\text{ N} \qquad\qquad F_{DE} = 561\text{ N } C \blacktriangleleft$$
$$\xrightarrow{+}\Sigma F_x = 0: \qquad C_x - F_{DE}\cos\alpha + 300\text{ N} = 0$$
$$C_x - (-561\text{ N})\cos 28.07° + 300\text{ N} = 0 \qquad C_x = -795\text{ N}$$
$$+\uparrow\Sigma F_y = 0: \qquad C_y - F_{DE}\sin\alpha - 480\text{ N} = 0$$
$$C_y - (-561\text{ N})\sin 28.07° - 480\text{ N} = 0 \qquad C_y = +216\text{ N}$$

From the signs obtained for C_x and C_y, the force components $\mathbf{C}_x$ and $\mathbf{C}_y$ exerted on member *BCD* are directed to the left and up, respectively. Thus, you have

$$\mathbf{C}_x = 795\text{ N}\leftarrow, \ \mathbf{C}_y = 216\text{ N}\uparrow \blacktriangleleft$$

REFLECT and THINK: Check the computations by considering the free body *ACE* (Fig. 3). For example,

$$+\circlearrowleft\Sigma M_A = (F_{DE}\cos\alpha)(300\text{ mm}) + (F_{DE}\sin\alpha)(100\text{ mm}) - C_x(220\text{ mm})$$
$$= (-561\cos\alpha)(300) + (-561\sin\alpha)(100) - (-795)(220) = 0$$

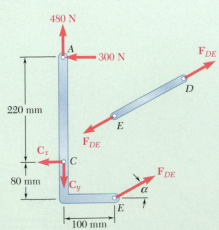

Fig. 3 Free-body diagrams of member *ACE* and *DE*.

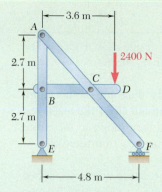

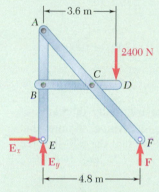

Fig. 1 Free-body diagram of entire frame.

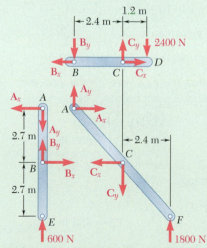

Fig. 2 Free-body diagrams of individual members.

Sample Problem 6.5

Determine the components of the forces acting on each member of the frame shown.

STRATEGY: The approach to this analysis is to consider the entire frame as a free body to determine the reactions, and then consider separate members. However, in this case, you will not be able to determine forces on one member without analyzing a second member at the same time.

MODELING and ANALYSIS: The external reactions involve only three unknowns, so compute the reactions by considering the free-body diagram of the entire frame (Fig. 1).

$$+\circlearrowleft\Sigma M_E = 0: \qquad -(2400\text{ N})(3.6\text{ m}) + F(4.8\text{ m}) = 0$$
$$F = +1800\text{ N} \qquad\qquad \mathbf{F} = 1800\text{ N}\uparrow \quad \blacktriangleleft$$
$$+\uparrow\Sigma F_y = 0: \qquad -2400\text{ N} + 1800\text{ N} + E_y = 0$$
$$E_y = +600\text{ N} \qquad\qquad \mathbf{E}_y = 600\text{ N}\uparrow \quad \blacktriangleleft$$
$$\xrightarrow{+}\Sigma F_x = 0: \qquad\qquad\qquad\qquad\qquad\qquad \mathbf{E}_x = 0 \quad \blacktriangleleft$$

Now dismember the frame. Since only two members are connected at each joint, force components are equal and opposite on each member at each joint (Fig. 2).

Free Body: Member BCD.

$$+\circlearrowleft\Sigma M_B = 0: \quad -(2400\text{ N})(3.6\text{ m}) + C_y(2.4\text{ m}) = 0 \quad C_y = +3600\text{ N} \quad \blacktriangleleft$$
$$+\circlearrowleft\Sigma M_C = 0: \quad -(2400\text{ N})(1.2\text{ m}) + B_y(2.4\text{ m}) = 0 \quad B_y = +1200\text{ N} \quad \blacktriangleleft$$
$$\xrightarrow{+}\Sigma F_x = 0: \quad -B_x + C_x = 0$$

Neither B_x nor C_x can be obtained by considering only member BCD; you need to look at member ABE. The positive values obtained for B_y and C_y indicate that the force components $\mathbf{B}_y$ and $\mathbf{C}_y$ are directed as assumed.

Free Body: Member ABE.

$$+\circlearrowleft\Sigma M_A = 0: \qquad\qquad B_x(2.7\text{ m}) = 0 \qquad\qquad B_x = 0 \quad \blacktriangleleft$$
$$\xrightarrow{+}\Sigma F_x = 0: \qquad\qquad +B_x - A_x = 0 \qquad\qquad A_x = 0 \quad \blacktriangleleft$$
$$+\uparrow\Sigma F_y = 0: \qquad\qquad -A_y + B_y + 600\text{ N} = 0$$
$$-A_y + 1200\text{ N} + 600\text{ N} = 0 \qquad A_y = +1800\text{ N} \quad \blacktriangleleft$$

Free Body: Member BCD. Returning now to member BCD, you have

$$\xrightarrow{+}\Sigma F_x = 0: \qquad -B_x + C_x = 0 \qquad 0 + C_x = 0 \qquad C_x = 0 \quad \blacktriangleleft$$

REFLECT and THINK: All unknown components have now been found. To check the results, you can verify that member ACF is in equilibrium.

$$+\circlearrowleft\Sigma M_C = (1800\text{ N})(2.4\text{ m}) - A_y(2.4\text{ m}) - A_x(2.7\text{ m})$$
$$= (1800\text{ N})(2.4\text{ m}) - (1800\text{ N})(2.4\text{ m}) - 0 = 0 \qquad \text{(checks)}$$

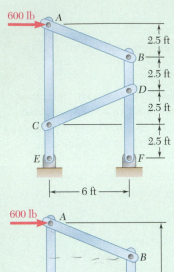

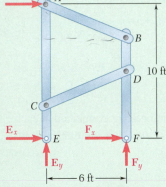

Fig. 1 Free-body diagram of entire frame.

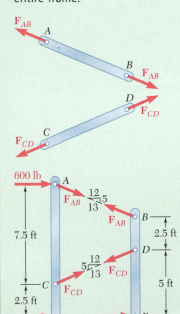

Fig. 2 Free-body diagrams of individual members.

Sample Problem 6.6

A 600-lb horizontal force is applied to pin A of the frame shown. Determine the forces acting on the two vertical members of the frame.

STRATEGY: Begin as usual with a free-body diagram of the entire frame, but this time you will not be able to determine all of the reactions. You will have to analyze a separate member and then return to the entire frame analysis in order to determine the remaining reaction forces.

MODELING and ANALYSIS: Choosing the entire frame as a free body (Fig. 1), you can write equilibrium equations to determine the two force components $\mathbf{E}_y$ and $\mathbf{F}_y$. However, these equations are not sufficient to determine $\mathbf{E}_x$ and $\mathbf{F}_x$.

$$+\circlearrowleft\Sigma M_E = 0: \qquad -(600\text{ lb})(10\text{ ft}) + F_y(6\text{ ft}) = 0$$
$$F_y = +1000\text{ lb} \qquad\qquad \mathbf{F}_y = 1000\text{ lb}\uparrow \quad \blacktriangleleft$$
$$+\uparrow\Sigma F_y = 0: \qquad E_y + F_y = 0$$
$$E_y = -1000\text{ lb} \qquad\qquad \mathbf{E}_y = 1000\text{ lb}\downarrow \quad \blacktriangleleft$$

To proceed with the solution, now consider the free-body diagrams of the various members (Fig. 2). In dismembering the frame, assume that pin A is attached to the multi-force member ACE so that the 600-lb force is applied to that member. Note that AB and CD are two-force members.

Free Body: Member ACE

$$+\uparrow\Sigma F_y = 0: \qquad -\tfrac{5}{13}F_{AB} + \tfrac{5}{13}F_{CD} - 1000\text{ lb} = 0$$
$$+\circlearrowleft\Sigma M_E = 0: \qquad -(600\text{ lb})(10\text{ ft}) - (\tfrac{12}{13}F_{AB})(10\text{ ft}) - (\tfrac{12}{13}F_{CD})(2.5\text{ ft}) = 0$$

Solving these equations simultaneously gives you

$$F_{AB} = -1040\text{ lb} \qquad F_{CD} = +1560\text{ lb} \quad \blacktriangleleft$$

The signs indicate that the sense assumed for F_{CD} was correct and the sense for F_{AB} was incorrect. Now summing x components, you have

$$\xrightarrow{+}\Sigma F_x = 0: \qquad 600\text{ lb} + \tfrac{12}{13}(-1040\text{ lb}) + \tfrac{12}{13}(+1560\text{ lb}) + E_x = 0$$
$$E_x = -1080\text{ lb} \qquad\qquad \mathbf{E}_x = 1080\text{ lb}\leftarrow \quad \blacktriangleleft$$

Free Body: Entire Frame.

Now that $\mathbf{E}_x$ is determined, you can return to the free-body diagram of the entire frame.

$$\xrightarrow{+}\Sigma F_x = 0: \qquad 600\text{ lb} - 1080\text{ lb} + F_x = 0$$
$$F_x = +480\text{ lb} \qquad\qquad \mathbf{F}_x = 480\text{ lb}\rightarrow \quad \blacktriangleleft$$

REFLECT and THINK: Check your computations by verifying that the equation $\Sigma M_B = 0$ is satisfied by the forces acting on member BDF.

$$+\circlearrowleft\Sigma M_B = -(\tfrac{12}{13}F_{CD})(2.5\text{ ft}) + (F_x)(7.5\text{ ft})$$
$$= -\tfrac{12}{13}(1560\text{ lb})(2.5\text{ ft}) + (480\text{ lb})(7.5\text{ ft})$$
$$= -3600\text{ lb·ft} + 3600\text{ lb·ft} = 0 \qquad \text{(checks)}$$

SOLVING PROBLEMS ON YOUR OWN

In this section, we analyzed **frames containing one or more multi-force members**. In the problems that follow, you will be asked to determine the external reactions exerted on the frame and the internal forces that hold together the members of the frame.

In solving problems involving frames containing one or more multi-force members, follow these steps.

1. Draw a free-body diagram of the entire frame. To the greatest extent possible, use this free-body diagram to calculate the reactions at the supports. (In Sample Prob. 6.6 only two of the four reaction components could be found from the free body of the entire frame.)

2. Dismember the frame, and draw a free-body diagram of each member.

3. First consider the two-force members. Equal and opposite forces apply to each two-force member at the points where it is connected to another member. If the two-force member is straight, these forces are directed along the axis of the member. If you cannot tell at this point whether the member is in tension or compression, *assume* that the member is in tension and *direct both of the forces away from the member*. Since these forces have the same unknown magnitude, give them both the *same name* and, to avoid any confusion later, *do not use a plus sign or a minus sign*.

4. Next consider the multi-force members. For each of these members, show all of the forces acting on the member, including *applied loads, reactions, and internal forces at connections*. Clearly indicate the magnitude and direction of any reaction or reaction component found earlier from the free-body diagram of the entire frame.

    a. Where a multi-force member is connected to a two-force member, apply a force to the multi-force member that is *equal and opposite* to the force drawn on the free-body diagram of the two-force member, *giving it the same name*.

    b. Where a multi-force member is connected to another multi-force member, use *horizontal and vertical components* to represent the internal forces at that point, since the directions and magnitudes of these forces are unknown. The direction you choose for each of the two force components exerted on the first multi-force member is arbitrary, but *you must apply equal and opposite force components of the same name* to the other multi-force member. Again, *do not use a plus sign or a minus sign*.

(continued)

5. Now determine the internal forces as well as any reactions that you have not already found.

    a. The free-body diagram of each multi-force member can provide you with *three equilibrium equations*.

    b. To simplify your solution, seek a way to write an equation involving a single unknown. If you can locate *a point where all but one of the unknown force components intersect,* you can obtain an equation in a single unknown by summing moments about that point. *If all unknown forces except one are parallel,* you can obtain an equation in a single unknown by summing force components in a direction perpendicular to the parallel forces.

    c. Since you arbitrarily chose the direction of each of the unknown forces, you cannot determine whether your guess was correct until the solution is complete. To do that, consider the *sign* of the value found for each of the unknowns: a *positive* sign means that the direction you selected was *correct;* a *negative* sign means that the direction is *opposite* to the direction you assumed.

6. To be more effective and efficient as you proceed through your solution, observe the following rules.

    a. If you can find an equation involving only one unknown, write that equation and *solve it for that unknown.* Immediately replace that unknown wherever it appears on other free-body diagrams by the value you have found. Repeat this process by seeking equilibrium equations involving only one unknown until you have found all of the internal forces and unknown reactions.

    b. If you cannot find an equation involving only one unknown, you may have to *solve a pair of simultaneous equations.* Before doing so, check that you have included the values of all of the reactions you obtained from the free-body diagram of the entire frame.

    c. The total number of equations of equilibrium for the entire frame and for the individual members *will be larger than the number of unknown forces and reactions.* After you have found all of the reactions and all of the internal forces, you can use the remaining equations to check the accuracy of your computations.

Problems

6.F1 For the frame and loading shown, draw the free-body diagram(s) needed to determine the force in member *BD* and the components of the reaction at *C*.

6.F2 For the frame and loading shown, draw the free-body diagram(s) needed to determine the components of all forces acting on member *ABC*.

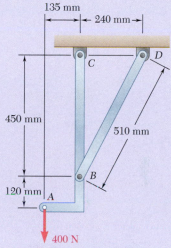

Fig. P6.F1

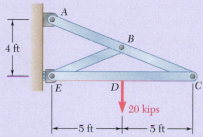

Fig. P6.F2

6.F3 Draw the free-body diagram(s) needed to determine all the forces exerted on member *AI* if the frame is loaded by a clockwise couple of magnitude 1200 lb·in. applied at point *D*.

6.F4 Knowing that the pulley has a radius of 0.5 m, draw the free-body diagram(s) needed to determine the components of the reactions at *A* and *E*.

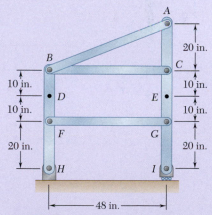

Fig. P6.F3

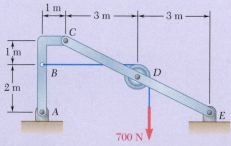

Fig. P6.F4

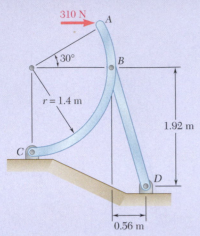

Fig. P6.75

6.75 and 6.76 Determine the force in member BD and the components of the reaction at C.

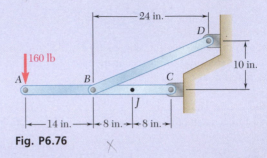

Fig. P6.76

6.77 For the frame and loading shown, determine the force acting on member ABC (*a*) at B, (*b*) at C.

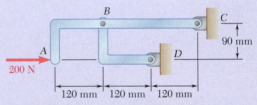

Fig. P6.77

6.78 Determine the components of all forces acting on member $ABCD$ of the assembly shown.

6.79 For the frame and loading shown, determine the components of all forces acting on member ABC.

6.80 Solve Prob. 6.79 assuming that the 18-kN load is replaced by a clockwise couple with a magnitude of 72 kN·m applied to member $CDEF$ at point D.

6.81 Determine the components of all forces acting on member $ABCD$ when $\theta = 0$.

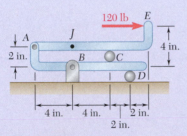

Fig. P6.78

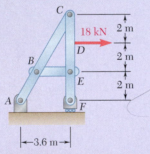

Fig. P6.79

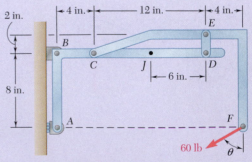

Fig. P6.81 and *P6.82*

6.82 Determine the components of all forces acting on member $ABCD$ when $\theta = 90°$.

6.83 Determine the components of the reactions at A and E, (a) if the 800-N load is applied as shown, (b) if the 800-N load is moved along its line of action and is applied at point D.

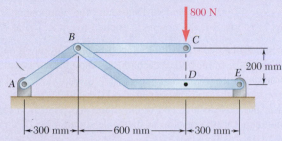

Fig. P6.83

6.84 Determine the components of the reactions at D and E if the frame is loaded by a clockwise couple of magnitude 150 N·m applied (a) at A, (b) at B.

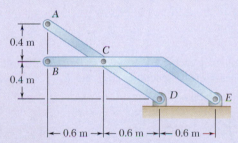

Fig. P6.84

6.85 Determine the components of the reactions at A and E if a 750-N force directed vertically downward is applied (a) at B, (b) at D.

6.86 Determine the components of the reactions at A and E if the frame is loaded by a clockwise couple with a magnitude of 36 N·m applied (a) at B, (b) at D.

6.87 Determine the components of the reactions at A and B, (a) if the 100-lb load is applied as shown, (b) if the 100-lb load is moved along its line of action and is applied at point F.

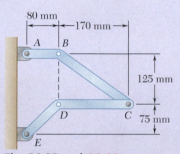

Fig. P6.85 and P6.86

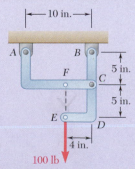

Fig. P6.87

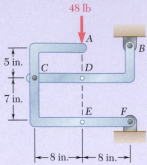

48 lb

5 in.

7 in.

C   D

A

B

E   F

←8 in.→←8 in.→

Fig. P6.88 and P6.89

6.88 The 48-lb load can be moved along the line of action shown and applied at A, D, or E. Determine the components of the reactions at B and F if the 48-lb load is applied (a) at A, (b) at D, (c) at E.

6.89 The 48-lb load is removed and a 288-lb·in. clockwise couple is applied successively at A, D, and E. Determine the components of the reactions at B and F if the couple is applied (a) at A, (b) at D, (c) at E.

6.90 (a) Show that, when a frame supports a pulley at A, an equivalent loading of the frame and of each of its component parts can be obtained by removing the pulley and applying at A two forces equal and parallel to the forces that the cable exerted on the pulley. (b) Show that, if one end of the cable is attached to the frame at a point B, a force of magnitude equal to the tension in the cable should also be applied at B.

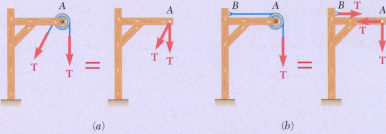

(a)   (b)

Fig. P6.90

✓ **6.91** Knowing that each pulley has a radius of 250 mm, determine the components of the reactions at D and E.

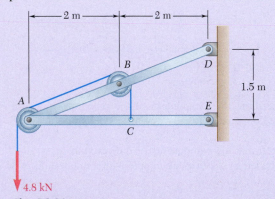

2 m   2 m

A   B   C   D   E

1.5 m

4.8 kN

Fig. P6.91

6.92 Knowing that the pulley has a radius of 75 mm, determine the components of the reactions at A and B.

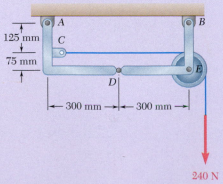

125 mm

75 mm

A   C   D   E   B

←300 mm→←300 mm→

240 N

Fig. P6.92

6.93 A 3-ft-diameter pipe is supported every 16 ft by a small frame like that shown. Knowing that the combined weight of the pipe and its contents is 500 lb/ft and assuming frictionless surfaces, determine the components (a) of the reaction at E, (b) of the force exerted at C on member CDE.

6.94 Solve Prob. 6.93 for a frame where $h = 6$ ft.

6.95 A trailer weighing 2400 lb is attached to a 2900-lb pickup truck by a ball-and-socket truck hitch at D. Determine (a) the reactions at each of the six wheels when the truck and trailer are at rest, (b) the additional load on each of the truck wheels due to the trailer.

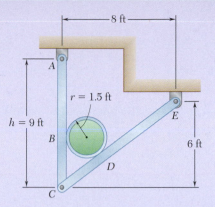

Fig. P6.93

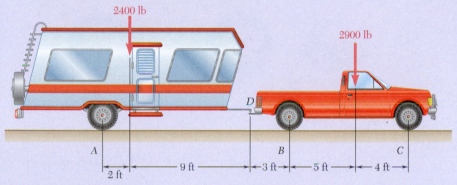

Fig. P6.95

6.96 In order to obtain a better weight distribution over the four wheels of the pickup truck of Prob. 6.95, a compensating hitch of the type shown is used to attach the trailer to the truck. The hitch consists of two bar springs (only one is shown in the figure) that fit into bearings inside a support rigidly attached to the truck. The springs are also connected by chains to the trailer frame, and specially designed hooks make it possible to place both chains in tension. (a) Determine the tension T required in each of the two chains if the additional load due to the trailer is to be evenly distributed over the four wheels of the truck. (b) What are the resulting reactions at each of the six wheels of the trailer-truck combination?

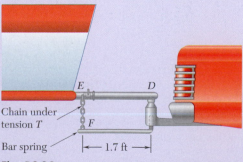

Fig. P6.96

6.97 The cab and motor units of the front-end loader shown are connected by a vertical pin located 2 m behind the cab wheels. The distance from C to D is 1 m. The center of gravity of the 300-kN motor unit is located at G_m, while the centers of gravity of the 100-kN cab and 75-kN load are located, respectively, at G_c and G_l. Knowing that the machine is at rest with its brakes released, determine (a) the reactions at each of the four wheels, (b) the forces exerted on the motor unit at C and D.

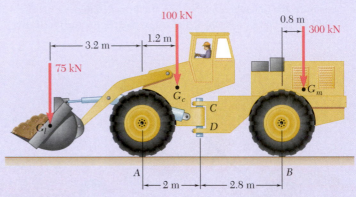

Fig. P6.97

6.98 Solve Prob. 6.97 assuming that the 75-kN load has been removed.

6.99 Knowing that P = 90 lb and Q = 60 lb, determine the components of all forces acting on member BCDE of the assembly shown.

6.100 Knowing that P = 60 lb and Q = 90 lb, determine the components of all forces acting on member BCDE of the assembly shown.

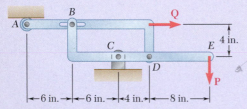

Fig. P6.99 and P6.100

6.101 and 6.102 For the frame and loading shown, determine the components of all forces acting on member ABE.

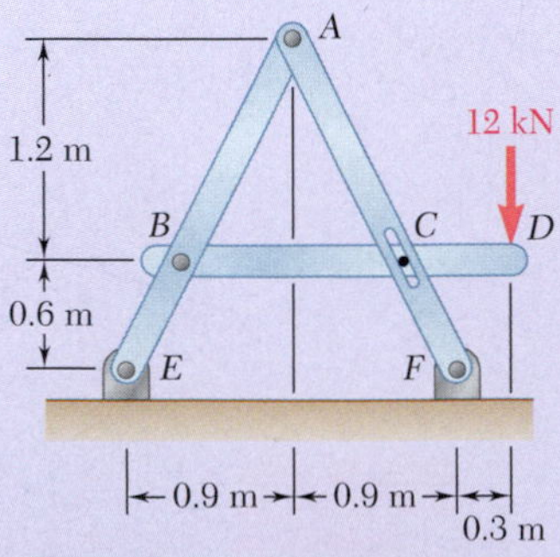

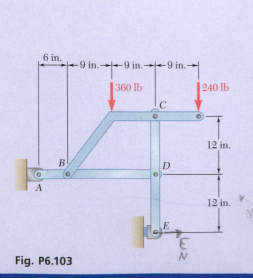

Fig. P6.103

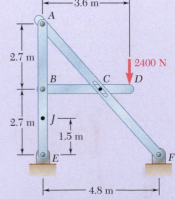

Fig. P6.101

Fig. P6.102

6.103 For the frame and loading shown, determine the components of all forces acting on member ABD.

6.104 Solve Prob. 6.103 assuming that the 360-lb load has been removed.

344

6.105 For the frame and loading shown, determine the components of the forces acting on member *DABC* at *B* and *D*.

6.106 Solve Prob. 6.105 assuming that the 6-kN load has been removed.

6.107 The axis of the three-hinge arch *ABC* is a parabola with vertex at *B*. Knowing that $P = 112$ kN and $Q = 140$ kN, determine (*a*) the components of the reaction at *A*, (*b*) the components of the force exerted at *B* on segment *AB*.

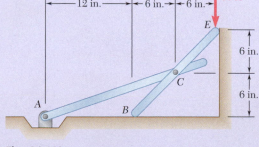

Fig. P6.105

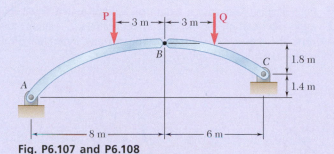
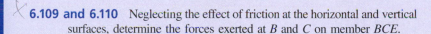

Fig. P6.107 and P6.108

6.108 The axis of the three-hinge arch *ABC* is a parabola with vertex at *B*. Knowing that $P = 140$ kN and $Q = 112$ kN, determine (*a*) the components of the reaction at *A*, (*b*) the components of the force exerted at *B* on segment *AB*.

6.109 and 6.110 Neglecting the effect of friction at the horizontal and vertical surfaces, determine the forces exerted at *B* and *C* on member *BCE*.

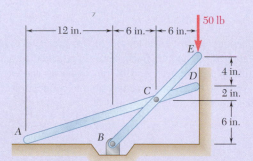

Fig. P6.109

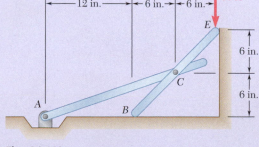

Fig. P6.110

6.111, 6.112, and 6.113 Members *ABC* and *CDE* are pin-connected at *C* and supported by four links. For the loading shown, determine the force in each link.

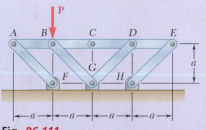

Fig. P6.111

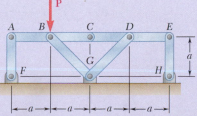

Fig. P6.112

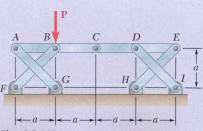

Fig. P6.113

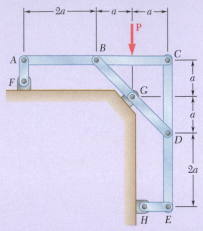

Fig. **P6.114**

6.114 Members *ABC* and *CDE* are pin-connected at *C* and supported by the four links *AF*, *BG*, *DG*, and *EH*. For the loading shown, determine the force in each link.

6.115 Solve Prob. 6.112 assuming that the force **P** is replaced by a clockwise couple of moment $\mathbf{M}_0$ applied to member *CDE* at *D*.

6.116 Solve Prob. 6.114 assuming that the force **P** is replaced by a clockwise couple of moment $\mathbf{M}_0$ applied at the same point.

6.117 Four beams, each with a length of 2*a*, are nailed together at their midpoints to form the support system shown. Assuming that only vertical forces are exerted at the connections, determine the vertical reactions at *A*, *D*, *E*, and *H*.

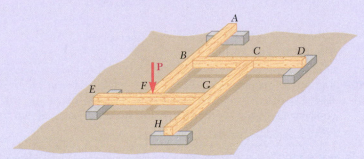

Fig. **P6.117**

6.118 Four beams, each with a length of 3*a*, are held together by single nails at *A*, *B*, *C*, and *D*. Each beam is attached to a support located at a distance *a* from an end of the beam as shown. Assuming that only vertical forces are exerted at the connections, determine the vertical reactions at *E*, *F*, *G*, and *H*.

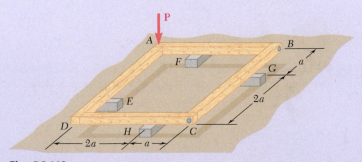

Fig. **P6.118**

6.119 through 6.121 Each of the frames shown consists of two L-shaped members connected by two rigid links. For each frame, determine the reactions at the supports and indicate whether the frame is rigid.

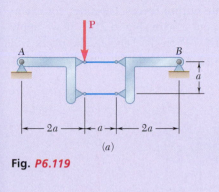

(a)

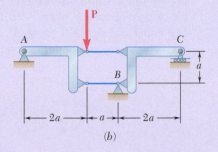

(b)

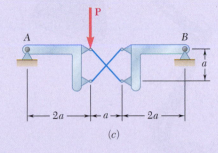

(c)

Fig. P6.119

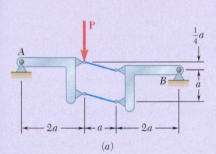

(a)

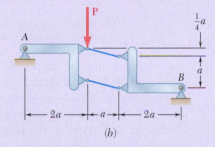

(b)

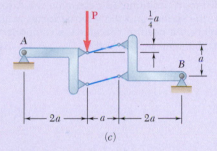

(c)

Fig. P6.120

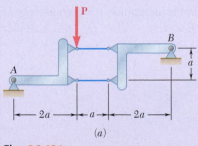

(a)

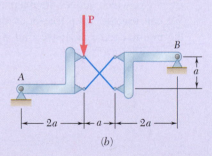

(b)

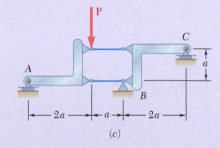

(c)

Fig. P6.121

Photo 6.7 This lamp can be placed in many different positions. To determine the forces in the springs and the internal forces at the joints, we need to consider the components of the lamp as free bodies.

6.4 MACHINES

Machines are structures designed to transmit and modify forces. Whether they are simple tools or include complicated mechanisms, their main purpose is to transform **input forces** into **output forces**. Consider, for example, a pair of cutting pliers used to cut a wire (Fig. 6.21a). If we apply two equal and opposite forces **P** and −**P** on the handles, the pliers will exert two equal and opposite forces **Q** and −**Q** on the wire (Fig. 6.21b).

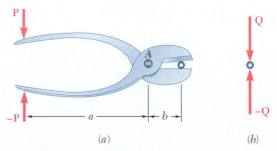

(a)   (b)

Fig. 6.21 (a) Input forces on the handles of a pair of cutting pliers; (b) output forces cut a wire.

To determine the magnitude Q of the output forces when we know the magnitude P of the input forces (or, conversely, to determine P when Q is known), we draw a free-body diagram of the pliers *alone* (i.e., without the wire), showing the input forces **P** and −**P** and the *reactions* −**Q** and **Q** that the wire exerts on the pliers (Fig. 6.22). However, since a pair of pliers forms a nonrigid structure, we must treat one of the component parts as a free body in order to determine the unknown forces. Consider Fig. 6.23a, for example. Taking moments about A, we obtain the relation $Pa = Qb$, which defines the magnitude Q in terms of P (or P in terms of Q). We can use the same free-body diagram to determine the components of the internal force at A; we find $A_x = 0$ and $A_y = P + Q$.

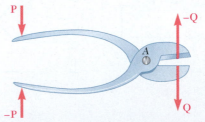

Fig. 6.22 To show a free-body diagram of the pliers in equilibrium, we include the input forces and the reactions to the output forces.

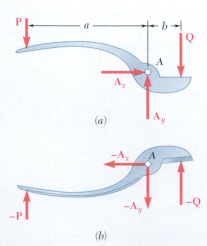

(a)

(b)

Fig. 6.23 Free-body diagrams of the members of the pliers, showing components of the internal forces at joint A.

In the case of more complicated machines, it is generally necessary to use several free-body diagrams and, possibly, to solve simultaneous equations involving various internal forces. You should choose the free bodies to include the input forces and the reactions to the output forces, and the total number of unknown force components involved should not exceed the number of available independent equations. It is advisable, before attempting to solve a problem, to determine whether the structure considered is determinate. There is no point, however, in discussing the rigidity of a machine, since a machine includes moving parts and thus *must* be nonrigid.

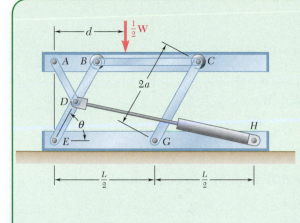

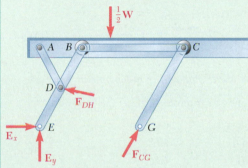

Fig. 1 Free-body diagram of machine.

Sample Problem 6.7

A hydraulic-lift table is used to raise a 1000-kg crate. The table consists of a platform and two identical linkages on which hydraulic cylinders exert equal forces. (Only one linkage and one cylinder are shown.) Members *EDB* and *CG* are each of length $2a$, and member *AD* is pinned to the midpoint of *EDB*. If the crate is placed on the table so that half of its weight is supported by the system shown, determine the force exerted by each cylinder in raising the crate for $\theta = 60°$, $a = 0.70$ m, and $L = 3.20$ m. Show that the result is independent of the distance d.

STRATEGY: The free-body diagram of the entire frame will involve more than three unknowns, so it alone can not be used to solve this problem. Instead, draw free-body diagrams of each component of the machine and work from them.

MODELING: The machine consists of the platform and the linkage. Its free-body diagram (Fig. 1) includes an input force $\mathbf{F}_{DH}$ exerted by the cylinder; the weight $W/2$, which is equal and opposite to the output force; and reactions at E and G, which are assumed to be directed as shown. Dismember the mechanism and draw a free-body diagram for each of its component parts (Fig. 2). Note that AD, BC, and CG are two-force members. Member CG has already been assumed to be in compression; now assume that AD and BC are in tension and direct the forces exerted on them as shown. Use equal and opposite vectors to represent the forces exerted by the two-force members on the platform, on member BDE, and on roller C.

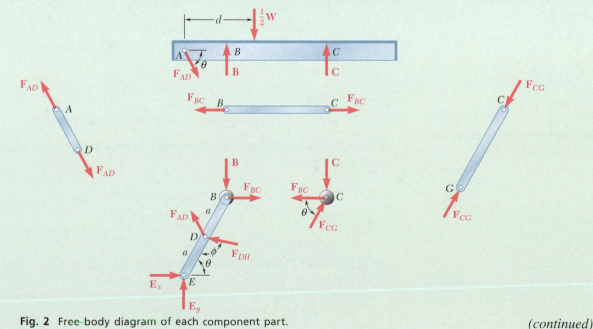

Fig. 2 Free-body diagram of each component part.

(continued)

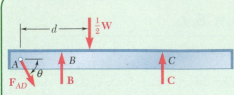

Fig. 3 Free-body diagram of platform *ABC*.

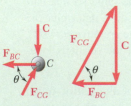

Fig. 4 Free-body diagram of roller *C* and its force triangle.

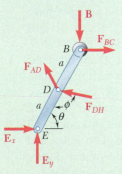

Fig. 5 Free-body diagram of member *BDE*.

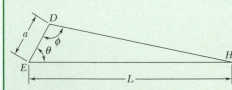

Fig. 6 Geometry of triangle *EDH*.

ANALYSIS:

Free Body: Platform *ABC* (Fig. 3).

$$\xrightarrow{+}\Sigma F_x = 0: \qquad F_{AD}\cos\theta = 0 \qquad F_{AD} = 0$$
$$+\uparrow\Sigma F_y = 0: \qquad B + C - \tfrac{1}{2}W = 0 \qquad B + C = \tfrac{1}{2}W \qquad (1)$$

Free-Body Roller *C* (Fig. 4). Draw a force triangle and obtain $F_{BC} = C\cot\theta$.

Free Body: Member *BDE* (Fig. 5). Recalling that $F_{AD} = 0$, you have

$$+\!\!\uparrow\!\!\Sigma M_E = 0: \quad F_{DH}\cos(\phi - 90°)a - B(2a\cos\theta) - F_{BC}(2a\sin\theta) = 0$$
$$F_{DH}a\sin\phi - B(2a\cos\theta) - (C\cot\theta)(2a\sin\theta) = 0$$
$$F_{DH}\sin\phi - 2(B + C)\cos\theta = 0$$

From Eq. (1), you obtain

$$F_{DH} = W\frac{\cos\theta}{\sin\phi} \qquad (2)$$

Note that *the result obtained is independent of d.* ◄

Applying first the law of sines to triangle *EDH* (Fig. 6), you have

$$\frac{\sin\phi}{EH} = \frac{\sin\theta}{DH} \qquad \sin\phi = \frac{EH}{DH}\sin\theta \qquad (3)$$

Now using the law of cosines, you get

$$(DH)^2 = a^2 + L^2 - 2aL\cos\theta$$
$$= (0.70)^2 + (3.20)^2 - 2(0.70)(3.20)\cos 60°$$
$$(DH)^2 = 8.49 \qquad DH = 2.91\text{ m}$$

Also note that

$$W = mg = (1000\text{ kg})(9.81\text{ m/s}^2) = 9810\text{ N} = 9.81\text{ kN}$$

Substituting for $\sin\phi$ from Eq. (3) into Eq. (2) and using the numerical data, your result is

$$F_{DH} = W\frac{DH}{EH}\cot\theta = (9.81\text{ kN})\frac{2.91\text{ m}}{3.20\text{ m}}\cot 60°$$

$$F_{DH} = 5.15\text{ kN} \quad ◄$$

REFLECT and THINK: Note that link *AD* ends up having zero force in this situation. However, this member still serves an important function, as it is necessary to enable the machine to support any horizontal load that might be exerted on the platform.

SOLVING PROBLEMS ON YOUR OWN

This section dealt with the analysis of *machines*. Since machines are designed to transmit or modify forces, they always contain moving parts. However, the machines considered here are always at rest, and you will be working with the set of *forces required to maintain the equilibrium of the machine.*

Known forces that act on a machine are called *input forces. A machine transforms the input forces into output forces,* such as the cutting forces applied by the pliers of Fig. 6.21. You will determine the output forces by finding the equal and opposite forces that should be applied to the machine to maintain its equilibrium.

In Sec. 6.3, you analyzed frames; you will use almost the same procedure to analyze machines by following these steps.

1. Draw a free-body diagram of the whole machine, and use it to determine as many as possible of the unknown forces exerted on the machine.

2. Dismember the machine and draw a free-body diagram of each member.

3. First consider the two-force members. Apply equal and opposite forces to each two-force member at the points where it is connected to another member. If you cannot tell at this point whether the member is in tension or in compression, *assume* that the member is in tension and *direct both of the forces away from the member.* Since these forces have the same unknown magnitude, *give them both the same name.*

4. Next consider the multi-force members. For each of these members, show all of the forces acting on it, including applied loads and forces, reactions, and internal forces at connections.

    a. Where a multi-force member is connected to a two-force member, apply to the multi-force member a force that is *equal and opposite* to the force drawn on the free-body diagram of the two-force member, *giving it the same name.*

    b. Where a multi-force member is connected to another multi-force member, use *horizontal and vertical components* to represent the internal forces at that point. The directions you choose for each of the two force components exerted on the first multi-force member are arbitrary, but *you must apply equal and opposite force components of the same name* to the other multi-force member.

5. Write equilibrium equations after you have completed the various free-body diagrams.

    a. To simplify your solution, you should, whenever possible, write and solve equilibrium equations involving single unknowns.

    b. Since you arbitrarily chose the direction of each of the unknown forces, you must determine at the end of the solution whether your guess was correct. To that effect, *consider the sign* of the value found for each of the unknowns. A *positive* sign indicates that your guess was correct, and a *negative* sign indicates that it was not.

6. Finally, check your solution by substituting the results obtained into an equilibrium equation that you have not previously used.

Problems

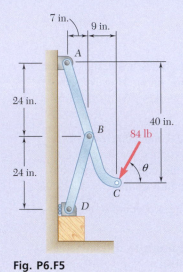

Fig. P6.F5

6.F5 An 84-lb force is applied to the toggle vise at C. Knowing that $\theta = 90°$, draw the free-body diagram(s) needed to determine the vertical force exerted on the block at D.

6.F6 For the system and loading shown, draw the free-body diagram(s) needed to determine the force **P** required for equilibrium.

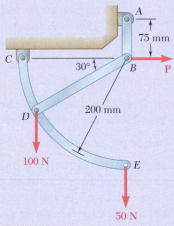

Fig. P6.F6

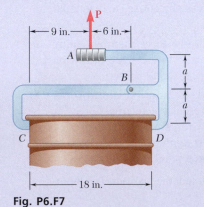

Fig. P6.F7

6.F7 A small barrel weighing 60 lb is lifted by a pair of tongs as shown. Knowing that $a = 5$ in., draw the free-body diagram(s) needed to determine the forces exerted at B and D on tong ABD.

6.F8 The position of member ABC is controlled by the hydraulic cylinder CD. Knowing that $\theta = 30°$, draw the free-body diagram(s) needed to determine the force exerted by the hydraulic cylinder on pin C, and the reaction at B.

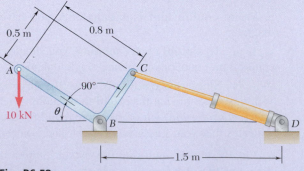

Fig. P6.F8

6.122 The shear shown is used to cut and trim electronic-circuit-board laminates. For the position shown, determine (a) the vertical component of the force exerted on the shearing blade at D, (b) the reaction at C.

6.123 A 100-lb force directed vertically downward is applied to the toggle vise at C. Knowing that link BD is 6 in. long and that a = 4 in., determine the horizontal force exerted on block E.

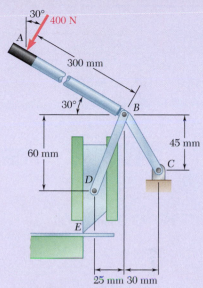

Fig. P6.122

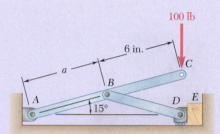

Fig. P6.123 and P6.124

6.124 A 100-lb force directed vertically downward is applied to the toggle vise at C. Knowing that link BD is 6 in. long and that a = 8 in., determine the horizontal force exerted on block E.

6.125 The control rod CE passes through a horizontal hole in the body of the toggle system shown. Knowing that link BD is 250 mm long, determine the force **Q** required to hold the system in equilibrium when β = 20°.

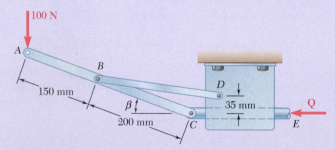

Fig. P6.125

6.126 Solve Prob. 6.125 when (a) β = 0, (b) β = 6°.

6.127 The press shown is used to emboss a small seal at E. Knowing that P = 250 N, determine (a) the vertical component of the force exerted on the seal, (b) the reaction at A.

6.128 The press shown is used to emboss a small seal at E. Knowing that the vertical component of the force exerted on the seal must be 900 N, determine (a) the required vertical force **P**, (b) the corresponding reaction at A.

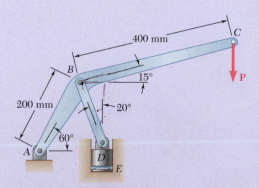

Fig. P6.127 and *P6.128*

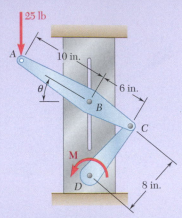

25 lb
A
10 in.
θ
6 in.
B
M
C
D
8 in.

Fig. P6.129 and P6.130

6.129 The pin at B is attached to member ABC and can slide freely along the slot cut in the fixed plate. Neglecting the effect of friction, determine the couple $\mathbf{M}$ required to hold the system in equilibrium when $\theta = 30°$.

6.130 The pin at B is attached to member ABC and can slide freely along the slot cut in the fixed plate. Neglecting the effect of friction, determine the couple $\mathbf{M}$ required to hold the system in equilibrium when $\theta = 60°$.

6.131 Arm ABC is connected by pins to a collar at B and to crank CD at C. Neglecting the effect of friction, determine the couple $\mathbf{M}$ required to hold the system in equilibrium when $\theta = 0$.

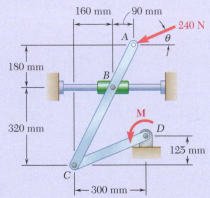

160 mm   90 mm
240 N
A   θ
180 mm
B
320 mm
M
D
125 mm
C
300 mm

Fig. P6.131 and P6.132

6.132 Arm ABC is connected by pins to a collar at B and to crank CD at C. Neglecting the effect of friction, determine the couple $\mathbf{M}$ required to hold the system in equilibrium when $\theta = 90°$.

6.133 The Whitworth mechanism shown is used to produce a quick-return motion of point D. The block at B is pinned to the crank AB and is free to slide in a slot cut in member CD. Determine the couple $\mathbf{M}$ that must be applied to the crank AB to hold the mechanism in equilibrium when (a) $\alpha = 0$, (b) $\alpha = 30°$.

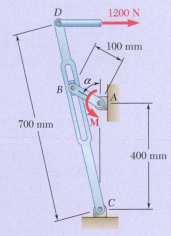

D
1200 N
100 mm
B
α
A
700 mm
M
400 mm
C

Fig. P6.133

6.134 Solve Prob. 6.133 when (a) $\alpha = 60°$, (b) $\alpha = 90°$.

6.135 and 6.136 Two rods are connected by a frictionless collar B. Knowing that the magnitude of the couple $\mathbf{M}_A$ is 500 lb·in., determine (a) the couple $\mathbf{M}_C$ required for equilibrium, (b) the corresponding components of the reaction at C.

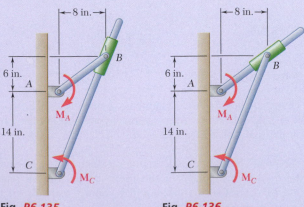

8 in.
B
6 in.
A
M_A
14 in.
C
M_C

Fig. P6.135

8 in.
B
6 in.
A
M_A
14 in.
C
M_C

Fig. P6.136

6.137 and 6.138 Rod *CD* is attached to the collar *D* and passes through a collar welded to end *B* of lever *AB*. Neglecting the effect of friction, determine the couple **M** required to hold the system in equilibrium when $\theta = 30°$.

6.139 Two hydraulic cylinders control the position of the robotic arm *ABC*. Knowing that in the position shown the cylinders are parallel, determine the force exerted by each cylinder when $P = 160$ N and $Q = 80$ N.

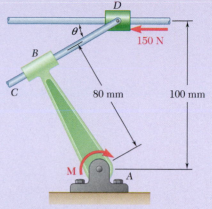

Fig. P6.137

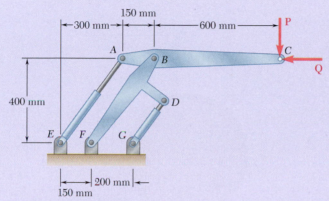

Fig. P6.139 and P6.140

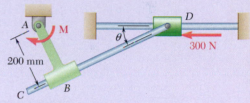

Fig. P6.138

6.140 Two hydraulic cylinders control the position of the robotic arm *ABC*. In the position shown, the cylinders are parallel and both are in tension. Knowing that $F_{AE} = 600$ N and $F_{DG} = 50$ N, determine the forces **P** and **Q** applied at *C* to arm *ABC*.

6.141 A 39-ft length of railroad rail of weight 44 lb/ft is lifted by the tongs shown. Determine the forces exerted at *D* and *F* on tong *BDF*.

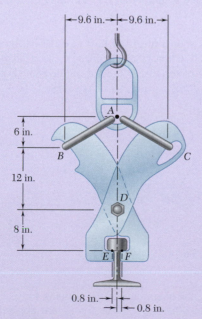

Fig. P6.141

6.142 A log weighing 800 lb is lifted by a pair of tongs as shown. Determine the forces exerted at E and F on tong DEF.

800 lb
1.5 in.
1.5 in.
3 in.
3 in.
A
B
C
D
2.5 in.
3.5 in.
E
12 in.
F
G
12 in.

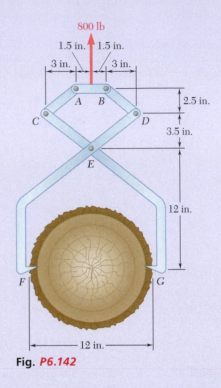

Fig. *P6.142*

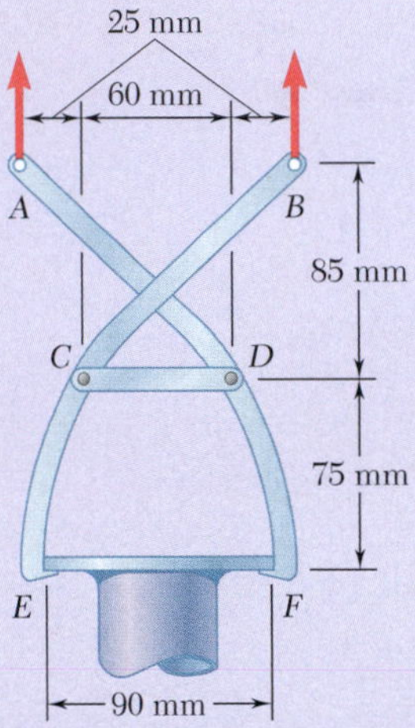

25 mm
60 mm
A
B
85 mm
C
D
75 mm
E
F
90 mm

Fig. P6.143

6.143 The tongs shown are used to apply a total upward force of 45 kN on a pipe cap. Determine the forces exerted at D and F on tong ADF.

6.144 If the toggle shown is added to the tongs of Prob. 6.143 and a single vertical force is applied at G, determine the forces exerted at D and F on tong ADF.

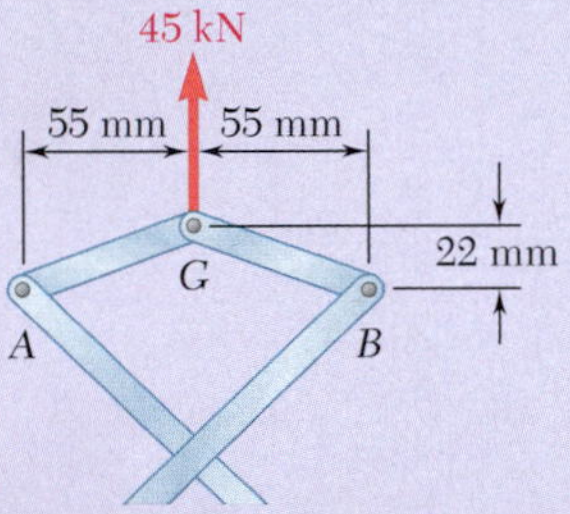

45 kN
55 mm
55 mm
22 mm
G
A
B

Fig. P6.144

6.145 The pliers shown are used to grip a 0.3-in.-diameter rod. Knowing that two 60-lb forces are applied to the handles, determine (*a*) the magnitude of the forces exerted on the rod, (*b*) the force exerted by the pin at A on portion AB of the pliers.

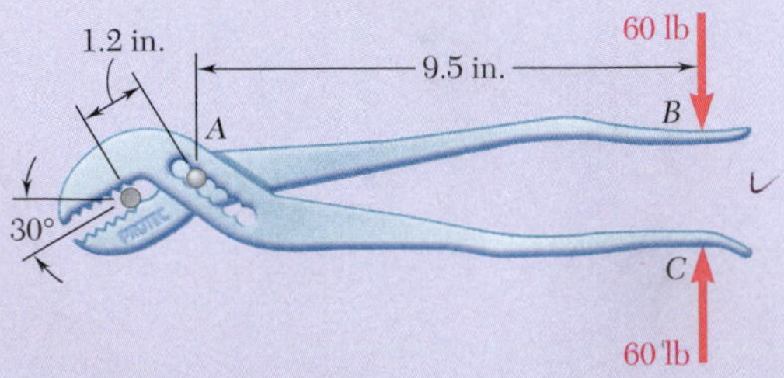

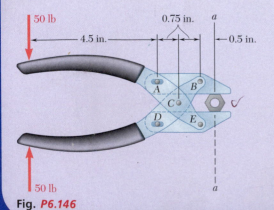

1.2 in.
9.5 in.
60 lb
A
B
30°
C
60 lb

Fig. P6.145

50 lb
0.75 in.
a
4.5 in.
0.5 in.
A
B
C
D
E
50 lb
a

Fig. *P6.146*

6.146 Determine the magnitude of the gripping forces exerted along line aa on the nut when two 50-lb forces are applied to the handles as shown. Assume that pins A and D slide freely in slots cut in the jaws.

6.147 In using the bolt cutter shown, a worker applies two 300-N forces to the handles. Determine the magnitude of the forces exerted by the cutter on the bolt.

6.148 Determine the magnitude of the gripping forces produced when two 300-N forces are applied as shown.

6.149 and 6.150 Determine the force **P** that must be applied to the toggle *CDE* to maintain bracket *ABC* in the position shown.

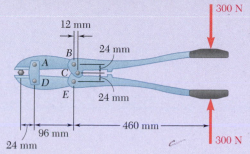

Fig. P6.147

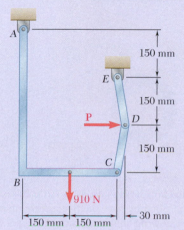

Fig. P6.149

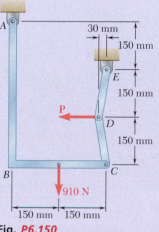

Fig. P6.150

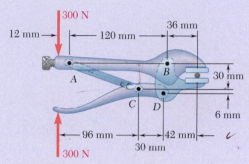

Fig. P6.148

6.151 Since the brace shown must remain in position even when the magnitude of **P** is very small, a single safety spring is attached at *D* and *E*. The spring *DE* has a constant of 50 lb/in. and an unstretched length of 7 in. Knowing that *l* = 10 in. and that the magnitude of **P** is 800 lb, determine the force **Q** required to release the brace.

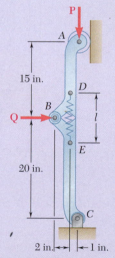

Fig. P6.151

6.152 The specialized plumbing wrench shown is used in confined areas (e.g., under a basin or sink). It consists essentially of a jaw *BC* pinned at *B* to a long rod. Knowing that the forces exerted on the nut are equivalent to a clockwise (when viewed from above) couple with a magnitude of 135 lb·in., determine (*a*) the magnitude of the force exerted by pin *B* on jaw *BC*, (*b*) the couple **M₀** that is applied to the wrench.

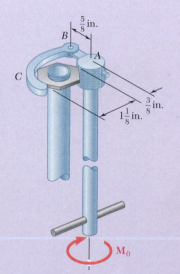

Fig. P6.152

357

6.153 The motion of the bucket of the front-end loader shown is controlled by two arms and a linkage that are pin-connected at *D*. The arms are located symmetrically with respect to the central, vertical, and longitudinal plane of the loader; one arm *AFJ* and its control cylinder *EF* are shown. The single linkage *GHDB* and its control cylinder *BC* are located in the plane of symmetry. For the position and loading shown, determine the force exerted (*a*) by cylinder *BC*, (*b*) by cylinder *EF*.

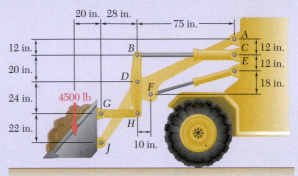

Fig. P6.153

6.154 The bucket of the front-end loader shown carries a 3200-lb load. The motion of the bucket is controlled by two identical mechanisms, only one of which is shown. Knowing that the mechanism shown supports one-half of the 3200-lb load, determine the force exerted (*a*) by cylinder *CD*, (*b*) by cylinder *FH*.

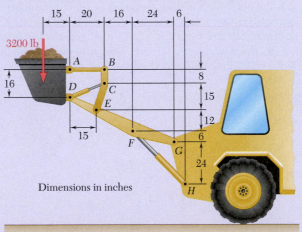

Dimensions in inches

Fig. P6.154

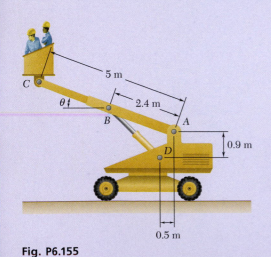

Fig. P6.155

6.155 The telescoping arm *ABC* is used to provide an elevated platform for construction workers. The workers and the platform together have a mass of 200 kg and have a combined center of gravity located directly above *C*. For the position when $\theta = 20°$, determine (*a*) the force exerted at *B* by the single hydraulic cylinder *BD*, (*b*) the force exerted on the supporting carriage at *A*.

6.156 The telescoping arm *ABC* of Prob. 6.155 can be lowered until end *C* is close to the ground, so that workers can easily board the platform. For the position when $\theta = -20°$, determine (*a*) the force exerted at *B* by the single hydraulic cylinder *BD*, (*b*) the force exerted on the supporting carriage at *A*.

6.157 The motion of the backhoe bucket shown is controlled by the hydraulic cylinders *AD*, *CG*, and *EF*. As a result of an attempt to dislodge
a portion of a slab, a 2-kip force **P** is exerted on the bucket teeth
at *J*. Knowing that $\theta = 45°$, determine the force exerted by each
cylinder.

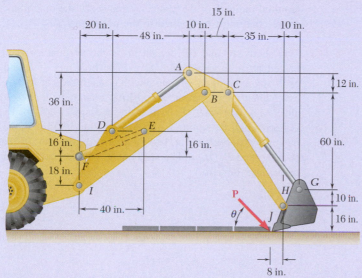

Fig. P6.157

6.158 Solve Prob. 6.157 assuming that the 2-kip force **P** acts horizontally
to the right ($\theta = 0$).

6.159 The gears *D* and *G* are rigidly attached to shafts that are held by
frictionless bearings. If $r_D = 90$ mm and $r_G = 30$ mm, determine
(*a*) the couple $\mathbf{M}_0$ that must be applied for equilibrium, (*b*) the reactions at *A* and *B*.

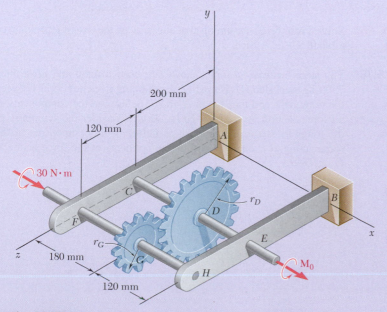

Fig. P6.159

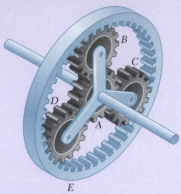

Fig. P6.160

6.160 In the planetary gear system shown, the radius of the central gear A is $a = 18$ mm, the radius of each planetary gear is b, and the radius of the outer gear E is $(a + 2b)$. A clockwise couple with a magnitude of $M_A = 10$ N·m is applied to the central gear A and a counterclockwise couple with a magnitude of $M_S = 50$ N·m is applied to the spider BCD. If the system is to be in equilibrium, determine (a) the required radius b of the planetary gears, (b) the magnitude M_E of the couple that must be applied to the outer gear E.

*6.161 Two shafts AC and CF, which lie in the vertical xy plane, are connected by a universal joint at C. The bearings at B and D do not exert any axial force. A couple with a magnitude of 500 lb·in. (clockwise when viewed from the positive x axis) is applied to shaft CF at F. At a time when the arm of the crosspiece attached to shaft CF is horizontal, determine (a) the magnitude of the couple that must be applied to shaft AC at A to maintain equilibrium, (b) the reactions at B, D, and E. (*Hint:* The sum of the couples exerted on the crosspiece must be zero.)

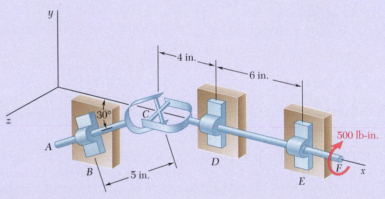

Fig. P6.161

*6.162 Solve Prob. 6.161 assuming that the arm of the crosspiece attached to shaft CF is vertical.

*6.163 The large mechanical tongs shown are used to grab and lift a thick 7500-kg steel slab HJ. Knowing that slipping does not occur between the tong grips and the slab at H and J, determine the components of all forces acting on member EFH. (*Hint:* Consider the symmetry of the tongs to establish relationships between the components of the force acting at E on EFH and the components of the force acting at D on DGJ.)

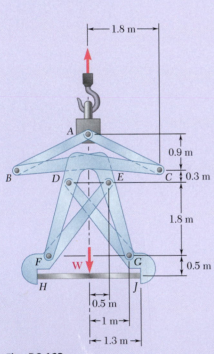

Fig. P6.163

Review and Summary

In this chapter, you studied ways to determine the **internal forces** holding together the various parts of a structure.

Analysis of Trusses

The first half of the chapter presented the analysis of **trusses**, i.e., structures consisting of *straight members connected at their extremities only*. Because the members are slender and unable to support lateral loads, all of the loads must be applied at the joints; thus, we can assume that a truss consists of *pins and two-force members* [Sec. 6.1A].

Simple Trusses

A truss is **rigid** if it is designed in such a way that it does not greatly deform or collapse under a small load. A triangular truss consisting of three members connected at three joints is clearly a rigid truss (Fig. 6.24*a*). The truss obtained by adding two new members to the first one and connecting them at a new joint (Fig. 6.24*b*) is also rigid. Trusses obtained by repeating this procedure are called **simple trusses**. We may check that, in a simple truss, the total number of members is $m = 2n - 3$, where n is the total number of joints [Sec. 6.1A].

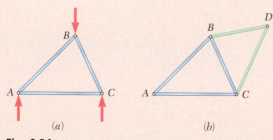

Fig. 6.24

Method of Joints

We can determine the forces in the various members of a simple truss by using the **method of joints** [Sec. 6.1B]. First, we obtain the reactions at the supports by considering the entire truss as a free body. Then we draw the free-body diagram of each pin, showing the forces exerted on the pin by the members or supports it connects. Since the members are straight two-force members, the force exerted by a member on the pin is directed along that member, and only the magnitude of the force is unknown. In the case of a simple truss, it is always possible to draw the free-body diagrams of the pins in such an order that only two unknown forces are included in each diagram. We obtain these forces from the corresponding two equilibrium equations or—if only three forces are involved—from the corresponding force triangle. If the force exerted by a member on a pin is directed toward that pin, the member is in **compression**; if it is directed away from the pin, the member is in **tension** [Sample Prob. 6.1]. The analysis of a truss is sometimes expedited by first recognizing **joints under special loading conditions** [Sec. 6.1C]. The method of joints also can be extended for the analysis of three-dimensional or **space trusses** [Sec. 6.1D].

Method of Sections

The **method of sections** is usually preferable to the method of joints when we want to determine the force in only one member—or very few members—of a truss [Sec. 6.2A]. To determine the force in member *BD* of the truss of Fig. 6.25*a*, for example, we *pass a section* through members *BD, BE,* and *CE;* remove these members; and use the portion *ABC* of the truss as a free body (Fig. 6.25*b*). Setting $\Sigma M_E = 0$, we determine the magnitude of force $\mathbf{F}_{BD}$ that represents the force in member *BD*. A positive sign indicates that the member is in *tension;* a negative sign indicates that it is in *compression* [Sample Probs. 6.2 and 6.3].

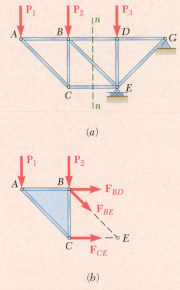

(a)

(b)

Fig. 6.25

Compound Trusses

The method of sections is particularly useful in the analysis of **compound trusses**, i.e., trusses that cannot be constructed from the basic triangular truss of Fig. 6.24*a* but are built by rigidly connecting several simple trusses [Sec. 6.2B]. If the component trusses are properly connected (e.g., one pin and one link, or three non-concurrent and unparallel links) and if the resulting structure is properly supported (e.g., one pin and one roller), the compound truss is **statically determinate, rigid, and completely constrained**. The following necessary—but not sufficient—condition is then satisfied: $m + r = 2n$, where *m* is the number of members, *r* is the number of unknowns representing the reactions at the supports, and *n* is the number of joints.

Frames and Machines

In the second part of the chapter, we analyzed **frames** and **machines**. These structures contain *multi-force members,* i.e., members acted upon by three or more forces. Frames are designed to support loads and are usually stationary, fully constrained structures. Machines are designed to transmit or modify forces and always contain moving parts [Sec. 6.3].

Analysis of a Frame

To analyze a frame, we first consider the entire frame to be a free body and write three equilibrium equations [Sec. 6.3A]. If the frame remains rigid when detached from its supports, the reactions involve only three unknowns and may be determined from these equations [Sample Probs. 6.4 and 6.5]. On the other hand, if the frame ceases to be rigid when detached from its supports, the reactions involve more than three unknowns, and we cannot determine them completely from the equilibrium equations of the frame [Sec. 6.3B; Sample Prob. 6.6].

Multi-force Members

We then dismember the frame and identify the various members as either two-force members or multi-force members; we assume pins form an integral part of one of the members they connect. We draw the free-body diagram of each of the multi-force members, noting that, when two multi-force members are connected to the same two-force member, they are acted upon by that member with *equal and opposite forces of unknown magnitude but known direction.* When two multi-force members are connected by a pin, they exert on each other *equal and opposite forces of unknown direction* that should be represented by *two unknown components*. We can then solve the equilibrium equations obtained from the free-body diagrams of the multi-force members for the various internal forces [Sample Probs. 6.4 and 6.5]. We also can use the equilibrium equations to complete the determination of the reactions at the supports [Sample Prob. 6.6]. Actually, if the frame is *statically determinate and rigid,* the free-body diagrams of the multi-force members could provide as many equations as there are unknown forces (including the reactions) [Sec. 6.3B]. However, as suggested previously, it is advisable to first consider the free-body diagram of the entire frame to minimize the number of equations that must be solved simultaneously.

Analysis of a Machine

To analyze a machine, we dismember it and, following the same procedure as for a frame, draw the free-body diagram of each multi-force member. The corresponding equilibrium equations yield the **output forces** exerted by the machine in terms of the **input forces** applied to it as well as the **internal forces** at the various connections [Sec. 6.4; Sample Prob. 6.7].

Review Problems

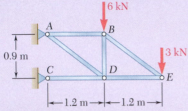

Fig. P6.164

6.164 Using the method of joints, determine the force in each member of the truss shown. State whether each member is in tension or compression.

6.165 Using the method of joints, determine the force in each member of the double-pitch roof truss shown. State whether each member is in tension or compression.

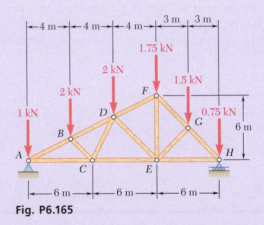

Fig. P6.165

6.166 A stadium roof truss is loaded as shown. Determine the force in members *AB*, *AG*, and *FG*.

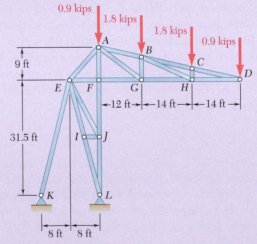

Fig. P6.166 and *P6.167*

6.167 A stadium roof truss is loaded as shown. Determine the force in members *AE*, *EF*, and *FJ*.

6.168 Determine the components of all forces acting on member *ABD* of the frame shown.

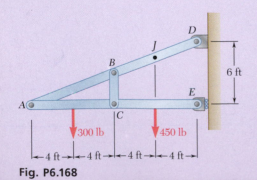

Fig. P6.168

6.169 Determine the components of the reactions at A and E if the frame is loaded by a clockwise couple of magnitude 36 N·m applied (*a*) at B, (*b*) at D.

6.170 Knowing that the pulley has a radius of 50 mm, determine the components of the reactions at B and E.

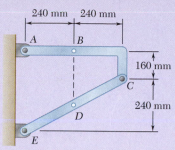

Fig. P6.169

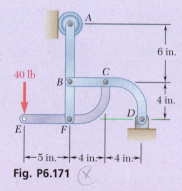

Fig. P6.170

6.171 For the frame and loading shown, determine the components of the forces acting on member CFE at C and F.

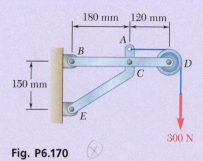

Fig. P6.171

6.172 For the frame and loading shown, determine the reactions at A, B, D, and E. Assume that the surface at each support is frictionless.

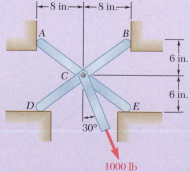

Fig. P6.172

6.173 Water pressure in the supply system exerts a downward force of 135 N on the vertical plug at A. Determine the tension in the fusible link DE and the force exerted on member BCE at B.

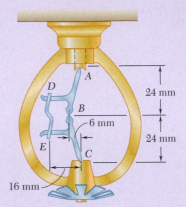

Fig. *P6.173*

6.174 A couple **M** with a magnitude of 1.5 kN·m is applied to the crank of the engine system shown. For each of the two positions shown, determine the force **P** required to hold the system in equilibrium.

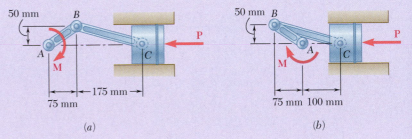

(a)   (b)

Fig. P6.174

6.175 The compound-lever pruning shears shown can be adjusted by placing pin A at various ratchet positions on blade ACE. Knowing that 300-lb vertical forces are required to complete the pruning of a small branch, determine the magnitude P of the forces that must be applied to the handles when the shears are adjusted as shown.

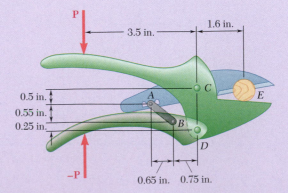

Fig. *P6.175*

7

Internal Forces and Moments

The Assut de l'Or Bridge in the City of Arts and Science in Valencia, Spain, is cable-stayed, where the bridge deck is supported by cables attached to the curved tower. The tower itself is partially supported by four anchor cables. The deck of the bridge consists of a system of beams that support the roadway.

Introduction

Objectives

- **Consider** the general state of internal member forces, which includes axial force, shearing force, and bending moment.

- **Apply** equilibrium analysis methods to obtain specific values, general expressions, and diagrams for shear and bending-moment in beams.

- **Examine** relations among load, shear, and bending-moment, and use these to obtain shear and bending-moment diagrams for beams.

- **Analyze** the tension forces in cables subjected to concentrated loads, loads uniformly distributed along the horizontal, and loads uniformly distributed along the cable itself.

Introduction

In previous chapters, we considered two basic problems involving structures: (1) determining the external forces acting on a structure (Chap. 4) and (2) determining the internal forces that hold together the various members forming a structure (Chap. 6). Now we consider the problem of determining the internal forces that hold together the parts of a given individual member.

We will first analyze the internal forces in the members of a frame, such as the crane considered in Fig. 6.1. Note that, whereas the internal forces in a straight two-force member can produce only **tension** or **compression** in that member, the internal forces in any other type of member usually produce **shear** and **bending** as well.

Most of this chapter is devoted to the analysis of the internal forces in two important types of engineering elements:

1. **Beams**, which are usually long, straight prismatic members designed to support loads applied at various points along it.
2. **Cables**, which are flexible members capable of withstanding only tension and are designed to support either concentrated or distributed loads. Cables are used in many engineering applications, such as suspension bridges and power transmission lines.

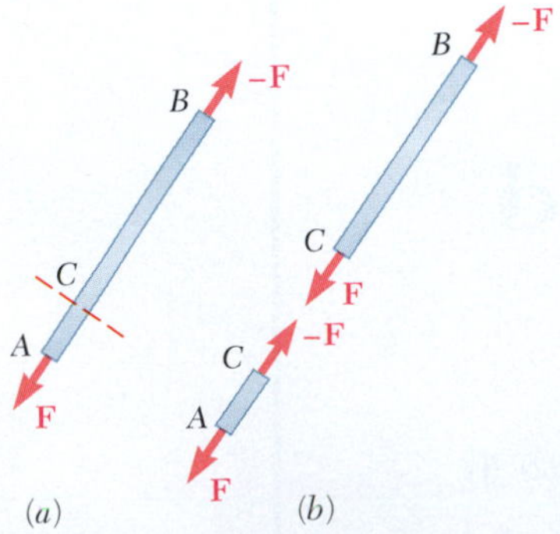

Fig. 7.1 A straight two-force member in tension. (a) External forces act at the ends of the member; (b) internal axial forces do not depend on the location of section C.

7.1 INTERNAL FORCES IN MEMBERS

Consider a straight two-force member AB (Fig. 7.1a). From Sec. 4.2A, we know that the forces $\mathbf{F}$ and $-\mathbf{F}$ acting at A and B, respectively, must be directed along AB in opposite sense and have the same magnitude F. Suppose we cut the member at C. To maintain equilibrium of the resulting free bodies AC and CB, we must apply to AC a force $-\mathbf{F}$ equal and opposite to **F** and to *CB* a force **F** equal and opposite to −**F** (Fig. 7.1*b*). These new forces are directed along *AB* in opposite sense and have the same magnitude *F*. Since the two parts *AC* and *CB* were in equilibrium before the member was cut, **internal forces** equivalent to these new forces must have existed in the member itself. We conclude that, in the case of a straight two-force member, the internal forces that the two portions of the member exert on each other are equivalent to **axial forces**. The common magnitude *F* of these forces does not depend upon the location of the section *C* and is referred to as the *force in member AB*. In the case shown in Fig. 7.1, the member is in tension and elongates under the action of the internal forces. In the case represented in Fig. 7.2, the member is in compression and decreases in length under the action of the internal forces.

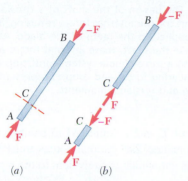

(*a*)   (*b*)

Fig. 7.2 A straight two-force member in compression. (*a*) External forces act at the ends; (*b*) internal axial forces are independent of the location of section *C*.

Next, consider a **multi-force member**. Take, for instance, member *AD* of the crane analyzed in Sec. 6.3A. This crane is shown again in Fig. 7.3*a*, and we drew the free-body diagram of member *AD* in Fig. 7.3*b*. Suppose we cut member *AD* at *J* and draw a free-body diagram for each of the portions *JD* and *AJ* (Fig. 7.3*c* and *d*). Considering the free body *JD*, we find that, to maintain its equilibrium, we need to apply at *J* a force **F** to balance the vertical component of **T**; a force **V** to balance the

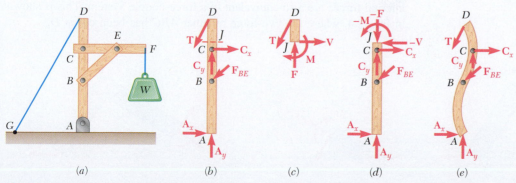

(*a*)   (*b*)   (*c*)   (*d*)   (*e*)

Fig. 7.3 (*a*) Crane from Chapter 6; (*b*) free-body diagram of multi-force member *AD*; (*c,d*) free-body diagrams of sections of member *AD* showing internal force-couple systems; (*e*) deformation of member *AD*.

Photo 7.1 The design of the shaft of a circular saw must account for the internal forces resulting from the forces applied to the teeth of the blade. At a given point in the shaft, these internal forces are equivalent to a force-couple system consisting of axial and shearing forces and couples representing the bending and torsional moments.

horizontal component of **T**; and a couple **M** to balance the moment of **T** about *J*. Again, we conclude that internal forces must have existed at *J* before member *AD* was cut, which is equivalent to the force-couple system shown in Fig. 7.3*c*.

According to Newton's third law, the internal forces acting on *AJ* must be equivalent to an equal and opposite force-couple system, as shown in Fig. 7.3*d*. Clearly, the action of the internal forces in member *AD is not limited to producing tension or compression,* as in the case of straight two-force members; the internal forces *also produce shear and bending.* The force **F** is an **axial force***;* the force **V** is called a **shearing force***;* and the moment **M** of the couple is known as the **bending moment at** *J*. Note that, when determining internal forces in a member, you should clearly indicate on which portion of the member the forces are supposed to act. The deformation that occurs in member *AD* is sketched in Fig. 7.3*e*. The actual analysis of such a deformation is part of the study of mechanics of materials.

Also note that, in a **two-force member that is not straight**, the internal forces are also equivalent to a force-couple system. This is shown in Fig. 7.4, where the two-force member *ABC* has been cut at *D*.

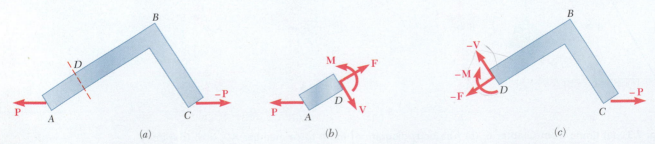

Fig. 7.4 (*a*) Free-body diagram of a two-force member that is not straight; (*b*, *c*) free-body diagrams of sections of member *ABC* showing internal force-couple systems.

Sample Problem 7.1

In the frame shown, determine the internal forces (*a*) in member *ACF* at point *J*, (*b*) in member *BCD* at point *K*. This frame was previously analyzed in Sample Prob. 6.5.

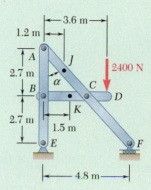

STRATEGY: After isolating each member, you can cut it at the given point and treat the resulting parts as objects in equilibrium. Analysis of the equilibrium equations, as we did before in Sample Problem 6.5, will determine the internal force-couple system.

MODELING: The reactions and the connection forces acting on each member of the frame were determined previously in Sample Prob. 6.5. The results are repeated in Fig. 1.

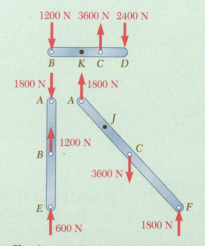

Fig. 1 Reactions and connection forces acting on each member of the frame.

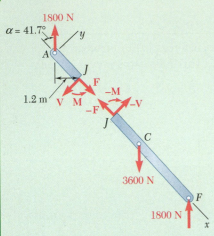

Fig. 2 Free-body diagrams of portion *AJ* and *FJ* of member *ACF*.

ANALYSIS:

a. Internal Forces at *J*.
Cut member *ACF* at point *J*, obtaining the two parts shown in Fig. 2. Represent the internal forces at *J* by an equivalent force-couple system, which can be determined by considering the equilibrium of either part. Considering the free body *AJ*, you have

$$+\curvearrowleft \Sigma M_J = 0: \qquad -(1800 \text{ N})(1.2 \text{ m}) + M = 0$$
$$M = +2160 \text{ N·m} \qquad\qquad \mathbf{M} = 2160 \text{ N·m} \nwarrow \quad \blacktriangleleft$$

$$+\searrow \Sigma F_x = 0: \qquad F - (1800 \text{ N}) \cos 41.7° = 0$$
$$F = +1344 \text{ N} \qquad\qquad \mathbf{F} = 1344 \text{ N} \searrow \quad \blacktriangleleft$$

$$+\nearrow \Sigma F_y = 0: \qquad -V + (1800 \text{ N}) \sin 41.7° = 0$$
$$V = +1197 \text{ N} \qquad\qquad \mathbf{V} = 1197 \text{ N} \swarrow \quad \blacktriangleleft$$

The internal forces at *J* are therefore equivalent to a couple **M**, an axial force **F**, and a shearing force **V**. The internal force-couple system acting on part *JCF* is equal and opposite.

b. Internal Forces at *K*.
Cut member *BCD* at *K*, obtaining the two parts shown in Fig. 3. Considering the free body *BK*, you obtain

$$+\curvearrowleft \Sigma M_K = 0: \qquad (1200 \text{ N})(1.5 \text{ m}) + M = 0$$
$$M = -1800 \text{ N·m} \qquad\qquad \mathbf{M} = 1800 \text{ N·m} \curvearrowright \quad \blacktriangleleft$$

$$\xrightarrow{+} \Sigma F_x = 0: \qquad F = 0 \qquad\qquad\qquad\qquad\quad \mathbf{F} = 0 \quad \blacktriangleleft$$

$$+\uparrow \Sigma F_y = 0: \qquad -1200 \text{ N} - V = 0$$
$$V = -1200 \text{ N} \qquad\qquad \mathbf{V} = 1200 \text{ N} \uparrow \quad \blacktriangleleft$$

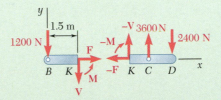

Fig. 3 Free-body diagrams of portion *BK* and *DK* of member *BCD*.

REFLECT and THINK: The mathematical techniques involved in solving a problem of this type are not new; they are simply applications of concepts presented in earlier chapters. However, the physical interpretation is new: we are now determining the internal forces and moments within a structural member. These are of central importance in the study of mechanics of materials.

SOLVING PROBLEMS
ON YOUR OWN

In this section, we discussed how to determine the internal forces in the member of a frame. The internal forces at a given point in a **straight two-force member** reduce to an axial force, but in all other cases, they are equivalent to a **force-couple system** consisting of an **axial force F**, a **shearing force V**, and a couple **M** representing the **bending moment** at that point.

To determine the internal forces at a given point J of the member of a frame, you should take the following steps.

1. Draw a free-body diagram of the entire frame, and use it to determine as many of the reactions at the supports as you can.

2. Dismember the frame and draw a free-body diagram of each of its members. Write as many equilibrium equations as are necessary to find all of the forces acting on the member on which point J is located.

3. Cut the member at point J and draw a free-body diagram of each resulting portion. Apply to each portion at point J the force components and couple representing the internal forces exerted by the other portion. These force components and couples are equal in magnitude and opposite in sense.

4. Select one of the two free-body diagrams you have drawn and use it to write three equilibrium equations for the corresponding portion of the member.

    a. Summing moments about J and equating them to zero yields the bending moment at point J.

    b. Summing components in directions parallel and perpendicular to the member at J and equating them to zero yields, respectively, the axial and shearing forces.

5. When recording your answers, be sure to specify the portion of the member you have used, since the forces and couples acting on the two portions have opposite senses.

The solutions of the problems in this section require you to determine the forces exerted on each other by the various members of a frame, so be sure to review the methods used in Chap. 6 to solve this type of problem. When frames involve pulleys and cables, for instance, remember that the forces exerted by a pulley on the member of the frame to which it is attached have the same magnitude and direction as the forces exerted by the cable on the pulley [Prob. 6.90].

Problems

7.1 and 7.2 Determine the internal forces (axial force, shearing force, and bending moment) at point *J* of the structure indicated.
  7.1 Frame and loading of Prob. 6.76
  7.2 Frame and loading of Prob. 6.78

7.3 Determine the internal forces at point *J* when $\alpha = 90°$.

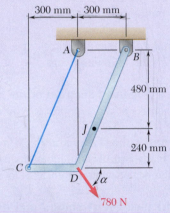

Fig. P7.3 and P7.4

7.4 Determine the internal forces at point *J* when $\alpha = 0$.

7.5 and 7.6 For the frame and loading shown, determine the internal forces at the point indicated:
  7.5 Point *J*
  7.6 Point *K*

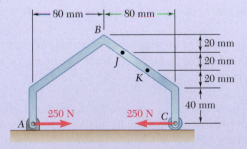

Fig. *P7.5* and *P7.6*

7.7 An archer aiming at a target is pulling with a 45-lb force on the bowstring. Assuming that the shape of the bow can be approximated by a parabola, determine the internal forces at point *J*.

7.8 For the bow of Prob. 7.7, determine the magnitude and location of the maximum (*a*) axial force, (*b*) shearing force, (*c*) bending moment.

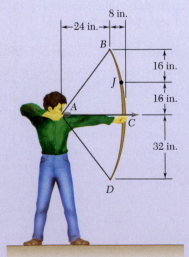

Fig. P7.7

7.9 A semicircular rod is loaded as shown. Determine the internal forces at point *J*.

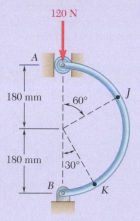

Fig. P7.9 and P7.10

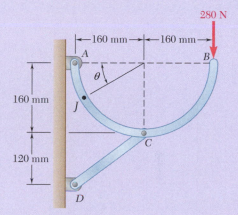

Fig. P7.11 and P7.12

7.10 A semicircular rod is loaded as shown. Determine the internal forces at point *K*.

7.11 A semicircular rod is loaded as shown. Determine the internal forces at point *J* knowing that $\theta = 30°$.

7.12 A semicircular rod is loaded as shown. Determine the magnitude and location of the maximum bending moment in the rod.

7.13 The axis of the curved member *AB* is a parabola with vertex at *A*. If a vertical load **P** of magnitude 450 lb is applied at *A*, determine the internal forces at *J* when *h* = 12 in., *L* = 40 in., and *a* = 24 in.

7.14 Knowing that the axis of the curved member *AB* is a parabola with vertex at *A*, determine the magnitude and location of the maximum bending moment.

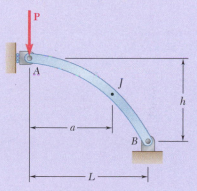

Fig. P7.13 and P7.14

7.15 Knowing that the radius of each pulley is 200 mm and neglecting friction, determine the internal forces at point *J* of the frame shown.

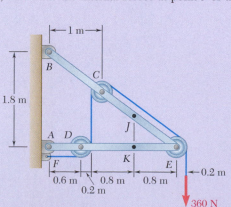

Fig. P7.15 and P7.16

7.16 Knowing that the radius of each pulley is 200 mm and neglecting friction, determine the internal forces at point *K* of the frame shown.

7.17 A 5-in.-diameter pipe is supported every 9 ft by a small frame consisting of two members as shown. Knowing that the combined weight of the pipe and its contents is 10 lb/ft and neglecting the effect of friction, determine the magnitude and location of the maximum bending moment in member *AC*.

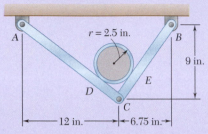

Fig. P7.17

7.18 For the frame of Prob. 7.17, determine the magnitude and location of the maximum bending moment in member *BC*.

7.19 Knowing that the radius of each pulley is 200 mm and neglecting friction, determine the internal forces at point *J* of the frame shown.

7.20 Knowing that the radius of each pulley is 200 mm and neglecting friction, determine the internal forces at point *K* of the frame shown.

7.21 and 7.22 A force **P** is applied to a bent rod that is supported by a roller and a pin and bracket. For each of the three cases shown, determine the internal forces at point *J*.

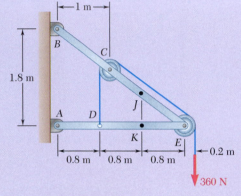

Fig. P7.19 and P7.20

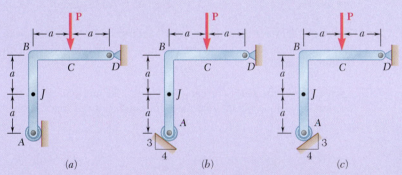

(a)   (b)   (c)

Fig. P7.21

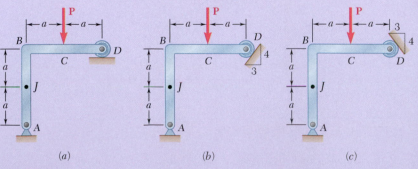

(a)   (b)   (c)

Fig. P7.22

7.23 A quarter-circular rod of weight W and uniform cross section is supported as shown. Determine the bending moment at point J when $\theta = 30°$.

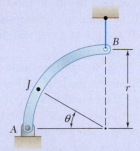

Fig. P7.23

7.24 For the rod of Prob. 7.23, determine the magnitude and location of the maximum bending moment.

7.25 A semicircular rod of weight W and uniform cross section is supported as shown. Determine the bending moment at point J when $\theta = 60°$.

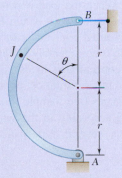

Fig. P7.25 and P7.26

7.26 A semicircular rod of weight W and uniform cross section is supported as shown. Determine the bending moment at point J when $\theta = 150°$.

7.27 and 7.28 A half section of pipe rests on a frictionless horizontal surface as shown. If the half section of pipe has a mass of 9 kg and a diameter of 300 mm, determine the bending moment at point J when $\theta = 90°$.

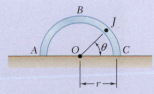

Fig. P7.27

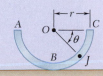

Fig. P7.28

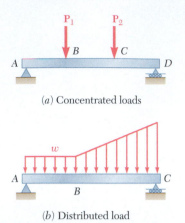

(*a*) Concentrated loads

w

(*b*) Distributed load

Fig. 7.5 A beam may be subjected to (*a*) concentrated loads or (*b*) distributed loads, or a combination of both.

7.2 BEAMS

A structural member designed to support loads applied at various points along the member is known as a **beam**. In most cases, the loads are perpendicular to the axis of the beam and cause only shear and bending in the beam. When the loads are not at a right angle to the beam, they also produce axial forces in the beam.

Beams are usually long, straight prismatic bars. Designing a beam for the most effective support of the applied loads is a two-part process: (1) determine the shearing forces and bending moments produced by the loads and (2) select the cross section best suited to resist these shearing forces and bending moments. Here we are concerned with the first part of the problem of beam design. The second part belongs to the study of mechanics of materials.

7.2A Various Types of Loading and Support

A beam can be subjected to **concentrated loads** P_1, P_2, ... that are expressed in newtons, pounds, or their multiples, kilonewtons and kips (Fig. 7.5*a*). We can also subject a beam to a **distributed load** *w*, expressed in N/m, kN/m, lb/ft, or kips/ft (Fig. 7.5*b*). In many cases, a beam is subjected to a combination of both types of load. When the load *w* per unit length has a constant value over part of the beam (as between *A* and *B* in Fig. 7.5*b*), the load is said to be **uniformly distributed** over that part of the beam. Determining the reactions at the supports is considerably simplified if we replace distributed loads by equivalent concentrated loads, as explained in Sec. 5.3A. However, you should not do this substitution, or at least perform it with care, when calculating internal forces (see Sample Prob. 7.3).

Beams are classified according to the way in which they are supported. Figure 7.6 shows several types of beams used frequently.

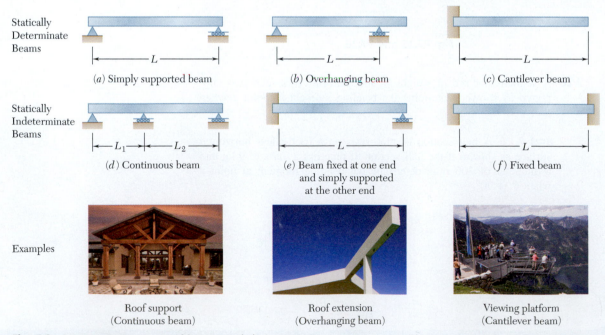

Statically Determinate Beams

(*a*) Simply supported beam   (*b*) Overhanging beam   (*c*) Cantilever beam

Statically Indeterminate Beams

(*d*) Continuous beam

(*e*) Beam fixed at one end and simply supported at the other end

(*f*) Fixed beam

Examples

Roof support (Continuous beam)

Roof extension (Overhanging beam)

Viewing platform (Cantilever beam)

Fig. 7.6 Some common types of beams and their supports.

The distance L between supports is called the **span**. Note that the reactions are determinate if the supports involve only three unknowns. If more unknowns are involved, the reactions are statically indeterminate, and the methods of statics are not sufficient to determine the reactions. In such a case, we must take into account the properties of the beam with regard to its resistance to bending. Beams supported by only two rollers are not shown here; they are partially constrained and move under certain types of loadings.

Sometimes two or more beams are connected by hinges to form a single continuous structure. Two examples of beams hinged at a point H are shown in Fig. 7.7. Here the reactions at the supports involve four unknowns and cannot be determined from the free-body diagram of the two-beam system. However, we can determine the reactions by considering the free-body diagram of each beam separately. Analysis of this situation involves six unknowns (including two force components at the hinge), and six equations are available.

7.2B Shear and Bending Moment in a Beam

Consider a beam AB subjected to various concentrated and distributed loads (Fig. 7.8a). We propose to determine the shearing force and bending moment at any point of the beam. In the example considered here, the beam is simply supported, but the method used could be applied to any type of statically determinate beam.

First we determine the reactions at A and B by choosing the entire beam as a free body (Fig. 7.8b). Setting $\Sigma M_A = 0$ and $\Sigma M_B = 0$, we obtain, respectively, $\mathbf{R}_B$ and $\mathbf{R}_A$.

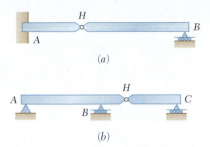

Fig. 7.7 Examples of two-beam systems connected by a hinge. In both cases, free-body diagrams of each individual beam enable you to determine the support reactions.

Photo 7.2 As a truck crosses a highway overpass, the internal forces vary in the beams of the overpass.

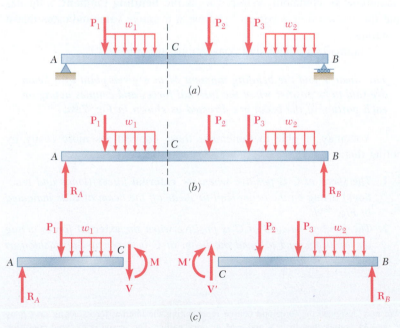

Fig. 7.8 (a) A simply supported beam AB; (b) free-body diagram of the beam; (c) free-body diagrams of portions AC and CB of the beam, showing internal shearing forces and couples.

To determine the internal forces at an arbitrary point C, we cut the beam at C and draw the free-body diagrams of the portions AC and CB (Fig. 7.8c). Using the free-body diagram of AC, we can determine the shearing force $\mathbf{V}$ at C by equating the sum of the vertical components of all forces acting on AC to zero. Similarly, we can find the bending moment $\mathbf{M}$ at C by equating the sum of the moments about C of all forces and couples acting on AC to zero. Alternatively, we could use the free-body diagram of CB[†] and determine the shearing force $\mathbf{V}'$ and the bending moment $\mathbf{M}'$ by equating the sum of the vertical components and the sum of the moments about C of all forces and couples acting on CB to zero. Although this choice of free bodies may make the computation of the numerical values of the shearing force and bending moment easier, it requires us to indicate on which portion of the beam the internal forces considered are acting. If we want to calculate and efficiently record the shearing force and bending moment at every point of the beam, we must devise a way to avoid having to specify which portion of the beam is used as a free body every time. Therefore, we shall adopt the following conventions.

In determining the shearing force in a beam, *we always assume* that the internal forces $\mathbf{V}$ and $\mathbf{V}'$ are directed as shown in Fig. 7.8c. A positive value obtained for their common magnitude V indicates that this assumption is correct and that the shearing forces are actually directed as shown. A negative value obtained for V indicates that the assumption is wrong and the shearing forces are directed in the opposite way. Thus, to define completely the shearing forces at a given point of the beam, we only need to record the magnitude V, together with a plus or minus sign. The scalar V is commonly referred to as the **shear** at the given point of the beam.

Similarly, *we always assume* that the internal couples $\mathbf{M}$ and $\mathbf{M}'$ are directed as shown in Fig. 7.8c. A positive value obtained for their magnitude M, commonly referred to as the **bending moment**, indicates that this assumption is correct, whereas a negative value indicates that it is wrong.

Summarizing these sign conventions, we state:

The shear V and the bending moment M at a given point of a beam are said to be positive when the internal forces and couples acting on each portion of the beam are directed as shown in Fig. 7.9a.

You may be able to remember these conventions more easily by noting that:

1. *The shear at C is positive when the* **external** *forces (loads and reactions) acting on the beam tend to shear off the beam at C as indicated in Fig. 7.9b.*
2. *The bending moment at C is positive when the* **external** *forces acting on the beam tend to bend the beam at C in a concave-up fashion as indicated in Fig. 7.9c.*

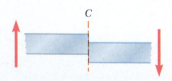

(*a*) Internal forces at section
(positive shear and positive bending moment)

(*b*) Effect of external forces
(positive shear)

C

(*c*) Effect of external forces
(positive bending moment)

Fig. 7.9 Figure for remembering the signs of shear and bending moment.

[†]We now designate the force and couple representing the internal forces acting on CB by $\mathbf{V}'$ and $\mathbf{M}'$, rather than by $-\mathbf{V}$ and $-\mathbf{M}$ as done earlier. The reason is to avoid confusion when applying the sign convention we are about to introduce.

It may also help to note that the situation described in Fig. 7.9, in which the values of both the shear and the bending moment are positive, is precisely the situation that occurs in the left half of a simply supported beam carrying a single concentrated load at its midpoint. This particular example is fully discussed in the following section.

7.2C Shear and Bending-Moment Diagrams

Now that we have clearly defined shear and bending moment in sense as well as in magnitude, we can easily record their values at any point along a beam by plotting these values against the distance x measured from one end of the beam. The graphs obtained in this way are called, respectively, the **shear diagram** and the **bending-moment diagram**.

As an example, consider a simply supported beam AB of span L subjected to a single concentrated load $\mathbf{P}$ applied at its midpoint D (Fig. 7.10a). We first determine the reactions at the supports from the free-body diagram of the entire beam (Fig. 7.10b); we find that the magnitude of each reaction is equal to $P/2$.

Next we cut the beam at a point C between A and D and draw the free-body diagrams of AC and CB (Fig. 7.10c). *Assuming that shear and bending moment are positive,* we direct the internal forces $\mathbf{V}$ and $\mathbf{V}'$ and the internal couples $\mathbf{M}$ and $\mathbf{M}'$ as indicated in Fig. 7.9a. Considering the free body AC, we set the sum of the vertical components and the sum of the moments about C of the forces acting on the free body to zero. From this, we find $V = +P/2$ and $M = +Px/2$. Therefore, both shear and bending moment are positive. (You can check this by observing that the reaction at A tends to shear off and to bend the beam at C as indicated in Fig. 7.9b and c.) Now let's plot V and M between A and D (Fig. 7.10e and f). The shear has a constant value $V = P/2$, whereas the bending moment increases linearly from $M = 0$ at $x = 0$ to $M = PL/4$ at $x = L/2$.

Proceeding along the beam, we cut it at a point E between D and B and consider the free body EB (Fig. 7.10d). As before, the sum of the vertical components and the sum of the moments about E of the forces acting on the free body are zero. We obtain $V = -P/2$ and $M = P(L - x)/2$. The shear is therefore negative and the bending moment is positive. (Again, you can check this by observing that the reaction at B bends the beam at E as indicated in Fig. 7.9c but tends to shear it off in a manner opposite to that shown in Fig. 7.9b.) We can now complete the shear and bending-moment diagrams of Fig. 7.10e and f. The shear has a constant value $V = -P/2$ between D and B, whereas the bending moment decreases linearly from $M = PL/4$ at $x = L/2$ to $M = 0$ at $x = L$.

Note that when a beam is subjected to concentrated loads only, the shear is of constant value between loads and the bending moment varies linearly between loads. However, when a beam is subjected to distributed loads, the shear and bending moment vary quite differently (see Sample Prob. 7.3).

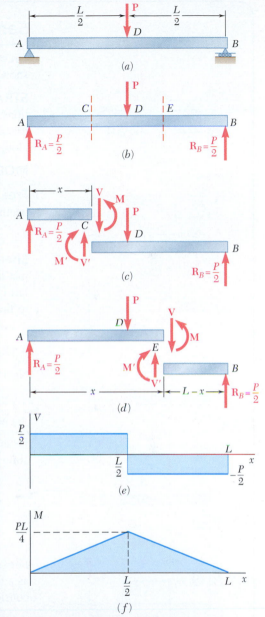

Fig. 7.10 (a) A beam supporting a single concentrated load at its midpoint; (b) free-body diagram of the beam; (c) free-body diagrams of parts of the beam after a cut at C; (d) free-body diagrams of parts of the beam after a cut at E; (e) shear diagram of the beam; (f) bending-moment diagram of the beam.

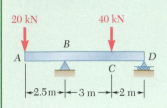

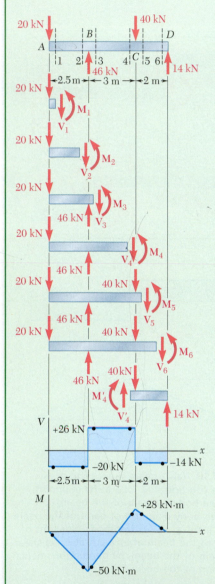

Fig. 1 Free-body diagrams of beam sections, and the resulting shear and bending-moment diagrams.

Sample Problem 7.2

Draw the shear and bending-moment diagrams for the beam and loading shown.

STRATEGY: Treat the entire beam as a free body to determine the reactions, then cut the beam just before and just after each external concentrated force (Fig. 1) to see how the shear and bending moment change along the length of the beam.

MODELING and ANALYSIS:

Free-Body, Entire Beam. From the free-body diagram of the entire beam, find the reactions at B and D:

$$\mathbf{R}_B = 46 \text{ kN} \uparrow \qquad \mathbf{R}_D = 14 \text{ kN} \uparrow$$

Shear and Bending Moment. First, determine the internal forces just to the right of the 20-kN load at A. Consider the stub of beam to the left of point 1 as a free body, and assume V and M are positive (according to the standard convention). Then you have

$$+\uparrow \Sigma F_y = 0: \qquad -20 \text{ kN} - V_1 = 0 \qquad V_1 = -20 \text{ kN}$$
$$+\curvearrowleft \Sigma M_1 = 0: \qquad (20 \text{ kN})(0 \text{ m}) + M_1 = 0 \qquad M_1 = 0$$

Next, consider the portion of the beam to the left of point 2 as a free body:

$$+\uparrow \Sigma F_y = 0: \qquad -20 \text{ kN} - V_2 = 0 \qquad V_2 = -20 \text{ kN}$$
$$+\curvearrowleft \Sigma M_2 = 0: \qquad (20 \text{ kN})(2.5 \text{ m}) + M_2 = 0 \qquad M_2 = -50 \text{ kN·m}$$

Determine the shear and bending moment at sections 3, 4, 5, and 6 in a similar way from the free-body diagrams. The results are

$$V_3 = +26 \text{ kN} \qquad M_3 = -50 \text{ kN·m}$$
$$V_4 = +26 \text{ kN} \qquad M_4 = +28 \text{ kN·m}$$
$$V_5 = -14 \text{ kN} \qquad M_5 = +28 \text{ kN·m}$$
$$V_6 = -14 \text{ kN} \qquad M_6 = 0$$

For several of the later cuts, the results are easier to obtain by considering as a free body the portion of the beam to the right of the cut. For example, consider the portion of the beam to the right of point 4. You have

$$+\uparrow \Sigma F_y = 0: \qquad V_4 - 40 \text{ kN} + 14 \text{ kN} = 0 \qquad V_4 = +26 \text{ kN}$$
$$+\curvearrowleft \Sigma M_4 = 0: \qquad -M_4 + (14 \text{ kN})(2 \text{ m}) = 0 \qquad M_4 = +28 \text{ kN·m}$$

Shear and Bending-Moment Diagrams. Now plot the six points shown on the shear and bending-moment diagrams. As indicated in Sec. 7.2C, the shear is of constant value between concentrated loads, and the bending moment varies linearly. You therefore obtain the shear and bending-moment diagrams shown in Fig. 1.

REFLECT and THINK: The calculations are pretty similar for each new choice of free body. However, moving along the beam, the shear changes magnitude whenever you pass a transverse force and the graph of the bending moment changes slope at these points.

Sample Problem 7.3

Draw the shear and bending-moment diagrams for the beam AB. The distributed load of 40 lb/in. extends over 12 in. of the beam from A to C, and the 400-lb load is applied at E.

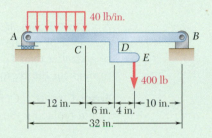

STRATEGY: Again, consider the entire beam as a free body to find the reactions. Then cut the beam within each region of continuous load. This will enable you to determine continuous functions for shear and bending moment, which you can then plot on a graph.

MODELING and ANALYSIS:

Free-Body, Entire Beam. Determine the reactions by considering the entire beam as a free body (Fig. 1).

$+\curvearrowleft \Sigma M_A = 0$: $B_y(32 \text{ in.}) - (480 \text{ lb})(6 \text{ in.}) - (400 \text{ lb})(22 \text{ in.}) = 0$
  $B_y = +365 \text{ lb}$ $\mathbf{B_y} = 365 \text{ lb} \uparrow$

$+\curvearrowleft \Sigma M_B = 0$: $(480 \text{ lb})(26 \text{ in.}) + (400 \text{ lb})(10 \text{ in.}) - A(32 \text{ in.}) = 0$
  $A = +515 \text{ lb}$ $\mathbf{A} = 515 \text{ lb} \uparrow$

$\xrightarrow{+} \Sigma F_x = 0$: $B_x = 0$ $\mathbf{B_x} = 0$

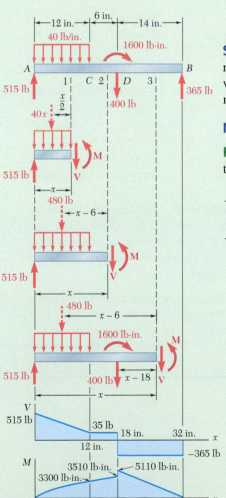

Fig. 2 Free-body diagrams of beam sections, and the resulting shear and bending-moment diagrams.

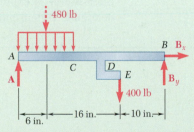

Fig. 1 Free-body diagram of entire beam.

Now, replace the 400-lb load by an equivalent force-couple system acting on the beam at point D and cut the beam at several points (Fig. 2).

Shear and Bending Moment. *From A to C.* Determine the internal forces at a distance x from point A by considering the portion of the beam to the left of point 1. Replace that part of the distributed load acting on the free body by its resultant. You get

$$+\uparrow\Sigma F_y = 0: \qquad 515 - 40x - V = 0 \qquad\qquad V = 515 - 40x$$
$$+\curvearrowleft\Sigma M_1 = 0: \qquad -515x + 40x(\tfrac{1}{2}x) + M = 0 \qquad M = 515x - 20x^2$$

Note that V and M are not numerical values, but they are expressed as functions of x. The free-body diagram shown can be used for all values of x smaller than 12 in., so the expressions obtained for V and M are valid throughout the region $0 < x < 12$ in.

From C to D. Consider the portion of the beam to the left of point 2. Again replacing the distributed load by its resultant, you have

$$+\uparrow\Sigma F_y = 0: \quad 515 - 480 - V = 0 \qquad\qquad V = 35 \text{ lb}$$
$$+\curvearrowleft\Sigma M_2 = 0: \quad -515x + 480(x - 6) + M = 0 \quad M = (2880 + 35x) \text{ lb·in.}$$

These expressions are valid in the region 12 in. $< x <$ 18 in.

From D to B. Use the portion of the beam to the left of point 3 for the region 18 in. $< x <$ 32 in. Thus,

$$+\uparrow\Sigma F_y = 0: \qquad 515 - 480 - 400 - V = 0 \qquad V = -365 \text{ lb}$$
$$+\curvearrowleft\Sigma M_3 = 0: \qquad -515x + 480(x - 6) - 1600 + 400(x - 18) + M = 0$$
$$M = (11,\!680 - 365x) \text{ lb·in.}$$

Shear and Bending-Moment Diagrams. Plot the shear and bending-moment diagrams for the entire beam. Note that the couple of moment 1600 lb·in. applied at point D introduces a discontinuity into the bending-moment diagram. Also note that the bending-moment diagram under the distributed load is not straight but is slightly curved.

REFLECT and THINK: Shear and bending-moment diagrams typically feature various kinds of curves and discontinuities. In such cases, it is often useful to express V and M as functions of location x as well as to determine certain numerical values.

SOLVING PROBLEMS
ON YOUR OWN

In this section, you saw how to determine the **shear** V and the **bending moment** M at any point in a beam. You also learned to draw the **shear diagram** and the **bending-moment diagram** for the beam by plotting, respectively, V and M against the distance x measured along the beam.

A. Determining the shear and bending moment in a beam. To determine the shear V and the bending moment M at a given point C of a beam, take the following steps.

1. Draw a free-body diagram of the entire beam, and use it to determine the reactions at the beam supports.

2. Cut the beam at point C, and using the original loading, select one of the two resulting portions of the beam.

3. Draw the free-body diagram of the portion of the beam you have selected. Show:

   a. The loads and the reactions exerted on that portion of the beam, replacing each distributed load by an equivalent concentrated load, as explained in Sec. 5.3A.

   b. The shearing force and the bending moment representing the internal forces at C. To facilitate recording the shear V and the bending moment M after determining them, follow the convention indicated in Figs. 7.8 and 7.9. Thus, if you are using the portion of the beam located to the *left of C*, apply at C a *shearing force* **V** *directed downward* and a *bending moment* **M** *directed counterclockwise*. If you are using the portion of the beam located to the *right of C*, apply at C a *shearing force* **V′** *directed upward* and a *bending moment* **M′** *directed clockwise* [Sample Prob. 7.2].

4. Write the equilibrium equations for the portion of the beam you have selected. Solve the equation $\Sigma F_y = 0$ for V and the equation $\Sigma M_C = 0$ for M.

5. Record the values of V and M with the sign obtained for each of them. A positive sign for V means that the shearing forces exerted at C on each of the two portions of the beam are directed as shown in Figs. 7.8 and 7.9; a negative sign means they have the opposite sense. Similarly, a positive sign for M means that the bending couples at C are directed as shown in these figures, and a negative sign means that they have the opposite sense. In addition, a positive sign for M means that the concavity of the beam at C is directed upward, and a negative sign means that it is directed downward.

B. Drawing the shear and bending-moment diagrams for a beam. Obtain these diagrams by plotting, respectively, V and M against the distance x measured along the beam. However, in most cases, you need to compute the values of V and M at only a few points.

1. For a beam supporting only concentrated loads, note [Sample Prob. 7.2] that

    a. The shear diagram consists of segments of horizontal lines. Thus, to draw the shear diagram of the beam, you need to compute V only just to the left or just to the right of the points where the loads or reactions are applied.

    b. The bending-moment diagram consists of segments of oblique straight lines. Thus, to draw the bending-moment diagram of the beam, you need to compute M only at the points where the loads or reactions are applied.

2. For a beam supporting uniformly distributed loads, note [Sample Prob. 7.3] that under each of the distributed loads:

    a. The shear diagram consists of a segment of an oblique straight line. Thus, you need to compute V only where the distributed load begins and where it ends.

    b. The bending-moment diagram consists of an arc of parabola. In most cases, you need to compute M only where the distributed load begins and where it ends.

3. For a beam with a more complicated loading, you need to consider the free-body diagram of a portion of the beam of arbitrary length x and determine V and M as functions of x. This procedure may have to be repeated several times, since V and M are often represented by different functions in various parts of the beam [Sample Prob. 7.3].

4. When a couple is applied to a beam, the shear has the same value on both sides of the point of application of the couple, but the bending-moment diagram shows a discontinuity at that point, rising or falling by an amount equal to the magnitude of the couple. Note that a couple can either be applied directly to the beam or result from the application of a load on a member rigidly attached to the beam [Sample Prob. 7.3].

Problems

7.29 through 7.32 For the beam and loading shown, (*a*) draw the shear and bending-moment diagrams, (*b*) determine the maximum absolute values of the shear and bending moment.

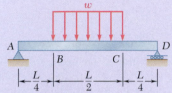

Fig. P7.29

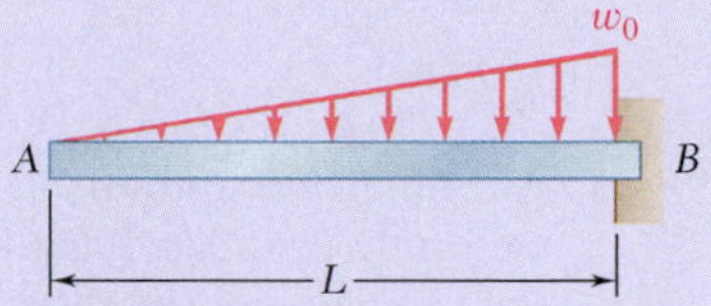

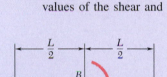

Fig. P7.30

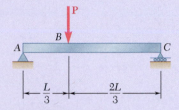

Fig. P7.31

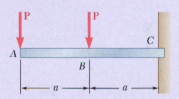

Fig. P7.32

7.33 and 7.34 For the beam and loading shown, (*a*) draw the shear and bending-moment diagrams, (*b*) determine the maximum absolute values of the shear and bending moment.

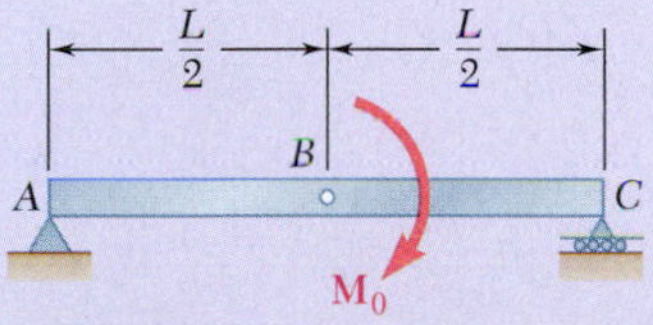

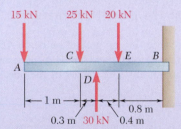

Fig. *P7.33*

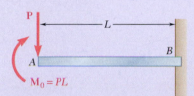

Fig. *P7.34*

7.35 and 7.36 For the beam and loading shown, (*a*) draw the shear and bending-moment diagrams, (*b*) determine the maximum absolute values of the shear and bending moment.

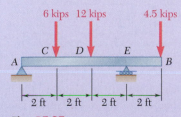

Fig. P7.35

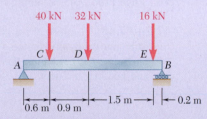

Fig. P7.36

7.37 and 7.38 For the beam and loading shown, (*a*) draw the shear and bending-moment diagrams, (*b*) determine the maximum absolute values of the shear and bending moment.

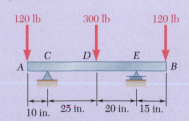

Fig. *P7.37*

Fig. *P7.38*

387

7.39 through 7.42 For the beam and loading shown, (a) draw the shear and bending-moment diagrams, (b) determine the maximum absolute values of the shear and bending moment.

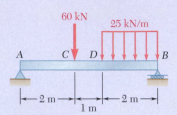

Fig. P7.39

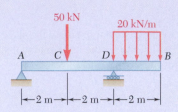

Fig. P7.40

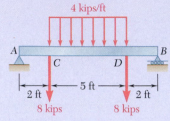

Fig. P7.41

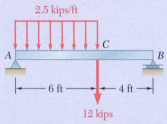

Fig. P7.42

7.43 Assuming the upward reaction of the ground on beam *AB* to be uniformly distributed and knowing that $P = wa$, (a) draw the shear and bending-moment diagrams, (b) determine the maximum absolute values of the shear and bending moment.

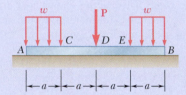

Fig. *P7.43*

7.44 Solve Prob. 7.43 knowing that $P = 3wa$.

7.45 Assuming the upward reaction of the ground on beam *AB* to be uniformly distributed, (a) draw the shear and bending-moment diagrams, (b) determine the maximum absolute values of the shear and bending moment.

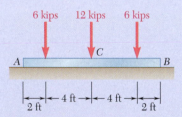

Fig. P7.45

7.46 Solve Prob. 7.45 assuming that the 12-kip load has been removed.

7.47 and 7.48 Assuming the upward reaction of the ground on beam *AB* to be uniformly distributed, (a) draw the shear and bending-moment diagrams, (b) determine the maximum absolute values of the shear and bending moment.

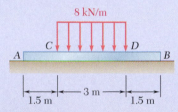

Fig. P7.47

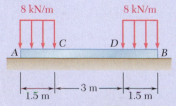

Fig. P7.48

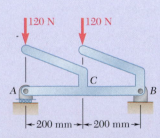

Fig. P7.49

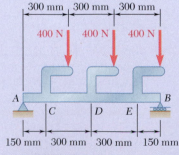

Fig. P7.50

7.49 and 7.50 Draw the shear and bending-moment diagrams for the beam *AB*, and determine the maximum absolute values of the shear and bending moment.

7.51 and 7.52 Draw the shear and bending-moment diagrams for the beam *AB*, and determine the maximum absolute values of the shear and bending moment.

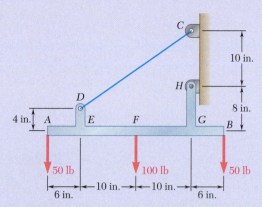

Fig. P7.51

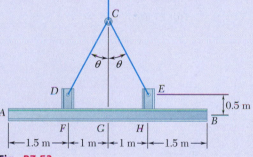

Fig. P7.52

7.53 Two small channel sections *DF* and *EH* have been welded to the uniform beam *AB* of weight $W = 3$ kN to form the rigid structural member shown. This member is being lifted by two cables attached at *D* and *E*. Knowing that $\theta = 30°$ and neglecting the weight of the channel sections, (*a*) draw the shear and bending-moment diagrams for beam *AB*, (*b*) determine the maximum absolute values of the shear and bending moment in the beam.

7.54 Solve Prob. 7.53 when $\theta = 60°$.

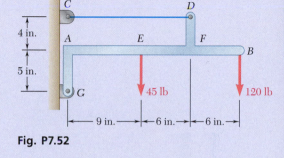

Fig. P7.53

7.55 For the structural member of Prob. 7.53, determine (*a*) the angle θ for which the maximum absolute value of the bending moment in beam *AB* is as small as possible, (*b*) the corresponding value of $|M|_{max}$. (*Hint:* Draw the bending-moment diagram and then equate the absolute values of the largest positive and negative bending moments obtained.)

7.56 For the beam of Prob. 7.43, determine (*a*) the ratio $k = P/wa$ for which the maximum absolute value of the bending moment in the beam is as small as possible, (*b*) the corresponding value of $|M|_{max}$. (See hint for Prob. 7.55.)

7.57 Determine (*a*) the distance a for which the maximum absolute value of the bending moment in beam *AB* is as small as possible, (*b*) the corresponding value of $|M|_{max}$. (See hint for Prob. 7.55.)

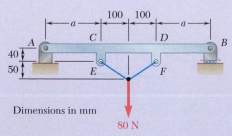

Dimensions in mm

Fig. P7.57

7.58 For the beam and loading shown, determine (*a*) the distance *a* for which the maximum absolute value of the bending moment in the beam is as small as possible, (*b*) the corresponding value of $|M|_{max}$. (See hint for Prob. 7.55.)

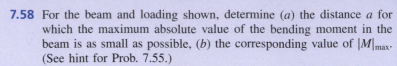

Fig. P7.58

7.59 A uniform beam is to be picked up by crane cables attached at *A* and *B*. Determine the distance *a* from the ends of the beam to the points where the cables should be attached if the maximum absolute value of the bending moment in the beam is to be as small as possible. (*Hint:* Draw the bending-moment diagram in terms of *a*, *L*, and the weight per unit length *w*, and then equate the absolute values of the largest positive and negative bending moments obtained.)

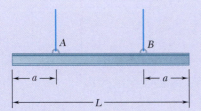

Fig. P7.59

7.60 Knowing that $P = Q = 150$ lb, determine (*a*) the distance *a* for which the maximum absolute value of the bending moment in beam *AB* is as small as possible, (*b*) the corresponding value of $|M|_{max}$. (See hint for Prob. 7.55.)

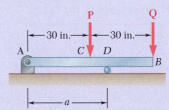

Fig. P7.60

7.61 Solve Prob. 7.60 assuming that $P = 300$ lb and $Q = 150$ lb.

*7.62 In order to reduce the bending moment in the cantilever beam *AB*, a cable and counterweight are permanently attached at end *B*. Determine the magnitude of the counterweight for which the maximum absolute value of the bending moment in the beam is as small as possible and the corresponding value of $|M|_{max}$. Consider (*a*) the case when the distributed load is permanently applied to the beam, (*b*) the more general case when the distributed load may either be applied or removed.

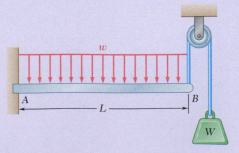

Fig. P7.62

7.3 RELATIONS AMONG LOAD, SHEAR, AND BENDING MOMENT

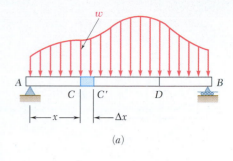

If a beam carries more than two or three concentrated loads or if it carries a distributed load, the method outlined in Sec. 7.2 for plotting shear and bending-moment diagrams is likely to be quite cumbersome. However, constructing a shear diagram and, especially, a bending-moment diagram, are much easier if we take into consideration some relations among load, shear, and bending moment.

Consider a simply supported beam AB carrying a distributed load w per unit length (Fig. 7.11a). Let C and C' be two points of the beam at a distance Δx from each other. We denote the shear and bending moment at C by V and M, respectively, and we assume they are positive. We denote the shear and bending moment at C' by $V + \Delta V$ and $M + \Delta M$.

Let us now detach the portion of beam CC' and draw its free-body diagram (Fig. 7.11b). The forces exerted on the free body include a load with a magnitude of $w\,\Delta x$ (indicated by a dashed arrow to distinguish it from the original distributed load from which it is derived) and internal forces and couples at C and C'. Since we assumed both shear and bending moment are positive, the forces and couples are directed as shown in the figure.

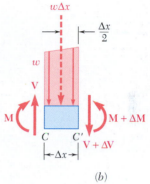

Fig. 7.11 (a) A simply supported beam carrying a distributed load; (b) free-body diagram of a portion CC' of the beam.

Relations Between Load and Shear. Because the free body CC' is in equilibrium, we set the sum of the vertical components of the forces acting on it to zero:

$$V - (V + \Delta V) - w\,\Delta x = 0$$
$$\Delta V = -w\,\Delta x$$

Dividing both sides of this equation by Δx and then letting Δx approach zero, we obtain

$$\frac{dV}{dx} = -w \tag{7.1}$$

Equation (7.1) indicates that, for a beam loaded as shown in Fig. 7.11a, the slope dV/dx of the shear curve is negative and the numerical value of the slope at any point is equal to the load per unit length at that point.

Integrating (7.1) between arbitrary points C and D, we have

$$V_D - V_C = -\int_{x_C}^{x_D} w\,dx \tag{7.2}$$

or

$$V_D - V_C = -(\text{area under load curve between } C \text{ and } D) \tag{7.2'}$$

Note that we could also obtain this result by considering the equilibrium of the portion of beam CD, since the area under the load curve represents the total load applied between C and D.

Equation (7.1) *is not valid* at a point where a concentrated load is applied; the shear curve is discontinuous at such a point, as we saw in

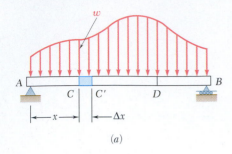

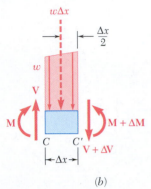

Fig. 7.11 (*repeated*)

Sec. 7.2. Similarly, formulas (7.2) and (7.2′) cease to be valid when concentrated loads are applied between C and D, since they do not take into account the sudden change in shear caused by a concentrated load. Formulas (7.2) and (7.2′), therefore, should be applied only between successive concentrated loads.

Relations Between Shear and Bending Moment. Returning to the free-body diagram of Fig. 7.11b, we can set the sum of the moments about C' to be zero, obtaining

$$(M + \Delta M) - M - V\,\Delta x + w\Delta x\frac{\Delta x}{2} = 0$$

$$\Delta M = V\,\Delta x - \tfrac{1}{2}w(\Delta x)^2$$

Dividing both sides of this equation by Δx and then letting Δx approach zero, we have

$$\frac{dM}{dx} = V \qquad (7.3)$$

Equation (7.3) indicates that the slope dM/dx of the bending-moment curve is equal to the value of the shear. This is true at any point where the shear has a well-defined value, i.e., at any point where no concentrated load is applied. Formula (7.3) also shows that the shear is zero at points where the bending moment is maximum. This property simplifies the determination of points where the beam is likely to fail under bending.

Integrating Eq. (7.3) between arbitrary points C and D, we obtain

$$M_D - M_C = \int_{x_C}^{x_D} V\,dx \qquad (7.4)$$

$$M_D - M_C = \text{area under shear curve between } C \text{ and } D \qquad (7.4')$$

Note that the area under the shear curve should be considered positive where the shear is positive and negative where the shear is negative. Formulas (7.4) and (7.4′) are valid even when concentrated loads are applied between C and D, as long as the shear curve has been drawn correctly. The formulas cease to be valid, however, if a *couple* is applied at a point between C and D, since they do not take into account the sudden change in bending moment caused by a couple (see Sample Prob. 7.7).

In most engineering applications, you need to know the value of the bending moment at only a few specific points. Once you have drawn the shear diagram and determined M at one end of the beam, you can obtain the value of the bending moment at any given point by computing the area under the shear curve and using formula (7.4′). For instance, since $M_A = 0$ for the beam of Fig. 7.12, you can determine the maximum value of the bending moment for that beam simply by measuring the area of the shaded triangle in the shear diagram as

$$M_{\max} = \frac{1}{2}\frac{L}{2}\frac{wL}{2} = \frac{wL^2}{8}$$

In this example, the load curve is a horizontal straight line, the shear curve is an oblique straight line, and the bending-moment curve

Concept Application 7.1

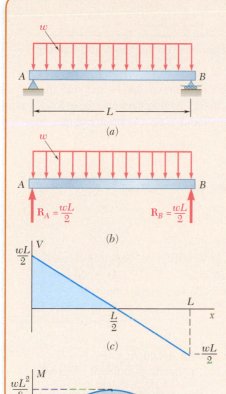

Consider a simply supported beam AB with a span of L carrying a uniformly distributed load w (Fig. 7.12a). From the free-body diagram of the entire beam, we determine the magnitude of the reactions at the supports: $R_A = R_B = wL/2$ (Fig. 7.12b). Then we draw the shear diagram. Close to end A of the beam, the shear is equal to R_A; that is, to $wL/2$, as we can check by considering a very small portion of the beam as a free body. Using formula (7.2), we can then determine the shear V at any distance x from A as

$$V - V_A = -\int_0^x w \, dx = -wx$$

$$V = V_A - wx = \frac{wL}{2} - wx = w\left(\frac{L}{2} - x\right)$$

The shear curve is thus an oblique straight line that crosses the x axis at $x = L/2$ (Fig. 7.12c). Now consider the bending moment. We first observe that $M_A = 0$. The value M of the bending moment at any distance x from A then can be obtained from Eq. (7.4), as

$$M - M_A = \int_0^x V \, dx$$

$$M = \int_0^x w\left(\frac{L}{2} - x\right) dx = \frac{w}{2}(Lx - x^2)$$

The bending-moment curve is a parabola. The maximum value of the bending moment occurs when $x = L/2$, since V (and thus dM/dx) is zero for that value of x. Substituting $x = L/2$ in the last equation, we obtain $M_{\max} = wL^2/8$.

Fig. 7.12 (a) A simply supported beam carrying a uniformly distributed load; (b) free-body diagram of the beam to determine the reactions at the supports; (c) the shear curve is an oblique straight line; (d) the bending-moment diagram is a parabola.

is a parabola. If the load curve had been an oblique straight line (first degree), the shear curve would have been a parabola (second degree), and the bending-moment curve would have been a cubic (third degree). The equations of the shear and bending-moment curves are always, respectively, one and two degrees higher than the equation of the load curve. Thus, once you have computed a few values of the shear and bending moment, you should be able to sketch the shear and bending-moment diagrams without actually determining the functions $V(x)$ and $M(x)$. The sketches will be more accurate if you make use of the fact that, at any point where the curves are continuous, the slope of the shear curve is equal to $-w$ and the slope of the bending-moment curve is equal to V.

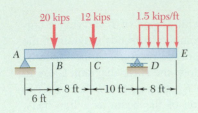

Sample Problem 7.4

Draw the shear and bending-moment diagrams for the beam and loading shown.

STRATEGY: The beam supports two concentrated loads and one distributed load. You can use the equations in this section between these loads and under the distributed load, but you should expect certain changes in the diagrams at the load points.

MODELING and ANALYSIS:

Free-Body, Entire Beam. Consider the entire beam as a free body and determine the reactions (Fig. 1):

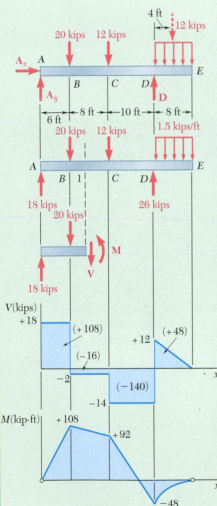

$+\uparrow\Sigma M_A = 0$:

$$D(24 \text{ ft}) - (20 \text{ kips})(6 \text{ ft}) - (12 \text{ kips})(14 \text{ ft}) - (12 \text{ kips})(28 \text{ ft}) = 0$$
$$D = +26 \text{ kips} \qquad \mathbf{D} = 26 \text{ kips} \uparrow$$

$+\uparrow\Sigma F_y = 0$: $\quad A_y - 20 \text{ kips} - 12 \text{ kips} + 26 \text{ kips} - 12 \text{ kips} = 0$
$$A_y = +18 \text{ kips} \qquad \mathbf{A}_y = 18 \text{ kips} \uparrow$$

$\xrightarrow{}\Sigma F_x = 0$: $\quad A_x = 0 \qquad\qquad\qquad\qquad \mathbf{A}_x = 0$

Note that the bending moment is zero at both A and E; thus, you know two points (indicated by small circles) on the bending-moment diagram.

Shear Diagram. Since $dV/dx = -w$, the slope of the shear diagram is zero (i.e., the shear is constant between concentrated loads and reactions). To find the shear at any point, divide the beam into two parts and consider either part as a free body. For example, using the portion of the beam to the left of point 1 (Fig. 1), you can obtain the shear between B and C:

$+\uparrow\Sigma F_y = 0$: $\qquad +18 \text{ kips} - 20 \text{ kips} - V = 0 \qquad V = -2 \text{ kips}$

You can also find that the shear is $+12$ kips just to the right of D and zero at end E. Since the slope $dV/dx = -w$ is constant between D and E, the shear diagram between these two points is a straight line.

Bending-Moment Diagram. Recall that the area under the shear curve between two points is equal to the change in bending moment between the same two points. For convenience, compute the area of each portion of the shear diagram and indicate it on the diagram (Fig. 1). Since you know the bending moment M_A at the left end is zero, you have

$M_B - M_A = +108$	$M_B = +108 \text{ kip·ft}$
$M_C - M_B = -16$	$M_C = +92 \text{ kip·ft}$
$M_D - M_C = -140$	$M_D = -48 \text{ kip·ft}$
$M_E - M_D = +48$	$M_E = 0$

Fig. 1 Free-body diagrams of beam, free-body diagram of section to left of cut, shear diagram, bending-moment diagram.

Since you know M_E is zero, this gives you a check of the calculations.

Between the concentrated loads and reactions, the shear is constant; thus, the slope dM/dx is constant. Therefore, you can draw the bending-moment diagram by connecting the known points with straight lines.

Between D and E, where the shear diagram is an oblique straight line, the bending-moment diagram is a parabola.

From the V and M diagrams, note that $V_{max} = 18$ kips and $M_{max} = 108$ kip·ft.

REFLECT and THINK: As expected, the values of shear and slopes of the bending-moment curves show abrupt changes at the points where concentrated loads act. Useful for design, these diagrams make it easier to determine the maximum values of shear and bending moment for a beam and its loading.

Sample Problem 7.5

Draw the shear and bending-moment diagrams for the beam and loading shown and determine the location and magnitude of the maximum bending moment.

STRATEGY: The load is a distributed load over part of the beam with no concentrated loads. You can use the equations in this section in two parts: for the load and no load regions. From the discussion in this section, you can expect the shear diagram will show an oblique line under the load, followed by a horizontal line. The bending-moment diagram should show a parabola under the load and an oblique line under the rest of the beam.

MODELING and ANALYSIS:

Free-Body, Entire Beam.   Consider the entire beam as a free body (Fig. 1) to obtain the reactions

$$\mathbf{R}_A = 80 \text{ kN} \uparrow \qquad \mathbf{R}_C = 40 \text{ kN} \uparrow$$

Shear Diagram.   The shear just to the right of A is $V_A = +80$ kN. Because the change in shear between two points is equal to *minus* the area under the load curve between these points, you can obtain V_B by writing

$$V_B - V_A = -(20 \text{ kN/m})(6 \text{ m}) = -120 \text{ kN}$$
$$V_B = -120 + V_A = -120 + 80 = -40 \text{ kN}$$

Since the slope $dV/dx = -w$ is constant between A and B, the shear diagram between these two points is represented by a straight line. Between B and C, the area under the load curve is zero; therefore,

$$V_C - V_B = 0 \qquad V_C = V_B = -40 \text{ kN}$$

and the shear is constant between B and C (Fig. 1).

Bending-Moment Diagram.   The bending moment at each end of the beam is zero. In order to determine the maximum bending moment, you need to locate the section D of the beam where $V = 0$. You have

$$V_D - V_A = -wx$$
$$0 - 80 \text{ kN} = -(20 \text{ kN/m})x$$

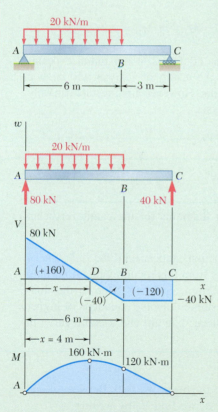

Fig. 1   Free-body diagram of beam, shear diagram, bending-moment diagram.

Solving for x: $x = 4$ m ◀

The maximum bending moment occurs at point D, where we have $dM/dx = V = 0$. Calculate the areas of the various portions of the shear diagram and mark them (in parentheses) on the diagram (Fig. 1). Since the area of the shear diagram between two points is equal to the change in bending moment between those points, you can write

$$M_D - M_A = +160 \text{ kN·m} \qquad M_D = +160 \text{ kN·m}$$
$$M_B - M_D = -40 \text{ kN·m} \qquad M_B = +120 \text{ kN·m}$$
$$M_C - M_B = -120 \text{ kN·m} \qquad M_C = 0$$

The bending-moment diagram consists of an arc of parabola followed by a segment of straight line; the slope of the parabola at A is equal to the value of V at that point.

The maximum bending moment is

$$M_{max} = M_D = +160 \text{ kN·m} ◀$$

REFLECT and THINK: The analysis conforms to our initial expectations. It is often useful to predict what the results of analysis will be as a way of checking against large-scale errors. However, final results can only depend on detailed modeling and analysis.

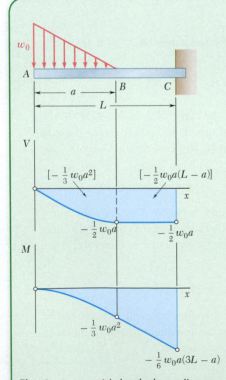

Fig. 1 Beam with load, shear diagram, bending-moment diagram.

Sample Problem 7.6

Sketch the shear and bending-moment diagrams for the cantilever beam shown.

STRATEGY: Because no support reactions appear until the right end of the beam, you can rely on the equations from this section without needing to use free-body diagrams and equilibrium equations. Due to the non-uniform load, you should expect the results to involve equations of higher degree with a parabolic curve in the shear diagram and a cubic curve in the bending-moment diagram.

MODELING and ANALYSIS:

Shear Diagram. At the free end of the beam, $V_A = 0$. Between A and B, the area under the load curve is $\frac{1}{2}w_0a$; we find V_B by writing

$$V_B - V_A = -\tfrac{1}{2}w_0a \qquad V_B = -\tfrac{1}{2}w_0a$$

Between B and C, the beam is not loaded; thus, $V_C = V_B$. At A, we have $w = w_0$, and according to Eq. (7.1), the slope of the shear curve is $dV/dx = -w_0$. At B, the slope is $dV/dx = 0$. Between A and B, the loading decreases linearly, and the shear diagram is parabolic (Fig. 1). Between B and C, $w = 0$ and the shear diagram is a horizontal line.

Bending-Moment Diagram. Note that $M_A = 0$ at the free end of the beam. You can compute the area under the shear curve, obtaining

$$M_B - M_A = -\tfrac{1}{3}w_0a^2 \qquad M_B = -\tfrac{1}{3}w_0a^2$$
$$M_C - M_B = -\tfrac{1}{2}w_0a(L - a)$$
$$M_C = -\tfrac{1}{6}w_0a(3L - a)$$

You can complete the sketch of the bending-moment diagram by recalling that $dM/dx = V$. The result is that between A and B the diagram is represented by a cubic curve with zero slope at A and between B and C the diagram is represented by a straight line.

REFLECT and THINK: Although not strictly required for the solution of this problem, determining the support reactions would serve as an excellent check of the final values of the shear and bending-moment diagrams.

Sample Problem 7.7

The simple beam AC is loaded by a couple of magnitude T applied at point B. Draw the shear and bending-moment diagrams for the beam.

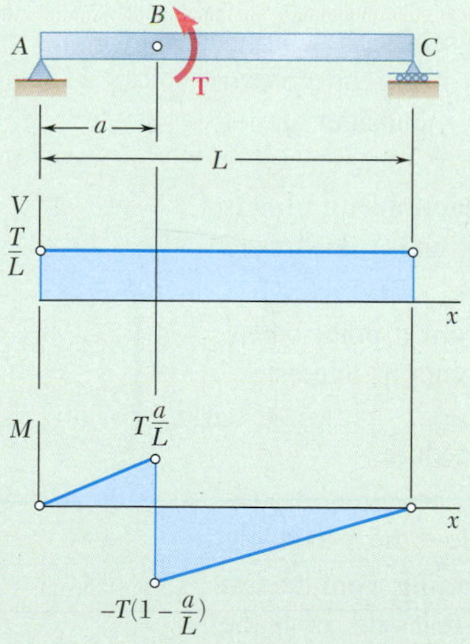

STRATEGY: The load supported by the beam is a concentrated couple. Since the only vertical forces are those associated with the support reactions, you should expect the shear diagram to be of constant value. However, the bending-moment diagram will have a discontinuity at B due to the couple.

MODELING and ANALYSIS:

Free-Body, Entire Beam. Consider the entire beam as a free body and determine the reactions:

$$\mathbf{R}_A = \frac{T}{L}\uparrow \qquad \mathbf{R}_B = \frac{T}{L}\downarrow$$

Shear and Bending-Moment Diagrams (Fig. 1). The shear at any section is constant and equal to T/L. Since a couple is applied at B, the bending-moment diagram is discontinuous at B; because the couple is counterclockwise, the bending moment *decreases* suddenly by an amount equal to T. You can demonstrate this by taking a section to the immediate right of B and applying equilibrium to solve for the bending moment at this location.

REFLECT and THINK: You can generalize the effect of a couple applied to a beam. At the point where the couple is applied, the bending-moment diagram increases by the value of the couple if it is clockwise and decreases by the value of the couple if it is counterclockwise.

Fig. 1 Beam with load, shear diagram, bending-moment diagram.

SOLVING PROBLEMS ON YOUR OWN

In this section, we described how to use the relations among load, shear, and bending moment to simplify the drawing of shear and bending-moment diagrams. These relations are

$$\frac{dV}{dx} = -w \tag{7.1}$$

$$\frac{dM}{dx} = V \tag{7.3}$$

$$V_D - V_C = -(\text{area under load curve between } C \text{ and } D) \tag{7.2'}$$

$$M_D - M_C = (\text{area under shear curve between } C \text{ and } D) \tag{7.4'}$$

Taking these relations into account, you can use the following procedure to draw the shear and bending-moment diagrams for a beam.

1. Draw a free-body diagram of the entire beam, and use it to determine the reactions at the beam supports.

2. Draw the shear diagram. This can be done as in the preceding section by cutting the beam at various points and considering the free-body diagram of one of the two resulting portions of the beam [Sample Prob. 7.3]. You can, however, consider one of the following alternative procedures.

    a. The shear V at any point of the beam is the sum of the reactions and loads to the left of that point; an upward force is counted as positive, and a downward force is counted as negative.

    b. For a beam carrying a distributed load, you can start from a point where you know V and use Eq. (7.2′) repeatedly to find V at all other points of interest.

3. Draw the bending-moment diagram, using the following procedure.

    a. Compute the area under each portion of the shear curve, assigning a positive sign to areas above the x axis and a negative sign to areas below the x axis.

    b. Apply Eq. (7.4′) repeatedly [Sample Probs. 7.4 and 7.5], starting from the left end of the beam, where $M = 0$ (except if a couple is applied at that end, or if the beam is a cantilever beam with a fixed left end).

    c. Where a couple is applied to the beam, be careful to show a discontinuity in the bending-moment diagram by *increasing* the value of M at that point by an amount equal to the magnitude of the couple if the couple is *clockwise*, or *decreasing* the value of M by that amount if the couple is *counterclockwise* [Sample Prob. 7.7].

4. Determine the location and magnitude of $|M|_{max}$. The maximum absolute value of the bending moment occurs at one of the points where $dM/dx = 0$ [according to Eq. (7.3), that is at a point where V is equal to zero or changes sign]. You should

   a. Determine from the shear diagram the value of $|M|$ where V changes sign; this will occur under a concentrated load [Sample Prob. 7.4].

   b. Determine the points where $V = 0$ and the corresponding values of $|M|$; this will occur under a distributed load. To find the distance x between point C where the distributed load starts and point D where the shear is zero, use Eq. (7.2'). For V_C, use the known value of the shear at point C; for V_D, use zero and express the area under the load curve as a function of x [Sample Prob. 7.5].

5. You can improve the quality of your drawings by keeping in mind that, at any given point according to Eqs. (7.1) and (7.3), the slope of the V curve is equal to $-w$ and the slope of the M curve is equal to V.

6. Finally, for beams supporting a distributed load expressed as a function $w(x)$, remember that you can obtain the shear V by integrating the function $-w(x)$, and you can obtain the bending moment M by integrating $V(x)$ [Eqs. (7.2) and (7.4)].

Problems

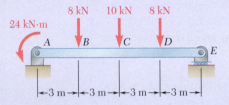

Fig. P7.69

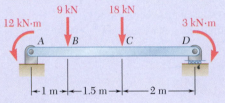

Fig. P7.70

7.63 Using the method of Sec. 7.3, solve Prob. 7.29.

7.64 Using the method of Sec. 7.3, solve Prob. 7.30.

7.65 Using the method of Sec. 7.3, solve Prob. 7.31.

7.66 Using the method of Sec. 7.3, solve Prob. 7.32.

7.67 Using the method of Sec. 7.3, solve Prob. 7.33.

7.68 Using the method of Sec. 7.3, solve Prob. 7.34.

7.69 and 7.70 For the beam and loading shown, (*a*) draw the shear and bending-moment diagrams, (*b*) determine the maximum absolute values of the shear and bending moment.

7.71 Using the method of Sec. 7.3, solve Prob. 7.39.

7.72 Using the method of Sec. 7.3, solve Prob. 7.40.

7.73 Using the method of Sec. 7.3, solve Prob. 7.41.

7.74 Using the method of Sec. 7.3, solve Prob. 7.42.

7.75 and 7.76 For the beam and loading shown, (*a*) draw the shear and bending-moment diagrams, (*b*) determine the maximum absolute values of the shear and bending moment.

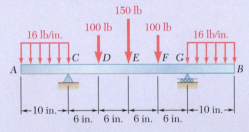

Fig. *P7.76*

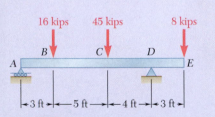

Fig. *P7.75*

7.77 and 7.78 For the beam and loading shown, (*a*) draw the shear and bending-moment diagrams, (*b*) determine the magnitude and location of the maximum absolute value of the bending moment.

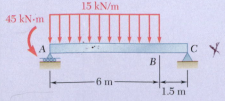

Fig. P7.77

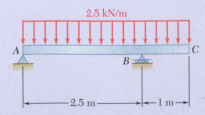

Fig. P7.78

7.79 and 7.80 For the beam and loading shown, (a) draw the shear and bending-moment diagrams, (b) determine the magnitude and location of the maximum absolute value of the bending moment.

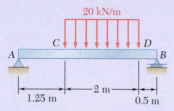

Fig. P7.79

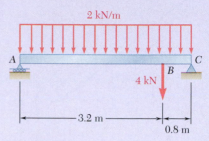

Fig. P7.80

7.81 and 7.82 For the beam and loading shown, (a) draw the shear and bending-moment diagrams, (b) determine the magnitude and location of the maximum absolute value of the bending moment.

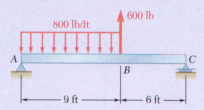

Fig. P7.81

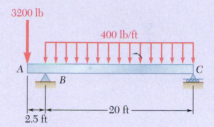

Fig. P7.82

7.83 (a) Draw the shear and bending-moment diagrams for beam AB, (b) determine the magnitude and location of the maximum absolute value of the bending moment.

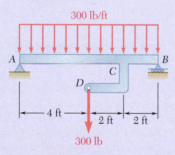

Fig. P7.83

7.84 Solve Prob. 7.83 assuming that the 300-lb force applied at D is directed upward.

7.85 and 7.86 For the beam and loading shown, (a) write the equations of the shear and bending-moment curves, (b) determine the magnitude and location of the maximum bending moment.

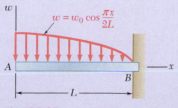

Fig. P7.85

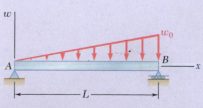

Fig. P7.86

7.87 and 7.88 For the beam and loading shown, (a) write the equations of the shear and bending-moment curves, (b) determine the magnitude and location of the maximum bending moment.

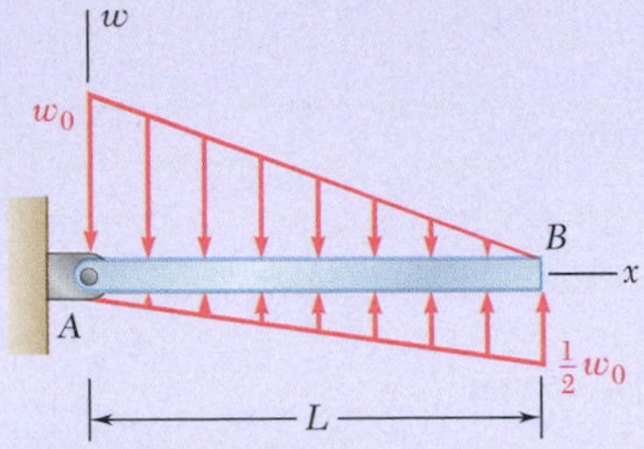

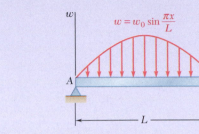

Fig. P7.87

$w = w_0 \sin \dfrac{\pi x}{L}$

Fig. *P7.88*

*7.89 The beam AB is subjected to the uniformly distributed load shown and to two unknown forces $\mathbf{P}$ and $\mathbf{Q}$. Knowing that it has been experimentally determined that the bending moment is $+800$ N·m at D and $+1300$ N·m at E, (a) determine $\mathbf{P}$ and $\mathbf{Q}$, (b) draw the shear and bending-moment diagrams for the beam.

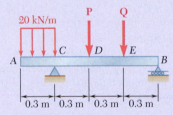

Fig. P7.89

*7.90 Solve Prob. 7.89 assuming that the bending moment was found to be $+650$ N·m at D and $+1450$ N·m at E.

*7.91 The beam AB is subjected to the uniformly distributed load shown and to two unknown forces $\mathbf{P}$ and $\mathbf{Q}$. Knowing that it has been experimentally determined that the bending moment is $+6.10$ kip·ft at D and $+5.50$ kip·ft at E, (a) determine $\mathbf{P}$ and $\mathbf{Q}$, (b) draw the shear and bending-moment diagrams for the beam.

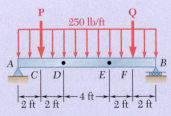

Fig. P7.91

*7.92 Solve Prob. 7.91 assuming that the bending moment was found to be $+5.96$ kip·ft at D and $+6.84$ kip·ft at E.

*7.4 CABLES

Cables are used in many engineering applications, such as suspension bridges, transmission lines, aerial tramways, guy wires for high towers, etc. Cables may be divided into two categories, according to their loading: (1) supporting concentrated loads and (2) supporting distributed loads.

7.4A Cables with Concentrated Loads

Consider a cable attached to two fixed points A and B and supporting n vertical concentrated loads $\mathbf{P}_1$, $\mathbf{P}_2$, . . . , $\mathbf{P}_n$ (Fig. 7.13a). We assume that the cable is *flexible*, i.e., that its resistance to bending is small and can be neglected. We further assume that the *weight of the cable is negligible* compared with the loads supported by the cable. We can therefore approximate any portion of cable between successive loads as a two-force member. Thus, the internal forces at any point in the cable reduce to a *force of tension directed along the cable.*

We assume that each of the loads lies in a given vertical line, i.e., that the horizontal distance from support A to each of the loads is known. We also assume that we know the horizontal and vertical distances between the supports. With these assumptions, we want to determine the shape of the cable (i.e., the vertical distance from support A to each of the points C_1, C_2, . . . , C_n) and also the tension T in each portion of the cable.

We first draw the free-body diagram of the entire cable (Fig. 7.13b). Since we do not know the slopes of the portions of cable attached at A and B, we represent the reactions at A and B by two components each. Thus, four unknowns are involved, and the three equations of equilibrium are not sufficient to determine the reactions. (Clearly, a cable is not a rigid body; thus, the equilibrium equations represent *necessary but not sufficient conditions.* See Sec. 6.3B.) We must therefore obtain an additional equation by considering the equilibrium of a portion of the cable. This is possible if we know the coordinates x and y of a point D of the cable.

We draw the free-body diagram of the portion of cable AD (Fig. 7.14a). From the equilibrium condition $\Sigma M_D = 0$, we obtain an additional relation between the scalar components A_x and A_y and can determine

Photo 7.3 The weight of the chairlift cables is negligible compared to the weights of the chairs and skiers, so we can use the methods of this section to determine the force at any point in the cable.

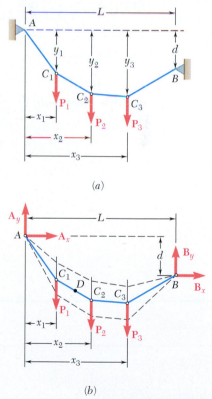

(a)

(b)

Fig. 7.13 (a) A cable supporting vertical concentrated loads; (b) free-body diagram of the entire cable.

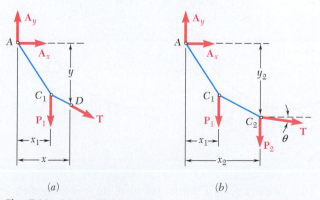

(a)

(b)

Fig. 7.14 (a) Free-body diagram of the portion of cable AD; (b) free-body diagram of the portion of cable AC_2.

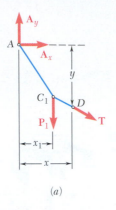

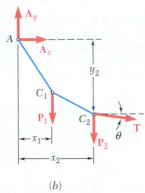

(a)

(b)

Fig. 7.14 *(repeated)*

the reactions at A and B. However, the problem remains indeterminate if we do not know the coordinates of D unless we are given some other relation between A_x and A_y (or between B_x and B_y). The cable might hang in any of various possible ways, as indicated by the dashed lines in Fig. 7.13b.

Once we have determined A_x and A_y, we can find the vertical distance from A to any point of the cable. Considering point C_2, for example, we draw the free-body diagram of the portion of cable AC_2 (Fig. 7.14b). From $\Sigma F_{C_2} = 0$, we obtain an equation that we can solve for y_2. From $\Sigma F_x = 0$ and $\Sigma F_y = 0$, we obtain the components of force $\mathbf{T}$ representing the tension in the portion of cable to the right of C_2. Note that $T \cos \theta = -A_x$; that is, *the horizontal component of the tension force is the same at any point of the cable.* It follows that the tension T is maximum when $\cos \theta$ is minimum, i.e., in the portion of cable that has the largest angle of inclination θ. Clearly, this portion of cable must be adjacent to one of the two supports of the cable.

7.4B Cables with Distributed Loads

Consider a cable attached to two fixed points A and B and carrying a *distributed load* (Fig. 7.15a). We just saw that for a cable supporting concentrated loads, the internal force at any point is a force of tension directed along the cable. By contrast, in the case of a cable carrying a distributed load, the cable hangs in the shape of a curve, and the internal force at a point D is a force of tension $\mathbf{T}$ *directed along the tangent to the curve.* Here we examine how to determine the tension at any point of a cable supporting a given distributed load. In the following sections, we will determine the shape of the cable for two common types of distributed loads.

Considering the most general case of distributed load, we draw the free-body diagram of the portion of cable extending from the lowest point C to a given point D of the cable (Fig. 7.15b). The three forces acting on the free body are the tension force $\mathbf{T}_0$ at C, which is horizontal; the tension force $\mathbf{T}$ at D, which is directed along the tangent to the cable at D; and the resultant $\mathbf{W}$ of the distributed load supported by the portion of cable CD. Drawing the corresponding force triangle (Fig. 7.15c), we obtain the relations

$$T \cos \theta = T_0 \qquad T \sin \theta = W \qquad (7.5)$$

$$T = \sqrt{T_0^2 + W^2} \qquad \tan \theta = \frac{W}{T_0} \qquad (7.6)$$

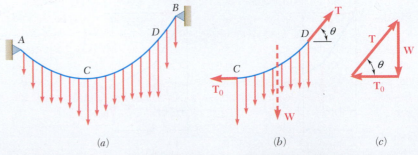

(a) (b) (c)

Fig. 7.15 (a) A cable carrying a distributed load; (b) free-body diagram of the portion of the cable CD; (c) force triangle for the free-body diagram in part (b).

From the relations in Eqs. (7.5), we see that the horizontal component of the tension force **T** is the same at any point. Furthermore, the vertical component of **T** at any point is equal to the magnitude W of the load when measured from the lowest point (C) to the point in question (D). Relations in Eq. (7.6) show that the tension T is minimum at the lowest point and maximum at one of the two support points.

7.4C Parabolic Cables

Now suppose that cable AB carries a load *uniformly distributed along the horizontal* (Fig. 7.16a). We can approximate the load on the cables of a suspension bridge in this way, since the weight of the cables is small compared with the uniform weight of the roadway. We denote the load per unit length by w (*measured horizontally*) and express it in N/m or lb/ft. Choosing coordinate axes with the origin at the lowest point C of the cable, we find that the magnitude W of the total load carried by the portion of cable extending from C to the point D with coordinates x and y is $W = wx$. The relations in Eqs. (7.6) defining the magnitude and direction of the tension force at D become

$$T = \sqrt{T_0^2 + w^2 x^2} \qquad \tan \theta = \frac{wx}{T_0} \qquad (7.7)$$

Moreover, the distance from D to the line of action of the resultant **W** is equal to half of the horizontal distance from C to D (Fig. 7.16b). Summing moments about D, we have

$$+\curvearrowleft \Sigma M_D = 0: \qquad wx\frac{x}{2} - T_0 y = 0$$

Solving for y, we have

Equation of parabolic cable

$$y = \frac{wx^2}{2T_0} \qquad (7.8)$$

This is the equation of a *parabola* with a vertical axis and its vertex at the origin of coordinates. Thus, the curve formed by cables loaded uniformly along the horizontal is a parabola.[‡]

When the supports A and B of the cable have the same elevation, the distance L between the supports is called the **span** of the cable and the vertical distance h from the supports to the lowest point is called the **sag** of the cable (Fig. 7.17a). If you know the span and sag of a cable and if the load w per unit horizontal length is given, you can find the minimum tension T_0 by substituting $x = L/2$ and $y = h$ in Eq. (7.8). Equations (7.7) then yield the tension and the slope at any point of the cable and Eq. (7.8) defines the shape of the cable.

[‡]Cables hanging under their own weight are not loaded uniformly along the horizontal and do not form parabolas. However, the error introduced by assuming a parabolic shape for cables hanging under their own weight is small when the cable is sufficiently taut. In the next section, we give a complete discussion of cables hanging under their own weight.

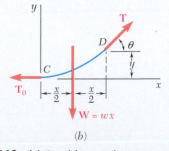

Photo 7.4 The main cables of suspension bridges, like the Golden Gate Bridge above, may be assumed to carry a loading uniformly distributed along the horizontal.

(a)

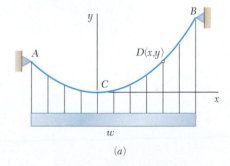

(b)

Fig. 7.16 (a) A cable carrying a uniformly distributed load along the horizontal; (b) free-body diagram of the portion of cable CD.

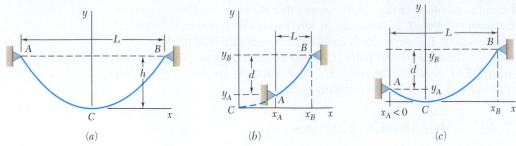

Fig. 7.17 (a) The shape of a parabolic cable is determined by its span L and sag h; (b, c) span and vertical distance between supports for cables with supports at different elevations.

When the supports have different elevations, the position of the lowest point of the cable is not known, and we must determine the coordinates x_A, y_A and x_B, y_B of the supports. To do this, we note that the coordinates of A and B satisfy Eq. (7.8) and that

$$x_B - x_A = L \quad \text{and} \quad y_B - y_A = d$$

where L and d denote, respectively, the horizontal and vertical distances between the two supports (Fig. 7.17b and c).

We can obtain the length of the cable from its lowest point C to its support B from the formula

$$s_B = \int_0^{x_B} \sqrt{1 + \left(\frac{dy}{dx}\right)^2}\, dx \tag{7.9}$$

Differentiating Eq. (7.8), we obtain the derivative $dy/dx = wx/T_0$. Substituting this into Eq. (7.9) and using the binomial theorem to expand the radical in an infinite series, we have

$$s_B = \int_0^{x_B} \sqrt{1 + \frac{w^2 x^2}{T_0^2}}\, dx = \int_0^{x_B} \left(1 + \frac{w^2 x^2}{2T_0^2} - \frac{w^4 x^4}{8T_0^4} + \cdots \right) dx$$

$$s_B = x_B \left(1 + \frac{w^2 x_B^2}{6T_0^2} - \frac{w^4 x_B^4}{40T_0^4} + \cdots \right)$$

Then, since $w x_B^2 / 2T_0 = y_B$, we obtain

$$s_B = x_B \left[1 + \frac{2}{3}\left(\frac{y_B}{x_B}\right)^2 - \frac{2}{5}\left(\frac{y_B}{x_B}\right)^4 + \cdots \right] \tag{7.10}$$

This series converges for values of the ratio y_B/x_B less than 0.5. In most cases, this ratio is much smaller, and only the first two terms of the series need be computed.

Sample Problem 7.8

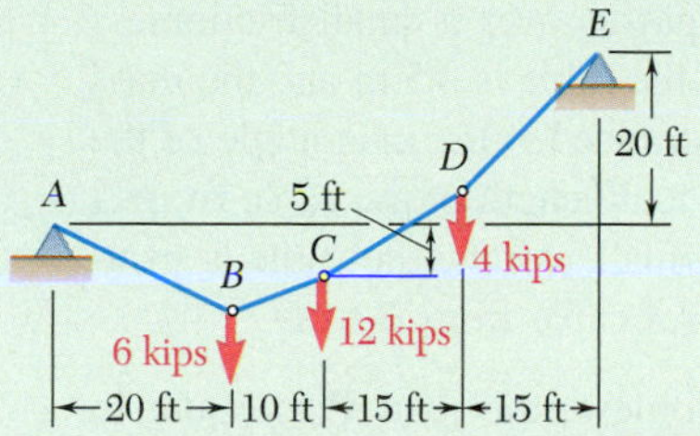

The cable *AE* supports three vertical loads from the points indicated. If point *C* is 5 ft below the left support, determine (*a*) the elevation of points *B* and *D*, (*b*) the maximum slope and the maximum tension in the cable.

STRATEGY: To solve for the support reactions at *A*, consider a free-body diagram of the entire cable as well as one that takes a section at *C*, since you know the coordinates of this point. Taking subsequent sections at *B* and *D* will then enable you to determine their elevations. The resulting cable geometry establishes the maximum slope, which is where the maximum tension in the cable occurs.

MODELING and ANALYSIS:

Free Body, Entire Cable. Determine the reaction components $\mathbf{A}_x$ and $\mathbf{A}_y$ as

$+\curvearrowleft \Sigma M_E = 0$:

$$A_x(20 \text{ ft}) - A_y(60 \text{ ft}) + (6 \text{ kips})(40 \text{ ft}) + (12 \text{ kips})(30 \text{ ft}) + (4 \text{ kips})(15 \text{ ft}) = 0$$

$$20A_x - 60A_y + 660 = 0$$

Free Body, ABC. Consider the portion *ABC* of the cable as a free body (Fig. 1). Then you have

$+\curvearrowleft \Sigma M_C = 0$: $\quad -A_x(5 \text{ ft}) - A_y(30 \text{ ft}) + (6 \text{ kips})(10 \text{ ft}) = 0$

$$-5A_x - 30A_y + 60 = 0$$

Solving the two equations simultaneously, you obtain

$$A_x = -18 \text{ kips} \qquad \mathbf{A}_x = 18 \text{ kips} \leftarrow$$
$$A_y = +5 \text{ kips} \qquad \mathbf{A}_y = 5 \text{ kips} \uparrow$$

a. Elevation of Points *B* and *D*:

Free Body, AB. Considering the portion of cable *AB* as a free body, you obtain

$+\curvearrowleft \Sigma M_B = 0$: $\quad (18 \text{ kips})y_B - (5 \text{ kips})(20 \text{ ft}) = 0$

$$y_B = 5.56 \text{ ft below } A \quad \blacktriangleleft$$

Free Body, ABCD. Using the portion of cable *ABCD* as a free body gives you

$+\curvearrowleft \Sigma M_D = 0$:

$$-(18 \text{ kips})y_D - (5 \text{ kips})(45 \text{ ft}) + (6 \text{ kips})(25 \text{ ft}) + (12 \text{ kips})(15 \text{ ft}) = 0$$

$$y_D = 5.83 \text{ ft above } A \quad \blacktriangleleft$$

b. Maximum Slope and Maximum Tension.

Note that the maximum slope occurs in portion *DE*. Since the horizontal component of the tension is constant and equal to 18 kips, you have

$$\tan \theta = \frac{14.17}{15 \text{ ft}} \qquad \theta = 43.4° \quad \blacktriangleleft$$

$$T_{\max} = \frac{18 \text{ kips}}{\cos \theta} \qquad T_{\max} = 24.8 \text{ kips} \quad \blacktriangleleft$$

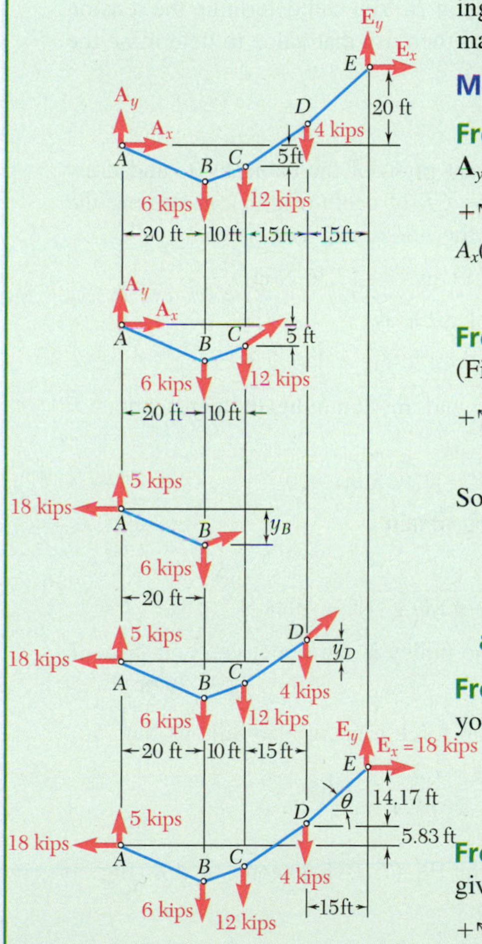

Fig. 1 Free-body diagrams of cable system.

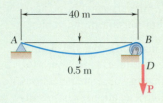

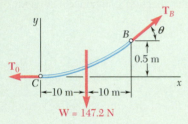

Fig. 1 Free-body diagram of cable portion *CB*.

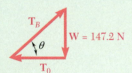

Fig. 2 Force triangle for cable portion *CB*.

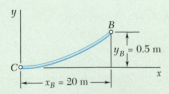

Fig. 3 Dimensions used to determine length of cable.

Sample Problem 7.9

A light cable is attached to a support at *A*, passes over a small frictionless pulley at *B*, and supports a load **P**. The sag of the cable is 0.5 m and the mass per unit length of the cable is 0.75 kg/m. Determine (*a*) the magnitude of the load **P**, (*b*) the slope of the cable at *B*, (*c*) the total length of the cable from *A* to *B*. Since the ratio of the sag to the span is small, assume the cable is parabolic. Also, neglect the weight of the portion of cable from *B* to *D*.

STRATEGY: Because the pulley is frictionless, the load **P** is equal in magnitude to the tension in the cable at *B*. You can determine the tension using the methods of this section and then use that value to determine the slope and length of the cable.

MODELING and ANALYSIS:

a. Load P. Denote the lowest point of the cable by *C* and draw the free-body diagram of the portion *CB* of cable (Fig. 1). Assuming the load is uniformly distributed along the horizontal, you have

$$w = (0.75 \text{ kg/m})(9.81 \text{ m/s}^2) = 7.36 \text{ N/m}$$

The total load for the portion *CB* of cable is

$$W = wx_B = (7.36 \text{ N/m})(20 \text{ m}) = 147.2 \text{ N}$$

This load acts halfway between *C* and *B*. Summing moments about *B* gives you

$$+\curvearrowleft \Sigma M_B = 0: \qquad (147.2 \text{ N})(10 \text{ m}) - T_0(0.5 \text{ m}) = 0 \qquad T_0 = 2944 \text{ N}$$

From the force triangle (Fig. 2), you obtain

$$T_B = \sqrt{T_0^2 + W^2}$$
$$= \sqrt{(2944 \text{ N})^2 + (147.2 \text{ N})^2} = 2948 \text{ N}$$

Since the tension on each side of the pulley is the same, you end up with

$$P = T_B = 2948 \text{ N} \quad \blacktriangleleft$$

b. Slope of Cable at B. The force triangle also tells us that

$$\tan \theta = \frac{W}{T_0} = \frac{147.2 \text{ N}}{2944 \text{ N}} = 0.05$$

$$\theta = 2.9° \quad \blacktriangleleft$$

c. Length of Cable. Applying Eq. (7.10) between *C* and *B* (Fig. 3) gives you

$$s_B = x_B \left[1 + \frac{2}{3}\left(\frac{y_B}{x_B}\right)^2 + \cdots \right]$$

$$= (20 \text{ m})\left[1 + \frac{2}{3}\left(\frac{0.5 \text{ m}}{20 \text{ m}}\right)^2 + \cdots \right] = 20.00833 \text{ m}$$

The total length of the cable between *A* and *B* is twice this value. Thus,

$$\text{Length} = 2s_B = 40.0167 \text{ m} \quad \blacktriangleleft$$

REFLECT and THINK: Notice that the length of the cable is only very slightly more than the length of the span between *A* and *B*. This means that the cable must be very taut, which is consistent with the relatively large value of load **P** (compared to the weight of the cable).

SOLVING PROBLEMS
ON YOUR OWN

In the problems of this section, you will apply the equations of equilibrium to *cables that lie in a vertical plane*. We assume that a cable cannot resist bending, so the force of tension in the cable is always directed along the cable.

A. In the first part of this lesson, we considered **cables subjected to concentrated loads.** Since we assume the weight of the cable is negligible, the cable is straight between loads.

Your solution will consist of the following steps.

1. Draw a free-body diagram of the entire cable showing the loads and the horizontal and vertical components of the reaction at each support. Use this free-body diagram to write the corresponding equilibrium equations.

2. You will have four unknown components and only three equations of equilibrium (see Fig. 7.13). You must therefore find an additional piece of information, such as the *position* of a point on the cable or the *slope* of the cable at a given point.

3. After you have identified the point of the cable where the additional information exists, cut the cable at that point, and draw a free-body diagram of one of the two resulting portions of the cable.

   a. If you know the position of the point where you have cut the cable, set $\Sigma M = 0$ about that point for the new free body. This will yield the additional equation required to solve for the four unknown components of the reactions [Sample Prob. 7.8].

   b. If you know the slope of the portion of the cable you have cut, set $\Sigma F_x = 0$ and $\Sigma F_y = 0$ for the new free body. This will yield two equilibrium equations that, together with the original three, you can solve for the four reaction components and for the tension in the cable where it has been cut.

4. To find the elevation of a given point of the cable and the slope and tension at that point once you have found the reactions at the supports, you should cut the cable at that point and draw a free-body diagram of one of the two resulting portions of the cable. Setting $\Sigma M = 0$ about the given point yields its elevation. Writing $\Sigma F_x = 0$ and $\Sigma F_y = 0$ yields the components of the tension force from which you can find its magnitude and direction.

5. For a cable supporting vertical loads only, *the horizontal component of the tension force is the same at any point. It follows that, for such a cable, the maximum tension occurs in the steepest portion of the cable.*

(*continued*)

B. In the second portion of this section, we considered **cables carrying a load that is uniformly distributed along the horizontal.** The shape of the cable is then parabolic.

Your solution will use one or more of the following concepts.

1. Place the origin of coordinates at the lowest point of the cable and direct the x and y axes to the right and upward, respectively. Then *the equation of the parabola* is

$$y = \frac{wx^2}{2T_0} \tag{7.8}$$

The minimum cable tension occurs at the origin, where the cable is horizontal. The maximum tension is at the support where the slope is maximum.

2. If the supports of the cable have the same elevation, the sag h of the cable is the vertical distance from the lowest point of the cable to the horizontal line joining the supports. To solve a problem involving such a parabolic cable, use Eq. (7.8) for one of the supports; this equation can be solved for one unknown.

3. If the supports of the cable have different elevations, you will have to write Eq. (7.8) for each of the supports (see Fig. 7.17).

4. To find the length of the cable from the lowest point to one of the supports, you can use Eq. (7.10). In most cases, you will need to compute only the first two terms of the series.

Problems

7.93 Three loads are suspended as shown from the cable *ABCDE*. Knowing that $d_C = 4$ m, determine (*a*) the components of the reaction at *E*, (*b*) the maximum tension in the cable.

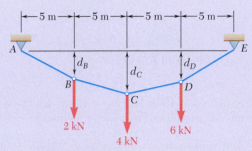

Fig. P7.93 and P7.94

7.94 Knowing that the maximum tension in cable *ABCDE* is 25 kN, determine the distance d_C.

7.95 If $d_C = 8$ ft, determine (*a*) the reaction at *A*, (*b*) the reaction at *E*.

7.96 If $d_C = 4.5$ ft, determine (*a*) the reaction at *A*, (*b*) the reaction at *E*.

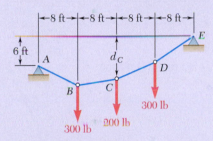

Fig. P7.95 and P7.96

7.97 Knowing that $d_C = 3$ m, determine (*a*) the distances d_B and d_D, (*b*) the reaction at *E*.

7.98 Determine (*a*) distance d_C for which portion *DE* of the cable is horizontal, (*b*) the corresponding reactions at *A* and *E*.

7.99 Knowing that $d_C = 15$ ft, determine (*a*) the distances d_B and d_D, (*b*) the maximum tension in the cable.

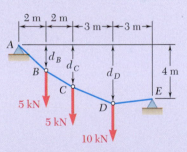

Fig. P7.97 and P7.98

7.100 Determine (*a*) the distance d_C for which portion *BC* of the cable is horizontal, (*b*) the corresponding components of the reaction at *E*.

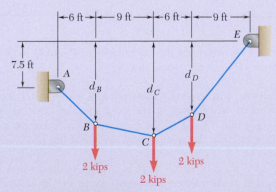

Fig. P7.99 and P7.100

411

7.101 Knowing that $m_B = 70$ kg and $m_C = 25$ kg, determine the magnitude of the force **P** required to maintain equilibrium.

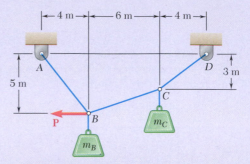

Fig. P7.101 and P7.102

7.102 Knowing that $m_B = 18$ kg and $m_C = 10$ kg, determine the magnitude of the force **P** required to maintain equilibrium.

7.103 Cable *ABC* supports two loads as shown. Knowing that $b = 21$ ft, determine (*a*) the required magnitude of the horizontal force **P**, (*b*) the corresponding distance *a*.

7.104 Cable *ABC* supports two loads as shown. Determine the distances *a* and *b* when a horizontal force **P** of magnitude 200 lb is applied at *C*.

7.105 If $a = 3$ m, determine the magnitudes of **P** and **Q** required to maintain the cable in the shape shown.

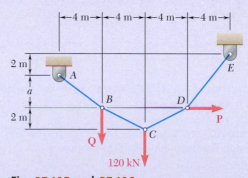

Fig. P7.105 and P7.106

7.106 If $a = 4$ m, determine the magnitudes of **P** and **Q** required to maintain the cable in the shape shown.

7.107 An electric wire having a mass per unit length of 0.6 kg/m is strung between two insulators at the same elevation that are 60 m apart. Knowing that the sag of the wire is 1.5 m, determine (*a*) the maximum tension in the wire, (*b*) the length of the wire.

7.108 The total mass of cable *ACB* is 20 kg. Assuming that the mass of the cable is distributed uniformly along the horizontal, determine (*a*) the sag *h*, (*b*) the slope of the cable at *A*.

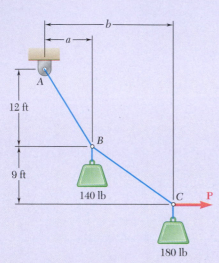

Fig. P7.103 and P7.104

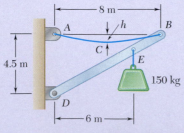

Fig. P7.108

7.109 The center span of the George Washington Bridge, as originally constructed, consisted of a uniform roadway suspended from four cables. The uniform load supported by each cable was w = 9.75 kips/ft along the horizontal. Knowing that the span L is 3500 ft and that the sag h is 316 ft, determine for the original configuration (a) the maximum tension in each cable, (b) the length of each cable.

7.110 The center span of the Verrazano-Narrows Bridge consists of two uniform roadways suspended from four cables. The design of the bridge allows for the effect of extreme temperature changes that cause the sag of the center span to vary from h_w = 386 ft in winter to h_s = 394 ft in summer. Knowing that the span is L = 4260 ft, determine the change in length of the cables due to extreme temperature changes.

7.111 Each cable of the Golden Gate Bridge supports a load w = 11.1 kips/ft along the horizontal. Knowing that the span L is 4150 ft and that the sag h is 464 ft, determine (a) the maximum tension in each cable, (b) the length of each cable.

7.112 Two cables of the same gauge are attached to a transmission tower at B. Since the tower is slender, the horizontal component of the resultant of the forces exerted by the cables at B is to be zero. Knowing that the mass per unit length of the cables is 0.4 kg/m, determine (a) the required sag h, (b) the maximum tension in each cable.

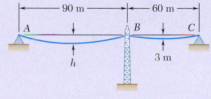

Fig. P7.112

7.113 A 76-m length of wire having a mass per unit length of 2.2 kg/m is used to span a horizontal distance of 75 m. Determine (a) the approximate sag of the wire, (b) the maximum tension in the wire. [*Hint:* Use only the first two terms of Eq. (7.10).]

7.114 A cable of length $L + \Delta$ is suspended between two points that are at the same elevation and a distance L apart. (a) Assuming that Δ is small compared to L and that the cable is parabolic, determine the approximate sag in terms of L and Δ. (b) If L = 100 ft and Δ = 4 ft, determine the approximate sag. [*Hint:* Use only the first two terms of Eq. (7.10).]

7.115 The total mass of cable AC is 25 kg. Assuming that the mass of the cable is distributed uniformly along the horizontal, determine the sag h and the slope of the cable at A and C.

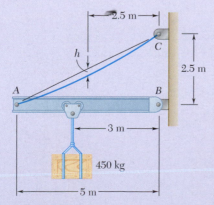

Fig. P7.115

7.116 Cable *ACB* supports a load uniformly distributed along the horizontal as shown. The lowest point *C* is located 9 m to the right of *A*. Determine (*a*) the vertical distance *a*, (*b*) the length of the cable, (*c*) the components of the reaction at *A*.

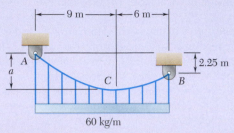

Fig. P7.116

7.117 Each cable of the side spans of the Golden Gate Bridge supports a load $w = 10.2$ kips/ft along the horizontal. Knowing that for the side spans the maximum vertical distance *h* from each cable to the chord *AB* is 30 ft and occurs at midspan, determine (*a*) the maximum tension in each cable, (*b*) the slope at *B*.

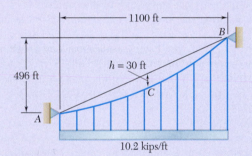

Fig. P7.117

7.118 A steam pipe weighing 45 lb/ft that passes between two buildings 40 ft apart is supported by a system of cables as shown. Assuming that the weight of the cable system is equivalent to a uniformly distributed loading of 5 lb/ft, determine (*a*) the location of the lowest point *C* of the cable, (*b*) the maximum tension in the cable.

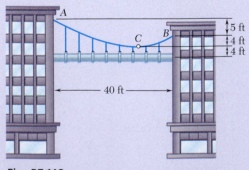

Fig. P7.118

*7.119 A cable AB of span L and a simple beam $A'B'$ of the same span are subjected to identical vertical loadings as shown. Show that the magnitude of the bending moment at a point C' in the beam is equal to the product T_0h, where T_0 is the magnitude of the horizontal component of the tension force in the cable and h is the vertical distance between point C and the chord joining the points of support A and B.

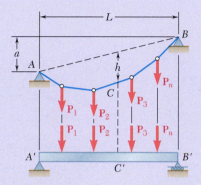

Fig. P7.119

7.120 through 7.123 Making use of the property established in Prob. 7.119, solve the problem indicated by first solving the corresponding beam problem.

7.120 Prob. 7.94.
7.121 Prob. 7.97a.
7.122 Prob. 7.99a.
7.123 Prob. 7.100a.

*7.124 Show that the curve assumed by a cable that carries a distributed load $w(x)$ is defined by the differential equation $d^2y/dx^2 = w(x)/T_0$, where T_0 is the tension at the lowest point.

*7.125 Using the property indicated in Prob. 7.124, determine the curve assumed by a cable of span L and sag h carrying a distributed load $w = w_0 \cos(\pi x/L)$, where x is measured from midspan. Also determine the maximum and minimum values of the tension in the cable.

*7.126 If the weight per unit length of the cable AB is $w_0/\cos^2\theta$, prove that the curve formed by the cable is a circular arc. (*Hint:* Use the property indicated in Prob. 7.124.)

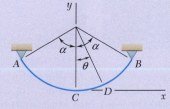

Fig. P7.126

*7.5   CATENARY CABLES

Let us now consider a cable *AB* carrying a load that is **uniformly distributed along the cable itself** (Fig. 7.18*a*). Cables hanging under their own weight are loaded in this way. We denote the load per unit length by *w* (*measured along the cable*) and express it in N/m or lb/ft. The magnitude *W* of the total load carried by a portion of cable with a length of *s*, extending

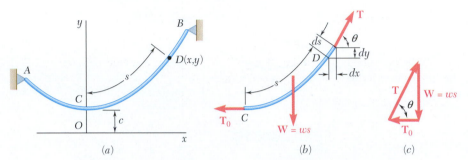

Fig. 7.18 (*a*) A cable carrying a load uniformly distributed along the cable; (*b*) free-body diagram of a portion of the cable *CD*; (*c*) force triangle for part (*b*).

from the lowest point *C* to some point *D*, is $W = ws$. Substituting this value for *W* in formula (7.6), we obtain the tension at *D*, as

$$T = \sqrt{T_0^2 + w^2 s^2}$$

In order to simplify the subsequent computations, we introduce the constant $c = T_0/w$. This gives us

$$T_0 = wc \qquad W = ws \qquad T = w\sqrt{c^2 + s^2} \tag{7.11}$$

The free-body diagram of the portion of cable *CD* is shown in Fig. 7.18*b*. However, we cannot use this diagram directly to obtain the equation of the curve assumed by the cable, because we do not know the horizontal distance from *D* to the line of action of the resultant **W** of the load. To obtain this equation, we note that the horizontal projection

(*a*) High-voltage power lines

(*b*) A spider's web

(*c*) The Gateway Arch

Photo 7.5 Catenary cables occur in nature as well as in engineered structures. (*a*) High-voltage power lines, common all across the country and in much of the world, support only their own weight. (*b*) Catenary cables can be as delicate as the silk threads of a spider's web. (*c*) The Gateway to the West Arch in St. Louis is an inverted catenary arch cast in concrete (which is in compression instead of tension).

of a small element of cable of length ds is $dx = ds \cos \theta$. Observing from Fig. 7.18c that $\cos \theta = T_0/T$ and using Eq. (7.11), we have

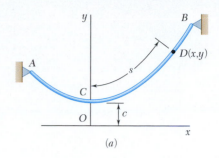

(a)

$$dx = ds \cos \theta = \frac{T_0}{T} ds = \frac{wc \, ds}{w\sqrt{c^2 + s^2}} = \frac{ds}{\sqrt{1 + s^2/c^2}}$$

Selecting the origin O of the coordinates at a distance c directly below C (Fig. 7.18a) and integrating from $C(0, c)$ to $D(x, y)$, we obtain†

$$x = \int_0^s \frac{ds}{\sqrt{1 + s^2/c^2}} = c\left[\sinh^{-1} \frac{s}{c} \right]_0^s = c \sinh^{-1} \frac{s}{c}$$

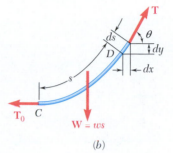

(b)

This equation, which relates the length s of the portion of cable CD and the horizontal distance x, can be written in the form

Length of catenary cable

$$s = c \sinh \frac{x}{c} \qquad (7.15)$$

We can now obtain the relation between the coordinates x and y by writing $dy = dx \tan \theta$. Observing from Fig. 7.18c that $\tan \theta = W/T_0$ and using (7.11) and (7.15), we have

(c)

Fig. 7.18 (continued).

$$dy = dx \tan \theta = \frac{W}{T_0} dx = \frac{s}{c} dx = \sinh \frac{x}{c} dx$$

Integrating from $C(0, c)$ to $D(x, y)$ and using Eqs. (7.12) and (7.13), we obtain

$$y - c = \int_0^x \sinh \frac{x}{c} dx = c\left[\cosh \frac{x}{c} \right]_0^x = c\left(\cosh \frac{x}{c} - 1 \right)$$

$$y - c = c \cosh \frac{x}{c} - c$$

†This integral appears in all standard integral tables. The function

$$z = \sinh^{-1} u$$

(read "arc hyperbolic sine u")is the inverse of the function $u = \sinh z$ (read "hyperbolic sine z"). This function and the function $v = \cosh z$ (read "hyperbolic cosine z") are defined as

$$u = \sinh z = \frac{1}{2}(e^z - e^{-z}) \qquad v = \cosh z = \frac{1}{2}(e^z + e^{-z})$$

Numerical values of the functions $\sinh z$ and $\cosh z$ are listed in tables of hyperbolic functions and also may be computed on most calculators, either directly or from the definitions. Refer to any calculus text for a complete description of the properties of these functions. In this section, we use only the following properties, which are easy to derive from the definitions:

$$\frac{d \sinh z}{dz} = \cosh z \qquad \frac{d \cosh z}{dz} = \sinh z \qquad (7.12)$$

$$\sinh 0 = 0 \qquad \cosh 0 = 1 \qquad (7.13)$$

$$\cosh^2 z - \sinh^2 z = 1 \qquad (7.14)$$

which reduces to

Equation of catenary cable

$$y = c \cosh \frac{x}{c} \qquad (7.16)$$

This is the equation of a **catenary** with vertical axis. The ordinate c of the lowest point C is called the *parameter* of the catenary. By squaring both sides of Eqs. (7.15) and (7.16), subtracting, and taking Eq. (7.14) into account, we obtain the following relation between y and s:

$$y^2 - s^2 = c^2 \qquad (7.17)$$

Solving Eq. (7.17) for s^2 and carrying into the last of the relations in Eqs. (7.11), we write these relations as

$$T_0 = wc \qquad W = ws \qquad T = wy \qquad (7.18)$$

The last relation indicates that the tension at any point D of the cable is proportional to the vertical distance from D to the horizontal line representing the x axis.

When the supports A and B of the cable have the same elevation, the distance L between the supports is called the *span* of the cable and the vertical distance h from the supports to the lowest point C is called the *sag* of the cable. These definitions are the same as those given for parabolic cables; note that, because of our choice of coordinate axes, the sag h is now

$$h = y_A - c \qquad (7.19)$$

Also note that some catenary problems involve transcendental equations, which must be solved by successive approximations (see Sample Prob. 7.10). When the cable is fairly taut, however, we can assume that the load is uniformly distributed *along the horizontal* and replace the catenary by a parabola. This greatly simplifies the solution of the problem, and the error introduced is small.

When the supports A and B have different elevations, the position of the lowest point of the cable is not known. We can then solve the problem in a manner similar to that indicated for parabolic cables by noting that the cable must pass through the supports and that $x_B - x_A = L$ and $y_B - y_A = d$, where L and d denote, respectively, the horizontal and vertical distances between the two supports.

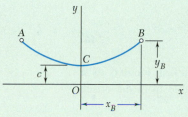

Fig. 1 Cable geometry.

Sample Problem 7.10

A uniform cable weighing 3 lb/ft is suspended between two points A and B as shown. Determine (a) the maximum and minimum values of the tension in the cable, (b) the length of the cable.

STRATEGY: This is a cable carrying only its own weight that is supported by its ends at the same elevation. You can use the analysis in this section to solve the problem.

MODELING and ANALYSIS:

Equation of Cable. Place the origin of coordinates at a distance c below the lowest point of the cable (Fig. 1). The equation of the cable is given by Eq. (7.16), as

$$y = c \cosh \frac{x}{c}$$

The coordinates of point B are

$$x_B = 250 \text{ ft} \qquad y_B = 100 + c$$

Substituting these coordinates into the equation of the cable, you obtain

$$100 + c = c \cosh \frac{250}{c}$$

$$\frac{100}{c} + 1 = \cosh \frac{250}{c}$$

Determine the value of c by substituting successive trial values, as shown in the following table.

c	$\dfrac{250}{c}$	$\dfrac{100}{c}$	$\dfrac{100}{c} + 1$	$\cosh \dfrac{250}{c}$
300	0.833	0.333	1.333	1.367
350	0.714	0.286	1.286	1.266
330	0.758	0.303	1.303	1.301
328	0.762	0.305	1.305	1.305

Taking $c = 328$, you have

$$y_B = 100 + c = 428 \text{ ft}$$

a. Maximum and Minimum Values of the Tension. Using Eqs. (7.18), you obtain

$$T_{\min} = T_0 = wc = (3 \text{ lb/ft})(328 \text{ ft}) \qquad T_{\min} = 984 \text{ lb} \blacktriangleleft$$
$$T_{\max} = T_B = wy_B = (3 \text{ lb/ft})(428 \text{ ft}) \qquad T_{\max} = 1284 \text{ lb} \blacktriangleleft$$

b. Length of Cable. You can find one-half of the length of the cable by solving Eq. (7.17). Hence,

$$y_B^2 - s_{CB}^2 = c^2 \qquad s_{CB}^2 = y_B^2 - c^2 = (428)^2 - (328)^2 \qquad s_{CB} = 275 \text{ ft}$$

The total length of the cable is therefore

$$s_{AB} = 2s_{CB} = 2(275 \text{ ft}) \qquad s_{AB} = 550 \text{ ft} \blacktriangleleft$$

REFLECT and THINK: The sag in the cable is one-fifth of the cable's span, so it is not very taut. The weight of the cable is $ws = (3 \text{ lb/ft})(550 \text{ ft}) = 1650 \text{ lb}$, while its maximum tension is only 1284 lb. This demonstrates that the total weight of a cable can exceed its maximum tension.

SOLVING PROBLEMS
ON YOUR OWN

In the last section of this chapter, we described how to solve problems involving a *cable carrying a load uniformly distributed along the cable*. The shape assumed by the cable is a catenary and is defined by

$$y = c \cosh \frac{x}{c} \qquad (7.16)$$

1. Keep in mind that the origin of coordinates for a catenary is located at a distance *c* directly below its lowest point. The length of the cable from the origin to any point is expressed as

$$s = c \sinh \frac{x}{c} \qquad (7.15)$$

2. You should first identify all of the known and unknown quantities. Then consider each of the equations listed in the text (Eqs. 7.15 through 7.19) and solve an equation that contains only one unknown. Substitute the value found into another equation, and solve that equation for another unknown.

3. If the sag *h* is given, use Eq. (7.19) to replace *y* by $h + c$ in Eq. (7.16) if *x* is known [Sample Prob. 7.10] or in Eq. (7.17) if *s* is known, and solve the resulting equation for the constant *c*.

4. Many of the problems you will encounter will involve the solution by trial and error of an equation involving a hyperbolic sine or cosine. You can make your work easier by keeping track of your calculations in a table, as in Sample Prob. 7.10, or by applying a numerical methods approach using a computer or calculator.

Problems

7.127 A 25-ft chain with a weight of 30 lb is suspended between two points at the same elevation. Knowing that the sag is 10 ft, determine (*a*) the distance between the supports, (*b*) the maximum tension in the chain.

7.128 A 500-ft-long aerial tramway cable having a weight per unit length of 2.8 lb/ft is suspended between two points at the same elevation. Knowing that the sag is 125 ft, find (*a*) the horizontal distance between the supports, (*b*) the maximum tension in the cable.

7.129 A 40-m cable is strung as shown between two buildings. The maximum tension is found to be 350 N, and the lowest point of the cable is observed to be 6 m above the ground. Determine (*a*) the horizontal distance between the buildings, (*b*) the total mass of the cable.

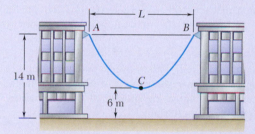

Fig. P7.129

7.130 A 50-m steel surveying tape has a mass of 1.6 kg. If the tape is stretched between two points at the same elevation and pulled until the tension at each end is 60 N, determine the horizontal distance between the ends of the tape. Neglect the elongation of the tape due to the tension.

7.131 A 20-m length of wire having a mass per unit length of 0.2 kg/m is attached to a fixed support at *A* and to a collar at *B*. Neglecting the effect of friction, determine (*a*) the force **P** for which $h = 8$ m, (*b*) the corresponding span *L*.

7.132 A 20-m length of wire having a mass per unit length of 0.2 kg/m is attached to a fixed support at *A* and to a collar at *B*. Knowing that the magnitude of the horizontal force applied to the collar is $P = 20$ N, determine (*a*) the sag *h*, (*b*) the span *L*.

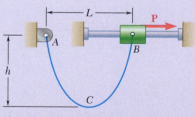

Fig. *P7.131*, *P7.132*, and P7.133

7.133 A 20-m length of wire having a mass per unit length of 0.2 kg/m is attached to a fixed support at *A* and to a collar at *B*. Neglecting the effect of friction, determine (*a*) the sag *h* for which $L = 15$ m, (*b*) the corresponding force **P**.

7.134 Determine the sag of a 30-ft chain that is attached to two points at the same elevation that are 20 ft apart.

7.135 A counterweight D is attached to a cable that passes over a small pulley at A and is attached to a support at B. Knowing that $L = 45$ ft and $h = 15$ ft, determine (a) the length of the cable from A to B, (b) the weight per unit length of the cable. Neglect the weight of the cable from A to D.

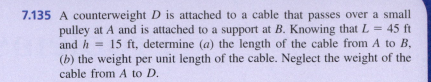
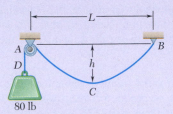

Fig. P7.135

7.136 A 90-m wire is suspended between two points at the same elevation that are 60 m apart. Knowing that the maximum tension is 300 N, determine (a) the sag of the wire, (b) the total mass of the wire.

7.137 A cable weighing 2 lb/ft is suspended between two points at the same elevation that are 160 ft apart. Determine the smallest allowable sag of the cable if the maximum tension is not to exceed 400 lb.

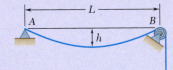

Fig. P7.138

7.138 A uniform cord 50 in. long passes over a pulley at B and is attached to a pin support at A. Knowing that $L = 20$ in. and neglecting the effect of friction, determine the smaller of the two values of h for which the cord is in equilibrium.

7.139 A motor M is used to slowly reel in the cable shown. Knowing that the mass per unit length of the cable is 0.4 kg/m, determine the maximum tension in the cable when $h = 5$ m.

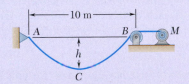

Fig. P7.139 and P7.140

7.140 A motor M is used to slowly reel in the cable shown. Knowing that the mass per unit length of the cable is 0.4 kg/m, determine the maximum tension in the cable when $h = 3$ m.

7.141 The cable ACB has a mass per unit length of 0.45 kg/m. Knowing that the lowest point of the cable is located at a distance $a = 0.6$ m below the support A, determine (a) the location of the lowest point C, (b) the maximum tension in the cable.

7.142 The cable ACB has a mass per unit length of 0.45 kg/m. Knowing that the lowest point of the cable is located at a distance $a = 2$ m below the support A, determine (a) the location of the lowest point C, (b) the maximum tension in the cable.

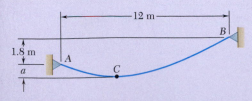

Fig. P7.141 and P7.142

7.143 A uniform cable weighing 3 lb/ft is held in the position shown by a horizontal force **P** applied at B. Knowing that $P = 180$ lb and $\theta_A = 60°$, determine (a) the location of point B, (b) the length of the cable.

7.144 A uniform cable weighing 3 lb/ft is held in the position shown by a horizontal force **P** applied at B. Knowing that $P = 150$ lb and $\theta_A = 60°$, determine (a) the location of point B, (b) the length of the cable.

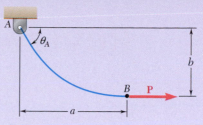

Fig. P7.143 and P7.144

7.145 To the left of point B, the long cable $ABDE$ rests on the rough horizontal surface shown. Knowing that the mass per unit length of the cable is 2 kg/m, determine the force **F** when $a = 3.6$ m.

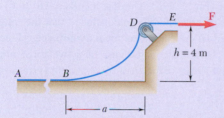

Fig. P7.145 and P7.146

7.146 To the left of point B, the long cable $ABDE$ rests on the rough horizontal surface shown. Knowing that the mass per unit length of the cable is 2 kg/m, determine the force **F** when $a = 6$ m.

*7.147 The 10-ft cable AB is attached to two collars as shown. The collar at A can slide freely along the rod; a stop attached to the rod prevents the collar at B from moving on the rod. Neglecting the effect of friction and the weight of the collars, determine the distance a.

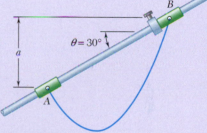

Fig. P7.147

*7.148 Solve Prob. 7.147 assuming that the angle θ formed by the rod and the horizontal is 45°.

7.149 Denoting the angle formed by a uniform cable and the horizontal by θ, show that at any point (a) $s = c \tan \theta$, (b) $y = c \sec \theta$.

*7.150 (a) Determine the maximum allowable horizontal span for a uniform cable with a weight per unit length of w if the tension in the cable is not to exceed a given value T_m. (b) Using the result of part a, determine the maximum span of a steel wire for which $w = 0.25$ lb/ft and $T_m = 8000$ lb.

*7.151 A cable has a mass per unit length of 3 kg/m and is supported as shown. Knowing that the span L is 6 m, determine the *two* values of the sag h for which the maximum tension is 350 N.

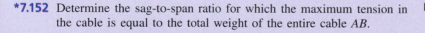

Fig. P7.151, P7.152 and P7.153

*7.152 Determine the sag-to-span ratio for which the maximum tension in the cable is equal to the total weight of the entire cable AB.

*7.153 A cable with a weight per unit length of w is suspended between two points at the same elevation that are a distance L apart. Determine (a) the sag-to-span ratio for which the maximum tension is as small as possible, (b) the corresponding values of θ_B and T_m.

Review and Summary

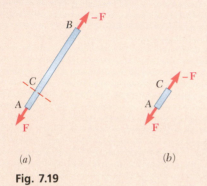

Fig. 7.19

In this chapter, you learned to determine the internal forces that hold together the various parts of a given member in a structure.

Forces in Straight Two-Force Members

Considering first a **straight two-force member** AB [Sec. 7.1], recall that such a member is subjected at A and B to equal and opposite forces $\mathbf{F}$ and $-\mathbf{F}$ directed along AB (Fig. 7.19a). Cutting member AB at C and drawing the free-body diagram of portion AC, we concluded that the internal forces existing at C in member AB are equivalent to an **axial force** $-\mathbf{F}$ equal and opposite to $\mathbf{F}$ (Fig. 7.19b). Note that, in the case of a two-force member that is not straight, the internal forces reduce to a force-couple system and not to a single force.

Forces in Multi-Force Members

Consider next a **multi-force member** AD (Fig. 7.20a). Cutting it at J and drawing the free-body diagram of portion JD, we concluded that the internal forces at J are equivalent to a force-couple system consisting of the **axial force** $\mathbf{F}$, the **shearing force** $\mathbf{V}$, and a couple $\mathbf{M}$ (Fig. 7.20b). The magnitude of the shearing force measures the **shear** at point J, and the moment of the couple is referred to as the **bending moment** at J. Since an equal and opposite force-couple system is obtained by considering the free-body diagram of portion AJ, it is necessary to specify which portion of member AD is used when recording the answers [Sample Prob. 7.1].

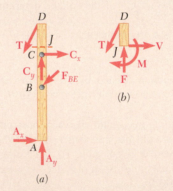

Fig. 7.20

Forces in Beams

Most of the chapter was devoted to the analysis of the internal forces in two important types of engineering structures: beams and cables. **Beams** are usually long, straight prismatic members designed to support loads applied at various points along the member. In general, the loads are perpendicular to the axis of the beam and produce only shear and bending in the beam. The loads may be either **concentrated** at specific points or **distributed** along the entire length or a portion of the beam. The beam itself may be supported in various ways; since only statically determinate beams are considered in this text, we limited our analysis to that of simply supported beams, overhanging beams, and cantilever beams [Sec. 7.2].

Shear and Bending Moment in a Beam

To obtain the shear V and bending moment M at a given point C of a beam, we first determine the reactions at the supports by considering the entire beam as a free body. We then cut the beam at C and use the free-body diagram of one of the two resulting portions to determine V and M. In order to avoid any confusion regarding the sense of the shearing force $\mathbf{V}$ and couple $\mathbf{M}$ (which act in opposite directions on the two portions of the beam), we adopted the sign convention illustrated in Fig. 7.21 [Sec. 7.2B]. Once we have determined the values of the shear and bending moment at a few selected points of the beam, it is usually possible to draw a **shear diagram** and a **bending-moment diagram** representing, respectively, the shear and bending moment at any point of the beam [Sec. 7.2C]. When a beam is subjected to concentrated loads only, the shear is of constant value between loads, and the bending moment varies linearly between loads [Sample Prob. 7.2]. When a beam is subjected to distributed loads, the shear and bending moment vary quite differently [Sample Prob. 7.3].

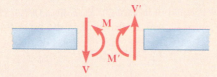

Internal forces at section
(positive shear and positive bending moment)

Fig. 7.21

Relations among Load, Shear, and Bending Moment

Construction of the shear and bending-moment diagrams is simplified by taking into account the following relations. Denoting the distributed load per unit length by w (assumed positive if directed downward), we have [Sec. 7.3]:

$$\frac{dV}{dx} = -w \tag{7.1}$$

$$\frac{dM}{dx} = V \tag{7.3}$$

In integrated form, these equations become

$$V_D - V_C = -(\text{area under load curve between } C \text{ and } D) \tag{7.2'}$$

$$M_D - M_C = \text{area under shear curve between } C \text{ and } D \tag{7.4'}$$

Equation (7.2′) makes it possible to draw the shear diagram of a beam from the curve representing the distributed load on that beam and the value of V at one end of the beam. Similarly, Eq. (7.4′) makes it possible to draw the bending-moment diagram from the shear diagram and the value of M at one end of the beam. However, discontinuities are introduced in the shear diagram by concentrated loads and in the bending-moment diagram by concentrated couples, none of which are accounted for in these equations [Sample Probs. 7.4 and 7.7]. Finally, we note from Eq. (7.3) that the points of the beam where the bending moment is maximum or minimum are also the points where the shear is zero [Sample Prob. 7.5].

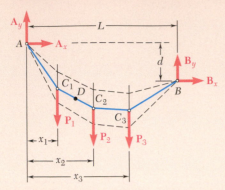

Fig. 7.22

Cables with Concentrated Loads

The second half of the chapter was devoted to the analysis of **flexible cables**. We first considered a cable of negligible weight supporting **concentrated loads** [Sec. 7.4A]. Using the entire cable AB as a free body (Fig. 7.22), we noted that the three available equilibrium equations were not sufficient to determine the four unknowns representing the reactions at supports A and B. However, if the coordinates of a point D of the cable are known, we can obtain an additional equation by considering the free-body diagram of portion AD or DB of the cable. Once we have determined the reactions at the supports, we can find the elevation of any point of the cable and the tension in any portion of the cable from the appropriate free-body diagram [Sample Prob. 7.8]. We noted that the horizontal component of the force $\mathbf{T}$ representing the tension is the same at any point of the cable.

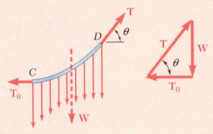

Fig. 7.23

Cables with Distributed Loads

We next considered cables carrying **distributed loads** [Sec. 7.4B]. Using as a free body a portion of cable CD extending from the lowest point C to an arbitrary point D of the cable (Fig. 7.23), we observed that the horizontal component of the tension force $\mathbf{T}$ at D is constant and equal to the tension T_0 at C, whereas its vertical component is equal to the weight W of the portion of cable CD. The magnitude and direction of $\mathbf{T}$ were obtained from the force triangle:

$$T = \sqrt{T_0^2 + W^2} \qquad \tan \theta = \frac{W}{T_0} \tag{7.6}$$

Parabolic Cable

In the case of a load uniformly distributed along the horizontal—as in a suspension bridge (Fig. 7.24)—the load supported by portion CD is $W = wx$, where w is the constant load per unit horizontal length [Sec. 7.4C]. We also found that the curve formed by the cable is a **parabola** with equation

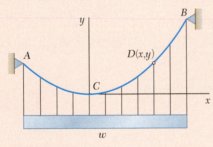

Fig. 7.24

$$y = \frac{wx^2}{2T_0} \tag{7.8}$$

and that the length of the cable can be found by using the expansion in series given in Eq. (7.10) [Sample Prob. 7.9].

Catenary

In the case of a load uniformly distributed along the cable itself—e.g., a cable hanging under its own weight (Fig. 7.25)—the load supported by portion CD is $W = ws$, where s is the length measured along the cable and w is the constant load per unit length [Sec. 7.5]. Choosing the origin O of the coordinate axes at a distance $c = T_0/w$ below C, we derived the relations

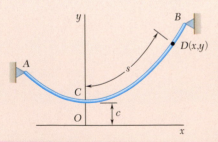

Fig. 7.25

$$s = c \sinh \frac{x}{c} \tag{7.15}$$

$$y = c \cosh \frac{x}{c} \tag{7.16}$$

$$y^2 - s^2 = c^2 \tag{7.17}$$

$$T_0 = wc \qquad W = ws \qquad T = wy \tag{7.18}$$

These equations can be used to solve problems involving cables hanging under their own weight [Sample Prob. 7.10]. Equation (7.16), which defines the shape of the cable, is the equation of a **catenary**.

Review Problems

7.154 and *7.155* Knowing that the turnbuckle has been tightened until the tension in wire *AD* is 850 N, determine the internal forces at the point indicated:

7.154 Point *J*

7.155 Point *K*

7.156 Two members, each consisting of a straight and a quarter-circular portion of rod, are connected as shown and support a 75-lb load at *A*. Determine the internal forces at point *J*.

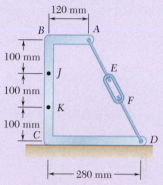

Fig. P7.154 and *P7.155*

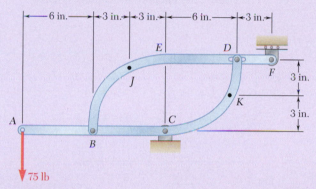

Fig. P7.156

7.157 Knowing that the radius of each pulley is 150 mm, that $\alpha = 20°$, and neglecting friction, determine the internal forces at (*a*) point *J*, (*b*) point *K*.

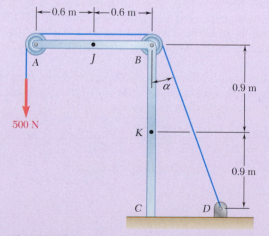

Fig. P7.157

7.158 For the beam shown, determine (*a*) the magnitude *P* of the two upward forces for which the maximum absolute value of the bending moment in the beam is as small as possible, (*b*) the corresponding value of $|M|_{max}$.

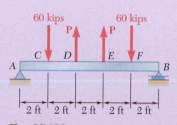

Fig. P7.158

427

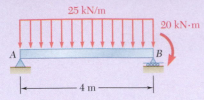

Fig. P7.159

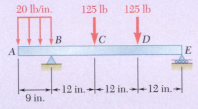

Fig. P7.160

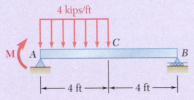

Fig. P7.161

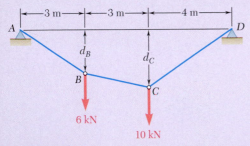

Fig. P7.163

7.159 For the beam and loading shown, (a) draw the shear and bending-moment diagrams, (b) determine the magnitude and location of the maximum absolute value of the bending moment.

7.160 For the beam and loading shown, (a) draw the shear and bending-moment diagrams, (b) determine the maximum absolute values of the shear and bending moment.

7.161 For the beam shown, draw the shear and bending-moment diagrams, and determine the magnitude and location of the maximum absolute value of the bending moment, knowing that (a) $M = 0$, (b) $M = 24$ kip·ft.

7.162 The beam AB, which lies on the ground, supports the parabolic load shown. Assuming the upward reaction of the ground to be uniformly distributed, (a) write the equations of the shear and bending-moment curves, (b) determine the maximum bending moment.

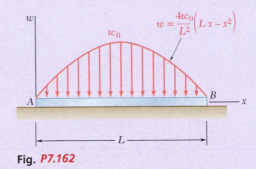

Fig. P7.162

7.163 Two loads are suspended as shown from the cable $ABCD$. Knowing that $d_B = 1.8$ m, determine (a) the distance d_C, (b) the components of the reaction at D, (c) the maximum tension in the cable.

7.164 A wire having a mass per unit length of 0.65 kg/m is suspended from two supports at the same elevation that are 120 m apart. If the sag is 30 m, determine (a) the total length of the wire, (b) the maximum tension in the wire.

7.165 A 10-ft rope is attached to two supports A and B as shown. Determine (a) the span of the rope for which the span is equal to the sag, (b) the corresponding angle θ_B.

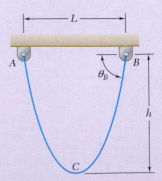

Fig. P7.165

8

Friction

The tractive force that a railroad locomotive can develop depends upon the frictional resistance between the drive wheels and the rails. When the potential exists for wheel slip to occur, such as when a train travels upgrade over wet rails, sand is deposited on top of the railhead to increase this friction.

Objectives

- **Examine** the laws of dry friction and the associated coefficients and angles of friction.
- **Consider** the equilibrium of rigid bodies where dry friction at contact surfaces is modeled.
- **Apply** the laws of friction to analyze problems involving wedges and square-threaded screws.
- **Study** engineering applications of the laws of friction, such as in modeling axle, disk, wheel, and belt friction.

Introduction

In the previous chapters, we assumed that surfaces in contact are either *frictionless* or *rough*. If they are frictionless, the force each surface exerts on the other is normal to the surfaces, and the two surfaces can move freely with respect to each other. If they are rough, tangential forces can develop that prevent the motion of one surface with respect to the other.

This view is a simplified one. Actually, no perfectly frictionless surface exists. When two surfaces are in contact, tangential forces, called **friction forces**, always develop if you attempt to move one surface with respect to the other. However, these friction forces are limited in magnitude and do not prevent motion if you apply sufficiently large forces. Thus, the distinction between frictionless and rough surfaces is a matter of degree. You will see this more clearly in this chapter, which is devoted to the study of friction and its applications to common engineering situations.

There are two types of friction: **dry friction**, sometimes called *Coulomb friction*, and **fluid friction** or *viscosity*. Fluid friction develops

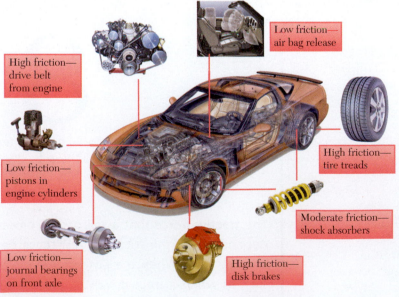

Low friction—
air bag release

High friction—
drive belt
from engine

High friction—
tire treads

Low friction—
pistons in
engine cylinders

Moderate friction—
shock absorbers

Low friction—
journal bearings
on front axle

High friction—
disk brakes

Photo 8.1 Examples of friction in an automobile. Depending upon the application, the degree of friction is controlled by design engineers.

between layers of fluid moving at different velocities. This is of great importance in analyzing problems involving the flow of fluids through pipes and orifices or dealing with bodies immersed in moving fluids. It is also basic for the analysis of the motion of *lubricated mechanisms.* Such problems are considered in texts on fluid mechanics. The present study is limited to dry friction, i.e., to situations involving rigid bodies that are in contact along *unlubricated* surfaces.

In the first section of this chapter, we examine the equilibrium of various rigid bodies and structures, assuming dry friction at the surfaces of contact. Afterward, we consider several specific engineering applications where dry friction plays an important role: wedges, square-threaded screws, journal bearings, thrust bearings, rolling resistance, and belt friction.

8.1 THE LAWS OF DRY FRICTION

We can illustrate the laws of dry friction by the following experiment. Place a block of weight **W** on a horizontal plane surface (Fig. 8.1*a*). The forces acting on the block are its weight **W** and the reaction of the surface. Since the weight has no horizontal component, the reaction of the surface also has no horizontal component; the reaction is therefore *normal* to the surface and is represented by **N** in Fig. 8.1*a*. Now suppose that you apply a horizontal force **P** to the block (Fig. 8.1*b*). If **P** is small, the block does not move; some other horizontal force must therefore exist, which balances **P**. This other force is the **static-friction force F**, which is actually the resultant of a great number of forces acting over the entire surface of contact between the block and the plane. The nature of these forces is not known exactly, but we generally assume that these forces are due to the irregularities of the surfaces in contact and, to a certain extent, to molecular attraction.

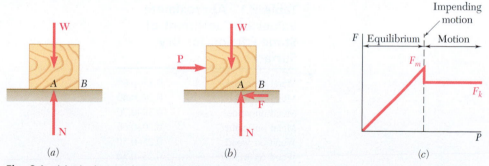

Fig. 8.1 (a) Block on a horizontal plane, friction force is zero; (b) a horizontally applied force **P** produces an opposing friction force **F**; (c) graph of **F** with increasing **P**.

If you increase the force **P**, the friction force **F** also increases, continuing to oppose **P**, until its magnitude reaches a certain *maximum value F_m* (Fig. 8.1*c*). If **P** is further increased, the friction force cannot balance it anymore, and the block starts sliding. As soon as the block has started in motion, the magnitude of **F** drops from F_m to a lower value F_k. This happens because less interpenetration occurs between the irregularities of the surfaces in contact when these surfaces move with respect to each other. From then on, the block keeps sliding with increasing velocity while the friction force, denoted by F_k and called the **kinetic-friction force**, remains approximately constant.

Note that, as the magnitude F of the friction force increases from 0 to F_m, the point of application A of the resultant $\mathbf{N}$ of the normal forces of contact moves to the right. In this way, the couples formed by $\mathbf{P}$ and $\mathbf{F}$ and by $\mathbf{W}$ and $\mathbf{N}$, respectively, remain balanced. If $\mathbf{N}$ reaches B before F reaches its maximum value F_m, the block starts to tip about B before it can start sliding (see Sample Prob. 8.4).

8.1A Coefficients of Friction

Experimental evidence shows that the maximum value F_m of the static-friction force is proportional to the normal component N of the reaction of the surface. We have

Static friction

$$F_m = \mu_s N \tag{8.1}$$

where μ_s is a constant called the **coefficient of static friction**. Similarly, we can express the magnitude F_k of the kinetic-friction force in the form

Kinetic friction

$$F_k = \mu_k N \tag{8.2}$$

where μ_k is a constant called the **coefficient of kinetic friction**. The coefficients of friction μ_s and μ_k do not depend upon the area of the surfaces in contact. Both coefficients, however, depend strongly on the *nature* of the surfaces in contact. Since they also depend upon the exact condition of the surfaces, their value is seldom known with an accuracy greater than 5%. Approximate values of coefficients of static friction for various combinations of dry surfaces are given in Table 8.1. The corresponding values of the coefficient of kinetic friction are about 25% smaller. Since coefficients of friction are dimensionless quantities, the values given in Table 8.1 can be used with both SI units and U.S. customary units.

Table 8.1 Approximate Values of Coefficient of Static Friction for Dry Surfaces

Metal on metal	0.15–0.60
Metal on wood	0.20–0.60
Metal on stone	0.30–0.70
Metal on leather	0.30–0.60
Wood on wood	0.25–0.50
Wood on leather	0.25–0.50
Stone on stone	0.40–0.70
Earth on earth	0.20–1.00
Rubber on concrete	0.60–0.90

From this discussion, it appears that four different situations can occur when a rigid body is in contact with a horizontal surface:

1. The forces applied to the body do not tend to move it along the surface of contact; there is no friction force (Fig. 8.2a).
2. The applied forces tend to move the body along the surface of contact but are not large enough to set it in motion. We can find the static-friction force $\mathbf{F}$ that has developed by solving the equations of equilibrium for the body. Since there is no evidence that $\mathbf{F}$ has reached its maximum

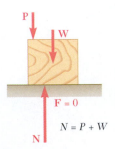

(a) No friction ($P_x = 0$)

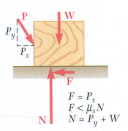

(b) No motion ($P_x < F_m$)

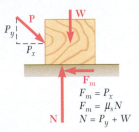

(c) Motion impending $\longrightarrow$ ($P_x = F_m$)

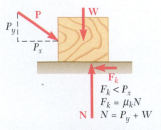

(d) Motion $\longrightarrow$ ($P_x > F_k$)

Fig. 8.2 (a) Applied force is vertical, friction force is zero; (b) horizontal component of applied force is less than F_m, no motion occurs; (c) horizontal component of applied force equals F_m, motion is impending; (d) horizontal component of applied force is greater than F_k, forces are unbalanced and motion continues.

value, the equation $F_m = \mu_s N$ *cannot be used* to determine the friction force (Fig. 8.2b).

3. The applied forces are such that the body is just about to slide. We say that *motion is impending*. The friction force **F** has reached its maximum value F_m and, together with the normal force **N**, balances the applied forces. Both the equations of equilibrium and the equation $F_m = \mu_s N$ can be used. Note that the friction force has a sense opposite to the sense of impending motion (Fig. 8.2c).

4. The body is sliding under the action of the applied forces, and the equations of equilibrium no longer apply. However, **F** is now equal to $\mathbf{F}_k$, and we can use the equation $F_k = \mu_k N$. The sense of $\mathbf{F}_k$ is opposite to the sense of motion (Fig. 8.2d).

8.1B Angles of Friction

It is sometimes convenient to replace the normal force **N** and the friction force **F** by their resultant **R**. Let's see what happens when we do that.

Consider again a block of weight **W** resting on a horizontal plane surface. If no horizontal force is applied to the block, the resultant **R** reduces to the normal force **N** (Fig. 8.3a). However, if the applied force **P** has a horizontal component $\mathbf{P}_x$ that tends to move the block, force **R** has a horizontal component **F** and, thus, forms an angle ϕ with the normal to the surface (Fig. 8.3b). If you increase $\mathbf{P}_x$ until motion becomes impending, the angle between **R** and the vertical grows and reaches a maximum value (Fig. 8.3c). This value is called the **angle of static friction** and is denoted by ϕ_s. From the geometry of Fig. 8.3c, we note that

Angle of static friction

$$\tan \phi_s = \frac{F_m}{N} = \frac{\mu_s N}{N}$$

$$\tan \phi_s = \mu_s \qquad (8.3)$$

If motion actually takes place, the magnitude of the friction force drops to F_k; similarly, the angle between **R** and **N** drops to a lower value ϕ_k, which is called the **angle of kinetic friction** (Fig. 8.3d). From the geometry of Fig. 8.3d, we have

Angle of kinetic friction

$$\tan \phi_k = \frac{F_k}{N} = \frac{\mu_k N}{N}$$

$$\tan \phi_k = \mu_k \qquad (8.4)$$

Another example shows how the angle of friction can be used to advantage in the analysis of certain types of problems. Consider a block resting on a board and subjected to no other force than its weight **W** and the reaction **R** of the board. The board can be given any desired inclination. If the board is horizontal, the force **R** exerted by the board on the block is perpendicular to the board and balances the weight **W** (Fig. 8.4a). If the board is given a small angle of inclination θ, force **R** deviates from the perpendicular to the board by angle θ and continues to balance **W** (Fig. 8.4b). The reaction **R** now has a normal component **N** with a magnitude of $N = W \cos \theta$ and a tangential component **F** with a magnitude of $F = W \sin \theta$.

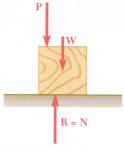

(a) No friction

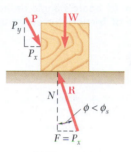

(b) No motion

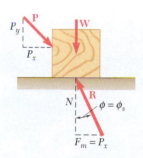

(c) Motion impending $\longrightarrow$

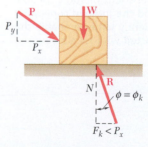

(d) Motion $\longrightarrow$

Fig. 8.3 (a) Applied force is vertical, friction force is zero; (b) applied force is at an angle, its horizontal component balanced by the horizontal component of the surface resultant; (c) impending motion, the horizontal component of the applied force equals the maximum horizontal component of the resultant; (d) motion, the horizontal component of the resultant is less than the horizontal component of the applied force.

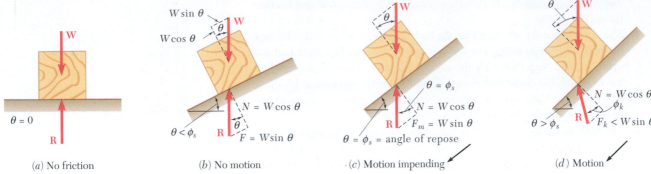

Fig. 8.4 (a) Block on horizontal board, friction force is zero; (b) board's angle of inclination is less than angle of static friction, no motion; (c) board's angle of inclination equals angle of friction, motion is impending; (d) angle of inclination is greater than angle of friction, forces are unbalanced and motion occurs.

If we keep increasing the angle of inclination, motion soon becomes impending. At that time, the angle between **R** and the normal reaches its maximum value $\theta = \phi_s$ (Fig. 8.4c). The value of the angle of inclination corresponding to impending motion is called the **angle of repose**. Clearly, the angle of repose is equal to the angle of static friction ϕ_s. If we further increase the angle of inclination θ, motion starts and the angle between **R** and the normal drops to the lower value ϕ_k (Fig. 8.4d). The reaction **R** is not vertical anymore, and the forces acting on the block are unbalanced.

8.1C   Problems Involving Dry Friction

Many engineering applications involve dry friction. Some are simple situations, such as variations on the block sliding on a plane just described. Others involve more complicated situations, as in Sample Prob. 8.3. Many problems deal with the stability of rigid bodies in accelerated motion and will be studied in dynamics. Also, several common machines and mechanisms can be analyzed by applying the laws of dry friction, including wedges, screws, journal and thrust bearings, and belt transmissions. We will study these applications in the following sections.

The methods used to solve problems involving dry friction are the same that we used in the preceding chapters. If a problem involves only a motion of translation with no possible rotation, we can usually treat the body under consideration as a particle and use the methods of Chap. 2. If the problem involves a possible rotation, we must treat the body as a rigid body and use the methods of Chap. 4. If the structure considered is made of several parts, we must apply the principle of action and reaction, as we did in Chap. 6.

If the body being considered is acted upon by more than three forces (including the reactions at the surfaces of contact), the reaction at each surface is represented by its components **N** and **F**, and we solve the problem using the equations of equilibrium. If only three forces act on the body under consideration, it may be more convenient to represent each reaction by the single force **R** and solve the problem by using a force triangle.

Most problems involving friction fall into one of the following three groups.

1. All applied forces are given, and we know the coefficients of friction; we are to determine whether the body being considered remains at rest or slides. The friction force **F** *required to maintain equilibrium* is

Photo 8.2 The coefficient of static friction between a crate and the inclined conveyer belt must be sufficiently large to enable the crate to be transported without slipping.

unknown (its magnitude is *not* equal to $\mu_s N$) and needs to be determined, together with the normal force **N**, by drawing a free-body diagram and solving the equations of equilibrium (Fig. 8.5*a*). We then compare the value found for the magnitude F of the friction force with the maximum value $F_m = \mu_s N$. If F is smaller than or equal to F_m, the body remains at rest. If the value found for F is larger than F_m, equilibrium cannot be maintained and motion takes place; the actual magnitude of the friction force is then $F_k = \mu_k N$.

2. All applied forces are given, and we know the motion is impending; we are to determine the value of the coefficient of static friction. Here again, we determine the friction force and the normal force by drawing a free-body diagram and solving the equations of equilibrium (Fig. 8.5*b*). Since we know that the value found for F is the maximum value F_m, we determine the coefficient of friction by solving the equation $F_m = \mu_s N$.

3. The coefficient of static friction is given, and we know that the motion is impending in a given direction; we are to determine the magnitude or the direction of one of the applied forces. The friction force should be shown in the free-body diagram with a *sense opposite to that of the impending motion* and with a magnitude $F_m = \mu_s N$ (Fig. 8.5*c*). We can then write the equations of equilibrium and determine the desired force.

As noted previously, when only three forces are involved, it may be more convenient to represent the reaction of the surface by a single force **R** and to solve the problem by drawing a force triangle. Such a solution is used in Sample Prob. 8.2.

When two bodies A and B are in contact (Fig. 8.6*a*), the forces of friction exerted, respectively, by A on B and by B on A are equal and opposite (Newton's third law). In drawing the free-body diagram of one of these bodies, it is important to include the appropriate friction force with its correct sense. Observe the following rule: *The sense of the friction force acting on A is opposite to that of the motion (or impending motion) of A as observed from B* (Fig. 8.6*b*). (It is therefore the same as the motion of B as observed from A.) The sense of the friction force acting on B is determined in a similar way (Fig. 8.6*c*). Note that the motion of A as observed from B is a *relative motion*. For example, if body A is fixed and body B moves, body A has a relative motion with respect to B. Also, if both B and A are moving down but B is moving faster than A, then body A is observed, from B, to be moving up.

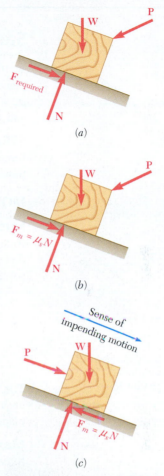

(a)

(b)

(c)

Fig. 8.5 Three types of friction problems: (*a*) given the forces and coefficient of friction, will the block slide or stay? (*b*) given the forces and that motion is pending, determine the coefficient of friction; (*c*) given the coefficient of friction and that motion is impending, determine the applied force.

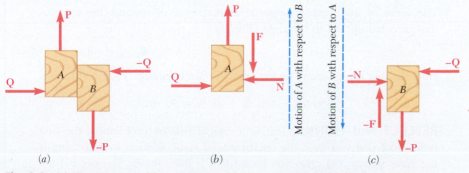

(a)   (b)   (c)

Fig. 8.6 (*a*) Two blocks held in contact by forces; (*b*) free-body diagram for block A, including direction of friction force; (*c*) free-body diagram for block B, including direction of friction force.

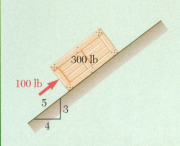

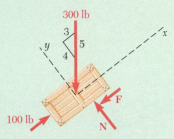

Fig. 1 Free-body diagram of crate showing assumed direction of friction force.

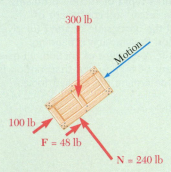

Fig. 2 Free-body diagram of crate showing actual friction force.

Sample Problem 8.1

A 100-lb force acts as shown on a 300-lb crate placed on an inclined plane. The coefficients of friction between the crate and the plane are $\mu_s = 0.25$ and $\mu_k = 0.20$. Determine whether the crate is in equilibrium, and find the value of the friction force.

STRATEGY: This is a friction problem of the first type: You know the forces and the friction coefficients and want to determine if the crate moves. You also want to find the friction force.

MODELING and ANALYSIS

Force Required for Equilibrium. First determine the value of the friction force *required to maintain equilibrium*. Assuming that **F** is directed down and to the left, draw the free-body diagram of the crate (Fig. 1) and solve the equilibrium equations:

$$+\nearrow \Sigma F_x = 0: \qquad 100 \text{ lb} - \tfrac{3}{5}(300 \text{ lb}) - F = 0$$
$$F = -80 \text{ lb} \qquad \mathbf{F} = 80 \text{ lb} \nearrow$$

$$+\nwarrow \Sigma F_y = 0: \qquad N - \tfrac{4}{5}(300 \text{ lb}) = 0$$
$$N = +240 \text{ lb} \qquad \mathbf{N} = 240 \text{ lb} \nwarrow$$

The force **F** required to maintain equilibrium is an 80-lb force directed up and to the right; the tendency of the crate is thus to move down the plane.

Maximum Friction Force. The magnitude of the maximum friction force that may be developed between the crate and the plane is

$$F_m = \mu_s N \qquad F_m = 0.25(240 \text{ lb}) = 60 \text{ lb}$$

Since the value of the force required to maintain equilibrium (80 lb) is larger than the maximum value that may be obtained (60 lb), equilibrium is not maintained and *the crate will slide down the plane*.

Actual Value of Friction Force. The magnitude of the actual friction force is

$$F_{\text{actual}} = F_k = \mu_k N = 0.20(240 \text{ lb}) = 48 \text{ lb}$$

The sense of this force is opposite to the sense of motion; the force is thus directed up and to the right (Fig. 2):

$$\mathbf{F}_{\text{actual}} = 48 \text{ lb} \nearrow \quad \blacktriangleleft$$

Note that the forces acting on the crate are not balanced. Their resultant is

$$\tfrac{3}{5}(300 \text{ lb}) - 100 \text{ lb} - 48 \text{ lb} = 32 \text{ lb} \swarrow$$

REFLECT and THINK: This is a typical friction problem of the first type. Note that you used the coefficient of static friction to determine if the crate moves, but once you found that it does move, you needed the coefficient of kinetic friction to determine the friction force.

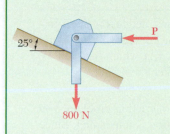

Sample Problem 8.2

A support block is acted upon by two forces as shown. Knowing that the coefficients of friction between the block and the incline are $\mu_s = 0.35$ and $\mu_k = 0.25$, determine the force **P** required to (a) start the block moving up the incline; (b) keep it moving up; (c) prevent it from sliding down.

STRATEGY: This problem involves practical variations of the third type of friction problem. You can approach the solutions through the concept of the angles of friction.

MODELING:

Free-Body Diagram. For each part of the problem, draw a free-body diagram of the block and a force triangle including the 800-N vertical force, the horizontal force **P**, and the force **R** exerted on the block by the incline. You must determine the direction of **R** in each separate case. Note that, since **P** is perpendicular to the 800-N force, the force triangle is a right triangle, which easily can be solved for **P**. In most other problems, however, the force triangle will be an oblique triangle and should be solved by applying the law of sines.

ANALYSIS:

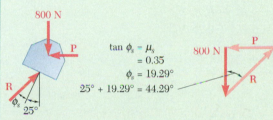

$$\tan \phi_s = \mu_s$$
$$= 0.35$$
$$\phi_s = 19.29°$$
$$25° + 19.29° = 44.29°$$

Fig. 1 Free-body diagram of block and its force triangle—motion impending up incline.

a. Force P to Start Block Moving Up. In this case, motion is impending up the incline, so the resultant is directed at the angle of static friction (Fig. 1). Note that the resultant is oriented to the left of the normal such that its friction component (not shown) is directed opposite the direction of impending motion.

$$P = (800 \text{ N}) \tan 44.29° \qquad \textbf{P = 780 N} \leftarrow \blacktriangleleft$$

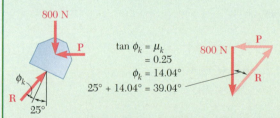

$$\tan \phi_k = \mu_k$$
$$= 0.25$$
$$\phi_k = 14.04°$$
$$25° + 14.04° = 39.04°$$

Fig. 2 Free-body diagram of block and its force triangle—motion continuing up incline.

b. Force P to Keep Block Moving Up. Motion is continuing, so the resultant is directed at the angle of kinetic friction (Fig. 2). Again, the resultant is oriented to the left of the normal such that its friction component is directed opposite the direction of motion.

$$P = (800 \text{ N}) \tan 39.04° \qquad \textbf{P = 649 N} \leftarrow \blacktriangleleft$$

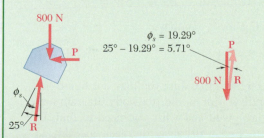

$$\phi_s = 19.29°$$
$$25° - 19.29° = 5.71°$$

Fig. 3 Free-body diagram of block and its force triangle—motion prevented down the slope.

c. Force P to Prevent Block from Sliding Down. Here, motion is impending down the incline, so the resultant is directed at the angle of static friction (Fig. 3). Note that the resultant is oriented to the right of the normal such that its friction component is directed opposite the direction of impending motion.

$$P = (800 \text{ N}) \tan 5.71° \qquad \textbf{P = 80.0 N} \leftarrow \blacktriangleleft$$

REFLECT and THINK: As expected, considerably more force is required to begin moving the block up the slope than is necessary to restrain it from sliding down the slope.

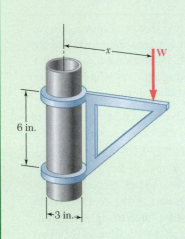

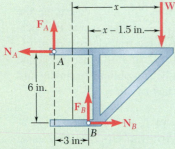

Fig. 1 Free-body diagram of bracket.

Sample Problem 8.3

The movable bracket shown may be placed at any height on the 3-in.-diameter pipe. If the coefficient of static friction between the pipe and bracket is 0.25, determine the minimum distance x at which the load **W** can be supported. Neglect the weight of the bracket.

STRATEGY: In this variation of the third type of friction problem, you know the coefficient of static friction and that motion is impending. Since the problem involves consideration of resistance to rotation, you should apply both moment equilibrium and force equilibrium.

MODELING:

Free-Body Diagram.   Draw the free-body diagram of the bracket (Fig. 1). When **W** is placed at the minimum distance x from the axis of the pipe, the bracket is just about to slip, and the forces of friction at A and B have reached their maximum values:

$$F_A = \mu_s N_A = 0.25\, N_A$$
$$F_B = \mu_s N_B = 0.25\, N_B$$

ANALYSIS:

Equilibrium Equations.

$$\xrightarrow{+} \Sigma F_x = 0: \qquad\qquad N_B - N_A = 0$$
$$N_B = N_A$$

$$+\uparrow \Sigma F_y = 0: \qquad F_A + F_B - W = 0$$
$$0.25 N_A + 0.25 N_B = W$$

Since N_B is equal to N_A,

$$0.50 N_A = W$$
$$N_A = 2W$$

$$+\,\curvearrowleft \Sigma M_B = 0: \qquad N_A(6\text{ in.}) - F_A(3\text{ in.}) - W(x - 1.5\text{ in.}) = 0$$
$$6 N_A - 3(0.25 N_A) - Wx + 1.5W = 0$$
$$6(2W) - 0.75(2W) - Wx + 1.5W = 0$$

Dividing through by W and solving for x, you have

$$x = 12\text{ in.} \quad \blacktriangleleft$$

REFLECT and THINK: In a problem like this, you may not figure out how to approach the solution until you draw the free-body diagram and examine what information you are given and what you need to find. In this case, since you are asked to find a distance, the need to evaluate moment equilibrium should be clear.

Sample Problem 8.4

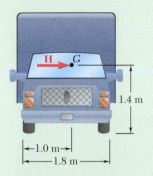

An 8400-kg truck is traveling on a level horizontal curve, resulting in an effective lateral force **H** (applied at the center of gravity G of the truck). Treating the truck as a rigid system with the center of gravity shown, and knowing that the distance between the outer edges of the tires is 1.8 m, determine (*a*) the maximum force **H** before tipping of the truck occurs, (*b*) the minimum coefficient of static friction between the tires and roadway such that slipping does not occur before tipping.

STRATEGY: For the direction of **H** shown, the truck would tip about the outer edge of the right tire. At the verge of tip, the normal force and friction force are zero at the left tire, and the normal force at the right tire is at the outer edge. You can apply equilibrium to determine the value of **H** necessary for tip and the required friction force such that slipping does not occur.

MODELING: Draw the free-body diagram of the truck (Fig. 1), which reflects impending tip about point B. Obtain the weight of the truck by multiplying its mass of 8400 kg by $g = 9.81$ m/s²; that is, $W = 82\ 400$ N or 82.4 kN.

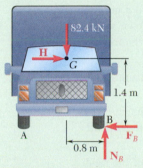

Fig. 1 Free-body diagram of truck.

ANALYSIS:

Free Body: Truck (Fig. 1).

$$+\curvearrowright \Sigma M_B = 0: \qquad (82.4\text{ kN})(0.8\text{ m}) - H(1.4\text{ m}) = 0$$
$$H = +47.1\text{ kN} \qquad\qquad H = 47.1\text{ kN} \rightarrow \quad \blacktriangleleft$$

$$\xrightarrow{+}\Sigma F_x = 0: \qquad 47.1\text{ kN} - F_B = 0$$
$$F_B = +47.1\text{ kN}$$

$$+\uparrow \Sigma F_y = 0: \qquad N_B - 82.4\text{ kN} = 0$$
$$N_B = +82.4\text{ kN}$$

Minimum Coefficient of Static Friction.   The magnitude of the maximum friction force that can be developed is

$$F_m = \mu_s N_B = \mu_s\ (82.4\text{ kN})$$

Setting this equal to the friction force required, $F_B = 47.1$ kN, gives

$$\mu_s\ (82.4\text{ kN}) = 47.1\text{ kN} \qquad \mu_s = 0.572 \quad \blacktriangleleft$$

REFLECT and THINK: Recall from physics that **H** represents the force due to the centripetal acceleration of the truck (of mass m), and its magnitude is

$$H = m(v^2/\rho)$$

where
> v = velocity of the truck
> ρ = radius of curvature

In this problem, if the truck were traveling around a curve of 100-m radius (measured to G), the velocity at which it would begin to tip would be 23.7 m/s (or 85.2 km/h). You will learn more about this aspect in your study of dynamics.

SOLVING PROBLEMS
ON YOUR OWN

In this section, you studied and applied the **laws of dry friction**. Previously, you had encountered only (*a*) frictionless surfaces that could move freely with respect to each other or (*b*) rough surfaces that allowed no motion relative to each other.

A. In solving problems involving dry friction, keep the following ideas in mind.

1. The reaction R exerted by a surface on a free body can be resolved into a normal component **N** and a tangential component **F**. The tangential component is known as the **friction force**. When a body is in contact with a fixed surface, the direction of the friction force **F** is opposite to that of the actual or impending motion of the body.

    a. No motion will occur as long as F does not exceed the maximum value $F_m = \mu_s N$, where μ_s is the **coefficient of static friction**.

    b. Motion will occur if a value of F larger than F_m is required to maintain equilibrium. As motion takes place, the actual value of F drops to $F_k = \mu_k N$, where μ_k is the **coefficient of kinetic friction** [Sample Prob. 8.1].

    c. Motion may also occur at a value of F smaller than F_m if tipping of the rigid body is a possibility [Sample Prob. 8.4}

2. When only three forces are involved, you might prefer an alternative approach to the analysis of friction [Sample Prob. 8.2]. The reaction **R** is defined by its magnitude R and the angle ϕ it forms with the normal to the surface. No motion occurs as long as ϕ does not exceed the maximum value ϕ_s, where $\tan \phi_s = \mu_s$. Motion does occur if a value of ϕ larger than ϕ_s is required to maintain equilibrium, and the actual value of ϕ drops to ϕ_k, where $\tan \phi_k = \mu_k$.

3. When two bodies are in contact, you must determine the sense of the actual or impending relative motion at the point of contact. On each of the two bodies, a friction force **F** is in a direction opposite to that of the actual or impending motion of the body as seen from the other body (see Fig. 8.6).

B. Methods of solution. The first step in your solution is to draw a free-body diagram of the body under consideration, resolving the force exerted on each surface where friction exists into a normal component **N** and a friction force **F**. If several bodies are involved, draw a free-body diagram for each of them, labeling and directing the forces at each surface of contact, as described for analyzing frames in Chap. 6.

The problem you have to solve may fall in one of the following three categories.

1. You know all the applied forces and the coefficients of friction, and you must determine whether equilibrium is maintained. In this situation, the friction force is unknown and *cannot be assumed to be equal* to $\mu_s N$.

    a. Write the equations of equilibrium to determine N and F.

    b. Calculate the maximum allowable friction force, $F_m = \mu_s N$. If $F \leq F_m$, equilibrium is maintained. If $F \geq F_m$, motion occurs, and the magnitude of the friction force is $F_k = \mu_k N$ [Sample Prob. 8.1].

2. You know all the applied forces, and you must find the smallest allowable value of μ_s for which equilibrium is maintained. Assume that motion is impending, and determine the corresponding value of μ_s.

    a. Write the equations of equilibrium to determine N and F.

    b. Since motion is impending, $F = F_m$. Substitute the values found for N and F into the equation $F_m = \mu_s N$ and solve for μ_s [Sample Prob. 8.4].

3. The motion of the body is impending and μ_s is known; you must find some unknown quantity, such as a distance, an angle, the magnitude of a force, or the direction of a force.

    a. Assume a possible motion of the body and, on the free-body diagram, draw the friction force in a direction opposite to that of the assumed motion.

    b. Since motion is impending, $F = F_m = \mu_s N$. Substituting the known value for μ_s, you can express F in terms of N on the free-body diagram, thus eliminating one unknown.

    c. Write and solve the equilibrium equations for the unknown you seek [Sample Prob. 8.3].

Problems

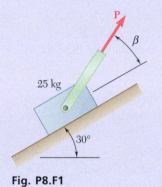

Fig. P8.F1

FREE-BODY PRACTICE PROBLEMS

8.F1 Knowing that the coefficient of friction between the 25-kg block and the incline is $\mu_s = 0.25$, draw the free-body diagram needed to determine both the smallest value of P required to start the block moving up the incline and the corresponding value of β.

8.F2 Two blocks A and B are connected by a cable as shown. Knowing that the coefficient of static friction at all surfaces of contact is 0.30 and neglecting the friction of the pulleys, draw the free-body diagrams needed to determine the smallest force **P** required to move the blocks.

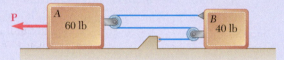

Fig. P8.F2

8.F3 A cord is attached to and partially wound around a cylinder with a weight of W and radius r that rests on an incline as shown. Knowing that $\theta = 30°$, draw the free-body diagram needed to determine both the tension in the cord and the smallest allowable value of the coefficient of static friction between the cylinder and the incline for which equilibrium is maintained.

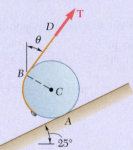

Fig. P8.F3

8.F4 A 40-kg packing crate must be moved to the left along the floor without tipping. Knowing that the coefficient of static friction between the crate and the floor is 0.35, draw the free-body diagram needed to determine both the largest allowable value of α and the corresponding magnitude of the force **P**.

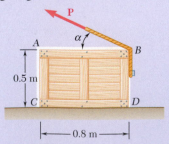

Fig. P8.F4

8.1 Determine whether the block shown is in equilibrium, and find the magnitude and direction of the friction force when $P = 150$ N.

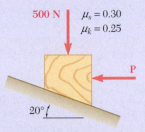

500 N $\mu_s = 0.30$
$\mu_k = 0.25$

P

20°

Fig. P8.1 and P8.2

8.2 Determine whether the block shown is in equilibrium, and find the magnitude and direction of the friction force when $P = 400$ N.

8.3 Determine whether the block shown is in equilibrium, and find the magnitude and direction of the friction force when $P = 120$ lb.

8.4 Determine whether the block shown is in equilibrium, and find the magnitude and direction of the friction force when $P = 80$ lb.

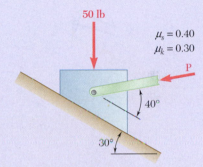

50 lb $\mu_s = 0.40$
$\mu_k = 0.30$

P

40°

30°

Fig. P8.3, P8.4, and P8.5

8.5 Determine the smallest value of P required to (*a*) start the block up the incline, (*b*) keep it moving up.

8.6 The 20-lb block A hangs from a cable as shown. Pulley C is connected by a short link to block E, which rests on a horizontal rail. Knowing that the coefficient of static friction between block E and the rail is $\mu_s = 0.35$ and neglecting the weight of block E and the friction in the pulleys, determine the maximum allowable value of θ if the system is to remain in equilibrium.

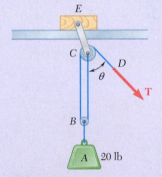

E

C   D

θ

T

B

A   20 lb

Fig. P8.6

8.7 The 10-kg block is attached to link AB and rests on a moving belt. Knowing that $\mu_s = 0.30$ and $\mu_k = 0.25$ and neglecting the weight of the link, determine the magnitude of the horizontal force **P** that should be applied to the belt to maintain its motion (*a*) to the left as shown, (*b*) to the right.

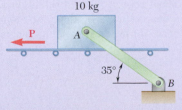

10 kg

P   A

35°   B

Fig. P8.7

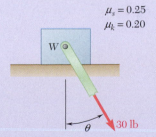

$\mu_s = 0.25$
$\mu_k = 0.20$

W

θ 30 lb

Fig. P8.8

8.8 Considering only values of θ less than 90°, determine the smallest value of θ required to start the block moving to the right when (*a*) $W = 75$ lb, (*b*) $W = 100$ lb.

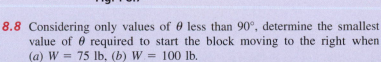

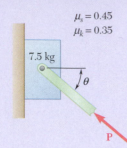

$\mu_s = 0.45$
$\mu_k = 0.35$

7.5 kg

θ

P

Fig. P8.9 and P8.10

8.9 Knowing that $\theta = 40°$, determine the smallest force **P** for which equilibrium of the 7.5-kg block is maintained.

8.10 Knowing that $P = 100$ N, determine the range of values of θ for which equilibrium of the 7.5-kg block is maintained.

8.11 The 50-lb block A and the 25-lb block B are supported by an incline that is held in the position shown. Knowing that the coefficient of static friction is 0.15 between the two blocks and zero between block B and the incline, determine the value of θ for which motion is impending.

8.12 The 50-lb block A and the 25-lb block B are supported by an incline that is held in the position shown. Knowing that the coefficient of static friction is 0.15 between all surfaces of contact, determine the value of θ for which motion is impending.

8.13 Three 4-kg packages A, B, and C are placed on a conveyor belt that is at rest. Between the belt and both packages A and C, the coefficients of friction are $\mu_s = 0.30$ and $\mu_k = 0.20$; between package B and the belt, the coefficients are $\mu_s = 0.10$ and $\mu_k = 0.08$. The packages are placed on the belt so that they are in contact with each other and at rest. Determine which, if any, of the packages will move and the friction force acting on each package.

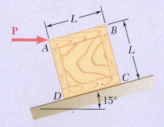

Fig. P8.11 and P8.12

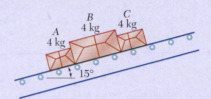

Fig. P8.13

8.14 Solve Prob. 8.13 assuming that package B is placed to the right of both packages A and C.

8.15 A uniform crate with a mass of 30 kg must be moved up along the 15° incline without tipping. Knowing that force **P** is horizontal, determine (*a*) the largest allowable coefficient of static friction between the crate and the incline, (*b*) the corresponding magnitude of force **P**.

8.16 A worker slowly moves a 50-kg crate to the left along a loading dock by applying a force **P** at corner B as shown. Knowing that the crate starts to tip about edge E of the loading dock when $a = 200$ mm, determine (*a*) the coefficient of kinetic friction between the crate and the loading dock, (*b*) the corresponding magnitude P of the force.

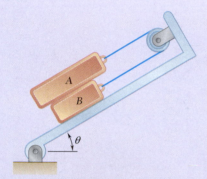

Fig. P8.15

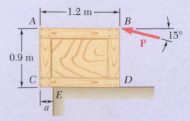

Fig. P8.16

8.17 A half-section of pipe weighing 200 lb is pulled by a cable as shown. The coefficient of static friction between the pipe and the floor is 0.40. If $\alpha = 30°$, determine (a) the tension T required to move the pipe, (b) whether the pipe will slide or tip.

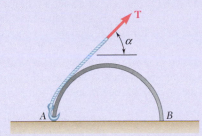

Fig. P8.17

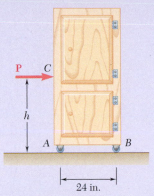

Fig. P8.18

8.18 A 120-lb cabinet is mounted on casters that can be locked to prevent their rotation. The coefficient of static friction between the floor and each caster is 0.30. Assuming that the casters at both A and B are locked, determine (a) the force **P** required to move the cabinet to the right, (b) the largest allowable value of h if the cabinet is not to tip over.

8.19 Wire is being drawn at a constant rate from a spool by applying a vertical force **P** to the wire as shown. The spool and the wire wrapped on the spool have a combined weight of 20 lb. Knowing that the coefficients of friction at both A and B are $\mu_s = 0.40$ and $\mu_k = 0.30$, determine the required magnitude of force **P**.

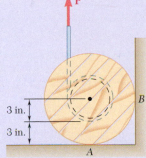

Fig. P8.19

8.20 Solve Prob. 8.19 assuming that the coefficients of friction at B are zero.

8.21 The cylinder shown has a weight W and radius r. Express in terms of W and r the magnitude of the largest couple **M** that can be applied to the cylinder if it is not to rotate, assuming the coefficient of static friction to be (a) zero at A and 0.30 at B, (b) 0.25 at A and 0.30 at B.

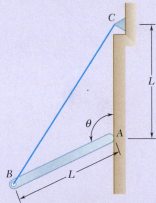

Fig. P8.23

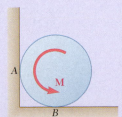

Fig. P8.21 and P.22

8.22 The cylinder shown has a weight W and radius r, and the coefficient of static friction μ_s is the same at A and B. Determine the magnitude of the largest couple **M** that can be applied to the cylinder if it is not to rotate.

8.23 and 8.24 End A of a slender, uniform rod with a length of L and weight W bears on a surface as shown, while end B is supported by a cord BC. Knowing that the coefficients of friction are $\mu_s = 0.40$ and $\mu_k = 0.30$, determine (a) the largest value of θ for which motion is impending, (b) the corresponding value of the tension in the cord.

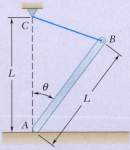

Fig. P8.24

445

Fig. P8.25 and *P8.26*

8.25 A 6.5-m ladder *AB* leans against a wall as shown. Assuming that the coefficient of static friction μ_s is zero at *B*, determine the smallest value of μ_s at *A* for which equilibrium is maintained.

8.26 A 6.5-m ladder *AB* leans against a wall as shown. Assuming that the coefficient of static friction μ_s is the same at *A* and *B*, determine the smallest value of μ_s for which equilibrium is maintained.

8.27 The press shown is used to emboss a small seal at *E*. Knowing that the coefficient of static friction between the vertical guide and the embossing die *D* is 0.30, determine the force exerted by the die on the seal.

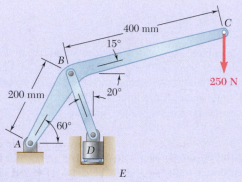

Fig. P8.27

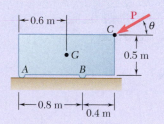

Fig. *P8.28*

8.28 The machine base shown has a mass of 75 kg and is fitted with skids at *A* and *B*. The coefficient of static friction between the skids and the floor is 0.30. If a force **P** with a magnitude of 500 N is applied at corner *C*, determine the range of values of θ for which the base will not move.

8.29 The 50-lb plate *ABCD* is attached at *A* and *D* to collars that can slide on the vertical rod. Knowing that the coefficient of static friction is 0.40 between both collars and the rod, determine whether the plate is in equilibrium in the position shown when the magnitude of the vertical force applied at *E* is (*a*) *P* = 0, (*b*) *P* = 20 lb.

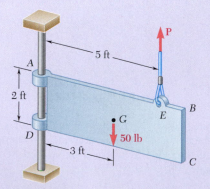

Fig. P8.29 and P8.30

8.30 In Prob. 8.29, determine the range of values of the magnitude *P* of the vertical force applied at *E* for which the plate will move downward.

8.31 A window sash weighing 10 lb is normally supported by two 5-lb sash weights. Knowing that the window remains open after one sash cord has broken, determine the smallest possible value of the coefficient of static friction. (Assume that the sash is slightly smaller than the frame and will bind only at points *A* and *D*.)

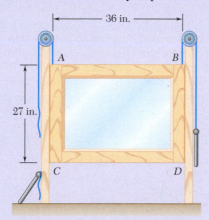

Fig. P8.31

8.32 A 500-N concrete block is to be lifted by the pair of tongs shown. Determine the smallest allowable value of the coefficient of static friction between the block and the tongs at *F* and *G*.

8.33 A pipe with a diameter of 60 mm is gripped by the stillson wrench shown. Portions *AB* and *DE* of the wrench are rigidly attached to each other, and portion *CF* is connected by a pin at *D*. If the wrench is to grip the pipe and be self-locking, determine the required minimum coefficients of friction at *A* and *C*.

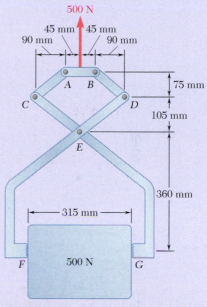

Fig. P8.32

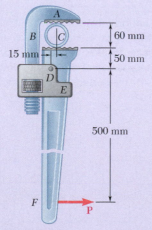

Fig. P8.33

8.34 A safety device used by workers climbing ladders fixed to high structures consists of a rail attached to the ladder and a sleeve that can slide on the flange of the rail. A chain connects the worker's belt to the end of an eccentric cam that can be rotated about an axle attached to the sleeve at *C*. Determine the smallest allowable common value of the coefficient of static friction between the flange of the rail, the pins at *A* and *B*, and the eccentric cam if the sleeve is not to slide down when the chain is pulled vertically downward.

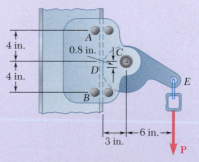

Fig. P8.34

447

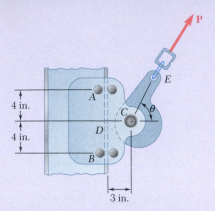

Fig. P8.35

8.35 To be of practical use, the safety sleeve described in Prob. 8.34 must be free to slide along the rail when pulled upward. Determine the largest allowable value of the coefficient of static friction between the flange of the rail and the pins at A and B if the sleeve is to be free to slide when pulled as shown in the figure. Assume (*a*) $\theta = 60°$, (*b*) $\theta = 50°$, (*c*) $\theta = 40°$.

8.36 Two 10-lb blocks A and B are connected by a slender rod of negligible weight. The coefficient of static friction is 0.30 between all surfaces of contact, and the rod forms an angle $\theta = 30°$ with the vertical. (*a*) Show that the system is in equilibrium when $P = 0$. (*b*) Determine the largest value of P for which equilibrium is maintained.

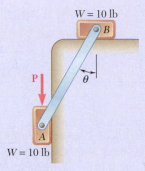

Fig. P8.36

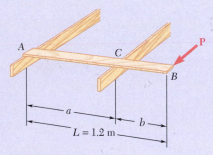

Fig. P8.37

8.37 A 1.2-m plank with a mass of 3 kg rests on two joists. Knowing that the coefficient of static friction between the plank and the joists is 0.30, determine the magnitude of the horizontal force required to move the plank when (*a*) $a = 750$ mm, (*b*) $a = 900$ mm.

8.38 Two identical uniform boards, each with a weight of 40 lb, are temporarily leaned against each other as shown. Knowing that the coefficient of static friction between all surfaces is 0.40, determine (*a*) the largest magnitude of the force **P** for which equilibrium will be maintained, (*b*) the surface at which motion will impend.

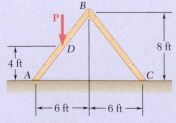

Fig. *P8.38*

8.39 Two rods are connected by a collar at B. A couple $\mathbf{M}_A$ with a magnitude of 15 N·m is applied to rod AB. Knowing that the coefficient of static friction between the collar and the rod is 0.30, determine the largest couple $\mathbf{M}_C$ for which equilibrium will be maintained.

8.40 In Prob. 8.39, determine the smallest couple $\mathbf{M}_C$ for which equilibrium will be maintained.

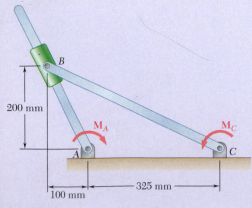

Fig. P8.39

8.41 A 10-ft beam, weighing 1200 lb, is to be moved to the left onto the platform as shown. A horizontal force **P** is applied to the dolly, which is mounted on frictionless wheels. The coefficients of friction between all surfaces are $\mu_s = 0.30$ and $\mu_k = 0.25$, and initially, $x = 2$ ft. Knowing that the top surface of the dolly is slightly higher than the platform, determine the force **P** required to start moving the beam. (*Hint:* The beam is supported at A and D.)

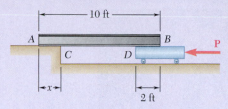

Fig. P8.41

8.42 (*a*) Show that the beam of Prob. 8.41 *cannot* be moved if the top surface of the dolly is slightly *lower* than the platform. (*b*) Show that the beam *can* be moved if two 175-lb workers stand on the beam at B, and determine how far to the left the beam can be moved.

8.43 Two 8-kg blocks A and B resting on shelves are connected by a rod of negligible mass. Knowing that the magnitude of a horizontal force **P** applied at C is slowly increased from zero, determine the value of P for which motion occurs and what that motion is when the coefficient of static friction between all surfaces is (*a*) $\mu_s = 0.40$, (*b*) $\mu_s = 0.50$.

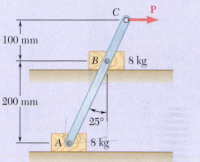

Fig. P8.43

8.44 A slender steel rod with a length of 225 mm is placed inside a pipe as shown. Knowing that the coefficient of static friction between the rod and the pipe is 0.20, determine the largest value of θ for which the rod will not fall into the pipe.

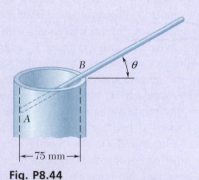

Fig. P8.44

8.45 In Prob. 8.44, determine the smallest value of θ for which the rod will not fall out of the pipe.

8.46 Two slender rods of negligible weight are pin-connected at C and attached to blocks A and B, each with a weight W. Knowing that $\theta = 80°$ and that the coefficient of static friction between the blocks and the horizontal surface is 0.30, determine the largest value of P for which equilibrium is maintained.

8.47 Two slender rods of negligible weight are pin-connected at C and attached to blocks A and B, each with a weight W. Knowing that $P = 1.260W$ and that the coefficient of static friction between the blocks and the horizontal surface is 0.30, determine the range of values of θ between 0 and 180° for which equilibrium is maintained.

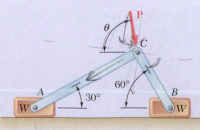

Fig. P8.46 and P8.47

Photo 8.3 Wedges are used as shown to split tree trunks because the normal forces exerted by a wedge on the wood are much larger than the force required to insert the wedge.

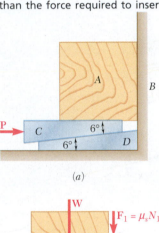

(a)

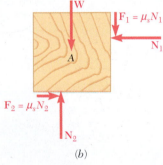

(b)

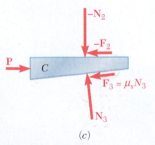

(c)

Fig. 8.7 (a) A wedge C used to raise a block A; (b) free-body diagram of block A; (c) free-body diagram of wedge C. Note the directions of the friction forces.

8.2 WEDGES AND SCREWS

Friction is a key element in analyzing the function and operation of several types of simple machines. Here we examine the wedge and the screw, which are both extensions of the inclined plane we analyzed in Sect. 8.1.

8.2A Wedges

Wedges are simple machines used to raise large stone blocks and other heavy loads. These loads are raised by applying to the wedge a force usually considerably smaller than the weight of the load. In addition, because of the friction between the surfaces in contact, a properly shaped wedge remains in place after being forced under the load. In this way, you can use a wedge advantageously to make small adjustments in the position of heavy pieces of machinery.

Consider the block A shown in Fig. 8.7a. This block rests against a vertical wall B, and we want to raise it slightly by forcing a wedge C between block A and a second wedge D. We want to find the minimum value of the force **P** that we must apply to wedge C to move the block. We assume that we know the weight **W** of the block, which is either given in pounds or determined in newtons from the mass of the block expressed in kilograms.

We have drawn the free-body diagrams of block A and wedge C in Fig. 8.7b and c. The forces acting on the block include its weight and the normal and friction forces at the surfaces of contact with wall B and wedge C. The magnitudes of the friction forces $\mathbf{F}_1$ and $\mathbf{F}_2$ are equal, respectively, to $\mu_s N_1$ and $\mu_s N_2$, because the motion of the block must be started. It is important to show the friction forces with their correct sense. Since the block will move upward, the force $\mathbf{F}_1$ exerted by the wall on the block must be directed downward. On the other hand, since wedge C moves to the right, the relative motion of A with respect to C is to the left, and the force $\mathbf{F}_2$ exerted by C on A must be directed to the right.

Now consider the free body C in Fig. 8.7c. The forces acting on C include the applied force **P** and the normal and friction forces at the surfaces of contact with A and D. The weight of the wedge is small compared with the other forces involved and can be neglected. The forces exerted by A on C are equal and opposite to the forces $\mathbf{N}_2$ and $\mathbf{F}_2$ exerted by C on A, so we denote them, respectively, by $-\mathbf{N}_2$ and $-\mathbf{F}_2$; the friction force $-\mathbf{F}_2$ therefore must be directed to the left. We check that the force $\mathbf{F}_3$ exerted by D is also directed to the left.

We can reduce the total number of unknowns involved in the two free-body diagrams to four if we express the friction forces in terms of the normal forces. Then, since block A and wedge C are in equilibrium, we obtain four equations that we can solve to obtain the magnitude of **P**. Note that, in the example considered here, it is more convenient to replace each pair of normal and friction forces by their resultant. Each free body is then subjected to only three forces, and we can solve the problem by drawing the corresponding force triangles (see Sample Prob. 8.5).

8.2B Square-Threaded Screws

Square-threaded screws are frequently part of jacks, presses, and other mechanisms. Their analysis is similar to the analysis of a block sliding along an inclined plane. (Screws are also commonly used as fasteners, but the threads on these screws are shaped differently.)

Consider the jack shown in Fig. 8.8. The screw carries a load **W** and is supported by the base of the jack. Contact between the screw and the base takes place along a portion of their threads. By applying a force **P** on the handle, the screw can be made to turn and to raise the load **W**.

In Fig. 8.9a, we have unwrapped the thread of the base and shown it as a straight line. We obtained the correct slope by horizontally drawing the product $2\pi r$, where r is the mean radius of the thread, and vertically drawing the **lead** L of the screw, i.e., the distance through which the screw advances in one turn. The angle θ this line forms with the horizontal is the **lead angle**. Since the force of friction between two surfaces in contact does not depend upon the area of contact, we can assume a much smaller than actual area of contact between the two threads, which allows us to represent the screw as the block shown in Fig. 8.9a. Note that, in this analysis of the jack, we neglect the small friction force between cap and screw.

The free-body diagram of the block includes the load **W**, the reaction **R** of the base thread, and a horizontal force **Q**, which has the same effect as the force **P** exerted on the handle. The force **Q** should have the same moment as **P** about the axis of the screw, so its magnitude should be $Q = Pa/r$. We can obtain the value of force **Q**, and thus that of force **P** required to raise load **W**, from the free-body diagram shown in Fig. 8.9a. The friction angle is taken to be equal to ϕ_s, since presumably the load is raised through a succession of short strokes. In mechanisms providing for the continuous rotation of a screw, it may be desirable to distinguish between the force required to start motion (using ϕ_s) and that required to maintain motion (using ϕ_k).

If the friction angle ϕ_s is larger than the lead angle θ, the screw is said to be *self-locking*; it will remain in place under the load. To lower the load, we must then apply the force shown in Fig. 8.9b. If ϕ_s is smaller than θ, the screw will unwind under the load; it is then necessary to apply the force shown in Fig. 8.9c to maintain equilibrium.

The lead of a screw should not be confused with its **pitch**. The *lead* is defined as the distance through which the screw advances in one turn; the *pitch* is the distance measured between two consecutive threads. Lead and pitch are equal in the case of *single-threaded* screws, but they are different in the case of *multiple-threaded* screws, i.e., screws having several independent threads. It is easily verified that for double-threaded screws the lead is twice as large as the pitch; for triple-threaded screws, it is three times as large as the pitch; etc.

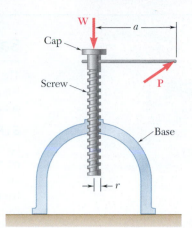

Fig. 8.8 A screw as part of a jack carrying a load **W**.

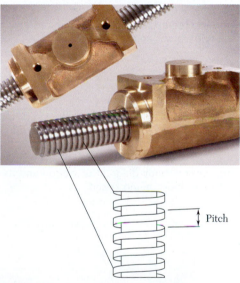

Photo 8.4 An example of a square-threaded screw, fitted to a sleeve, as might be used in an industrial application.

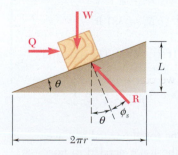

(a) Impending motion upward

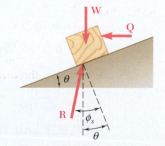

(b) Impending motion downward with $\phi_s > \theta$

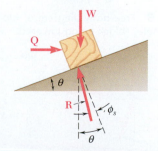

(c) Impending motion downward with $\phi_s < \theta$

Fig. 8.9 Block-and-incline analysis of a screw. We can represent the screw as a block, because the force of friction does not depend on the area of contact between two surfaces.

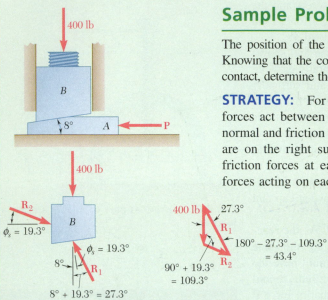

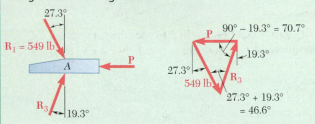

Fig. 1 Free-body diagram of block and its force triangle—block being raised.

Fig. 2 Free-body diagram of wedge and its force triangle—block being raised.

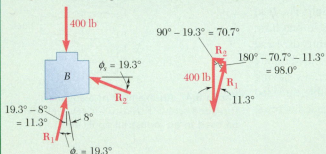

Fig. 3 Free-body diagram of block and its force triangle—block being lowered.

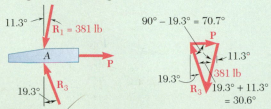

Fig. 4 Free-body diagram of wedge and its force triangle—block being lowered.

Sample Problem 8.5

The position of the machine block B is adjusted by moving the wedge A. Knowing that the coefficient of static friction is 0.35 between all surfaces of contact, determine the force $\mathbf{P}$ required to (a) raise block B, (b) lower block B.

STRATEGY: For both parts of the problem, normal forces and friction forces act between the wedge and the block. In part (a), you also have normal and friction forces at the left surface of the block; for part (b), they are on the right surface of the block. If you combine the normal and friction forces at each surface into resultants, you have a total of three forces acting on each body and can use force triangles to solve.

MODELING: For each part, draw the free-body diagrams of block B and wedge A together with the corresponding force triangles. Then use the law of sines to find the desired forces. Note that, since $\mu_s = 0.35$, the angle of friction is

$$\phi_s = \tan^{-1} 0.35 = 19.3°$$

ANALYSIS: a. Force P to raise block

Free Body: Block B (Fig. 1). The friction force on block B due to wedge A is to the left, so the resultant $\mathbf{R}_1$ is at an angle equal to the slope of the wedge plus the angle of friction.

$$\frac{R_1}{\sin 109.3°} = \frac{400 \text{ lb}}{\sin 43.4°} \qquad R_1 = 549 \text{ lb}$$

Free Body: Wedge A (Fig. 2). The friction forces on wedge A are to the right.

$$\frac{P}{\sin 46.6°} = \frac{549 \text{ lb}}{\sin 70.7°} \qquad \mathbf{P = 423 \text{ lb}} \leftarrow \quad ◄$$

b. Force P to lower block

Free Body: Block B (Fig. 3). Now the friction force on block B due to wedge A is to the right, so the resultant $\mathbf{R}_1$ is at an angle equal to the angle of friction minus the slope of the wedge.

$$\frac{R_1}{\sin 70.7°} = \frac{400 \text{ lb}}{\sin 98.0°} \qquad R_1 = 381 \text{ lb}$$

Free Body: Wedge A (Fig. 4). The friction forces on wedge A are to the left.

$$\frac{P}{\sin 30.6°} = \frac{381 \text{ lb}}{\sin 70.7°} \qquad \mathbf{P = 206 \text{ lb}} \rightarrow \quad ◄$$

REFLECT and THINK: The force needed to lower the block is much less than the force needed to raise the block, which makes sense.

Sample Problem 8.6

A clamp is used to hold two pieces of wood together as shown. The clamp has a double square thread with a mean diameter of 10 mm and a pitch of 2 mm. The coefficient of friction between threads is $\mu_s = 0.30$. If a maximum couple of 40 N·m is applied in tightening the clamp, determine (a) the force exerted on the pieces of wood, (b) the couple required to loosen the clamp.

STRATEGY: If you represent the screw by a block, as in the analysis of this section, you can determine the incline of the screw from the geometry given in the problem, and you can find the force applied to the block by setting the moment of that force equal to the applied couple.

MODELING and ANALYSIS:

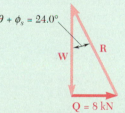

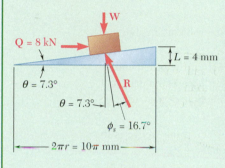

Fig. 1 Free-body diagram of block and its force triangle—clamp being tightened.

a. Force Exerted by Clamp.

The mean radius of the screw is $r = 5$ mm. Since the screw is double-threaded, the lead L is equal to twice the pitch: $L = 2(2$ mm$) = 4$ mm. Obtain the lead angle θ and the friction angle ϕ_s from

$$\tan \theta = \frac{L}{2\pi r} = \frac{4 \text{ mm}}{10\pi \text{ mm}} = 0.1273 \qquad \theta = 7.3°$$

$$\tan \phi_s = \mu_s = 0.30 \qquad\qquad \phi_s = 16.7°$$

You can find the force **Q** that should be applied to the block representing the screw by setting its moment Qr about the axis of the screw equal to the applied couple.

$$Q(5 \text{ mm}) = 40 \text{ N·m}$$

$$Q = \frac{40 \text{ N·m}}{5 \text{ mm}} = \frac{40 \text{ N·m}}{5 \times 10^{-3} \text{ m}} = 8000 \text{ N} = 8 \text{ kN}$$

Now you can draw the free-body diagram and the corresponding force triangle for the block (Fig. 1). Solve the triangle to find the magnitude of the force **W** exerted on the pieces of wood.

$$W = \frac{Q}{\tan(\theta + \phi_s)} = \frac{8 \text{ kN}}{\tan 24.0°}$$

$$W = 17.97 \text{ kN} \quad \blacktriangleleft$$

b. Couple Required to Loosen Clamp.

You can obtain the force **Q** required to loosen the clamp and the corresponding couple from the free-body diagram and force triangle shown in Fig. 2.

$$Q = W \tan (\phi_s - \theta) = (17.97 \text{ kN}) \tan 9.4°$$
$$= 2.975 \text{ kN}$$

$$\text{Couple} = Qr = (2.975 \text{ kN})(5 \text{ mm})$$
$$= (2.975 \times 10^3 \text{ N})(5 \times 10^{-3} \text{ m}) = 14.87 \text{ N·m}$$

$$\text{Couple} = 14.87 \text{ N·m} \quad \blacktriangleleft$$

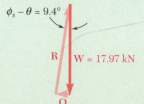

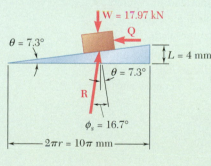

Fig. 2 Free-body diagram of block and its force triangle—clamp being loosened.

REFLECT and THINK: In practice, you often have to determine the force effectively acting on a screw by setting the moment of that force about the axis of the screw equal to an applied couple. However, the rest of the analysis is mostly an application of dry friction. Also note that the couple required to loosen a screw is not the same as the couple required to tighten it.

SOLVING PROBLEMS
ON YOUR OWN

In this section, you saw how to apply the laws of friction to the solution of problems involving **wedges** and **square-threaded screws**.

1. Wedges. Keep the following steps in mind when solving a problem involving a wedge.

   a.   First draw a free-body diagram of the wedge and of each of the other bodies involved. Carefully note the sense of the relative motion of all surfaces of contact and show each friction force acting in *a direction opposite* to the direction of that relative motion.

   b.   Show the maximum static friction force F_m at each surface if the wedge is to be inserted or removed, since motion will be impending in each of these cases.

   c.   The reaction R and the angle of friction, rather than the normal force and the friction force, are most useful in many applications. You can then draw one or more force triangles and determine the unknown quantities either graphically or by trigonometry [Sample Prob. 8.5].

2. Square-Threaded Screws.   The analysis of a square-threaded screw is equivalent to the analysis of a block sliding on an incline. To draw the appropriate incline, you need to unwrap the thread of the screw and represent it as a straight line [Sample Prob. 8.6]. When solving a problem involving a square-threaded screw, keep the following steps in mind.

   a.   Do not confuse the pitch of a screw with the lead of a screw. The **pitch** of a screw is the distance between two consecutive threads, whereas the **lead** of a screw is the distance the screw advances in one full turn. The lead and the pitch are equal only in single-threaded screws. In a double-threaded screw, the lead is twice the pitch.

   b.   The couple required to tighten a screw is different from the couple required to loosen it. Also, screws used in jacks and clamps are usually *self-locking;* that is, the screw will remain stationary as long as no couple is applied to it, and a couple must be applied to the screw to loosen it [Sample Prob. 8.6].

Problems

8.48 The machine part *ABC* is supported by a frictionless hinge at *B* and a 10° wedge at *C*. Knowing that the coefficient of static friction is 0.20 at both surfaces of the wedge, determine (*a*) the force **P** required to move the wedge to the left, (*b*) the components of the corresponding reaction at *B*.

8.49 Solve Prob. 8.48 assuming that the wedge is moved to the right.

8.50 and 8.51 Two 8° wedges of negligible weight are used to move and position the 800-kg block. Knowing that the coefficient of static friction is 0.30 at all surfaces of contact, determine the smallest force **P** that should be applied as shown to one of the wedges.

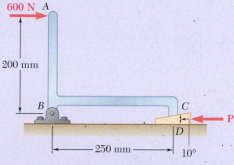

Fig. P8.48 and P8.49

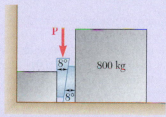

Fig. P8.50

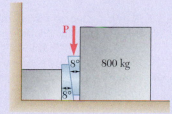

Fig. P8.51

8.52 The elevation of the end of the steel beam supported by a concrete floor is adjusted by means of the steel wedges *E* and *F*. The base plate *CD* has been welded to the lower flange of the beam, and the end reaction of the beam is known to be 18 kips. The coefficient of static friction is 0.30 between two steel surfaces and 0.60 between steel and concrete. If horizontal motion of the beam is prevented by the force **Q**, determine (*a*) the force **P** required to raise the beam, (*b*) the corresponding force **Q**.

8.53 Solve Prob. 8.52 assuming that the end of the beam is to be lowered.

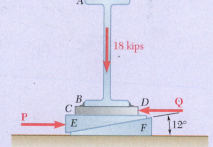

Fig. P8.52

8.54 Block *A* supports a pipe column and rests as shown on wedge *B*. Knowing that the coefficient of static friction at all surfaces of contact is 0.25 and that *θ* = 45°, determine the smallest force **P** required to raise block *A*.

8.55 Block *A* supports a pipe column and rests as shown on wedge *B*. Knowing that the coefficient of static friction at all surfaces of contact is 0.25 and that *θ* = 45°, determine the smallest force **P** for which equilibrium is maintained.

8.56 Block *A* supports a pipe column and rests as shown on wedge *B*. The coefficient of static friction at all surfaces of contact is 0.25. If **P** = 0, determine (*a*) the angle *θ* for which sliding is impending, (*b*) the corresponding force exerted on the block by the vertical wall.

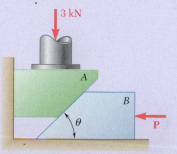

Fig. P8.54, P8.55, and P8.56

8.57 A wedge A of negligible weight is to be driven between two 100-lb blocks B and C resting on a horizontal surface. Knowing that the coefficient of static friction between all surfaces of contact is 0.35, determine the smallest force P required to start moving the wedge (*a*) if the blocks are equally free to move, (*b*) if block C is securely bolted to the horizontal surface.

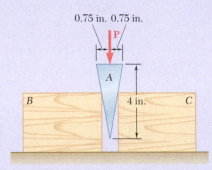

Fig. P8.57

8.58 A 15° wedge is forced into a saw cut to prevent binding of the circular saw. The coefficient of static friction between the wedge and the wood is 0.25. Knowing that a horizontal force P with a magnitude of 30 lb was required to insert the wedge, determine the magnitude of the forces exerted on the board by the wedge after insertion.

8.59 A 12° wedge is used to spread a split ring. The coefficient of static friction between the wedge and the ring is 0.30. Knowing that a force P with a magnitude of 120 N was required to insert the wedge, determine the magnitude of the forces exerted on the ring by the wedge after insertion.

8.60 The spring of the door latch has a constant of 1.8 lb/in. and in the position shown exerts a 0.6-lb force on the bolt. The coefficient of static friction between the bolt and the strike plate is 0.40; all other surfaces are well lubricated and may be assumed frictionless. Determine the magnitude of the force P required to start closing the door.

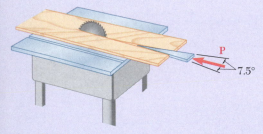

Fig. *P8.58*

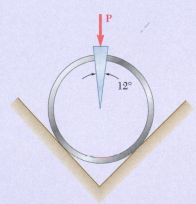

Fig. P8.59

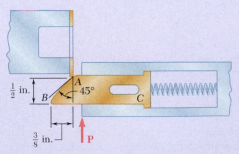

Fig. P8.60

8.61 In Prob. 8.60, determine the angle that the face of the bolt near B should form with line BC if the force P required to close the door is to be the same for both the position shown and the position when B is almost at the strike plate.

8.62 A 5° wedge is to be forced under a 1400-lb machine base at A. Knowing that the coefficient of static friction at all surfaces is 0.20, (a) determine the force $\mathbf{P}$ required to move the wedge, (b) indicate whether the machine base will move.

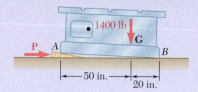

8.63 Solve Prob. 8.62 assuming that the wedge is to be forced under the machine base at B instead of A.

Fig. P8.62

8.64 A 15° wedge is forced under a 50-kg pipe as shown. The coefficient of static friction at all surfaces is 0.20. (a) Show that slipping will occur between the pipe and the vertical wall. (b) Determine the force $\mathbf{P}$ required to move the wedge.

8.65 A 15° wedge is forced under a 50-kg pipe as shown. Knowing that the coefficient of static friction at both surfaces of the wedge is 0.20, determine the largest coefficient of static friction between the pipe and the vertical wall for which slipping will occur at A.

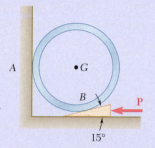

Fig. P8.64 and P8.65

*8.66 A 200-N block rests as shown on a wedge of negligible weight. The coefficient of static friction μ_s is the same at both surfaces of the wedge, and friction between the block and the vertical wall may be neglected. For $P = 100$ N, determine the value of μ_s for which motion is impending. (*Hint:* Solve the equation obtained by trial and error.)

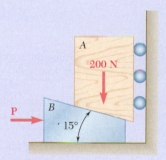

Fig. P8.66

*8.67 Solve Prob. 8.66 assuming that the rollers are removed and that μ_s is the coefficient of friction at all surfaces of contact.

8.68 Derive the following formulas relating the load $\mathbf{W}$ and the force $\mathbf{P}$ exerted on the handle of the jack discussed in Sec. 8.2B. (a) $P = (Wr/a) \tan(\theta + \phi_s)$ to raise the load; (b) $P = (Wr/a) \tan(\phi_s - \theta)$ to lower the load if the screw is self-locking; (c) $P = (Wr/a) \tan(\theta - \phi_s)$ to hold the load if the screw is not self-locking.

8.69 The square-threaded worm gear shown has a mean radius of 2 in. and a lead of 0.5 in. The large gear is subjected to a constant clockwise couple of 9.6 kip·in. Knowing that the coefficient of static friction between the two gears is 0.12, determine the couple that must be applied to shaft AB in order to rotate the large gear counterclockwise. Neglect friction in the bearings at A, B, and C.

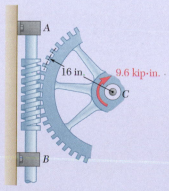

8.70 In Prob. 8.69, determine the couple that must be applied to shaft AB in order to rotate the large gear clockwise.

Fig. P8.69

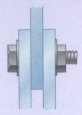

Fig. P8.71

8.71 High-strength bolts are used in the construction of many steel structures. For a 1-in.-nominal-diameter bolt, the required minimum bolt tension is 51 kips. Assuming the coefficient of friction to be 0.30, determine the required couple that should be applied to the bolt and nut. The mean diameter of the thread is 0.94 in., and the lead is 0.125 in. Neglect friction between the nut and washer, and assume the bolt to be square-threaded.

8.72 The position of the automobile jack shown is controlled by a screw *ABC* that is single-threaded at each end (right-handed thread at *A*, left-handed thread at *C*). Each thread has a pitch of 2.5 mm and a mean diameter of 9 mm. If the coefficient of static friction is 0.15, determine the magnitude of the couple **M** that must be applied to raise the automobile.

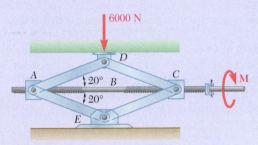

Fig. P8.72

8.73 For the jack of Prob. 8.72, determine the magnitude of the couple **M** that must be applied to lower the automobile.

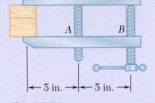

Fig. P8.74

8.74 The vise shown consists of two members connected by two double-threaded screws with a mean radius of 0.25 in. and pitch of 0.08 in. The lower member is threaded at *A* and *B* ($\mu_s = 0.35$), but the upper member is not threaded. It is desired to apply two equal and opposite forces of 120 lb on the blocks held between the jaws. (*a*) What screw should be adjusted first? (*b*) What is the maximum couple applied in tightening the second screw?

8.75 The ends of two fixed rods *A* and *B* are each made in the form of a single-threaded screw with a mean radius of 6 mm and pitch of 2 mm. Rod *A* has a right-handed thread, and rod *B* has a left-handed thread. The coefficient of static friction between the rods and the threaded sleeve is 0.12. Determine the magnitude of the couple that must be applied to the sleeve in order to draw the rods closer together.

Fig. P8.75

8.76 Assuming that in Prob. 8.75 a right-handed thread is used on *both* rods *A* and *B*, determine the magnitude of the couple that must be applied to the sleeve in order to rotate it.

*8.3 FRICTION ON AXLES, DISKS, AND WHEELS

Journal bearings are used to provide lateral support to rotating shafts and axles. **Thrust bearings** are used to provide axial support to shafts and axles. If the journal bearing is fully lubricated, the frictional resistance depends upon the speed of rotation, the clearance between axle and bearing, and the viscosity of the lubricant. As indicated in Sec. 8.1, such problems are studied in fluid mechanics. However, we can apply the methods of this chapter to the study of axle friction when the bearing is not lubricated or only partially lubricated. In this case, we can assume that the axle and the bearing are in direct contact along a single straight line.

8.3A Journal Bearings and Axle Friction

Consider two wheels, each with a weight of **W**, rigidly mounted on an axle supported symmetrically by two journal bearings (Fig. 8.10a). If the wheels rotate, we find that, to keep them rotating at constant speed, it is

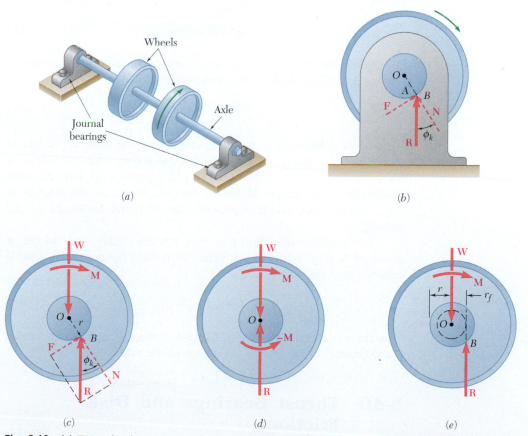

Fig. 8.10 (a) Two wheels supported by two journal bearings; (b) point of contact when the axle is rotating; (c) free-body diagram of one wheel and corresponding half axle; (d) frictional resistance produces a couple that opposes the couple maintaining the axle in motion; (e) graphical analysis with circle of friction.

necessary to apply a couple **M** to each of them. The free-body diagram in Fig. 8.10c represents one of the wheels and the corresponding half axle in projection on a plane perpendicular to the axle. The forces acting on the free body include the weight **W** of the wheel, the couple **M** required to maintain its motion, and a force **R** representing the reaction of the bearing. This force is vertical, equal, and opposite to **W**, but it does not pass through the center O of the axle; **R** is located to the right of O at a distance such that its moment about O balances the moment **M** of the couple. Therefore, when the axle rotates, contact between the axle and bearing does not take place at the lowest point A. Instead, contact takes place at point B (Fig. 8.10b) or, rather, along a straight line intersecting the plane of the figure at B.

Physically, the location of contact is explained by the fact that, when the wheels are set in motion, the axle "climbs" in the bearings until slippage occurs. After sliding back slightly, the axle settles more or less in the position shown. This position is such that the angle between the reaction **R** and the normal to the surface of the bearing is equal to the angle of kinetic friction ϕ_k. The distance from O to the line of action of **R** is thus $r \sin \phi_k$, where r is the radius of the axle. Setting $\Sigma M_O = 0$ for the forces acting on the free body (the wheel), we obtain the magnitude of the couple **M** required to overcome the frictional resistance of one of the bearings:

$$M = Rr \sin \phi_k \tag{8.5}$$

For small values of the angle of friction, we can replace $\sin \phi_k$ by $\tan \phi_k$; that is, by μ_k. This gives us the approximate formula

$$M \approx Rr\mu_k \tag{8.6}$$

In the solution of certain problems, it may be more convenient to let the line of action of **R** pass through O, as it does when the axle does not rotate. In such a case, you need to add a couple $-\mathbf{M}$, with the same magnitude as the couple **M** but of opposite sense, to the reaction **R** (Fig. 8.10d). This couple represents the frictional resistance of the bearing.

If a graphical solution is preferred, you can readily draw the line of action of **R** (Fig. 8.10e) if you note that it must be tangent to a circle centered at O and with a radius

$$r_f = r \sin \phi_k \approx r\mu_k \tag{8.7}$$

This circle is called the **circle of friction** of the axle and bearing, and it is independent of the loading conditions of the axle.

8.3B Thrust Bearings and Disk Friction

Two types of thrust bearings are commonly used to provide axial support to rotating shafts and axles: (1) **end bearings** and (2) **collar bearings** (Fig. 8.11). In the case of collar bearings, friction forces develop between

(a) End bearing   (b) Collar bearing

Fig. 8.11 In thrust bearings, an axial force keeps the rotating axle in contact with the support bearing.

the two ring-shaped areas in contact. In the case of end bearings, friction takes place over full circular areas or over ring-shaped areas when the end of the shaft is hollow. Friction between circular areas, called **disk friction**, also occurs in other mechanisms, such as disk clutches.

To obtain a formula for the most general case of disk friction, let us consider a rotating hollow shaft. A couple **M** keeps the shaft rotating at constant speed, while an axial force **P** maintains it in contact with a fixed bearing (Fig. 8.12). Contact between the shaft and the bearing takes place over a ring-shaped area with an inner radius of R_1 and an outer radius of R_2. Assuming that the pressure between the two surfaces in contact is uniform, we find that the magnitude of the normal force $\Delta \mathbf{N}$ exerted on an element of area ΔA is $\Delta N = P\,\Delta A/A$, where $A = \pi\,(R_2^2 - R_1^2)$ and that the magnitude of the friction force $\Delta \mathbf{F}$ acting on ΔA is $\Delta F = \mu_k\,\Delta N$. Let's use r to denote the distance from the axis of the shaft to the element of area ΔA. Then the magnitude ΔM of the moment of $\Delta \mathbf{F}$ about the axis of the shaft is

$$\Delta M = r\,\Delta F = \frac{r\mu_k P\,\Delta A}{\pi(R_2^2 - R_1^2)}$$

Equilibrium of the shaft requires that the moment **M** of the couple applied to the shaft be equal in magnitude to the sum of the moments of the

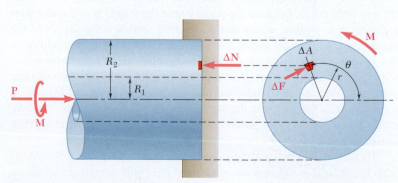

Fig. 8.12 Geometry of the frictional contact surface in a thrust bearing.

friction forces $\Delta\mathbf{F}$ opposing the motion of the shaft. Replacing ΔA by the infinitesimal element $dA = r\,d\theta\,dr$ used with polar coordinates and integrating over the area of contact, the expression for the magnitude of the couple $\mathbf{M}$ required to overcome the frictional resistance of the bearing is

$$M = \frac{\mu_k P}{\pi(R_2^2 - R_1^2)} \int_0^{2\pi} \int_{R_1}^{R_2} r^2\,dr\,d\theta$$

$$= \frac{\mu_k P}{\pi(R_2^2 - R_1^2)} \int_0^{2\pi} \tfrac{1}{3}(R_2^3 - R_1^3)\,d\theta$$

$$M = \tfrac{2}{3}\mu_k P \frac{R_2^3 - R_1^3}{R_2^2 - R_1^2} \tag{8.8}$$

When contact takes place over a full circle with a radius of R, formula (8.8) reduces to

$$M = \tfrac{2}{3}\mu_k PR \tag{8.9}$$

This value of M is the same value we would obtain if contact between shaft and bearing took place at a single point located at a distance $2R/3$ from the axis of the shaft.

The largest couple that can be transmitted by a disk clutch without causing slippage is given by a formula similar to Eq. (8.9), where μ_k has been replaced by the coefficient of static friction μ_s.

8.3C   Wheel Friction and Rolling Resistance

The wheel is one of the most important inventions of our civilization. Among many other uses, with a wheel we can move heavy loads with relatively little effort. Because the point where the wheel is in contact with the ground at any given instant has no relative motion with respect to the ground, use of the wheel avoids the large friction forces that would arise if the load were in direct contact with the ground. However, some resistance to the wheel's motion does occur. This resistance has two distinct causes. It is due to (1) a combined effect of axle friction and friction at the rim and (2) the fact that the wheel and the ground deform, causing contact between wheel and ground to take place over an area rather than at a single point.

To understand better the first cause of resistance to the motion of a wheel, consider a railroad car supported by eight wheels mounted on axles and bearings. We assume the car is moving to the right at constant speed along a straight horizontal track. The free-body diagram of one of the wheels is shown in Fig. 8.13a. The forces acting on the free body include the load **W** supported by the wheel and the normal reaction **N** of the track. Since **W** passes through the center O of the axle, we represent the frictional resistance of the bearing by a counterclockwise couple **M** (see Sec. 8.3A). Then to keep the free body in equilibrium, we must add two equal and opposite forces **P** and **F**, forming a clockwise couple of moment $-$**M**. The force **F** is the friction force exerted by the track on the wheel, and **P** represents the force that should be applied to the wheel to keep it rolling at constant speed. Note that the forces **P** and **F** would not exist if there were no friction between the wheel and the track. The couple **M** representing the axle friction would then be zero; the wheel would slide on the track without turning in its bearing.

The couple **M** and the forces **P** and **F** also reduce to zero when there is no axle friction. For example, a wheel that is not held in bearings but rolls freely and at constant speed on horizontal ground (Fig. 8.13b) is subjected to only two forces: its own weight **W** and the normal reaction **N** of the ground. No friction force acts on the wheel regardless of the value of the coefficient of friction between wheel and ground. Thus, a wheel rolling freely on horizontal ground should keep rolling indefinitely.

Experience, however, indicates that a free wheel does slow down and eventually come to rest. This is due to the second type of resistance mentioned at the beginning of this section, known as **rolling resistance**. Under the load **W**, both the wheel and the ground deform slightly, causing the contact between wheel and ground to take place over a certain area. Experimental evidence shows that the resultant of the forces exerted by the ground on the wheel over this area is a force **R** applied at a point B, which is not located directly under the center O of the wheel but slightly in front of it (Fig. 8.13c). To balance the moment of **W** about B and to keep the wheel rolling at constant speed, it is necessary to apply a horizontal force **P** at the center of the wheel. Setting $\Sigma M_B = 0$, we obtain

$$Pr = Wb \tag{8.10}$$

where r = radius of wheel
  b = horizontal distance between O and B

The distance b is commonly called the **coefficient of rolling resistance**. Note that b is not a dimensionless coefficient, since it represents a length; b is usually expressed in inches or in millimeters. The value of b depends upon several parameters in a manner that has not yet been clearly established. Values of the coefficient of rolling resistance vary from about 0.01 in. or 0.25 mm for a steel wheel on a steel rail to 5.0 in. or 125 mm for the same wheel on soft ground.

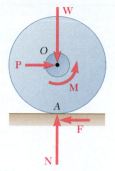

(a) Effect of axle friction

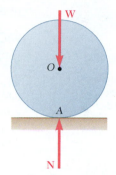

(b) Free wheel

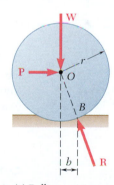

(c) Rolling resistance

Fig. 8.13 (a) Free-body diagram of a rolling wheel, showing the effect of axle friction; (b) free-body diagram of a free wheel, not connected to an axle; (c) free-body diagram of a rolling wheel, showing the effect of rolling resistance.

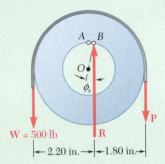

Fig. 1 Free-body diagram of pulley—smallest vertical force to raise the load.

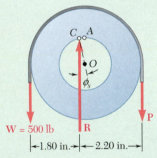

Fig. 2 Free-body diagram of pulley—smallest vertical force to hold load.

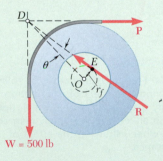

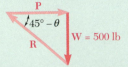

Fig. 3 Free-body diagram of pulley and force triangle—smallest horizontal force to raise load.

Sample Problem 8.7

A pulley with a diameter of 4 in. can rotate about a fixed shaft with a diameter of 2 in. The coefficient of static friction between the pulley and shaft is 0.20. Determine (*a*) the smallest vertical force **P** required to start raising a 500-lb load, (*b*) the smallest vertical force **P** required to hold the load, (*c*) the smallest horizontal force **P** required to start raising the same load.

STRATEGY: You can use the radius of the circle of friction to position the reaction of the pulley in each scenario and then apply the principles of equilibrium.

MODELING and ANALYSIS:

a. Vertical Force P Required to Start Raising the Load. When the forces in both parts of the rope are equal, contact between the pulley and shaft takes place at *A* (Fig. 1). When **P** is increased, the pulley rolls around the shaft slightly and contact takes place at *B*. Draw the free-body diagram of the pulley when motion is impending. The perpendicular distance from the center *O* of the pulley to the line of action of **R** is

$$r_f = r \sin \phi_s \approx r\mu_s \qquad r_f \approx (1 \text{ in.})0.20 = 0.20 \text{ in.}$$

Summing moments about *B*, you obtain

$$+\circlearrowleft \Sigma M_B = 0: \qquad (2.20 \text{ in.})(500 \text{ lb}) - (1.80 \text{ in.})P = 0$$
$$P = 611 \text{ lb} \qquad\qquad \mathbf{P} = 611 \text{ lb} \downarrow \quad \blacktriangleleft$$

b. Vertical Force P to Hold the Load. As the force **P** is decreased, the pulley rolls around the shaft, and contact takes place at *C* (Fig. 2). Considering the pulley as a free body and summing moments about *C*, you find

$$+\circlearrowleft \Sigma M_C = 0: \qquad (1.80 \text{ in.})(500 \text{ lb}) - (2.20 \text{ in.})P = 0$$
$$P = 409 \text{ lb} \qquad\qquad \mathbf{P} = 409 \text{ lb} \downarrow \quad \blacktriangleleft$$

c. Horizontal Force P to Start Raising the Load. Since the three forces **W**, **P**, and **R** are not parallel, they must be concurrent (Fig. 3). The direction of **R** is thus determined from the fact that its line of action must pass through the point of intersection *D* of **W** and **P** and must be tangent to the circle of friction. Recall that the radius of the circle of friction is $r_f = 0.20$ in., so you can calculate the angle marked θ in Fig. 3 as

$$\sin \theta = \frac{OE}{OD} = \frac{0.20 \text{ in.}}{(2 \text{ in.})\sqrt{2}} = 0.0707 \quad \theta = 4.1°$$

From the force triangle, you can determine

$$P = W \cot (45° - \theta) = (500 \text{ lb}) \cot 40.9°$$
$$= 577 \text{ lb} \qquad\qquad \mathbf{P} = 577 \text{ lb} \rightarrow \quad \blacktriangleleft$$

REFLECT and THINK: Many elementary physics problems treat pulleys as frictionless, but when you do take friction into account, the results can be quite different, depending on the direction of motion, the directions of the forces involved, and especially the coefficient of friction.

SOLVING PROBLEMS
ON YOUR OWN

In this section, we described several additional engineering applications of the laws of friction.

1. Journal bearings and axle friction. In journal bearings, the reaction does not pass through the center of the shaft or axle that is being supported. The distance from the center of the shaft or axle to the line of action of the reaction (Fig. 8.10) is defined by

$$r_f = r \sin \phi_k \approx r\mu_k$$

if motion is actually taking place.

It is defined by

$$r_f = r \sin \phi_s \approx r\mu_s$$

if motion is impending.

Once you have determined the line of action of the reaction, you can draw a free-body diagram and use the corresponding equations of equilibrium to complete the solution [Sample Prob. 8.7]. In some problems, it is useful to observe that the line of action of the reaction must be tangent to a circle with a radius of $r_f \approx r\mu_k$ or $r_f \approx r\mu_s$, which is known as the **circle of friction** [Sample Prob. 8.7, part *c*].

2. Thrust bearings and disk friction. In a thrust bearing, the magnitude of the couple required to overcome frictional resistance is equal to the sum of the moments of the *kinetic* friction forces exerted on the end of the shaft [Eqs. (8.8) and (8.9)].

An example of disk friction is the **disk clutch**. It is analyzed in the same way as a thrust bearing, except that to determine the largest couple that can be transmitted you must compute the sum of the moments of the maximum *static* friction forces exerted on the disk.

3. Wheel friction and rolling resistance. The rolling resistance of a wheel is caused by deformations of both the wheel and the ground. The line of action of the reaction **R** of the ground on the wheel intersects the ground at a horizontal distance b from the center of the wheel. The distance b is known as the **coefficient of rolling resistance** and is expressed in inches or millimeters.

4. In problems involving both rolling resistance and axle friction, the free-body diagram should show that the line of action of the reaction **R** of the ground on the wheel is tangent to the friction circle of the axle and intersects the ground at a horizontal distance from the center of the wheel equal to the coefficient of rolling resistance.

Problems

8.77 A lever of negligible weight is loosely fitted onto a 75-mm-diameter fixed shaft. It is observed that the lever will just start rotating if a 3-kg mass is added at *C*. Determine the coefficient of static friction between the shaft and the lever.

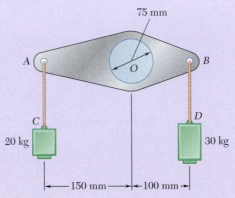

Fig. P8.77

8.78 A hot-metal ladle and its contents weigh 130 kips. Knowing that the coefficient of static friction between the hooks and the pinion is 0.30, determine the tension in cable *AB* required to start tipping the ladle.

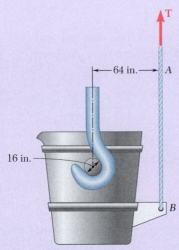

Fig. P8.78

8.79 and 8.80 The double pulley shown is attached to a 10-mm-radius shaft that fits loosely in a fixed bearing. Knowing that the coefficient of static friction between the shaft and the poorly lubricated bearing is 0.40, determine the magnitude of the force **P** required to start raising the load.

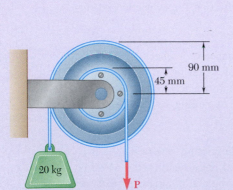

Fig. P8.79 and P8.81

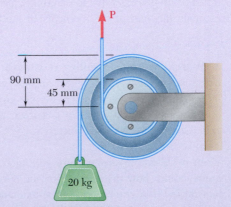

Fig. P8.80 and P8.82

8.81 and 8.82 The double pulley shown is attached to a 10-mm-radius shaft that fits loosely in a fixed bearing. Knowing that the coefficient of static friction between the shaft and the poorly lubricated bearing is 0.40, determine the magnitude of the smallest force **P** required to maintain equilibrium.

8.83 The block and tackle shown are used to raise a 150-lb load. Each of the 3-in.-diameter pulleys rotates on a 0.5-in.-diameter axle. Knowing that the coefficient of static friction is 0.20, determine the tension in each portion of the rope as the load is slowly raised.

8.84 The block and tackle shown are used to lower a 150-lb load. Each of the 3-in.-diameter pulleys rotates on a 0.5-in.-diameter axle. Knowing that the coefficient of static friction is 0.20, determine the tension in each portion of the rope as the load is slowly lowered.

8.85 A scooter is to be designed to roll down a 2 percent slope at a constant speed. Assuming that the coefficient of kinetic friction between the 25-mm-diameter axles and the bearings is 0.10, determine the required diameter of the wheels. Neglect the rolling resistance between the wheels and the ground.

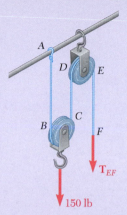
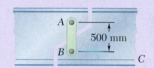

Fig. *P8.83* and P8.84

8.86 The link arrangement shown is frequently used in highway bridge construction to allow for expansion due to changes in temperature. At each of the 60-mm-diameter pins *A* and *B*, the coefficient of static friction is 0.20. Knowing that the vertical component of the force exerted by *BC* on the link is 200 kN, determine (*a*) the horizontal force that should be exerted on beam *BC* to just move the link, (*b*) the angle that the resulting force exerted by beam *BC* on the link will form with the vertical.

Fig. P8.86

8.87 and 8.88 A lever *AB* of negligible weight is loosely fitted onto a 2.5-in.-diameter fixed shaft. Knowing that the coefficient of static friction between the fixed shaft and the lever is 0.15, determine the force **P** required to start the lever rotating counterclockwise.

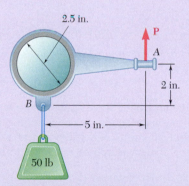

Fig. *P8.87* and P8.89

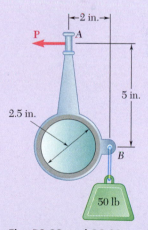

Fig. P8.88 and P8.90

8.89 and 8.90 A lever *AB* of negligible weight is loosely fitted onto a 2.5-in.-diameter fixed shaft. Knowing that the coefficient of static friction between the fixed shaft and the lever is 0.15, determine the force **P** required to start the lever rotating clockwise.

8.91 A loaded railroad car has a mass of 30 Mg and is supported by eight 800-mm-diameter wheels with 125-mm-diameter axles. Knowing that the coefficients of friction are $\mu_s = 0.020$ and $\mu_k = 0.015$, determine the horizontal force required (*a*) to start the car moving, (*b*) to keep the car moving at a constant speed. Neglect rolling resistance between the wheels and the rails.

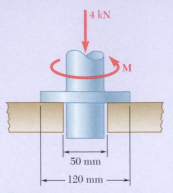

4 kN

M

50 mm

120 mm

Fig. P8.92

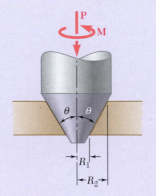

−Q

20 in.

Q

18 in.

Fig. P8.93

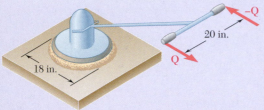

P

M

θ   θ

R_1

R_2

Fig. *P8.96*

P

Fig. P8.100

8.92 Knowing that a couple of magnitude 30 N·m is required to start the vertical shaft rotating, determine the coefficient of static friction between the annular surfaces of contact.

8.93 A 50-lb electric floor polisher is operated on a surface for which the coefficient of kinetic friction is 0.25. Assuming that the normal force per unit area between the disk and the floor is uniformly distributed, determine the magnitude Q of the horizontal forces required to prevent motion of the machine.

*8.94 The frictional resistance of a thrust bearing decreases as the shaft and bearing surfaces wear out. It is generally assumed that the wear is directly proportional to the distance traveled by any given point of the shaft and thus to the distance r from the point to the axis of the shaft. Assuming then that the normal force per unit area is inversely proportional to r, show that the magnitude M of the couple required to overcome the frictional resistance of a worn-out end bearing (with contact over the full circular area) is equal to 75 percent of the value given by Eq. (8.9) for a new bearing.

*8.95 Assuming that bearings wear out as indicated in Prob. 8.94, show that the magnitude M of the couple required to overcome the frictional resistance of a worn-out collar bearing is

$$M = \tfrac{1}{2}\,\mu_k\,P(R_1 + R_2)$$

where P = magnitude of the total axial force
R_1, R_2 = inner and outer radii of collar

*8.96 Assuming that the pressure between the surfaces of contact is uniform, show that the magnitude M of the couple required to overcome frictional resistance for the conical bearing shown is

$$M = \frac{2}{3}\frac{\mu_k P}{\sin\theta}\frac{R_2^3 - R_1^3}{R_2^2 - R_1^2}$$

8.97 Solve Prob. 8.93 assuming that the normal force per unit area between the disk and the floor varies linearly from a maximum at the center to zero at the circumference of the disk.

8.98 Determine the horizontal force required to move a 2500-lb automobile with 23-in.-diameter tires along a horizontal road at a constant speed. Neglect all forms of friction except rolling resistance, and assume the coefficient of rolling resistance to be 0.05 in.

8.99 Knowing that a 6-in.-diameter disk rolls at a constant velocity down a 2 percent incline, determine the coefficient of rolling resistance between the disk and the incline.

8.100 A 900-kg machine base is rolled along a concrete floor using a series of steel pipes with outside diameters of 100 mm. Knowing that the coefficient of rolling resistance is 0.5 mm between the pipes and the base and 1.25 mm between the pipes and the concrete floor, determine the magnitude of the force **P** required to slowly move the base along the floor.

8.101 Solve Prob. 8.85 including the effect of a coefficient of rolling resistance of 1.75 mm.

8.102 Solve Prob. 8.91 including the effect of a coefficient of rolling resistance of 0.5 mm.

8.4 BELT FRICTION

Another common application of dry friction concerns belts, which serve many different purposes in engineering, such as transmitting a torque from a lawn mower engine to its wheels. Some of the same analysis affects the design of band brakes and the operation of ropes and pulleys.

Consider a flat belt passing over a fixed cylindrical drum (Fig. 8.14*a*). We want to determine the relation between the values T_1 and T_2 of the tension in the two parts of the belt when the belt is just about to slide toward the right.

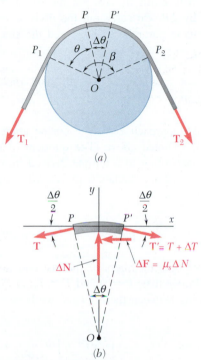

(a)

(b)

Fig. 8.14 (a) Tensions at the ends of a belt passing over a drum; (b) free-body diagram of an element of the belt, indicating the condition that the belt is about to slip to the right.

First we detach from the belt a small element PP' subtending an angle $\Delta\theta$. Denoting the tension at P by T and the tension at P' by $T + \Delta T$, we draw the free-body diagram of the element of the belt (Fig. 8.14*b*). Besides the two forces of tension, the forces acting on the free body are the normal component $\Delta\mathbf{N}$ of the reaction of the drum and the friction force $\Delta\mathbf{F}$. Since we assume motion is impending, we have $\Delta F = \mu_s \Delta N$. Note that if $\Delta\theta$ approaches zero, the magnitudes ΔN and ΔF and the *difference* ΔT between the tension at P and the tension at P' also approach zero; the value T of the tension at P, however, remains unchanged. This observation helps in understanding our choice of notation.

Choosing the coordinate axes shown in Fig. 8.14*b*, we can write the equations of equilibrium for the element PP' as

$$\Sigma F_x = 0: \quad (T + \Delta T)\cos\frac{\Delta\theta}{2} - T\cos\frac{\Delta\theta}{2} - \mu_s\Delta N = 0 \qquad \textbf{(8.11)}$$

$$\Sigma F_y = 0: \quad \Delta N - (T + \Delta T) \sin \frac{\Delta \theta}{2} - T \sin \frac{\Delta \theta}{2} = 0 \qquad \textbf{(8.12)}$$

Solving Eq. (8.12) for ΔN and substituting into Eq. (8.11), we obtain after reductions

$$\Delta T \cos \frac{\Delta \theta}{2} - \mu_s (2T + \Delta T) \sin \frac{\Delta \theta}{2} = 0$$

Now we divide both terms by $\Delta \theta$. For the first term, we do this simply by dividing ΔT by $\Delta \theta$. We carry out the division of the second term by dividing the terms in parentheses by 2 and the sine by $\Delta \theta / 2$. The result is

$$\frac{\Delta T}{\Delta \theta} \cos \frac{\Delta \theta}{2} - \mu_s \left(T + \frac{\Delta T}{2} \right) \frac{\sin(\Delta \theta / 2)}{\Delta \theta / 2} = 0$$

If we now let $\Delta \theta$ approach zero, the cosine approaches one and $\Delta T / 2$ approaches zero, as noted above. The quotient of $\sin (\Delta \theta / 2)$ over $\Delta \theta / 2$ approaches one, according to a lemma derived in all calculus textbooks. Since the limit as $\Delta \theta$ approaches 0 of $\Delta T / \Delta \theta$ is equal to the derivative $dT/d\theta$ by definition, we get

$$\frac{dT}{d\theta} - \mu_s T = 0 \qquad \frac{dT}{T} = \mu_s \, d\theta$$

Now we integrate both members of the last equation from P_1 to P_2 (see Fig. 8.14a). At P_1, we have $\theta = 0$ and $T = T_1$; at P_2, we have $\theta = \beta$ and $T = T_2$. Integrating between these limits, we have

$$\int_{T_1}^{T_2} \frac{dT}{T} = \int_0^\beta \mu_s \, d\theta$$

$$\ln T_2 - \ln T_1 = \mu_s \beta$$

Noting that the left-hand side is equal to the natural logarithm of the quotient of T_2 and T_1, this reduces to

$$\ln \frac{T_2}{T_1} = \mu_s \beta \qquad \textbf{(8.13)}$$

We can also write this relation in the form

Belt friction, impending slip

$$\frac{T_2}{T_1} = e^{\mu_s \beta} \qquad \textbf{(8.14)}$$

The formulas we have derived apply equally well to problems involving flat belts passing over fixed cylindrical drums and to problems

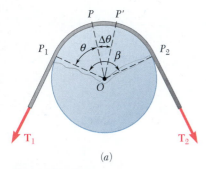

Fig. 8.14a *(repeated)*

(*a*)

Photo 8.5 A sailor wraps a rope around the smooth post (called a *bollard*) in order to control the rope using much less force than the tension in the taut part of the rope.

involving ropes wrapped around a post or capstan. They also can be used to solve problems involving band brakes. (In this situation, the drum is about to rotate, but the band remains fixed.) The formulas also can be applied to problems involving belt drives. In these problems, both the pulley and the belt rotate; our concern is then to find whether the belt will slip; i.e., whether it will move with respect to the pulley.

Formulas (8.13) and (8.14) should be used only if the belt, rope, or brake is *about to slip*. Generally, it is easier to use Eq. (8.14) if you need to find T_1 or T_2; it is preferable to use Eq. (8.13) if you need to find either μ_s or the angle of contact β. Note that T_2 is always larger than T_1. T_2 therefore represents the tension in that part of the belt or rope that *pulls*, whereas T_1 is the tension in the part that *resists*. Also observe that the angle of contact β must be expressed in *radians*. The angle of contact β may be larger than 2π; for example, if a rope is wrapped n times around a post, β is equal to $2\pi n$.

If the belt, rope, or brake is actually slipping, you should use formulas similar to Eqs. (8.13) and (8.14) involving the coefficient of kinetic friction μ_k to find the difference in forces. If the belt, rope, or brake is not slipping and is not about to slip, none of these formulas can be used.

The belts used in belt drives are often V-shaped. In the V belt shown in Fig. 8.15a, contact between belt and pulley takes place along the sides of the groove. Again, we can obtain the relation between the values T_1 and T_2 of the tension in the two parts of the belt when the belt is just about to slip by drawing the free-body diagram of an element of the belt (Fig. 8.15b and c). Formulas similar to Eqs. (8.11) and (8.12) are derived, but the magnitude of the total friction force acting on the element is now $2\,\Delta F$, and the sum of the y components of the normal forces is $2\,\Delta N \sin(\alpha/2)$. Proceeding as previoulsy, we obtain

$$\ln \frac{T_2}{T_1} = \frac{\mu_s \beta}{\sin(\alpha/2)} \tag{8.15}$$

or

$$\frac{T_2}{T_1} = e^{\mu_s \beta/\sin(\alpha/2)} \tag{8.16}$$

(a)

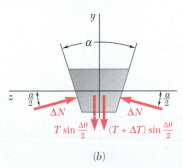

(b)

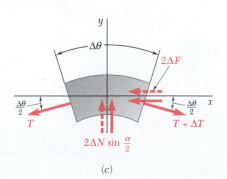

(c)

Fig. 8.15 (a) A V belt lying in the groove of a pulley; (b) free-body diagram of a cross-sectional element of the belt; (c) free-body diagram of a short length of belt.

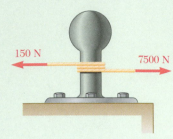

150 N ⟵   ⟶ 7500 N

Sample Problem 8.8

A hawser (a thick docking rope) thrown from a ship to a pier is wrapped two full turns around a bollard. The tension in the hawser is 7500 N; by exerting a force of 150 N on its free end, a dockworker can just keep the hawser from slipping. (*a*) Determine the coefficient of friction between the hawser and the bollard. (*b*) Determine the tension in the hawser that could be resisted by the 150-N force if the hawser were wrapped three full turns around the bollard.

STRATEGY: You are given the difference in forces and the angle of contact through which the friction acts. You can insert these data in the equations of belt friction to determine the coefficient of friction, and then you can use the result to determine the ratio of forces in the second situation.

MODELING and ANALYSIS:

a. Coefficient of Friction. Since slipping of the hawser is impending, we use Eq. (8.13):

$$\ln \frac{T_2}{T_1} = \mu_s \beta$$

Since the hawser is wrapped two full turns around the bollard, you have

$$\beta = 2(2\pi \text{ rad}) = 12.57 \text{ rad}$$
$$T_1 = 150 \text{ N} \quad T_2 = 7500 \text{ N}$$

Therefore,

$$\mu_s \beta = \ln \frac{T_2}{T_1}$$
$$\mu_s(12.57 \text{ rad}) = \ln \frac{7500 \text{ N}}{150 \text{ N}} = \ln 50 = 3.91$$
$$\mu_s = 0.311 \qquad\qquad\qquad \mu_s = 0.311 \blacktriangleleft$$

Photo 8.6 Dockworker mooring a ship using a hawser wrapped around a bollard.

b. Hawser Wrapped Three Turns Around Bollard. Using the value of μ_s obtained in part *a*, you now have (Fig. 1)

$$\beta = 3(2\pi \text{ rad}) = 18.85 \text{ rad}$$
$$T_1 = 150 \text{ N} \quad \mu_s = 0.311$$

Substituting these values into Eq. (8.14), you obtain

$$\frac{T_2}{T_1} = e^{\mu_s \beta}$$
$$\frac{T_2}{150 \text{ N}} = e^{(0.311)(18.85)} = e^{5.862} = 351.5$$
$$T_2 = 52\,725 \text{ N}$$

$$T_2 = 52.7 \text{ kN} \blacktriangleleft$$

REFLECT and THINK: You can see how the use of a simple post or pulley can have an enormous effect of the magnitude of a force. This is why such systems are commonly used to control, load, and unload container ships in a harbor.

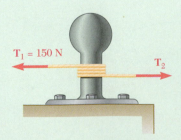

$T_1 = 150 \text{ N}$ ⟵   ⟶ T_2

Fig. 1 Hawser wrapped three turns around bollard.

Sample Problem 8.9

A flat belt connects pulley A, which drives a machine tool, to pulley B, which is attached to the shaft of an electric motor. The coefficients of friction are $\mu_s = 0.25$ and $\mu_k = 0.20$ between both pulleys and the belt. Knowing that the maximum allowable tension in the belt is 600 lb, determine the largest torque that the belt can exert on pulley A.

STRATEGY: The key to solving this problem is to identify the pulley where slippage would first occur, and then find the corresponding belt tensions when slippage is impending. The resistance to slippage depends upon the angle of contact β between pulley and belt, as well as upon the coefficient of static friction μ_s. Since μ_s is the same for both pulleys, slippage occurs first on pulley B, for which β is smaller (Fig. 1).

MODELING and ANALYSIS:

Pulley B. Using Eq. (8.14) with $T_2 = 600$ lb, $\mu_s = 0.25$, and $\beta = 120°$ $= 2\pi/3$ rad (Fig. 2), you obtain

$$\frac{T_2}{T_1} = e^{\mu_s \beta} \qquad \frac{600 \text{ lb}}{T_1} = e^{0.25(2\pi/3)} = 1.688$$

$$T_1 = \frac{600 \text{ lb}}{1.688} = 355.4 \text{ lb}$$

Pulley A. Draw the free-body diagram of pulley A (Fig. 3). The couple $\mathbf{M}_A$ is applied to the pulley using the machine tool to which it is attached and is equal and opposite to the torque exerted by the belt. Setting the sum of the moments equal to zero gives

$$+\circlearrowleft \Sigma M_A = 0: \quad M_A - (600 \text{ lb})(8 \text{ in.}) + (355.4 \text{ lb})(8 \text{ in.}) = 0$$
$$M_A = 1957 \text{ lb·in.} \qquad\qquad \boxed{M_A = 163.1 \text{ lb·ft}} \;\blacktriangleleft$$

REFLECT and THINK: You may check that the belt does not slip on pulley A by computing the value of μ_s required to prevent slipping at A and verify that it is smaller than the actual value of μ_s. From Eq. (8.13), you have

$$\mu_s \beta = \ln \frac{T_2}{T_1} = \ln \frac{600 \text{ lb}}{355.4 \text{ lb}} = 0.524$$

Since $\beta = 240° = 4\pi/3$ rad,

$$\frac{4\pi}{3}\mu_s = 0.524 \qquad \mu_s = 0.125 < 0.25$$

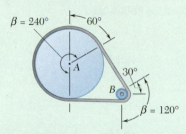

Fig. 1 Angles of contact for the pulleys.

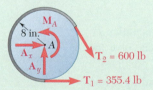

Fig. 2 Belt tensions at pulley B.

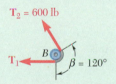

Fig. 3 Free-body diagram of pulley A.

SOLVING PROBLEMS
ON YOUR OWN

In the preceding section, you studied **belt friction**. The problems you will solve include belts passing over fixed drums, band brakes in which the drum rotates when the band remains fixed, and belt drives.

1. Problems involving belt friction fall into one of the following two categories.

    a. Problems in which slipping is impending. You can use one of the following formulas involving the *coefficient of static friction* μ_s.

$$\ln\frac{T_2}{T_1} = \mu_s\beta \tag{8.13}$$

or

$$\frac{T_2}{T_1} = e^{\mu_s\beta} \tag{8.14}$$

    b. Problems in which slipping is occurring. You can obtain the formulas to be used from Eqs. (8.13) and (8.14) by replacing μ_s with the *coefficient of kinetic friction* μ_k.

2. As you start solving a belt-friction problem, remember these conventions:

    a. The angle β must be expressed in radians. In a belt-and-drum problem, this is the angle subtending the arc of the drum on which the belt is wrapped.

    b. The larger tension is always denoted by T_2 and the smaller tension is denoted by T_1.

    c. The larger tension occurs at the end of the belt which is in the direction of the motion, or impending motion, of the belt relative to the drum.

3. In each of the problems you will be asked to solve, three of the four quantities T_1, T_2, β, and μ_s (or μ_k) will either be given or readily found, and you will then solve the appropriate equation for the fourth quantity. You will encounter two kinds of problems.

    a. Find μ_s between belt and drum, knowing that slipping is impending. From the given data, determine T_1, T_2, and β; substitute these values into Eq. (8.13) and solve for μ_s [Sample Prob. 8.8, part *a*]. Follow the same procedure to find the smallest value of μ_s for which slipping will not occur.

    b. Find the magnitude of a force or couple applied to the belt or drum, knowing that slipping is impending. The given data should include μ_s and β. If it also includes T_1 or T_2, use Eq. (8.14) to find the other tension. If neither T_1 nor T_2 is known but some other data is given, use the free-body diagram of the belt-drum system to write an equilibrium equation that you can solve simultaneously with Eq. (8.14) for T_1 and T_2. You then will be able to find the magnitude of the specified force or couple from the free-body diagram of the system. Follow the same procedure to determine the largest value of a force or couple that can be applied to the belt or drum if no slipping is to occur [Sample Prob. 8.9].

Problems

8.103 A rope having a weight per unit length of 0.4 lb/ft is wound $2\frac{1}{2}$ times around a horizontal rod. Knowing that the coefficient of static friction between the rope and the rod is 0.30, determine the minimum length x of rope that should be left hanging if a 100-lb load is to be supported.

8.104 A hawser is wrapped two full turns around a bollard. By exerting an 80-lb force on the free end of the hawser, a dockworker can resist a force of 5000 lb on the other end of the hawser. Determine (a) the coefficient of static friction between the hawser and the bollard, (b) the number of times the hawser should be wrapped around the bollard if a 20,000-lb force is to be resisted by the same 80-lb force.

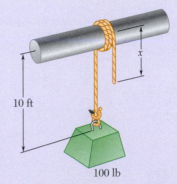

Fig. P8.103

8.105 Two cylinders are connected by a rope that passes over two fixed rods as shown. Knowing that the coefficient of static friction between the rope and the rods is 0.40, determine the range of the mass m of cylinder D for which equilibrium is maintained.

8.106 Two cylinders are connected by a rope that passes over two fixed rods as shown. Knowing that for cylinder D upward motion impends when $m = 20$ kg, determine (a) the coefficient of static friction between the rope and the rods, (b) the corresponding tension in portion BC of the rope.

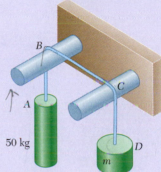

Fig. P8.105 and P8.106

8.107 Knowing that the coefficient of static friction is 0.25 between the rope and the horizontal pipe and 0.20 between the rope and the vertical pipe, determine the range of values of P for which equilibrium is maintained.

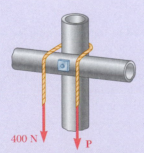

400 N   P

Fig. P8.107 and P8.108

8.108 Knowing that the coefficient of static friction is 0.30 between the rope and the horizontal pipe and that the smallest value of P for which equilibrium is maintained is 80 N, determine (a) the largest value of P for which equilibrium is maintained, (b) the coefficient of static friction between the rope and the vertical pipe.

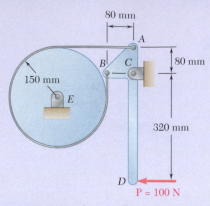

80 mm

A

B C

150 mm

E

80 mm

320 mm

D

P = 100 N

Fig. P8.109

8.109 A band belt is used to control the speed of a flywheel as shown. Determine the magnitude of the couple being applied to the flywheel, knowing that the coefficient of kinetic friction between the belt and the flywheel is 0.25 and that the flywheel is rotating clockwise at a constant speed. Show that the same result is obtained if the flywheel rotates counterclockwise.

8.110 The setup shown is used to measure the output of a small turbine. When the flywheel is at rest, the reading of each spring scale is 14 lb. If a 105-lb·in. couple must be applied to the flywheel to keep it rotating clockwise at a constant speed, determine (*a*) the reading of each scale at that time, (*b*) the coefficient of kinetic friction. Assume that the length of the belt does not change.

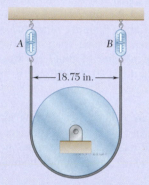

A   B

18.75 in.

Fig. P8.110 and P8.111

8.111 The setup shown is used to measure the output of a small turbine. The coefficient of kinetic friction is 0.20, and the reading of each spring scale is 16 lb when the flywheel is at rest. Determine (*a*) the reading of each scale when the flywheel is rotating clockwise at a constant speed, (*b*) the couple that must be applied to the flywheel. Assume that the length of the belt does not change.

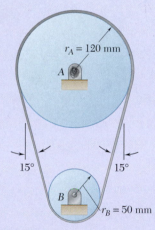

$r_A = 120$ mm

A

15°   15°

B

$r_B = 50$ mm

Fig. P8.112

8.112 A flat belt is used to transmit a couple from drum *B* to drum *A*. Knowing that the coefficient of static friction is 0.40 and that the allowable belt tension is 450 N, determine the largest couple that can be exerted on drum *A*.

8.113 A flat belt is used to transmit a couple from pulley *A* to pulley *B*. The radius of each pulley is 60 mm, and a force of magnitude *P* = 900 N is applied as shown to the axle of pulley *A*. Knowing that the coefficient of static friction is 0.35, determine (*a*) the largest couple that can be transmitted, (*b*) the corresponding maximum value of the tension in the belt.

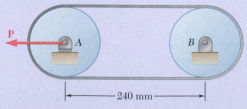

P

A   B

240 mm

Fig. P8.113

8.114 Solve Prob. 8.113 assuming that the belt is looped around the pulleys in a figure eight.

476

476

8.115 The speed of the brake drum shown is controlled by a belt attached to the control bar AD. A force $\mathbf{P}$ with a magnitude of 25 lb is applied to the control bar at A. Determine the magnitude of the couple being applied to the drum knowing that the coefficient of kinetic friction between the belt and the drum is 0.25, that $a = 4$ in., and that the drum is rotating at a constant speed (a) counterclockwise, (b) clockwise.

8.116 The speed of the brake drum shown is controlled by a belt attached to the control bar AD. Knowing that $a = 4$ in., determine the maximum value of the coefficient of static friction for which the brake is not self-locking when the drum rotates counterclockwise.

8.117 The speed of the brake drum shown is controlled by a belt attached to the control bar AD. Knowing that the coefficient of static friction is 0.30 and that the brake drum is rotating counterclockwise, determine the minimum value of a for which the brake is not self-locking.

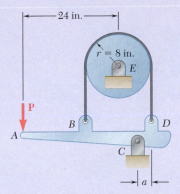

Fig. *P8.115*, *P8.116*, and P8.117

8.118 Bucket A and block C are connected by a cable that passes over drum B. Knowing that drum B rotates slowly counterclockwise and that the coefficients of friction at all surfaces are $\mu_s = 0.35$ and $\mu_k = 0.25$, determine the smallest combined mass m of the bucket and its contents for which block C will (a) remain at rest, (b) start moving up the incline, (c) continue moving up the incline at a constant speed.

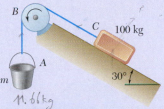

11.66kg

Fig. P8.118

8.119 Solve Prob. 8.118 assuming that drum B is frozen and cannot rotate.

8.120 and 8.122 A cable is placed around three parallel pipes. Knowing that the coefficients of friction are $\mu_s = 0.25$ and $\mu_k = 0.20$, determine (a) the smallest weight W for which equilibrium is maintained, (b) the largest weight W that can be raised if pipe B is slowly rotated counterclockwise while pipes A and C remain fixed.

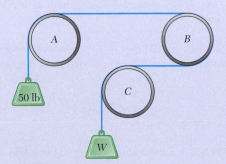

Fig. P8.120 and P8.121

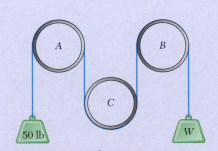

Fig. *P8.122* and *P8.123*

8.121 and 8.123 A cable is placed around three parallel pipes. Two of the pipes are fixed and do not rotate; the third pipe is slowly rotated. Knowing that the coefficients of friction are $\mu_s = 0.25$ and $\mu_k = 0.20$, determine the largest weight W that can be raised (a) if only pipe A is rotated counterclockwise, (b) if only pipe C is rotated clockwise.

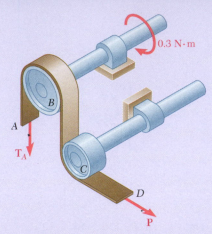

Fig. P8.124

8.124 A recording tape passes over the 20-mm-radius drive drum B and under the idler drum C. Knowing that the coefficients of friction between the tape and the drums are $\mu_s = 0.40$ and $\mu_k = 0.30$ and that drum C is free to rotate, determine the smallest allowable value of P if slipping of the tape on drum B is not to occur.

8.125 Solve Prob. 8.124 assuming that the idler drum C is frozen and cannot rotate.

8.126 The strap wrench shown is used to grip the pipe firmly without marring the external surface of the pipe. Knowing that the coefficient of static friction is the same for all surfaces of contact, determine the smallest value of μ_s for which the wrench will be self-locking when $a = 200$ mm, $r = 30$ mm, and $\theta = 65°$.

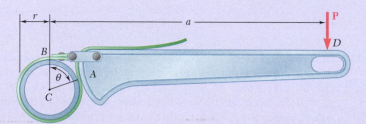

Fig. P8.126

8.127 Solve Prob. 8.126 assuming that $\theta = 75°$.

8.128 The 10-lb bar AE is suspended by a cable that passes over a 5-in.-radius drum. Vertical motion of end E of the bar is prevented by the two stops shown. Knowing that $\mu_s = 0.30$ between the cable and the drum, determine (a) the largest counterclockwise couple $\mathbf{M}_0$ that can be applied to the drum if slipping is not to occur, (b) the corresponding force exerted on end E of the bar.

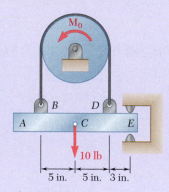

Fig. P8.128

8.129 Solve Prob. 8.128 assuming that a clockwise couple $\mathbf{M}_0$ is applied to the drum.

8.130 Prove that Eqs. (8.13) and (8.14) are valid for any shape of surface provided that the coefficient of friction is the same at all points of contact.

8.131 Complete the derivation of Eq. (8.15), which relates the tension in both parts of a V belt.

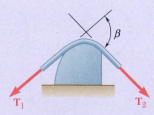

Fig. P8.130

8.132 Solve Prob. 8.112 assuming that the flat belt and drums are replaced by a V belt and V pulleys with $\alpha = 36°$. (The angle α is as shown in Fig. 8.15a.)

8.133 Solve Prob. 8.113 assuming that the flat belt and pulleys are replaced by a V belt and V pulleys with $\alpha = 36°$. (The angle α is as shown in Fig. 8.15a.)

Review and Summary

This chapter was devoted to the study of **dry friction**, i.e., to problems involving rigid bodies in contact along unlubricated surfaces.

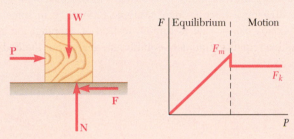

Fig. 8.16

Static and Kinetic Friction

If we apply a horizontal force P to a block resting on a horizontal surface [Sec. 8.1], we note that at first the block does not move. This shows that a **friction force F** must have developed to balance **P** (Fig. 8.16). As the magnitude of **P** increases, the magnitude of **F** also increases until it reaches a maximum value F_m. If **P** is further increased, the block starts sliding, and the magnitude of **F** drops from F_m to a lower value F_k. Experimental evidence shows that F_m and F_k are proportional to the normal component N of the reaction of the surface. We have

$$F_m = \mu_s N \qquad F_k = \mu_k N \qquad \text{(8.1, 8.2)}$$

where μ_s and μ_k are called, respectively, the **coefficient of static friction** and the **coefficient of kinetic friction**. These coefficients depend on the nature and the condition of the surfaces in contact. Approximate values of the coefficients of static friction are given in Table 8.1.

Angles of Friction

It is sometimes convenient to replace the normal force **N** and the friction force **F** by their resultant **R** (Fig. 8.17). As the friction force increases and reaches its maximum value $F_m = \mu_s N$, the angle ϕ that **R** forms with the normal to the surface increases and reaches a maximum value ϕ_s, which is called the **angle of static friction**. If motion actually takes place, the magnitude of **F** drops to F_k; similarly, the angle ϕ drops to a lower value ϕ_k, which is called the **angle of kinetic friction**. As shown in Sec. 8.1B, we have

$$\tan \phi_s = \mu_s \qquad \tan \phi_k = \mu_k \qquad \text{(8.3, 8.4)}$$

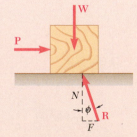

Fig. 8.17

Problems Involving Friction

When solving equilibrium problems involving friction, you should keep in mind that the magnitude F of the friction force is equal to $F_m = \mu_s N$ only if *the body is about to slide* [Sec. 8.1C]. *If motion is not impending,* you should treat F and N as independent unknowns to be determined from the equilibrium equations (Fig. 8.18a). You should also check that the value of F required to maintain equilibrium is not larger than F_m; if it were, the body would move, and the magnitude of the friction force would be $F_k = \mu_k N$ [Sample Prob. 8.1]. On the other hand, *if motion is known to be impending*, F has reached its maximum value $F_m = \mu_s N$ (Fig. 8.18b), and you should substitute this expression for F in the equilibrium equations [Sample Prob. 8.3]. When only three forces are involved in a free-body diagram, including the reaction **R** of the surface in contact with the body, it is usually more convenient to solve the problem by drawing a force triangle [Sample Prob. 8.2]. In some problems, impending motion can be due to tipping instead of slipping; the assessment of this condition requires a moment equilibrium analysis of the body [Sample Prob. 8.4].

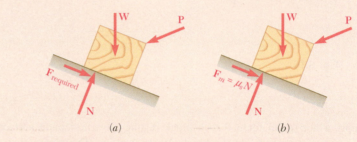

Fig. 8.18

When a problem involves the analysis of the forces exerted on each other by *two bodies A and B*, it is important to show the friction forces with their correct sense. The correct sense for the friction force exerted by B on A, for instance, is opposite to that of the *relative motion* (or impending motion) of A with respect to B [Fig. 8.6].

Wedges and Screws

In the later sections of this chapter, we considered several specific engineering applications where dry friction plays an important role. In the case of **wedges**, which are simple machines used to raise heavy loads [Sec. 8.2A], we must draw two or more free-body diagrams, taking care to show each friction force with its correct sense [Sample Prob. 8.5]. The analysis of **square-threaded screws**, which are frequently used in jacks, presses, and other mechanisms, is reduced to the analysis of a block sliding on an incline by unwrapping the thread of the screw and showing it as a straight line [Sec. 8.2B]. This is shown again in Fig. 8.19, where r denotes the *mean radius* of the thread, L is the *lead* of the screw (i.e., the distance through which the screw advances in one turn), **W** is the load, and Qr is equal to the couple exerted on the screw. We noted in the case of multiple-threaded screws that the lead L of the screw is *not* equal to its pitch, which is the distance measured between two consecutive threads.

Other engineering applications considered in this chapter were **journal bearings** and **axle friction** [Sec. 8.3A], **thrust bearings** and **disk friction** [Sec. 8.3B], **wheel friction** and **rolling resistance** [Sec. 8.3C], and **belt friction** [Sec. 8.4].

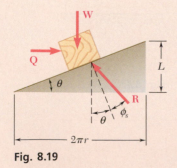

Fig. 8.19

Belt Friction

In solving a problem involving a flat belt passing over a fixed cylinder, it is important to first determine the direction in which the belt slips or is about to slip. If the drum is rotating, the motion or impending motion of the belt should be determined *relative* to the rotating drum. For instance, if the belt shown in Fig. 8.20 is about to slip to the right relative to the drum, the friction

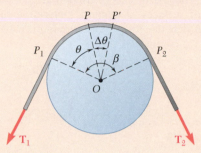

Fig. 8.20

forces exerted by the drum on the belt are directed to the left, and the tension is larger in the right-hand portion of the belt than in the left-hand portion. Denoting the larger tension by T_2, the smaller tension by T_1, the coefficient of static friction by μ_s, and the angle (in radians) subtended by the belt by β, we derived in Sec. 8.4 the formulas

$$\ln \frac{T_2}{T_1} = \mu_s \beta \qquad\qquad (8.13)$$

$$\frac{T_2}{T_1} = e^{\mu_s \beta} \qquad\qquad (8.14)$$

that we used in solving Sample Probs. 8.8 and 8.9. If the belt actually slips on the drum, the coefficient of static friction μ_s should be replaced by the coefficient of kinetic friction μ_k in both of these formulas.

Review Problems

Fig. P8.134

8.134 and 8.135 The coefficients of friction are $\mu_s = 0.40$ and $\mu_k = 0.30$ between all surfaces of contact. Determine the smallest force **P** required to start the 30-kg block moving if cable AB (a) is attached as shown, (b) is removed.

Fig. *P8.135*

8.136 A 120-lb cabinet is mounted on casters that can be locked to prevent their rotation. The coefficient of static friction between the floor and each caster is 0.30. If $h = 32$ in., determine the magnitude of the force **P** required to move the cabinet to the right (a) if all casters are locked, (b) if the casters at B are locked and the casters at A are free to rotate, (c) if the casters at A are locked and the casters at B are free to rotate.

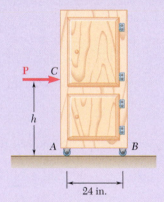

24 in.

Fig. P8.136

8.137 A slender rod with a length of L is lodged between peg C and the vertical wall, and supports a load **P** at end A. Knowing that the coefficient of static friction between the peg and the rod is 0.15 and neglecting friction at the roller, determine the range of values of the ratio L/a for which equilibrium is maintained.

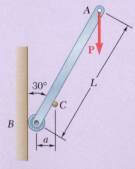

Fig. P8.137

8.138 The hydraulic cylinder shown exerts a force of 3 kN directed to the right on point B and to the left on point E. Determine the magnitude of the couple $\mathbf{M}$ required to rotate the drum clockwise at a constant speed.

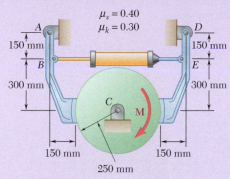

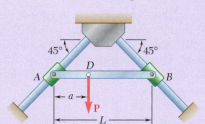

Fig. P8.138

8.139 A rod DE and a small cylinder are placed between two guides as shown. The rod is not to slip downward, however large the force $\mathbf{P}$ may be; i.e., the arrangement is said to be self-locking. Neglecting the weight of the cylinder, determine the minimum allowable coefficients of static friction at A, B, and C.

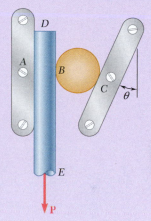

Fig. *P8.139*

8.140 Bar AB is attached to collars that can slide on the inclined rods shown. A force $\mathbf{P}$ is applied at point D located at a distance a from end A. Knowing that the coefficient of static friction μ_s between each collar and the rod upon which it slides is 0.30 and neglecting the weights of the bar and of the collars, determine the smallest value of the ratio a/L for which equilibrium is maintained.

Fig. P8.140

8.141 Two 10° wedges of negligible weight are used to move and position the 400-lb block. Knowing that the coefficient of static friction is 0.25 at all surfaces of contact, determine the smallest force $\mathbf{P}$ that should be applied as shown to one of the wedges.

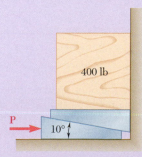

Fig. P8.141

8.142 A 10° wedge is used to split a section of a log. The coefficient of static friction between the wedge and the log is 0.35. Knowing that a force **P** with a magnitude of 600 lb was required to insert the wedge, determine the magnitude of the forces exerted on the wood by the wedge after insertion.

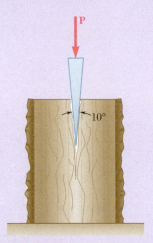

Fig. *P8.142*

8.143 In the gear-pulling assembly shown, the square-threaded screw *AB* has a mean radius of 15 mm and a lead of 4 mm. Knowing that the coefficient of static friction is 0.10, determine the couple that must be applied to the screw in order to produce a force of 3 kN on the gear. Neglect friction at end *A* of the screw.

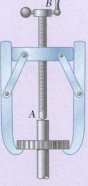

Fig. P8.143

8.144 A lever of negligible weight is loosely fitted onto a 30-mm-radius fixed shaft as shown. Knowing that a force **P** of magnitude 275 N will just start the lever rotating clockwise, determine (*a*) the coefficient of static friction between the shaft and the lever, (*b*) the smallest force **P** for which the lever does not start rotating counterclockwise.

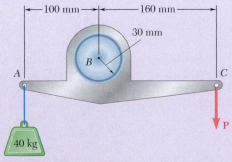

Fig. P8.144

8.145 In the pivoted motor mount shown, the weight **W** of the 175-lb motor is used to maintain tension in the drive belt. Knowing that the coefficient of static friction between the flat belt and drums *A* and *B* is 0.40 and neglecting the weight of platform *CD*, determine the largest couple that can be transmitted to drum *B* when the drive drum *A* is rotating clockwise.

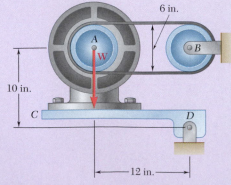

Fig. *P8.145*

9

Distributed Forces: Moments of Inertia

The strength of structural members used in the construction of buildings depends to a large extent on the properties of their cross sections. This includes the second moments of area, or moments of inertia, of these cross sections.

Introduction

Objectives

- **Describe** the second moment, or moment of inertia, of an area.
- **Determine** the rectangular and polar moments of inertia of areas and their corresponding radii of gyration by integration.
- **Develop** the parallel-axis theorem and apply it to determine the moments of inertia of composite areas.
- **Introduce** the product of inertia and apply it to analyze the transformation of moments of inertia when coordinate axes are rotated.
- **Describe** the moment of inertia of a mass with respect to an axis.
- **Apply** the parallel-axis theorem to facilitate mass moment of inertia computations.
- **Analyze** the transformation of mass moments of inertia when coordinate axes are rotated.

Introduction

In Chap. 5, we analyzed various systems of forces distributed over an area or volume. The three main types of forces considered were (1) weights of homogeneous plates of uniform thickness (Secs. 5.1 and 5.2); (2) distributed loads on beams and submerged surfaces (Sec. 5.3); and (3) weights of homogeneous three-dimensional bodies (Sec. 5.4). In all of these cases, the distributed forces were proportional to the elemental areas or volumes associated with them. Therefore, we could obtain the resultant of these forces by summing the corresponding areas or volumes, and we determined the moment of the resultant about any given axis by computing the first moments of the areas or volumes about that axis.

In the first part of this chapter, we consider distributed forces $\Delta \mathbf{F}$ where the magnitudes depend not only upon the elements of area ΔA on which these forces act but also upon the distance from ΔA to some given axis. More precisely, we assume the magnitude of the force per unit area $\Delta F/\Delta A$ varies linearly with the distance to the axis. Forces of this type arise in the study of the bending of beams and in problems involving submerged nonrectangular surfaces.

Starting with the assumption that the elemental forces involved are distributed over an area A and vary linearly with the distance y to the x axis, we will show that the magnitude of their resultant $\mathbf{R}$ depends upon the first moment Q_x of the area A. However, the location of the point where $\mathbf{R}$ is applied depends upon the *second moment*, or *moment of inertia*, I_x of the same area with respect to the x axis. You will see how to compute the moments of inertia of various areas with respect to given x and y axes. We also introduce the *polar moment of inertia* J_O of an area. To facilitate these computations, we establish a relation between the moment of inertia I_x

of an area A with respect to a given x axis and the moment of inertia $I_{x'}$ of the same area with respect to the parallel centroidal x' axis (a relation known as the parallel-axis theorem). You will also study the transformation of the moments of inertia of a given area when the coordinate axes are rotated.

In the second part of this chapter, we will explain how to determine the moments of inertia of various *masses* with respect to a given axis. Moments of inertia of masses are common in dynamics problems involving the rotation of a rigid body about an axis. To facilitate the computation of mass moments of inertia, we introduce another version of the parallel-axis theorem. Finally, we will analyze the transformation of moments of inertia of masses when the coordinate axes are rotated.

9.1 MOMENTS OF INERTIA OF AREAS

In the first part of this chapter, we consider distributed forces $\Delta \mathbf{F}$ whose magnitudes ΔF are proportional to the elements of area ΔA on which the forces act and, at the same time, vary linearly with the distance from ΔA to a given axis.

9.1A Second Moment, or Moment of Inertia, of an Area

Consider a beam with a uniform cross section that is subjected to two equal and opposite couples: one applied at each end of the beam. Such a beam is said to be in **pure bending**. The internal forces in any section of the beam are distributed forces whose magnitudes $\Delta F = ky \, \Delta A$ vary linearly with the distance y between the element of area ΔA and an axis passing through the centroid of the section. (This statement can be derived in a course on mechanics of materials.) This axis, represented by the x axis in Fig. 9.1, is known as the **neutral axis** of the section. The forces on one side of the neutral axis are forces of compression, whereas those on the other side are forces of tension. On the neutral axis itself, the forces are zero.

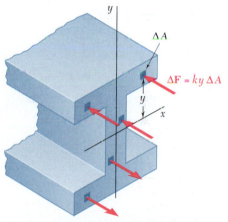

Fig. 9.1 Representative forces on a cross section of a beam subjected to equal and opposite couples at each end.

The magnitude of the resultant $\mathbf{R}$ of the elemental forces $\Delta \mathbf{F}$ that act over the entire section is

$$R = \int ky \, dA = k \int y \, dA$$

You might recognize this last integral as the **first moment** Q_x of the section about the x axis; it is equal to $\bar{y}A$ and is thus equal to zero, since the centroid of the section is located on the x axis. The system of forces $\Delta \mathbf{F}$ thus reduces to a couple. The magnitude M of this couple (bending moment) must be equal to the sum of the moments $\Delta M_x = y \, \Delta F = ky^2 \, \Delta A$ of the elemental forces. Integrating over the entire section, we obtain

$$M = \int ky^2 \, dA = k \int y^2 \, dA$$

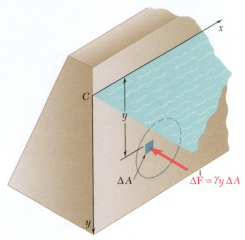

Fig. 9.2 Vertical circular gate, submerged under water, used to close the outlet of a reservoir.

This last integral is known as the **second moment**, or **moment of inertia**,[†] of the beam section with respect to the x axis and is denoted by I_x. We obtain it by multiplying each element of area dA by the *square of its distance* from the x axis and integrating over the beam section. Since each product $y^2\,dA$ is positive, regardless of the sign of y, or zero (if y is zero), the integral I_x is always positive.

Another example of a second moment, or moment of inertia, of an area is provided by the following problem from hydrostatics. A vertical circular gate used to close the outlet of a large reservoir is submerged under water as shown in Fig. 9.2. What is the resultant of the forces exerted by the water on the gate, and what is the moment of the resultant about the line of intersection of the plane of the gate and the water surface (x axis)?

If the gate were rectangular, we could determine the resultant of the forces due to water pressure from the pressure curve, as we did in Sec. 5.3B. Since the gate is circular, however, we need to use a more general method. Denoting the depth of an element of area ΔA by y and the specific weight of water by γ, the pressure at an element is $p = \gamma y$, and the magnitude of the elemental force exerted on ΔA is $\Delta F = p\,\Delta A = \gamma y\,\Delta A$. The magnitude of the resultant of the elemental forces is thus

$$R = \int \gamma y\,dA = \gamma \int y\,dA$$

We can obtain this by computing the first moment $Q_x = \int y\,dA$ of the area of the gate with respect to the x axis. The moment M_x of the resultant must be equal to the sum of the moments $\Delta M_x = y\,\Delta F = \gamma y^2\,\Delta A$ of the elemental forces. Integrating over the area of the gate, we have

$$M_x = \int \gamma y^2\,dA = \gamma \int y^2\,dA$$

Here again, the last integral represents the second moment, or moment of inertia, I_x of the area with respect to the x axis.

9.1B Determining the Moment of Inertia of an Area by Integration

We just defined the second moment, or moment of inertia, I_x of an area A with respect to the x axis. In a similar way, we can also define the moment of inertia I_y of the area A with respect to the y axis (Fig. 9.3a):

Moments of inertia of an area

$$I_x = \int y^2\,dA \qquad I_y = \int x^2\,dA \tag{9.1}$$

[†]The term *second moment* is more proper than the term *moment of inertia*, which logically should be used only to denote integrals of mass (see Sec. 9.5). In engineering practice, however, moment of inertia is used in connection with areas as well as masses.

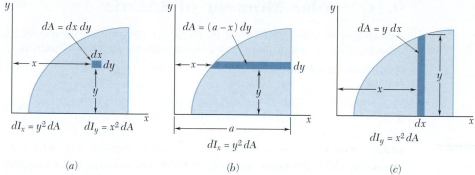

Fig. 9.3 (a) Rectangular moments of inertia dI_x and dI_y of an area dA; (b) calculating I_x with a horizontal strip; (c) calculating I_y with a vertical strip.

We can evaluate these integrals, which are known as the **rectangular moments of inertia** of the area A, more easily if we choose dA to be a thin strip parallel to one of the coordinate axes. To compute I_x, we choose the strip parallel to the x axis, so that all points of the strip are at the same distance y from the x axis (Fig. 9.3b). We obtain the moment of inertia dI_x of the strip by multiplying the area dA of the strip by y^2. To compute I_y, we choose the strip parallel to the y axis, so that all points of the strip are at the same distance x from the y axis (Fig. 9.3c). Then the moment of inertia dI_y of the strip is $x^2\,dA$.

Moment of Inertia of a Rectangular Area.

As an example, let us determine the moment of inertia of a rectangle with respect to its base (Fig. 9.4). Dividing the rectangle into strips parallel to the x axis, we have

$$dA = b\,dy \qquad dI_x = y^2 b\,dy$$

$$I_x = \int_0^h by^2\,dy = \frac{1}{3}bh^3 \tag{9.2}$$

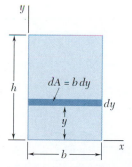

Fig. 9.4 Calculating the moment of inertia of a rectangular area with respect to its base.

Computing I_x and I_y Using the Same Elemental Strips.

We can use Eq. (9.2) to determine the moment of inertia dI_x with respect to the x axis of a rectangular strip that is parallel to the y axis, such as the strip shown in Fig. 9.3c. Setting $b = dx$ and $h = y$ in formula (9.2), we obtain

$$dI_x = \frac{1}{3}y^3\,dx$$

We also have

$$dI_y = x^2\,dA = x^2 y\,dx$$

Thus, we can use the same element to compute the moments of inertia I_x and I_y of a given area (Fig. 9.5).

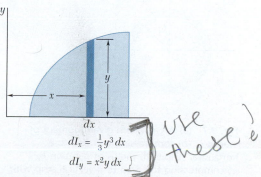

Fig. 9.5 Using the same strip element of a given area to calculate I_x and I_y.

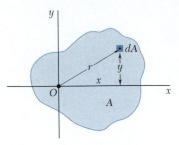

Fig. 9.6 Distance r used to evaluate the polar moment of inertia of area A.

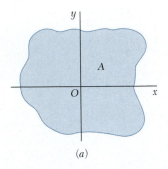

(a)

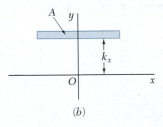

(b)

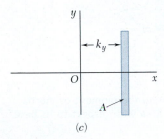

(c)

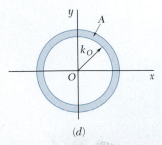

(d)

Fig. 9.7 (a) Area A with given moment of inertia I_x; (b) compressing the area to a horizontal strip with radius of gyration k_x; (c) compressing the area to a vertical strip with radius of gyration k_y; (d) compressing the area to a circular ring with polar radius of gyration k_O.

9.1C Polar Moment of Inertia

An integral of great importance in problems concerning the torsion of cylindrical shafts and in problems dealing with the rotation of slabs is

Polar moment of inertia

$$J_O = \int r^2 \, dA \qquad (9.3)$$

where r is the distance from O to the element of area dA (Fig. 9.6). This integral is called the **polar moment of inertia** of the area A with respect to the "pole" O.

We can compute the polar moment of inertia of a given area from the rectangular moments of inertia I_x and I_y of the area if these quantities are already known. Indeed, noting that $r^2 = x^2 + y^2$, we have

$$J_O = \int r^2 dA = \int (x^2 + y^2) dA = \int y^2 dA + \int x^2 dA$$

that is,

$$J_O = I_x + I_y \qquad (9.4)$$

9.1D Radius of Gyration of an Area

Consider an area A that has a moment of inertia I_x with respect to the x axis (Fig. 9.7a). Imagine that we concentrate this area into a thin strip parallel to the x axis (Fig. 9.7b). If the concentrated area A is to have the same moment of inertia with respect to the x axis, the strip should be placed at a distance k_x from the x axis, where k_x is defined by the relation

$$I_x = k_x^2 A$$

Solving for k_x, we have

Radius of gyration

$$k_x = \sqrt{\frac{I_x}{A}} \qquad (9.5)$$

The distance k_x is referred to as the **radius of gyration** of the area with respect to the x axis. In a similar way, we can define the radii of gyration k_y and k_O (Fig. 9.7c and d); we have

$$I_y = k_y^2 A \qquad k_y = \sqrt{\frac{I_y}{A}} \qquad (9.6)$$

$$J_O = k_O^2 A \qquad k_O = \sqrt{\frac{J_O}{A}} \qquad (9.7)$$

If we rewrite Eq. (9.4) in terms of the radii of gyration, we find that

$$k_O^2 = k_x^2 + k_y^2 \qquad (9.8)$$

Concept Application 9.1

For the rectangle shown in Fig. 9.8, compute the radius of gyration k_x with respect to its base. Using formulas (9.5) and (9.2), you have

$$k_x^2 = \frac{I_x}{A} = \frac{\frac{1}{3}bh^3}{bh} = \frac{h^2}{3} \qquad k_x = \frac{h}{\sqrt{3}}$$

The radius of gyration k_x of the rectangle is shown in Fig. 9.8. Do not confuse it with the ordinate $\bar{y} = h/2$ of the centroid of the area. The radius of gyration k_x depends upon the *second moment* of the area, whereas the ordinate $\bar{y}$ is related to the *first moment* of the area.

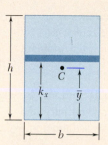

Fig. 9.8 Radius of gyration of a rectangle with respect to its base.

Sample Problem 9.1

Determine the moment of inertia of a triangle with respect to its base.

STRATEGY: To find the moment of inertia with respect to the base, it is expedient to use a differential strip of area parallel to the base. Use the geometry of the situation to carry out the integration.

MODELING: Draw a triangle with a base b and height h, choosing the x axis to coincide with the base (Fig. 1). Choose a differential strip parallel to the x axis to be dA. Since all portions of the strip are at the same distance from the x axis, you have

$$dI_x = y^2\, dA \qquad dA = l\, dy$$

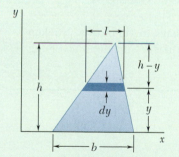

Fig. 1 Triangle with differential strip element parallel to its base.

ANALYSIS: Using similar triangles, you have

$$\frac{l}{b} = \frac{h-y}{h} \qquad l = b\frac{h-y}{h} \qquad dA = b\frac{h-y}{h}dy$$

Integrating dI_x from $y = 0$ to $y = h$, you obtain

$$I_x = \int y^2\, dA = \int_0^h y^2 b\frac{h-y}{h}\,dy = \frac{b}{h}\int_0^h (hy^2 - y^3)\, dy$$

$$= \frac{b}{h}\left[h\frac{y^3}{3} - \frac{y^4}{4}\right]_0^h \qquad\qquad I_x = \frac{bh^3}{12} \blacktriangleleft$$

REFLECT and THINK: This problem also could have been solved using a differential strip perpendicular to the base by applying Eq. (9.2) to express the moment of inertia of this strip. However, because of the geometry of this triangle, you would need two integrals to complete the solution.

Sample Problem 9.2

(a) Determine the centroidal polar moment of inertia of a circular area by direct integration. (b) Using the result of part (a), determine the moment of inertia of a circular area with respect to a diameter.

STRATEGY: Since the area is circular, you can evaluate part (a) by using an annular differential area. For part (b), you can use symmetry and Eq. (9.4) to solve for the moment of inertia with respect to a diameter.

MODELING and ANALYSIS:

a. Polar Moment of Inertia.   Choose an annular differential element of area to be dA (Fig. 1). Since all portions of the differential area are at the same distance from the origin, you have

$$dJ_O = u^2\, dA \quad dA = 2\pi u\, du$$

$$J_O = \int dJ_O = \int_0^r u^2 (2\pi u\, du) = 2\pi \int_0^r u^3\, du$$

$$J_O = \frac{\pi}{2} r^4 \quad \blacktriangleleft$$

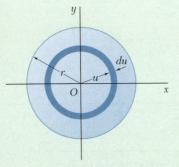

Fig. 1 Circular area with an annular differential element.

b. Moment of Inertia with Respect to a Diameter.   Because of the symmetry of the circular area, $I_x = I_y$. Then from Eq. (9.4), you have

$$J_O = I_x + I_y = 2I_x \quad \frac{\pi}{2} r^4 = 2I_x \quad I_{\text{diameter}} = I_x = \frac{\pi}{4} r^4 \quad \blacktriangleleft$$

REFLECT and THINK: Always look for ways to simplify a problem by the use of symmetry. This is especially true for situations involving circles or spheres.

Sample Problem 9.3

(a) Determine the moment of inertia of the shaded region shown with respect to each of the coordinate axes. (Properties of this region were considered in Sample Prob. 5.4.) (b) Using the results of part (a), determine the radius of gyration of the shaded area with respect to each of the coordinate axes.

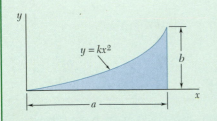

STRATEGY: You can determine the moments of inertia by using a single differential strip of area; a vertical strip will be more convenient. You can calculate the radii of gyration from the moments of inertia and the area of the region.

MODELING: Referring to Sample Prob. 5.4, you can find the equation of the curve and the total area using

$$y = \frac{b}{a^2}x^2 \qquad A = \tfrac{1}{3}ab$$

ANALYSIS:

a. Moments of Inertia.

Moment of Inertia I_x. Choose a vertical differential element of area for dA (Fig. 1). Since all portions of this element are *not* at the same distance from the x axis, you must treat the element as a thin rectangle. The moment of inertia of the element with respect to the x axis is then

$$dI_x = \tfrac{1}{3}y^3\,dx = \frac{1}{3}\left(\frac{b}{a^2}x^2\right)^3 dx = \frac{1}{3}\frac{b^3}{a^6}x^6\,dx$$

$$I_x = \int dI_x = \int_0^a \frac{1}{3}\frac{b^3}{a^6}x^6\,dx = \left[\frac{1}{3}\frac{b^3}{a^6}\frac{x^7}{7}\right]_0^a$$

$$\boxed{I_x = \frac{ab^3}{21}} \quad \blacktriangleleft$$

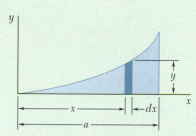

Fig. 1 Subject area with vertical differential strip element.

Moment of Inertia I_y. Use the same vertical differential element of area. Since all portions of the element are at the same distance from the y axis, you have

$$dI_y = x^2\,dA = x^2(y\,dx) = x^2\left(\frac{b}{a^2}x^2\right)dx = \frac{b}{a^2}x^4\,dx$$

$$I_y = \int dI_y = \int_0^a \frac{b}{a^2}x^4\,dx = \left[\frac{b}{a^2}\frac{x^5}{5}\right]_0^a$$

$$\boxed{I_y = \frac{a^3b}{5}} \quad \blacktriangleleft$$

b. Radii of Gyration k_x and k_y.

From the definition of radius of gyration, you have

$$k_x^2 = \frac{I_x}{A} = \frac{ab^3/21}{ab/3} = \frac{b^2}{7} \qquad \boxed{k_x = \sqrt{\tfrac{1}{7}}\,b} \quad \blacktriangleleft$$

and

$$k_y^2 = \frac{I_y}{A} = \frac{a^3b/5}{ab/3} = \tfrac{3}{5}a^2 \qquad \boxed{k_y = \sqrt{\tfrac{3}{5}}\,a} \quad \blacktriangleleft$$

REFLECT and THINK: This problem demonstrates how you can calculate I_x and I_y using the same strip element. However, the general mathematical approach in each case is distinctly different.

SOLVING PROBLEMS ON YOUR OWN

In this section, we introduced the **rectangular and polar moments of inertia of areas** and the corresponding **radii of gyration**. Although the problems you are about to solve may appear more appropriate for a calculus class than for one in mechanics, we hope that our introductory comments have convinced you of the relevance of moments of inertia to your study of a variety of engineering topics.

1. Calculating the rectangular moments of inertia I_x and I_y. We defined these quantities as

$$I_x = \int y^2\, dA \qquad I_y = \int x^2\, dA \tag{9.1}$$

where dA is a differential element of area $dx\, dy$. The moments of inertia are the **second moments of the area**; it is for that reason that I_x, for example, depends on the perpendicular distance y to the area dA. As you study Sec. 9.1, you should recognize the importance of carefully defining the shape and the orientation of dA. Furthermore, you should note the following points.

a. You can obtain the moments of inertia of most areas by means of a single integration. You can use the expressions given in Figs. 9.3b and c and Fig. 9.5 to calculate I_x and I_y. Regardless of whether you use a single or a double integration, be sure to show the element dA that you have chosen on your sketch.

b. The moment of inertia of an area is always positive, regardless of the location of the area with respect to the coordinate axes. The reason is that the moment of inertia is obtained by integrating the product of dA and the *square* of distance. (Note how this differs from the first moment of the area.) Only when an area is *removed* (as in the case for a hole) does its moment of inertia enter in your computations with a minus sign.

c. As a partial check of your work, observe that the moments of inertia are equal to an area times the square of a length. Thus, every term in an expression for a moment of inertia must be a length to the fourth power.

2. Computing the polar moment of inertia J_O. We defined J_O as

$$J_O = \int r^2\, dA \tag{9.3}$$

where $r^2 = x^2 + y^2$. If the given area has circular symmetry (as in Sample Prob. 9.2), it is possible to express dA as a function of r and to compute J_O with a single integration. When the area lacks circular symmetry, it is usually easier first to calculate I_x and I_y and then to determine J_O from

$$J_O = I_x + I_y \tag{9.4}$$

Lastly, if the equation of the curve that bounds the given area is expressed in polar coordinates, then $dA = r\, dr\, d\theta$, and you need to perform a double integration to compute the integral for J_O [see Prob. 9.27].

3. Determining the radii of gyration k_x and k_y and the polar radius of gyration k_O. These quantities are defined in Sec. 9.1D. You should realize that they can be determined only after you have computed the area and the appropriate moments of inertia. It is important to remember that k_x is measured in the y direction, whereas k_y is measured in the x direction; you should carefully study Sec. 9.1D until you understand this point.

Problems

9.1 through 9.4 Determine by direct integration the moment of inertia of the shaded area with respect to the y axis.

9.5 through 9.8 Determine by direct integration the moment of inertia of the shaded area with respect to the x axis.

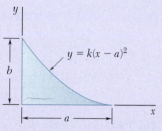

$y = k(x - a)^2$

Fig. P9.1 and *P9.5*

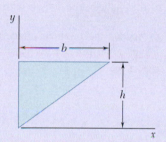

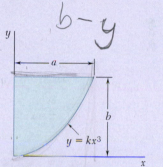

$y = kx^{1/3}$

Fig. P9.2 and P9.6

Fig. P9.3 and *P9.7*

$y = kx^3$

Fig. P9.4 and P9.8

9.9 through 9.11 Determine by direct integration the moment of inertia of the shaded area with respect to the x axis.

9.12 through *9.14* Determine by direct integration the moment of inertia of the shaded area with respect to the y axis.

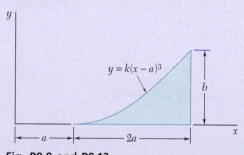

$y = k(x - a)^3$

Fig. P9.9 and P9.12

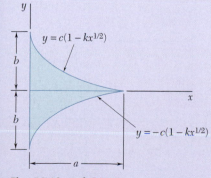

$y = c(1 - kx^{1/2})$

$y = -c(1 - kx^{1/2})$

Fig. P9.10 and *P9.13*

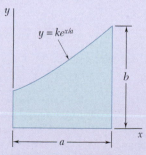

$y = ke^{x/a}$

Fig. P9.11 and *P9.14*

9.15 and 9.16 Determine the moment of inertia and the radius of gyration of the shaded area shown with respect to the x axis.

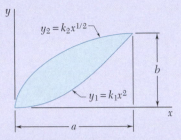

Fig. P9.15 and P9.17

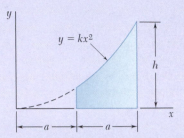

Fig. P9.16 and P9.18

9.17 and 9.18 Determine the moment of inertia and the radius of gyration of the shaded area shown with respect to the y axis.

9.19 Determine the moment of inertia and the radius of gyration of the shaded area shown with respect to the x axis.

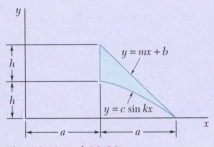

Fig. *P9.19* and *P9.20*

9.20 Determine the moment of inertia and the radius of gyration of the shaded area shown with respect to the y axis.

9.21 and 9.22 Determine the polar moment of inertia and the polar radius of gyration of the shaded area shown with respect to point P.

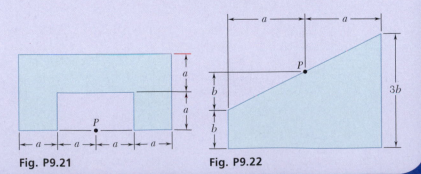

Fig. P9.21

Fig. P9.22

9.23 and 9.24 Determine the polar moment of inertia and the polar radius of gyration of the shaded area shown with respect to point P.

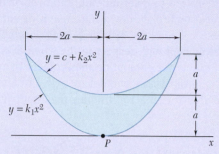

Fig. P9.23

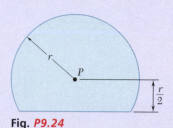

Fig. *P9.24*

9.25 (*a*) Determine by direct integration the polar moment of inertia of the annular area shown with respect to point O. (*b*) Using the result of part *a*, determine the moment of inertia of the given area with respect to the x axis.

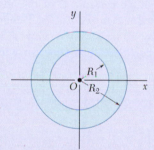

Fig. P9.25 and P9.26

9.26 (*a*) Show that the polar radius of gyration k_O of the annular area shown is approximately equal to the mean radius $R_m = (R_1 + R_2)/2$ for small values of the thickness $t = R_2 - R_1$. (*b*) Determine the percentage error introduced by using R_m in place of k_O for the following values of t/R_m: 1, $\frac{1}{2}$, and $\frac{1}{10}$.

9.27 Determine the polar moment of inertia and the polar radius of gyration of the shaded area shown with respect to point O.

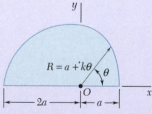

Fig. *P9.27*

9.28 Determine the polar moment of inertia and the polar radius of gyration of the isosceles triangle shown with respect to point O.

*9.29 Using the polar moment of inertia of the isosceles triangle of Prob. 9.28, show that the centroidal polar moment of inertia of a circular area of radius r is $\pi r^4/2$. (*Hint:* As a circular area is divided into an increasing number of equal circular sectors, what is the approximate shape of each circular sector?)

*9.30 Prove that the centroidal polar moment of inertia of a given area A cannot be smaller than $A^2/2\pi$. (*Hint:* Compare the moment of inertia of the given area with the moment of inertia of a circle that has the same area and the same centroid.)

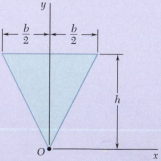

Fig. P9.28

9.2 PARALLEL-AXIS THEOREM AND COMPOSITE AREAS

In practice, we often need to determine the moment of inertia of a complicated area that can be broken down into a sum of simple areas. However, in doing these calculations, we have to determine the moment of inertia of each simple area with respect to the same axis. In this section, we first derive a formula for computing the moment of inertia of an area with respect to a centroidal axis parallel to a given axis. Then we show how you can use this formula for finding the moment of inertia of a composite area.

9.2A The Parallel-Axis Theorem

Consider the moment of inertia I of an area A with respect to an axis AA' (Fig. 9.9). We denote the distance from an element of area dA to AA' by y. This gives us

$$I = \int y^2 \, dA$$

Let us now draw through the centroid C of the area an axis BB' parallel to AA'; this axis is called a *centroidal axis*. Denoting the distance from the element dA to BB' by y', we have $y = y' + d$, where d is the distance between the axes AA' and BB'. Substituting for y in the previous integral, we obtain

$$I = \int y^2 \, dA = \int (y' + d)^2 \, dA$$

$$= \int y'^2 \, dA + 2d \int y' \, dA + d^2 \int dA$$

The first integral represents the moment of inertia $\bar{I}$ of the area with respect to the centroidal axis BB'. The second integral represents the first moment of the area with respect to BB', but since the centroid C of the area is located on this axis, the second integral must be zero. The last integral is equal to the total area A. Therefore, we have

Parallel-axis theorem

$$I = \bar{I} + Ad^2 \tag{9.9}$$

This formula states that the moment of inertia I of an area with respect to any given axis AA' is equal to the moment of inertia $\bar{I}$ of the area with respect to a centroidal axis BB' parallel to AA' *plus* the product of the area A and the square of the distance d between the two axes. This theorem is known as the **parallel-axis theorem**. Substituting $k^2 A$ for I and $\bar{k}^2 A$ for $\bar{I}$, we can also express this theorem as

$$k^2 = \bar{k}^2 + d^2 \tag{9.10}$$

A similar theorem relates the polar moment of inertia J_O of an area about a point O to the polar moment of inertia $\bar{J}_C$ of the same area about its centroid C. Denoting the distance between O and C by d, we have

$$J_O = \bar{J}_C + Ad^2 \quad \text{or} \quad k_O^2 = \bar{k}_C^2 + d^2 \tag{9.11}$$

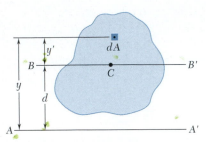

Fig. 9.9 The moment of inertia of an area A with respect to an axis AA' can be determined from its moment of inertia with respect to the centroidal axis BB' by a calculation involving the distance d between the axes.

Concept Application 9.2

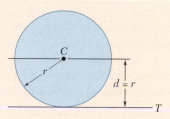

Fig. 9.10 Finding the moment of inertia of a circle with respect to a line tangent to it.

As an application of the parallel-axis theorem, let us determine the moment of inertia I_T of a circular area with respect to a line tangent to the circle (Fig. 9.10). We found in Sample Prob. 9.2 that the moment of inertia of a circular area about a centroidal axis is $\bar{I} = \frac{1}{4}\pi r^4$. Therefore, we have

$$I_T = \bar{I} + Ad^2 = \frac{1}{4}\pi r^4 + (\pi r^2)r^2 = \frac{5}{4}\pi r^4$$

Concept Application 9.3

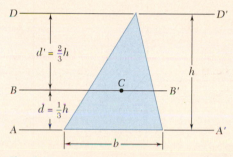

Fig. 9.11 Finding the centroidal moment of inertia of a triangle from the moment of inertia about a parallel axis.

We can also use the parallel-axis theorem to determine the centroidal moment of inertia of an area when we know the moment of inertia of the area with respect to a parallel axis. Consider, for instance, a triangular area (Fig. 9.11). We found in Sample Prob. 9.1 that the moment of inertia of a triangle with respect to its base AA' is equal to $\frac{1}{12}bh^3$. Using the parallel-axis theorem, we have

$$I_{AA'} = \bar{I}_{BB'} + Ad^2$$
$$\bar{I}_{BB'} = I_{AA'} - Ad^2 = \frac{1}{12}bh^3 - \frac{1}{2}bh(\frac{1}{3}h)^2 = \frac{1}{36}bh^3$$

Note that we *subtracted* the product Ad^2 from the given moment of inertia in order to obtain the centroidal moment of inertia of the triangle. That is, this product is *added* when transferring *from* a centroidal axis to a parallel axis, but it is *subtracted* when transferring *to* a centroidal axis. In other words, the moment of inertia of an area is always smaller with respect to a centroidal axis than with respect to any parallel axis.

Returning to Fig. 9.11, we can obtain the moment of inertia of the triangle with respect to the line DD' (which is drawn through a vertex) by writing

$$I_{DD'} = \bar{I}_{BB'} + Ad'^2 = \frac{1}{36}bh^3 + \frac{1}{2}bh(\frac{2}{3}h)^2 = \frac{1}{4}bh^3$$

Note that we could not have obtained $I_{DD'}$ directly from $I_{AA'}$. We can apply the parallel-axis theorem only if one of the two parallel axes passes through the centroid of the area.

9.2B Moments of Inertia of Composite Areas

Consider a composite area A made of several component areas A_1, A_2, A_3, ... The integral representing the moment of inertia of A can be subdivided into integrals evaluated over A_1, A_2, A_3, ... Therefore, we can obtain the moment of inertia of A with respect to a given axis by adding the moments of inertia of the areas A_1, A_2, A_3, . . . with respect to the same axis.

Photo 9.1 Figure 9.13 tabulates data for a small sample of the rolled-steel shapes that are readily available. Shown above are examples of wide-flange shapes that are commonly used in the construction of buildings.

Figure 9.12 shows several common geometric shapes along with formulas for the moments of inertia of each one. Before adding the moments of inertia of the component areas, however, you may have to use the parallel-axis theorem to transfer each moment of inertia to the desired axis. Sample Probs. 9.4 and 9.5 illustrate the technique.

Properties of the cross sections of various structural shapes are given in Fig. 9.13. As we noted in Sec. 9.1A, the moment of inertia of a beam

Rectangle	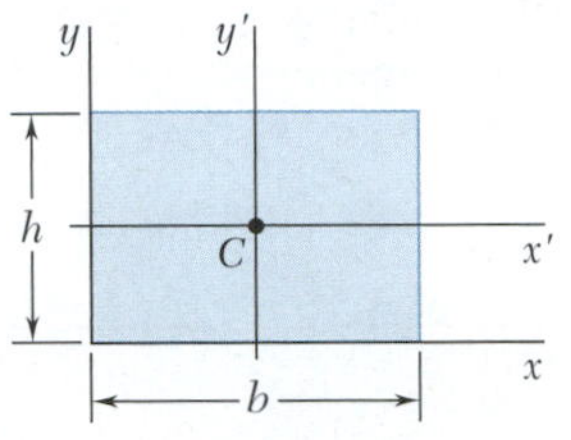	$\overline{I}_{x'} = \frac{1}{12}bh^3$ $\overline{I}_{y'} = \frac{1}{12}b^3h$ $I_x = \frac{1}{3}bh^3$ $I_y = \frac{1}{3}b^3h$ $J_C = \frac{1}{12}bh(b^2 + h^2)$
Triangle	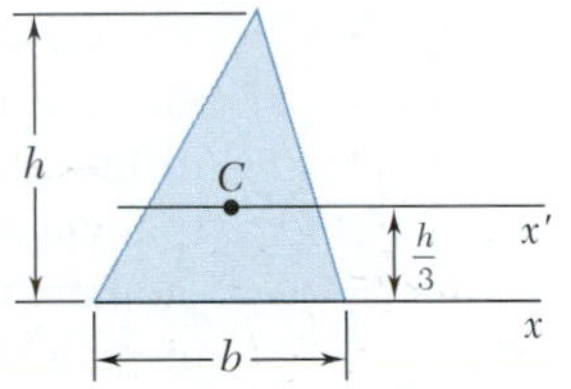	$\overline{I}_{x'} = \frac{1}{36}bh^3$ $I_x = \frac{1}{12}bh^3$
Circle	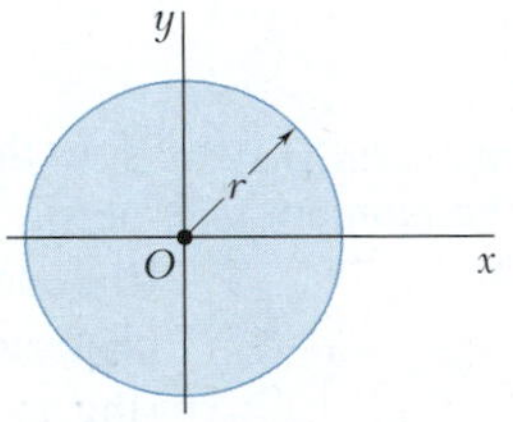	$\overline{I}_x = \overline{I}_y = \frac{1}{4}\pi r^4$ $J_O = \frac{1}{2}\pi r^4$
Semicircle	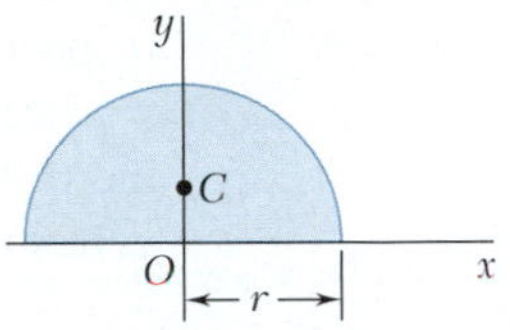	$I_x = I_y = \frac{1}{8}\pi r^4$ $J_O = \frac{1}{4}\pi r^4$
Quarter circle	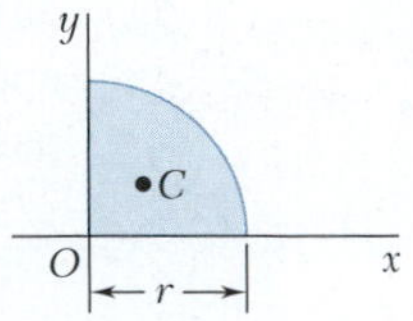	$I_x = I_y = \frac{1}{16}\pi r^4$ $J_O = \frac{1}{8}\pi r^4$
Ellipse	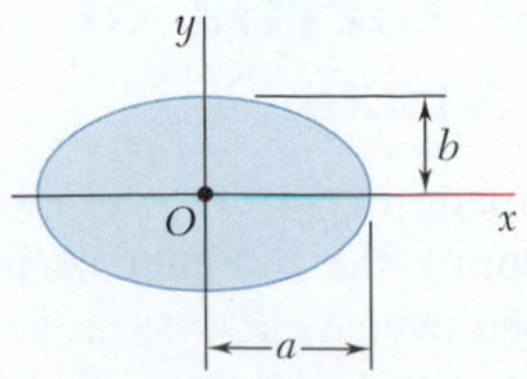	$\overline{I}_x = \frac{1}{4}\pi ab^3$ $\overline{I}_y = \frac{1}{4}\pi a^3b$ $J_O = \frac{1}{4}\pi ab(a^2 + b^2)$

Fig. 9.12 Moments of inertia of common geometric shapes.

		Area in²	Depth in.	Width in.	Axis X–X			Axis Y–Y		
	Designation				$\bar{I}_x$, in⁴	$\bar{k}_x$, in.	$\bar{y}$, in.	$\bar{I}_y$, in⁴	$\bar{k}_y$, in.	$\bar{x}$, in.
W Shapes (Wide-Flange Shapes)	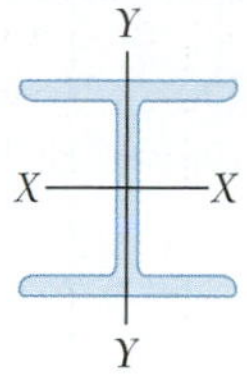 W18 × 76†	22.3	18.2	11.0	1330	7.73		152	2.61	
	W16 × 57	16.8	16.4	7.12	758	6.72		43.1	1.60	
	W14 × 38	11.2	14.1	6.77	385	5.87		26.7	1.55	
	W8 × 31	9.12	8.00	8.00	110	3.47		37.1	2.02	
S Shapes (American Standard Shapes)	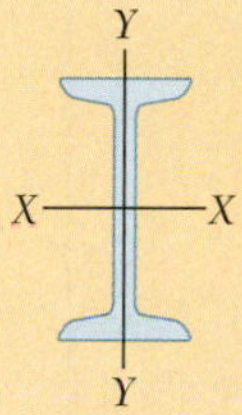 S18 × 54.7†	16.0	18.0	6.00	801	7.07		20.7	1.14	
	S12 × 31.8	9.31	12.0	5.00	217	4.83		9.33	1.00	
	S10 × 25.4	7.45	10.0	4.66	123	4.07		6.73	0.950	
	S6 × 12.5	3.66	6.00	3.33	22.0	2.45		1.80	0.702	
C Shapes (American Standard Channels)	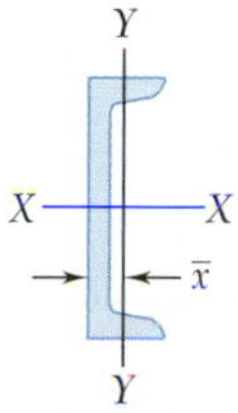 C12 × 20.7†	6.08	12.0	2.94	129	4.61		3.86	0.797	0.698
	C10 × 15.3	4.48	10.0	2.60	67.3	3.87		2.27	0.711	0.634
	C8 × 11.5	3.37	8.00	2.26	32.5	3.11		1.31	0.623	0.572
	C6 × 8.2	2.39	6.00	1.92	13.1	2.34		0.687	0.536	0.512
Angles	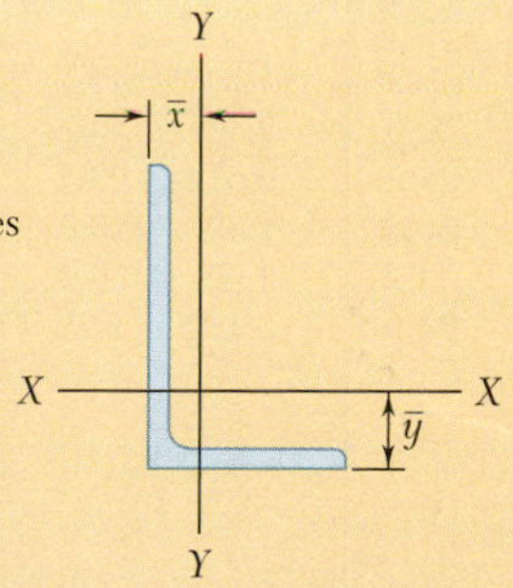 L6 × 6 × 1‡	11.0			35.4	1.79	1.86	35.4	1.79	1.86
	L4 × 4 × ½	3.75			5.52	1.21	1.18	5.52	1.21	1.18
	L3 × 3 × ¼	1.44			1.23	0.926	0.836	1.23	0.926	0.836
	L6 × 4 × ½	4.75			17.3	1.91	1.98	6.22	1.14	0.981
	L5 × 3 × ½	3.75			9.43	1.58	1.74	2.55	0.824	0.746
	L3 × 2 × ¼	1.19			1.09	0.953	0.980	0.390	0.569	0.487

Fig. 9.13A Properties of rolled-steel shapes (U.S. customary units).*

*Courtesy of the American Institute of Steel Construction, Chicago, Illinois

†Nominal depth in inches and weight in pounds per foot

‡Depth, width, and thickness in inches

section about its neutral axis is closely related to the computation of the bending moment in that section of the beam. Thus, determining moments of inertia is a prerequisite to the analysis and design of structural members.

Note that the radius of gyration of a composite area is *not* equal to the sum of the radii of gyration of the component areas. In order to determine the radius of gyration of a composite area, you must first compute the moment of inertia of the area.

		Designation	Area mm²	Depth mm	Width mm	Axis X–X			Axis Y–Y		
						$\overline{I}_x$ 10⁶ mm⁴	$\overline{k}_x$ mm	$\overline{y}$ mm	$\overline{I}_y$ 10⁶ mm⁴	$\overline{k}_y$ mm	$\overline{x}$ mm
W Shapes (Wide-Flange Shapes) 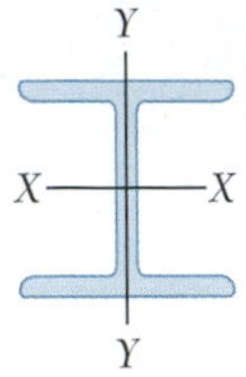		W460 × 113†	14 400	462	279	554	196		63.3	66.3	
		W410 × 85	10 800	417	181	316	171		17.9	40.6	
		W360 × 57.8	7230	358	172	160	149		11.1	39.4	
		W200 × 46.1	5880	203	203	45.8	88.1		15.4	51.3	
S Shapes (American Standard Shapes) 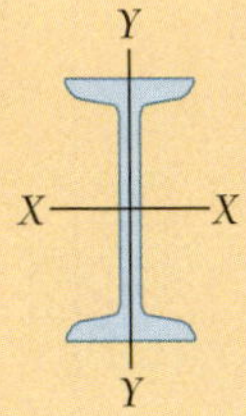		S460 × 81.4†	10 300	457	152	333	180		8.62	29.0	
		S310 × 47.3	6010	305	127	90.3	123		3.88	25.4	
		S250 × 37.8	4810	254	118	51.2	103		2.80	24.1	
		S150 × 18.6	2360	152	84.6	9.16	62.2		0.749	17.8	
C Shapes (American Standard Channels) 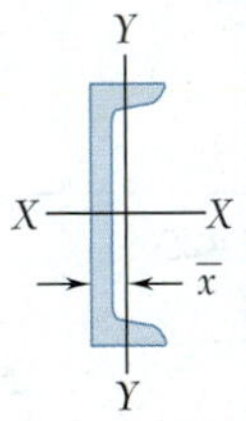		C310 × 30.8†	3920	305	74.7	53.7	117		1.61	20.2	17.7
		C250 × 22.8	2890	254	66.0	28.0	98.3		0.945	18.1	16.1
		C200 × 17.1	2170	203	57.4	13.5	79.0		0.545	15.8	14.5
		C150 × 12.2	1540	152	48.8	5.45	59.4		0.286	13.6	13.0
Angles 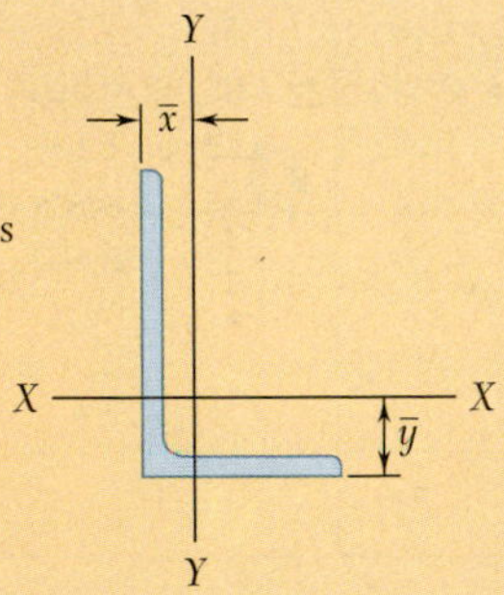		L152 × 152 × 25.4‡	7100			14.7	45.5	47.2	14.7	45.5	47.2
		L102 × 102 × 12.7	2420			2.30	30.7	30.0	2.30	30.7	30.0
		L76 × 76 × 6.4	929			0.512	23.5	21.2	0.512	23.5	21.2
		L152 × 102 × 12.7	3060			7.20	48.5	50.3	2.59	29.0	24.9
		L127 × 76 × 12.7	2420			3.93	40.1	44.2	1.06	20.9	18.9
		L76 × 51 × 6.4	768			0.454	24.2	24.9	0.162	14.5	12.4

Fig. 9.13B Properties of rolled-steel shapes (SI units).

†Nominal depth in millimeters and mass in kilograms per meter

‡Depth, width, and thickness in millimeters

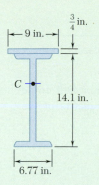

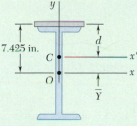

Fig. 1 Origin of coordinates placed at centroid of wide-flange shape.

Sample Problem 9.4

The strength of a W14 × 38 rolled-steel beam is increased by attaching a 9 × 3/4-in. plate to its upper flange as shown. Determine the moment of inertia and the radius of gyration of the composite section with respect to an axis that is parallel to the plate and passes through the centroid C of the section.

STRATEGY: This problem involves finding the moment of inertia of a composite area with respect to its centroid. You should first determine the location of this centroid. Then, using the parallel-axis theorem, you can determine the moment of inertia relative to this centroid for the overall section from the centroidal moment of inertia for each component part.

MODELING and ANALYSIS: Place the origin O of coordinates at the centroid of the wide-flange shape, and compute the distance $\overline{Y}$ to the centroid of the composite section by using the methods of Chap. 5 (Fig. 1). Refer to Fig. 9.13A for the area of the wide-flange shape. The area and the y coordinate of the centroid of the plate are

$$A = (9 \text{ in.})(0.75 \text{ in.}) = 6.75 \text{ in}^2$$
$$\overline{y} = \tfrac{1}{2}(14.1 \text{ in.}) + \tfrac{1}{2}(0.75 \text{ in.}) = 7.425 \text{ in.}$$

Section	Area, in^2	$\overline{y}$, in.	$\overline{y}A$, in^3
Plate	6.75	7.425	50.12
Wide-flange shape	11.2	0	0
	$\Sigma A = 17.95$		$\Sigma \overline{y}A = 50.12$

$$\overline{Y}\Sigma A = \Sigma \overline{y}A \qquad \overline{Y}(17.95) = 50.12 \qquad \overline{Y} = 2.792 \text{ in.}$$

Moment of Inertia. Use the parallel-axis theorem to determine the moments of inertia of the wide-flange shape and the plate with respect to the x' axis. This axis is a centroidal axis for the composite section but *not* for either of the elements considered separately. You can obtain the value of $\overline{I}_x$ for the wide-flange shape from Fig. 9.13A.

For the wide-flange shape,

$$I_{x'} = \overline{I}_x + A\overline{Y}^2 = 385 + (11.2)(2.792)^2 = 472.3 \text{ in}^4$$

For the plate,

$$I_{x'} = \overline{I}_x + Ad^2 = (\tfrac{1}{12})(9)(\tfrac{3}{4})^3 + (6.75)(7.425 - 2.792)^2 = 145.2 \text{ in}^4$$

For the composite area,

$$I_{x'} = 472.3 + 145.2 = 617.5 \text{ in}^4 \qquad I_{x'} = 618 \text{ in}^4 \blacktriangleleft$$

Radius of Gyration. From the moment of inertia and area just calculated, you obtain

$$k_{x'}^2 = \frac{I_{x'}}{A} = \frac{617.5 \text{ in}^4}{17.95 \text{ in}^2} \qquad k_{x'} = 5.87 \text{ in.} \blacktriangleleft$$

REFLECT and THINK: This is a common type of calculation for many different situations. It is often helpful to list data in a table to keep track of the numbers and identify which data you need.

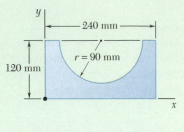

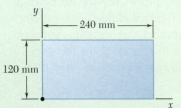

Fig. 1 Modeling given area by subtracting a half circle from a rectangle.

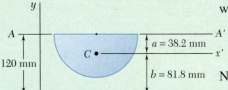

Fig. 2 Centroid location of the half circle.

Sample Problem 9.5

Determine the moment of inertia of the shaded area with respect to the x axis.

STRATEGY: You can obtain the given area by subtracting a half circle from a rectangle (Fig. 1). Then compute the moments of inertia of the rectangle and the half circle separately.

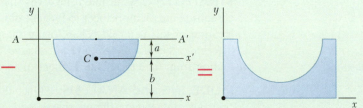

MODELING and ANALYSIS:

Moment of Inertia of Rectangle. Referring to Fig. 9.12, you have

$$I_x = \tfrac{1}{3}bh^3 = \tfrac{1}{3}(240 \text{ mm})(120 \text{ mm})^3 = 138.2 \times 10^6 \text{ mm}^4$$

Moment of Inertia of Half Circle. Refer to Fig. 5.8 and determine the location of the centroid C of the half circle with respect to diameter AA'. As shown in Fig. 2, you have

$$a = \frac{4r}{3\pi} = \frac{(4)(90 \text{ mm})}{3\pi} = 38.2 \text{ mm}$$

The distance b from the centroid C to the x axis is

$$b = 120 \text{ mm} - a = 120 \text{ mm} - 38.2 \text{ mm} = 81.8 \text{ mm}$$

Referring now to Fig. 9.12, compute the moment of inertia of the half circle with respect to diameter AA' and then compute the area of the half circle.

$$I_{AA'} = \tfrac{1}{8}\pi r^4 = \tfrac{1}{8}\pi(90 \text{ mm})^4 = 25.76 \times 10^6 \text{ mm}^4$$
$$A = \tfrac{1}{2}\pi r^2 = \tfrac{1}{2}\pi(90 \text{ mm})^2 = 12.72 \times 10^3 \text{ mm}^2$$

Next, using the parallel-axis theorem, obtain the value of $\bar{I}_{x'}$ as

$$I_{AA'} = \bar{I}_{x'} + Aa^2$$
$$25.76 \times 10^6 \text{ mm}^4 = \bar{I}_{x'} + (12.72 \times 10^3 \text{ mm}^2)(38.2 \text{ mm})^2$$
$$\bar{I}_{x'} = 7.20 \times 10^6 \text{ mm}^4$$

Again using the parallel-axis theorem, obtain the value of I_x as

$$I_x = \bar{I}_{x'} + Ab^2 = 7.20 \times 10^6 \text{ mm}^4 + (12.72 \times 10^3 \text{ mm}^2)(81.8 \text{ mm})^2$$
$$= 92.3 \times 10^6 \text{ mm}^4$$

Moment of Inertia of Given Area. Subtracting the moment of inertia of the half circle from that of the rectangle, you obtain

$$I_x = 138.2 \times 10^6 \text{ mm}^4 - 92.3 \times 10^6 \text{ mm}^4$$
$$I_x = 45.9 \times 10^6 \text{ mm}^4 \quad \blacktriangleleft$$

REFLECT and THINK: Figures 5.8 and 9.12 are useful references for locating centroids and moments of inertia of common areas; don't forget to use them.

SOLVING PROBLEMS
ON YOUR OWN

In this section, we introduced the **parallel-axis theorem** and showed how to use it to simplify the computation of moments and polar moments of inertia of composite areas. The areas that you will consider in the following problems will consist of common shapes and rolled-steel shapes. You will also use the parallel-axis theorem to locate the point of application (the center of pressure) of the resultant of the hydrostatic forces acting on a submerged plane area.

1. Applying the parallel-axis theorem. In Sec. 9.2, we derived the parallel-axis theorem

$$I = \bar{I} + Ad^2 \tag{9.9}$$

which states that the moment of inertia I of an area A with respect to a given axis is equal to the sum of the moment of inertia $\bar{I}$ of that area with respect to a *parallel centroidal axis* and the product Ad^2, where d is the distance between the two axes. It is important that you remember the following points as you use the parallel-axis theorem.

  a. You can obtain the centroidal moment of inertia $\bar{I}$ of an area A by subtracting the product Ad^2 from the moment of inertia I of the area with respect to a parallel axis. It follows that the moment of inertia $\bar{I}$ is *smaller* than the moment of inertia I of the same area with respect to any parallel axis.

  b. You can apply the parallel-axis theorem only if one of the two axes involved is a centroidal axis. Therefore, as we noted in Concept Application 9.3, to compute the moment of inertia of an area with respect to a *noncentroidal axis* when the moment of inertia of the area is known with respect to *another noncentroidal axis*, it is necessary to first compute the moment of inertia of the area with respect to a centroidal axis parallel to the two given axes.

2. Computing the moments and polar moments of inertia of composite areas. Sample Probs. 9.4 and 9.5 illustrate the steps you should follow to solve problems of this type. As with all composite-area problems, you should show on your sketch the common shapes or rolled-steel shapes that constitute the various elements of the given area, as well as the distances between the centroidal axes of the elements and the axes about which the moments of inertia are to be computed. In addition, it is important to note the following points.

  a. The moment of inertia of an area is always positive, regardless of the location of the axis with respect to which it is computed. As pointed out in the comments for the preceding section, only when an area is *removed* (as in the case of a hole) should you enter its moment of inertia in your computations with a minus sign.

(continued)

b. The moments of inertia of a semiellipse and a quarter ellipse can be determined by dividing the moment of inertia of an ellipse by 2 and 4, respectively. Note, however, that the moments of inertia obtained in this manner are *with respect to the axes of symmetry of the ellipse*. To obtain the *centroidal* moments of inertia of these shapes, use the parallel-axis theorem. This remark also applies to a semicircle and to a quarter circle. Also note that the expressions given for these shapes in Fig. 9.12 are *not* centroidal moments of inertia.

c. To calculate the polar moment of inertia of a composite area, you can use either the expressions given in Fig. 9.12 for J_O or the relationship

$$J_O = I_x + I_y \tag{9.4}$$

depending on the shape of the given area.

d. Before computing the centroidal moments of inertia of a given area, you may find it necessary to first locate the centroid of the area using the methods of Chap. 5.

3. Locating the point of application of the resultant of a system of hydrostatic forces. In Sec. 9.1, we found that

$$R = \gamma \int y \, dA = \gamma \bar{y} A$$

$$M_x = \gamma \int y^2 \, dA = \gamma I_x$$

where $\bar{y}$ is the distance from the x axis to the centroid of the submerged plane area. Since **R** is equivalent to the system of elemental hydrostatic forces, it follows that

$$\Sigma M_x: \qquad y_P R = M_x$$

where y_P is the depth of the point of application of **R**. Then

$$y_P(\gamma \bar{y} A) = \gamma I_x \quad \text{or} \quad y_P = \frac{I_x}{\bar{y} A}$$

In closing, we encourage you to carefully study the notation used in Fig. 9.13 for the rolled-steel shapes, as you will likely encounter it again in subsequent engineering courses.

for integration :

$$dI_y = x^2 y \, dx$$

$$dI_x = \frac{y^3}{3} \, dx$$

Problems

9.31 and 9.32 Determine the moment of inertia and the radius of gyration of the shaded area with respect to the x axis.

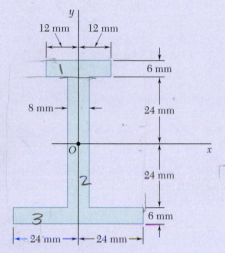

Fig. P9.31 and P9.33

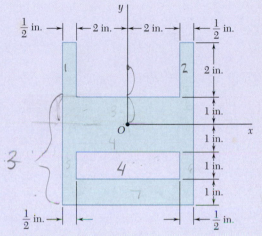

Fig. P9.32 and P9.34

9.33 and 9.34 Determine the moment of inertia and the radius of gyration of the shaded area with respect to the y axis.

9.35 and 9.36 Determine the moments of inertia of the shaded area shown with respect to the x and y axes.

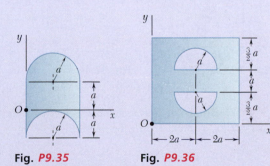

Fig. P9.35   **Fig. P9.36**

9.37 The centroidal polar moment of inertia $\bar{J}_C$ of the 24-in^2 shaded area is 600 in^4. Determine the polar moments of inertia J_B and J_D of the shaded area knowing that $J_D = 2J_B$ and $d = 5$ in.

9.38 Determine the centroidal polar moment of inertia $\bar{J}_C$ of the 25-in^2 shaded area knowing that the polar moments of inertia of the area with respect to points A, B, and D are, respectively, $J_A = 281$ in^4, $J_B = 810$ in^4, and $J_D = 1578$ in^4.

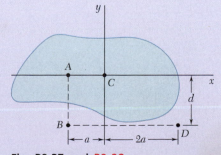

Fig. P9.37 and P9.38

9.39 Determine the shaded area and its moment of inertia with respect to the centroidal axis parallel to AA' knowing that $d_1 = 25$ mm and $d_2 = 10$ mm and that its moments of inertia with respect to AA' and BB' are 2.2×10^6 mm^4 and 4×10^6 mm^4, respectively.

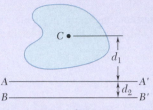

Fig. P9.39 and P9.40

9.40 Knowing that the shaded area is equal to 6000 mm^2 and that its moment of inertia with respect to AA' is 18×10^6 mm^4, determine its moment of inertia with respect to BB' for $d_1 = 50$ mm and $d_2 = 10$ mm.

9.41 through 9.44 Determine the moments of inertia $\bar{I}_x$ and $\bar{I}_y$ of the area shown with respect to centroidal axes respectively parallel and perpendicular to side AB.

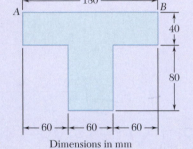

Fig. P9.41

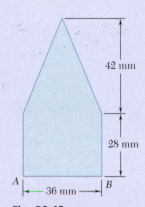

Fig. P9.42

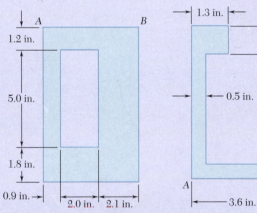

Fig. P9.43   **Fig. P9.44**

9.45 and 9.46 Determine the polar moment of inertia of the area shown with respect to (a) point O, (b) the centroid of the area.

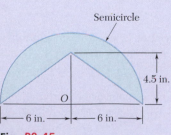

Fig. P9.45

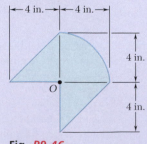

Fig. P9.46

9.47 and 9.48 Determine the polar moment of inertia of the area shown with respect to (a) point O, (b) the centroid of the area.

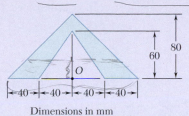

Dimensions in mm

Fig. P9.47

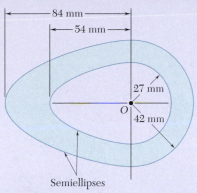

Semiellipses

Fig. P9.48

9.49 Two channels and two plates are used to form the column section shown. For $b = 200$ mm, determine the moments of inertia and the radii of gyration of the combined section with respect to the centroidal x and y axes.

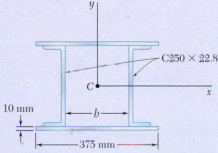

Fig. P9.49

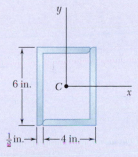

Fig. P9.50

9.50 Two L6 $\times$ 4 $\times \frac{1}{2}$-in. angles are welded together to form the section shown. Determine the moments of inertia and the radii of gyration of the combined section with respect to the centroidal x and y axes.

9.51 Four L3 $\times$ 3 $\times \frac{1}{4}$-in. angles are welded to a rolled W section as shown. Determine the moments of inertia and the radii of gyration of the combined section with respect to the centroidal x and y axes.

9.52 Two 20-mm steel plates are welded to a rolled S section as shown. Determine the moments of inertia and the radii of gyration of the combined section with respect to the centroidal x and y axes.

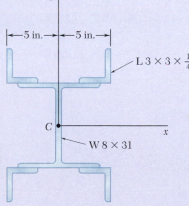

Fig. P9.51

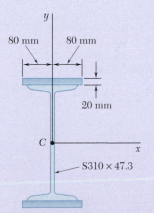

Fig. P9.52

9.53 A channel and a plate are welded together as shown to form a section that is symmetrical with respect to the y axis. Determine the moments of inertia of the combined section with respect to its centroidal x and y axes.

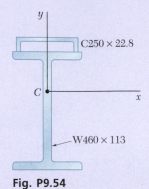

Fig. P9.54

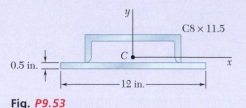

Fig. P9.53

9.54 The strength of the rolled W section shown is increased by welding a channel to its upper flange. Determine the moments of inertia of the combined section with respect to its centroidal x and y axes.

9.55 Two L76 × 76 × 6.4-mm angles are welded to a C250 × 22.8 channel. Determine the moments of inertia of the combined section with respect to centroidal axes respectively parallel and perpendicular to the web of the channel.

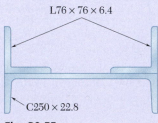

Fig. P9.55

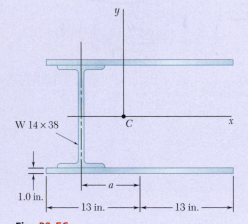

Fig. P9.56

9.56 Two steel plates are welded to a rolled W section as indicated. Knowing that the centroidal moments of inertia $\bar{I}_x$ and $\bar{I}_y$ of the combined section are equal, determine (a) the distance a, (b) the moments of inertia with respect to the centroidal x and y axes.

9.57 and 9.58 The panel shown forms the end of a trough that is filled with water to the line AA'. Referring to Sec. 9.1A, determine the depth of the point of application of the resultant of the hydrostatic forces acting on the panel (the center of pressure).

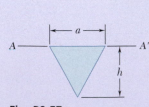

Fig. P9.57

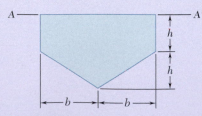

Fig. P9.58

9.59 and *9.60 The panel shown forms the end of a trough that is filled with water to the line AA'. Referring to Sec. 9.1A, determine the depth of the point of application of the resultant of the hydrostatic forces acting on the panel (the center of pressure).

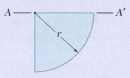

Fig. P9.59

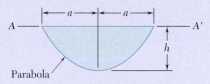

Parabola

Fig. P9.60

9.61 A vertical trapezoidal gate that is used as an automatic valve is held shut by two springs attached to hinges located along edge AB. Knowing that each spring exerts a couple of magnitude 1470 N·m, determine the depth d of water for which the gate will open.

9.62 The cover for a 0.5-m-diameter access hole in a water storage tank is attached to the tank with four equally spaced bolts as shown. Determine the additional force on each bolt due to the water pressure when the center of the cover is located 1.4 m below the water surface.

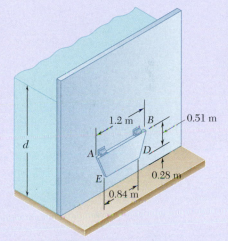

Fig. *P9.61*

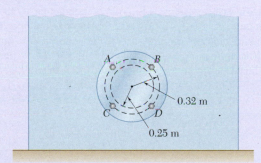

Fig. *P9.62*

*9.63 Determine the x coordinate of the centroid of the volume shown. (*Hint:* The height y of the volume is proportional to the x coordinate; consider an analogy between this height and the water pressure on a submerged surface.)

*9.64 Determine the x coordinate of the centroid of the volume shown; this volume was obtained by intersecting an elliptic cylinder with an oblique plane. (See hint of Prob. 9.63.)

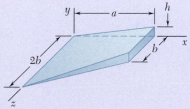

Fig. P9.63

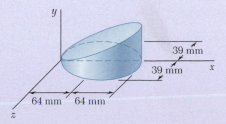

Fig. P9.64

*9.65 Show that the system of hydrostatic forces acting on a submerged plane area A can be reduced to a force $\mathbf{P}$ at the centroid C of the area and two couples. The force $\mathbf{P}$ is perpendicular to the area and has a magnitude of $P = \gamma A \bar{y} \sin \theta$, where γ is the specific weight of the liquid. The couples are $\mathbf{M}_{x'} = (\gamma \bar{I}_{x'} \sin \theta)\mathbf{i}$ and $\mathbf{M}_{y'} = (\gamma \bar{I}_{x'y'} \sin \theta)\mathbf{j}$, where $\bar{I}_{x'y'} = \int x'y'\,dA$ (see Sec. 9.3). Note that the couples are independent of the depth at which the area is submerged.

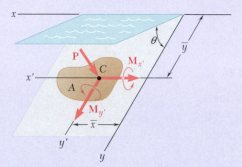

Fig. P9.65

*9.66 Show that the resultant of the hydrostatic forces acting on a submerged plane area A is a force $\mathbf{P}$ perpendicular to the area and of magnitude $P = \gamma A \bar{y} \sin \theta = \bar{p} A$, where γ is the specific weight of the liquid and $\bar{p}$ is the pressure at the centroid C of the area. Show that $\mathbf{P}$ is applied at a point C_P, called the center of pressure, whose coordinates are $x_p = I_{xy}/A\bar{y}$ and $y_p = I_x/A\bar{y}$, where $I_{xy} = \int xy\,dA$ (see Sec. 9.3). Show also that the difference of ordinates $y_p - \bar{y}$ is equal to $\bar{k}_{x'}^2/\bar{y}$ and thus depends upon the depth at which the area is submerged.

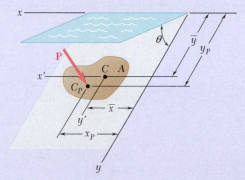

Fig. P9.66

*9.3 TRANSFORMATION OF MOMENTS OF INERTIA

The moments of inertia of an area can have different values depending on what axes we use to calculate them. It turns out that it is often important to determine the maximum and minimum values of the moments of inertia, which means finding the particular orientation of axes that produce these values. The first step in calculating moments of inertia with regard to rotated axes is to determine a new kind of second moment, called the product of inertia. In this section, we illustrate the procedures for this.

9.3A Product of Inertia

The **product of inertia** of an area A with respect to the x and y axes is defined by the integral

Product of inertia

$$I_{xy} = \int xy \, dA \qquad (9.12)$$

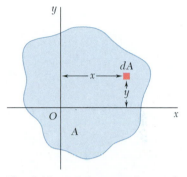

Fig. 9.14 An element of area dA with coordinates x and y.

We calculate it by multiplying each element dA of an area A by its coordinates x and y and integrating over the area (Fig. 9.14). Unlike the moments of inertia I_x and I_y, the product of inertia I_{xy} can be positive, negative, or zero. We will see shortly that the product of inertia is necessary for transforming moments of inertia with respect to a different set of axes; in a course on mechanics of materials, you will find other applications of this quantity.

When one or both of the x and y axes are axes of symmetry for the area A, the product of inertia I_{xy} is zero. Consider, for example, the channel section shown in Fig. 9.15. Since this section is symmetrical with respect to the x axis, we can associate with each element dA of coordinates x and y an element dA' of coordinates x and $-y$. Clearly, the contributions to I_{xy} of any pair of elements chosen in this way cancel out, and the integral of Eq. (9.12) reduces to zero.

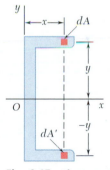

Fig. 9.15 If an area has an axis of symmetry, its product of inertia is zero.

We can derive a parallel-axis theorem for products of inertia similar to the one established in Sec. 9.2 for moments of inertia. Consider an area A and a system of rectangular coordinates x and y (Fig. 9.16). Through the centroid C of the area, with coordinates $\bar{x}$ and $\bar{y}$, we draw two centroidal axes x' and y' that are parallel, respectively, to the x and y axes. We denote the coordinates of an element of area dA with respect to the original axes by x and y, and the coordinates of the same element with respect to the centroidal axes by x' and y'. This gives us

$$x = x' + \bar{x} \quad \text{and} \quad y = y' + \bar{y}$$

Substituting into Eq. (9.12), we obtain the expression for the product of inertia I_{xy} as

$$I_{xy} = \int xy \, dA = \int (x' + \bar{x})(y' + \bar{y}) \, dA$$

$$= \int x'y' \, dA + \bar{y} \int x' \, dA + \bar{x} \int y' \, dA + \bar{x}\bar{y} \int dA$$

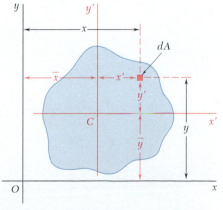

Fig. 9.16 An element of area dA with respect to x and y axes and the centroidal axes x' and y' for area A.

The first integral represents the product of inertia $\bar{I}_{xy}$ of the area A with respect to the centroidal axes x' and y'. The next two integrals represent first moments of the area with respect to the centroidal axes; they reduce to zero, since the centroid C is located on these axes. The last integral is equal to the total area A. Therefore, we have

Parallel-axis theorem for products of inertia

$$I_{xy} = \bar{I}_{x'y'} + \bar{x}\bar{y}A \tag{9.13}$$

9.3B Principal Axes and Principal Moments of Inertia

Consider an area A with coordinate axes x and y (Fig. 9.17) and assume that we know the moments and product of inertia of the area A. We have

$$I_x = \int y^2\,dA \qquad I_y = \int x^2\,dA \qquad I_{xy} = \int xy\,dA \tag{9.14}$$

We propose to determine the moments and product of inertia $I_{x'}$, $I_{y'}$, and $I_{x'y'}$ of A with respect to new axes x' and y' that we obtain by rotating the original axes about the origin through an angle θ.

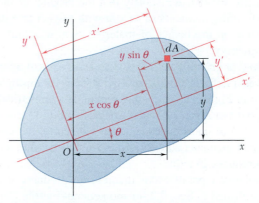

Fig. 9.17 An element of area dA with respect to x and y axes and a set of x' and y' axes rotated about the origin through an angle θ.

We first note that the relations between the coordinates x', y' and x, y of an element of area dA are

$$x' = x \cos\theta + y \sin\theta \qquad\qquad y' = y \cos\theta - x \sin\theta$$

Substituting for y' in the expression for $I_{x'}$, we obtain

$$I_{x'} = \int (y')^2\,dA = \int (y \cos\theta - x \sin\theta)^2\,dA$$

$$= \cos^2\theta \int y^2\,dA - 2 \sin\theta \cos\theta \int xy\,dA + \sin^2\theta \int x^2\,dA$$

Using the relations in Eq. (9.14), we have

$$I_{x'} = I_x \cos^2 \theta - 2I_{xy} \sin \theta \cos \theta + I_y \sin^2 \theta \qquad \textbf{(9.15)}$$

Similarly, we obtain for $I_{y'}$ and $I_{x'y'}$ the expressions

$$I_{y'} = I_x \sin^2 \theta + 2I_{xy} \sin \theta \cos \theta + I_y \cos^2 \theta \qquad \textbf{(9.16)}$$

$$I_{x'y'} = (I_x - I_y) \sin \theta \cos \theta + I_{xy}(\cos^2 \theta - \sin^2 \theta) \qquad \textbf{(9.17)}$$

Recalling the trigonometric relations

$$\sin 2\theta = 2 \sin \theta \cos \theta \qquad \cos 2\theta = \cos^2 \theta - \sin^2 \theta$$

and

$$\cos^2 \theta = \frac{1 + \cos 2\theta}{2} \qquad \sin^2 \theta = \frac{1 - \cos 2\theta}{2}$$

we can write Eqs. (9.15), (9.16), and (9.17) as

$$I_{x'} = \frac{I_x + I_y}{2} + \frac{I_x - I_y}{2} \cos 2\theta - I_{xy} \sin 2\theta \qquad \textbf{(9.18)}$$

$$I_{y'} = \frac{I_x + I_y}{2} - \frac{I_x - I_y}{2} \cos 2\theta + I_{xy} \sin 2\theta \qquad \textbf{(9.19)}$$

$$I_{x'y'} = \frac{I_x - I_y}{2} \sin 2\theta + I_{xy} \cos 2\theta \qquad \textbf{(9.20)}$$

Now, adding Eqs. (9.18) and (9.19), we observe that

$$I_{x'} + I_{y'} = I_x + I_y \qquad \textbf{(9.21)}$$

We could have anticipated this result, since both members of Eq. (9.21) are equal to the polar moment of inertia J_O.

   Equations (9.18) and (9.20) are the parametric equations of a circle. This means that, if we choose a set of rectangular axes and plot a point M of abscissa $I_{x'}$ and ordinate $I_{x'y'}$ for any given value of the parameter θ, all of the points will lie on a circle. To establish this property algebraically, we can eliminate θ from Eqs. (9.18) and (9.20) by transposing $(I_x + I_y)/2$ in Eq. (9.18), squaring both sides of Eqs. (9.18) and (9.20), and adding. The result is

$$\left(I_{x'} - \frac{I_x + I_y}{2}\right)^2 + I_{x'y'}^2 = \left(\frac{I_x - I_y}{2}\right)^2 + I_{xy}^2 \qquad \textbf{(9.22)}$$

Setting

$$I_{ave} = \frac{I_x + I_y}{2} \quad \text{and} \quad R = \sqrt{\left(\frac{I_x - I_y}{2}\right)^2 + I_{xy}^2} \qquad \textbf{(9.23)}$$

we can write the identity equation (9.22) in the form

$$(I_{x'} - I_{ave})^2 + I_{x'y'}^2 = R^2 \qquad \textbf{(9.24)}$$

This is the equation of a circle of radius R centered at the point C whose x and y coordinates are I_{ave} and 0, respectively (Fig. 9.18a).

   Note that Eqs. (9.19) and (9.20) are parametric equations of the same circle. Furthermore, because of the symmetry of the circle about the

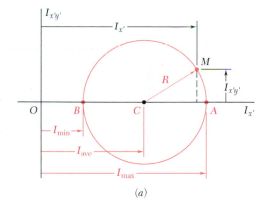

(a)

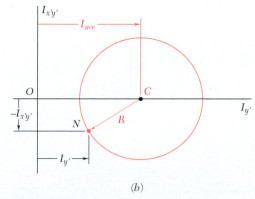

(b)

Fig. 9.18 Plots of $I_{x'y'}$ versus (a) $I_{x'}$ and (b) $I_{y'}$ for different values of the parameter θ are identical circles. The circle in part (a) indicates the average, maximum, and minimum values of the moment of inertia.

horizontal axis, we would obtain the same result if we plot a point N of coordinates $I_{y'}$ and $-I_{x'y'}$ (Fig. 9.18b) instead of plotting M. We will use this property in Sec. 9.4.

The two points A and B where this circle intersects the horizontal axis (Fig. 9.18a) are of special interest: Point A corresponds to the maximum value of the moment of inertia $I_{x'}$, whereas point B corresponds to its minimum value. In addition, both points correspond to a zero value of the product of inertia $I_{x'y'}$. Thus, we can obtain the values θ_m of the parameter θ corresponding to the points A and B by setting $I_{x'y'} = 0$ in Eq. (9.20). The result is[†]

$$\tan 2\theta_m = \frac{2I_{xy}}{I_x - I_y} \qquad (9.25)$$

This equation defines two values $(2\theta_m)$ that are $180°$ apart and thus two values (θ_m) that are $90°$ apart. One of these values corresponds to point A in Fig. 9.18a and to an axis through O in Fig. 9.17 with respect to which the moment of inertia of the given area is maximum. The other value corresponds to point B and to an axis through O with respect to which the moment of inertia of the area is minimum. These two perpendicular axes are called the **principal axes of the area about** O. The corresponding values $I_{\max}$ and $I_{\min}$ of the moment of inertia are called the **principal moments of inertia of the area about** O. Since we obtained the two values θ_m defined by Eq. (9.25) by setting $I_{x'y'} = 0$ in Eq. (9.20), it is clear that the product of inertia of the given area with respect to its principal axes is zero.

Note from Fig. 9.18a that

$$I_{\max} = I_{\text{ave}} + R \qquad I_{\min} = I_{\text{ave}} - R \qquad (9.26)$$

Using the values for I_{ave} and R from formulas (9.23), we obtain

$$I_{\max,\,\min} = \frac{I_x + I_y}{2} \pm \sqrt{\left(\frac{I_x - I_y}{2}\right)^2 + I_{xy}^2} \qquad (9.27)$$

Unless you can tell by inspection which of the two principal axes corresponds to $I_{\max}$ and which corresponds to $I_{\min}$, you must substitute one of the values of θ_m into Eq. (9.18) in order to determine which of the two corresponds to the maximum value of the moment of inertia of the area about O.

Referring to Sec. 9.3A, note that, if an area possesses an axis of symmetry through a point O, this axis must be a principal axis of the area about O. On the other hand, a principal axis does not need to be an axis of symmetry; whether or not an area possesses any axes of symmetry, it will always have two principal axes of inertia about any point O.

The properties we have established hold for any point O located inside or outside the given area. If we choose the point O to coincide with the centroid of the area, any axis through O is a centroidal axis; the two principal axes of the area about its centroid are referred to as the **principal centroidal axes of the area**.

[†]We can also obtain this relation by differentiating $I_{x'}$ in Eq. (9.18) and setting $dI_{x'}/d\theta = 0$.

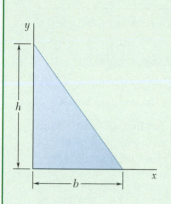

Sample Problem 9.6

Determine the product of inertia of the right triangle shown (*a*) with respect to the *x* and *y* axes and (*b*) with respect to centroidal axes parallel to the *x* and *y* axes.

STRATEGY: You can approach this problem by using a vertical differential strip element. Because each point of the strip is at a different distance from the *x* axis, it is necessary to describe this strip mathematically using the parallel-axis theorem. Once you have completed the solution for the product of inertia with respect to the *x* and *y* axes, a second application of the parallel-axis theorem yields the product of inertia with respect to the centroidal axes.

MODELING and ANALYSIS:

a. Product of Inertia I_{xy}. Choose a vertical rectangular strip as the differential element of area (Fig. 1). Using a differential version of the parallel-axis theorem, you have

$$dI_{xy} = dI_{x'y'} + \bar{x}_{el}\bar{y}_{el}\, dA$$

The element is symmetrical with respect to the x' and y' axes, so $dI_{x'y'} = 0$. From the geometry of the triangle, you can express the variables in terms of *x* and *y*.

$$y = h\left(1 - \frac{x}{b}\right) \qquad dA = y\, dx = h\left(1 - \frac{x}{b}\right) dx$$

$$\bar{x}_{el} = x \qquad \bar{y}_{el} = \tfrac{1}{2}y = \tfrac{1}{2}h\left(1 - \frac{x}{b}\right)$$

Integrating dI_{xy} from $x = 0$ to $x = b$ gives you I_{xy}:

$$I_{xy} = \int dI_{xy} = \int \bar{x}_{el}\bar{y}_{el}\, dA = \int_0^b x(\tfrac{1}{2})h^2\left(1 - \frac{x}{b}\right)^2 dx$$

$$= h^2 \int_0^b \left(\frac{x}{2} - \frac{x^2}{b} + \frac{x^3}{2b^2}\right) dx = h^2\left[\frac{x^2}{4} - \frac{x^3}{3b} + \frac{x^4}{8b^2}\right]_0^b$$

$$I_{xy} = \tfrac{1}{24}b^2h^2 \quad \blacktriangleleft$$

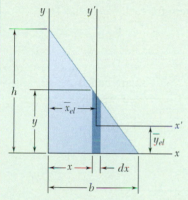

Fig. 1 Using a vertical rectangular strip as the differential element.

b. Product of Inertia $\bar{I}_{x''y''}$. The coordinates of the centroid of the triangle relative to the *x* and *y* axes are (Fig. 2 and Fig. 5.8A)

$$\bar{x} = \frac{1}{3}b \qquad \bar{y} = \frac{1}{3}h$$

Using the expression for I_{xy} obtained in part *a*, apply the parallel-axis theorem again:

$$I_{xy} = \bar{I}_{x''y''} + \bar{x}\,\bar{y}A$$
$$\tfrac{1}{24}b^2h^2 = \bar{I}_{x''y''} + (\tfrac{1}{3}b)(\tfrac{1}{3}h)(\tfrac{1}{2}bh)$$
$$\bar{I}_{x''y''} = \tfrac{1}{24}b^2h^2 - \tfrac{1}{18}b^2h^2$$

$$\bar{I}_{x''y''} = -\tfrac{1}{72}b^2h^2 \quad \blacktriangleleft$$

REFLECT and THINK: An equally effective alternative strategy would be to use a horizontal strip element. Again, you would need to use the parallel-axis theorem to describe this strip, since each point in the strip would be a different distance from the *y* axis.

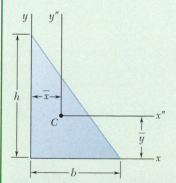

Fig. 2 Centroid of the triangular area.

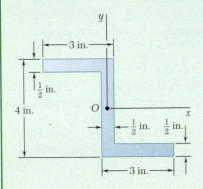

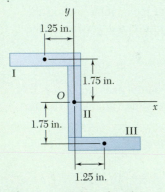

Fig. 1 Modeling the given area as three rectangles.

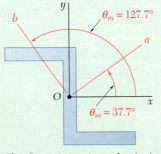

Fig. 2 Orientation of principal axes.

Sample Problem 9.7

For the section shown, the moments of inertia with respect to the x and y axes have been computed and are known to be

$$I_x = 10.38 \text{ in}^4 \qquad I_y = 6.97 \text{ in}^4$$

Determine (a) the orientation of the principal axes of the section about O, (b) the values of the principal moments of inertia of the section about O.

STRATEGY: The first step is to compute the product of inertia with respect to the x and y axes, treating the section as a composite area of three rectangles. Then you can use Eq. (9.25) to find the principal axes and Eq. (9.27) to find the principal moments of inertia.

MODELING and ANALYSIS: Divide the area into three rectangles as shown (Fig. 1). Note that the product of inertia $I_{x'y'}$ with respect to centroidal axes parallel to the x and y axes is zero for each rectangle. Thus, using the parallel-axis theorem

$$I_{xy} = I_{x'y'} + \bar{x}\bar{y}A$$

you find that I_{xy} reduces to $\bar{x}\bar{y}A$ for each rectangle.

Rectangle	Area, in^2	$\bar{x}$, in.	$\bar{y}$, in.	$\bar{x}\bar{y}A$, in^4
I	1.5	−1.25	+1.75	−3.28
II	1.5	0	0	0
III	1.5	+1.25	−1.75	−3.28
				$\Sigma\bar{x}\bar{y}A = -6.56$

$$I_{xy} = \Sigma\bar{x}\,\bar{y}A = -6.56 \text{ in}^4$$

a. Principal Axes. Since you know the magnitudes of I_x, I_y, and I_{xy}, you can use Eq. (9.25) to determine the values of θ_m (Fig. 2):

$$\tan 2\theta_m = -\frac{2I_{xy}}{I_x - I_y} = -\frac{2(-6.56)}{10.38 - 6.97} = +3.85$$

$$2\theta_m = 75.4° \text{ and } 255.4°$$

$$\theta_m = 37.7° \quad \text{and} \quad \theta_m = 127.7° \quad \blacktriangleleft$$

b. Principal Moments of Inertia. Using Eq. (9.27), you have

$$I_{max,min} = \frac{I_x + I_y}{2} \pm \sqrt{\left(\frac{I_x - I_y}{2}\right)^2 + I_{xy}^2}$$

$$= \frac{10.38 + 6.97}{2} \pm \sqrt{\left(\frac{10.38 - 6.97}{2}\right)^2 + (-6.56)^2}$$

$$I_{max} = 15.45 \text{ in}^4 \quad I_{min} = 1.897 \text{ in}^4 \quad \blacktriangleleft$$

REFLECT and THINK: Note that the elements of the area of the section are more closely distributed about the b axis than about the a axis. Therefore, you can conclude that $I_a = I_{max} = 15.45 \text{ in}^4$ and $I_b = I_{min} = 1.897 \text{ in}^4$. You can verify this conclusion by substituting $\theta = 37.7°$ into Eqs. (9.18) and (9.19).

SOLVING PROBLEMS
ON YOUR OWN

In the problems for this section, you will continue your work with **moments of inertia** and use various techniques for computing **products of inertia**. Although the problems are generally straightforward, several items are worth noting.

1. Calculating the product of inertia I_{xy} by integration. We defined this quantity as

$$I_{xy} = \int xy \, dA \tag{9.12}$$

and stated that its value can be positive, negative, or zero. You can compute the product of inertia directly from this equation using double integration, or you can find it by using single integration as shown in Sample Prob. 9.6. When applying single integration and using the parallel-axis theorem, it is important to remember that in the equation

$$dI_{xy} = dI_{x'y'} + \bar{x}_{el}\bar{y}_{el} dA$$

$\bar{x}_{el}$ and $\bar{y}_{el}$ are the coordinates of the centroid of the element of area dA. Thus, if dA is not in the first quadrant, one or both of these coordinates is negative.

2. Calculating the products of inertia of composite areas. You can easily compute these quantities from the products of inertia of their component parts by using the parallel-axis theorem, as

$$I_{xy} = \bar{I}_{x'y'} + \bar{x}\bar{y}A \tag{9.13}$$

The proper technique to use for problems of this type is illustrated in Sample Probs. 9.6 and 9.7. In addition to the usual rules for composite-area problems, it is essential that you remember the following points.

   a. If either of the centroidal axes of a component area is an axis of symmetry for that area, the product of inertia $\bar{I}_{x'y'}$ for that area is zero. Thus, $\bar{I}_{x'y'}$ is zero for component areas such as circles, semicircles, rectangles, and isosceles triangles, which possess an axis of symmetry parallel to one of the coordinate axes.

   b. Pay careful attention to the signs of the coordinates $\bar{x}$ and $\bar{y}$ of each component area when you use the parallel-axis theorem [Sample Prob. 9.7].

3. Determining the moments of inertia and the product of inertia for rotated coordinate axes. In Sec. 9.3B, we derived Eqs. (9.18), (9.19), and (9.20) from which you can compute the moments of inertia and the product of inertia for coordinate axes that have been rotated about the origin O. To apply these equations, you must know a set of values I_x, I_y, and I_{xy} for a given orientation of the axes, and you must remember that θ is positive for counterclockwise rotations of the axes and negative for clockwise rotations of the axes.

4. Computing the principal moments of inertia. We showed in Sec. 9.3B that a particular orientation of the coordinate axes exists for which the moments of inertia attain their maximum and minimum values, I_{max} and I_{min}, and for which the product of inertia is zero. Equation (9.27) can be used to compute these values that are known as the **principal moments of inertia** of the area about O. The corresponding axes are referred to as the **principal axes** of the area about O, and their orientation is defined by Eq. (9.25). To determine which of the principal axes corresponds to I_{max} and which corresponds to I_{min}, you can either follow the procedure outlined in the text after Eq. (9.27) or observe about which of the two principal axes the area is more closely distributed; that axis corresponds to I_{min} [Sample Prob. 9.7].

Problems

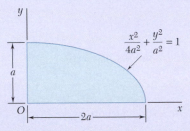

Fig. P9.67

9.67 through 9.70 Determine by direct integration the product of inertia of the given area with respect to the x and y axes.

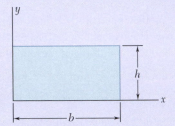

Fig. P9.68

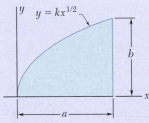

Fig. P9.69

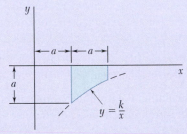

Fig. *P9.70*

9.71 through 9.74 Using the parallel-axis theorem, determine the product of inertia of the area shown with respect to the centroidal x and y axes.

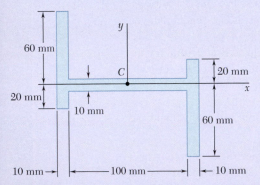

Fig. P9.71

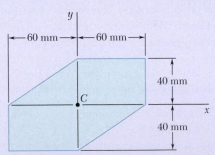

Fig. P9.72

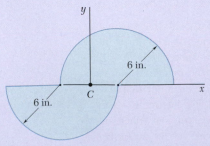

Fig. *P9.73*

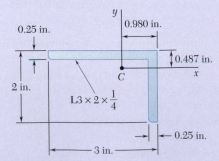

Fig. P9.74

9.75 through 9.78 Using the parallel-axis theorem, determine the product of inertia of the area shown with respect to the centroidal x and y axes.

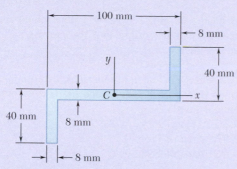

Fig. P9.75

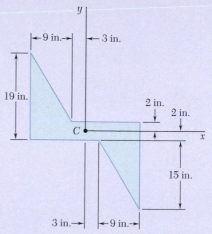

Fig. P9.76

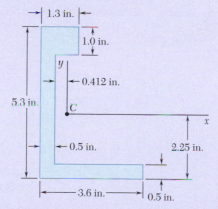

Fig. *P9.77*

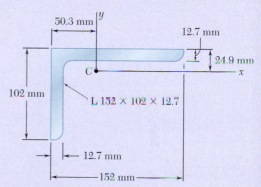

Fig. P9.78

9.79 Determine for the quarter ellipse of Prob. 9.67 the moments of inertia and the product of inertia with respect to new axes obtained by rotating the x and y axes about O (*a*) through 45° counterclockwise, (*b*) through 30° clockwise.

9.80 Determine the moments of inertia and the product of inertia of the area of Prob. 9.72 with respect to new centroidal axes obtained by rotating the x and y axes 30° counterclockwise.

9.81 Determine the moments of inertia and the product of inertia of the area of Prob. 9.73 with respect to new centroidal axes obtained by rotating the x and y axes 60° counterclockwise.

9.82 Determine the moments of inertia and the product of inertia of the area of Prob. 9.75 with respect to new centroidal axes obtained by rotating the x and y axes 45° clockwise.

9.83 Determine the moments of inertia and the product of inertia of the L3 × 2 × $\frac{1}{4}$-in. angle cross section of Prob. 9.74 with respect to new centroidal axes obtained by rotating the x and y axes 30° clockwise.

9.84 Determine the moments of inertia and the product of inertia of the L152 × 102 × 12.7-mm angle cross section of Prob. 9.78 with respect to new centroidal axes obtained by rotating the x and y axes 30° clockwise.

9.85 For the quarter ellipse of Prob. 9.67, determine the orientation of the principal axes at the origin and the corresponding values of the moments of inertia.

9.86 through 9.88 For the area indicated, determine the orientation of the principal axes at the origin and the corresponding values of the moments of inertia.

  9.86 Area of Prob. 9.72
  9.87 Area of Prob. 9.73
  9.88 Area of Prob. 9.75

9.89 and 9.90 For the angle cross section indicated, determine the orientation of the principal axes at the origin and the corresponding values of the moments of inertia.

  9.89 The L3 × 2 × $\frac{1}{4}$-in. angle cross section of Prob. 9.74
  9.90 The L152 × 102 × 12.7-mm angle cross section of Prob. 9.78

*9.4 MOHR'S CIRCLE FOR MOMENTS OF INERTIA

The circle introduced in the preceding section to illustrate the relations between the moments and products of inertia of a given area with respect to axes passing through a fixed point O was first introduced by the German engineer Otto Mohr (1835–1918) and is known as **Mohr's circle**. Here, we show that, if we know the moments and product of inertia of an area A with respect to two rectangular x and y axes that pass through a point O, we can use Mohr's circle to graphically determine (a) the principal axes and principal moments of inertia of the area about O and (b) the moments and product of inertia of the area with respect to any other pair of rectangular axes x' and y' through O.

Consider a given area A and two rectangular coordinate axes x and y (Fig. 9.19a). Assuming that we know the moments of inertia I_x and I_y and the product of inertia I_{xy}, we can represent them on a diagram by plotting a point X with coordinates I_x and I_{xy} and a point Y with coordinates I_y and $-I_{xy}$ (Fig. 9.19b). If I_{xy} is positive, as assumed in Fig. 9.19a, then point X is located above the horizontal axis and point Y is located below, as shown in Fig. 9.19b. If I_{xy} is negative, X is located below the horizontal axis and Y is located above. Joining X and Y with a straight line, we denote the point of intersection of line XY with the horizontal axis by C. Then we draw the circle of center C and diameter XY. Noting that the abscissa of C and the radius of the circle are respectively equal to the quantities I_{ave} and R defined by formula (9.23), we conclude that the circle obtained is Mohr's circle for the given area about point O. Thus, the abscissas of the points A and B where the circle intersects the horizontal axis represent, respectively, the principal moments of inertia I_{max} and I_{min} of the area.

Also note that, since tan $(XCA) = 2I_{xy}/(I_x - I_y)$, the angle XCA is equal in magnitude to one of the angles $2\theta_m$ that satisfy Eq. (9.25). Thus, the angle θ_m, which in Fig. 9.19a defines the principal axis Oa corresponding

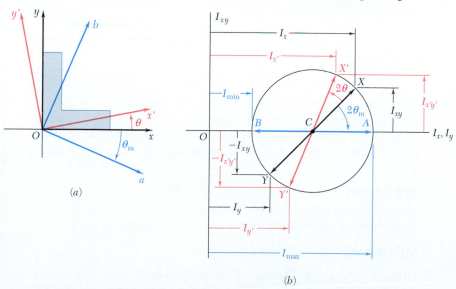

(a)

(b)

Fig. 9.19 (a) An area A with principal axes Oa and Ob and axes Ox' and Oy' obtained by rotation through an angle θ; (b) Mohr's circle used to calculate angles and moments of inertia.

to point A in Fig. 9.19b, is equal to half of the angle XCA of Mohr's circle. In addition, if $I_x > I_y$ and $I_{xy} > 0$, as in the case considered here, the rotation that brings CX into CA is clockwise. Also, under these conditions, the angle θ_m obtained from Eq. (9.25) is negative; thus, the rotation that brings Ox into Oa is also clockwise. We conclude that the senses of rotation in both parts of Fig. 9.19 are the same. If a clockwise rotation through $2\theta_m$ is required to bring CX into CA on Mohr's circle, a clockwise rotation through θ_m will bring Ox into the corresponding principal axis Oa in Fig. 9.19a.

Since Mohr's circle is uniquely defined, we can obtain the same circle by considering the moments and product of inertia of the area A with respect to the rectangular axes x' and y' (Fig. 9.19a). The point X' with coordinates $I_{x'}$ and $I_{x'y'}$ and the point Y' with coordinates $I_{y'}$ and $-I_{x'y'}$ are thus located on Mohr's circle, and the angle $X'CA$ in Fig. 9.19b must be equal to twice the angle $x'Oa$ in Fig. 9.19a. Since, as noted previously, the angle XCA is twice the angle xOa, it follows that the angle XCX' in Fig. 9.19b is twice the angle xOx' in Fig. 9.19a. The diameter $X'Y'$, which defines the moments and product of inertia $I_{x'}$, $I_{y'}$, and $I_{x'y'}$ of the given area with respect to rectangular axes x' and y' forming an angle θ with the x and y axes, can be obtained by rotating through an angle 2θ the diameter XY, which corresponds to the moments and product of inertia I_x, I_y, and I_{xy}. Note that the rotation that brings the diameter XY into the diameter $X'Y'$ in Fig. 9.19b has the same sense as the rotation that brings the x and y axes into the x' and y' axes in Fig. 9.19a.

Finally, also note that the use of Mohr's circle is not limited to graphical solutions, i.e., to solutions based on the careful drawing and measuring of the various parameters involved. By merely sketching Mohr's circle and using trigonometry, you can easily derive the various relations required for a numerical solution of a given problem (see Sample Prob. 9.8).

Sample Problem 9.8

For the section shown, the moments and product of inertia with respect to the x and y axes are

$$I_x = 7.20 \times 10^6 \text{ mm}^4 \quad I_y = 2.59 \times 10^6 \text{ mm}^4 \quad I_{xy} = -2.54 \times 10^6 \text{ mm}^4$$

Using Mohr's circle, determine (a) the principal axes of the section about O, (b) the values of the principal moments of inertia of the section about O, and (c) the moments and product of inertia of the section with respect to the x' and y' axes that form an angle of 60° with the x and y axes.

STRATEGY: You should carefully draw Mohr's circle and use the geometry of the circle to determine the orientation of the principal axes. Then complete the analysis for the requested moments of inertia.

MODELING:

Drawing Mohr's Circle. First plot point X with coordinates $I_x = 7.20$, $I_{xy} = -2.54$, and plot point Y with coordinates $I_y = 2.59$, $-I_{xy} = +2.54$. Join X and Y with a straight line to define the center C

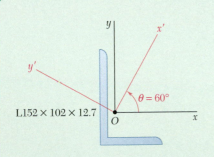

L152 × 102 × 12.7

$\theta = 60°$

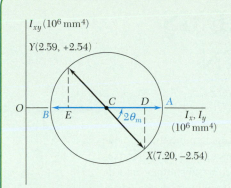

Fig. 1 Mohr's circle.

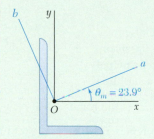

Fig. 2 Orientation of the principal axes.

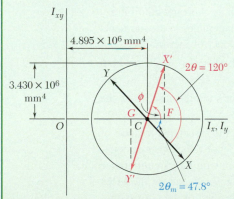

Fig. 3 Using Mohr's circle to determine the moments and product of inertia with respect to x' and y' axes.

of Mohr's circle (Fig. 1). You can measure the abscissa of C, which represents I_{ave}, and the radius R of the circle either directly or using

$$I_{ave} = OC = \tfrac{1}{2}(I_x + I_y) = \tfrac{1}{2}(7.20 \times 10^6 + 2.59 \times 10^6) = 4.895 \times 10^6 \text{ mm}^4$$

$$CD = \tfrac{1}{2}(I_x - I_y) = \tfrac{1}{2}(7.20 \times 10^6 - 2.59 \times 10^6) = 2.305 \times 10^6 \text{ mm}^4$$

$$R = \sqrt{(CD)^2 + (DX)^2} = \sqrt{(2.305 \times 10^6)^2 + (2.54 \times 10^6)^2}$$
$$= 3.430 \times 10^6 \text{ mm}^4$$

ANALYSIS:

a. Principal Axes.
The principal axes of the section correspond to points A and B on Mohr's circle, and the angle through which you should rotate CX to bring it into CA defines $2\theta_m$. You obtain

$$\tan 2\theta_m = \frac{DX}{CD} = \frac{2.54}{2.305} = 1.102 \qquad 2\theta_m = 47.8° \nwarrow \qquad \theta_m = 23.9° \nwarrow \quad \blacktriangleleft$$

Thus, the principal axis Oa corresponding to the maximum value of the moment of inertia is obtained by rotating the x axis through 23.9° counterclockwise; the principal axis Ob corresponding to the minimum value of the moment of inertia can be obtained by rotating the y axis through the same angle (Fig. 2).

b. Principal Moments of Inertia.
The principal moments of inertia are represented by the abscissas of A and B. The results are

$$I_{max} = OA = OC + CA = I_{ave} + R = (4.895 + 3.430)10^6 \text{ mm}^4$$

$$I_{max} = 8.33 \times 10^6 \text{ mm}^4 \quad \blacktriangleleft$$

$$I_{min} = OB = OC - BC = I_{ave} - R = (4.895 - 3.430)10^6 \text{ mm}^4$$

$$I_{min} = 1.47 \times 10^6 \text{ mm}^4 \quad \blacktriangleleft$$

c. Moments and Product of Inertia with Respect to the x' and y' Axes.
On Mohr's circle, you obtain the points X' and Y', which correspond to the x' and y' axes, by rotating CX and CY through an angle $2\theta = 2(60°) = 120°$ counterclockwise (Fig. 3). The coordinates of X' and Y' yield the desired moments and product of inertia. Noting that the angle that CX' forms with the horizontal axis is $\phi = 120° - 47.8° = 72.2°$, you have

$$I_{x'} = OF = OC + CF = 4.895 \times 10^6 \text{ mm}^4 + (3.430 \times 10^6 \text{ mm}^4) \cos 72.2°$$
$$I_{x'} = 5.94 \times 10^6 \text{ mm}^4 \quad \blacktriangleleft$$

$$I_{y'} = OG = OC - GC = 4.895 \times 10^6 \text{ mm}^4 - (3.430 \times 10^6 \text{ mm}^4) \cos 72.2°$$
$$I_{y'} = 3.85 \times 10^6 \text{ mm}^4 \quad \blacktriangleleft$$

$$I_{x'y'} = FX' = (3.430 \times 10^6 \text{ mm}^4) \sin 72.2°$$
$$I_{x'y'} = 3.27 \times 10^6 \text{ mm}^4 \quad \blacktriangleleft$$

REFLECT and THINK: This problem illustrates typical calculations with Mohr's circle. The technique is a useful one to learn and remember.

SOLVING PROBLEMS
ON YOUR OWN

In the problems for this section, you will use **Mohr's circle** to determine the moments and products of inertia of a given area for different orientations of the coordinate axes. Although in some cases using Mohr's circle may not be as direct as substituting into the appropriate equations [Eqs. (9.18) through (9.20)], this method of solution has the advantage of providing a visual representation of the relationships among the variables involved. Also, Mohr's circle shows all of the values of the moments and products of inertia that are possible for a given problem.

Using Mohr's circle. We presented the underlying theory in Sec. 9.3B, and we discussed the application of this method in Sec. 9.4 and in Sample Prob. 9.8. In the same problem, we presented the steps you should follow to determine the **principal axes**, the **principal moments of inertia**, and the **moments and product of inertia with respect to a specified orientation of the coordinates axes**. When you use Mohr's circle to solve problems, it is important that you remember the following points.

   a. Mohr's circle is completely defined by the quantities R and I_{ave}, which represent, respectively, the radius of the circle and the distance from the origin O to the center C of the circle. You can obtain these quantities from Eqs. (9.23) if you know the moments and product of inertia for a given orientation of the axes. However, Mohr's circle can be defined by other combinations of known values [Probs. 9.103, 9.106, and 9.107]. For these cases, it may be necessary to first make one or more assumptions, such as choosing an arbitrary location for the center when I_{ave} is unknown, assigning relative magnitudes to the moments of inertia (for example, $I_x > I_y$), or selecting the sign of the product of inertia.

   b. Point X of coordinates (I_x, I_{xy}) and point Y of coordinates $(I_y, -I_{xy})$ are both located on Mohr's circle and are diametrically opposite.

   c. Since moments of inertia must be positive, all of Mohr's circle must lie to the right of the I_{xy} axis; it follows that $I_{ave} > R$ for all cases.

   d. As the coordinate axes are rotated through an angle θ, the associated rotation of the diameter of Mohr's circle is equal to 2θ and is in the same sense (clockwise or counterclockwise). We strongly suggest that you label the known points on the circumference of the circle with the appropriate capital letter, as was done in Fig. 9.19b and for the Mohr circles of Sample Prob. 9.8. This will enable you to determine the sign of the corresponding product of inertia for each value of θ and which moment of inertia is associated with each of the coordinate axes [Sample Prob. 9.8, parts a and c].

Although we have introduced Mohr's circle within the specific context of the study of moments and products of inertia, the Mohr circle technique also applies to the solution of analogous but physically different problems in mechanics of materials. This multiple use of a specific technique is not unique, and as you pursue your engineering studies, you will encounter several methods of solution that can be applied to a variety of problems.

Problems

9.91 Using Mohr's circle, determine for the quarter ellipse of Prob. 9.67 the moments of inertia and the product of inertia with respect to new axes obtained by rotating the x and y axes about O (*a*) through 45° counterclockwise, (*b*) through 30° clockwise.

9.92 Using Mohr's circle, determine the moments of inertia and the product of inertia of the area of Prob. 9.72 with respect to new centroidal axes obtained by rotating the x and y axes 30° counterclockwise.

9.93 Using Mohr's circle, determine the moments of inertia and the product of inertia of the area of Prob. 9.73 with respect to new centroidal axes obtained by rotating the x and y axes 60° counterclockwise.

9.94 Using Mohr's circle, determine the moments of inertia and the product of inertia of the area of Prob. 9.75 with respect to new centroidal axes obtained by rotating the x and y axes 45° clockwise.

9.95 Using Mohr's circle, determine the moments of inertia and the product of inertia of the L3 × 2 × $\frac{1}{4}$-in. angle cross section of Prob. 9.74 with respect to new centroidal axes obtained by rotating the x and y axes 30° clockwise.

9.96 Using Mohr's circle, determine the moments of inertia and the product of inertia of the L152 × 102 × 12.7-mm angle cross section of Prob. 9.78 with respect to new centroidal axes obtained by rotating the x and y axes 30° clockwise.

9.97 For the quarter ellipse of Prob. 9.67, use Mohr's circle to determine the orientation of the principal axes at the origin and the corresponding values of the moments of inertia.

9.98 through 9.102 Using Mohr's circle, determine for the area indicated the orientation of the principal centroidal axes and the corresponding values of the moments of inertia.

    9.98 Area of Prob. 9.72
    9.99 Area of Prob. 9.76
    9.100 Area of Prob. 9.73
    9.101 Area of Prob. 9.74
    9.102 Area of Prob. 9.77

(The moments of inertia $\bar{I}_x$ and $\bar{I}_y$ of the area of Prob. 9.102 were determined in Prob. 9.44)

9.103 The moments and product of inertia of an L4 × 3 × $\frac{1}{4}$-in. angle cross section with respect to two rectangular axes x and y through C are, respectively, $\bar{I}_x = 1.33$ in^4, $\bar{I}_y = 2.75$ in^4, and $\bar{I}_{xy} < 0$, with the minimum value of the moment of inertia of the area with respect to any axis through C being $\bar{I}_{min} = 0.692$ in^4. Using Mohr's circle, determine (*a*) the product of inertia $\bar{I}_{xy}$ of the area, (*b*) the orientation of the principal axes, (*c*) the value of $\bar{I}_{max}$.

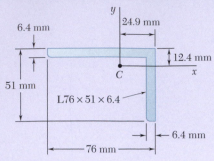

Fig. P9.104

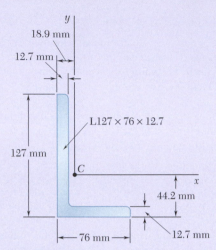

Fig. P9.105

9.104 and 9.105 Using Mohr's circle, determine the orientation of the principal centroidal axes and the corresponding values of the moments of inertia for the cross section of the rolled-steel angle shown. (Properties of the cross sections are given in Fig. 9.13.)

*9.106 For a given area, the moments of inertia with respect to two rectangular centroidal x and y axes are $\bar{I}_x = 1200$ in^4 and $\bar{I}_y = 300$ in^4, respectively. Knowing that, after rotating the x and y axes about the centroid 30° counterclockwise, the moment of inertia relative to the rotated x axis is 1450 in^4, use Mohr's circle to determine (*a*) the orientation of the principal axes, (*b*) the principal centroidal moments of inertia.

9.107 It is known that for a given area $\bar{I}_y = 48 \times 10^6$ mm^4 and $\bar{I}_{xy} = -20 \times 10^6$ mm^4, where the x and y axes are rectangular centroidal axes. If the axis corresponding to the maximum product of inertia is obtained by rotating the x axis 67.5° counterclockwise about C, use Mohr's circle to determine (*a*) the moment of inertia $\bar{I}_x$ of the area, (*b*) the principal centroidal moments of inertia.

9.108 Using Mohr's circle, show that for any regular polygon (such as a pentagon) (*a*) the moment of inertia with respect to every axis through the centroid is the same, (*b*) the product of inertia with respect to every pair of rectangular axes through the centroid is zero.

9.109 Using Mohr's circle, prove that the expression $I_{x'}I_{y'} - I_{x'y'}^2$ is independent of the orientation of the x' and y' axes, where $I_{x'}$, $I_{y'}$, and $I_{x'y'}$ represent the moments and product of inertia, respectively, of a given area with respect to a pair of rectangular axes x' and y' through a given point O. Also show that the given expression is equal to the square of the length of the tangent drawn from the origin of the coordinate system to Mohr's circle.

9.110 Using the invariance property established in the preceding problem, express the product of inertia I_{xy} of an area A with respect to a pair of rectangular axes through O in terms of the moments of inertia I_x and I_y of A and the principal moments of inertia I_{min} and I_{max} of A about O. Use the formula obtained to calculate the product of inertia I_{xy} of the L3 × 2 × $\frac{1}{4}$-in. angle cross section shown in Fig. 9.13A, knowing that its maximum moment of inertia is 1.257 in^4.

9.5   MASS MOMENTS OF INERTIA

So far in this chapter, we have examined moments of inertia of areas. In the rest of this chapter, we consider moments of inertia associated with the masses of bodies. This will be an important concept in dynamics when studying the rotational motion of a rigid body about an axis.

9.5A   Moment of Inertia of a Simple Mass

Consider a small mass Δm mounted on a rod of negligible mass that can rotate freely about an axis AA' (Fig. 9.20a). If we apply a couple to the system, the rod and mass (assumed to be initially at rest) start rotating about AA'. We will study the details of this motion later in dynamics. At present, we wish to indicate only that the time required for the system to reach a given speed of rotation is proportional to the mass Δm and to the square of the distance r. The product $r^2\,\Delta m$ thus provides a measure of the **inertia** of the system; i.e., a measure of the resistance the system offers when we try to set it in motion. For this reason, the product $r^2\,\Delta m$ is called the **moment of inertia** of the mass Δm with respect to axis AA'.

Now suppose a body of mass m is to be rotated about an axis AA' (Fig. 9.20b). Dividing the body into elements of mass Δm_1, Δm_2, etc., we find that the body's resistance to being rotated is measured by the sum $r_1^2\,\Delta m_1 + r_2^2\,\Delta m_2 + \ldots$. This sum defines the moment of inertia of the body with respect to axis AA'. Increasing the number of elements, we find that the moment of inertia is equal, in the limit, to the integral

Moment of inertia of a mass

$$I = \int r^2\, dm \qquad\qquad \textbf{(9.28)}$$

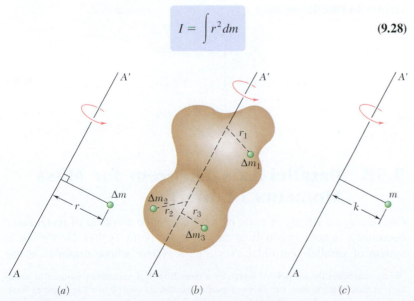

| (a) | (b) | (c) |

Fig. 9.20   (a) An element of mass Δm at a distance r from an axis AA'; (b) the moment of inertia of a rigid body is the sum of the moments of inertia of many small masses; (c) the moment of inertia is unchanged if all the mass is concentrated at a point at a distance from the axis equal to the radius of gyration.

We define the **radius of gyration** k of the body with respect to axis AA' by the relation

Radius of gyration of a mass

$$I = k^2 m \quad \text{or} \quad k = \sqrt{\frac{I}{m}} \tag{9.29}$$

The radius of gyration k represents the distance at which the entire mass of the body should be concentrated if its moment of inertia with respect to AA' is to remain unchanged (Fig. 9.20c). Whether it stays in its original shape (Fig. 9.20b) or is concentrated as shown in Fig. 9.20c, the mass m reacts in the same way to a rotation (or *gyration*) about AA'.

If SI units are used, the radius of gyration k is expressed in meters and the mass m in kilograms, so the unit for the moment of inertia of a mass is kg·m². If U.S. customary units are used, the radius of gyration is expressed in feet and the mass in slugs (i.e., in lb·s²/ft), so the derived unit for the moment of inertia of a mass is lb·ft·s².[†]

We can express the moment of inertia of a body with respect to a coordinate axis in terms of the coordinates x, y, z of the element of mass dm (Fig. 9.21). Noting, for example, that the square of the distance r from the element dm to the y axis is $z^2 + x^2$, the moment of inertia of the body with respect to the y axis is

$$I_y = \int r^2 \, dm = \int (z^2 + x^2) \, dm$$

We obtain similar expressions for the moments of inertia with respect to the x and z axes.

Moments of inertia with respect to coordinate axes

$$I_x = \int (y^2 + z^2) \, dm$$

$$I_y = \int (z^2 + x^2) \, dm \tag{9.30}$$

$$I_z = \int (x^2 + y^2) \, dm$$

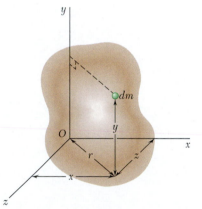

Fig. 9.21 An element of mass dm in an x, y, z coordinate system.

9.5B Parallel-Axis Theorem for Mass Moments of Inertia

Consider again a body of mass m and let $Oxyz$ be a system of rectangular coordinates whose origin is at the arbitrary point O. Let $Gx'y'z'$ be a system of parallel centroidal axes; i.e., a system whose origin is at the

[†]When converting the moment of inertia of a mass from U.S. customary units to SI units, keep in mind that the base unit (pound) used in the derived unit (lb·ft·s²) is a unit of force (*not* of mass). Therefore, it should be converted into newtons. We have

$$1 \text{ lb·ft·s}^2 = (4.45 \text{ N})(0.3048 \text{ m})(1 \text{ s})^2 = 1.356 \text{ N·m·s}^2$$

or since $1 \text{ N} = 1 \text{ kg·m/s}^2$

$$1 \text{ lb·ft·s}^2 = 1.356 \text{ kg·m}^2$$

Photo 9.2 The rotational behavior of this crankshaft depends upon its mass moment of inertia with respect to its axis of rotation, as you will see in a dynamics course.

center of gravity G of the body and whose axes x', y', z' are parallel to the x, y, and z axes, respectively (Fig. 9.22). (Note that we use the term centroidal here to define axes passing through the center of gravity G of the body, regardless of whether or not G coincides with the centroid of the volume of the body.) We denote by $\bar{x}, \bar{y}, \bar{z}$ the coordinates of G with respect to $Oxyz$. Then we have the following relations between the coordinates x, y, z of the element dm with respect to $Oxyz$ and its coordinates x', y', z' with respect to the centroidal axes $Gx'y'z'$:

$$x = x' + \bar{x} \quad y = y' + \bar{y} \quad z = z' + \bar{z} \tag{9.31}$$

Referring to Eqs. (9.30), we can express the moment of inertia of the body with respect to the x axis as

$$I_x = \int (y^2 + z^2)\, dm = \int [(y' + \bar{y})^2 + (z' + \bar{z})^2]\, dm$$

$$= \int (y'^2 + z'^2)\, dm + 2\bar{y} \int y'\, dm + 2\bar{z} \int z'\, dm + (\bar{y}^2 + \bar{z}^2) \int dm$$

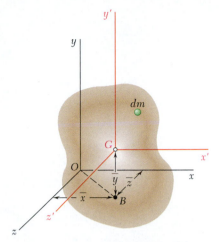

Fig. 9.22 A body of mass m with an arbitrary rectangular coordinate system at O and a parallel centroidal coordinate system at G. Also shown is an element of mass dm.

The first integral in this expression represents the moment of inertia $\bar{I}_{x'}$ of the body with respect to the centroidal axis x'. The second and third integrals represent the first moment of the body with respect to the $z'x'$ and $x'y'$ planes, respectively, and since both planes contain G, these two integrals are zero. The last integral is equal to the total mass m of the body. Therefore, we have

$$I_x = \bar{I}_{x'} + m(\bar{y}^2 + \bar{z}^2) \tag{9.32}$$

Similarly,

$$I_y = \bar{I}_{y'} + m(\bar{z}^2 + \bar{x}^2) \quad I_z = \bar{I}_{z'} + m(\bar{x}^2 + \bar{y}^2) \tag{9.32'}$$

We easily verify from Fig. 9.22 that the sum $\bar{z}^2 + \bar{x}^2$ represents the square of the distance OB between the y and y' axes. Similarly, $\bar{y}^2 + \bar{z}^2$ and $\bar{x}^2 + \bar{y}^2$ represent the squares of the distance between the x and x' axes and the z and z' axes, respectively. We denote the distance between an arbitrary axis AA' and a parallel centroidal axis BB' by d (Fig. 9.23). Then the general relation between the moment of inertia I of the body with respect to AA' and its moment of inertia $\bar{I}$ with respect to BB', known as the parallel-axis theorem for mass moments of inertia, is

Parallel-axis theorem for mass moments of inertia

$$I = \bar{I} + md^2 \tag{9.33}$$

Expressing the moments of inertia in terms of the corresponding radii of gyration, we can also write

$$k^2 = \bar{k}^2 + d^2 \tag{9.34}$$

where k and $\bar{k}$ represent the radii of gyration of the body about AA' and BB', respectively.

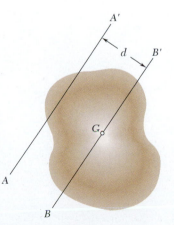

Fig. 9.23 We use d to denote the distance between an arbitrary axis AA' and a parallel centroidal axis BB'.

9.5C Moments of Inertia of Thin Plates

Now imagine a thin plate of uniform thickness t, made of a homogeneous material of density ρ (density = mass per unit volume). The mass moment of inertia of the plate with respect to an axis AA' *contained in the plane* of the plate (Fig. 9.24a) is

$$I_{AA',\,\text{mass}} = \int r^2\,dm$$

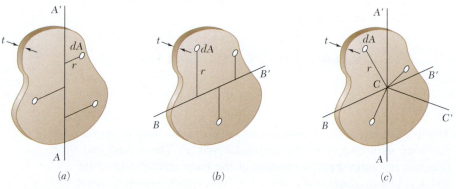

Fig. 9.24 (a) A thin plate with an axis AA' in the plane of the plate; (b) an axis BB' in the plane of the plate and perpendicular to AA'; (c) an axis CC' perpendicular to the plate and passing through the intersection of AA' and BB'.

Since $dm = \rho t\,dA$, we have

$$I_{AA',\,\text{mass}} = \rho t \int r^2\,dA$$

However, r represents the distance of the element of area dA to the axis AA'. Therefore, the integral is equal to the moment of inertia of the area of the plate with respect to AA'.

$$I_{AA',\,\text{mass}} = \rho t I_{AA',\,\text{area}} \tag{9.35}$$

Similarly, for an axis BB' that is contained in the plane of the plate and is perpendicular to AA' (Fig. 9.24b), we have

$$I_{BB',\,\text{mass}} = \rho t I_{BB',\,\text{area}} \tag{9.36}$$

Consider now the axis CC', which is *perpendicular* to the plate and passes through the point of intersection C of AA' and BB' (Fig. 9.24c). This time we have

$$I_{CC',\,\text{mass}} = \rho t J_{C,\,\text{area}} \tag{9.37}$$

where J_C is the polar moment of inertia of the area of the plate with respect to point C.

Recall the relation $J_C = I_{AA'} + I_{BB'}$ between the polar and rectangular moments of inertia of an area. We can use this to write the relation between the mass moments of inertia of a thin plate as

$$I_{CC'} = I_{AA'} + I_{BB'} \tag{9.38}$$

Rectangular Plate. In the case of a rectangular plate of sides a and b (Fig. 9.25), we obtain the mass moments of inertia with respect to axes through the center of gravity of the plate as

$$I_{AA',\,mass} = \rho t I_{AA',\,area} = \rho t(\tfrac{1}{12}a^3 b)$$
$$I_{BB',\,mass} = \rho t I_{BB',\,area} = \rho t(\tfrac{1}{12}ab^3)$$

Since the product ρabt is equal to the mass m of the plate, we can also write the mass moments of inertia of a thin rectangular plate as

$$I_{AA'} = \tfrac{1}{12}ma^2 \qquad I_{BB'} = \tfrac{1}{12}mb^2 \qquad\qquad \textbf{(9.39)}$$
$$I_{CC'} = I_{AA'} + I_{BB'} = \tfrac{1}{12}m(a^2 + b^2) \qquad\qquad \textbf{(9.40)}$$

Circular Plate. In the case of a circular plate, or disk, of radius r (Fig. 9.26), Eq. (9.35) becomes

$$I_{AA',\,mass} = \rho t I_{AA',\,area} = \rho t(\tfrac{1}{4}\pi r^4)$$

In this case, the product $\rho \pi r^2 t$ is equal to the mass m of the plate, and $I_{AA'} = I_{BB'}$. Therefore, we can write the mass moments of inertia of a circular plate as

$$I_{AA'} = I_{BB'} = \tfrac{1}{4}mr^2 \qquad\qquad \textbf{(9.41)}$$
$$I_{CC'} = I_{AA'} + I_{BB'} = \tfrac{1}{2}mr^2 \qquad\qquad \textbf{(9.42)}$$

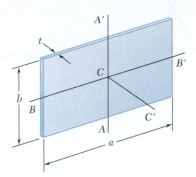

Fig. 9.25 A thin rectangular plate of sides a and b.

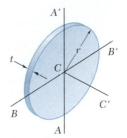

Fig. 9.26 A thin circular plate of radius r.

9.5D Determining the Moment of Inertia of a Three-Dimensional Body by Integration

We obtain the moment of inertia of a three-dimensional body by evaluating the integral $I = \int r^2\, dm$. If the body is made of a homogeneous material with a density ρ, the element of mass dm is equal to $\rho\, dV$, and we have $I = \rho \int r^2\, dV$. This integral depends only upon the shape of the body. Thus, in order to compute the moment of inertia of a three-dimensional body, it is generally necessary to perform a triple, or at least a double, integration.

However, if the body possesses two planes of symmetry, it is usually possible to determine the body's moment of inertia with a single integration. We do this by choosing as the element of mass dm a thin slab that is perpendicular to the planes of symmetry. In the case of bodies of revolution, for example, the element of mass is a thin disk (Fig. 9.27). Using formula (9.42), we can express the moment of inertia of the disk with respect to the axis of revolution as indicated in Fig. 9.27. Its moment of inertia with respect to each of the other two coordinate axes is obtained by using formula (9.41) and the parallel-axis theorem. Integration of these expressions yields the desired moment of inertia of the body.

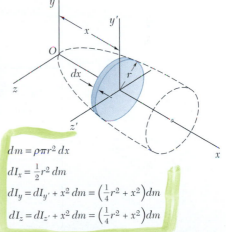

$$dm = \rho \pi r^2\, dx$$
$$dI_x = \tfrac{1}{2}r^2\, dm$$
$$dI_y = dI_{y'} + x^2\, dm = \left(\tfrac{1}{4}r^2 + x^2\right)dm$$
$$dI_z = dI_{z'} + x^2\, dm = \left(\tfrac{1}{4}r^2 + x^2\right)dm$$

Fig. 9.27 Using a thin disk to determine the moment of inertia of a body of revolution.

9.5E Moments of Inertia of Composite Bodies

Figure 9.28 lists the moments of inertia of a few common shapes. For a body consisting of several of these simple shapes in combination, you can obtain the moment of inertia of the body with respect to a given axis by first computing the moments of inertia of its component parts about the desired axis and then adding them together. As was the case for areas, the radius of gyration of a composite body *cannot* be obtained by adding the radii of gyration of its component parts.

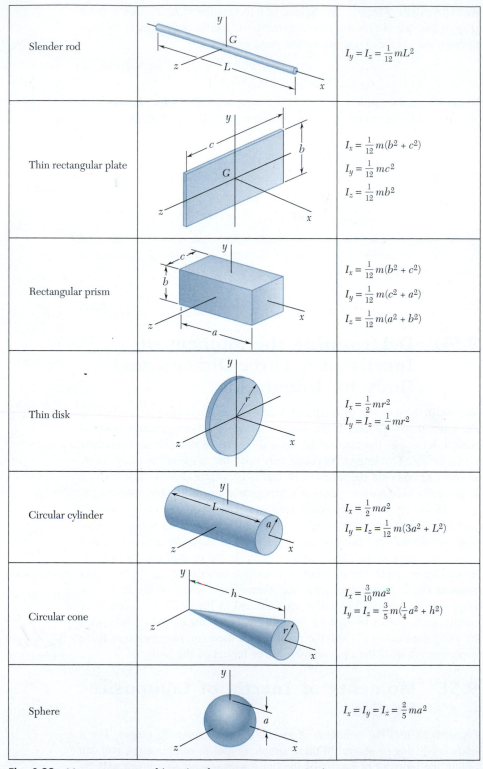

Slender rod		$I_y = I_z = \frac{1}{12} mL^2$
Thin rectangular plate		$I_x = \frac{1}{12} m(b^2 + c^2)$ $I_y = \frac{1}{12} mc^2$ $I_z = \frac{1}{12} mb^2$
Rectangular prism		$I_x = \frac{1}{12} m(b^2 + c^2)$ $I_y = \frac{1}{12} m(c^2 + a^2)$ $I_z = \frac{1}{12} m(a^2 + b^2)$
Thin disk		$I_x = \frac{1}{2} mr^2$ $I_y = I_z = \frac{1}{4} mr^2$
Circular cylinder		$I_x = \frac{1}{2} ma^2$ $I_y = I_z = \frac{1}{12} m(3a^2 + L^2)$
Circular cone		$I_x = \frac{3}{10} ma^2$ $I_y = I_z = \frac{3}{5} m(\frac{1}{4} a^2 + h^2)$
Sphere		$I_x = I_y = I_z = \frac{2}{5} ma^2$

Fig. 9.28 Mass moments of inertia of common geometric shapes.

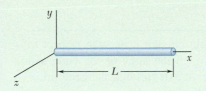

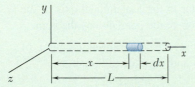

Fig. 1 Differential element of mass.

Sample Problem 9.9

Determine the moment of inertia of a slender rod of length L and mass m with respect to an axis that is perpendicular to the rod and passes through one end.

STRATEGY: Approximating the rod as a one-dimensional body enables you to solve the problem by a single integration.

MODELING and ANALYSIS: Choose the differential element of mass shown in Fig. 1 and express it as a mass per unit length.

$$dm = \frac{m}{L}\,dx$$

$$I_y = \int x^2\,dm = \int_0^L x^2 \frac{m}{L}\,dx = \left[\frac{m}{L}\frac{x^3}{3}\right]_0^L \qquad I_y = \tfrac{1}{3}mL^2 \;\blacktriangleleft$$

REFLECT and THINK: This problem could also have been solved by starting with the moment of inertia for a slender rod with respect to its centroid, as given in Fig. 9.28, and using the parallel-axis theorem to obtain the moment of inertia with respect to an end of the rod.

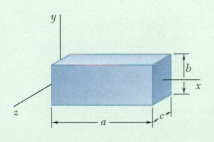

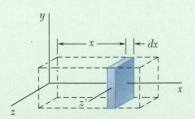

Fig. 1 Differential element of mass.

Sample Problem 9.10

For the homogeneous rectangular prism shown, determine the moment of inertia with respect to the z axis.

STRATEGY: You can approach this problem by choosing a differential element of mass perpendicular to the long axis of the prism; find its moment of inertia with respect to a centroidal axis parallel to the z axis; and then apply the parallel-axis theorem.

MODELING and ANALYSIS: Choose as the differential element of mass the thin slab shown in Fig. 1. Then

$$dm = \rho bc\,dx$$

Referring to Sec. 9.5C, the moment of inertia of the element with respect to the z' axis is

$$dI_{z'} = \tfrac{1}{12}b^2\,dm$$

Applying the parallel-axis theorem, you can obtain the mass moment of inertia of the slab with respect to the z axis.

$$dI_z = dI_{z'} + x^2\,dm = \tfrac{1}{12}b^2\,dm + x^2\,dm = (\tfrac{1}{12}b^2 + x^2)\rho bc\,dx$$

Integrating from $x = 0$ to $x = a$ gives you

$$I_z = \int dI_z = \int_0^a (\tfrac{1}{12}b^2 + x^2)\rho bc\,dx = \rho abc(\tfrac{1}{12}b^2 + \tfrac{1}{3}a^2)$$

Since the total mass of the prism is $m = \rho abc$, you can write

$$I_z = m(\tfrac{1}{12}b^2 + \tfrac{1}{3}a^2) \qquad I_z = \tfrac{1}{12}m(4a^2 + b^2) \;\blacktriangleleft$$

REFLECT and THINK: Note that if the prism is thin, b is small compared to a, and the expression for I_z reduces to $\tfrac{1}{3}ma^2$, which is the result obtained in Sample Prob. 9.9 when $L = a$.

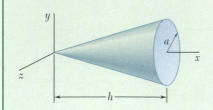

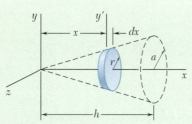

Fig. 1 Differential element of mass.

Sample Problem 9.11

Determine the moment of inertia of a right circular cone with respect to (*a*) its longitudinal axis, (*b*) an axis through the apex of the cone and perpendicular to its longitudinal axis, (*c*) an axis through the centroid of the cone and perpendicular to its longitudinal axis.

STRATEGY: For parts (a) and (b), choose a differential element of mass in the form of a thin circular disk perpendicular to the longitudinal axis of the cone. You can solve part (c) by an application of the parallel-axis theorem.

MODELING and ANALYSIS: Choose the differential element of mass shown in Fig. 1. Express the radius and mass of this disk as

$$r = a\frac{x}{h} \qquad dm = \rho\pi r^2 \, dx = \rho\pi\frac{a^2}{h^2}x^2 \, dx$$

a. Moment of Inertia I_x.

Using the expression derived in Sec. 9.5C for a thin disk, compute the mass moment of inertia of the differential element with respect to the *x* axis.

$$dI_x = \tfrac{1}{2}r^2 \, dm = \tfrac{1}{2}\left(a\frac{x}{h}\right)^2\left(\rho\pi\frac{a^2}{h^2}x^2 \, dx\right) = \tfrac{1}{2}\rho\pi\frac{a^4}{h^4}x^4 \, dx$$

Integrating from $x = 0$ to $x = h$ gives you

$$I_x = \int dI_x = \int_0^h \tfrac{1}{2}\rho\pi\frac{a^4}{h^4}x^4 \, dx = \tfrac{1}{2}\rho\pi\frac{a^4}{h^4}\frac{h^5}{5} = \tfrac{1}{10}\rho\pi a^4 h$$

Since the total mass of the cone is $m = \tfrac{1}{3}\rho\pi a^2 h$, you can write this as

$$I_x = \tfrac{1}{10}\rho\pi a^4 h = \tfrac{3}{10}a^2(\tfrac{1}{3}\rho\pi a^2 h) = \tfrac{3}{10}ma^2 \quad I_x = \tfrac{3}{10}ma^2 \blacktriangleleft$$

b. Moment of Inertia I_y.

Use the same differential element. Applying the parallel-axis theorem and using the expression derived in Sec. 9.5C for a thin disk, you have

$$dI_y = dI_{y'} + x^2 \, dm = \tfrac{1}{4}r^2 \, dm + x^2 \, dm = (\tfrac{1}{4}r^2 + x^2) \, dm$$

Substituting the expressions for *r* and *dm* into this equation yields

$$dI_y = \left(\frac{1}{4}\frac{a^2}{h^2}x^2 + x^2\right)\left(\rho\pi\frac{a^2}{h^2}x^2 \, dx\right) = \rho\pi\frac{a^2}{h^2}\left(\frac{a^2}{4h^2} + 1\right)x^4 \, dx$$

$$I_y = \int dI_y = \int_0^h \rho\pi\frac{a^2}{h^2}\left(\frac{a^2}{4h^2} + 1\right)x^4 \, dx = \rho\pi\frac{a^2}{h^2}\left(\frac{a^2}{4h^2} + 1\right)\frac{h^5}{5}$$

Introducing the total mass of the cone *m*, you can rewrite I_y as

$$I_y = \tfrac{3}{5}(\tfrac{1}{4}a^2 + h^2)\tfrac{1}{3}\rho\pi a^2 h \qquad I_y = \tfrac{3}{5}m(\tfrac{1}{4}a^2 + h^2) \blacktriangleleft$$

c. Moment of Inertia $\bar{I}_{y''}$.

Apply the parallel-axis theorem to obtain

$$I_y = \bar{I}_{y''} + m\bar{x}^2$$

Solve for $\bar{I}_{y''}$ and recall from Fig. 5.21 that $\bar{x} = \tfrac{3}{4}h$ (Fig. 2). The result is

$$\bar{I}_{y''} = I_y - m\bar{x}^2 = \tfrac{3}{5}m(\tfrac{1}{4}a^2 + h^2) - m(\tfrac{3}{4}h)^2$$

$$\bar{I}_{y''} = \tfrac{3}{20}m(a^2 + \tfrac{1}{4}h^2) \blacktriangleleft$$

REFLECT and THINK: The parallel-axis theorem for masses can be just as useful as the version for areas. Don't forget to use the reference figures for centroids of volumes when needed.

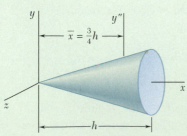

Fig. 2 Centroid of a right circular cone.

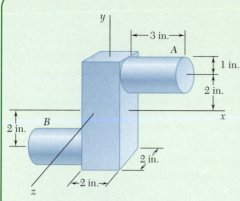

Fig. 1 Geometry of each component.

Sample Problem 9.12

A steel forging consists of a $6 \times 2 \times 2$-in. rectangular prism and two cylinders with a diameter of 2 in. and length of 3 in. as shown. Determine the moments of inertia of the forging with respect to the coordinate axes. The specific weight of steel is 490 lb/ft^3.

STRATEGY: Compute the moments of inertia of each component from Fig. 9.28 using the parallel-axis theorem when necessary. Note that all lengths should be expressed in feet to be consistent with the units for the given specific weight.

MODELING and ANALYSIS:

Computation of Masses.
Prism

$$V = (2 \text{ in.})(2 \text{ in.})(6 \text{ in.}) = 24 \text{ in}^3$$

$$W = \frac{(24 \text{ in}^3)(490 \text{ lb/ft}^3)}{1728 \text{ in}^3/\text{ft}^3} = 6.81 \text{ lb}$$

$$m = \frac{6.81 \text{ lb}}{32.2 \text{ ft/s}^2} = 0.211 \text{ lb·s}^2/\text{ft}$$

Each Cylinder

$$V = \pi(1 \text{ in.})^2(3 \text{ in.}) = 9.42 \text{ in}^3$$

$$W = \frac{(9.42 \text{ in}^3)(490 \text{ lb/ft}^3)}{1728 \text{ in}^3/\text{ft}^3} = 2.67 \text{ lb}$$

$$m = \frac{2.67 \text{ lb}}{32.2 \text{ ft/s}^2} = 0.0829 \text{ lb·s}^2/\text{ft}$$

Moments of Inertia (Fig. 1).
Prism

$$I_x = I_z = \tfrac{1}{12}(0.211 \text{ lb·s}^2/\text{ft})[(\tfrac{6}{12} \text{ ft})^2 + (\tfrac{2}{12} \text{ ft})^2] = 4.88 \times 10^{-3} \text{ lb·ft·s}^2$$
$$I_y = \tfrac{1}{12}(0.211 \text{ lb·s}^2/\text{ft})[(\tfrac{2}{12} \text{ ft})^2 + (\tfrac{2}{12} \text{ ft})^2] = 0.977 \times 10^{-3} \text{ lb·ft·s}^2$$

Each Cylinder

$$I_x = \tfrac{1}{2}ma^2 + m\bar{y}^2 = \tfrac{1}{2}(0.0829 \text{ lb·s}^2/\text{ft})(\tfrac{1}{12} \text{ ft})^2$$
$$+ (0.0829 \text{ lb·s}^2/\text{ft})(\tfrac{2}{12} \text{ ft})^2 = 2.59 \times 10^{-3} \text{ lb·ft·s}^2$$

$$I_y = \tfrac{1}{12}m(3a^2 + L^2) = m\bar{x}^2 = \tfrac{1}{12}(0.0829 \text{ lb·s}^2/\text{ft})[3(\tfrac{1}{12} \text{ ft})^2 + (\tfrac{3}{12} \text{ ft})^2]$$
$$+ (0.0829 \text{ lb·s}^2/\text{ft})(\tfrac{2.5}{12} \text{ ft})^2 = 4.17 \times 10^{-3} \text{ lb·ft·s}^2$$

$$I_z = \tfrac{1}{12}m(3a^2 + L^2) + m(\bar{x}^2 + \bar{y}^2) = \tfrac{1}{12}(0.0829 \text{ lb·s}^2/\text{ft})[3(\tfrac{1}{12} \text{ ft})^2 + (\tfrac{3}{12} \text{ ft})^2]$$
$$+ (0.0829 \text{ lb·s}^2/\text{ft})[(\tfrac{2.5}{12} \text{ ft})^2 + (\tfrac{2}{12} \text{ ft})^2] = 6.48 \times 10^{-3} \text{ lb·ft·s}^2$$

Entire Body. Adding the values obtained for the prism and two cylinders, you have

$$I_x = 4.88 \times 10^{-3} + 2(2.59 \times 10^{-3}) \qquad \boxed{I_x = 10.06 \times 10^{-3} \text{ lb·ft·s}^2} \blacktriangleleft$$
$$I_y = 0.977 \times 10^{-3} + 2(4.17 \times 10^{-3}) \qquad \boxed{I_y = 9.32 \times 10^{-3} \text{ lb·ft·s}^2} \blacktriangleleft$$
$$I_z = 4.88 \times 10^{-3} + 2(6.48 \times 10^{-3}) \qquad \boxed{I_z = 17.84 \times 10^{-3} \text{ lb·ft·s}^2} \blacktriangleleft$$

REFLECT and THINK: The results indicate this forging has more resistance to rotation about the z axis (largest moment of inertia) than about the x or y axes. This makes intuitive sense, because more of the mass is farther from the z axis than from the x or y axes.

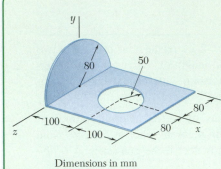

Dimensions in mm

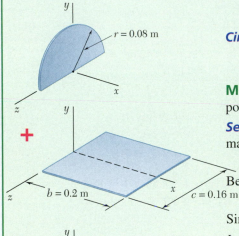

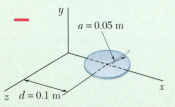

Fig. 1 Modeling the machine part as a combination of simple geometric shapes.

Sample Problem 9.13

A thin steel plate that is 4 mm thick is cut and bent to form the machine part shown. The density of the steel is 7850 kg/m³. Determine the moments of inertia of the machine part with respect to the coordinate axes.

STRATEGY: The machine part consists of a semicircular plate and a rectangular plate from which a circular plate has been removed (Fig. 1). After calculating the moments of inertia for each part, add those of the semicircular plate and the rectangular plate, then subtract those of the circular plate to determine the moments of inertia for the entire machine part.

MODELING and ANALYSIS:

Computation of Masses.   *Semicircular Plate*

$$V_1 = \tfrac{1}{2}\pi r^2 t = \tfrac{1}{2}\pi(0.08 \text{ m})^2(0.004 \text{ m}) = 40.21 \times 10^{-6} \text{ m}^3$$
$$m_1 = \rho V_1 = (7.85 \times 10^3 \text{ kg/m}^3)(40.21 \times 10^{-6} \text{ m}^3) = 0.3156 \text{ kg}$$

Rectangular Plate

$$V_2 = (0.200 \text{ m})(0.160 \text{ m})(0.004 \text{ m}) = 128 \times 10^{-6} \text{ m}^3$$
$$m_2 = \rho V_2 = (7.85 \times 10^3 \text{ kg/m}^3)(128 \times 10^{-6} \text{ m}^3) = 1.005 \text{ kg}$$

Circular Plate

$$V_3 = \pi a^2 t = \pi(0.050 \text{ m})^2(0.004 \text{ m}) = 31.42 \times 10^{-6} \text{ m}^3$$
$$m_3 = \rho V_3 = (7.85 \times 10^3 \text{ kg/m}^3)(31.42 \times 10^{-6} \text{ m}^3) = 0.2466 \text{ kg}$$

Moments of Inertia.   Compute the moments of inertia of each component, using the method presented in Sec. 9.5C.

Semicircular Plate.   Observe from Fig. 9.28 that, for a circular plate of mass m and radius r,

$$I_x = \tfrac{1}{2}mr^2 \quad I_y = I_z = \tfrac{1}{4}mr^2$$

Because of symmetry, halve these values for a semicircular plate. Thus,

$$I_x = \tfrac{1}{2}(\tfrac{1}{2}mr^2) \quad I_y = I_z = \tfrac{1}{2}(\tfrac{1}{4}mr^2)$$

Since the mass of the semicircular plate is $m_1 = \tfrac{1}{2}m$, you have

$$I_x = \tfrac{1}{2}m_1 r^2 = \tfrac{1}{2}(0.3156 \text{ kg})(0.08 \text{ m})^2 = 1.010 \times 10^{-3}\,\text{kg·m}^2$$
$$I_y = I_z = \tfrac{1}{4}(\tfrac{1}{2}mr^2) = \tfrac{1}{4}m_1 r^2 = \tfrac{1}{4}(0.3156 \text{ kg})(0.08 \text{ m})^2 = 0.505 \times 10^{-3}\,\text{kg·m}^2$$

Rectangular Plate

$$I_x = \tfrac{1}{12}m_2 c^2 = \tfrac{1}{12}(1.005 \text{ kg})(0.16 \text{ m})^2 = 2.144 \times 10^{-3}\,\text{kg·m}^2$$
$$I_z = \tfrac{1}{3}m_2 b^2 = \tfrac{1}{3}(1.005 \text{ kg})(0.2 \text{ m})^2 = 13.400 \times 10^{-3}\,\text{kg·m}^2$$
$$I_y = I_x + I_z = (2.144 + 13.400)(10^{-3}) = 15.544 \times 10^{-3}\,\text{kg·m}^2$$

Circular Plate

$$I_x = \tfrac{1}{4}m_3 a^2 = \tfrac{1}{4}(0.2466 \text{ kg})(0.05 \text{ m})^2 = 0.154 \times 10^{-3}\,\text{kg·m}^2$$
$$I_y = \tfrac{1}{2}m_3 a^2 + m_3 d^2$$
$$= \tfrac{1}{2}(0.2466 \text{ kg})(0.05 \text{ m})^2 + (0.2466 \text{ kg})(0.1 \text{ m})^2 = 2.774 \times 10^{-3}\,\text{kg·m}^2$$
$$I_z = \tfrac{1}{4}m_3 a^2 + m_3 d^2 = \tfrac{1}{4}(0.2466 \text{ kg})(0.05 \text{ m})^2 + (0.2466 \text{ kg})(0.1 \text{ m})^2$$
$$= 2.620 \times 10^{-3}\,\text{kg·m}^2$$

Entire Machine Part

$$I_x = (1.010 + 2.144 - 0.154)(10^{-3}) \text{ kg·m}^2 \qquad I_x = 3.00 \times 10^{-3}\,\text{kg·m}^2 \;\blacktriangleleft$$
$$I_y = (0.505 + 15.544 - 2.774)(10^{-3}) \text{ kg·m}^2 \quad I_y = 13.28 \times 10^{-3}\,\text{kg·m}^2 \;\blacktriangleleft$$
$$I_z = (0.505 + 13.400 - 2.620)(10^{-3}) \text{ kg·m}^2 \quad I_z = 11.29 \times 10^{-3}\,\text{kg·m}^2 \;\blacktriangleleft$$

SOLVING PROBLEMS ON YOUR OWN

In this section, we introduced the **mass moment of inertia** and the **radius of gyration** of a three-dimensional body with respect to a given axis [Eqs. (9.28) and (9.29)]. We also derived a **parallel-axis theorem** for use with mass moments of inertia and discussed the computation of the mass moments of inertia of thin plates and three-dimensional bodies.

1. Computing mass moments of inertia. You can calculate the mass moment of inertia I of a body with respect to a given axis directly from the definition given in Eq. (9.28) for simple shapes [Sample Prob. 9.9]. In most cases, however, it is necessary to divide the body into thin slabs, compute the moment of inertia of a typical slab with respect to the given axis—using the parallel-axis theorem if necessary—and integrate the resulting expression.

2. Applying the parallel-axis theorem. In Sec. 9.5B, we derived the parallel-axis theorem for mass moments of inertia as

$$I = \bar{I} + md^2 \tag{9.33}$$

This theorem states that the moment of inertia I of a body of mass m with respect to a given axis is equal to the sum of the moment of inertia $\bar{I}$ of that body with respect to a parallel centroidal axis and the product md^2, where d is the distance between the two axes. When you calculate the moment of inertia of a three-dimensional body with respect to one of the coordinate axes, you can replace d^2 by the sum of the squares of distances measured along the other two coordinate axes [Eqs. (9.32) and (9.32′)].

3. Avoiding unit-related errors. To avoid errors, you must be consistent in your use of units. Thus, all lengths should be expressed in meters or feet, as appropriate, and for problems using U.S. customary units, masses should be given in lb·s^2/ft. In addition, we strongly recommend that you include units as you perform your calculations [Sample Probs. 9.12 and 9.13].

4. Calculating the mass moment of inertia of thin plates. We showed in Sec. 9.5C that you can obtain the mass moment of inertia of a thin plate with respect to a given axis by multiplying the corresponding moment of inertia of the area of the plate by the density ρ and the thickness t of the plate [Eqs. (9.35) through (9.37)]. Note that, since the axis CC' in Fig. 9.24c is perpendicular to the plate, $I_{CC',\text{mass}}$ is associated with the *polar* moment of inertia $J_{C,\text{area}}$.

Instead of calculating the moment of inertia of a thin plate with respect to a specified axis directly, you may sometimes find it convenient to first compute its moment of inertia with respect to an axis parallel to the specified axis and to then apply the parallel-axis theorem. Furthermore, to determine the moment of inertia of a thin plate with respect to an axis perpendicular to the plate, you may wish to first determine its moments of inertia with respect to two perpendicular in-plane axes and to then use Eq. (9.38). Finally, remember that the mass of a plate consists of area A, thickness t, and density ρ, as $m = \rho t A$.

(continued)

5. Determining the moment of inertia of a body by direct single integration. We discussed in Sec. 9.5D and illustrated in Sample Probs. 9.10 and 9.11 how you can use single integration to compute the moment of inertia of a body that can be divided into a series of thin, parallel slabs. For such cases, you will often need to express the mass of the body in terms of the body's density and dimensions. Assuming that the body has been divided, as in the sample problems, into thin slabs perpendicular to the x axis, you will need to express the dimensions of each slab as functions of the variable x.

   a. In the special case of a body of revolution, the elemental slab is a thin disk, and you can use the equations given in Fig. 9.27 to determine the moments of inertia of the body [Sample Prob. 9.11].

   b. In the general case, when the body is not a solid of revolution, the differential element is not a disk but a thin slab of a different shape. You cannot use the equations of Fig. 9.27 in this case. See, for example, Sample Prob. 9.10, where the element was a thin, rectangular slab. For more complex configurations, you may want to use one or more of the following equations, which are based on Eqs. (9.32) and (9.32′) of Sec. 9.5B.

$$dI_x = dI_{x'} + (\bar{y}_{el}^2 + \bar{z}_{el}^2)\, dm$$
$$dI_y = dI_{y'} + (\bar{z}_{el}^2 + \bar{x}_{el}^2)\, dm$$
$$dI_z = dI_{z'} + (\bar{x}_{el}^2 + \bar{y}_{el}^2)\, dm$$

Here, the primes denote the centroidal axes of each elemental slab and $\bar{x}_{el}$, $\bar{y}_{el}$, and $\bar{z}_{el}$ represent the coordinates of its centroid. Determine the centroidal moments of inertia of the slab in the manner described earlier for a thin plate: Refer to Fig. 9.12, calculate the corresponding moments of inertia of the area of the slab, and multiply the result by the density ρ and the thickness t of the slab. Also, assuming that the body has been divided into thin slabs perpendicular to the x axis, remember that you can obtain $dI_{x'}$ by adding $dI_{y'}$ and $dI_{z'}$ instead of computing it directly. Finally, using the geometry of the body, express the result obtained in terms of the single variable x, and integrate in x.

6. Computing the moment of inertia of a composite body. As stated in Sec. 9.5E, the moment of inertia of a composite body with respect to a specified axis is equal to the sum of the moments of its components with respect to that axis. Sample Probs. 9.12 and 9.13 illustrate the appropriate method of solution. Also remember that the moment of inertia of a component is negative only if the component is *removed* (as in the case of a hole).

Although the composite-body problems in this section are relatively straightforward, you will have to work carefully to avoid computational errors. In addition, if some of the moments of inertia that you need are not given in Fig. 9.28, you will have to derive your own formulas, using the techniques described in this section.

Problems

9.111 A thin plate with a mass m is cut in the shape of an equilateral triangle of side a. Determine the mass moment of inertia of the plate with respect to (a) the centroidal axes AA' and BB', (b) the centroidal axis CC' that is perpendicular to the plate.

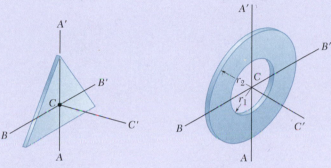

Fig. P9.111   **Fig. P9.112**

9.112 A ring with a mass m is cut from a thin uniform plate. Determine the mass moment of inertia of the ring with respect to (a) the axis AA', (b) the centroidal axis CC' that is perpendicular to the plane of the ring.

9.113 A thin, semielliptical plate has a mass m. Determine the mass moment of inertia of the plate with respect to (a) the centroidal axis BB', (b) the centroidal axis CC' that is perpendicular to the plate.

9.114 The parabolic spandrel shown was cut from a thin, uniform plate. Denoting the mass of the spandrel by m, determine its mass moment of inertia with respect to (a) the axis BB', (b) the axis DD' that is perpendicular to the spandrel. (*Hint:* See Sample Prob. 9.3.)

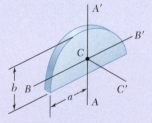

Fig. P9.113

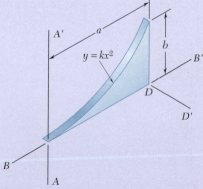

Fig. P9.114

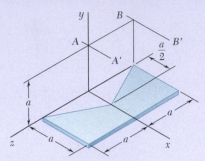

Fig. P9.115 and P9.116

9.115 A piece of thin, uniform sheet metal is cut to form the machine component shown. Denoting the mass of the component by m, determine its mass moment of inertia with respect to (a) the x axis, (b) the y axis.

9.116 A piece of thin, uniform sheet metal is cut to form the machine component shown. Denoting the mass of the component by m, determine its mass moment of inertia with respect to (a) the axis AA', (b) the axis BB', where the AA' and BB' axes are parallel to the x axis and lie in a plane parallel to and at a distance a above the xz plane.

9.117 A thin plate with a mass m has the trapezoidal shape shown. Determine the mass moment of inertia of the plate with respect to (a) the x axis, (b) the y axis.

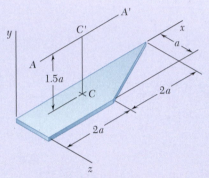

Fig. P9.117 and P9.118

9.118 A thin plate with a mass m has the trapezoidal shape shown. Determine the mass moment of inertia of the plate with respect to (a) the centroidal axis CC' that is perpendicular to the plate, (b) the axis AA' that is parallel to the x axis and is located at a distance $1.5a$ from the plate.

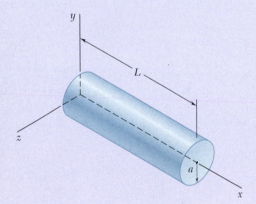

Fig. P9.119

9.119 Determine by direct integration the mass moment of inertia with respect to the z axis of the right circular cylinder shown, assuming that it has a uniform density and a mass m.

9.120 The area shown is revolved about the x axis to form a homogeneous solid of revolution of mass m. Using direct integration, express the mass moment of inertia of the solid with respect to the x axis in terms of m and h.

9.121 The area shown is revolved about the x axis to form a homogeneous solid of revolution of mass m. Determine by direct integration the mass moment of inertia of the solid with respect to (a) the x axis, (b) the y axis. Express your answers in terms of m and the dimensions of the solid.

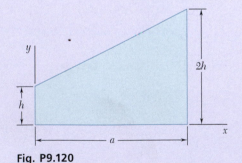

Fig. P9.120

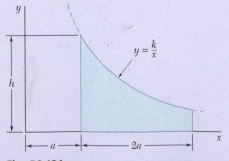

Fig. P9.121

9.122 Determine by direct integration the mass moment of inertia with respect to the x axis of the tetrahedron shown, assuming that it has a uniform density and a mass m.

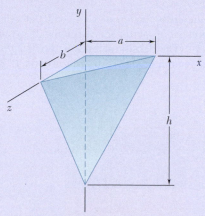

Fig. P9.122 and *P9.123*

9.123 Determine by direct integration the mass moment of inertia with respect to the y axis of the tetrahedron shown, assuming that it has a uniform density and a mass m.

9.124 Determine by direct integration the mass moment of inertia and the radius of gyration with respect to the x axis of the paraboloid shown, assuming that it has a uniform density and a mass m.

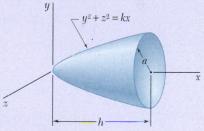

Fig. P9.124

9.125 A thin, rectangular plate with a mass m is welded to a vertical shaft AB as shown. Knowing that the plate forms an angle θ with the y axis, determine by direct integration the mass moment of inertia of the plate with respect to (a) the y axis, (b) the z axis.

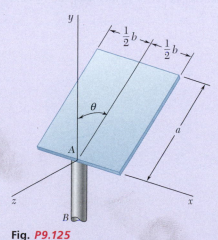

Fig. *P9.125*

*9.126 A thin steel wire is bent into the shape shown. Denoting the mass per unit length of the wire by m', determine by direct integration the mass moment of inertia of the wire with respect to each of the coordinate axes.

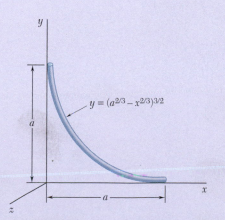

Fig. P9.126

543

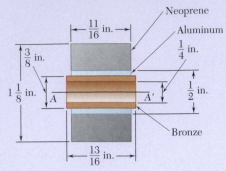

Fig. P9.127

9.127 Shown is the cross section of an idler roller. Determine its mass moment of inertia and its radius of gyration with respect to the axis AA'. (The specific weight of bronze is 0.310 lb/in^3; of aluminum, 0.100 lb/in^3; and of neoprene, 0.0452 lb/in^3.)

9.128 Shown is the cross section of a molded flat-belt pulley. Determine its mass moment of inertia and its radius of gyration with respect to the axis AA'. (The density of brass is 8650 kg/m^3, and the density of the fiber-reinforced polycarbonate used is 1250 kg/m^3.)

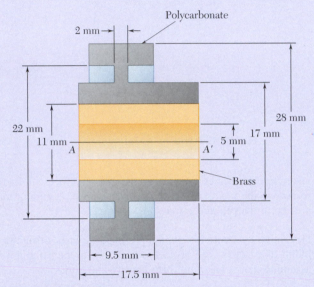

Fig. P9.128

9.129 The machine part shown is formed by machining a conical surface into a circular cylinder. For $b = \frac{1}{2}h$, determine the mass moment of inertia and the radius of gyration of the machine part with respect to the y axis.

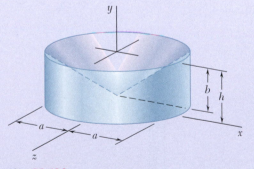

Fig. P9.129

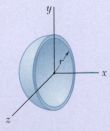

Fig. P9.130

9.130 Knowing that the thin hemispherical shell shown has a mass m and thickness t, determine the mass moment of inertia and the radius of gyration of the shell with respect to the x axis. (*Hint:* Consider the shell as formed by removing a hemisphere of radius r from a hemisphere of radius $r + t$; then neglect the terms containing t^2 and t^3 and keep those terms containing t.)

9.131 A square hole is centered in and extends through the aluminum machine component shown. Determine (a) the value of a for which the mass moment of inertia of the component is maximum with respect to the axis AA' that bisects the top surface of the hole, (b) the corresponding values of the mass moment of inertia and the radius of gyration with respect to the axis AA'. (The specific weight of aluminum is 0.100 lb/in³.)

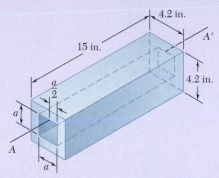

Fig. P9.131

9.132 The cups and the arms of an anemometer are fabricated from a material of density ρ. Knowing that the mass moment of inertia of a thin, hemispherical shell with a mass m and thickness t with respect to its centroidal axis GG' is $5ma^2/12$, determine (a) the mass moment of inertia of the anemometer with respect to the axis AA', (b) the ratio of a to l for which the centroidal moment of inertia of the cups is equal to 1 percent of the moment of inertia of the cups with respect to the axis AA'.

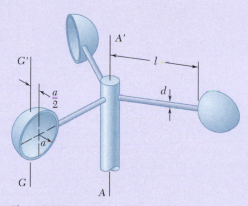

Fig. P9.132

9.133 After a period of use, one of the blades of a shredder has been worn to the shape shown and is of mass 0.18 kg. Knowing that the mass moments of inertia of the blade with respect to the AA' and BB' axes are 0.320 g·m² and 0.680 g·m², respectively, determine (a) the location of the centroidal axis GG', (b) the radius of gyration with respect to axis GG'.

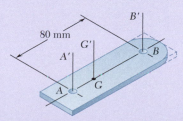

Fig. P9.133

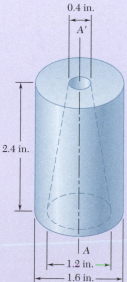

9.134 Determine the mass moment of inertia of the 0.9-lb machine component shown with respect to the axis AA'.

Fig. P9.134

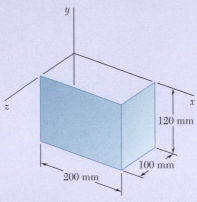

Fig. P9.135

9.135 and 9.136 A 2-mm thick piece of sheet steel is cut and bent into the machine component shown. Knowing that the density of steel is 7850 kg/m^3, determine the mass moment of inertia of the component with respect to each of the coordinate axes.

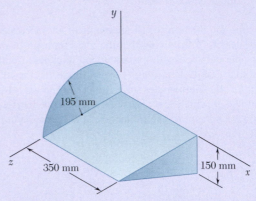

Fig. P9.136

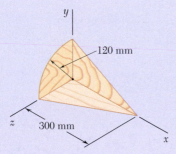

Fig. P9.137

9.137 A subassembly for a model airplane is fabricated from three pieces of 1.5-mm plywood. Neglecting the mass of the adhesive used to assemble the three pieces, determine the mass moment of inertia of the subassembly with respect to each of the coordinate axes. (The density of the plywood is 780 kg/m^3.)

9.138 A section of sheet steel 0.03 in. thick is cut and bent into the sheet metal machine component shown. Determine the mass moment of inertia of the component with respect to each of the coordinate axes. (The specific weight of steel is 490 lb/ft^3.)

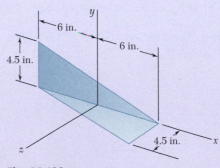

Fig. P9.138

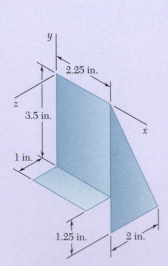

Fig. P9.139

9.139 A framing anchor is formed of 0.05-in.-thick galvanized steel. Determine the mass moment of inertia of the anchor with respect to each of the coordinate axes. (The specific weight of galvanized steel is 470 lb/ft^3.)

*9.140 A farmer constructs a trough by welding a rectangular piece of 2-mm-thick sheet steel to half of a steel drum. Knowing that the density of steel is 7850 kg/m³ and that the thickness of the walls of the drum is 1.8 mm, determine the mass moment of inertia of the trough with respect to each of the coordinate axes. Neglect the mass of the welds.

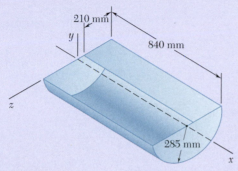

210 mm

840 mm

y

285 mm

z

x

Fig. *P9.140*

9.141 The machine element shown is fabricated from steel. Determine the mass moment of inertia of the assembly with respect to (*a*) the *x* axis, (*b*) the *y* axis, (*c*) the *z* axis. (The density of steel is 7850 kg/m³.)

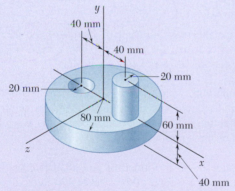

y

40 mm

40 mm

20 mm

20 mm

80 mm

60 mm

z

x

40 mm

Fig. P9.141

9.142 Determine the mass moments of inertia and the radii of gyration of the steel machine element shown with respect to the *x* and *y* axes. (The density of steel is 7850 kg/m³.)

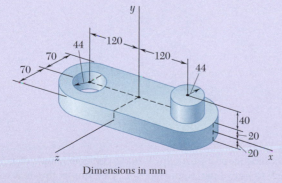

y

44

70

70

120

120

44

40

20

20

z

x

Dimensions in mm

Fig. P9.142

9.143 Determine the mass moment of inertia of the steel machine element shown with respect to the *x* axis. (The specific weight of steel is 490 lb/ft³.)

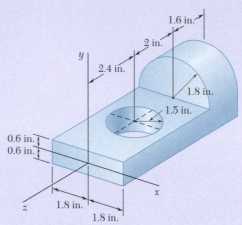

Fig. P9.143 and *P.9.144*

9.144 Determine the mass moment of inertia of the steel machine element shown with respect to the *y* axis. (The specific weight of steel is 490 lb/ft³.)

9.145 Determine the mass moment of inertia of the steel fixture shown with respect to (*a*) the *x* axis, (*b*) the *y* axis, (*c*) the *z* axis. (The density of steel is 7850 kg/m³.)

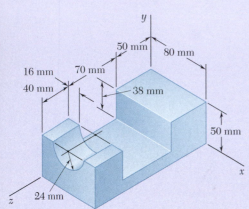

Fig. P9.145

9.146 Aluminum wire with a weight per unit length of 0.033 lb/ft is used to form the circle and the straight members of the figure shown. Determine the mass moment of inertia of the assembly with respect to each of the coordinate axes.

9.147 The figure shown is formed of ⅛-in.-diameter steel wire. Knowing that the specific weight of the steel is 490 lb/ft³, determine the mass moment of inertia of the wire with respect to each of the coordinate axes.

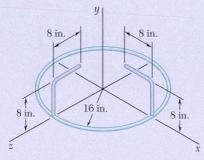

Fig. *P9.146*

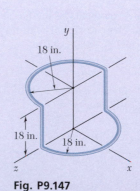

Fig. P9.147

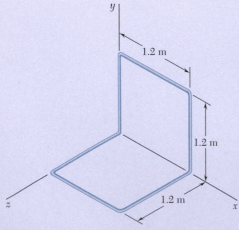

Fig. P9.148

9.148 A homogeneous wire with a mass per unit length of 0.056 kg/m is used to form the figure shown. Determine the mass moment of inertia of the wire with respect to each of the coordinate axes.

*9.6 ADDITIONAL CONCEPTS OF MASS MOMENTS OF INERTIA

In this final section of the chapter, we present several concepts involving mass moments of inertia that are analogous to material presented in Sec. 9.4 involving moments of inertia of areas. These ideas include mass products of inertia, principal axes of inertia, and principal moments of inertia for masses, which are necessary for the study of the dynamics of rigid bodies in three dimensions.

9.6A Mass Products of Inertia

In this section, you will see how to determine the moment of inertia of a body with respect to an arbitrary axis OL through the origin (Fig. 9.29) if its moments of inertia with respect to the three coordinate axes, as well as certain other quantities defined here, have already been determined.

The moment of inertia I_{OL} of the body with respect to OL is equal to $\int p^2\, dm$, where p denotes the perpendicular distance from the element of mass dm to the axis OL. If we denote the unit vector along OL by $\boldsymbol{\lambda}$ and the position vector of the element dm by $\mathbf{r}$, the perpendicular distance p is equal to $r \sin \theta$, which is the magnitude of the vector product $\boldsymbol{\lambda} \times \mathbf{r}$. We therefore have

$$I_{OL} = \int p^2\, dm = \int |\boldsymbol{\lambda} \times \mathbf{r}|^2\, dm \tag{9.43}$$

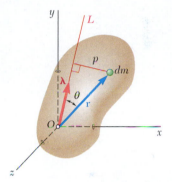

Fig. 9.29 An element of mass dm of a body and its perpendicular distance to an arbitrary axis OL through the origin.

Expressing $|\boldsymbol{\lambda} \times \mathbf{r}|^2$ in terms of the rectangular components of the vector product, we have

$$I_{OL} = \int [(\lambda_x y - \lambda_y x)^2 + (\lambda_y z - \lambda_z y)^2 + (\lambda_z x - \lambda_x z)^2]\, dm$$

Here, the components λ_x, λ_y, λ_z of the unit vector $\boldsymbol{\lambda}$ represent the direction cosines of the axis OL, and the components x, y, z of $\mathbf{r}$ represent the coordinates of the element of mass dm. Expanding the squares and rearranging the terms, we obtain

$$I_{OL} = \lambda_x^2 \int (y^2 + z^2)\, dm + \lambda_y^2 \int (z^2 + x^2)\, dm + \lambda_z^2 \int (x^2 + y^2)\, dm$$
$$- 2\lambda_x \lambda_y \int xy\, dm - 2\lambda_y \lambda_z \int yz\, dm - 2\lambda_z \lambda_x \int zx\, dm \tag{9.44}$$

Referring to Eqs. (9.30), note that the first three integrals in Eq. (9.44) represent, respectively, the moments of inertia I_x, I_y, and I_z of the body with respect to the coordinate axes. The last three integrals in Eq. (9.44), which involve products of coordinates, are called the **products of inertia** of the body with respect to the x and y axes, the y and z axes, and the z and x axes, respectively.

Mass products of inertia

$$I_{xy} = \int xy\, dm \qquad I_{yz} = \int yz\, dm \qquad I_{zx} = \int zx\, dm \tag{9.45}$$

Rewriting Eq. (9.44) in terms of the integrals defined in Eqs. (9.30) and (9.45), we have

$$I_{OL} = I_x\lambda_x^2 + I_y\lambda_y^2 + I_z\lambda_z^2 - 2I_{xy}\lambda_x\lambda_y - 2I_{yz}\lambda_y\lambda_z - 2I_{zx}\lambda_z\lambda_x \tag{9.46}$$

The definition of the products of inertia of a mass given in Eqs. (9.45) is an extension of the definition of the product of inertia of an area (Sec. 9.3). Mass products of inertia reduce to zero under the same conditions of symmetry as do products of inertia of areas, and the parallel-axis theorem for mass products of inertia is expressed by relations similar to the formula derived for the product of inertia of an area. Substituting the expressions for x, y, and z given in Eqs. (9.31) into Eqs. (9.45), we find that

Parallel-axis theorem for mass products of inertia

$$\begin{aligned} I_{xy} &= \bar{I}_{x'y'} + m\bar{x}\bar{y} \\ I_{yz} &= \bar{I}_{y'z'} + m\bar{y}\bar{z} \\ I_{zx} &= \bar{I}_{z'x'} + m\bar{z}\bar{x} \end{aligned} \tag{9.47}$$

Here $\bar{x}$, $\bar{y}$, $\bar{z}$ are the coordinates of the center of gravity G of the body and $\bar{I}_{x'y'}$, $\bar{I}_{y'z'}$, $\bar{I}_{z'x'}$ denote the products of inertia of the body with respect to the centroidal axes x', y', and z' (see Fig. 9.22).

9.6B   Principal Axes and Principal Moments of Inertia

Let us assume that we have determined the moment of inertia of the body considered in the preceding section with respect to a large number of axes OL through the fixed point O. Suppose that we plot a point Q on each axis OL at a distance $OQ = 1/\sqrt{I_{OL}}$ from O. The locus of the points Q forms a surface (Fig. 9.30). We can obtain the equation of that surface by substituting $1/(OQ)^2$ for I_{OL} in Eq. (9.46) and then multiplying both sides of the equation by $(OQ)^2$. Observing that

$$(OQ)\lambda_x = x \qquad (OQ)\lambda_y = y \qquad (OQ)\lambda_z = z$$

where x, y, z denote the rectangular coordinates of Q, we have

$$I_x x^2 + I_y y^2 + I_z z^2 - 2I_{xy}xy - 2I_{yz}yz - 2I_{zx}zx = 1 \tag{9.48}$$

This is the equation of a *quadric surface*. Since the moment of inertia I_{OL} is different from zero for every axis OL, no point Q can be at an infinite distance from O. Thus, the quadric surface obtained is an *ellipsoid*. This ellipsoid, which defines the moment of inertia of the body with respect to any axis through O, is known as the **ellipsoid of inertia** of the body at O.

Observe that, if we rotate the axes in Fig. 9.30, the coefficients of the equation defining the ellipsoid change, since they are equal to the moments and products of inertia of the body with respect to the rotated coordinate axes. However, the *ellipsoid itself remains unaffected*, since its

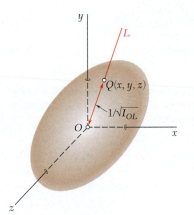

Fig. 9.30 The ellipsoid of inertia defines the moment of inertia of a body with respect to any axis through O.

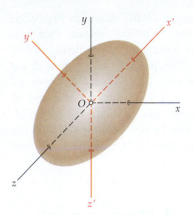

Fig. 9.31 Principal axes of inertia x', y', z' of the body at O.

shape depends only upon the distribution of mass in the given body. Suppose that we choose as coordinate axes the principal axes x', y', and z' of the ellipsoid of inertia (Fig. 9.31). The equation of the ellipsoid with respect to these coordinate axes is known to be of the form

$$I_{x'}x'^2 + I_{y'}y'^2 + I_{z'}z'^2 = 1 \qquad (9.49)$$

which does not contain any products of the coordinates. Comparing Eqs. (9.48) and (9.49), we observe that the products of inertia of the body with respect to the x', y', and z' axes must be zero. The x', y', and z' axes are known as the **principal axes of inertia** of the body at O, and the coefficients $I_{x'}$, $I_{y'}$, and $I_{z'}$ are referred to as the **principal moments of inertia** of the body at O. Note that, given a body of arbitrary shape and a point O, it is always possible to find principal axes of inertia of the body at O; that is, axes with respect to which the products of inertia of the body are zero. Indeed, whatever the shape of the body, the moments and products of inertia of the body with respect to the x, y, and z axes through O define an ellipsoid, and this ellipsoid has principal axes that, by definition, are the principal axes of inertia of the body at O.

If the principal axes of inertia x', y', and z' are used as coordinate axes, the expression in Eq. (9.46) for the moment of inertia of a body with respect to an arbitrary axis through O reduces to

$$I_{OL} = I_{x'}\lambda_{x'}^2 + I_{y'}\lambda_{y'}^2 + I_{z'}\lambda_{z'}^2 \qquad (9.50)$$

The determination of the principal axes of inertia of a body of arbitrary shape is somewhat involved and is discussed in the next section. In many cases, however, these axes can be spotted immediately. Consider, for instance, the homogeneous cone of elliptical base shown in Fig. 9.32; this cone possesses two mutually perpendicular planes of symmetry OAA' and OBB'. From the definition of Eq. (9.45), we observe that if we choose the $x'y'$ and $y'z'$ planes to coincide with the two planes of symmetry, all of the products of inertia are zero. The x', y', and z' axes selected in this way are therefore the principal axes of inertia of the cone at O. In the

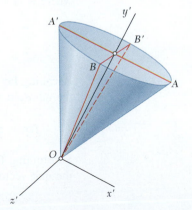

Fig. 9.32 A homogeneous cone with elliptical base has two mutually perpendicular planes of symmetry.

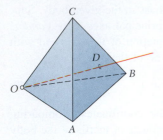

Fig. 9.33 A line drawn from a corner to the center of the opposite face of a homogeneous regular tetrahedron is a principal axis, since each 120° rotation of the body about this axis leaves its shape and mass distribution unchanged.

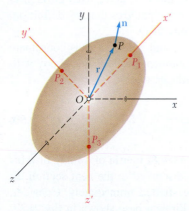

Fig. 9.34 The principal axes intersect an ellipsoid of inertia at points where the radius vectors are collinear with the unit normal vectors to the surface.

case of the homogeneous regular tetrahedron $OABC$ shown in Fig. 9.33, the line joining the corner O to the center D of the opposite face is a principal axis of inertia at O, and any line through O perpendicular to OD is also a principal axis of inertia at O. This property is apparent if we observe that rotating the tetrahedron through 120° about OD leaves its shape and mass distribution unchanged. It follows that the ellipsoid of inertia at O also remains unchanged under this rotation. The ellipsoid, therefore, is a body of revolution whose axis of revolution is OD, and the line OD, as well as any perpendicular line through O, must be a principal axis of the ellipsoid.

9.6C Principal Axes and Moments of Inertia for a Body of Arbitrary Shape

The method of analysis described in this section extends the analysis in the preceding section. However, generally speaking, you should use it only when the body under consideration has no obvious property of symmetry.

Consider the ellipsoid of inertia of a body at a given point O (Fig. 9.34). Let $\mathbf{r}$ be the radius vector of a point P on the surface of the ellipsoid, and let $\mathbf{n}$ be the unit vector along the normal to that surface at P. We observe that the only points where $\mathbf{r}$ and $\mathbf{n}$ are collinear are points P_1, P_2, and P_3, where the principal axes intersect the visible portion of the surface of the ellipsoid (along with the corresponding points on the other side of the ellipsoid).

Recall from calculus that the direction of the normal to a surface of equation $f(x, y, z) = 0$ at a point $P(x, y, z)$ is defined by the gradient ∇f of the function f at that point. To obtain the points where the principal axes intersect the surface of the ellipsoid of inertia, we must therefore express that $\mathbf{r}$ and ∇f are collinear,

$$\nabla f = (2K)\mathbf{r} \tag{9.51}$$

where K is a constant, $\mathbf{r} = x\mathbf{i} + y\mathbf{j} + z\mathbf{k}$, and

$$\nabla f = \frac{\partial f}{\partial x}\mathbf{i} + \frac{\partial f}{\partial y}\mathbf{j} + \frac{\partial f}{\partial z}\mathbf{k}$$

Recalling Eq. (9.48), we note that the function $f(x, y, z)$ corresponding to the ellipsoid of inertia is

$$f(x, y, z) = I_x x^2 + I_y y^2 + I_z z^2 - 2I_{xy}xy - 2I_{yz}yz - 2I_{zx}zx - 1$$

Substituting for $\mathbf{r}$ and ∇f into Eq. (9.51) and equating the coefficients of the unit vectors, we obtain

$$\begin{aligned} I_x x \quad - I_{xy}y - I_{zx}z &= Kx \\ -I_{xy}x + I_y y - I_{yz}z &= Ky \\ -I_{zx}x - I_{yz}y + I_z z &= Kz \end{aligned} \tag{9.52}$$

Dividing each term by the distance r from O to P, we obtain similar equations involving the direction cosines λ_x, λ_y, and λ_z:

$$
\begin{aligned}
I_x\lambda_x - I_{xy}\lambda_y - I_{zx}\lambda_z &= K\lambda_x \\
-I_{xy}\lambda_x + I_y\lambda_y - I_{yz}\lambda_z &= K\lambda_y \\
-I_{zx}\lambda_x - I_{yz}\lambda_y + I_z\lambda_z &= K\lambda_z
\end{aligned}
\tag{9.53}
$$

Transposing the right-hand members leads to the homogeneous linear equations, as

$$
\begin{aligned}
(I_x - K)\lambda_x - I_{xy}\lambda_y - I_{zx}\lambda_z &= 0 \\
-I_{xy}\lambda_x + (I_y - K)\lambda_y - I_{yz}\lambda_z &= 0 \\
-I_{zx}\lambda_x - I_{yz}\lambda_y + (I_z - K)\lambda_z &= 0
\end{aligned}
\tag{9.54}
$$

For this system of equations to have a solution different from $\lambda_x = \lambda_y = \lambda_z = 0$, its discriminant must be zero. Thus,

$$
\begin{vmatrix}
I_x - K & -I_{xy} & -I_{zx} \\
-I_{xy} & I_y - K & -I_{yz} \\
-I_{zx} & -I_{yz} & I_z - K
\end{vmatrix} = 0
\tag{9.55}
$$

Expanding this determinant and changing signs, we have

$$
K^3 - (I_x + I_y + I_z)K^2 + (I_xI_y + I_yI_z + I_zI_x - I_{xy}^2 - I_{yz}^2 - I_{zx}^2)K \\
- (I_xI_yI_z - I_xI_{yz}^2 - I_yI_{zx}^2 - I_zI_{xy}^2 - 2I_{xy}I_{yz}I_{zx}) = 0
\tag{9.56}
$$

This is a cubic equation in K, which yields three real, positive roots: K_1, K_2, and K_3.

To obtain the direction cosines of the principal axis corresponding to the root K_1, we substitute K_1 for K in Eqs. (9.54). Since these equations are now linearly dependent, only two of them may be used to determine λ_x, λ_y, and λ_z. We can obtain an additional equation, however, by recalling from Sec. 2.4A that the direction cosines must satisfy the relation

$$
\lambda_x^2 + \lambda_y^2 + \lambda_z^2 = 1
\tag{9.57}
$$

Repeating this procedure with K_2 and K_3, we obtain the direction cosines of the other two principal axes.

We now show that *the roots K_1, K_2, and K_3 of Eq. (9.56) are the principal moments of inertia of the given body.* Let us substitute for K in Eqs. (9.53) the root K_1, and for λ_x, λ_y, and λ_z the corresponding values $(\lambda_x)_1$, $(\lambda_y)_1$, and $(\lambda_z)_1$ of the direction cosines; the three equations are satisfied. We now multiply by $(\lambda_x)_1$, $(\lambda_y)_1$, and $(\lambda_z)_1$, respectively, each term in the first, second, and third equation and add the equations obtained in this way. The result is

$$
I_x^2(\lambda_x)_1^2 + I_y^2(\lambda_y)_1^2 + I_z^2(\lambda_z)_1^2 - 2I_{xy}(\lambda_x)_1(\lambda_y)_1 \\
- 2I_{yz}(\lambda_y)_1(\lambda_z)_1 - 2I_{zx}(\lambda_z)_1(\lambda_x)_1 = K_1[(\lambda_x)_1^2 + (\lambda_y)_1^2 + (\lambda_z)_1^2]
$$

Recalling Eq. (9.46), we observe that the left-hand side of this equation represents the moment of inertia of the body with respect to the principal axis corresponding to K_1; it is thus the principal moment of inertia corresponding to that root. On the other hand, recalling Eq. (9.57), we note that the right-hand member reduces to K_1. Thus, K_1 itself is the principal moment of inertia. In the same fashion, we can show that K_2 and K_3 are the other two principal moments of inertia of the body.

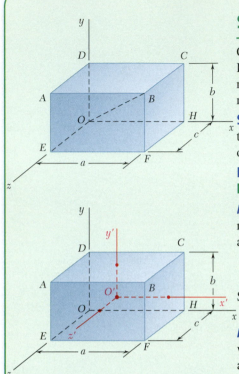

Fig. 1 Centroidal axes for the rectangular prism.

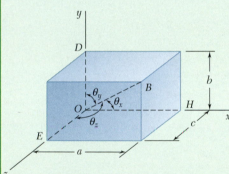

Fig. 2 Direction angles for *OB*.

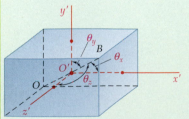

Fig. 3 Line *OB* passes through the centroid *O'*.

Sample Problem 9.14

Consider a rectangular prism with a mass of m and sides a, b, and c. Determine (*a*) the moments and products of inertia of the prism with respect to the coordinate axes shown, (*b*) its moment of inertia with respect to the diagonal *OB*.

STRATEGY: For part (a), you can introduce centroidal axes and apply the parallel-axis theorem. For part (b), determine the direction cosines of line *OB* from the given geometry and use either Eq. (9.46) or (9.50).

MODELING and ANALYSIS: a. Moments and Products of Inertia with Respect to the Coordinate Axes.

Moments of Inertia.   Introduce the centroidal axes x', y', and z' with respect to which the moments of inertia are given in Fig. 9.28, and then apply the parallel-axis theorem (Fig. 1). Thus,

$$I_x = \bar{I}_{x'} + m(\bar{y}^2 + \bar{z}^2) = \tfrac{1}{12}m(b^2 + c^2) + m(\tfrac{1}{4}b^2 + \tfrac{1}{4}c^2)$$

$$I_x = \tfrac{1}{3}m(b^2 + c^2) \blacktriangleleft$$

Similarly,

$$I_y = \tfrac{1}{3}m(c^2 + a^2) \qquad I_z = \tfrac{1}{3}m(a^2 + b^2) \blacktriangleleft$$

Products of Inertia.   Because of symmetry, the products of inertia with respect to the centroidal axes x', y', and z' are zero, and these axes are principal axes of inertia. Using the parallel-axis theorem, you have

$$I_{xy} = \bar{I}_{x'y'} + m\bar{x}\bar{y} = 0 + m(\tfrac{1}{2}a)(\tfrac{1}{2}b) \qquad I_{xy} = \tfrac{1}{4}mab \blacktriangleleft$$

Similarly,

$$I_{yz} = \tfrac{1}{4}mbc \qquad I_{zx} = \tfrac{1}{4}mca \blacktriangleleft$$

b. Moment of Inertia with Respect to *OB*.   Recall Eq. (9.46):

$$I_{OB} = I_x\lambda_x^2 + I_y\lambda_y^2 + I_z\lambda_z^2 - 2I_{xy}\lambda_x\lambda_y - 2I_{yz}\lambda_y\lambda_z - 2I_{zx}\lambda_z\lambda_x$$

where the direction cosines of *OB* are (Fig. 2)

$$\lambda_x = \cos\theta_x = \frac{OH}{OB} = \frac{a}{(a^2 + b^2 + c^2)^{1/2}}$$

$$\lambda_y = \frac{b}{(a^2 + b^2 + c^2)^{1/2}} \qquad \lambda_z = \frac{c}{(a^2 + b^2 + c^2)^{1/2}}$$

Substituting the values obtained in part (a) for the moments and products of inertia and for the direction cosines into the equation for I_{OB}, you obtain

$$I_{OB} = \frac{1}{a^2 + b^2 + c^2}\left[\tfrac{1}{3}m(b^2 + c^2)a^2 + \tfrac{1}{3}m(c^2 + a^2)b^2 + \tfrac{1}{3}m(a^2 + b^2)c^2\right.$$
$$\left. -\tfrac{1}{2}ma^2b^2 - \tfrac{1}{2}mb^2c^2 - \tfrac{1}{2}mc^2a^2\right]$$

$$I_{OB} = \frac{m}{6}\frac{a^2b^2 + b^2c^2 + c^2a^2}{a^2 + b^2 + c^2} \blacktriangleleft$$

REFLECT and THINK: You can also obtain the moment of inertia I_{OB} directly from the principal moments of inertia $\bar{I}_{x'}$, $\bar{I}_{y'}$, and $\bar{I}_{z'}$, since the line *OB* passes through the centroid *O'*. Since the x', y', and z' axes are principal axes of inertia (Fig. 3), use Eq. (9.50) to write

$$I_{OB} = \bar{I}_{x'}\lambda_x^2 + \bar{I}_{y'}\lambda_y^2 + \bar{I}_{z'}\lambda_z^2$$
$$= \frac{1}{a^2 + b^2 + c^2}\left[\frac{m}{12}(b^2 + c^2)a^2 + \frac{m}{12}(c^2 + a^2)b^2 + \frac{m}{12}(a^2 + b^2)c^2\right]$$

$$I_{OB} = \frac{m}{6}\frac{a^2b^2 + b^2c^2 + c^2a^2}{a^2 + b^2 + c^2} \blacktriangleleft$$

Sample Problem 9.15

If $a = 3c$ and $b = 2c$ for the rectangular prism of Sample Prob. 9.14, determine (*a*) the principal moments of inertia at the origin O, (*b*) the principal axes of inertia at O.

STRATEGY: Substituting the data into the results from Sample Prob. 9.14 gives you values you can use with Eq. (9.56) to determine the principal moments of inertia. You can then use these values to set up a system of equations for finding the direction cosines of the principal axes.

MODELING and ANALYSIS:

a. Principal Moments of Inertia at the Origin O.
Substituting $a = 3c$ and $b = 2c$ into the solution to Sample Prob. 9.14 gives you

$$I_x = \tfrac{5}{3}mc^2 \qquad I_y = \tfrac{10}{3}mc^2 \qquad I_z = \tfrac{13}{3}mc^2$$
$$I_{xy} = \tfrac{3}{2}mc^2 \qquad I_{yz} = \tfrac{1}{2}mc^2 \qquad I_{zx} = \tfrac{3}{4}mc^2$$

Substituting the values of the moments and products of inertia into Eq. (9.56) and collecting terms yields

$$K^3 - (\tfrac{28}{3}mc^2)K^2 + (\tfrac{3479}{144}m^2c^4)K - \tfrac{589}{54}m^3c^6 = 0$$

Now solve for the roots of this equation; from the discussion in Sec. 9.6C, it follows that these roots are the principal moments of inertia of the body at the origin.

$$K_1 = 0.568867mc^2 \qquad K_2 = 4.20885mc^2 \qquad K_3 = 4.55562mc^2$$
$$K_1 = 0.569mc^2 \qquad K_2 = 4.21mc^2 \qquad K_3 = 4.56mc^2 \qquad \blacktriangleleft$$

b. Principal Axes of Inertia at O.
To determine the direction of a principal axis of inertia, first substitute the corresponding value of K into two of the equations (9.54). The resulting equations, together with Eq. (9.57), constitute a system of three equations from which you can determine the direction cosines of the corresponding principal axis. Thus, for the first principal moment of inertia K_1, you have

$$(\tfrac{5}{3} - 0.568867)mc^2(\lambda_x)_1 - \tfrac{3}{2}mc^2(\lambda_y)_1 - \tfrac{3}{4}mc^2(\lambda_z)_1 = 0$$
$$-\tfrac{3}{2}mc^2(\lambda_x)_1 + (\tfrac{10}{3} - 0.568867)\,mc^2(\lambda_y)_1 - \tfrac{1}{2}mc^2(\lambda_z)_1 = 0$$
$$(\lambda_x)_1^2 + (\lambda_y)_1^2 + (\lambda_z)_1^2 = 1$$

Solving yields

$$(\lambda_x)_1 = 0.836600 \qquad (\lambda_y)_1 = 0.496001 \qquad (\lambda_z)_1 = 0.232557$$

The angles that the first principal axis of inertia forms with the coordinate axes are then

$$(\theta_x)_1 = 33.2° \qquad (\theta_y)_1 = 60.3° \qquad (\theta_z)_1 = 76.6° \qquad \blacktriangleleft$$

Using the same set of equations successively with K_2 and K_3, you can find that the angles associated with the second and third principal moments of inertia at the origin are, respectively,

$$(\theta_x)_2 = 57.8° \qquad (\theta_y)_2 = 146.6° \qquad (\theta_z)_2 = 98.0° \qquad \blacktriangleleft$$

and

$$(\theta_x)_3 = 82.8° \qquad (\theta_y)_3 = 76.1° \qquad (\theta_z)_3 = 164.3° \qquad \blacktriangleleft$$

SOLVING PROBLEMS
ON YOUR OWN

In this section, we defined the **mass products of inertia** I_{xy}, I_{yz}, and I_{zx} of a body and showed you how to determine the moments of inertia of that body with respect to an arbitrary axis passing through the origin O. You also saw how to determine at the origin O the **principal axes of inertia** of a body and the corresponding **principal moments of inertia**.

1. Determining the mass products of inertia of a composite body. You can express the mass products of inertia of a composite body with respect to the coordinate axes as the sums of the products of inertia of its component parts with respect to those axes. For each component part, use the parallel-axis theorem to write Eqs. (9.47)

$$I_{xy} = \bar{I}_{x'y'} + m\bar{x}\bar{y} \qquad I_{yz} = \bar{I}_{y'z'} + m\bar{y}\bar{z} \qquad I_{zx} = \bar{I}_{z'x'} + m\bar{z}\bar{x}$$

Here the primes denote the centroidal axes of each component part, and $\bar{x}$, $\bar{y}$, and $\bar{z}$ represent the coordinates of its center of gravity. Keep in mind that the mass products of inertia can be positive, negative, or zero, and be sure to take into account the signs of $\bar{x}$, $\bar{y}$, and $\bar{z}$.

   a. From the properties of symmetry of a component part, you can deduce that two or all three of its centroidal mass products of inertia are zero. For instance, you can verify for a thin plate parallel to the xy plane, a wire lying in a plane parallel to the xy plane, a body with a plane of symmetry parallel to the xy plane, and a body with an axis of symmetry parallel to the z axis that the products of inertia $\bar{I}_{y'z'}$ and $\bar{I}_{z'x'}$ are zero.

   For rectangular, circular, or semicircular plates with axes of symmetry parallel to the coordinate axes, straight wires parallel to a coordinate axis, circular and semicircular wires with axes of symmetry parallel to the coordinate axes, and rectangular prisms with axes of symmetry parallel to the coordinate axes, the products of inertia $\bar{I}_{x'y'}$, $\bar{I}_{y'z'}$, and $\bar{I}_{z'x'}$ are all zero.

   b. Mass products of inertia that are different from zero can be computed from Eqs. (9.45). Although, in general, you need a triple integration to determine a mass product of inertia, you can use a single integration if you can divide the given body into a series of thin, parallel slabs. The computations are then similar to those discussed in the preceding section for moments of inertia.

2. Computing the moment of inertia of a body with respect to an arbitrary axis OL. In Sec. 9.6A, we derived an expression for the moment of inertia I_{OL} that was given in Eq. (9.46). Before computing I_{OL}, you must first determine the mass moments and products of inertia of the body with respect to the given coordinate axes, as well as the direction cosines of the unit vector λ along OL.

3. Calculating the principal moments of inertia of a body and determining its principal axes of inertia. You saw in Sec. 9.6B that it is always possible to find an orientation of the coordinate axes for which the mass products of inertia are zero. These axes are referred to as the **principal axes of inertia**, and the corresponding moments of inertia are known as the **principal moments of inertia** of the body. In many cases, you can determine the principal axes of inertia of a body from its properties of symmetry. The procedure required to determine the principal moments and principal axes of inertia of a body with no obvious property of symmetry was discussed in Sec. 9.6C and was illustrated in Sample Prob. 9.15. It consists of the following steps.

   a. Expand the determinant in Eq. (9.55) and solve the resulting cubic equation. You can obtain the solution by trial and error or (preferably) with an advanced scientific calculator or appropriate computer software. The roots K_1, K_2, and K_3 of this equation are the principal moments of inertia of the body.

   b. To determine the direction of the principal axis corresponding to K_1, substitute this value for K in two of the equations (9.54) and solve these equations, together with Eq. (9.57), for the direction cosines of the principal axis corresponding to K_1.

   c. Repeat this procedure with K_2 and K_3 to determine the directions of the other two principal axes. As a check of your computations, you may wish to verify that the scalar product of any two of the unit vectors along the three axes you have obtained is zero and, thus, that these axes are perpendicular to each other.

Problems

9.149 Determine the mass products of inertia I_{xy}, I_{yz}, and I_{zx} of the steel fixture shown. (The density of steel is 7850 kg/m³.)

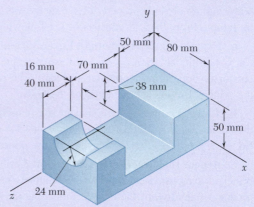

Fig. P9.149

9.150 Determine the mass products of inertia I_{xy}, I_{yz}, and I_{zx} of the steel machine element shown. (The density of steel is 7850 kg/m³.)

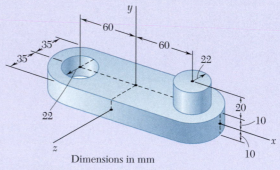

Dimensions in mm

Fig. P9.150

9.151 and 9.152 Determine the mass products of inertia I_{xy}, I_{yz}, and I_{zx} of the cast aluminum machine component shown. (The specific weight of aluminum is 0.100 lb/in³.)

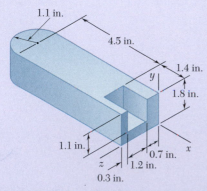

Fig. P9.151

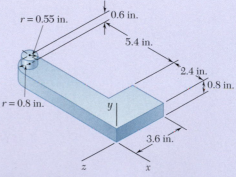

Fig. P9.152

9.153 through 9.156 A section of sheet steel 2 mm thick is cut and bent into the machine component shown. Knowing that the density of steel is 7850 kg/m^3, determine the mass products of inertia I_{xy}, I_{yz}, and I_{zx} of the component.

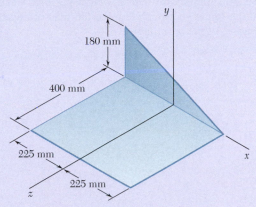

Fig. *P9.153*

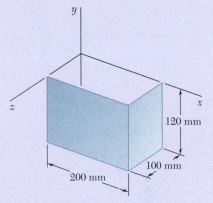

Fig. *P9.154*

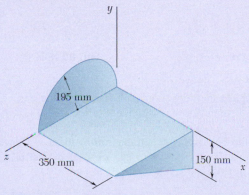

Fig. P9.155

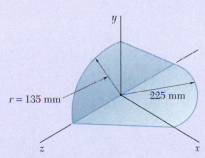

Fig. P9.156

9.157 The figure shown is formed of 1.5-mm-diameter aluminum wire. Knowing that the density of aluminum is 2800 kg/m^3, determine the mass products of inertia I_{xy}, I_{yz}, and I_{zx} of the wire figure.

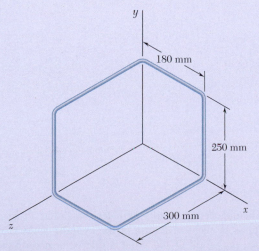

Fig. P9.157

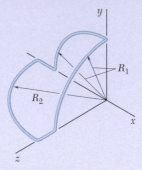

Fig. P9.158

9.158 Thin aluminum wire of uniform diameter is used to form the figure shown. Denoting the mass per unit length of the wire by m', determine the mass products of inertia I_{xy}, I_{yz}, and I_{zx} of the wire figure.

9.159 and 9.160 Brass wire with a weight per unit length w is used to form the figure shown. Determine the mass products of inertia I_{xy}, I_{yz}, and I_{zx} of the wire figure.

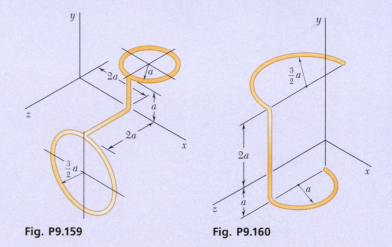

Fig. P9.159   Fig. P9.160

9.161 Complete the derivation of Eqs. (9.47) that expresses the parallel-axis theorem for mass products of inertia.

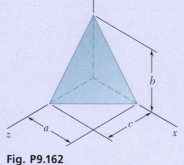

Fig. P9.162

9.162 For the homogeneous tetrahedron of mass m shown, (a) determine by direct integration the mass product of inertia I_{zx}, (b) deduce I_{yz} and I_{xy} from the result obtained in part a.

9.163 The homogeneous circular cone shown has a mass m. Determine the mass moment of inertia of the cone with respect to the line joining the origin O and point A.

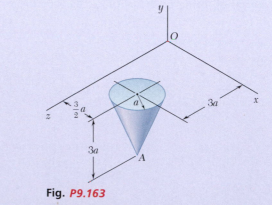

Fig. *P9.163*

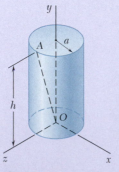

Fig. *P9.164*

9.164 The homogeneous circular cylinder shown has a mass m. Determine the mass moment of inertia of the cylinder with respect to the line joining the origin O and point A that is located on the perimeter of the top surface of the cylinder.

9.165 Shown is the machine element of Prob. 9.141. Determine its mass moment of inertia with respect to the line joining the origin O and point A.

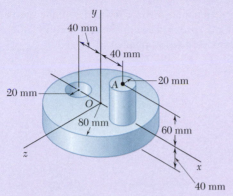

Fig. P9.165

9.166 Determine the mass moment of inertia of the steel fixture of Probs. 9.145 and 9.149 with respect to the axis through the origin that forms equal angles with the x, y, and z axes.

9.167 The thin, bent plate shown is of uniform density and weight W. Determine its mass moment of inertia with respect to the line joining the origin O and point A.

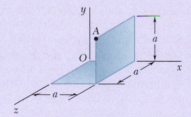

Fig. P9.167

9.168 A piece of sheet steel with thickness t and specific weight γ is cut and bent into the machine component shown. Determine the mass moment of inertia of the component with respect to the line joining the origin O and point A.

9.169 Determine the mass moment of inertia of the machine component of Probs. 9.136 and 9.155 with respect to the axis through the origin characterized by the unit vector $\boldsymbol{\lambda} = (-4\mathbf{i} + 8\mathbf{j} + \mathbf{k})/9$.

9.170 through 9.172 For the wire figure of the problem indicated, determine the mass moment of inertia of the figure with respect to the axis through the origin characterized by the unit vector $\boldsymbol{\lambda} = (-3\mathbf{i} - 6\mathbf{j} + 2\mathbf{k})/7$.

    9.170 Prob. 9.148
    9.171 Prob. 9.147
    9.172 Prob. 9.146

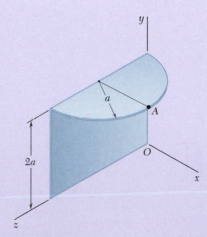

Fig. P9.168

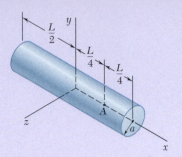

Fig. P9.173

9.173 For the homogeneous circular cylinder shown with radius a and length L, determine the value of the ratio a/L for which the ellipsoid of inertia of the cylinder is a sphere when computed (a) at the centroid of the cylinder, (b) at point A.

9.174 For the rectangular prism shown, determine the values of the ratios b/a and c/a so that the ellipsoid of inertia of the prism is a sphere when computed (a) at point A, (b) at point B.

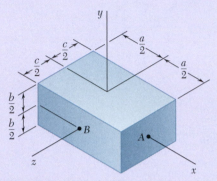

Fig. P9.174

9.175 For the right circular cone of Sample Prob. 9.11, determine the value of the ratio a/h for which the ellipsoid of inertia of the cone is a sphere when computed (a) at the apex of the cone, (b) at the center of the base of the cone.

9.176 Given an arbitrary body and three rectangular axes x, y, and z, prove that the mass moment of inertia of the body with respect to any one of the three axes cannot be larger than the sum of the mass moments of inertia of the body with respect to the other two axes. That is, prove that the inequality $I_x \leq I_y + I_z$ and the two similar inequalities are satisfied. Furthermore, prove that $I_y \geq \frac{1}{2}I_x$ if the body is a homogeneous solid of revolution, where x is the axis of revolution and y is a transverse axis.

9.177 Consider a cube with mass m and side a. (a) Show that the ellipsoid of inertia at the center of the cube is a sphere, and use this property to determine the moment of inertia of the cube with respect to one of its diagonals. (b) Show that the ellipsoid of inertia at one of the corners of the cube is an ellipsoid of revolution, and determine the principal moments of inertia of the cube at that point.

9.178 Given a homogeneous body of mass m and of arbitrary shape and three rectangular axes x, y, and z with origin at O, prove that the sum $I_x + I_y + I_z$ of the mass moments of inertia of the body cannot be smaller than the similar sum computed for a sphere of the same mass and the same material centered at O. Furthermore, using the result of Prob. 9.176, prove that, if the body is a solid of revolution where x is the axis of revolution, its mass moment of inertia I_y about a transverse axis y cannot be smaller than $3ma^2/10$, where a is the radius of the sphere of the same mass and the same material.

*9.179 The homogeneous circular cylinder shown has a mass m, and the diameter OB of its top surface forms 45° angles with the x and z axes. (*a*) Determine the principal mass moments of inertia of the cylinder at the origin O. (*b*) Compute the angles that the principal axes of inertia at O form with the coordinate axes. (*c*) Sketch the cylinder, and show the orientation of the principal axes of inertia relative to the x, y, and z axes.

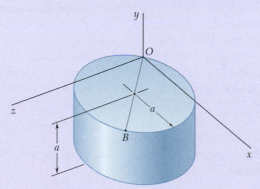

Fig. P9.179

9.180 through 9.184 For the component described in the problem indicated, determine (*a*) the principal mass moments of inertia at the origin, (*b*) the principal axes of inertia at the origin. Sketch the body and show the orientation of the principal axes of inertia relative to the x, y, and z axes.

    *9.180 Prob. 9.165
    *9.181 Probs. 9.145 and 9.149
    *9.182 Prob. 9.167
    *9.183 Prob. 9.168
    *9.184 Probs. 9.148 and 9.170

Review and Summary

In the first half of this chapter, we discussed how to determine the resultant **R** of forces $\Delta\mathbf{F}$ distributed over a plane area A when the magnitudes of these forces are proportional to both the areas ΔA of the elements on which they act and the distances y from these elements to a given x axis; we thus had $\Delta F = ky\,\Delta A$. We found that the magnitude of the resultant **R** is proportional to the first moment $Q_x = \int y\,dA$ of area A, whereas the moment of **R** about the x axis is proportional to the **second moment**, or **moment of inertia**, $I_x = \int y^2\,dA$ of A with respect to the same axis [Sec. 9.1A].

Rectangular Moments of Inertia

The **rectangular moments of inertia I_x and I_y of an area** [Sec. 9.1B] are obtained by evaluating the integrals

$$I_x = \int y^2\,dA \qquad I_y = \int x^2\,dA \tag{9.1}$$

We can reduce these computations to single integrations by choosing dA to be a thin strip parallel to one of the coordinate axes. We also recall that it is possible to compute I_x and I_y from the same elemental strip (Fig. 9.35) using the formula for the moment of inertia of a rectangular area [Sample Prob. 9.3].

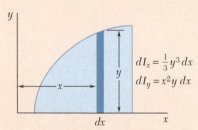

$dI_x = \frac{1}{3}y^3\,dx$

$dI_y = x^2 y\,dx$

Fig. 9.35

Polar Moment of Inertia

We defined the **polar moment of inertia of an area** A with respect to the pole O [Sec. 9.1C] as

$$J_O = \int r^2\,dA \tag{9.3}$$

where r is the distance from O to the element of area dA (Fig. 9.36). Observing that $r^2 = x^2 + y^2$, we established the relation

$$J_O = I_x + I_y \tag{9.4}$$

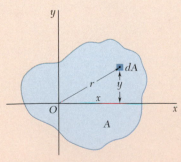

Fig. 9.36

Radius of Gyration

We defined the **radius of gyration of an area** A with respect to the x axis [Sec. 9.1D] as the distance k_x, where $I_x = k_x^2 A$. With similar definitions for the radii of gyration of A with respect to the y axis and with respect to O, we have

$$k_x = \sqrt{\frac{I_x}{A}} \qquad k_y = \sqrt{\frac{I_y}{A}} \qquad k_O = \sqrt{\frac{J_O}{A}} \qquad \textbf{(9.5–9.7)}$$

Parallel-Axis Theorem

The **parallel-axis theorem,** presented in Sec. 9.2A, states that the moment of inertia I of an area with respect to any given axis AA' (Fig. 9.37) is equal to the moment of inertia $\bar{I}$ of the area with respect to the centroidal axis BB' that is parallel to AA' *plus* the product of the area A and the square of the distance d between the two axes:

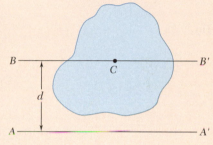

Fig. 9.37

$$I = \bar{I} + Ad^2 \qquad \textbf{(9.9)}$$

You can use this formula to determine the moment of inertia $\bar{I}$ of an area with respect to a centroidal axis BB' if you know its moment of inertia I with respect to a parallel axis AA'. In this case, however, the product Ad^2 should be *subtracted* from the known moment of inertia I.

A similar relation holds between the polar moment of inertia J_O of an area about a point O and the polar moment of inertia $\bar{J}_C$ of the same area about its centroid C. Letting d be the distance between O and C, we have

$$J_C = \bar{J}_C + Ad^2 \qquad \textbf{(9.11)}$$

Composite Areas

The parallel-axis theorem can be used very effectively to compute the **moment of inertia of a composite area** with respect to a given axis [Sec. 9.2B]. Considering each component area separately, we first compute the moment of inertia of each area with respect to its centroidal axis, using the data provided in Figs. 9.12 and 9.13 whenever possible. Then apply the parallel-axis theorem to determine the moment of inertia of each component area with respect to the desired axis, and add the values [Sample Probs. 9.4 and 9.5].

Product of Inertia

Section 9.3 was devoted to the transformation of the moments of inertia of an area under a rotation of the coordinate axes. First, we defined the **product of inertia of an area** A as

$$I_{xy} = \int xy\,dA \qquad \textbf{(9.12)}$$

and showed that $I_{xy} = 0$ if the area A is symmetrical with respect to either or both of the coordinate axes. We also derived the **parallel-axis theorem for products of inertia** as

$$I_{xy} = \bar{I}_{x'y'} + \bar{x}\bar{y}A \qquad \textbf{(9.13)}$$

where $\bar{I}_{x'y'}$ is the product of inertia of the area with respect to the centroidal axes x' and y' that are parallel to the x and y axes and $\bar{x}$ and $\bar{y}$ are the coordinates of the centroid of the area [Sec. 9.3A].

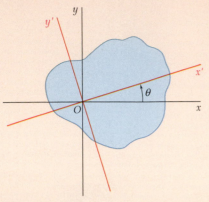

Fig. 9.38

Rotation of Axes

In Sec. 9.3B, we determined the moments and product of inertia $I_{x'}$, $I_{y'}$, and $I_{x'y'}$ of an area with respect to x' and y' axes obtained by rotating the original x and y coordinate axes counterclockwise through an angle θ (Fig. 9.38). We expressed $I_{x'}$, $I_{y'}$, and $I_{x'y'}$ in terms of the moments and product of inertia I_x, I_y, and I_{xy} computed with respect to the original x and y axes.

$$I_{x'} = \frac{I_x + I_y}{2} + \frac{I_x - I_y}{2}\cos 2\theta - I_{xy}\sin 2\theta \tag{9.18}$$

$$I_{y'} = \frac{I_x + I_y}{2} - \frac{I_x - I_y}{2}\cos 2\theta + I_{xy}\sin 2\theta \tag{9.19}$$

$$I_{x'y'} = \frac{I_x - I_y}{2}\sin 2\theta + I_{xy}\cos 2\theta \tag{9.20}$$

Principal Axes

We defined the **principal axes of the area about** O as the two axes perpendicular to each other with respect to which the moments of inertia of the area are maximum and minimum. The corresponding values of θ, denoted by θ_m, were obtained from

$$\tan 2\theta_m = -\frac{2I_{xy}}{I_x - I_y} \tag{9.25}$$

Principal Moments of Inertia

The corresponding maximum and minimum values of I are called the **principal moments of inertia** of the area about O:

$$I_{\text{max,min}} = \frac{I_x + I_y}{2} \pm \sqrt{\left(\frac{I_x - I_y}{2}\right)^2 + I_{xy}^2} \tag{9.27}$$

We also noted that the corresponding value of the product of inertia is zero.

Mohr's Circle

The transformation of the moments and product of inertia of an area under a rotation of axes can be represented graphically by drawing **Mohr's circle** [Sec. 9.4]. Given the moments and product of inertia I_x, I_y, and I_{xy} of the area with respect to the x and y coordinate axes, we plot points X (I_x, I_{xy}) and Y (I_y, $-I_{xy}$) and draw the line joining these two points (Fig. 9.39). This line is a diameter of Mohr's circle and thus defines this circle. As the coordinate axes are rotated through θ, the diameter rotates through *twice that angle*, and the coordinates of X' and Y' yield the new values $I_{x'}$, $I_{y'}$, and $I_{x'y'}$ of the moments and product of inertia of the area. Also, the angle θ_m and the coordinates of points A and B define the principal axes a and b and the principal moments of inertia of the area [Sample Prob. 9.8].

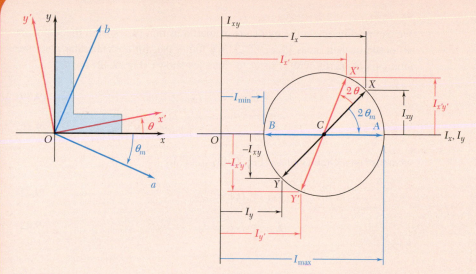

Fig. 9.39

Moments of Inertia of Masses

The second half of the chapter was devoted to determining **moments of inertia of masses**, which are encountered in dynamics problems involving the rotation of a rigid body about an axis. We defined the mass moment of inertia of a body with respect to an axis AA' (Fig. 9.40) as

$$I = \int r^2 \, dm \qquad \textbf{(9.28)}$$

where r is the distance from AA' to the element of mass [Sec. 9.5A]. We defined the **radius of gyration** of the body as

$$k = \sqrt{\frac{I}{m}} \qquad \textbf{(9.29)}$$

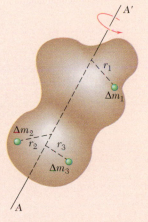

Fig. 9.40

The moments of inertia of a body with respect to the coordinate axes were expressed as

$$I_x = \int (y^2 + z^2) \, dm$$

$$I_y = \int (z^2 + x^2) \, dm \qquad \textbf{(9.30)}$$

$$I_z = \int (x^2 + y^2) \, dm$$

Parallel-Axis Theorem

We saw that the **parallel-axis theorem** also applies to mass moments of inertia [Sec. 9.5B]. Thus, the moment of inertia I of a body with respect to an arbitrary axis AA' (Fig. 9.41) can be expressed as

$$I = \bar{I} + md^2 \qquad \textbf{(9.33)}$$

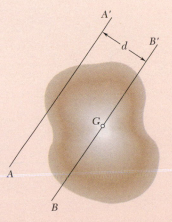

where $\bar{I}$ is the moment of inertia of the body with respect to the centroidal axis BB' that is parallel to the axis AA', m is the mass of the body, and d is the distance between the two axes.

Fig. 9.41

Moments of Inertia of Thin Plates

We can readily obtain the moments of inertia of thin plates from the moments of inertia of their areas [Sec. 9.5C]. We found that for a rectangular plate the moments of inertia with respect to the axes shown (Fig. 9.42) are

$$I_{AA'} = \tfrac{1}{12}ma^2 \qquad I_{BB'} = \tfrac{1}{12}mb^2 \tag{9.39}$$

$$I_{CC'} = I_{AA'} + I_{BB'} = \tfrac{1}{12}m(a^2 + b^2) \tag{9.40}$$

whereas for a circular plate (Fig. 9.43), they are

$$I_{AA'} = I_{BB'} = \tfrac{1}{4}mr^2 \tag{9.41}$$

$$I_{CC'} = I_{AA'} + I_{BB'} = \tfrac{1}{2}mr^2 \tag{9.42}$$

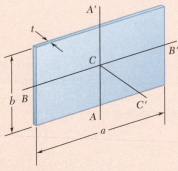

Fig. 9.42

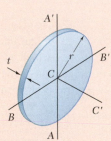

Fig. 9.43

Composite Bodies

When a body possesses two planes of symmetry, it is usually possible to use a single integration to determine its moment of inertia with respect to a given axis by selecting the element of mass dm to be a thin plate [Sample Probs. 9.10 and 9.11]. On the other hand, when a body consists of several common geometric shapes, we can obtain its moment of inertia with respect to a given axis by using the formulas given in Fig. 9.28 together with the parallel-axis theorem [Sample Probs. 9.12 and 9.13].

Moment of Inertia with Respect to an Arbitrary Axis

In the last section of the chapter, we described how to determine the moment of inertia of a body with respect to an arbitrary axis OL that is drawn through the origin O [Sec. 9.6A]. We denoted the components of the unit vector $\boldsymbol{\lambda}$ along OL by λ_x, λ_y, and λ_z (Fig. 9.44) and introduced the **products of inertia as**

$$I_{xy} = \int xy\,dm \qquad I_{yz} = \int yz\,dm \qquad I_{zx} = \int zx\,dm \tag{9.45}$$

We found that the moment of inertia of the body with respect to OL could be expressed as

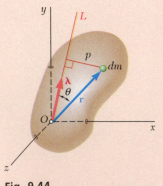

Fig. 9.44

$$I_{OL} = I_x\lambda_x^2 + I_y\lambda_y^2 + I_z\lambda_z^2 - 2I_{xy}\lambda_x\lambda_y - 2I_{yz}\lambda_y\lambda_z - 2I_{zx}\lambda_z\lambda_x \tag{9.46}$$

Ellipsoid of Inertia

By plotting a point Q along each axis OL at a distance $OQ = 1/\sqrt{I_{OL}}$ from O [Sec. 9.6B], we obtained the surface of an ellipsoid, known as the **ellipsoid of inertia** of the body at point O.

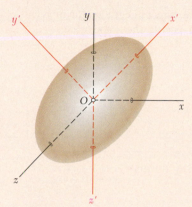

Fig. 9.45

Principal Axes and Principal Moments of Inertia

The principal axes x', y', and z' of this ellipsoid (Fig. 9.45) are the **principal axes of inertia** of the body; that is, the products of inertia $I_{x'y'}$, $I_{y'z'}$, and $I_{z'x'}$ of the body with respect to these axes are all zero. In many situations, you can deduce the principal axes of inertia of a body from its properties of symmetry. Choosing these axes to be the coordinate axes, we can then express I_{OL} as

$$I_{OL} = I_{x'}\lambda_{x'}^2 + I_{y'}\lambda_{y'}^2 + I_{z'}\lambda_{z'}^2 \qquad \text{(9.50)}$$

where $I_{x'}$, $I_{y'}$, and $I_{z'}$ are the **principal moments of inertia** of the body at O.

When the principal axes of inertia cannot be obtained by observation [Sec. 9.6B], it is necessary to solve the cubic equation

$$K^3 - (I_x + I_y + I_z)K^2 + (I_xI_y + I_yI_z + I_zI_x - I_{xy}^2 - I_{yz}^2 - I_{zx}^2)K$$
$$- (I_xI_yI_z - I_xI_{yz}^2 - I_yI_{zx}^2 - I_zI_{xy}^2 - 2I_{xy}I_{yz}I_{zx}) = 0 \qquad \text{(9.56)}$$

We found [Sec. 9.6C] that the roots K_1, K_2, and K_3 of this equation are the principal moments of inertia of the given body. The direction cosines $(\lambda_x)_1$, $(\lambda_y)_1$, and $(\lambda_z)_1$ of the principal axis corresponding to the principal moment of inertia K_1 are then determined by substituting K_1 into Eqs. (9.54) and by solving two of these equations and Eq. (9.57) simultaneously. The same procedure is then repeated using K_2 and K_3 to determine the direction cosines of the other two principal axes [Sample Prob. 9.15].

Review Problems

9.185 Determine by direct integration the moments of inertia of the shaded area with respect to the x and y axes.

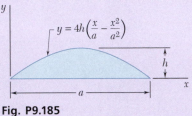

$$y = 4h\left(\frac{x}{a} - \frac{x^2}{a^2}\right)$$

Fig. P9.185

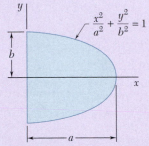

$$\frac{x^2}{a^2} + \frac{y^2}{b^2} = 1$$

Fig. P9.186

9.186 Determine the moment of inertia and the radius of gyration of the shaded area shown with respect to the y axis.

9.187 Determine the moment of inertia and the radius of gyration of the shaded area shown with respect to the x axis.

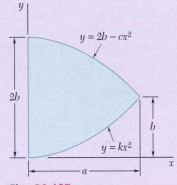

$$y = 2b - cx^2$$

$$y = kx^2$$

Fig. P9.187

9.188 Determine the moments of inertia $\bar{I}_x$ and $\bar{I}_y$ of the area shown with respect to centroidal axes respectively parallel and perpendicular to side AB.

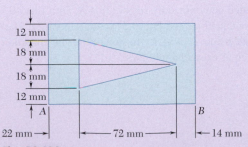

Fig. P9.188

9.189 Determine the polar moment of inertia of the area shown with respect to (a) point O, (b) the centroid of the area.

9.190 Two L4 × 4 × ½-in. angles are welded to a steel plate as shown. Determine the moments of inertia of the combined section with respect to centroidal axes respectively parallel and perpendicular to the plate.

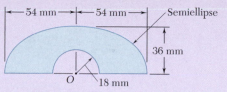

Fig. P9.189

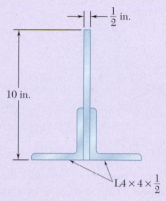

Fig. P9.190

9.191 Using the parallel-axis theorem, determine the product of inertia of the L5 × 3 × ½-in. angle cross section shown with respect to the centroidal x and y axes.

9.192 For the L5 × 3 × ½-in. angle cross section shown, use Mohr's circle to determine (a) the moments of inertia and the product of inertia with respect to new centroidal axes obtained by rotating the x and y axes 30° clockwise, (b) the orientation of the principal axes through the centroid and the corresponding values of the moments of inertia.

9.193 A thin plate with a mass m was cut in the shape of a parallelogram as shown. Determine the mass moment of inertia of the plate with respect to (a) the x axis, (b) the axis BB' that is perpendicular to the plate.

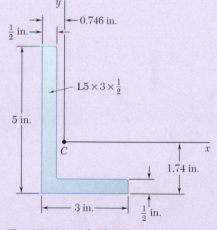

Fig. P9.191 and P9.192

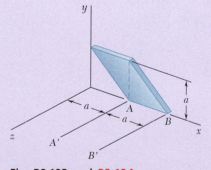

Fig. P9.193 and P9.194

9.194 A thin plate with mass m was cut in the shape of a parallelogram as shown. Determine the mass moment of inertia of the plate with respect to (a) the y axis, (b) the axis AA' that is perpendicular to the plate.

9.195 A 2-mm-thick piece of sheet steel is cut and bent into the machine component shown. Knowing that the density of steel is 7850 kg/m³, determine the mass moment of inertia of the component with respect to each of the coordinate axes.

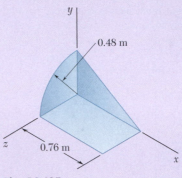

0.48 m

0.76 m

Fig. P9.195

9.196 Determine the mass moment of inertia of the steel machine element shown with respect to the z axis. (The specific weight of steel is 490 lb/ft³.)

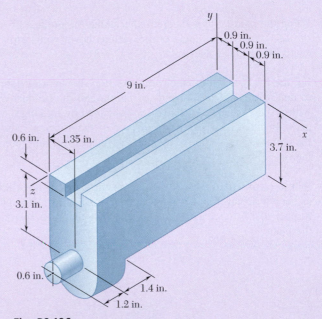

0.9 in.
0.9 in.
0.9 in.
9 in.
0.6 in.   1.35 in.
3.7 in.
3.1 in.
0.6 in.
1.4 in.
1.2 in.

Fig. P9.196

10

Method of Virtual Work

The method of virtual work is particularly effective when a simple relation can be found among the displacements of the points of application of the various forces involved. This is the case for the scissor lift platform being used by workers to gain access to a highway bridge under construction.

Objectives

- **Define** the work of a force, and consider the circumstances when a force does no work.
- **Examine** the principle of virtual work, and apply it to analyze the equilibrium of machines and mechanisms.
- **Apply** the concept of potential energy to determine the equilibrium position of a rigid body or a system of rigid bodies.
- **Evaluate** the mechanical efficiency of machines, and consider the stability of equilibrium.

*Introduction

In the preceding chapters, we solved problems involving the equilibrium of rigid bodies by expressing the balance of external forces acting on the bodies. We wrote the equations of equilibrium $\Sigma F_x = 0$, $\Sigma F_y = 0$, and $\Sigma M_A = 0$ and solved them for the desired unknowns. We now consider a different method, which turns out to be more effective for solving certain types of equilibrium problems. This method, based on the **principle of virtual work,** was first formally used by the Swiss mathematician Jean Bernoulli in the eighteenth century.

As you will see in Sec. 10.1B, the principle of virtual work considers a particle or rigid body (or more generally, a system of connected rigid bodies) that is in equilibrium under various external forces. The principle states that, if the body is given an arbitrary displacement from that position of equilibrium, the total work done by the external forces during the displacement is zero. This principle is particularly effective when applied to the solution of problems involving the equilibrium of machines or mechanisms consisting of several connected members.

In the second part of this chapter, we apply the method of virtual work in an alternative form based on the concept of **potential energy.** We will show in Sec. 10.2 that, if a particle, rigid body, or system of rigid bodies is in equilibrium, the derivative of its potential energy with respect to a variable defining its position must be zero.

You will also learn in this chapter to evaluate the mechanical efficiency of a machine (Sec. 10.1D) and to determine whether a given position of equilibrium is stable, unstable, or neutral (Sec. 10.2D).

*10.1 THE BASIC METHOD

The first step in explaining the method of virtual work is to define the terms displacement and work as they are used in mechanics. Then we can state the principle of virtual work and show how to apply it in practical situations. We also take the opportunity to define mechanical efficiency, which is a useful and important parameter for the design of real machines.

10.1A Work of a Force

Consider a particle that moves from a point A to a neighboring point A' (Fig. 10.1). If $\mathbf{r}$ denotes the position vector corresponding to point A, we denote the small vector joining A and A' by the differential $d\mathbf{r}$; we call the vector $d\mathbf{r}$ the **displacement** of the particle.

Now let us assume that a force $\mathbf{F}$ is acting on the particle. The **work** dU **of force F corresponding to the displacement** $d\mathbf{r}$ is defined as the quantity

Definition of work

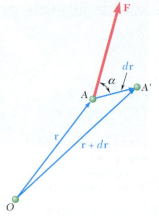

$$dU = \mathbf{F} \cdot d\mathbf{r} \qquad (10.1)$$

Fig. 10.1 The work of a force acting on a particle is the scalar product of the force and the particle's displacement.

That is, dU is the scalar product of the force $\mathbf{F}$ and the displacement $d\mathbf{r}$. Suppose we denote the magnitudes of the force by F, the displacement by ds, and the angle formed by $\mathbf{F}$ and $d\mathbf{r}$ by α. Then, recalling the definition of the scalar product of two vectors (Sec. 3.2A), we have

$$dU = F\,ds\,\cos\alpha \qquad (10.1')$$

Work is a scalar quantity, so it has a magnitude and a sign, but no direction. Note that work should be expressed in units obtained by multiplying units of length by units of force. Thus, if we use U.S. customary units, we should express work in ft·lb or in·lb. If we use SI units, we express work in N·m. This unit of work is called a **joule** (J).[†]

It follows from $(10.1')$ that work dU is positive if the angle α is acute and negative if α is obtuse. Three particular cases are of special interest.

- If the force $\mathbf{F}$ has the same direction as $d\mathbf{r}$, the work dU reduces to $F\,ds$.
- If $\mathbf{F}$ has a direction opposite to that of $d\mathbf{r}$, the work is $dU = -F\,ds$.
- Finally, if $\mathbf{F}$ is perpendicular to $d\mathbf{r}$, the work dU is zero.

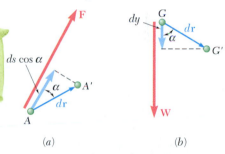

(a)   (b)

Fig. 10.2 (a) You can think of work as the product of a force and the component of displacement in the direction of the force. (b) This is useful for computing the work done by an object's weight.

We can also consider the work dU of a force $\mathbf{F}$ during a displacement $d\mathbf{r}$ to be the product of F and the component $ds\,\cos\alpha$ of the displacement $d\mathbf{r}$ along $\mathbf{F}$ (Fig. 10.2a). This view is particularly useful in computing the work done by the weight $\mathbf{W}$ of a body (Fig. 10.2b). The work of $\mathbf{W}$ is equal to the product of W and the vertical displacement dy of the center of gravity G of the body. If the displacement is downward, the work is positive; if the displacement is upward, the work is negative.

Some forces frequently encountered in statics do no work, such as forces applied to fixed points ($ds = 0$) or acting in a direction perpendicular to the displacement ($\cos\alpha = 0$). Among these forces are the reaction at a frictionless pin when the body supported rotates about the pin; the reaction at a frictionless surface when the body in contact moves along the surface; the reaction at a roller moving along its track; the weight of a body when its center of gravity moves horizontally; and the friction force

---

[†]The joule is the SI unit of *energy*, whether in mechanical form (work, potential energy, kinetic energy) or in chemical, electrical, or thermal form. Note that even though 1 N·m = 1 J, we must express the moment of a force in N·m, and not in joules, since the moment of a force is not a form of energy.

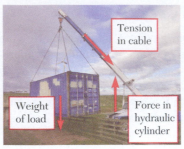

Photo 10.1 (*a*) In analyzing a crane, we might consider displacements associated with vertical movement of a container. (*b*) A force does work if it has a component in the direction of a displacement. (*c*) A force does no work if there is no displacement or if the force is perpendicular to a displacement.

acting on a wheel rolling without slipping (since at any instant the point of contact does not move). Examples of forces that do work are the weight of a body (except in the case considered previously), the friction force acting on a body sliding on a rough surface, and most forces applied on a moving body.

In certain cases, the sum of the work done by several forces is zero. Consider, for example, two rigid bodies *AC* and *BC* that are connected at *C* by a *frictionless pin* (Fig. 10.3*a*). Among the forces acting on *AC* is the force **F** exerted at *C* by *BC*. In general, the work of this force is not zero, but it is equal in magnitude and opposite in sign to the work of the force −**F** exerted by *AC* on *BC*, since these forces are equal and opposite and are applied to the same particle. Thus, when the total work done by all the forces acting on *AB* and *BC* is considered, the work of the two internal forces at *C* cancels out. We obtain a similar result if we consider a system consisting of two blocks connected by a *cord AB* that is not extensible (Fig. 10.3*b*). The work of the tension force **T** at *A* is equal in magnitude to the work of the tension force **T**′ at *B*, since these forces have the same magnitude and the points *A* and *B* move through the same distance; but in one case, the work is positive, and in the other, it is negative. Thus, the work of the internal forces again cancels out.

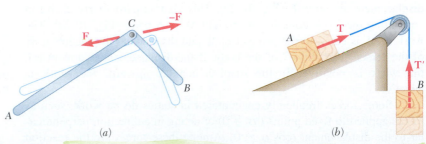

(*a*)   (*b*)

Fig. 10.3 (*a*) For a frictionless pin or (*b*) a cord that is not extensible, the total work done by the pairs of internal forces is zero.

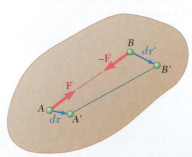

Fig. 10.4 As demonstrated here for an arbitrary pair of particles, the total work of the internal forces holding a rigid body together is zero.

We can show that the total work of the internal forces holding together the particles of a rigid body is zero. Consider two particles *A* and *B* of a rigid body and the two equal and opposite forces **F** and −**F** they exert on each other (Fig. 10.4). Although, in general, small displacements *d***r** and *d***r**′ of the two particles are different, the components of these displacements along *AB* must be equal; otherwise, the particles would not remain at the same distance from each other, so the body would not be rigid. Therefore, the work of **F** is equal in magnitude and opposite in sign to the work of −**F**, and their sum is zero.

In computing the work of the external forces acting on a rigid body, it is often convenient to determine the work of a couple without considering separately the work of each of the two forces forming the couple. Consider the two forces **F** and −**F** forming a couple of moment **M** and acting on a rigid body (Fig. 10.5). Any small displacement of the rigid body bringing A and B, respectively, into A' and B'' can be divided into two parts: one in which points A and B undergo equal displacements $d\mathbf{r}_1$, the other in which A' remains fixed while B' moves into B'' through a displacement $d\mathbf{r}_2$ with a magnitude of $ds_2 = r\,d\theta$. In the first part of the motion, the work of **F** is equal in magnitude and opposite in sign to the work of −**F**, and their sum is zero. In the second part of the motion, only force **F** works, and its work is $dU = F\,ds_2 = Fr\,d\theta$. But the product Fr is equal to the magnitude M of the moment of the couple. Thus, the work of a couple of moment **M** acting on a rigid body is

Work of a couple

$$dU = M\,d\theta \qquad (10.2)$$

where $d\theta$ is the small angle (expressed in radians) through which the body rotates. We again note that work should be expressed in units obtained by multiplying units of force by units of length.

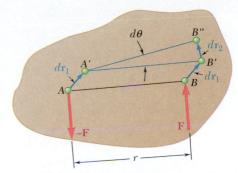

Fig. 10.5 The work of a couple acting on a rigid body is the moment of the couple times the angular rotation.

10.1B Principle of Virtual Work

Consider a particle acted upon by several forces $\mathbf{F}_1, \mathbf{F}_2, \ldots, \mathbf{F}_n$ (Fig. 10.6). We can imagine that the particle undergoes a small displacement from A to A'. This displacement is possible, but it does not necessarily take place. The forces may be balanced and the particle remains at rest, or the particle may move under the action of the given forces in a direction different from that of AA'. Since the considered displacement does not actually occur, it is called a **virtual displacement**, which is denoted by $\delta\mathbf{r}$. The symbol $\delta\mathbf{r}$ represents a differential of the first order; it is used to distinguish the virtual displacement from the displacement $d\mathbf{r}$ that would take place under actual motion. As you will see, we can use virtual displacements to determine whether the conditions of equilibrium of a particle are satisfied.

The work of each of the forces $\mathbf{F}_1, \mathbf{F}_2, \ldots, \mathbf{F}_n$ during the virtual displacement $\delta\mathbf{r}$ is called **virtual work**. The virtual work of all the forces acting on the particle of Fig. 10.6 is

$$\delta U = \mathbf{F}_1 \cdot \delta\mathbf{r} + \mathbf{F}_2 \cdot \delta\mathbf{r} + \ldots + \mathbf{F}_n \cdot \delta\mathbf{r}$$
$$= (\mathbf{F}_1 + \mathbf{F}_2 + \ldots + \mathbf{F}_n) \cdot \delta\mathbf{r}$$

or

$$\delta U = \mathbf{R} \cdot \delta\mathbf{r} \qquad (10.3)$$

where **R** is the resultant of the given forces. Thus, the total virtual work of the forces $\mathbf{F}_1, \mathbf{F}_2, \ldots, \mathbf{F}_n$ is equal to the virtual work of their resultant **R**.

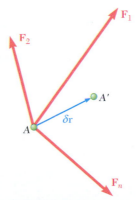

Fig. 10.6 Forces acting on a particle that goes through a virtual displacement.

The principle of virtual work for a particle states:

If a particle is in equilibrium, the total virtual work of the forces acting on the particle is zero for any virtual displacement of the particle.

This condition is necessary: if the particle is in equilibrium, the resultant **R** of the forces is zero, and it follows from Eq. (10.3) that the total virtual work δU is zero. The condition is also sufficient: if the total virtual work δU is zero for any virtual displacement, the scalar product $\mathbf{R} \cdot \delta\mathbf{r}$ is zero for any $\delta\mathbf{r}$, and the resultant **R** must be zero.

In the case of a rigid body, the principle of virtual work states:

If a rigid body is in equilibrium, the total virtual work of the external forces acting on the rigid body is zero for any virtual displacement of the body.

The condition is necessary: if the body is in equilibrium, all the particles forming the body are in equilibrium and the total virtual work of the forces acting on all the particles must be zero. However, we have seen in the preceding section that the total work of the internal forces is zero; therefore, the total work of the external forces also must be zero. The condition can also be proven to be sufficient.

The principle of virtual work can be extended to the case of a **system of connected rigid bodies**. If the system remains connected during the virtual displacement, **only the work of the forces external to the system need be considered**, since the total work of the internal forces at the various connections is zero.

10.1C Applying the Principle of Virtual Work

The principle of virtual work is particularly effective when applied to the solution of problems involving machines or mechanisms consisting of several connected rigid bodies. Consider, for instance, the toggle vise *ACB* of Fig. 10.7*a* used to compress a wooden block. Suppose we wish to determine the force exerted by the vise on the block when a given force **P** is applied at *C*, assuming there is no friction. Denoting the reaction of the block on the vise by **Q**, we draw the free-body diagram of the vise and

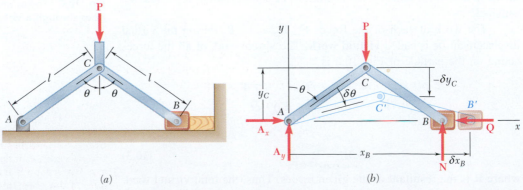

Fig. 10.7 (*a*) A toggle vise used to compress a wooden block, assuming no friction; (*b*) a virtual displacement of the vise.

consider the virtual displacement obtained by giving a positive increment $\delta\theta$ to angle θ (Fig. 10.7b). Choosing a system of coordinate axes with origin at A, we note that x_B increases as y_C decreases. This is indicated in the figure, where we indicate a positive increment δx_B and a negative increment $-\delta y_C$. The reactions $\mathbf{A}_x$, $\mathbf{A}_y$, and $\mathbf{N}$ do no work during the virtual displacement considered, so we need only compute the work done by $\mathbf{P}$ and $\mathbf{Q}$. Since $\mathbf{Q}$ and δx_B have opposite senses, the virtual work of $\mathbf{Q}$ is $\delta U_Q = -Q\ \delta x_B$. Since $\mathbf{P}$ and the increment shown ($-\delta y_C$) have the same sense, the virtual work of $\mathbf{P}$ is $\delta U_P = +P(-\delta y_C) = -P\ \delta y_C$. (We could have predicted the minus signs by simply noting that the forces $\mathbf{Q}$ and $\mathbf{P}$ are directed opposite to the positive x and y axes, respectively.) Expressing the coordinates x_B and y_C in terms of the angle θ and differentiating, we obtain

$$x_B = 2l \sin\theta \qquad\qquad y_C = l\cos\theta$$
$$\delta x_B = 2l\cos\theta\ \delta\theta \qquad \delta y_C = -l\sin\theta\ \delta\theta \qquad \textbf{(10.4)}$$

The total virtual work of the forces $\mathbf{Q}$ and $\mathbf{P}$ is thus

$$\delta U = \delta U_Q + \delta U_P = -Q\ \delta x_B - P\ \delta y_C$$
$$= -2Ql\cos\theta\ \delta\theta + Pl\sin\theta\ \delta\theta$$

Setting $\delta U = 0$, we obtain

$$2Ql\cos\theta\ \delta\theta = Pl\sin\theta\ \delta\theta \qquad\qquad \textbf{(10.5)}$$

and

$$Q = \frac{1}{2}P\tan\theta \qquad\qquad \textbf{(10.6)}$$

The superiority of the method of virtual work over the conventional equilibrium equations in the problem considered here is clear: by using the method of virtual work, we were able to eliminate all unknown reactions, whereas the equation $\Sigma M_A = 0$ would have eliminated only two of the unknown reactions. This property of the method of virtual work can be used in solving many problems involving machines and mechanisms.

If the virtual displacement considered is consistent with the constraints imposed by the supports and connections, all reactions and internal forces are eliminated and only the work of the loads, applied forces, and friction forces need be considered.

We can also use the method of virtual work to solve problems involving completely constrained structures, although the virtual displacements considered never actually take place. Consider, for example, the frame ACB shown in Fig. 10.8a. If point A is kept fixed while point B is given a horizontal virtual displacement (Fig. 10.8b), we need consider only the work of P and $\mathbf{B}_x$. We can thus determine the reaction component $\mathbf{B}_x$ in the same way as the force $\mathbf{Q}$ of the preceding example (Fig. 10.7b); we have

$$B_x = \frac{1}{2}P\tan\theta$$

By keeping B fixed and giving a horizontal virtual displacement to A, we can similarly determine the reaction component $\mathbf{A}_x$. Then we can determine the components $\mathbf{A}_y$ and $\mathbf{B}_y$ by rotating the frame ACB as a rigid body about B and A, respectively.

Photo 10.2 The method of virtual work is useful for determining the forces exerted by the hydraulic cylinders positioning the bucket lift. The reason is that a simple relation exists among the displacements of the points of application of the forces acting on the members of the lift.

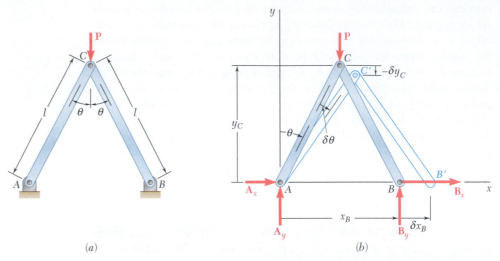

Fig. 10.8 (a) A completely constrained frame ACB; (b) a virtual displacement of the frame in order to determine B_x, keeping A fixed.

We can also use the method of virtual work to determine the configuration of a system in equilibrium under given forces. For example, we can obtain the value of the angle θ for which the linkage of Fig. 10.7 is in equilibrium under two given forces **P** and **Q** by solving Eq. (10.6) for tan θ.

Note, however, that the attractiveness of the method of virtual work depends to a large extent upon the existence of simple geometric relations between the various virtual displacements involved in the solution of a given problem. When no such simple relations exist, it is usually advisable to revert to the conventional method of Chap. 6.

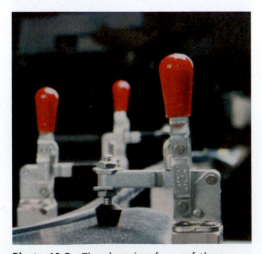

Photo 10.3 The clamping force of the toggle clamp shown can be expressed as a function of the force applied to the handle by first establishing the geometric relations among the members of the clamp and then applying the method of virtual work.

10.1D Mechanical Efficiency of Real Machines

In analyzing the toggle vise of Fig. 10.7, we assumed that no friction forces were involved. Thus, the virtual work consisted only of the work of the applied force **P** and of the reaction **Q**. However, the work of reaction **Q** is equal in magnitude and opposite in sign to the work of the force exerted by the vise on the block. Therefore, Equation (10.5) states that the **output work** $2Ql \cos \theta \, \delta\theta$ is equal to the **input work** $Pl \sin \theta \, \delta\theta$. A machine in which input and output work are equal is said to be an "ideal" machine. In a "real" machine, friction forces always do some work, and the output work is smaller than the input work.

Consider again the toggle vise of Fig. 10.7a. and now assume that a friction force **F** develops between the sliding block B and the horizontal plane (Fig. 10.9). Using the conventional methods of statics and summing moments about A, we find $N = P/2$. Denoting the coefficient of friction between block B and the horizontal plane by μ, we have $F = \mu N = \mu P/2$.

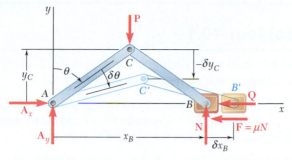

Fig. 10.9 A virtual displacement of the toggle vise with friction.

Recalling formulas (10.4), we find that the total virtual work of the forces **Q**, **P**, and **F** during the virtual displacement shown in Fig. 10.9 is

$$\delta U = -Q\,\delta x_B - P\,\delta y_C - F\,\delta x_B$$
$$= -2Ql\cos\theta\,\delta\theta + Pl\sin\theta\,\delta\theta - \mu Pl\cos\theta\,\delta\theta$$

Setting $\delta U = 0$, we obtain

$$2Ql\cos\theta\,\delta\theta = Pl\sin\theta\,\delta\theta - \mu Pl\cos\theta\,\delta\theta \qquad \textbf{(10.7)}$$

This equation states that the output work is equal to the input work minus the work of the friction force. Solving for Q, we have

$$Q = \frac{1}{2}P\,(\tan\theta - \mu) \qquad \textbf{(10.8)}$$

Note that $Q = 0$ when $\tan\theta = \mu$, that is, when θ is equal to the angle of friction ϕ, and that $Q < 0$ when $\theta < \phi$. Thus, we can use the toggle vise only for values of θ larger than the angle of friction.

We define the **mechanical efficiency** η of a machine as the ratio

Mechanical efficiency

$$\eta = \frac{\text{output work}}{\text{input work}} \qquad \textbf{(10.9)}$$

Clearly, the mechanical efficiency of an ideal machine is $\eta = 1$ when input and output work are equal, whereas the mechanical efficiency of a real machine is always less than 1.

In the case of the toggle vise we have just analyzed, we have

$$\eta = \frac{\text{output work}}{\text{input work}} = \frac{2Ql\cos\theta\,\delta\theta}{Pl\sin\theta\,\delta\theta} \qquad \textbf{(10.10)}$$

We can check that, in the absence of friction forces, we would have $\mu = 0$ and $\eta = 1$. In the general case when μ is different from zero, the efficiency η becomes zero for $\mu\cot\theta = 1$, that is, for $\tan\theta = \mu$ or $\theta = \tan^{-1}\mu = \phi$. We note again that the toggle vise can be used only for values of θ larger than the angle of friction ϕ.

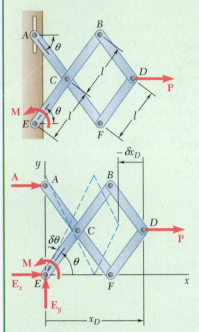

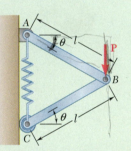

Fig. 1 Free-body diagram of mechanism showing a virtual displacement.

Sample Problem 10.1

Using the method of virtual work, determine the magnitude of the couple **M** required to maintain the equilibrium of the mechanism shown.

STRATEGY: For a virtual displacement consistent with the constraints, the reactions do no work, so you can focus solely on the force **P** and the moment **M**. You can solve for **M** in terms of **P** and the geometric parameters.

MODELING: Choose a coordinate system with origin at E (Fig. 1). Then

$$x_D = 3l \cos \theta \qquad\qquad \delta x_D = -3l \sin \theta \; \delta\theta$$

ANALYSIS: Principle of Virtual Work. Since the reactions **A**, $\mathbf{E}_x$, and $\mathbf{E}_y$ do no work during the virtual displacement, the total virtual work done by **M** and **P** must be zero. Notice that **P** acts in the positive x direction and **M** acts in the positive θ direction. You obtain

$$\delta U = 0: \qquad +M \, \delta\theta + P \, \delta x_D = 0$$
$$+M \, \delta\theta + P(-3l \sin \theta \; \delta\theta) = 0$$
$$M = 3Pl \sin \theta \quad \blacktriangleleft$$

REFLECT and THINK: This problem illustrates that the principle of virtual work can help determine a moment as well as a force in a straightforward computation.

Sample Problem 10.2

Determine the expressions for θ and for the tension in the spring that correspond to the equilibrium position of the mechanism. The unstretched length of the spring is h, and the spring constant is k. Neglect the weight of the mechanism.

STRATEGY: The tension in the spring is a force **F** exerted at C. Applying the principle of virtual work, you can obtain a relationship between **F** and the applied force **P**.

MODELING: With the coordinate system shown in Fig. 1,

$$y_B = l \sin \theta \qquad\qquad y_C = 2l \sin \theta$$
$$\delta y_B = l \cos \theta \; \delta\theta \qquad\qquad \delta y_C = 2l \cos \theta \; \delta\theta$$

The elongation of the spring is $s = y_C - h = 2l \sin \theta - h$. The magnitude of the force exerted at C by the spring is

$$F = ks = k(2l \sin \theta - h) \qquad\qquad \textbf{(1)}$$

ANALYSIS: Principle of Virtual Work. Since the reactions $\mathbf{A}_x$, $\mathbf{A}_y$, and **C** do no work, the total virtual work done by **P** and **F** must be zero.

$$\delta U = 0: \qquad P \, \delta y_B - F \, \delta y_C = 0$$
$$P(l \cos \theta \; \delta\theta) - k(2l \sin \theta - h)(2l \cos \theta \; \delta\theta) = 0$$
$$\sin \theta = \frac{P + 2kh}{4kl} \quad \blacktriangleleft$$

Substituting this expression into Eq. (1), you obtain $\quad F = \tfrac{1}{2}P \quad \blacktriangleleft$

REFLECT and THINK: You can verify these results by applying the appropriate equations of equilibrium.

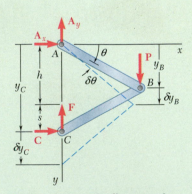

Fig. 1 Free-body diagram of mechanism showing a virtual displacement.

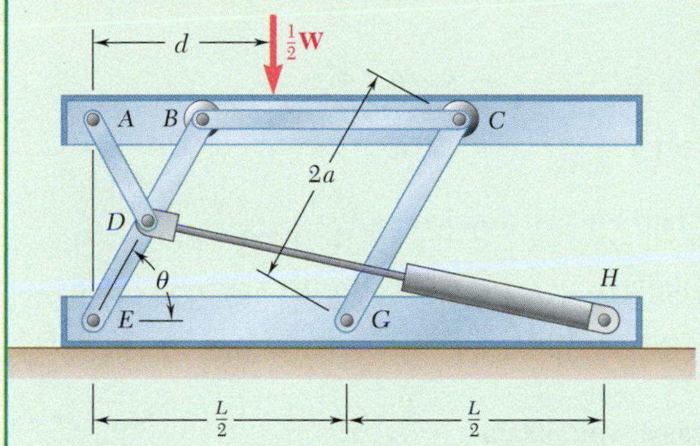

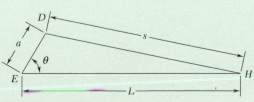

Sample Problem 10.3

A hydraulic-lift table is used to raise a 1000-kg crate. The table consists of a platform and two identical linkages on which hydraulic cylinders exert equal forces. (Only one linkage and one cylinder are shown.) Members *EDB* and *CG* are each of length $2a$, and member *AD* is pinned to the midpoint of *EDB*. If the crate is placed on the table so that half of its weight is supported by the system shown, determine the force exerted by each cylinder in raising the crate for $\theta = 60°$, $a = 0.70$ m, and $L = 3.20$ m. (This mechanism was previously considered in Sample Prob. 6.7.)

STRATEGY: The principle of virtual work allows you to find a relationship between the force applied by the cylinder and the weight without involving the reactions. However, you need a relationship between the virtual displacement and the change in angle θ, which is found from the law of cosines applied to the given geometry.

MODELING: The free body consists of the platform and the linkage (Fig. 1), with an input force $\mathbf{F}_{DH}$ exerted by the cylinder and an output force equal and opposite to $\frac{1}{2}\mathbf{W}$.

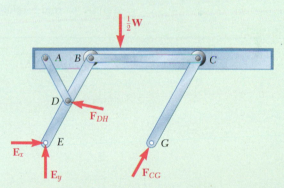

Fig. 1 Free-body diagram of the platform and linkage.

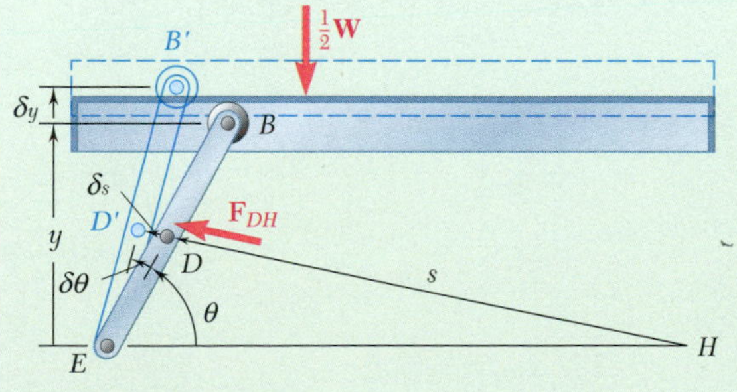

ANALYSIS: Principle of Virtual Work. First observe that the reactions at *E* and *G* do no work. Denoting the elevation of the platform above the base by y and the length *DH* of the cylinder-and-piston assembly by s (Fig. 2), you have

$$\delta U = 0: \qquad -\tfrac{1}{2}W\,\delta y + F_{DH}\,\delta s = 0 \qquad \textbf{(1)}$$

You can express the vertical displacement δy of the platform in terms of the angular displacement $\delta\theta$ of *EDB* as

$$y = (EB)\sin\theta = 2a\sin\theta$$
$$\delta y = 2a\cos\theta\,\delta\theta$$

To express δs similarly in terms of $\delta\theta$, first note that by the law of cosines (Fig. 3),

$$s^2 = a^2 + L^2 - 2aL\cos\theta$$

Fig. 2 Virtual displacement of the machine.

Fig. 3 Geometry associated with the cylinder-and-piston assembly.

Differentiating,

$$2s\,\delta s = -2aL(-\sin\theta)\,\delta\theta$$

$$\delta s = \frac{aL\sin\theta}{s}\,\delta\theta$$

Substituting for δy and δs into Eq. (1), you have

$$(-\tfrac{1}{2}W)2a\cos\theta\,\delta\theta + F_{DH}\frac{aL\sin\theta}{s}\,\delta\theta = 0$$

$$F_{DH} = W\frac{s}{L}\cot\theta$$

With the given numerical data, you obtain

$$W = mg = (1000\text{ kg})(9.81\text{ m/s}^2) = 9810\text{ N} = 9.81\text{ kN}$$

$$s^2 = a^2 + L^2 - 2aL\cos\theta$$

$$= (0.70)^2 + (3.20)^2 - 2(0.70)(3.20)\cos60° = 8.49$$

$$s = 2.91\text{ m}$$

$$F_{DH} = W\frac{s}{L}\cot\theta = (9.81\text{ kN})\frac{2.91\text{ m}}{3.20\text{ m}}\cot60°$$

$$F_{DH} = 5.15\text{ kN} \quad \blacktriangleleft$$

REFLECT and THINK: The principle of virtual work gives you a relationship between forces, but sometimes you need to review the geometry carefully to find a relationship between the displacements.

SOLVING PROBLEMS
ON YOUR OWN

In this section, we described how to use the **method of virtual work**, which is a different way of solving problems involving the equilibrium of rigid bodies.

The work done by a force during a displacement of its point of application or by a couple during a rotation is found, respectively, by using:

$$dU = F \, ds \cos \alpha \qquad\qquad (10.1)$$
$$dU = M \, d\theta \qquad\qquad (10.2)$$

Principle of virtual work. In its more general and more useful form, this principle can be stated as:

> **If a system of connected rigid bodies is in equilibrium, the total virtual work of the external forces applied to the system is zero for any virtual displacement of the system.**

As you apply the principle of virtual work, keep in mind the following points.

1. Virtual displacement. A machine or mechanism in equilibrium has no tendency to move. However, we can cause—or imagine—a small displacement. Since it does not actually occur, such a displacement is called a **virtual displacement**.

2. Virtual work. The work done by a force or couple during a virtual displacement is called **virtual work**.

3. You need consider only the forces that do work during the virtual displacement.

4. Forces that do no work during a virtual displacement that are consistent with the constraints imposed on the system are

    a. Reactions at supports

    b. Internal forces at connections

    c. Forces exerted by inextensible cords and cables

None of these forces need be considered when you use the method of virtual work.

5. Be sure to express the various virtual displacements involved in your computations in terms of a single virtual displacement. This is done in each of the three preceding sample problems, where the virtual displacements are all expressed in terms of $\delta\theta$.

6. Remember that the method of virtual work is effective only in those cases where the geometry of the system makes it relatively easy to relate the displacements involved.

Problems

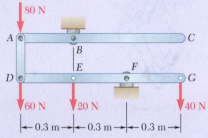

Fig. P10.1 and P10.3

10.1 Determine the vertical force **P** that must be applied at C to maintain the equilibrium of the linkage.

10.2 Determine the horizontal force **P** that must be applied at A to maintain the equilibrium of the linkage.

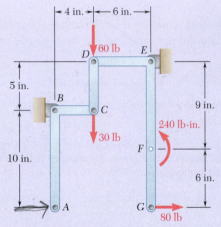

Fig. P10.2 and P10.4

10.3 and 10.4 Determine the couple **M** that must be applied to member ABC to maintain the equilibrium of the linkage.

10.5 A spring of constant 15 kN/m connects points C and F of the linkage shown. Neglecting the weight of the spring and linkage, determine the force in the spring and the vertical motion of point G when a vertical downward 120-N force is applied (*a*) at point C, (*b*) at points C and H.

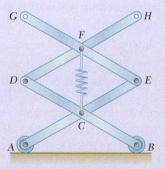

Fig. P10.5 and P10.6

10.6 A spring of constant 15 kN/m connects points C and F of the linkage shown. Neglecting the weight of the spring and linkage, determine the force in the spring and the vertical motion of point G when a vertical downward 120-N force is applied (*a*) at point E, (*b*) at points E and F.

10.7 The two-bar linkage shown is supported by a pin and bracket at B and a collar at D that slides freely on a vertical rod. Determine the force P required to maintain the equilibrium of the linkage.

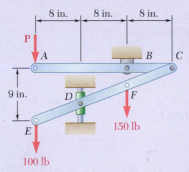

Fig. P10.7

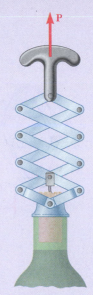

Fig. P10.8

10.8 Knowing that the maximum friction force exerted by the bottle on the cork is 60 lb, determine (a) the force P that must be applied to the corkscrew to open the bottle, (b) the maximum force exerted by the base of the corkscrew on the top of the bottle.

10.9 Rod AD is acted upon by a vertical force P at end A and by two equal and opposite horizontal forces of magnitude Q at points B and C. Derive an expression for the magnitude Q of the horizontal forces required for equilibrium.

10.10 and 10.11 The slender rod AB is attached to a collar A and rests on a small wheel at C. Neglecting the radius of the wheel and the effect of friction, derive an expression for the magnitude of the force Q required to maintain the equilibrium of the rod.

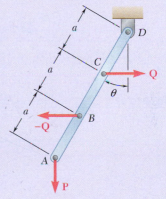

Fig. P10.9

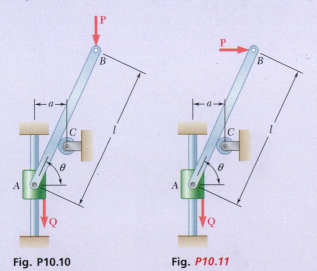

Fig. P10.10   **Fig. P10.11**

10.12 Knowing that the line of action of the force Q passes through point C, derive an expression for the magnitude of Q required to maintain equilibrium.

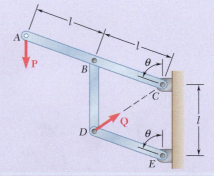

Fig. P10.12

10.13 Solve Prob. 10.12 assuming that the force P applied at point A acts horizontally to the left.

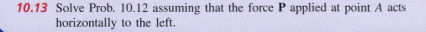

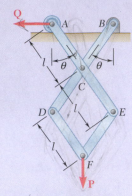

Fig. P10.14

10.14 The mechanism shown is acted upon by the force **P**; derive an expression for the magnitude of the force **Q** required to maintain equilibrium.

10.15 and 10.16 Derive an expression for the magnitude of the couple **M** required to maintain the equilibrium of the linkage shown.

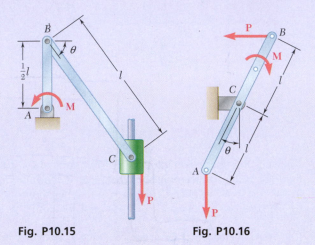

Fig. P10.15   **Fig. P10.16**

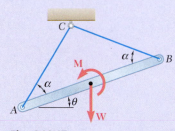

Fig. P10.17

10.17 A uniform rod AB with length l and weight W is suspended from two cords AC and BC of equal length. Derive an expression for the magnitude of the couple **M** required to maintain equilibrium of the rod in the position shown.

10.18 The pin at C is attached to member BCD and can slide along a slot cut in the fixed plate shown. Neglecting the effect of friction, derive an expression for the magnitude of the couple **M** required to maintain equilibrium when the force **P** that acts at D is directed (a) as shown, (b) vertically downward, (c) horizontally to the right.

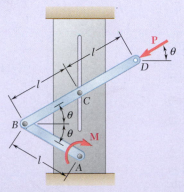

Fig. P10.18

10.19 For the linkage shown, determine the couple **M** required for equilibrium when $l = 1.8$ ft, $Q = 40$ lb, and $\theta = 65°$.

10.20 For the linkage shown, determine the force **Q** required for equilibrium when $l = 18$ in., $M = 600$ lb·in., and $\theta = 70°$.

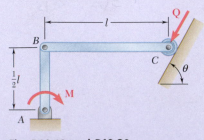

Fig. P10.19 and P10.20

10.21 A 4-kN force **P** is applied as shown to the piston of the engine system. Knowing that $AB = 50$ mm and $BC = 200$ mm, determine the couple **M** required to maintain the equilibrium of the system when (*a*) $\theta = 30°$, (*b*) $\theta = 150°$.

10.22 A couple **M** with a magnitude of 100 N·m is applied as shown to the crank of the engine system. Knowing that $AB = 50$ mm and $BC = 200$ mm, determine the force **P** required to maintain the equilibrium of the system when (*a*) $\theta = 60°$, (*b*) $\theta = 120°$.

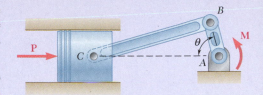

Fig. *P10.21* and *P10.22*

10.23 Rod *AB* is attached to a block at *A* that can slide freely in the vertical slot shown. Neglecting the effect of friction and the weights of the rods, determine the value of θ corresponding to equilibrium.

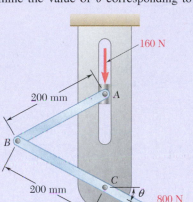

Fig. P10.23

10.24 Solve Prob. 10.23 assuming that the 800-N force is replaced by a 24-N·m clockwise couple applied at *D*.

10.25 Determine the value of θ corresponding to the equilibrium position of the rod of Prob. 10.10 when $l = 30$ in., $a = 5$ in., $P = 25$ lb, and $Q = 40$ lb.

10.26 Determine the values of θ corresponding to the equilibrium position of the rod of Prob. 10.11 when $l = 24$ in., $a = 4$ in., $P = 10$ lb, and $Q = 18$ lb.

10.27 Determine the value of θ corresponding to the equilibrium position of the mechanism of Prob. 10.12 when $P = 80$ N and $Q = 100$ N.

10.28 Determine the value of θ corresponding to the equilibrium position of the mechanism of Prob. 10.14 when $P = 270$ N and $Q = 960$ N.

10.29 Two rods *AC* and *CE* are connected by a pin at *C* and by a spring *AE*. The constant of the spring is k, and the spring is unstretched when $\theta = 30°$. For the loading shown, derive an equation in P, θ, l, and k that must be satisfied when the system is in equilibrium.

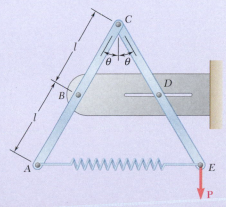

10.30 Two rods *AC* and *CE* are connected by a pin at *C* and by a spring *AE*. The constant of the spring is 1.5 lb/in., and the spring is unstretched when $\theta = 30°$. Knowing that $l = 10$ in. and neglecting the weight of the rods, determine the value of θ corresponding to equilibrium when $P = 40$ lb.

Fig. *P10.29* and P10.30

10.31 Solve Prob. 10.30 assuming that force **P** is moved to *C* and acts vertically downward.

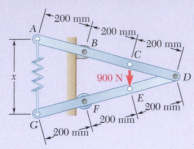

Fig. P10.32

10.32 Two bars AD and DG are connected by a pin at D and by a spring AG. Knowing that the spring is 300 mm long when unstretched and that the constant of the spring is 5 kN/m, determine the value of x corresponding to equilibrium when a 900-N load is applied at E as shown.

10.33 Solve Prob. 10.32 assuming that the 900-N vertical force is applied at C instead of E.

10.34 Two 5-kg bars AB and BC are connected by a pin at B and by a spring DE. Knowing that the spring is 150 mm long when unstretched and that the constant of the spring is 1 kN/m, determine the value of x corresponding to equilibrium.

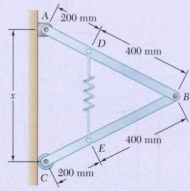

Fig. P10.34

10.35 A vertical force $\mathbf{P}$ with a magnitude of 150 N is applied to end E of cable CDE that passes over a small pulley D and is attached to the mechanism at C. The constant of the spring is $k = 4$ kN/m, and the spring is unstretched when $\theta = 0$. Neglecting the weight of the mechanism and the radius of the pulley, determine the value of θ corresponding to equilibrium.

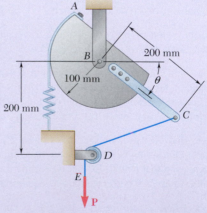

Fig. P10.35

10.36 A load **W** with a magnitude of 72 lb is applied to the mechanism at *C*. Neglecting the weight of the mechanism, determine the value of θ corresponding to equilibrium. The constant of the spring is $k = 20$ lb/in., and the spring is unstretched when $\theta = 0$.

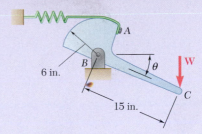

10.37 and P10.38 Knowing that the constant of spring *CD* is *k* and that the spring is unstretched when rod *ABC* is horizontal, determine the value of θ corresponding to equilibrium for the data indicated.

    10.37 $P = 300$ N, $l = 400$ mm, and $k = 5$ kN/m
    10.38 $P = 75$ lb, $l = 15$ in., and $k = 20$ lb/in.

Fig. P10.36

10.39 The lever *AB* is attached to the horizontal shaft *BC* that passes through a bearing and is welded to a fixed support at *C*. The torsional spring constant of the shaft *BC* is *K*; that is, a couple of magnitude *K* is required to rotate end *B* through 1 rad. Knowing that the shaft is untwisted when *AB* is horizontal, determine the value of θ corresponding to the position of equilibrium when $P = 100$ N, $l = 250$ mm, and $K = 12.5$ N·m/rad.

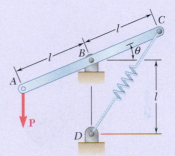

Fig. P10.37 and P10.38

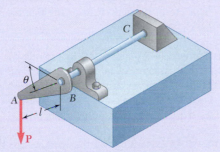

Fig. P10.39

10.40 Solve Prob. 10.39 assuming that $P = 350$ N, $l = 250$ mm, and $K = 12.5$ N·m/rad. Obtain answers in each of the following quadrants: $0 < \theta < 90°$, $270° < \theta < 360°$, and $360° < \theta < 450°$.

10.41 The position of boom *ABC* is controlled by the hydraulic cylinder *BD*. For the loading shown, determine the force exerted by the hydraulic cylinder on pin *B* when $\theta = 70°$.

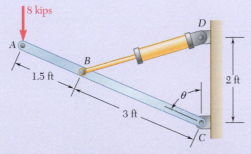

Fig. P10.41 and P10.42

10.42 The position of boom *ABC* is controlled by the hydraulic cylinder *BD*. For the loading shown, determine the largest allowable value of the angle θ if the maximum force that the cylinder can exert on pin *B* is 25 kips.

10.43 The position of member *ABC* is controlled by the hydraulic cylinder *CD*. For the loading shown, determine the force exerted by the hydraulic cylinder on pin *C* when $\theta = 55°$.

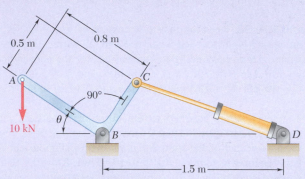

Fig. P10.43 and P10.44

10.44 The position of member *ABC* is controlled by the hydraulic cylinder *CD*. Determine angle θ, knowing that the hydraulic cylinder exerts a 15-kN force on pin *C*.

10.45 The telescoping arm *ABC* is used to provide an elevated platform for construction workers. The workers and the platform together weigh 500 lb, and their combined center of gravity is located directly above *C*. For the position when $\theta = 20°$, determine the force exerted on pin *B* by the single hydraulic cylinder *BD*.

10.46 Solve Prob. 10.45 assuming that the workers are lowered to a point near the ground so that $\theta = -20°$.

10.47 Denoting the coefficient of static friction between collar *C* and the vertical rod by μ_s, derive an expression for the magnitude of the largest couple **M** for which equilibrium is maintained in the position shown. Explain what happens if $\mu_s \geq \tan \theta$.

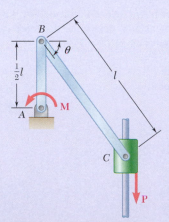

Fig. *P10.47* and P10.48

10.48 Knowing that the coefficient of static friction between collar *C* and the vertical rod is 0.40, determine the magnitude of the largest and smallest couple **M** for which equilibrium is maintained in the position shown, when $\theta = 35°$, $l = 600$ mm, and $P = 300$ N.

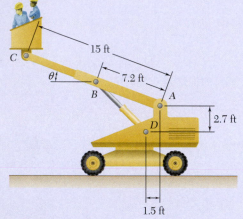

Fig. P10.45

10.49 A block with weight W is pulled up a plane forming an angle α with the horizontal by a force $\mathbf{P}$ directed along the plane. If μ is the coefficient of friction between the block and the plane, derive an expression for the mechanical efficiency of the system. Show that the mechanical efficiency cannot exceed $\frac{1}{2}$ if the block is to remain in place when the force $\mathbf{P}$ is removed.

10.50 Derive an expression for the mechanical efficiency of the jack discussed in Sec. 8.2B. Show that if the jack is to be self-locking, the mechanical efficiency cannot exceed $\frac{1}{2}$.

10.51 Denoting the coefficient of static friction between the block attached to rod ACE and the horizontal surface by μ_s, derive expressions in terms of P, μ_s, and θ for the largest and smallest magnitude of the force $\mathbf{Q}$ for which equilibrium is maintained.

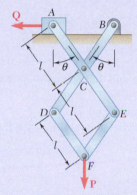

Fig. *P10.51* and P10.52

10.52 Knowing that the coefficient of static friction between the block attached to rod ACE and the horizontal surface is 0.15, determine the magnitude of the largest and smallest force $\mathbf{Q}$ for which equilibrium is maintained when $\theta = 30°$, $l = 0.2$ m, and $P = 40$ N.

10.53 Using the method of virtual work, determine separately the force and couple representing the reaction at A.

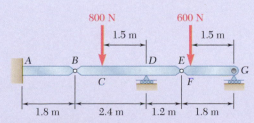

Fig. P10.53 and P10.54

10.54 Using the method of virtual work, determine the reaction at D.

10.55 Referring to Prob. 10.43 and using the value found for the force exerted by the hydraulic cylinder *CD*, determine the change in the length of *CD* required to raise the 10-kN load by 15 mm.

10.56 Referring to Prob. 10.45 and using the value found for the force exerted by the hydraulic cylinder *BD*, determine the change in the length of *BD* required to raise the platform attached at *C* by 2.5 in.

10.57 Determine the vertical movement of joint *D* if the length of member *BF* is increased by 1.5 in. (*Hint:* Apply a vertical load at joint *D*, and using the methods of Chap. 6, compute the force exerted by member *BF* on joints *B* and *F*. Then apply the method of virtual work for a virtual displacement resulting in the specified increase in length of member *BF*. This method should be used only for small changes in the lengths of members.)

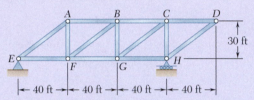

Fig. P10.57 and P10.58

10.58 Determine the horizontal movement of joint *D* if the length of member *BF* is increased by 1.5 in. (See the hint for Prob. 10.57.)

*10.2 WORK, POTENTIAL ENERGY, AND STABILITY

The concept of virtual work has another important connection with equilibrium, leading to criteria for conditions of stable, unstable, and neutral equilibrium. However, to explain this connection, we first need to introduce expressions for the work of a force during a finite displacement and then to define the concept of potential energy.

10.2A Work of a Force During a Finite Displacement

Consider a force $\mathbf{F}$ acting on a particle. In Sec. 10.1A, we defined the work of $\mathbf{F}$ corresponding to an infinitesimal displacement $d\mathbf{r}$ of the particle as

$$dU = \mathbf{F} \cdot d\mathbf{r} \tag{10.1}$$

We obtain the work of $\mathbf{F}$ corresponding to a finite displacement of the particle from A_1 to A_2 (Fig. 10.10a) that is denoted by $U_{1\rightarrow2}$ by integrating Eq. (10.1) along the curve described by the particle. Thus,

Work during a finite displacement

$$U_{1\rightarrow2} = \int_{A_1}^{A_2} \mathbf{F} \cdot d\mathbf{r} \tag{10.11}$$

Using the alternative expression

$$dU = F \, ds \cos \alpha \tag{10.1'}$$

given in Sec. 10.1 for the elementary work dU, we can also express the work $U_{1\rightarrow2}$ as

$$U_{1\rightarrow2} = \int_{s_1}^{s_2} (F \cos \alpha) \, ds \tag{10.11'}$$

Here, the variable of integration s measures the distance along the path traveled by the particle. We can represent the work $U_{1\rightarrow2}$ by the area under the curve obtained by plotting $F \cos \alpha$ against s (Fig. 10.10b). In the case of a force $\mathbf{F}$ of constant magnitude acting in the direction of motion, formula (10.11') yields $U_{1\rightarrow2} = F(s_2 - s_1)$.

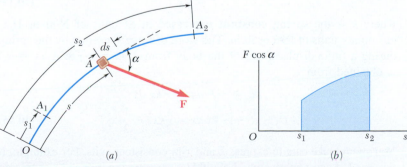

(a) (b)

Fig. 10.10 (a) A force acting on a particle moving along a path from A_1 to A_2; (b) the work done by the force in (a) equals the area under the graph of $F \cos \alpha$ versus s.

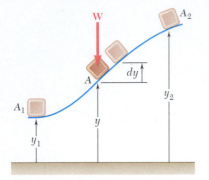

Fig. 10.11 The work done by the weight of a body equals the magnitude of the weight times the vertical displacement of its center of gravity.

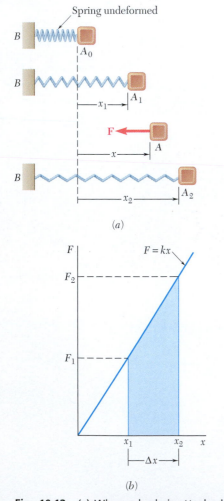

(a)

(b)

Fig. 10.12 (a) When a body is attached to a fixed point by a spring, the force on it is the product of the spring constant and the displacement from the undeformed position; (b) the work of the force equals the area under the graph of F versus x between x_1 and x_2.

Recall from Sec. 10.1 that the work of a couple of moment **M** during an infinitesimal rotation $d\theta$ of a rigid body is

$$dU = M\, d\theta \tag{10.2}$$

Therefore, we can express the work of the couple during a finite rotation of the body as

Work during a finite rotation

$$U_{1\to2} = \int_{\theta_1}^{\theta_2} M\, d\theta \tag{10.12}$$

In the case of a constant couple, formula (10.12) yields

$$U_{1\to2} = M(\theta_2 - \theta_1)$$

Work of a Weight. We stated in Sec. 10.1 that the work of a body's weight **W** during an infinitesimal displacement of the body is equal to the product of W and the vertical displacement of the body's center of gravity. With the y axis pointing upward, we obtain the work of **W** during a finite displacement of the body (Fig. 10.11) from

$$dU = -W\, dy$$

Integrating from A_1 to A_2, we have

$$U_{1\to2} = -\int_{y_1}^{y_2} W\, dy = Wy_1 - Wy_2 \tag{10.13}$$

or

$$U_{1\to2} = -W(y_2 - y_1) = -W\,\Delta y \tag{10.13'}$$

where Δy is the vertical displacement from A_1 to A_2. The work of the weight **W** is thus equal to **the product of W and the vertical displacement of the center of gravity of the body**. The work is *positive* when $\Delta y < 0$, that is, when the body moves down.

Work of the Force Exerted by a Spring. Consider a body A attached to a fixed point B by a spring. We assume that the spring is undeformed when the body is at A_0 (Fig. 10.12a). Experimental evidence shows that the magnitude of the force **F** exerted by the spring on a body A is proportional to the deflection x of the spring measured from position A_0. We have

$$F = kx \tag{10.14}$$

where k is the **spring constant** expressed in SI units of N/m or U.S. customary units of lb/ft or lb/in. The work of force **F** exerted by the spring during a finite displacement of the body from A_1 ($x = x_1$) to A_2 ($x = x_2$) is obtained from

$$dU = -F\, dx = -kx\, dx$$

$$U_{1\to2} = -\int_{x_1}^{x_2} kx\, dx = \tfrac{1}{2}kx_1^2 - \tfrac{1}{2}kx_2^2 \tag{10.15}$$

You should take care to express k and x in consistent units. For example, if you use U.S. customary units, k should be expressed in lb/ft and x expressed in feet, or k is given in lb/in. and x in inches. In the first case, the work is obtained in ft·lb; in the second case, it is in in·lb. We note that the work of the force **F** exerted by the spring on the body is *positive* when $x_2 < x_1$, that is, *when the spring is returning to its undeformed position.*

Since Eq. (10.14) is the equation of a straight line of slope k passing through the origin, we can obtain the work $U_{1\to2}$ of **F** during the displacement from A_1 to A_2 by evaluating the area of the trapezoid shown in Fig. 10.12b. This is done by computing the values F_1 and F_2 and multiplying the base Δx of the trapezoid by its mean height as $\frac{1}{2}(F_1 + F_2)$. Since the work of the force **F** exerted by the spring is positive for a negative value of Δx, we have

$$U_{1\to2} = -\frac{1}{2}(F_1 + F_2)\,\Delta x \qquad \textbf{(10.16)}$$

Equation (10.16) is usually more convenient to use than Eq. (10.15) and affords fewer chances of confusing the units involved.

10.2B Potential Energy

Let's consider again the body of Fig. 10.11. Using Eq. (10.13), we obtain the work of weight **W** during a finite displacement by subtracting the value of the function Wy corresponding to the second position of the body from its value corresponding to the first position. Thus, the work of **W** is independent of the actual path followed; it depends only upon the initial and final values of the function Wy. This function is called the **potential energy** of the body with respect to the force due to gravity **W** and is denoted by V_g. Thus,

$$U_{1\to2} = (V_g)_1 - (V_g)_2 \qquad \text{with } V_g = Wy \qquad \textbf{(10.17)}$$

Note that if $(V_g)_2 > (V_g)_1$, that is, *if the potential energy increases* during the displacement (as in the case considered here), *the work $U_{1\to2}$ is negative.* If, on the other hand, the work of **W** is positive, the potential energy decreases. Therefore, the potential energy V_g of the body provides a measure of *the work that can be done* by its weight **W**. Since only the *change* in potential energy—not the actual value of V_g—is involved in formula (10.17), we can add an arbitrary constant to the expression obtained for V_g. In other words, the level from which the elevation y is measured can be chosen arbitrarily. Note that potential energy is expressed in the same units as work, i.e., in joules (J) if SI units are used[†] and in ft·lb or in·lb if U.S. customary units are used.

Now consider the body of Fig. 10.12a. Using Eq. (10.15), we obtain the work of the elastic force **F** by subtracting the value of the function $\frac{1}{2}kx^2$ corresponding to the second position of the body from its value corresponding to the first position. This function, denoted by V_e, is called the **potential energy** of the body with respect to the **elastic force F**. We have

$$U_{1\to2} = (V_e)_1 - (V_e)_2 \qquad \text{with } V_e = \frac{1}{2}kx^2 \qquad \textbf{(10.18)}$$

Note that during the displacement considered, the work of force **F** exerted by the spring on the body is negative and the potential energy V_e increases. Also note that the expression obtained for V_e is valid only if the deflection of the spring is measured from its undeformed position.

We can use the concept of potential energy when forces other than gravity forces and elastic forces are involved. It remains valid as long as

[†]See footnote, p. 575.

the elementary work dU of the force considered is an *exact differential*. It is then possible to find a function V, called potential energy, such that

$$dU = -dV \tag{10.19}$$

Integrating Eq. (10.19) over a finite displacement, we obtain

Potential energy, general formulation

$$U_{1 \to 2} = V_1 - V_2 \tag{10.20}$$

This equation says that **the work of the force is independent of the path followed and is equal to minus the change in potential energy.** A force that satisfies Eq. (10.20) is called a **conservative force.**[†]

10.2C Potential Energy and Equilibrium

Applying the principle of virtual work is considerably simplified if we know the potential energy of a system. In the case of a virtual displacement, formula (10.19) becomes $\delta U = -\delta V$. Moreover, if the position of the system is defined by a single independent variable θ, we can write $\delta V = (dV/d\theta)\,\delta\theta$. Since $\delta\theta$ must be different from zero, the condition $\delta U = 0$ for the equilibrium of the system becomes

Equilibrium condition $$\frac{dV}{d\theta} = 0 \tag{10.21}$$

In terms of potential energy, therefore, the principle of virtual work states:

> **If a system is in equilibrium, the derivative of its total potential energy is zero.**

If the position of the system depends upon several independent variables (the system is then said to possess *several degrees of freedom*), the partial derivatives of V with respect to each of the independent variables must be zero.

Consider, for example, a structure made of two members AC and CB and carrying a load W at C. The structure is supported by a pin at A and a roller at B, and a spring BD connects B to a fixed point D (Fig. 10.13a). The constant of the spring is k, and we assume that the natural length of the spring is equal to AD, so that the spring is undeformed when B coincides with A. Neglecting friction forces and the weights of the members, we find that the only forces that do work during a virtual displacement of the structure are the weight $\mathbf{W}$ and the force $\mathbf{F}$ exerted by the spring at point B (Fig. 10.13b). Therefore, we can obtain the total potential energy of the system by adding the potential energy V_g corresponding to the gravity force $\mathbf{W}$ and the potential energy V_e corresponding to the elastic force $\mathbf{F}$.

Choosing a coordinate system with the origin at A and noting that the deflection of the spring measured from its undeformed position is $AB = x_B$, we have

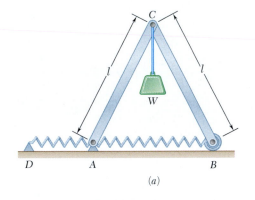

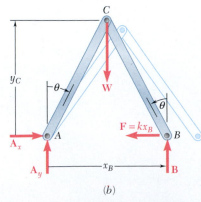

Fig. 10.13 (a) Structure carrying a load at C with a spring from B to D; (b) free-body diagram of the structure, and a virtual displacement.

$$V_e = \frac{1}{2}kx_B^2 \quad \text{and} \quad V_g = Wy_C$$

[†]A detailed discussion of conservative forces is given in Sec. 13.2B of *Dynamics*.

Expressing the coordinates x_B and y_C in terms of the angle θ, we have

$$x_B = 2l \sin \theta \qquad\qquad y_C = l \cos \theta$$
$$V_e = \tfrac{1}{2}k(2l \sin \theta)^2 \qquad V_g = W(l \cos \theta)$$
$$V = V_e + V_g = 2kl^2 \sin^2 \theta + Wl \cos \theta \qquad\qquad \textbf{(10.22)}$$

We obtain the positions of equilibrium of the system by setting the derivative of the potential energy V to zero, as

$$\frac{dV}{d\theta} = 4kl^2 \sin \theta \cos \theta - Wl \sin \theta = 0$$

or, factoring out $l \sin \theta$, as

$$\frac{dV}{d\theta} = l \sin \theta(4kl \cos \theta - W) = 0$$

There are therefore two positions of equilibrium corresponding to the values $\theta = 0$ and $\theta = \cos^{-1}(W/4kl)$, respectively.[†]

10.2D   Stability of Equilibrium

Consider the three uniform rods with a length of $2a$ and weight $\mathbf{W}$ shown in Fig. 10.14. Although each rod is in equilibrium, there is an important difference between the three cases considered. Suppose that each rod is slightly disturbed from its position of equilibrium and then released. Rod a moves back toward its original position; rod b keeps moving away from its original position; and rod c remains in its new position. In case a, the equilibrium of the rod is said to be **stable**; in case b, it is **unstable**; and in case c, it is **neutral**.

Recall from Sec. 10.2B that the potential energy V_g with respect to gravity is equal to Wy, where y is the elevation of the point of application of $\mathbf{W}$ measured from an arbitrary level. We observe that the potential energy of rod a is minimum in the position of equilibrium considered, that the potential energy of rod b is maximum, and that the potential energy of rod c is constant. Equilibrium is thus *stable, unstable,* or *neutral* according to whether the potential energy is *minimum, maximum,* or *constant* (Fig. 10.15).

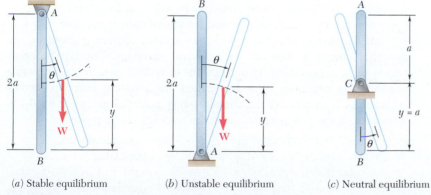

(a) Stable equilibrium   (b) Unstable equilibrium   (c) Neutral equilibrium

Fig. 10.14   (a) Rod supported from above, stable equilibrium; (b) rod supported from below, unstable equilibrium; (c) rod supported at its midpoint, neutral equilibrium.

[†]The second position does not exist if $W > 4kl$.

This result is quite general, as we now show. We first observe that a force always tends to do positive work and thus to decrease the potential energy of the system on which it is applied. Therefore, when a system is disturbed from its position of equilibrium, the forces acting on the system tend to bring it back to its original position if V is minimum (Fig. 10.15a) and to move it farther away if V is maximum (Fig. 10.15b). If V is constant (Fig. 10.15c), the forces do not tend to move the system either way.

Recall from calculus that a function is minimum or maximum according to whether its second derivative is positive or negative. Therefore, we can summarize the conditions for the equilibrium of a system with one degree of freedom (i.e., a system for which the position is defined by a single independent variable θ) as

$$\frac{dV}{d\theta} = 0 \quad \frac{d^2V}{d\theta^2} > 0 : \text{stable equilibrium}$$

(10.23)

$$\frac{dV}{d\theta} = 0 \quad \frac{d^2V}{d\theta^2} < 0 : \text{unstable equilibrium}$$

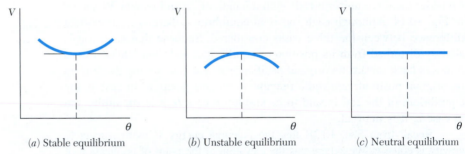

(a) Stable equilibrium   (b) Unstable equilibrium   (c) Neutral equilibrium

Fig. 10.15 Stable, unstable, and neutral equilibria correspond to potential energy values that are minimum, maximum, or constant, respectively.

If both the first and the second derivatives of V are zero, it is necessary to examine derivatives of a higher order to determine whether the equilibrium is stable, unstable, or neutral. The equilibrium is neutral if all derivatives are zero, since the potential energy V is then a constant. The equilibrium is stable if the first derivative found to be different from zero is of even order and positive. In all other cases, the equilibrium is unstable.

If the system of interest possesses *several degrees of freedom*, the potential energy V depends upon several variables. Thus, it becomes necessary to apply the theory of functions of several variables to determine whether V is minimum. It can be verified that a system with two degrees of freedom is stable, and the corresponding potential energy $V(\theta_1, \theta_2)$ is minimum, if the following relations are satisfied simultaneously:

$$\frac{\partial V}{\partial \theta_1} = \frac{\partial V}{\partial \theta_2} = 0$$

$$\left(\frac{\partial^2 V}{\partial \theta_1 \partial \theta_2} \right)^2 - \frac{\partial^2 V}{\partial \theta_1^2} \frac{\partial^2 V}{\partial \theta_2^2} < 0$$

(10.24)

$$\frac{\partial^2 V}{\partial \theta_1^2} > 0 \quad \text{or} \quad \frac{\partial^2 V}{\partial \theta_2^2} > 0$$

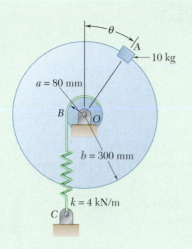

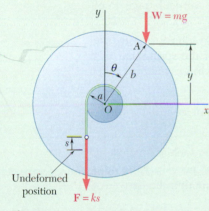

Fig. 1 Free-body diagram of rotated disk, showing only those forces that do work.

Sample Problem 10.4

A 10-kg block is attached to the rim of a 300-mm-radius disk as shown. Knowing that spring BC is unstretched when $\theta = 0$, determine the position or positions of equilibrium, and state in each case whether the equilibrium is stable, unstable, or neutral.

STRATEGY: The first step is to determine a potential energy function V for the system. You can find the positions of equilibrium by determining where the derivative of V is zero. You can find the types of stability by finding where V is maximum or minimum.

MODELING and ANALYSIS:

Potential Energy.   Denote the deflection of the spring from its undeformed position by s, and place the origin of coordinates at O (Fig. 1). You obtain

$$V_e = \tfrac{1}{2}ks^2 \quad V_g = Wy = mgy$$

Measuring θ in radians, you have

$$s = a\theta \qquad y = b\cos\theta$$

Substituting for s and y in the expressions for V_e and V_g gives you

$$V_e = \tfrac{1}{2}ka^2\theta^2 \quad V_g = mgb\cos\theta$$

$$V = V_e + V_g = \tfrac{1}{2}ka^2\theta^2 + mgb\cos\theta$$

Positions of Equilibrium.   Setting $dV/d\theta = 0$, you obtain

$$\frac{dV}{d\theta} = ka^2\theta - mgb\sin\theta = 0$$

$$\sin\theta = \frac{ka^2}{mgb}\theta$$

Now substitute $a = 0.08$ m, $b = 0.3$ m, $k = 4$ kN/m, and $m = 10$ kg. The result is

$$\sin\theta = \frac{(4\text{ kN/m})(0.08\text{ m})^2}{(10\text{ kg})(9.81\text{ m/s}^2)(0.3\text{ m})}\theta$$

$$\sin\theta = 0.8699\,\theta$$

where θ is expressed in radians. Solving by trial and error for θ, you find

$$\theta = 0 \qquad \text{and} \qquad \theta = 0.902\text{ rad}$$
$$\theta = 0 \qquad \text{and} \qquad \theta = 51.7° \quad \blacktriangleleft$$

Stability of Equilibrium.   The second derivative of the potential energy V with respect to θ is

$$\frac{d^2V}{d\theta^2} = ka^2 - mgb\cos\theta$$
$$= (4\text{ kN/m})(0.08\text{ m})^2 - (10\text{ kg})(9.81\text{ m/s}^2)(0.3\text{ m})\cos\theta$$
$$= 25.6 - 29.43\cos\theta$$

For $\theta = 0$, $\qquad \dfrac{d^2V}{d\theta^2} = 25.6 - 29.43\cos 0° = -3.83 < 0$

The equilibrium is unstable for $\theta = 0$ ◀

For $\theta = 51.7°$, $\qquad \dfrac{d^2V}{d\theta^2} = 25.6 - 29.43\cos 51.7° = +7.36 > 0$

The equilibrium is stable for $\theta = 51.7°$ ◀

REFLECT and THINK: If you just let the block-and-disk system fall on its own, it will come to rest at $\theta = 51.7°$. If you balance the system at $\theta = 0$, the slightest touch will put it in motion.

SOLVING PROBLEMS
ON YOUR OWN

In this section, we defined the **work of a force during a finite displacement** and the **potential energy** of a rigid body or a system of rigid bodies. You saw how to use the concept of potential energy to determine the **equilibrium position** of a rigid body or a system of rigid bodies.

1. The potential energy V of a system is the sum of the potential energies associated with the various forces acting on the system that do work as the system moves. In the problems of this section, you will determine the following energies.

   a. Potential energy of a weight. This is the potential energy due to *gravity*, $V_g = Wy$, where y is the elevation of the weight W measured from some arbitrary reference level. You can use the potential energy V_g with any vertical force **P** of constant magnitude directed downward; we write $V_g = Py$.

   b. Potential energy of a spring. This is the potential energy due to the *elastic* force exerted by a spring, $V_e = \frac{1}{2}kx^2$, where k is the constant of the spring and x is the deformation of the spring measured from its unstretched position.

Reactions at fixed supports, internal forces at connections, forces exerted by inextensible cords and cables, and other forces that do no work do not contribute to the potential energy of the system.

2. Express all distances and angles in terms of a single variable, such as an angle θ, when computing the potential energy V of a system. This is necessary because determining the equilibrium position of the system requires computing the derivative $dV/d\theta$.

3. When a system is in equilibrium, the first derivative of its potential energy is zero. Therefore,

   a. To determine a position of equilibrium of a system, first express its potential energy V in terms of the single variable θ, and then compute its derivative and solve the equation $dV/d\theta = 0$ for θ.

   b. To determine the force or couple required to maintain a system in a given position of equilibrium, substitute the known value of θ in the equation $dV/d\theta = 0$, and solve this equation for the desired force or couple.

4. Stability of equilibrium. The following rules generally apply:

   a. Stable equilibrium occurs when the potential energy of the system is *minimum*, that is, when $dV/d\theta = 0$ and $d^2V/d\theta^2 > 0$ (Figs. 10.14a and 10.15a).

   b. Unstable equilibrium occurs when the potential energy of the system is *maximum*, that is, when $dV/d\theta = 0$ and $d^2V/d\theta^2 < 0$ (Figs. 10.14b and 10.15b).

   c. Neutral equilibrium occurs when the potential energy of the system is *constant*; $dV/d\theta$, $d^2V/d\theta^2$, and all the successive derivatives of V are then equal to zero (Figs. 10.14c and 10.15c).

See page 600 for a discussion of the case when $dV/d\theta$, $d^2V/d\theta^2$, but *not all* of the successive derivatives of V are equal to zero.

Problems

10.59 Using the method of Sec. 10.2C, solve Prob. 10.29.

10.60 Using the method of Sec. 10.2C, solve Prob. 10.30.

10.61 Using the method of Sec. 10.2C, solve Prob. 10.31.

10.62 Using the method of Sec. 10.2C, solve Prob. 10.32.

10.63 Using the method of Sec. 10.2C, solve Prob. 10.34.

10.64 Using the method of Sec. 10.2C, solve Prob. 10.35.

10.65 Using the method of Sec. 10.2C, solve Prob. 10.37.

10.66 Using the method of Sec. 10.2C, solve Prob. 10.38.

10.67 Show that equilibrium is neutral in Prob. 10.1.

10.68 Show that equilibrium is neutral in Prob. 10.7.

10.69 Two uniform rods, each with a mass m, are attached to gears of equal radii as shown. Determine the positions of equilibrium of the system and state in each case whether the equilibrium is stable, unstable, or neutral.

10.70 Two uniform rods, AB and CD, are attached to gears of equal radii as shown. Knowing that $W_{AB} = 8$ lb and $W_{CD} = 4$ lb, determine the positions of equilibrium of the system and state in each case whether the equilibrium is stable, unstable, or neutral.

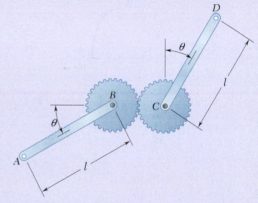

Fig. P10.69 and P10.70

10.71 Two uniform rods AB and CD, of the same length l, are attached to gears as shown. Knowing that rod AB weighs 3 lb and that rod CD weighs 2 lb, determine the positions of equilibrium of the system and state in each case whether the equilibrium is stable, unstable, or neutral.

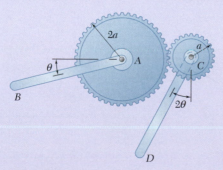

Fig. P10.71

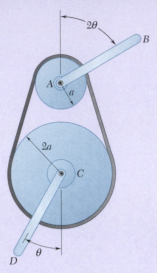

Fig. P10.72

10.72 Two uniform rods, each of mass m and length l, are attached to drums that are connected by a belt as shown. Assuming that no slipping occurs between the belt and the drums, determine the positions of equilibrium of the system and state in each case whether the equilibrium is stable, unstable, or neutral.

10.73 Using the method of Sec. 10.2C, solve Prob. 10.39. Determine whether the equilibrium is stable, unstable, or neutral. (*Hint:* The potential energy corresponding to the couple exerted by a torsion spring is $\frac{1}{2}K\theta^2$, where K is the torsional spring constant and θ is the angle of twist.)

10.74 In Prob. 10.40, determine whether each of the positions of equilibrium is stable, unstable, or neutral. (See hint for Prob. 10.73.)

10.75 A load **W** with a magnitude of 100 lb is applied to the mechanism at C. Knowing that the spring is unstretched when $\theta = 15°$, determine that value of θ corresponding to equilibrium and check that the equilibrium is stable.

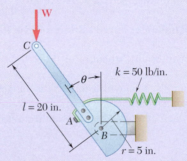

Fig. *P10.75* and *P10.76*

10.76 A load **W** with a magnitude of 100 lb is applied to the mechanism at C. Knowing that the spring is unstretched when $\theta = 30°$, determine that value of θ corresponding to equilibrium and check that the equilibrium is stable.

10.77 A slender rod AB with a weight W is attached to two blocks A and B that can move freely in the guides shown. Knowing that the spring is unstretched when $y = 0$, determine the value of y corresponding to equilibrium when $W = 80$ N, $l = 500$ mm, and $k = 600$ N/m.

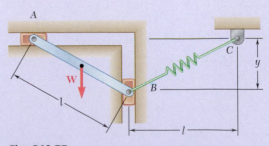

Fig. P10.77

10.78 A slender rod AB with a weight W is attached to two blocks A and B that can move freely in the guides shown. Knowing that both springs are unstretched when $y = 0$, determine the value of y corresponding to equilibrium when $W = 80$ N, $l = 500$ mm, and $k = 600$ N/m.

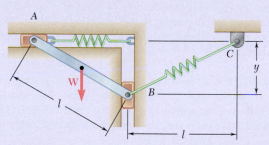

Fig. P10.78

10.79 A slender rod AB with a weight W is attached to two blocks A and B that can move freely in the guides shown. The constant of the spring is k, and the spring is unstretched when AB is horizontal. Neglecting the weight of the blocks, derive an equation in θ, W, l, and k that must be satisfied when the rod is in equilibrium.

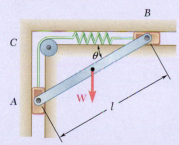

Fig. P10.79 and P10.80

10.80 A slender rod AB with a weight W is attached to two blocks A and B that can move freely in the guides shown. Knowing that the spring is unstretched when AB is horizontal, determine three values of θ corresponding to equilibrium when $W = 300$ lb, $l = 16$ in., and $k = 75$ lb/in. State in each case whether the equilibrium is stable, unstable, or neutral.

10.81 A spring AB of constant k is attached to two identical gears as shown. Knowing that the spring is undeformed when $\theta = 0$, determine two values of the angle θ corresponding to equilibrium when $P = 30$ lb, $a = 4$ in., $b = 3$ in., $r = 6$ in., and $k = 5$ lb/in. State in each case whether the equilibrium is stable, unstable, or neutral.

10.82 A spring AB of constant k is attached to two identical gears as shown. Knowing that the spring is undeformed when $\theta = 0$, and given that $a = 60$ mm, $b = 45$ mm, $r = 90$ mm, and $k = 6$ kN/m, determine (a) the range of values of P for which a position of equilibrium exists, (b) two values of θ corresponding to equilibrium if the value of P is equal to half the upper limit of the range found in part a.

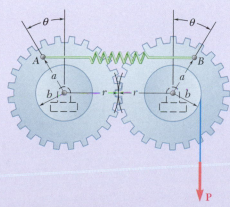

Fig. P10.81 and P10.82

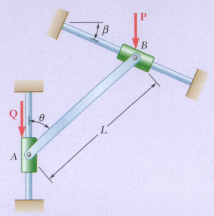

Fig. P10.83 and P10.84

10.83 A slender rod AB is attached to two collars A and B that can move freely along the guide rods shown. Knowing that $\beta = 30°$ and $P = Q = 400$ N, determine the value of the angle θ corresponding to equilibrium.

10.84 A slender rod AB is attached to two collars A and B that can move freely along the guide rods shown. Knowing that $\beta = 30°$, $P = 100$ N, and $Q = 25$ N, determine the value of the angle θ corresponding to equilibrium.

10.85 and 10.86 Cart B, which weighs 75 kN, rolls along a sloping track that forms an angle β with the horizontal. The spring constant is 5 kN/m, and the spring is unstretched when $x = 0$. Determine the distance x corresponding to equilibrium for the angle β indicated.

  10.85 Angle $\beta = 30°$
  10.86 Angle $\beta = 60°$

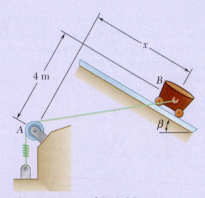

Fig. P10.85 and P10.86

10.87 and 10.88 Collar A can slide freely on the semicircular rod shown. Knowing that the constant of the spring is k and that the unstretched length of the spring is equal to the radius r, determine the value of θ corresponding to equilibrium when $W = 50$ lb, $r = 9$ in., and $k = 15$ lb/in.

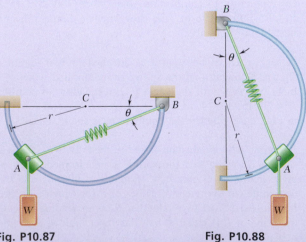

Fig. P10.87   **Fig. P10.88**

10.89 Two bars AB and BC of negligible weight are attached to a single spring of constant k that is unstretched when the bars are horizontal. Determine the range of values of the magnitude P of two equal and opposite forces $\mathbf{P}$ and $-\mathbf{P}$ for which the equilibrium of the system is stable in the position shown.

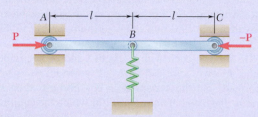

Fig. P10.89

10.90 A vertical bar AD is attached to two springs of constant k and is in equilibrium in the position shown. Determine the range of values of the magnitude P of two equal and opposite vertical forces $\mathbf{P}$ and $-\mathbf{P}$ for which the equilibrium position is stable if (a) $AB = CD$, (b) $AB = 2CD$.

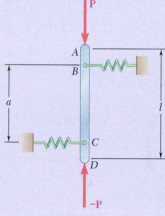

Fig. *P10.90*

10.91 Rod AB is attached to a hinge at A and to two springs, each of constant k. If $h = 25$ in., $d = 12$ in., and $W = 80$ lb, determine the range of values of k for which the equilibrium of the rod is stable in the position shown. Each spring can act in either tension or compression.

10.92 Rod AB is attached to a hinge at A and to two springs, each of constant k. If $h = 45$ in., $k = 6$ lb/in., and $W = 60$ lb, determine the smallest distance d for which the equilibrium of the rod is stable in the position shown. Each spring can act in either tension or compression.

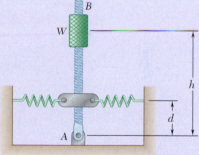

Fig. P10.91 and P10.92

10.93 and 10.94 Two bars are attached to a single spring of constant k that is unstretched when the bars are vertical. Determine the range of values of P for which the equilibrium of the system is stable in the position shown.

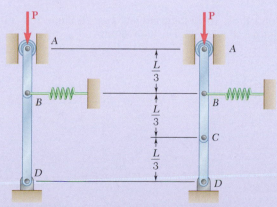

Fig. P10.93   Fig. P10.94

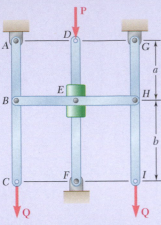

Fig. *P10.95* and *P10.96*

10.95 The horizontal bar *BEH* is connected to three vertical bars. The collar at *E* can slide freely on bar *DF*. Determine the range of values of *Q* for which the equilibrium of the system is stable in the position shown when $a = 24$ in., $b = 20$ in., and $P = 150$ lb.

10.96 The horizontal bar *BEH* is connected to three vertical bars. The collar at *E* can slide freely on bar *DF*. Determine the range of values of *P* for which the equilibrium of the system is stable in the position shown when $a = 150$ mm, $b = 200$ mm, and $Q = 45$ N.

*10.97 Bars *AB* and *BC*, each with a length *l* and of negligible weight, are attached to two springs, each of constant *k*. The springs are undeformed, and the system is in equilibrium when $\theta_1 = \theta_2 = 0$. Determine the range of values of *P* for which the equilibrium position is stable.

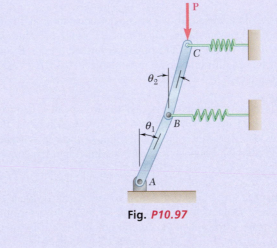

Fig. *P10.97*

*10.98 Solve Prob. 10.97 knowing that $l = 800$ mm and $k = 2.5$ kN/m.

*10.99 Two rods of negligible weight are attached to drums of radius *r* that are connected by a belt and spring of constant *k*. Knowing that the spring is undeformed when the rods are vertical, determine the range of values of *P* for which the equilibrium position $\theta_1 = \theta_2 = 0$ is stable.

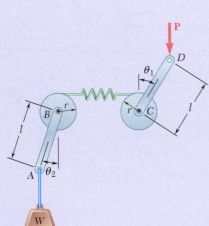

Fig. *P10.99*

*10.100 Solve Prob. 10.99 knowing that $k = 20$ lb/in., $r = 3$ in., $l = 6$ in., and (*a*) $W = 15$ lb, (*b*) $W = 60$ lb.

Review and Summary

Work of a Force

The first section of this chapter was devoted to the **principle of virtual work** and to its direct application to the solution of equilibrium problems. We first defined the **work of a force F corresponding to the small displacement** $d\mathbf{r}$ [Sec. 10.1A] as the quantity

$$dU = \mathbf{F} \cdot d\mathbf{r} \tag{10.1}$$

obtained by forming the scalar product of the force $\mathbf{F}$ and the displacement $d\mathbf{r}$ (Fig. 10.16). Denoting the magnitudes of the force and of the displacement by F and ds, respectively, and the angle formed by $\mathbf{F}$ and $d\mathbf{r}$ by α, we have

$$dU = F\, ds \cos \alpha \tag{10.19}$$

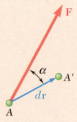

Fig. 10.16

The work dU is positive if $\alpha < 90°$, zero if $\alpha = 90°$, and negative if $\alpha > 90°$. We also found that the **work of a couple of moment M** acting on a rigid body is

$$dU = M\, d\theta \tag{10.2}$$

where $d\theta$ is the small angle expressed in radians through which the body rotates.

Virtual Displacement

Considering a particle located at A and acted upon by several forces $\mathbf{F}_1$, $\mathbf{F}_2, \ldots, \mathbf{F}_n$ [Sec. 10.1B], we imagined that the particle moved to a new position A' (Fig. 10.17). Since this displacement does not actually take place, we refer to it to as a **virtual displacement** denoted by $\delta\mathbf{r}$. The corresponding work of the forces is called **virtual work** and is denoted by δU. We have

$$\delta U = \mathbf{F}_1 \cdot \delta\mathbf{r} + \mathbf{F}_2 \cdot \delta\mathbf{r} + \ldots + \mathbf{F}_n \cdot \delta\mathbf{r}$$

Principle of Virtual Work

The **principle of virtual work** states that **if a particle is in equilibrium, the total virtual work δU of the forces acting on the particle is zero for any virtual displacement of the particle.**

The principle of virtual work can be extended to the case of rigid bodies and systems of rigid bodies. Since it involves only forces that do work, its application provides a useful alternative to the use of the equilibrium equations in the solution of many engineering problems. It is particularly effective in the case of machines and mechanisms consisting of connected rigid bodies, since the work of the reactions at the supports is zero and the work of the internal forces at the pin connections cancels out [Sec. 10.1C; Sample Probs. 10.1, 10.2, and 10.3].

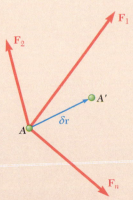

Fig. 10.17

Mechanical Efficiency

In the case of real machines, however [Sec. 10.1D], the work of the friction forces should be taken into account with the result that the **output work is less than the input work.** We defined the **mechanical efficiency** of a machine as the ratio

$$\eta = \frac{\text{output work}}{\text{input work}} \tag{10.9}$$

We noted that, for an ideal machine (no friction), $\eta = 1$, whereas for a real machine, $\eta < 1$.

Work of a Force over a Finite Displacement

In the second section of this chapter, we considered the work of forces corresponding to finite displacements of their points of application. We obtained the work $U_{1\to2}$ of the force **F** corresponding to a displacement of the particle A from A_1 to A_2 (Fig. 10.18) by integrating the right-hand side of Eqs. (10.1) or (10.1′) along the curve described by the particle [Sec. 10.2A]. Thus,

$$U_{1\to2} = \int_{A_1}^{A_2} \mathbf{F} \cdot d\mathbf{r} \tag{10.11}$$

or

$$U_{1\to2} = \int_{s_1}^{s_2} (F \cos \alpha)\, ds \tag{10.11′}$$

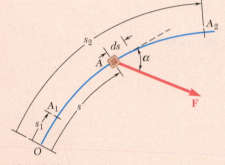

Fig. 10.18

Similarly, we expressed the work of a couple of moment **M** corresponding to a finite rotation from θ_1 to θ_2 of a rigid body as

$$U_{1\to2} = \int_{\theta_1}^{\theta_2} M\, d\theta \tag{10.12}$$

Work of a Weight

We obtained the **work of the weight W of a body** as its center of gravity moves from the elevation y_1 to y_2 (Fig. 10.19) by setting $F = W$ and $\alpha = 180°$ in Eq. (10.11′) as

$$U_{1\to2} = -\int_{y_1}^{y_2} W\, dy = Wy_1 - Wy_2 \tag{10.13}$$

The work of **W** is therefore positive when the elevation y decreases.

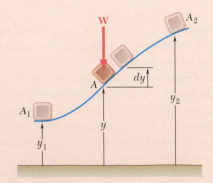

Fig. 10.19

Work of the Force Exerted by a Spring

The **work of the force F exerted by a spring** on a body A as the spring is stretched from x_1 to x_2 (Fig. 10.20) can be obtained by setting $F = kx$, where k is the constant of the spring, and $\alpha = 180°$ in Eq. (10.11'). Hence,

$$U_{1\to2} = -\int_{x_1}^{x_2} kx\,dx = \frac{1}{2}kx_1^2 - \frac{1}{2}kx_2^2 \qquad (10.15)$$

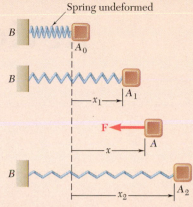

Spring undeformed

Fig. 10.20

The work of $\mathbf{F}$ is therefore positive when the spring is returning to its undeformed position.

Potential Energy

When the work of a force $\mathbf{F}$ is independent of the path actually followed between A_1 and A_2, the force is said to be a **conservative force**, and we can express its work as

$$U_{1\to2} = V_1 - V_2 \qquad (10.20)$$

Here V is the **potential energy** associated with $\mathbf{F}$, and V_1 and V_2 represent the values of V at A_1 and A_2, respectively [Sec. 10.2B]. We found the potential energies associated, respectively, with the force of gravity $\mathbf{W}$ and the elastic force $\mathbf{F}$ exerted by a spring to be

$$V_g = Wy \quad \text{and} \quad V_e = \frac{1}{2}kx^2 \qquad (10.17, 10.18)$$

Alternative Expression for the Principle of Virtual Work

When the position of a mechanical system depends upon a single independent variable θ, the potential energy of the system is a function $V(\theta)$ of that variable, and it follows from Eq. (10.20) that $\delta U = -\delta V = -(dV/d\delta)\,\delta\theta$. The condition $\delta U = 0$ required by the principle of virtual work for the equilibrium of the system thus can be replaced by the condition

$$\frac{dV}{d\theta} = 0 \qquad (10.21)$$

When all the forces involved are conservative, it may be preferable to use Eq. (10.21) rather than apply the principle of virtual work directly [Sec. 10.2C; Sample Prob. 10.4].

Stability of Equilibrium

This alternative approach presents another advantage, since it is possible to determine from the sign of the second derivative of V whether the equilibrium of the system is *stable, unstable,* or *neutral* [Sec. 10.2D]. If $d^2V/d\theta^2 > 0$, V is *minimum* and the equilibrium is *stable*; if $d^2V/d\theta^2 < 0$, V is *maximum* and the equilibrium is *unstable*; if $d^2V/d\theta^2 = 0$, it is necessary to examine derivatives of a higher order.

Review Problems

10.101 Determine the vertical force **P** that must be applied at G to maintain the equilibrium of the linkage.

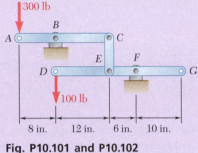
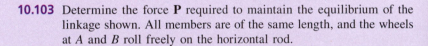
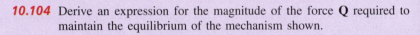

Fig. P10.101 and P10.102

10.102 Determine the couple **M** that must be applied to member $DEFG$ to maintain the equilibrium of the linkage.

10.103 Determine the force **P** required to maintain the equilibrium of the linkage shown. All members are of the same length, and the wheels at A and B roll freely on the horizontal rod.

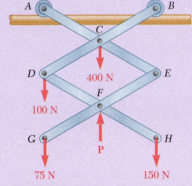

Fig. P10.103

10.104 Derive an expression for the magnitude of the force **Q** required to maintain the equilibrium of the mechanism shown.

10.105 Derive an expression for the magnitude of the couple **M** required to maintain the equilibrium of the linkage shown.

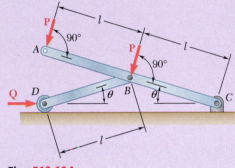

Fig. P10.104

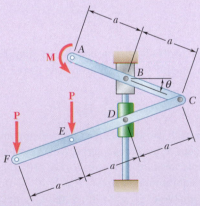
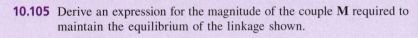

Fig. P10.105

10.106 A vertical load **W** is applied to the linkage at B. The constant of the spring is k, and the spring is unstretched when AB and BC are horizontal. Neglecting the weight of the linkage, derive an equation in θ, W, l, and k that must be satisfied when the linkage is in equilibrium.

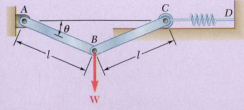

Fig. P10.106

10.107 A force **P** with a magnitude of 240 N is applied to end E of cable CDE, which passes under pulley D and is attached to the mechanism at C. Neglecting the weight of the mechanism and the radius of the pulley, determine the value of θ corresponding to equilibrium. The constant of the spring is $k = 4$ kN/m, and the spring is unstretched when $\theta = 90°$.

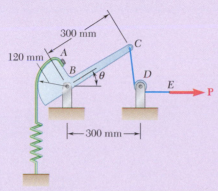

Fig. P10.107

10.108 Two identical rods ABC and DBE are connected by a pin at B and by a spring CE. Knowing that the spring is 4 in. long when unstretched and that the constant of the spring is 8 lb/in., determine the distance x corresponding to equilibrium when a 24-lb load is applied at E as shown.

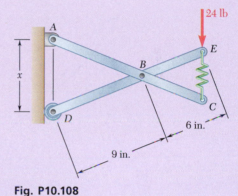

Fig. P10.108

10.109 Solve Prob. 10.108 assuming that the 24-lb load is applied at C instead of E.

10.110 Two uniform rods each with a mass m and length l are attached to gears as shown. For the range $0 \le \theta \le 180°$, determine the positions of equilibrium of the system, and state in each case whether the equilibrium is stable, unstable, or neutral.

10.111 A homogeneous hemisphere with a radius r is placed on an incline as shown. Assuming that friction is sufficient to prevent slipping between the hemisphere and the incline, determine the angle θ corresponding to equilibrium when $\beta = 10°$.

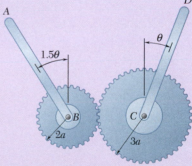

Fig. P10.110

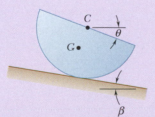

Fig. P10.111 and P10.112

10.112 A homogeneous hemisphere with a radius r is placed on an incline as shown. Assuming that friction is sufficient to prevent slipping between the hemisphere and the incline, determine (a) the largest angle β for which a position of equilibrium exists, (b) the angle θ corresponding to equilibrium when the angle β is equal to half the value found in part a.

Fundamentals of Engineering Examination

Engineers are required to be licensed when their work directly affects the public health, safety, and welfare. The intent is to ensure that engineers have met minimum qualifications involving competence, ability, experience, and character. The licensing process involves an initial exam, called the *Fundamentals of Engineering Examination;* professional experience; and a second exam, called the *Principles and Practice of Engineering.* Those who successfully complete these requirements are licensed as a *Professional Engineer.* The exams are developed under the auspices of the *National Council of Examiners for Engineering and Surveying.*

The first exam, the *Fundamentals of Engineering Examination,* can be taken just before or after graduation from a four-year accredited engineering program. The exam stresses subject material in a typical undergraduate engineering program, including statics. The topics included in the exam cover much of the material in this book. The following is a list of the main topic areas, with references to the appropriate sections in this book. Also included are problems that can be solved to review this material.

Concurrent Force Systems (2.1–2.2; 2.4)
Problems: 2.31, 2.35, 2.36, 2.37, 2.77, 2.83, 2.92, 2.94, 2.97

Vector Forces (3.1–3.2)
Problems: 3.17, 3.18, 3.26, 3.33, 3.37, 3.39

Equilibrium in Two Dimensions (2.3; 4.1–4.2)
Problems: 4.1, 4.13, 4.14, 4.17, 4.31, 4.33, 4.67, 4.77

Equilibrium in Three Dimensions (2.5; 4.3)
Problems: 4.99, 4.101, 4.103, 4.108, 4.115, 4.117, 4.127, 4.129, 4.135

Centroids of Areas and Volumes (5.1–5.2; 5.4)
Problems: 5.9, 5.16, 5.30, 5.35, 5.41, 5.55, 5.62, 5.96, 5.102, 5.103, 5.125

Analysis of Trusses (6.1–6.2)
Problems: 6.3, 6.4, 6.32, 6.43, 6.44, 6.53

Equilibrium of Two-Dimensional Frames (6.3)
Problems: 6.75, 6.81, 6.85, 6.93, 6.94

Shear and Bending Moment (7.1–7.3)
Problems: 7.22, 7.30, 7.36, 7.41, 7.45, 7.49, 7.71, 7.79

Friction (8.1–8.2; 8.4)
Problems: 8.11, 8.18, 8.19, 8.30, 8.49, 8.52, 8.103, 8.104, 8.105

Moments of Inertia (9.1–9.4)
Problems: 9.6, 9.31, 9.32, 9.33, 9.72, 9.74, 9.80, 9.83, 9.98, 9.103

Answers to Problems

2.1 1391 N ∡ 47.8°.

2.2 906 lb ∡ 26.6°.

2.4 8.03 kips ▷ 3.8°.

2.5 (*a*) 101.4 N. (*b*) 196.6 N.

2.6 (*a*) 853 lb. (*b*) 567 lb.

2.8 (*a*) T_{AC} = 2.60 kN. (*b*) R = 4.26 kN.

2.9 (*a*) $\mathbf{T}_{AC}$ = 2.66 kN ⤸ 34.3°.

2.10 (*a*) 37.1°. (*b*) 73.2 N.

2.11 (*a*) 392 lb. (*b*) 346 lb.

2.13 (*a*) 368 lb →. (*b*) 213 lb.

2.14 (*a*) 21.1 N ↓. (*b*) 45.3 N.

2.15 414 lb ⤸ 72.0°.

2.16 1391 N ∡ 47.8°.

2.17 8.03 kips ▷ 3.8°.

2.19 104.4 N ⤡ 86.7°.

2.21 (29 lb) 21.0 lb, 20.0 lb; (50 lb)−14.00 lb, 48.0 lb; (51 lb) 24.0 lb, −45.0 lb.

2.23 (80 N) 61.3 N, 51.4 N; (120 N) 41.0 N, 112.8 N; (150 N)−122.9 N, 86.0 N.

2.24 (40 lb) 20.0 lb, −34.6 lb; (50 lb) −38.3 lb, −32.1 lb; (60 lb) 54.4 lb, 25.4 lb.

2.26 (*a*) 523 lb. (*b*) 428 lb.

2.27 (*a*) 621 N. (*b*) 160.8 N.

2.28 (*a*) 610 lb. (*b*) 500 lb.

2.29 (*a*) 2190 N. (*b*) 2060 N.

2.31 38.6 lb ∡ 36.6°.

2.32 251 N ⤡ 85.3°.

2.34 654 N ⤸ 21.5°.

2.35 309 N ▷ 86.6°.

2.36 474 N ⤸ 32.5°.

2.37 203 lb ∡ 8.46°.

2.39 (*a*) 21.7°. (*b*) 229 N.

2.40 (*a*) 26.5 N. (*b*) 623 N.

2.42 (*a*) 56.3°. (*b*) 204 lb.

2.43 (*a*) 352 lb. (*b*) 261 lb.

2.44 (*a*) 5.22 kN. (*b*) 3.45 kN

2.46 (*a*) 305 N. (*b*) 514 N.

2.48 (*a*) 1244 lb. (*b*) 115.4 lb.

2.49 T_{CA} = 134.6 N; T_{CB} = 110.4 N.

2.50 179.3 N < *P* < 669 N.

2.51 F_A = 1303 lb; F_B = 420 lb.

2.53 F_C = 6.40 kN; F_D = 4.80 kN.

2.54 F_B = 15.00 kN; F_C = 8.00 kN.

2.55 (*a*) T_{ACB} = 269 lb. (*b*) T_{CD} = 37.0 lb.

2.57 (*a*) α = 35.0°; T_{AC} = 4.91 kN; T_{BC} = 3.44 kN, (*b*) α = 55.0°; T_{AC} = T_{BC} = 3.66 kN.

2.58 (*a*) 784 N. (*b*) α = 71.0°.

2.59 (*a*) α = 5.00°. (*b*) 104.6 lb.

2.61 1.250 m.

2.62 75.6 mm.

2.63 (*a*) 10.98 lb. (*b*) 30.0 lb.

2.65 27.4° ≤ α ≤ 222.6°.

2.67 (*a*) 300 lb. (*b*) 300 lb. (*c*) 200 lb. (*d*) 200 lb. (*e*) 150.0 lb.

2.68 (*a*) 200 lb. (*b*) 150.0 lb.

2.69 (*a*) 1293 N. (*b*) 2220 N.

2.71 (*a*) 220 N, 544 N, 126.8 N. (*b*) 68.5°, 25.0°, 77.8°.

2.72 (*a*) −237 N, 258 N, 282 N. (*b*) 121.8°, 55.0°, 51.1°.

2.73 (*a*) −175.8 N, −257 N, 251 N. (*b*) 116.1°, 130.0°, 51.1°.

2.74 (*a*) 350 N, −169.0 N, 93.8 N. (*b*) 28.9°, 115.0°, 76.4°.

2.75 (*a*) −20.5 lb, 43.3 lb, −14.33 lb. (*b*) 114.2°, 30.0°, 106.7°.

2.77 (*a*) −1861 lb, 3360 lb, 677 lb. (*b*) 118.5°, 30.5°, 80.0°.

2.79 (*a*) 770 N; 71.8°; 110.5°; 28.0°.

2.81 (*a*) 140.3°. (*b*) F_x = 79.9 lb, F_z = 120.1 lb; F = 226 lb.

2.82 (*a*) 118.2°. (*b*) F_x = 36.0 lb, F_y = −90.0 lb; F = 110.0 lb.

2.84 (*a*) F_x = 507 N, F_y = 919 N, F_z = 582 N. (*b*) 61.0°.

2.85 240 N; −255 N; 160.0 N.

2.87 −1.260 kips; 1.213 kips; 0.970 kips.

2.88 −0.820 kips; 0.978 kips; −0.789 kips.

2.89 192.0 N; 288 N; −216 N.

2.91 515 N; θ_x = 70.2°; θ_y = 27.6°; θ_z = 71.5°

2.92 515 N; θ_x = 79.8°; θ_y = 33.4°; θ_z = 58.6°.

2.94 913 lb; θ_x = 50.6°; θ_y = 117.6°; θ_z = 51.8°.

2.95 748 N; θ_x = 120.1°; θ_y = 52.5°; θ_z = 128.0°.

2.96 T_{AB} = 490 N; T_{AD} = 515 N.

2.97 130.0 lb.

2.99 13.98 kN.

2.101 926 N ↑.

2.103 T_{DA} = 14.42 lb; T_{DB} = T_{DC} = 13.00 lb.

2.104 T_{DA} = 14.42 lb; T_{DB} = T_{DC} = 13.27 lb.

2.106 T_{AB} = 571 lb; T_{AC} = 830 lb; T_{AD} = 528 lb.

2.107 960 N.

2.108 0 ≤ *Q* < 300 N.

2.109 845 N.

2.110 768 N.

2.112 2000 lb.

2.113 T_{AB} = 30.8 lb; T_{AC} = 62.5 lb.

2.115 T_{AB} = 510 N; T_{AC} = 56.2 N; T_{AD} = 536 N.

2.116 T_{AB} = 1340 N; T_{AC} = 1025 N; T_{AD} = 915 N.

2.117 T_{AB} = 1431 N; T_{AC} = 1560 N; T_{AD} = 183.0 N.

2.118 T_{AB} = 1249 N; T_{AC} = 490 N; T_{AD} = 1647 N.

2.119 T_{AB} = 974 lb; T_{AC} = 531 lb; T_{AD} = 533 lb.

2.121 378 N.

2.123 T_{BAC} = 76.7 lb; T_{AD} = 26.9 lb; T_{AE} = 49.2 lb.

2.124 (*a*) 305 lb. (*b*) T_{BAC} = 117.0 lb; T_{AD} = 40.9 lb.

2.125 (*a*) 1155 N. (*b*) 1012 N.

2.127 21.8 kN ⤸ 73.4°.

2.128 (102 lb) −48.0 lb, 90.0 lb; (106 lb) 56.0 lb, 90.0 lb; (200 lb) −160.0 lb, −120.0 lb

2.130 (*a*) 172.7 lb. (*b*) 231 lb.

2.131 (*a*) 312 N. (*b*) 144.0 N.

2.133 (*a*) 56.4 lb; −103.9 lb; −20.5 lb. (*b*) 62.0°, 150.0°, 99.8°.

2.135 940 N; 65.7°, 28.2°, 76.4°.

2.136 *P* = 131.2 N; *Q* = 29.6 N.

2.137 (*a*) 125.0 lb. (*b*) 45.0 lb.

CHAPTER 3

3.1 (a) 196.2 N·m ↓. (b) 199.0 N ↘ 59.5°.

3.2 (a) 196.2 N·m ↓. (b) 321 N ⬈ 35.0°. (c) 231 N ↑ at point D.

3.4 (a) 41.7 N·m ↑. (b) 147.4 N ∡ 45.0°.

3.5 (a) 41.7 N·m ↑. (b) 334 N. (c) 176.8 N ∡ 58.0°.

3.6 115.7 lb·in.

3.7 115.7 lb·in.

3.9 (a) 292 N·m ↓. (b) 292 N·m ↓.

3.11 116.2 lb·ft ↑.

3.12 128.2 lb·ft ↑.

3.13 140.0 N·m ↑.

3.17 (a) $\lambda = -0.677\mathbf{i} - 0.369\mathbf{j} - 0.636\mathbf{k}$.
(b) $\lambda = -0.0514\mathbf{i} + 0.566\mathbf{j} + 0.823\mathbf{k}$.

3.18 1.184 m.

3.20 (a) $9\mathbf{i} + 22\mathbf{j} + 21\mathbf{k}$. (b) $22\mathbf{i} + 11\mathbf{k}$. (c) 0.

3.22 $(2400 \text{ lb·ft})\mathbf{j} + (1440 \text{ lb·ft})\mathbf{k}$.

3.23 $(7.50 \text{ N·m})\mathbf{i} - (6.00 \text{ N·m})\mathbf{j} - (10.39 \text{ N·m})\mathbf{k}$.

3.25 $(-25.4 \text{ lb·ft})\mathbf{i} - (12.60 \text{ lb·ft})\mathbf{j} - (12.60 \text{ lb·ft})\mathbf{k}$.

3.26 $(1200 \text{ N·m})\mathbf{i} - (1500 \text{ N·m})\mathbf{j} - (900 \text{ N·m})\mathbf{k}$.

3.27 7.37 ft.

3.28 100.8 mm.

3.29 144.8 mm.

3.30 5.17 ft.

3.32 2.36 m.

3.33 1.491 m.

3.35 $\mathbf{P·Q} = -5$; $\mathbf{P·S} = +5$; $\mathbf{Q·S} = -38$.

3.37 77.9°.

3.39 (a) 59.0°. (b) 144.0 lb.

3.40 (a) 70.5°. (b) 60.0 lb

3.41 (a) 52.9°. (b) 326 N.

3.43 26.8°.

3.44 33.3°.

3.45 (a) 67.0. (b) 111.0.

3.46 2.

3.47 $M_x = 78.9 \text{ kN·m}$, $M_y = 13.15 \text{ kN·m}$, $M_z = -9.86 \text{ kN·m}$.

3.48 3.04 kN.

3.49 $\phi = 24.6°$; $d = 34.6$ in.

3.51 1.252 m.

3.52 1.256 m.

3.53 283 lb.

3.55 +207 lb·ft.

3.57 −90.0 N·m.

3.58 −111.0 N·m.

3.59 +2.28 N·m.

3.60 −9.50 N·m.

3.61 $a P/\sqrt{2}$.

3.64 13.06 in.

3.65 12.69 in.

3.67 0.249 m.

3.68 0.1198 m.

3.70 (a) 7.33 N·m ↑. (b) 91.6 mm.

3.71 6.19 N·m ↓.

3.73 1.125 in.

3.74 (a) 26.7 N. (b) 50.0 N. (c) 23.5 N.

3.76 $M = 604$ lb·in.; $\theta_x = 72.8°$, $\theta_y = 27.3°$, $\theta_z = 110.5°$.

3.77 $M = 1170$ lb·in.; $\theta_x = 81.2°$, $\theta_y = 13.70°$, $\theta_z = 100.4°$.

3.78 $M = 3.22$ N·m; $\theta_x = 90.0°$, $\theta_y = 53.1°$, $\theta_z = 36.9°$.

3.79 $M = 2.72$ N·m; $\theta_x = 134.9°$, $\theta_y = 58.0°$; $\theta_z = 61.9°$.

3.80 $M = 2150$ lb·ft; $\theta_x = 113.0°$, $\theta_y = 92.7°$, $\theta_z = 23.2°$.

3.82 (a) $\mathbf{F}_A = 560$ lb ↖ 20.0°; $\mathbf{M}_A = 7720$ lb·ft ↓.
(b) $\mathbf{F}_B = 560$ lb ↖ 20.0°; $\mathbf{M}_B = 4290$ lb·ft ↓.

3.83 $\mathbf{F}_A = 389$ N ↖ 60.0°; $\mathbf{F}_C = 651$ N ↖ 60.0°.

3.84 (a) $\mathbf{F} = 30.0$ lb ↓; $\mathbf{M} = 150.0$ lb.in. ↑. (b) $\mathbf{B} = 50.0$ lb ← ; $\mathbf{C} = 50.0$ lb →.

3.86 $\mathbf{F}_A = 168.0$ N ⬈ 50.0°; $\mathbf{F}_C = 192.0$ N ⬈ 50.0°.

3.87 $\mathbf{F} = 900$ N ↓; $x = 50.0$ mm.

3.89 (a) $\mathbf{F} = 48.0$ lb ∡ 65.0°; $\mathbf{M} = 490$ lb.in. ↓.
(b) $\mathbf{F} = 48.0$ lb ∡ 65.0° applied 17.78 in. to the left of B.

3.90 (a) 48.0 N intersecting line AB 144.0 mm to the right of A.
(b) 77.7° or −15.72°.

3.91 $(0.227 \text{ lb})\mathbf{i} + (0.1057 \text{ lb})\mathbf{k}$; 63.6 in. to the right of B.

3.93 $\mathbf{F} = -(250 \text{ kN})\mathbf{j}$; $\mathbf{M} = (15.00 \text{ kN·m})\mathbf{i} + (7.50 \text{ kN·m})\mathbf{k}$.

3.95 $\mathbf{F} = -(122.9 \text{ N})\mathbf{j} - (86.0 \text{ N})\mathbf{k}$; $\mathbf{M} = (22.6 \text{ N·m})\mathbf{i} + (15.49 \text{ N·m})\mathbf{j} - (22.1 \text{ N·m})\mathbf{k}$.

3.96 $\mathbf{F} = (5.00 \text{ N})\mathbf{i} + (150.0 \text{ N})\mathbf{j} - (90.0 \text{ N})\mathbf{k}$; $\mathbf{M} = (77.4 \text{ N·m})\mathbf{i} + (61.5 \text{ N·m})\mathbf{j} + (106.8 \text{ N·m})\mathbf{k}$.

3.97 $\mathbf{F} = (36.0 \text{ lb})\mathbf{i} - (28.0 \text{ lb})\mathbf{j} - (6.00 \text{ lb})\mathbf{k}$; $\mathbf{M} = -(157.0 \text{ lb·ft})\mathbf{i} + (22.5 \text{ lb·ft})\mathbf{j} - (240 \text{ lb·ft})\mathbf{k}$.

3.98 $\mathbf{F} = -(28.5 \text{ N})\mathbf{i} + (106.3 \text{ N})\mathbf{k}$; $\mathbf{M} = (12.35 \text{ N·m})\mathbf{i} - (19.16 \text{ N·m})\mathbf{j} - (5.13 \text{ N·m})\mathbf{k}$.

3.99 $\mathbf{F} = -(128.0 \text{ lb})\mathbf{i} - (256 \text{ lb})\mathbf{j} + (32.0 \text{ lb})\mathbf{k}$; $\mathbf{M} = (4.10 \text{ kip·ft})\mathbf{i} + (16.38 \text{ kip·ft})\mathbf{k}$.

3.101 (a) Loading a: 500 N ↓; 1000 N·m ↓.
Loading b: 500 N ↑; 500 N·m ↑.
Loading c: 500 N ↓; 500 N·m ↓.
Loading d: 500 N ↓; 1100 N·m ↓.
Loading e: 500 N ↓; 1000 N·m ↓.
Loading f: 500 N ↓; 200 N·m ↓.
Loading g: 500 N ↓; 2300 N·m ↑.
Loading h: 500 N ↓; 600 N·m ↑.
(b) Loadings a and e are equivalent.

3.102 Equivalent to case f of problem 3.101.

3.104 Equivalent force-couple system at D.

3.105 (a) 2.00 ft to the right of C. (b) 2.31 ft to the right of C.

3.106 (a) 39.6 in. to the right of D. (b) 33.1 in.

3.108 44.7 lb ↘ 26.6°; 10.61 in. to the left of C and 5.30 in. below C.

3.110 (a) 224 N ↙ 63.4°. (b) 130.0 mm to the left of B and 260 mm below B.

3.111 (a) 269 N ↙ 68.2°. (b) 120.0 mm to the left of B and 300 mm below B.

3.113 773 lb ⬈ 79.0°; 9.54 ft to the right of A.

3.114 (a) 29.9 lb ↘ 23.0°. (b) AB: 10.30 in. to the left of B; BC: 4.36 in. below B.

3.115 (a) 60.2 lb·in. ↑. (b) 200 lb·in.↑. (C) 20.0 lb·in. ↓.

3.116 (a) 0.365 m above G. (b) 0.227 m to the right of G.

3.117 (a) 0.299 m above G. (b) 0.259 m to the right of G.

3.118 (a) $\mathbf{R} = F ⬈ \tan^{-1}(a^2/2bx)$; $\mathbf{M} = 2Fb^2(x - x^3/a^2)/\sqrt{a^4 + 4b^2x^2}$ ↑.
(b) 0.369 m.

3.119 $\mathbf{R} = -(300 \text{ N})\mathbf{i} - (240 \text{ N})\mathbf{j} + (25.0 \text{ N})\mathbf{k}$;
$\mathbf{M} = -(3.00 \text{ N·m})\mathbf{i} + (13.50 \text{ N·m})\mathbf{j} + (9.00 \text{ N·m})\mathbf{k}$.

3.120 $\mathbf{R} = (420 \text{ N})\mathbf{j} - (339 \text{ N})\mathbf{k}$; $\mathbf{M} = (1.125 \text{ N·m})\mathbf{i} + (163.9 \text{ N·m})\mathbf{j} - (109.9 \text{ N·m})\mathbf{k}$.

3.122 (a) 60.0°. (b) $(20.0 \text{ lb})\mathbf{i} - (34.6 \text{ lb})\mathbf{j}$; $(520 \text{ lb·in.})\mathbf{i}$.

3.124 $\mathbf{R} = -(420 \text{ N})\mathbf{i} - (50.0 \text{ N})\mathbf{j} - (250 \text{ N})\mathbf{k}$; $\mathbf{M} = (30.8 \text{ N·m})\mathbf{j} - (22.0 \text{ N·m})\mathbf{k}$.

3.125 (a) $\mathbf{B} = -(75.0 \text{ N})\mathbf{k}$, $\mathbf{C} = -(25.0 \text{ N})\mathbf{i} + (37.5 \text{ N})\mathbf{k}$.
(b) $R_y = 0$, $R_z = -37.5$ N. (c) when the slot is vertical.

3.126 $\mathbf{A} = (1.600 \text{ lb})\mathbf{i} - (36.0 \text{ lb})\mathbf{j} + (2.00 \text{ lb})\mathbf{k}$,
$\mathbf{B} = -(9.60 \text{ lb})\mathbf{i} + (36.0 \text{ lb})\mathbf{j} + (2.00 \text{ lb})\mathbf{k}$.

3.127 1035 N; 2.57 m from OG and 3.05 m from OE.

3.128 2.32 m from OG and 1.165 m from OE.

3.129 405 lb; 12.60 ft to the right of AB and 2.94 ft below BC.

3.130 $a = 0.722$ ft; $b = 20.6$ ft.

3.133 (a) $P\sqrt{3}$; $\theta_x = \theta_y = \theta_z = 54.7°$. (b) −a
(c) Axis of the wrench is diagonal OA.

3.134 (a) P; $\theta_x = 90.0°$, $\theta_y = 90.0°$, $\theta_z = 0$. (b) $5a/2$.
(c) Axis of the wrench is parallel to the z-axis at $x = a$, $y = -a$.

3.136 (a) $-(21.0\ \text{lb})\ \mathbf{j}$. (b) 0.571 in. (c) At $x = 0$, $z = 1.667$ in; and is parallel to the y axis.

3.137 (a) $-(84.0\ \text{N})\mathbf{j} - (80.0\ \text{N})\mathbf{k}$. (b) 0.477 m. (c) $x = 0.526$ m, $y = 0$, $z = -0.1857$ m.

3.140 (a) $3P\ (2\mathbf{i} - 20\mathbf{j} - \mathbf{k})/25$. (b) $-0.0988a$.
(c) $x = 2.00a$, $y = 0$, $z = -1.990a$.

3.141 $\mathbf{R} = (20.0\ \text{N})\mathbf{i} + (30.0\ \text{N})\mathbf{j} - (10.00\ \text{N})\mathbf{k}$; $y = -0.540$ m, $z = -0.420$ m.

3.143 $\mathbf{F}_A = (M/b)\mathbf{i} + R\ [1 + (a/b)]\mathbf{k}$; $\mathbf{F}_B = -(M/b)\mathbf{i} - (a\ R/b)\mathbf{k}$.

3.147 (a) 20.5 N·m ↘. (b) 68.4 mm.

3.148 760 N·m ↘.

3.150 43.6°.

3.151 23.0 N·m.

3.153 $\mathbf{M} = 4.50$ N·m; $\theta_x = 90.0°$, $\theta_y = 177.1°$, $\theta_z = 87.1°$.

3.154 $\mathbf{F} = 260$ lb ⬈67.4°; $\mathbf{M}_c = 200$ lb.in ↓.

3.156 (a) 135.0 mm. (b) $\mathbf{F}_2 = (42.0\ \text{N})\mathbf{i} + (42.0\ \text{N})\mathbf{j} - (49.0\ \text{N})\mathbf{k}$;
$\mathbf{M}_2 = -(25.9\ \text{N·m})\mathbf{i} + (21.2\ \text{N·m})\mathbf{j}$

3.158 (a) $\mathbf{B} = (2.50\ \text{lb})\mathbf{i}$, $\mathbf{C} = (0.1000\ \text{lb})\mathbf{i} - (2.47\ \text{lb})\mathbf{j} - (0.700\ \text{lb})\mathbf{k}$.
(b) $\mathbf{R}_y = -2.47$ lb; $\mathbf{M}_x = 1.360$ lb·ft.

CHAPTER 4

4.1 42.0 N ↑.

4.2 0.264 m.

4.4 (a) 245 lb ↑. (b) 140.0 lb.

4.5 (a) 34.0 kN ↑. (b) 4.96 kN ↑.

4.6 (a) 81.1 kN. (b) 134.1 kN ↑.

4.7 (a) $\mathbf{A} = 20.0$ lb ↓; $\mathbf{B} = 150.0$ lb ↑. (b) $\mathbf{A} = 10.00$ lb ↓; $\mathbf{B} = 140.0$ lb ↑.

4.9 1.250 kN $\le Q \le 27.5$ kN.

4.12 6.00 kips $\le P \le 27.0$ kips.

4.13 150.0 mm $\le d \le 400$ mm.

4.14 2.00 in. $\le a \le 10.00$ in.

4.15 (a) 600 N. (b) 1253 N ∡69.8°.

4.17 (a) 80.8 lb ↓. (b) 216 lb ∡22.0°.

4.18 232 lb.

4.19 (a) 2.00 kN. (b) 2.32 kN ∡46.4°.

4.22 (a) 400 N. (b) 458 N ∡49.1°.

4.23 (a) $\mathbf{A} = 44.7$ lb ↘ 26.6°; $\mathbf{B} = 30.0$ lb ↑.
(b) $\mathbf{A} = 30.2$ lb ↘ 41.4°; $\mathbf{B} = 34.6$ lb ↘ 60.0°.

4.24 (a) $\mathbf{A} = 20.0$ lb ↑; $\mathbf{B} = 50.0$ lb ↘ 36.9°.
(b) $\mathbf{A} = 23.1$ lb ∡60.0°; $\mathbf{B} = 59.6$ lb ↘ 30.2°.

4.25 (a) 190.9 N. (b) 142.3 N ∡18.43°.

4.26 (a) 324 N. (b) 270 N →.

4.28 (a) $\mathbf{A} = 225$ N ↑; $\mathbf{C} = 641$ N ⬈ 20.6°.
(b) $\mathbf{A} = 365$ N ∡60.0°; $\mathbf{C} = 884$ N ⬈ 22.0°.

4.31 $T = 2P/3$; $\mathbf{C} = 0.577P$ →.

4.32 $T = 0.586P$; $\mathbf{C} = 0.414P$ →.

4.33 (a) 117.0 lb. (b) 129.8 lb ↖ 56.3°.

4.34 (a) 195.0 lb. (b) 225 lb ↖ 45.0°.

4.35 (a) 1432 N. (b) 1100 N ↑. (c) 1400 N ←.

4.36 $T_{BE} = 196.2$ N; $\mathbf{A} = 73.6$ N →; $\mathbf{D} = 73.6$ N ←.

4.39 (a) 600 N. (b) $\mathbf{A} = 4.00$ kN ←; $\mathbf{B} = 4.00$ kN →.

4.40 (a) 105.1 N. (b) $\mathbf{A} = 147.2$ N ↑; $\mathbf{B} = 105.1$ N ←.

4.41 (a) $\mathbf{A} = 20.2$ lb ↑; $\mathbf{B} = 30.0$ lb ↘ 60.0°. (b) 16.21 lb ↓.

4.42 5.44 lb $\le P \le 17.23$ lb.

4.43 (a) $\mathbf{E} = 8.80$ kips ↑; $\mathbf{M}_E = 36.0$ kip·ft ↓.
(b) $\mathbf{E} = 4.80$ kips ↑; $\mathbf{M}_E = 51.0$ kip·ft ↓.

4.45 $T_{max} = 2240$ N; $T_{min} = 1522$ N.

4.46 $\mathbf{C} = 1951$ N ↘88.5°; $\mathbf{M}_C = 75.0$ N·m ↓.

4.47 1.232 kN $\le T \le 1.774$ kN.

4.48 (a) $\mathbf{D} = 20.0$ lb ↓; $\mathbf{M}_D = 20.0$ lb·ft ↘.
(b) $\mathbf{D} = 10.00$ lb ↓; $\mathbf{M}_D = 30.0$ lb·ft ↓.

4.50 (a) $\mathbf{A} = 78.5$ N ↑; $\mathbf{M}_A = 125.6$ N·m ↘.
(b) $\mathbf{A} = 111.0$ N ↑; $\mathbf{M}_A = 125.6$ N·m ↘.
(c) $\mathbf{A} = 157.0$ N ↑; $\mathbf{M}_A = 251$ N·m ↘.

4.51 $\theta = \sin^{-1}(2M\cot\alpha/Wl)$.

4.52 $\theta = \tan^{-1}(Q/3P)$.

4.53 (a) $T = (W/2)/(1 - \tan\theta)$. (b) 39.8°.

4.54 (a) $\theta = 2\cos^{-1}\left[\frac{1}{4}\left(\frac{W}{P} \pm \sqrt{\frac{W^2}{P^2} + 8}\right)\right]$. (b) 65.1°.

4.55 (a) $\theta = 2\sin^{-1}(W/2P)$. (b) 29.0°.

4.57 141.1°.

4.59 (1) completely constrained; determinate; $\mathbf{A} = \mathbf{C} = 196.2$ N ↑.
(2) completely constrained; determinate; $\mathbf{B} = 0$,
$\mathbf{C} = \mathbf{D} = 196.2$ N ↑.
(3) completely constrained; indeterminate; $\mathbf{A}_x = 294$ N →;
$\mathbf{D}_x = 294$ ←.
(4) improperly constrained; indeterminate; no equilibrium.
(5) partially constrained; determinate; equilibrium;
$\mathbf{C} = \mathbf{D} = 196.2$ N ↑.
(6) completely constrained; determinate;
$\mathbf{B} = 294$ N →, $\mathbf{D} = 491$ N ↘ 53.1°.
(7) partially constrained; no equilibrium.
(8) completely constrained; indeterminate; $\mathbf{B} = 196.2$ N ↑,
$\mathbf{D}_y = 196.2$ N ↑.

4.61 $T = 289$ lb; $\mathbf{A} = 577$ lb ∡ 60.0°.

4.62 $\mathbf{A} = 400$ N ↑; $\mathbf{B} = 500$ N ↖ 53.1°.

4.63 $a \ge 138.6$ mm.

4.65 $\mathbf{B} = 501$ N ↘ 56.3°; $\mathbf{C} = 324$ N ↖ 31.0°.

4.66 $\mathbf{A} = 82.5$ lb ∡14.04°; $T = 100.0$ lb.

4.67 $\mathbf{B} = 888$ N ↖41.3°; $\mathbf{D} = 943$ N ↘ 45.0°.

4.69 (a) 499 N. (b) 457 N ↘ 26.6°.

4.71 (a) 5.63 kips. (b) 4.52 kips ⬈4.76°.

4.72 (a) 24.9 lb ⬈30.0°. (b) 15.34 lb ∡30.0°.

4.73 $\mathbf{A} = 778$ N ↓; $\mathbf{C} = 1012$ N ↘ 77.9°.

4.75 $\mathbf{A} = 170.0$ N ↘ 33.9°; $\mathbf{C} = 160.0$ N ∡28.1°.

4.77 $T = 100.0$ lb; $\mathbf{B} = 111.1$ lb ↖ 30.3°.

4.78 (a) 400 N. (b) 458 N ∡49.1°.

4.79 (a) $2P$ ↘ 60.0°. (b) $1.239P$ ↖ 36.2°.

4.80 (a) $1.55\ P$ ↘ 30.0°. (b) $1.086\ P$ ∡22.9°.

4.81 $\mathbf{A} = 163.1$ N ↖74.1°; $\mathbf{B} = 258$ N ↘ 65.0°.

4.83 60.0 mm.

4.84 $\tan\theta = 2\tan\beta$.

4.85 (a) 49.1°. (b) $\mathbf{A} = 45.3$ N ←; $\mathbf{B} = 90.6$ N ∡60.0°.

4.86 32.5°.

4.88 (a) 225 mm. (b) 23.1 N. (c) 12.21 N →.

4.90 (a) 59.4°. (b) $\mathbf{A} = 8.45$ lb →; $\mathbf{B} = 13.09$ lb ↘ 49.8°.

4.91 $\mathbf{A} = (120.0\ \text{N})\mathbf{j} + (133.3\ \text{N})\mathbf{k}$; $\mathbf{D} = (60.0\ \text{N})\mathbf{j} + (166.7\ \text{N})\mathbf{k}$.

4.93 (a) 96.0 lb. (b) $\mathbf{A} = (2.40\ \text{lb})\mathbf{j}$; $\mathbf{B} = (214\ \text{lb})\mathbf{j}$.

4.94 $\mathbf{A} = (22.9\ \text{lb})\mathbf{i} + (8.50\ \text{lb})\mathbf{j}$; $\mathbf{B} = (22.9\ \text{lb})\mathbf{i} + (25.5\ \text{lb})\mathbf{j}$;
$\mathbf{C} = -(45.8\ \text{lb})\mathbf{i}$.

4.95 (a) 78.5 N. (b) $\mathbf{A} = -(27.5\ \text{N})\mathbf{i} + (58.9\ \text{N})\mathbf{j}$; $\mathbf{B} = (106.0\ \text{N})\mathbf{i} + (58.9\ \text{N})\mathbf{j}$.

4.97 $T_A = 21.0$ lb; $T_B = T_C = 17.50$ lb.

4.99 (a) 121.9 N. (b) -46.2 N. (c) 100.9 N.

4.100 (a) 95.6 N. (b) -7.36 N. (c) 88.3 N.

4.101 $T_A = 23.5$ N; $T_C = 11.77$ N; $T_D = 105.9$ N.

4.102 (a) 0.480 in. (b) $T_A = 23.5$ N; $T_C = 0$; $T_D = 117.7$ N.

4.103 (a) $T_A = 6.00$ lb; $T_B = T_C = 9.00$ lb. (b) 15.00 in.

4.105 $T_{BD} = 1100$ lb; $T_{BE} = 1100$ lb; $\mathbf{A} = (1200\ \text{lb})\mathbf{i} - (560\ \text{lb})\mathbf{j}$.

4.106 $T_{BD} = 780$ N; $T_{BE} = 390$ N; $\mathbf{A} = -(195.0\ \text{N})\mathbf{i} + (1170\ \text{N})\mathbf{j} + (130.0\ \text{N})\mathbf{k}$.

4.107 $T_{BD} = 525$ N; $T_{BE} = 105.0$ N; $\mathbf{A} = -(105.0\ \text{N})\mathbf{i} + (840\ \text{N})\mathbf{j} + (140.0\ \text{N})\mathbf{k}$.

4.108 $T_{AD} = 2.60$ kN; $T_{AE} = 2.80$ kN; $\mathbf{C} = (1.800$ kN$)\mathbf{j} + (4.80$ kN$)\mathbf{k}$.

4.109 $T_{AD} = 5.20$ kN; $T_{AE} = 5.60$ kN; $\mathbf{C} = (9.60$ kN$)\mathbf{k}$.

4.110 $T_{BD} = T_{BE} = 176.8$ lb; $\mathbf{C} = -(50.0$ lb$)\mathbf{j} + (216.5$ lb$)\mathbf{k}$.

4.113 $F_{CD} = 19.62$ N; $\mathbf{A} = -(19.22$ N$)\mathbf{i} + (45.1$ N$)\mathbf{j}$;
$\mathbf{B} = (49.1$ N$)\mathbf{j}$.

4.115 $\mathbf{A} = -(56.3$ lb$)\mathbf{i}$; $\mathbf{B} = -(56.2$ lb$)\mathbf{i} + (150.0$ lb$)\mathbf{j} - (75.0$ lb$)\mathbf{k}$;
$F_{CE} = 202$ lb.

4.116 (a) 116.6 lb. (b) $\mathbf{A} = -(72.7$ lb$)\mathbf{j} - (38.1$ lb$)\mathbf{k}$;
$\mathbf{B} = (37.5$ lb$)\mathbf{j}$.

4.117 (a) 345 N. (b) $\mathbf{A} = (114.4$ N$)\mathbf{i} + (377$ N$)\mathbf{j} + (141.5$ N$)\mathbf{k}$;
$\mathbf{B} = (113.2$ N$)\mathbf{j} + (185.5$ N$)\mathbf{k}$.

4.119 $F_{CD} = 19.62$ N; $\mathbf{B} = -(19.22$ N$)\mathbf{i} + (94.2$ N$)\mathbf{j}$;
$\mathbf{M}_B = -(40.6$ N·m$)\mathbf{i} - (17.30$ N·m$)\mathbf{j}$.

4.120 $\mathbf{A} = -(112.5$ lb$)\mathbf{i} + (150.0$ lb$)\mathbf{j} - (75.0$ lb$)\mathbf{k}$;
$\mathbf{M}_A = (600$ lb·ft$)\mathbf{i} + (225$ lb·ft$)\mathbf{j}$; $F_{CE} = 202$ lb.

4.121 (a) 5.00 lb. (b) $\mathbf{C} = -(5.00$ lb$)\mathbf{i} + (6.00$ lb$)\mathbf{j} - (5.00$ lb$)\mathbf{k}$;
$\mathbf{M}_C = (8.00$ lb·in.$)\mathbf{j} - (12.00$ lb·in$)\mathbf{k}$.

4.122 $T_{CF} = 200$ N; $T_{DE} = 450$ N; $\mathbf{A} = (160.0$ N$)\mathbf{i} + (270$ N$)\mathbf{k}$;
$\mathbf{M}_A = -(16.20$ N·m$)\mathbf{i}$.

4.123 $T_{BD} = 2.18$ kN; $T_{BE} = 3.96$ kN; $T_{CD} = 1.500$ kN.

4.124 $T_{BD} = 0$; $T_{BE} = 3.96$ kN; $T_{CD} = 3.00$ kN.

4.127 $\mathbf{A} = (120.0$ lb$)\mathbf{j} - (150.0$ lb$)\mathbf{k}$; $\mathbf{B} = (180.0$ lb$)\mathbf{i} + (150.0$ lb$)\mathbf{k}$;
$\mathbf{C} = -(180.0$ lb$)\mathbf{i} + (120.0$ lb$)\mathbf{j}$.

4.128 $\mathbf{A} = (20.0$ lb$)\mathbf{j} + (25.0$ lb$)\mathbf{k}$; $\mathbf{B} = (30.0$ lb$)\mathbf{i} - (25.0$ lb$)\mathbf{k}$;
$\mathbf{C} = -(30.0$ lb$)\mathbf{i} - (20.0$ lb$)\mathbf{j}$.

4.129 $T_{BE} = 975$ N; $T_{CF} = 600$ N; $T_{DG} = 625$ N;
$\mathbf{A} = (2100$ N$)\mathbf{i} + (175.0$ N$)\mathbf{j} - (375$ N$)\mathbf{k}$.

4.131 $T_B = -0.366\,P$; $T_C = 1.219\,P$; $T_D = -0.853\,P$;
$\mathbf{F} = -0.345\,P\mathbf{i} + P\mathbf{j} - 0.862\,P\mathbf{k}$.

4.133 360 N.

4.135 85.3 lb.

4.136 181.7 lb.

4.137 $(45.0$ lb$)\mathbf{j}$.

4.138 343 N.

4.140 (a) $x = 4.00$ ft, $y = 8.00$ ft. (b) 10.73 lb.

4.141 (a) $x = 0$, $y = 16.00$ ft. (b) 11.31 lb.

4.142 (a) 1761 lb↑. (b) 689 lb ↑.

4.143 (a) 150.0 lb. (b) 225 lb ⬚ 32.3°.

4.145 (a) 130.0 N. (b) 224 ⬚ 2.05°.

4.146 $T = 80.0$ N; $\mathbf{A} = 160.0$ N ⬚ 30.0°; $\mathbf{C} = 160.0$ N ⬚ 30.0°.

4.148 $\mathbf{A} = 680$ N ∠ 28.1°; $\mathbf{B} = 600$ N ←.

4.149 $\mathbf{A} = 63.6$ lb ⬚ 45.0°; $\mathbf{C} = 87.5$ lb ⬚ 59.0°.

4.151 $T_A = 5.63$ lb; $T_B = 16.88$ lb; $T_C = 22.5$ lb.

4.153 (a) $\mathbf{A} = 0.745\,P$ ∠ 63.4°; $\mathbf{C} = 0.471\,P$ ⬚ 45.0°.
(b) $\mathbf{A} = 0.812\,P$ ∠ 60.0°; $\mathbf{C} = 0.503\,P$ ⬚ 36.2°.
(c) $\mathbf{A} = 0.448\,P$ ⬚ 60.0°; $\mathbf{C} = 0.652\,P$ ∠ 69.9°.
(d) improperly constrained; no equilibrium.

CHAPTER 5

5.1 $\bar{X} = 42.2$ mm, $\bar{Y} = 24.2$ mm.

5.2 $\bar{X} = 1.045$ in., $\bar{Y} = 3.59$ in.

5.3 $\bar{X} = 2.84$ mm, $\bar{Y} = 24.8$ mm.

5.4 $\bar{X} = 52.0$ mm, $\bar{Y} = 65.0$ mm.

5.5 $\bar{X} = 3.27$ in., $\bar{Y} = 2.82$ in.

5.6 $\bar{X} = -10.00$ mm, $\bar{Y} = 87.5$ mm.

5.9 $\bar{X} = \bar{Y} = 16.75$ mm.

5.10 $\bar{X} = 10.11$ in., $\bar{Y} = 3.88$ in.

5.11 $\bar{X} = 30.0$ mm, $\bar{Y} = 64.8$ mm.

5.13 $\bar{X} = 3.20$ in., $\bar{Y} = 2.00$ in.

5.14 $\bar{X} = 0$, $\bar{Y} = 1.372$ m.

5.16 $\bar{Y} = \left(\dfrac{2}{3}\right)\left(\dfrac{r_2^3 - r_1^3}{r_2^2 - r_1^2}\right)\left(\dfrac{2\cos\alpha}{\pi - 2\alpha}\right)$.

5.17 $\bar{Y} = (r_1 + r_2)(\cos\alpha)/(\pi - 2\alpha)$.

5.19 0.520.

5.20 459 N.

5.21 0.235 in^3 for A_1, -0.235 in^3 for A_2.

5.23 (a) $b(c^2 - y^2)/2$. (b) $y = 0$; $Q_x = bc^2/2$.

5.24 $\bar{X} = 40.9$ mm, $\bar{Y} = 25.3$ mm.

5.26 $\bar{X} = 3.38$ in., $\bar{Y} = 2.93$ in.

5.29 (a) 125.3 N. (b) 137.0 N ∠ 56.7°.

5.30 120.0 mm.

5.31 99.5 mm.

5.32 (a) $0.513a$. (b) $0.691a$.

5.34 $\bar{x} = \frac{2}{3}a$, $\bar{y} = \frac{2}{3}h$.

5.35 $\bar{x} = a/2$, $\bar{y} = 2h/5$.

5.37 $\bar{x} = a(3 - 4\sin\alpha)/6\,(1 - \alpha)$, $\bar{y} = 0$.

5.39 $\bar{x} = 2a/3(4 - \pi)$, $\bar{y} = 2b/3(4 - \pi)$.

5.40 $\bar{x} = a/4$, $\bar{y} = 3b/10$.

5.41 $\bar{x} = 3a/5$, $\bar{y} = 12b/35$.

5.43 $\bar{x} = 17a/130$, $\bar{y} = 11b/26$.

5.44 $\bar{x} = a$, $\bar{y} = 17b/35$.

5.45 $2a\,/5$.

5.46 $-2\sqrt{2}r/3\pi$.

5.48 $\bar{x} = -9.27a$, $\bar{y} = 3.09a$.

5.49 $\bar{x} = L/\pi$, $\bar{y} = \pi a/8$.

5.51 $\bar{x} = \bar{y} = 1.027$ in.

5.52 (a) $V = 401 \times 10^3$ mm^3; $A = 34.1 \times 10^3$ mm^2.
(b) $V = 492 \times 10^3$ mm^3; $A = 41.9 \times 10^3$ mm^2.

5.53 (a) $V = 248$ in^3; $A = 547$ in^2.
(b) $V = 72.3$ in^3; $A = 169.6$ in^2.

5.54 (a) $V = 2.26 \times 10^6$ mm^3; $A = 116.3 \times 10^3$ mm^2.
(b) $V = 1.471 \times 10^6$ mm^3; $A = 116.3 \times 10^3$ mm^2.

5.55 $V = 3470$ mm^3; $A = 2320$ mm^2.

5.58 308 in^2.

5.60 31.9 liters.

5.62 $V = 3.96$ in^3, $W = 1.211$ lb.

5.63 14.52 in^2.

5.64 0.0305 kg.

5.66 (a) $\mathbf{R} = 6000$ N ↓, $\bar{x} = 3.60$ m.
(b) $\mathbf{A} = 6000$ N ↑, $\mathbf{M}_A = 21.6$ kN·m ↱.

5.67 (a) $\mathbf{R} = 7.60$ kN ↓, $\bar{x} = 2.57$ m.
(b) $\mathbf{A} = 4.35$ kN ↑, $\mathbf{B} = 3.25$ kN ↑.

5.69 $\mathbf{A} = 900$ lb↑; $\mathbf{M}_A = 9200$ lb·in. ↱.

5.70 $\mathbf{B} = 1360$ lb↑; $\mathbf{C} = 2360$ lb↑.

5.71 $\mathbf{A} = 105.0$ N ↑; $\mathbf{B} = 270$ N ↑.

5.73 $\mathbf{A} = 3.00$ kN ↑; $\mathbf{M}_A = 12.60$ kN·m ↱.

5.74 (a) 0.536 m. (b) $\mathbf{A} = \mathbf{B} = 761$ N ↑.

5.76 $\mathbf{B} = 3770$ lb↑; $\mathbf{C} = 429$ lb↑.

5.77 (a) 900 lb/ft. (b) 7200 lb↑.

5.78 $w_A = 10.00$ kN/m; $w_B = 50$ kN/m.

5.80 (a) $\mathbf{H} = 10.11$ kips →, $\mathbf{V} = 37.8$ kips↑.
(b) 10.48 ft to the right of A.
(c) $\mathbf{R} = 10.66$ kips ⬚ 18.43°.

5.81 (a) $\mathbf{H} = 44.1$ kN →, $\mathbf{V} = 228$ kN ↑.
(b) 1.159 m to the right of A.
(c) $\mathbf{R} = 59.1$ kN ⬚ 41.6°.

5.82 6.98%.

5.84 12.00 in.

5.85 4.00 in.

5.87 $\mathbf{T} = 6.72$ kN ← ; $\mathbf{A} = 141.2$ kN ←.

5.88 $\mathbf{A} = 1197$ N ⬚ 53.1°; $\mathbf{B} = 1511$ N ⬚ 53.1°.

5.89 3570 N.

5.90 6.00 ft.

5.92 0.683 m.

5.93 0.0711 m.

5.94 208 lb.

5.96 (a) 0.548 L. (b) $2\sqrt{3}$.

5.97 (a) b/10 to the left of base of cone.
(b) 0.1136b to the right of base of cone.

5.98 (a) −0.402 a. (b) h/a = 2/5 or 2/3.

5.99 27.8 mm above base of cone.

5.100 18.28 mm.

5.102 −0.0656 in.

5.103 2.57 in.

5.104 −19.02 mm.

5.106 $\bar{X} = 125.0$ mm, $\bar{Y} = 167.0$ mm, $\bar{Z} = 33.5$ mm.

5.107 $\bar{X} = 0.295$ m, $\bar{Y} = 0.423$ m, $\bar{Z} = 1.703$ m.

5.109 $\bar{X} = \bar{Z} = 4.21$ in., $\bar{Y} = 7.03$ in.

5.110 $\bar{X} = 180.2$ mm, $\bar{Y} = 38.0$ mm, $\bar{Z} = 193.5$ mm.

5.111 $\bar{X} = 17.00$ in., $\bar{Y} = 15.68$ in., $\bar{Z} = 14.16$ in.

5.113 $\bar{X} = 46.5$ mm, $\bar{Y} = 27.2$ mm, $\bar{Z} = 30.0$ mm.

5.114 $\bar{X} = 0.909$ m, $\bar{Y} = 0.1842$ m, $\bar{Z} = 0.884$ m.

5.116 $\bar{X} = 0.410$ m, $\bar{Y} = 0.510$ m, $\bar{Z} = 0.1500$ m.

5.117 $\bar{X} = 0$, $\bar{Y} = 10.05$ in., $\bar{Z} = 5.15$ in.

5.118 $\bar{X} = 61.1$ mm from the end of the handle.

5.119 $\bar{Y} = 0.526$ in. above the base.

5.121 $\bar{Y} = 421$ mm above the floor.

5.122 $(\bar{x}_1) = 21a/88$; $(\bar{x}_2) = 27a/40$.

5.123 $(\bar{x}_1) = 21h/88$; $(\bar{x}_2) = 27h/40$.

5.124 $(\bar{x}_1) = 2h/9$; $(\bar{x}_2) = 2\,h/3$.

5.125 $\bar{x} = 2.34$ m; $\bar{y} = \bar{z} = 0$.

5.128 $\bar{x} = 1.297a$; $\bar{y} = \bar{z} = 0$.

5.129 $\bar{x} = z = 0$; $\bar{y} = 0.374b$.

5.132 (a) $\bar{x} = \bar{z} = 0$, $\bar{y} = -121.9$ mm. (b) $\bar{x} = \bar{z} = 0$,
$\bar{y} = -90.2$ mm.

5.134 $\bar{x} = 0$, $\bar{y} = 5h/16$, $\bar{z} = -b/4$.

5.135 $\bar{x} = a/2$, $\bar{y} = 8h/25$, $\bar{z} = b/2$.

5.136 $V = 688$ ft^3; $\bar{x} = 15.91$ ft.

5.137 $\bar{X} = 5.67$ in., $\bar{Y} = 5.17$ in.

5.138 $\bar{X} = 92.0$ mm, $\bar{Y} = 23.3$ mm.

5.139 (a) 5.09 lb. (b) 9.48 lb ⟋ 57.5°.

5.141 $\bar{x} = 2L/5$, $\bar{y} = 12h/25$.

5.143 **A** = 2860 lb↑; **B** = 740 lb↑.

5.144 $w_{BC} = 2810$ N/m; $w_{DE} = 3150$ N/m.

5.146 $-(2h^2 - 3b^2)/2\ (4h - 3b)$.

5.148 $\bar{X} = \bar{Z} = 0$, $\bar{Y} = 83.3$ mm above the base.

CHAPTER 6

6.1 $F_{AB} = 900$ lb T; $F_{AC} = 780$ lb C; $F_{BC} = 720$ lb T.

6.2 $F_{AB} = 1.700$ kN T; $F_{AC} = 2.00$ kN C; $F_{BC} = 2.50$ kN T.

6.3 $F_{AB} = 720$ lb T; $F_{AC} = 1200$ lb C; $F_{BC} = 780$ lb C.

6.4 $F_{AB} = F_{BC} = 0$; $F_{AD} = F_{CF} = 7.00$ kN C; $F_{BD} = F_{BF} = 34.0$ kN C; $F_{BE} = 8.00$ kN T; $F_{DE} = F_{EF} = 30.0$ kN T.

6.6 $F_{AC} = 80.0$ kN T; $F_{CE} = 45.0$ kN C; $F_{DE} = 51.0$ kN C; $F_{BD} = 51.0$ kN C; $F_{CD} = 48.0$ kN T; F_{BC} 19.00 kN C.

6.8 $F_{AB} = 20.0$ kN T; $F_{AD} = 20.6$ kN C; $F_{BC} = 30.0$ kN T; $F_{BD} = 11.18$ kN C; $F_{CD} = 10.00$ kN T.

6.9 $F_{AB} = F_{DE} = 8.00$ kN C; $F_{AF} = F_{FG} = F_{GH} = F_{EH} = 6.93$ kN T; $F_{BC} = F_{CD} = F_{BG} = F_{DG} = 4.00$ kN C; $F_{BF} = F_{DH} = F_{CG} = 4.00$ kN T.

6.11 $F_{AB} = F_{FH} = 1500$ lb C; $F_{AC} = F_{CE} = F_{EG} = F_{GH} = 1200$ lb T; $F_{BC} = F_{FG} = 0$; $F_{BD} = F_{DF} = 1200$ lb C; $F_{BE} = F_{EF} = 60.0$ lb C; $F_{DE} = 72.0$ lb T.

6.12 $F_{AB} = F_{FH} = 1500$ lb C; $F_{AC} = F_{CE} = F_{EG} = F_{GH} = 1200$ lb T; $F_{BC} = F_{FG} = 0$; $F_{BD} = F_{DF} = 1000$ lb C; $F_{BE} = F_{EF} = 500$ lb C; $F_{DE} = 600$ lb T.

6.13 $F_{AB} = 6.24$ kN C; $F_{AC} = 2.76$ kN T; $F_{BC} = 2.50$ kN C; $F_{BD} = 4.16$ kN C; $F_{CD} = 1.867$ kN T; $F_{CE} = 2.88$ kN T; $F_D = 3.75$ kN C; $F_{DF} = 0$; $F_{EF} = 1.200$ kN C.

6.15 $F_{AB} = F_{FG} = 7.50$ kips C; $F_{AC} = F_{EG} = 4.50$ kips T; $F_{BC} = F_{EF} = 7.50$ kips T; $F_{BD} = F_{DF} = 9.00$ kips C; $F_{CD} = F_{DE} = 0$; $F_{CE} = 9.00$ kips T.

6.17 $F_{AB} = 47.2$ kN C; $F_{AC} = 44.6$ kN T; $F_{BC} = 10.50$ kN C; $F_{BD} = 47.2$ kN C; $F_{CD} = 17.50$ kN T; $F_{CE} = 30.6$ kN T; $F_{DE} = 0$.

6.18 $F_{AB} = 2250$ N C; $F_{AC} = 1200$ N T; $F_{BC} = 750$ N T; $F_{BD} = 1700$ N C; $F_{BE} = 400$ N C; $F_{CE} = 850$ N C; $F_{CF} = 1600$ N T; $F_{DE} = 1500$ N T; $F_{EF} = 2250$ N T.

6.19 $F_{AB} = F_{FH} = 7.50$ kips C; $F_{AC} = F_{GH} = 4.50$ kips T; $F_{BC} = F_{FG} = 4.00$ kips T; $F_{BD} = F_{DF} = 6.00$ kips C; $F_{BE} = F_{EF} = 2.50$ kips T; $F_{CE} = F_{EG} = 4.50$ kips T; $F_{DE} = 0$.

6.21 $F_{AB} = 9.90$ kN C; $F_{AC} = 7.83$ kN T; $F_{BC} = 0$; $F_{BD} = 7.07$ kN C; $F_{BE} = 2.00$ kN C; $F_{CE} = 7.83$ kN T; $F_{DE} = 1.000$; kN T; $F_{DF} = 5.03$ kN C; $F_{DG} = 0.559$ kN C; $F_{EG} = 5.59$ kN T.

6.22 $F_{AB} = 3610$ lb C; $F_{AC} = 4110$ lb T; $F_{BC} = 768$ lb C; $F_{BD} = 3840$ lb C; $F_{CD} = 1371$ lb T; $F_{CE} = 2740$ lb T; $F_{DE} = 1536$ lb C.

6.23 $F_{DF} = 4060$ lb C; $F_{DG} = 1371$ lb T; $F_{EG} = 2740$ lb T; $F_{FG} = 768$ lb C; $F_{FH} = 4290$ lb C; $F_{GH} = 4110$ lb T.

6.24 $F_{AB} = F_{DF} = 2.29$ kN T; $F_{AC} = F_{EF} = 2.29$ kN C; $F_{BC} = F_{DE} = 0.600$ kN C; $F_{BD} = 2.21$ kN T; $F_{BE} = F_{EH} = 0$; $F_{CE} = 2.21$ kN C; $F_{CH} = F_{EJ} = 1.200$ kN C.

6.27 $F_{AB} = F_{BC} = F_{CD} = 36.0$ kips T; $F_{AE} = 57.6$ kips T; $F_{AF} = 45.0$ kips C; $F_{BF} = F_{BG} = F_{CG} = F_{CH} = 0$; $F_{DH} = F_{FG} = F_{GH} = 39.0$ kips C; $F_{EF} = 36.0$ kips C.

6.28 $F_{AB} = 128.0$ kN T; $F_{AC} = 136.7$ kN C; $F_{BD} = F_{DF} = F_{FH} = 128.0$ kN T; $F_{CE} = F_{EG} = 136.7$ kN C; $F_{GH} = 192.7$ kN C; $F_{BC} = F_{BE} = F_{DE} = F_{DG} = F_{FG} = 0$.

6.29 Truss of Prob. 6.33a is the only simple truss.

6.30 Trusses of Prob. 6.32b and Prob. 6.33b are simple trusses.

6.32 (a) AI, BJ, CK, DI, EI, FK, GK.
(b) FK, IO.

6.34 (a) BC, HI, IJ, JK. (b) BF, BG, CG, CH.

6.35 $F_{AB} = F_{AD} = 244$ lb C; $F_{AC} = 1040$ lb T; $F_{BC} = F_{CD} = 500$ lb C; $F_{BD} = 280$ lb T.

6.36 $F_{AB} = F_{AD} = 861$ N C; $F_{AC} = 676$ N C; $F_{BC} = F_{CD} = 162.5$ N T; $F_{BD} = 244$ N T.

6.37 $F_{AB} = F_{AD} = 2810$ N T; $F_{AC} = 5510$ N C; $F_{BC} = F_{CD} = 1325$ N T; $F_{BD} = 1908$ N C.

6.38 $F_{AB} = F_{AC} = 1061$ lb C; $F_{AD} = 2500$ lb T; $F_{BC} = 2100$ lb T; $F_{BD} = F_{CD} = 1250$ lb C; $F_{BE} = F_{CE} = 1250$ lb C; $F_{DE} = 1500$ lb T.

6.39 $F_{AB} = 840$ N C; $F_{AC} = 110.6$ N C; $F_{AD} = 394$ N C; $F_{AE} = 0$; $F_{BC} = 160.0$ N T; $F_{BE} = 200$ N T; $F_{CD} = 225$ N T; $F_{CE} = 233$ N C; $F_{DE} = 120.0$ N T.

6.40 $F_{AB} = F_{AE} = F_{BC} = 0$; $F_{AC} = 995$ N T; $F_{AD} = 1181$ N C; $F_{BE} = 600$ N T; $F_{CD} = 375$ N T; $F_{CE} = 700$ N C; $F_{DE} = 360$ N T.

6.43 $F_{DF} = 5.45$ kN C; $F_{DG} = 1.000$ kN T; $F_{EG} = 4.65$ kN T.

6.44 $F_{GI} = 4.65$ kN T; $F_{HI} = 1.800$ kN C; $F_{HJ} = 4.65$ kN C.

6.45 $F_{BD} = 36.0$ kips C; $F_{CD} = 45.0$ kips C.

6.46 $F_{DF} = 60.0$ kips C; $F_{DG} = 15.00$ kips C.

6.49 $F_{CD} = 20.0$ kN C; $F_{DF} = 52.0$ kN C.

6.50 $F_{CE} = 36.0$ kN T; $F_{EF} = 15.00$ kN C.

6.51 $F_{DE} = 25.0$ kips T; $F_{DF} = 13.00$ kips C.

6.52 $F_{EG} = 16.00$ kips T; $F_{EF} = 6.40$ kips C.
6.53 $F_{DF} = 91.4$ kN T; $F_{DE} = 38.6$ kN C.
6.54 $F_{CD} = 64.2$ kN T; $F_{CE} = 92.1$ kN C.
6.55 $F_{CE} = 7.20$ kN T; $F_{DE} = 1.047$ kN C; $F_{DF} = 6.39$ kN C.
6.56 $F_{EG} = 3.46$ kN T; $F_{GH} = 3.78$ kN C; $F_{HJ} = 3.55$ kN C.
6.59 $F_{AD} = 3.38$ kips C; $F_{CD} = 0$; $F_{CE} = 14.03$ kips T.
6.60 $F_{DG} = 18.75$ kips C; $F_{FG} = 14.03$ kips T; $F_{FH} = 17.43$ kips T.
6.61 $F_{DG} = 3.75$ kN T; $F_{FI} = 3.75$ kN C.
6.62 $F_{GJ} = 11.25$ kN T; $F_{IK} = 11.25$ kN C.
6.65 (*a*) *CJ*. (*b*) 1.026 kN T.
6.66 (*a*) *IO*. (*b*) 2.05 kN T.
6.67 $F_{BE} = 10.00$ kips T; $F_{DE} = 0$; $F_{EF} = 5.00$ kips T.
6.68 $F_{BE} = 2.50$ kips T; $F_{DE} = 1.500$ kips C; $F_{DG} = 2.50$ kips T.
6.69 (*a*) improperly constrained. (*b*) completely constrained, determinate. (*c*) completely constrained, indeterminate.
6.70 (*a*) completely constrained, determinate. (*b*) partially constrained. (*c*) improperly constrained.
6.71 (*a*) completely constrained, determinate. (*b*) completely constrained, indeterminate. (*c*) improperly constrained.
6.72 (a) partially constrained. (*b*) completely constrained. determinate. (*c*) completely constrained, indeterminate.
6.75 $F_{BD} = 375$ N C; $\mathbf{C}_x = 205$ N $\leftarrow$; $\mathbf{C}_y = 360$ N $\downarrow$.
6.76 $F_{BD} = 780$ lb T; $\mathbf{C}_x = 720$ lb $\leftarrow$, $\mathbf{C}_y = 140.0$ lb $\downarrow$.
6.77 (*a*) 125.0 N $\searrow$ 36.9°. (*b*) 125.0 N $\nearrow$ 36.9°.
6.78 $\mathbf{A}_x = 120.0$ lb $\rightarrow$, $\mathbf{A}_y = 30.0$ lb $\uparrow$; $\mathbf{B}_x = 120.0$ lb $\leftarrow$, $\mathbf{B}_y = 80.0$ lb $\downarrow$; $\mathbf{C} = 30.0$ lb $\downarrow$; $\mathbf{D} = 80.0$ lb $\uparrow$.
6.79 $\mathbf{A}_x = 18.00$ kN $\leftarrow$, $\mathbf{A}_y = 20.0$ kN $\downarrow$; $\mathbf{B} = 9.00$ kN $\rightarrow$; $\mathbf{C}_x = 9.00$ kN $\rightarrow$, $\mathbf{C}_y = 20.0$ kN $\uparrow$.
6.80 $\mathbf{A} = 20.0$ kN $\downarrow$, $\mathbf{B} = 18.00$ kN $\leftarrow$; $\mathbf{C}_x = 18.00$ kN $\rightarrow$, $\mathbf{C}_y = 20.0$ kN $\uparrow$.
6.81 $\mathbf{A} = 150.0$ lb $\rightarrow$; $\mathbf{B}_x = 150.0$ lb $\leftarrow$, $\mathbf{B}_y = 60.0$ lb $\uparrow$; $\mathbf{C} = 20.0$ lb $\uparrow$; $\mathbf{D} = 80.0$ lb $\downarrow$.
6.83 (*a*) $\mathbf{A}_x = 2700$ N $\rightarrow$, $\mathbf{A}_y = 200$ N $\uparrow$; $\mathbf{E}_x = 2700$ N $\leftarrow$, $\mathbf{E}_y = 600$ N $\uparrow$.
(*b*) $\mathbf{A}_x = 300$ N $\rightarrow$, $\mathbf{A}_y = 200$ N $\uparrow$; $\mathbf{E}_x = 300$ N $\leftarrow$, $\mathbf{E}_y = 600$ N $\uparrow$.
6.85 (*a*) $\mathbf{A}_x = 300$ N $\leftarrow$, $\mathbf{A}_y = 660$ N $\uparrow$; $\mathbf{E}_x = 300$ N $\rightarrow$, $\mathbf{E}_y = 90.0$ N $\uparrow$. (*b*) $\mathbf{A}_x = 300$ N $\leftarrow$, $\mathbf{A}_y = 150.0$ N $\uparrow$; $\mathbf{E}_x = 300$ N $\rightarrow$, $\mathbf{E}_y = 600$ N $\uparrow$.
6.87 (*a*) $\mathbf{A}_x = 80.0$ lb $\leftarrow$, $\mathbf{A}_y = 40.0$ lb $\uparrow$; $\mathbf{B}_x = 80.0$ lb $\rightarrow$, $\mathbf{B}y = 60.0$ lb $\uparrow$. (*b*) $\mathbf{A}_x = 0$, $\mathbf{A}_y = 40.0$ lb $\uparrow$; $\mathbf{B}_x = 0$, $\mathbf{B}_y = 60.0$ lb $\uparrow$.
6.88 (*a*) and (*c*) $\mathbf{B}_x = 32.0$ lb $\rightarrow$, $\mathbf{B}_y = 10.00$ lb $\uparrow$; $\mathbf{F}_x = 32.0$ lb $\leftarrow$, $\mathbf{F}_y = 38.0$ lb $\uparrow$. (*b*) $\mathbf{B}_x = 32.0$ lb $\leftarrow$, $\mathbf{B}_y = 34.0$ lb $\uparrow$; $\mathbf{F}_x = 32.0$ lb $\rightarrow$, $\mathbf{F}y = 14.00$ lb $\uparrow$.
6.89 (*a*) and (*c*) $\mathbf{B}_x = 24.0$ lb $\leftarrow$, $\mathbf{B}_y = 7.50$ lb $\downarrow$; $\mathbf{F}_x = 24.0$ lb $\rightarrow$, $\mathbf{F}_y = 7.50$ lb $\uparrow$. (*b*) $\mathbf{B}_x = 24.0$ lb $\leftarrow$, $\mathbf{B}_y = 10.50$ lb $\uparrow$; $\mathbf{F}_x = 24.0$ lb $\rightarrow$, $\mathbf{F}_y = 10.50$ lb $\downarrow$.
6.91 $\mathbf{D}_x = 13.60$ kN $\rightarrow$, $\mathbf{D}_y = 7.50$ kN $\uparrow$; $\mathbf{E}_x = 13.60$ kN $\leftarrow$, $\mathbf{E}_y = 2.70$ kN $\downarrow$.
6.92 $\mathbf{A}_x = 45.0$ N $\leftarrow$, $\mathbf{A}_y = 30.0$ N $\downarrow$; $\mathbf{B}_x = 45.0$ N $\rightarrow$, $\mathbf{B}_y = 270$ N $\uparrow$.
6.93 (*a*) $\mathbf{E}_x = 2.00$ kips $\leftarrow$, $\mathbf{E}_y = 2.25$ kips $\uparrow$. (*b*) $\mathbf{C}_x = 4.00$ kips $\leftarrow$, $\mathbf{C}_y = 5.75$ kips $\uparrow$.
6.94 (*a*) $\mathbf{E}_x = 3.00$ kips $\leftarrow$, $\mathbf{E}_y = 1.500$ kips $\uparrow$. (*b*) $\mathbf{C}_x = 3.00$ kips $\leftarrow$, $\mathbf{C}_y = 6.50$ kips $\uparrow$.
6.95 (*a*) $\mathbf{A} = 982$ lb $\uparrow$; $\mathbf{B} = 935$ lb $\uparrow$; $\mathbf{C} = 733$ lb $\uparrow$. (*b*) $\Delta B = +291$ lb; $\Delta C = -72.7$ lb.
6.96 (*a*) 572 lb. (*b*) $\mathbf{A} = 1070$ lb $\uparrow$; $\mathbf{B} = 709$ lb $\uparrow$; $\mathbf{C} = 870$ lb $\uparrow$.
6.99 $\mathbf{B} = 152.0$ lb $\downarrow$; $\mathbf{C}_x = 60.0$ lb $\leftarrow$, $\mathbf{C}_y = 200$ lb $\uparrow$; $\mathbf{D}_x = 60.0$ lb $\rightarrow$, 42.0 lb $\uparrow$.
6.100 $\mathbf{B} = 108.0$ lb $\downarrow$; $\mathbf{C}_x = 90.0$ lb $\leftarrow$, $\mathbf{C}_y = 150.0$ lb $\uparrow$; $\mathbf{D}_x = 90.0$ lb $\rightarrow$, $\mathbf{D}_y = 18.00$ lb $\uparrow$.

6.101 $\mathbf{A}_x = 13.00$ kN $\leftarrow$, $\mathbf{A}_y = 4.00$ kN $\downarrow$; $\mathbf{B}_x = 36.0$ kN $\rightarrow$. $\mathbf{B}_y = 6.00$ kN $\uparrow$; $\mathbf{E}_x = 23.0$ kN $\leftarrow$, $\mathbf{E}_y = 2.00$ kN $\downarrow$.
6.102 $\mathbf{A}_x = 2025$ N $\leftarrow$, $\mathbf{A}_y = 1800$ kN $\downarrow$; $\mathbf{B}_x = 4050$ N $\rightarrow$, $\mathbf{B}_y = 1200$ N $\uparrow$; $\mathbf{E}_x = 2025$ N $\leftarrow$, $\mathbf{E}_y = 600$ N $\uparrow$.
6.103 $\mathbf{A}_x = 1110$ lb $\leftarrow$, $\mathbf{A}_y = 600$ lb $\uparrow$; $\mathbf{B}_x = 1110$ lb $\leftarrow$, $\mathbf{B}_y = 800$ lb $\downarrow$; $\mathbf{D}_x = 2220$ lb $\rightarrow$, $\mathbf{D}_y = 200$ lb $\uparrow$.
6.104 $\mathbf{A}_x = 660$ lb $\leftarrow$, $\mathbf{A}_y = 240$ lb $\uparrow$; $\mathbf{B}_x = 660$ lb $\leftarrow$, $\mathbf{B}_y = 320$ lb $\downarrow$; $\mathbf{D}_x = 1320$ lb $\rightarrow$, $\mathbf{D}_y = 80.0$ lb $\uparrow$.
6.107 (*a*) $\mathbf{A}_x = 200$ kN $\rightarrow$, $\mathbf{A}_y = 122.0$ kN $\uparrow$. (*b*) $\mathbf{B}_x = 200$ kN $\leftarrow$, $\mathbf{B}_y = 10.00$ kN $\downarrow$.
6.108 (*a*) $\mathbf{A}_x = 205$ kN $\rightarrow$, $\mathbf{A}_y = 134.5$ kN $\uparrow$. (*b*) $\mathbf{B}_x = 205$ kN $\leftarrow$, $\mathbf{B}_y = 5.50$ kN $\uparrow$.
6.109 $\mathbf{B} = 98.5$ lb $\measuredangle$ 24.0°; $\mathbf{C} = 90.6$ lb $\searined$ 6.34°.
6.110 $\mathbf{B} = 25.0$ lb $\uparrow$; $\mathbf{C} = 79.1$ lb $\searrow$ 18.43°.
6.112 $F_{AF} = P/4$ C; $F_{BG} = F_{DG} = P/\sqrt{2}$ C; $F_{EH} = P/4$ T.
6.113 $F_{AG} = \sqrt{2}P/6$ C; $F_{BF} = 2\sqrt{2}P/3$ C; $F_{DI} = \sqrt{2}P/3$ C; $F_{EH} = \sqrt{2}P/6$ T.
6.115 $F_{AF} = M_0/4a$ C; $F_{BG} = F_{DG} = M_0/\sqrt{2}a$ T; $F_{EH} = 3M_0/4a$ C.
6.116 $F_{AF} = M_0/6a$ T; $F_{BG} = \sqrt{2}M_0/6a$ T; $F_{DG} = \sqrt{2}M_0/3a$ T; $F_{EH} = M_0/6a$ C.
6.117 $\mathbf{A} = P/15$ $\uparrow$; $\mathbf{D} = 2P/15$ $\uparrow$; $\mathbf{E} = 8P/15$ $\uparrow$; $\mathbf{H} = 4P/15$ $\uparrow$.
6.118 $\mathbf{E} = P/5$ $\downarrow$; $\mathbf{F} = 8P/5$ $\uparrow$; $\mathbf{G} = 4P/5$ $\downarrow$; $\mathbf{H} = 2P/5$ $\uparrow$.
6.120 (*a*) $\mathbf{A} = 2.06P$ $\measuredangle$ 14.04°; $\mathbf{B} = 2.06$ $\searrow$ 14.04°; frame is rigid. (*b*) Frame is not rigid. (*c*) $\mathbf{A} = 1.25P$ $\searrow$ 36.9°. $\mathbf{B} = 1.031P$ $\measuredangle$ 14.04°; frame is rigid.
6.122 (*a*) 2860 N $\downarrow$. (*b*) 2700 N $\nearrow$ 68.5°.
6.123 564 lb $\rightarrow$.
6.124 275 lb $\rightarrow$.
6.125 764 N $\leftarrow$.
6.127 (*a*) 746 N $\downarrow$. (*b*) 565 N $\searrow$ 61.3°.
6.129 832 lb·in. $\nearrow$.
6.130 360 lb·in. $\nearrow$.
6.131 195.0 kN·m $\downarrow$.
6.132 40.5 kN·m $\nearrow$.
6.133 (*a*) 160.8 N·m $\nearrow$. (*b*) 155.9 N·m $\nearrow$.
6.134 (*a*) 117.8 N·m $\nearrow$. (*b*) 47.9 N·m $\nearrow$.
6.137 18.43 N·m $\downarrow$.
6.138 208 N·m $\downarrow$.
6.139 $F_{AE} = 800$ N T; $F_{DG} = 100.0$ N C.
6.140 $\mathbf{P} = 120.0$ N $\downarrow$; $\mathbf{Q} = 110.0$ N $\leftarrow$.
6.141 $\mathbf{F} = 3290$ lb $\searrow$ 15.12°; $\mathbf{D} = 4550$ lb $\leftarrow$.
6.143 $\mathbf{D} = 30.0$ kN $\leftarrow$; $\mathbf{F} = 37.5$ kN $\searrow$ 36.9°.
6.144 $\mathbf{D} = 150.0$ kN $\leftarrow$; $\mathbf{F} = 96.4$ kN $\searrow$ 13.50°.
6.145 (*a*) 475 lb. (*b*) 528 lb $\searrow$ 63.3°.
6.147 44.8 kN.
6.148 8.45 kN.
6.149 140.0 N.
6.151 315 lb.
6.152 (*a*) 312 lb. (*b*) 135.0 lb·in. $\downarrow$.
6.153 (*a*) 4.91 kips C. (*b*) 10.69 kips C.
6.154 (*a*) 2.86 kips C. (*b*) 9.43 kips C.
6.155 (*a*) 9.29 kN $\searrow$ 44.4°. (*b*) 8.04 kN $\searrow$ 34.4°.
6.159 (*a*) (90.0 N·m)$\mathbf{i}$. (*b*) $\mathbf{A} = 0$; $\mathbf{M}_A = -(48.0$ N·m$)\mathbf{i}$, $\mathbf{B} = 0$; $\mathbf{M}_B = -(72.0$ N·m$)\mathbf{i}$.
6.160 (*a*) 27.0 mm. (*b*) 40.0 N·m $\downarrow$.
6.163 $\mathbf{E}_x = 100.0$ kN $\rightarrow$, $\mathbf{E}_y = 154.9$ kN $\uparrow$; $\mathbf{F}_x = 26.5$ kN $\rightarrow$. $\mathbf{F}_y = 118.1$ kN $\downarrow$; $\mathbf{H}_x = 126.5$ kN $\leftarrow$, $\mathbf{H}_y = 36.8$ kN $\downarrow$.
6.164 $F_{AB} = 4.00$ kN T; $F_{AD} = 15.00$ kN T; $F_{BD} = 9.00$ kN C; $F_{BE} = 5.00$ kN T; $F_{CD} = 16.00$ kN C; $F_{DE} = 4.00$ kN C.

6.165 $F_{AB} = 7.83$ kN C; $F_{AC} = 7.00$ kN T; $F_{BC} = 1.886$ kN C;
$F_{BD} = 6.34$ kN C; $F_{CD} = 1.491$ kN T; $F_{CE} = 5.00$ kN T;
$F_{DE} = 2.83$ kN C; $F_{DF} = 3.35$ kN C; $F_{EF} = 2.75$ kN T;
$F_{EG} = 1.061$ kN C; $F_{EH} = 3.75$ kN T; $F_{FG} = 4.24$ kN C;
$F_{GH} = 5.30$ kN C.

6.166 $F_{AB} = 8.20$ kips T; $F_{AG} = 4.50$ kips T; $F_{FG} = 11.60$ kips C.

6.168 $\mathbf{A}_x = 900$ lb $\leftarrow$; $\mathbf{A}_y = 75.0$ lb$\uparrow$; $\mathbf{B} = 825$ lb$\downarrow$;
$\mathbf{D}_x = 900$ lb $\rightarrow$; $\mathbf{D}_y = 750$ lb$\uparrow$.

6.170 $\mathbf{B}_x = 700$ N $\leftarrow$, $\mathbf{B}_y = 200$ N $\downarrow$; $\mathbf{E}_x = 700$ N $\rightarrow$,
$\mathbf{E}_y = 500$ N $\uparrow$.

6.171 $\mathbf{C}_x = 78.0$ lb $\rightarrow$, $\mathbf{C}_y = 28.0$ lb$\uparrow$; $\mathbf{F}_x = 78.0$ lb $\leftarrow$,
$\mathbf{F}_y = 12.00$ lb$\uparrow$.

6.172 $\mathbf{A} = 327$ lb $\rightarrow$; $\mathbf{B} = 827$ lb $\leftarrow$; $\mathbf{D} = 621$ lb$\uparrow$; $\mathbf{E} = 246$ lb$\uparrow$.

6.174 (a) 21.0 kN $\leftarrow$. (b) 52.5 kN $\leftarrow$.

CHAPTER 7

7.1 $\mathbf{F} = 720$ lb $\rightarrow$; $\mathbf{V} = 140.0$ lb$\uparrow$; $\mathbf{M} = 1120$ lb·in. $\uparrow$ (On JC).

7.2 $\mathbf{F} = 120.0$ lb $\leftarrow$; $\mathbf{V} = 30.0$ lb$\downarrow$; $\mathbf{M} = 120.0$ lb·in. $\uparrow$.

7.3 $\mathbf{F} = 125.0$ N $\measuredangle 67.4°$; $\mathbf{V} = 300$ N $\measuredangle 22.6°$;
$\mathbf{M} = 156.0$ N·m. $\downarrow$.

7.4 $\mathbf{F} = 2330$ N $\measuredangle 67.4°$; $\mathbf{V} = 720$ N $\measuredangle 22.6°$; $\mathbf{M} = 374$ N·m. $\downarrow$.

7.7 $\mathbf{F} = 23.6$ lb $\measuredangle 76.0°$; $\mathbf{V} = 29.1$ lb $\measuredangle 14.04°$;
$\mathbf{M} = 540$ lb·in. $\downarrow$.

7.8 (a) 30.0 lb at C. (b) 33.5 lb at B and D. (c) 960 lb·in. at C.

7.9 $\mathbf{F} = 103.9$ N $\measuredangle 60.0°$; $\mathbf{V} = 60.0$ N $\measuredangle 30.0°$;
$\mathbf{M} = 18.71$ N·m $\downarrow$ (On AJ).

7.10 $\mathbf{F} = 60.0$ N $\measuredangle 30.0°$; $\mathbf{V} = 103.9$ N $\measuredangle 60.0°$;
$\mathbf{M} = 10.80$ N·m $\uparrow$ (On BK).

7.11 $\mathbf{F} = 194.6$ N $\measuredangle 60.0°$; $\mathbf{V} = 257$ N $\measuredangle 30.0°$;
$\mathbf{M} = 24.7$ N·m $\downarrow$ (On AJ).

7.12 45.2 N·m for $\theta = 82.9°$.

7.15 $\mathbf{F} = 250$ N $\measuredangle 36.9°$; $\mathbf{V} = 120.0$ N $\measuredangle 53.1$;
$\mathbf{M} = 120.0$ N·m $\uparrow$ (On BJ).

7.16 $\mathbf{F} = 560$ N $\leftarrow$; $\mathbf{V} = 90.0$ N $\downarrow$; $\mathbf{M} = 72.0$ N·m $\downarrow$ (On AK).

7.17 150.0 lb·in. at D.

7.18 105.0 lb·in. at E.

7.19 $\mathbf{F} = 200$ N $\measuredangle 36.9°$; $\mathbf{V} = 120.0$ N $\measuredangle 53.1°$;
$\mathbf{M} = 120.0$ N·m $\uparrow$ (On BJ).

7.20 $\mathbf{F} = 520$ N $\leftarrow$; $\mathbf{V} = 120.0$ N $\downarrow$; $\mathbf{M} = 96.0$ N·m$\downarrow$ (On AK).

7.23 $0.0557 \, Wr$ (On AJ).

7.24 $0.1009 \, Wr$ for $\theta = 57.3°$.

7.25 $0.289 \, Wr$ (On BJ).

7.26 $0.417 \, Wr$ (On BJ).

7.29 (b) $|V|_{max} = wL/4$; $|M|_{max} = 3wL^2/32$.

7.30 (b) $|V|_{max} = w_0L/2$; $|M|_{max} = w_0L^2/6$.

7.31 (b) $|V|_{max} = 2P/3$; $|M|_{max} = 2PL/9$.

7.32 (b) $|V|_{max} = 2P$; $|M|_{max} = 3Pa$.

7.35 (b) $|V|_{max} = 40.0$ kN; $|M|_{max} = 55.0$ kN·m.

7.36 (b) $|V|_{max} = 50.5$ kN; $|M|_{max} = 39.8$ kN·m.

7.39 (b) $|V|_{max} = 64.0$ kN; $|M|_{max} = 92.0$ kN·m.

7.40 (b) $|V|_{max} = 40.0$ kN; $|M|_{max} = 40.0$ kN·m.

7.41 (b) $|V|_{max} = 18.00$ kips; $|M|_{max} = 48.5$ kip·ft.

7.42 (b) $|V|_{max} = 15.30$ kips; $|M|_{max} = 46.8$ kip·ft.

7.45 (b) $|V|_{max} = 6.00$ kips; $|M|_{max} = 12.00$ kip·ft.

7.46 (b) $|V|_{max} = 4.00$ kips; $|M|_{max} = 6.00$ kip·ft.

7.47 (b) $|V|_{max} = 6.00$ kN; $|M|_{max} = 9.00$ kN·m.

7.48 (b) $|V|_{max} = 6.00$ kN; $|M|_{max} = 9.00$ kN·m.

7.49 $|V|_{max} = 180.0$ N; $|M|_{max} = 36.0$ N·m.

7.50 $|V|_{max} = 800$ N; $|M|_{max} = 180.0$ N·m.

7.51 $|V|_{max} = 90.0$ lb; $|M|_{max} = 1400$ lb·in.

7.52 $|V|_{max} = 165.0$ lb; $|M|_{max} = 1625$ lb·in.

7.55 (a) 54.5°. (b) 675 N·m.

7.56 (a) 1.236. (b) $0.1180 \, wa^2$.

7.57 (a) 40.0 mm. (b) 1.600 N·m.

7.58 (a) 0.840 m. (b) 1.680 N·m.

7.59 $0.207 \, L$.

7.62 (a) $0.414 \, wL$; $0.0858 \, wL^2$. (b) $0.250 \, wL$; $0.250 \, wL^2$.

7.69 (a) $|V|_{max} = 15.00$ kN; $|M|_{max} = 42.0$ kN·m.

7.70 (b) $|V|_{max} = 17.00$ kN; $|M|_{max} = 17.00$ kN·m.

7.77 (b) 75.0 kN·m, 4.00 m from A.

7.78 (b) 1.378 kN·m, 1.050 m from A.

7.79 (b) 26.4 kN·m, 2.05 m from A.

7.80 (b) 5.76 kN·m, 2.40 m from A.

7.81 (b) 14.40 kip·ft, 6.00 ft from A.

7.82 (b) 16.20 kip·ft, 13.50 ft from A.

7.86 (a) $V = (w_0/6L)(L^2 - 3x^2)$; $M = (w_0/6L)(L^2x - x^3)$.
(b) $0.0642 \, w_0 L^2 A \measuredangle = 0.577L$.

7.87 (a) $V = (w_0 L/4)[3(x/L)^2 - 4(x/L) + 1]$;
$M = (w_0L^2/4) \, [(x/L)^3 - 2(x/L)^2 + (x/L)]$.
(b) $w_0L^2/27$, at $x = L/3$.

7.89 (a) $\mathbf{P} = 4.00$ kN $\downarrow$; $\mathbf{Q} = 6.00$ kN $\downarrow$. (b) $M_C = -900$ N·m.

7.90 (a) $\mathbf{P} = 2.50$ kN $\downarrow$; $\mathbf{Q} = 7.50$ kN $\downarrow$. (b) $M_C = -900$ N·m.

7.91 (a) $\mathbf{P} = 1.350$ kips$\downarrow$; $\mathbf{Q} = 0.450$ kips$\downarrow$. (b) $V_{max} = 2.70$ kips
at A; $M_{max} = 6.345$ kip·ft, 5.40 ft from A.

7.92 (a) $\mathbf{P} = 0.540$ kips $\downarrow$; $\mathbf{Q} = 1.860$ kips$\downarrow$.
(b) $V_{max} = 3.14$ kips at B; $M_{max} = 7.00$ kip·ft, 6.88 ft from A.

7.93 (a) $\mathbf{E}_x = 10.00$ kN $\rightarrow$, $\mathbf{E}_y = 7.00$ kN $\uparrow$. (b) 12.21 kN.

7.94 1.667 m.

7.95 (a) 838 lb $\measuredangle 17.35°$. (b) 971 lb $\measuredangle 34.5°$.

7.96 (a) 2670 lb $\measuredangle 2.10°$. (b) 2810 lb $\measuredangle 18.65°$.

7.97 (a) $d_B = 1.733$ m; $d_D = 4.20$ m. (b) 21.5 kN $\measuredangle 3.81°$.

7.98 (a) 2.80 m. (b) $\mathbf{A} = 32.0$ kN $\measuredangle 38.7°$; $\mathbf{E} = 25.0$ kN $\rightarrow$.

7.101 196.2 N.

7.102 157.0 N.

7.103 (a) 240 lb. (b) 9.00 ft.

7.104 $a = 7.50$ ft; $b = 17.50$ ft

7.107 (a) 1775 N. (b) 60.1 m.

7.109 (a) 50,200 kips. (b) 3580 ft.

7.110 3.75 ft.

7.111 (a) 56,400 kips. (b) 4284 ft.

7.112 (a) 6.75 m. (b) $T_{AB} = 615$ N; $T_{BC} = 600$ N.

7.114 (a) $\sqrt{3L\Delta/8}$. (b) 12.25 ft.

7.115 $h = 27.6$ mm; $\theta_A = 25.5°$; $\theta_C = 27.6°$.

7.116 (a) 4.05 m. (b) 16.41 m. (c) $A_x = 5890$ N $\leftarrow$, $A_y = 5300$ N $\uparrow$.

7.117 (a) 58,900 kips, (b) 29.2°.

7.118 (a) 16.00 ft to the left of B. (b) 2000 lb.

7.125 $Y = h[1 - \cos(\pi x/L)]$; $T_{min} = w_0L^2/h\pi^2$;
$T_{max} = (w_0L/\pi)\sqrt{(L^2/h^2\pi^2) + 1}$

7.127 (a) 12.36 ft. (b) 15.38 lb.

7.128 (a) 412 ft. (b) 875 lb.

7.129 (a) 35.6 m. (b) 49.2 kg.

7.130 49.86 ft.

7.133 (a) 5.89 m. (b) 10.89 N $\rightarrow$.

7.134 10.05 ft.

7.135 (a) 56.3 ft. (b) 2.36 lb/ft.

7.136 (a) 30.2 m. (b) 56.6 kg.

7.139 31.8 N.

7.140 29.8 N.

7.143 (a) $a = 79.0$ ft; $b = 60.0$ ft. (b) 103.9 ft.

7.144 (a) $a = 65.8$ ft; $b = 50.0$ ft. (b) 86.6 ft.

7.145 119.1 N $\rightarrow$.

7.146 177.6 N $\rightarrow$.

7.147 3.50 ft.

7.148 5.71 ft.

7.151 0.394 m and 10.97 m.

7.152 0.1408.

7.153 (a) 0.338. (b) 56.5°; 0.755 wL.

7.154 (On AJ) $\mathbf{F}$ = 750 N ↑; $\mathbf{V}$ = 400 N ←; $\mathbf{M}$ = 130.0 N·m ⤴.

7.156 (On BJ) $\mathbf{F}$ = 12.50 lb ⤢ 30.0°; $\mathbf{V}$ = 21.7 lb ⤡ 60.0°;
$\mathbf{M}$ = 75.0 lb·in.⤵.

7.157 (a) (On AJ) $\mathbf{F}$ = 500 N ←; $\mathbf{V}$ = 500 N ↑; $\mathbf{M}$ = 300 N·m ⤵.
(b) (On AK) $\mathbf{F}$ = 970 N ↑; $\mathbf{V}$ = 171.0 N ←; $\mathbf{M}$ = 446 N·m ⤵.

7.158 (a) 40.0 kips. (b) 40.0 kip·ft.

7.161 (a) 18.00 kip·ft, 3.00 ft from A.
(b) 34.1 kip·ft, 2.25 ft from A.

7.163 (a) 2.28 m. (b) $\mathbf{D}_x$ = 13.67 kN →; $\mathbf{D}_y$ = 7.80 kN ↑.
(c) 15.94 kN.

7.164 (a) 138.1 m. (b) 602 N.

7.165 (a) 4.22 ft. (b) 80.3°.

CHAPTER 8

8.1 Block is in equilibrium, $\mathbf{F}$ = 30.1 N ⤡ 20.0°.

8.2 Block moves up, $\mathbf{F}$ = 151.7 N ⤧ 20.0°.

8.3 Block moves, $\mathbf{F}$ = 36.1 lb ⤧ 30.0°.

8.4 Block is in equilibrium, $\mathbf{F}$ = 36.3 lb ⤧ 30.0°.

8.5 (a) 83.2 lb. (b) 66.3 lb.

8.7 (a) 29.7 N ←. (b) 20.9 N →.

8.9 74.5 N.

8.10 $17.91° \le \theta \le 66.4°$.

8.11 31.0°.

8.12 46.4°.

8.13 Package C does not move; $\mathbf{F}_C$ = 10.16 N ↗.
Package A and B move; $\mathbf{F}_A$ = 7.58 N ↗; $\mathbf{F}_B$ = 3.03 N ↗.

8.14 All packages move; $\mathbf{F}_A = \mathbf{F}_C$ = 7.58 N ↗; $\mathbf{F}_B$ = 3.03 N ↗.

8.17 (a) 75.0 lb. (b) Pipe will slide.

8.18 (a) P = 36.0 lb →. (b) h_{max} = 40.0 in.

8.19 P = 8.34 lb.

8.20 P = 7.50 lb.

8.21 (a) 0.300 Wr. (b) 0.349 Wr.

8.22 $M = Wr\mu_s (1 + \mu_s)/(1 + \mu_s^2)$.

8.23 (a) 136.4°. (b) 0.928 W.

8.25 0.208.

8.27 664 N ↓.

8.29 (a) Plate in equilibrium. (b) Plate moves downward.

8.30 $10.00 \text{ lb} < P < 36.7 \text{ lb}$.

8.32 0.860.

8.34 0.0533.

8.35 (a) 1.333. (b) 1.192. (c) 0.839.

8.36 (b) 2.69 lb.

8.37 (a) 2.94 N. (b) 4.41 N.

8.39 30.6 N·m ⤴.

8.40 18.90 N·m ⤴.

8.41 135.0 lb.

8.43 (a) System slides; P = 62.8 N. (b) System rotates about B;
P = 73.2 N.

8.44 35.8°.

8.45 20.5°.

8.46 1.225 W.

8.47 $46.4° \le \theta \le 52.4°$ and $67.6° \le \theta \le 79.4°$.

8.48 (a) 283 N ←. (b) $\mathbf{B}_x$ = 413 N ←; $\mathbf{B}_y$ = 480 N ↓.

8.49 (a) 107.0 N ←. (b) $\mathbf{B}_x$ = 611 N ←; $\mathbf{B}_y$ = 480 N ↓.

8.52 (a) 15.26 kips. (b) 5.40 kips.

8.53 (a) 6.88 kips. (b) 5.40 kips.

8.54 9.86 kN ←.

8.55 9.13 N ←.

8.56 (a) 28.1°. (b) 728 N ⤢ 14.04°.

8.57 (a) 50.4 lb ↓. (b) 50.4 lb ↓.

8.59 143.4 N.

8.60 1.400 lb.

8.62 (a) 197.0 lb →. (b) Base will not move.

8.63 (a) 280 lb ←. (b) Base moves.

8.64 (b) 283 N ←.

8.65 0.442.

8.66 0.1103.

8.67 0.1013.

8.71 693 lb·ft.

8.72 35.8 N·m.

8.73 9.02 N·m.

8.74 (a) Screw A. (b) 14.06 lb·in.

8.77 0.226.

8.78 4.70 kips.

8.79 450 N.

8.80 412 N.

8.81 334 N.

8.82 376 N.

8.84 T_{AB} = 77.5 lb; T_{CD} = 72.5 lb. T_{EF} = 67.8 lb.

8.86 (a) 4.80 kN. (b) 1.375°.

8.88 22.0 lb ←.

8.89 1.948 lb ↓.

8.90 18.01 lb ←.

8.92 0.1670.

8.93 3.75 lb.

8.98 10.87 lb.

8.99 0.0600 in.

8.100 154.4 N.

8.101 300 mm.

8.102 (a) 1.288 kN. (b) 1.058 kN.

8.103 2.34 ft.

8.104 (a) 0.329. (b) 2.67 turns.

8.105 $14.23 \text{ kg} \le m \le 175.7 \text{ kg}$.

8.106 (a) 0.292. (b) 310 N.

8.109 31.8 N·m ⤴.

8.110 (a) T_A = 8.40 lb; T_B = 19.60 lb. (b) 0.270.

8.111 (a) T_A = 11.13 lb; T_B = 20.9 lb. (b) 91.3 lb·in. ⤵.

8.112 35.1 N·m.

8.113 (a) 27.0 N·m. (b) 675 N.

8.114 (a) 39.0 N·m. (b) 844 N.

8.117 4.49 in.

8.118 (a) 11.66 kg. (b) 38.6 kg. (c) 34.4 kg.

8.119 (a) 9.46 kg. (b) 167.2 kg. (c) 121.0 kg.

8.120 (a) 10.39 lb. (b) 58.5 lb.

8.121 (a) 28.9 lb. (b) 28.9 lb.

8.124 5.97 N.

8.125 9.56 N.

8.126 0.350.

8.128 (a) 30.3 lb·in. ⤴. (b) 3.78 lb ↓.

8.129 (a) 17.23 lb·in. ⤵. (b) 2.15 lb ↑.

8.133 (a) 51.0 N·m. (b) 875 N.

8.134 (a) 353 N ←. (b) 196.2 N ←.

8.136 (a) 136.0 lb →. (b) 30.0 lb →. (c) 12.86 lb →.

8.137 $6.35 \le L/a \le 10.81$.

8.138 151.5 N·m.

8.140 0.225.

8.141 313 lb →.

8.143 6.44 N·m.

8.144 (a) 0.238. (b) 218 N ↓.

CHAPTER 9

9.1 $a^3b/30$.

9.2 $3a^3b/10$.

9.3 $b^3h/12$.

9.4 $a^3b/6$.

9.6 $ab^3/6$.

9.8 $3ab^3/10$.

9.9 $ab^3/15$.

9.10 $ab^3/15$.

9.11 $0.1056\ ab^3$.

9.12 $3.43\ a^3b$.

9.15 $3a^3/35$; $b\sqrt{9/35}$.

9.16 $0.0945ah^3$; $0.402h$.

9.17 $3a^3b/35$; $a\sqrt{9/35}$.

9.18 $31a^3h/20$; $a\sqrt{93/35}$.

9.21 $20a^4$; $1.826a$.

9.22 $4ab(a^2 + 4b^2)/3$; $\sqrt{(a^2 + 4b^2)/3}$.

9.23 $64a^4/15$; $1.265a$.

9.25 $(\pi/2)(R_2^4 - R_1^4)$; $(\pi/4)(R_2^4 - R_1^4)$.

9.26 (b) for $t/R_m = 1$, $-10.56\ \%$; for $t/R_m = 1/2$, -2.99%; for $t/R_m = 1/10$, $-0.1250\ \%$.

9.28 $bh(12h^2 + b^2)/48$; $\sqrt{(12h^2 + b^2)/24}$.

9.31 $390 \times 10^3\ \text{mm}^4$; $21.9\ \text{mm}$.

9.32 $46.0\ \text{in}^4$; $1.599\ \text{in}$.

9.33 $64.3 \times 10^3\ \text{mm}^4$; $8.87\ \text{mm}$.

9.34 $46.5\ \text{in}^4$; $1.607\ \text{in}$.

9.37 $J_B = 1800\ \text{in}^4$; $J_D = 3600\ \text{in}^4$.

9.39 $3000\ \text{mm}^2$; $325 \times 10^3\ \text{mm}^4$.

9.40 $24.6 \times 10^6\ \text{mm}^4$.

9.41 $\bar{I}_x = 13.89 \times 10^6\ \text{mm}^4$; $\bar{I}_y = 20.9 \times 10^6\ \text{mm}^4$.

9.42 $\bar{I}_x = 479 \times 10^3\ \text{mm}^4$; $\bar{I}_y = 149.7 \times 10^3\ \text{mm}^4$.

9.43 $\bar{I}_x = 191.3\ \text{in}^4$; $\bar{I}_y = 75.2\ \text{in}^4$.

9.44 $\bar{I}_x = 18.13\ \text{in}^4$; $\bar{I}_y = 4.51\ \text{in}^4$.

9.47 (a) $11.57 \times 10^6\ \text{mm}^4$. (b) $7.81 \times 10^6\ \text{mm}^4$.

9.48 (a) $12.16 \times 10^6\ \text{mm}^4$. (b) $9.73 \times 10^6\ \text{mm}^4$.

9.49 $\bar{I}_x = 186.7 \times 10^6\ \text{mm}^4$; $\bar{k}_x = 118.6\ \text{mm}$; $\bar{I}_y = 167.7 \times 10^6\ \text{mm}^4$. $\bar{k}_y = 112.4\ \text{mm}$.

9.50 $\bar{I}_x = 44.5\ \text{in}^4$; $\bar{k}_x = 2.16\ \text{in.}$; $\bar{I}_y = 27.7\ \text{in}^4$; $\bar{k}_y = 1.709\ \text{in}$.

9.51 $\bar{I}_x = 250\ \text{in}^4$; $\bar{k}_x = 4.10\ \text{in.}$; $\bar{I}_y = 141.9\ \text{in}^4$; $\bar{k}_y = 3.09\ \text{in}$.

9.52 $\bar{I}_x = 260 \times 10^6\ \text{mm}^4$; $\bar{k}_x = 144.6\ \text{mm}$; $\bar{I}_y = 17.53\ \text{mm}^4$; $\bar{k}_y = 37.6\ \text{mm}$.

9.54 $\bar{I}_x = 745 \times 10^6\ \text{mm}^4$; $\bar{I}_y = 91.3 \times 10^6\ \text{mm}^4$.

9.55 $\bar{I}_x = 3.55 \times 10^6\ \text{mm}^4$; $\bar{I}_y = 49.8 \times 10^6\ \text{mm}^4$.

9.57 $h/2$.

9.58 $15h/14$.

9.59 $3\pi r/16$.

9.60 $4h/7$.

9.63 $5a/8$.

9.64 $80.0\ \text{mm}$.

9.67 $a^4/2$.

9.68 $b^2h^2/4$.

9.69 $a^2b^2/6$.

9.71 $-1.760 \times 10^6\ \text{mm}^4$.

9.72 $2.40 \times 10^6\ \text{mm}^4$.

9.74 $-0.380\ \text{in}^4$.

9.75 $471 \times 10^3\ \text{mm}^4$.

9.76 $-9010\ \text{in}^4$.

9.78 $2.54 \times 10^6\ \text{mm}^4$.

9.79 (a) $\bar{I}_{x'} = 0.482a^4$; $\bar{I}_{y'} = 1.482a^4$; $\bar{I}_{x'y'} = -0.589a^4$. (b) $\bar{I}_{x'} = 1.120a^4$; $\bar{I}_{y'} = 0.843a^4$; $\bar{I}_{x'y'} = 0.760a^4$.

9.80 $\bar{I}_{x'} = 2.12 \times 10^6\ \text{mm}^4$; $\bar{I}_{y'} = 8.28 \times 10^6\ \text{mm}^4$; $\bar{I}_{x'y'} = -0.532 \times 10^6\ \text{mm}^4$.

9.81 $\bar{I}_{x'} = 1033\ \text{in}^4$; $\bar{I}_{y'} = 2020\ \text{in}^4$; $\bar{I}_{x'y'} = -873\ \text{in}^4$.

9.83 $\bar{I}_{x'} = 0.236\ \text{in}^4$; $\bar{I}_{y'} = 1.244\ \text{in}^4$; $\bar{I}_{x'y'} = 0.1132\ \text{in}^4$.

9.85 $20.2°$ and $110.2°$; $1.754a^4$; $0.209a^4$.

9.86 $25.1°$ and $115.1°$; $\bar{I}_{\max} = 8.32 \times 10^6\ \text{mm}^4$; $\bar{I}_{\min} = 2.08 \times 10^6\ \text{mm}^4$.

9.87 $29.7°$ and $119.7°$; $2530\ \text{in}^4$; $524\ \text{in}^4$.

9.89 $-23.7°$ and $66.3°$; $1.257\ \text{in}^4$; $0.224\ \text{in}^4$.

9.91 (a) $\bar{I}_{x'} = 0.482a^4$; $\bar{I}_{y'} = 1.482a^4$; $\bar{I}_{x'y'} = -0.589a^4$. (b) $\bar{I}_{x'} = 1.120a^4$; $\bar{I}_{y'} = 0.843a^4$; $0.760a^4$.

9.92 $\bar{I}_{x'} = 2.12 \times 10^6\ \text{mm}^4$; $\bar{I}_{y'} = 8.28 \times 10^6\ \text{mm}^4$; $\bar{I}_{x'y'} = -0.532 \times 10^6\ \text{mm}^4$.

9.93 $\bar{I}_{x'} = 1033\ \text{in}^4$; $\bar{I}_{y'} = 2020\ \text{in}^4$; $\bar{I}_{x'y'} = -873\ \text{in}^4$.

9.95 $\bar{I}_{x'} = 0.236\ \text{in}^4$; $\bar{I}_{y'} = 1.244\ \text{in}^4$; $\bar{I}_{x'y'} = 0.1132\ \text{in}^4$.

9.97 $20.2°$; $1.754a^4$; $0.209a^4$.

9.98 $23.9°$; $8.33 \times 10^6\ \text{mm}^4$; $1.465 \times 10^6\ \text{mm}^4$.

9.99 $33.4°$; $22.1 \times 10^3\ \text{in}^4$; $2490\ \text{in}^4$.

9.100 $29.7°$; $2530\ \text{in}^4$; $524\ \text{in}^4$.

9.103 (a) $-1.146\ \text{in}^4$. (b) $29.1°$ clockwise. (c) $3.39\ \text{in}^4$.

9.104 $23.8°$ clockwise; $0.524 \times 10^6\ \text{mm}^4$; $0.0917 \times 10^6\ \text{mm}^4$.

9.105 $19.54°$ counterclockwise; $4.34 \times 10^6\ \text{mm}^4$; $0.647 \times 10^6\ \text{mm}^4$.

9.106 (a) $25.3°$. (b) $1459\ \text{in}^4$; $40.5\ \text{in}^4$.

9.107 (a) $88.0 \times 10^6\ \text{mm}^4$. (b) $96.3 \times 10^6\ \text{mm}^4$; $39.7 \times 10^6\ \text{mm}^4$.

9.111 (a) $\bar{I}_{AA'} = \bar{I}_{BB'} = ma^2/24$. (b) $ma^2/12$.

9.112 (a) $m\ (r_1^2 + r_2^2)/4$. (b) $m\ (r_1^2 + r_2^2)/2$.

9.113 (a) $0.0699\ mb^2$. (b) $m(a^2 + 0.279\ b^2)/4$.

9.114 (a) $mb^2/7$. (b) $m(7a^2 + 10b^2)/70$.

9.117 (a) $5ma^2/18$. (b) $3.61\ ma^2$.

9.118 (a) $0.994\ ma^2$. (b) $2.33\ ma^2$.

9.119 $m(3a^2 + 4L^2)/12$.

9.120 $1.329\ mh^2$.

9.121 (a) $0.241\ mh^2$. (b) $m(3a^2 + 0.1204\ h^2)$.

9.122 $m(b^2 + h^2)/10$.

9.124 $ma^2/3$; $a/\sqrt{3}$.

9.126 $I_x = I_y = ma^2/4$; $I_z = ma^2/2$.

9.127 $1.160 \times 10^{-6}\ \text{lb·ft·s}^2$; $0.341\ \text{in}$.

9.128 $837 \times 10^{-9}\ \text{kg·m}^2$; $6.92\ \text{mm}$.

9.130 $2\ mr^2/3$; $0.816r$.

9.131 (a) $2.30\ \text{in.}$ (b) $20.6 \times 10^{-3}\ \text{lb·ft·s}^2$; $2.27\ \text{in}$.

9.132 (a) $\pi pl^2\ [6a^2t(5a^2/3l^2 + 2a/l + 1) + d^2l/4]$. (b) 0.1851.

9.133 (a) $27.5\ \text{mm}$ to the right of A. (b) $32.0\ \text{mm}$.

9.135 $I_x = 7.11 \times 10^{-3}\ \text{kg·m}^2$; $I_y = 16.96 \times 10^{-3}\ \text{kg·m}^2$; $I_z = 15.27 \times 10^{-3}\ \text{kg·m}^2$.

9.136 $I_x = 175.5 \times 10^{-3}\ \text{kg·m}^2$; $I_y = 309.10^{-3}\ \text{kg·m}^2$; $I_z = 154.4 \times 10^{-3}\ \text{kg·m}^2$.

9.138 $I_x = 334 \times 10^{-6}\ \text{lb·ft·s}^2$; $I_z = 1.356 \times 10^{-3}\ \text{lb·ft·s}^2$.

9.139 $I_x = 344 \times 10^{-6}\ \text{lb·ft·s}^2$; $I_y = 132.1 \times 10^{-6}\ \text{lb·ft·s}^2$; $I_z = 453 \times 10^{-6}\ \text{lb·ft·s}^2$.

9.141 (a) $13.99 \times 10^{-3}\ \text{kg·m}^2$. (b) $20.6 \times 10^{-3}\ \text{kg·m}^2$. (c) $14.30 \times 10^{-3}\ \text{kg·m}^2$.

9.142 $I_x = 28.3 \times 10^{-3}\ \text{kg·m}^2$; $I_y = 183.8 \times 10^{-3}\ \text{kg·m}^2$; $k_x = 42.9\ \text{mm}$; $k_y = 109.3\ \text{mm}$.

9.143 $30.5 \times 10^{-3}\ \text{lb·ft·s}^2$.

9.145 (a) $26.4 \times 10^{-3}\ \text{kg·m}^2$. (b) $31.2 \times 10^{-3}\ \text{kg·m}^2$. (c) $8.58 \times 10^{-3}\ \text{kg·m}^2$.

9.147 $I_x = 0.0392\ \text{lb·ft·s}^2$; $I_y = 0.0363\ \text{lb·ft·s}^2$; $I_z = 0.0304\ \text{lb·ft·s}^2$.

9.148 $I_x = 0.323\ \text{kg·m}^2$; $I_y = I_z = 0.419\ \text{kg·m}^2$.

9.149 $I_{xy} = 2.50 \times 10^{-3}\ \text{kg·m}^2$; $I_{yz} = 4.06 \times 10^{-3}\ \text{kg·m}^2$; $I_{zx} = 8.81 \times 10^{-3}\ \text{kg·m}^2$.

9.150 $I_{xy} = 286 \times 10^{-6}\ \text{kg·m}^2$; $I_{yz} = I_{zx} = 0$.

9.151 $I_{xy} = -1.726 \times 10^{-3}\ \text{lb·ft·s}^2$; $I_{yz} = 0.507 \times 10^{-3}\ \text{lb·ft·s}^2$; $I_{zx} = -2.12 \times 10^{-3}\ \text{lb·ft·s}^2$.

9.152 $I_{xy} = -538 \times 10^{-6}\ \text{lb·ft·s}^2$; $I_{yz} = -171.4 \times 10^{-6}\ \text{lb·ft·s}^2$; $I_{zx} = 1120 \times 10^{-6}\ \text{lb·ft·s}^2$.

9.155 $I_{xy} = -8.04 \times 10^{-3}$ kg·m^2; $I_{yz} = 12.90 \times 10^{-3}$ kg·m^2;
$I_{zx} = 94.0 \times 10^{-3}$ kg·m^2.

9.156 $I_{xy} = 0$; $I_{yz} = 48.3 \times 10^{-6}$ kg·m^2;
$I_{zx} = -4.43 \times 10^{-3}$ kg·m^2.

9.157 $I_{xy} = 47.9 \times 10^{-6}$ kg·m^2; $I_{yz} = 102.1 \times 10^{-6}$ kg·m^2;
$I_{zx} = 64.1 \times 10^{-6}$ kg·m^2.

9.158 $I_{xy} = -m' R_1^3/2$; $I_{yz} = m' R_1^3/2$; $I_{zx} = -m' R_2^3/2$.

9.159 $I_{xy} = wa^3(1 - 5\pi)g$; $I_{yz} = -11\pi \, wa^3/g$;
$I_{zx} = 4wa^3(1 + 2\pi)/g$.

9.160 $I_{xy} = -11wa^3/g$; $I_{yz} = wa^3(\pi + 6)/2g$; $I_{zx} = -wa^3/4g$.

9.162 (a) $mac/20$. (b) $I_{xy} = mab/20$; $I_{yz} = mbc/20$.

9.165 18.17×10^{-3} kg·m^2.

9.166 11.81×10^{-3} kg·m^2.

9.167 $5Wa^2/18g$.

9.168 $4.41 \, \gamma ta^4/g$.

9.169 281×10^{-3} kg·m^2.

9.170 0.354 kg·m^2.

9.173 (a) $1/\sqrt{3}$. (b) $\sqrt{7/12}$.

9.174 (a) $b/a = 2$; $c/a = 2$. (b) $b/a = 1$; $c/a = 0.5$.

9.175 (a) 2. (b) $\sqrt{2/3}$.

9.179 (a) $K_1 = 0.363ma^2$; $K_2 = 1.583ma^2$; $K_3 = 1.720ma^2$.
(b) $(\theta_x)_1 = (\theta_z)_1 = 49.7°$; $(\theta_y)_1 = 113.7°$; $(\theta_x)_3 = 45.0°$
$(\theta_y)_2 = 90.0°$, $(\theta_z)_2 = 135.0°$; $(\theta_x)_3 = (\theta_z)_3 = 73.5°$, $(\theta_y)_3 = 23.7°$.

9.180 (a) $K_1 = 14.30 \times 10^{-3}$ kg·m^2; $K_2 = 13.96 \times 10^{-3}$ kg·m^2;
$K_3 = 20.6 \times 10^{-3}$ kg·m^2.
(b) $(\theta_x)_1 = (\theta_y)_1 = 90.0°$; $(\theta_z)_1 = 0°$; $(\theta_x)_2 = 3.42°$,
$(\theta_y)_2 = 86.6°$.
$(\theta_z)_2 = 90.0°$; $(\theta_x)_3 = 93.4°$, $(\theta_y)_3 = 3.43°$, $(\theta_z)_3 = 90.0°$

9.182 (a) $K_1 = 0.1639Wa^2/g$; $K_2 = 1.054Wa^2/g$; $K_3 = 1.115Wa^2/g$.
(b) $(\theta_x)_1 = 36.7°$, $(\theta_y)_1 = 71.6°$; $(\theta_z)_1 = 59.5°$; $(\theta_x)_2 = 74.9°$,
$(\theta_y)_2 = 54.5°$, $(\theta_z)_2 = 140.5°$; $(\theta_x)_3 = 57.5°$, $(\theta_y)_3 = 138.8°$,
$(\theta_z)_3 = 112.4°$

9.183 (a) $K_1 = 2.26\gamma ta^4/g$; $K_2 = 17.27\gamma ta^4/g$; $K_3 = 19.08\gamma ta^4/g$.
(b) $(\theta_x)_1 = 85.0°$, $(\theta_y)_1 = 36.8°$, $(\theta_z)_1 = 53.7°$; $(\theta_x)_2 = 81.7°$,
$(\theta_y)_2 = 54.7°$; $(\theta_z)_2 = 143.4°$; $(\theta_x)_3 = 9.70°$, $(\theta_y)_3 = 99.0°$,
$(\theta_z)_3 = 86.3°$.

9.185 $I_x = 16ah^3/105$; $I_y = ha^3/5$.

9.186 $\pi a^3 b/8$; $a/2$.

9.188 $\bar{I}_x = 1.874 \times 10^6$ mm^4; $\bar{I}_y = 5.82 \times 10^6$ mm^4.

9.189 (a) 3.13×10^6 mm^4. (b) 2.41×10^6 mm^4.

9.191 -2.81 in^4.

9.193 (a) $ma^2/3$. (b) $3ma^2/2$.

9.195 $I_x = 0.877$ kg·m^2; $I_y = 1.982$ kg·m^2;
$I_z = 1.652$ kg·m^2.

9.196 0.0442 lb·ft·s^2.

CHAPTER 10

10.1 65.0 N ↓.

10.2 132.0 lb →.

10.3 39.0 N·m ↲.

10.4 1320 lb·in. ↰.

10.5 (a) 60.0 N C, 8.00 mm↓. (b) 300 N C, 40.0 mm↓.

10.6 (a) 120.0 N C, 16.00 mm↓. (b) 300 N C, 40.0 mm↓.

10.9 $Q = 3 P \tan \theta$.

10.10 $Q = P[(l/a)] \cos^3 \theta - 1]$.

10.12 $Q = 2 P \sin \theta/\cos (\theta/2)$.

10.14 $Q = (3P/2) \tan \theta$.

10.15 $M = Pl/2 \tan \theta$.

10.16 $M = Pl(\sin \theta + \cos \theta)$.

10.17 $M = \frac{1}{2}Wl \tan \alpha \sin \theta$.

10.18 (a) $M = Pl \sin 2\theta$. (b) $M = 3Pl \cos \theta$. (c) $M = Pl \sin \theta$.

10.19 85.2 lb·ft ↲.

10.20 22.8 lb ⟋ 70.0°.

10.23 38.7°.

10.24 68.0°.

10.27 36.4°.

10.28 67.1°.

10.30 25.0°.

10.31 39.7° and 69.0°.

10.32 390 mm.

10.33 330 mm.

10.35 38.7°.

10.36 52.4°.

10.37 22.6°.

10.38 51.1°.

10.39 59.0°.

10.40 78.7°, 324°, 379°.

10.43 12.03 kN ↘.

10.44 20.4°.

10.45 2370 lb ↖.

10.46 2550 lb ↖.

10.48 300 N·m, 81.8 N·m.

10.49 $\eta = 1/(1 + \mu \cot \alpha)$.

10.50 $\eta = \tan \theta/\tan (\theta + \phi_s)$.

10.52 37.6 N, 31.6 N.

10.53 $\mathbf{A} = 250$ N ↑; $\mathbf{M}_A = 450$ N·m ↰.

10.54 1050 N ↑.

10.57 0.833 in.↓.

10.58 0.625 in. →.

10.60 25.0°.

10.61 39.7° and 69.0°.

10.62 390 mm.

10.69 $\theta = -45.0°$, unstable; $\theta = 135.0°$, stable.

10.70 $\theta = -63.4°$, unstable; $\theta = 116.6°$, stable.

10.71 $\theta = 90.0°$ and $\theta = 270°$, unstable; $\theta = 22.0°$ and
$\theta = 158.0°$, stable.

10.72 $\theta = 0$ and $\theta = 180.0°$, unstable; $\theta = 75.5°$ and
$\theta = 284°$, stable.

10.73 59.0°, stable.

10.74 78.7°, stable; 324°, unstable; 379°, stable.

10.77 357 mm.

10.78 252 mm.

10.80 9.39° and 90.0°, stable; 34.2°, unstable.

10.81 17.11°, stable; 72.9°, unstable.

10.83 49.1°.

10.86 16.88 m.

10.87 54.8°.

10.88 37.4°.

10.89 $P < kl/2$.

10.91 $k > 6.94$ lb/in.

10.92 15.00 in.

10.93 $P < 2kL/9$.

10.94 $P < kL/18$.

10.96 $P < 160.0$ N.

10.98 $P < 764$ N.

10.100 (a) P < 10.00 lb. (b) P < 20.0 lb.

10.101 60.0 lb↓.

10.102 600 lb·in. ↲.

10.103 500 N ↑.

10.105 $M = 7Pa \cos \theta$

10.107 19.40°.

10.108 7.13 in.

10.110 $\theta = 0$, unstable; $\theta = 137.8°$, stable.

10.112 (a) 22.0°. (b) 30.6°.

Photo Credits

CHAPTER 1
Opener: ©Renato Bordoni/Alamy; **Photo 1.1:** NASA; **Photo 1.2:** NASA/JPL-Caltech.

CHAPTER 2
Opener: ©Getty Images RF; **Photo 2.1:** ©DGB/Alamy; **Photo 2.2:** ©Michael Doolittle/Alamy; **Photo 2.3:** ©Flirt/SuperStock; **Photo 2.4:** ©WIN-Initiative/Neleman/Getty Images.

CHAPTER 3
Opener: ©St Petersburg Times/ZumaPress/Newscom; **Fig. 3.11a:** ©Gavela Montes Productions/Getty Images RF; **Fig. 3.11b:** ©Image Source/Getty Images RF; **Fig. 3.11c:** ©Valery Voennyy/Alamy RF; **Fig. 3.11d:** ©Monty Rakusen/Getty Images RF; **Photo 3.1:** ©McGraw-Hill Education/Photo by Lucinda Dowell; **Photo 3.2:** ©Steve Hix; **Photo 3.3:** ©Jose Luis Pelaez/Getty Images; **Photo 3.4:** ©Images-USA/Alamy RF; **Photo 3.5:** ©Dana White/PhotoEdit.

CHAPTER 4
Opener: ©View Stock/Getty Images RF; **Photos 4.1, 4.2:** ©McGraw-Hill Education/Photos by Lucinda Dowell; **Fig. 4.1(Rocker bearing):** Courtesy Godden Collection. National Information Service for Earthquake Engineering, University of California, Berkeley; **Fig. 4.1(Links):** Courtesy Michigan Department of Transportation; **Fig. 4.1(Slider and rod):** ©McGraw-Hill Education/Photo by Lucinda Dowell; **Fig. 4.1(Pin support):** Courtesy Michigan Department of Transportation; **Fig. 4.1(Cantilever support):** ©Richard Ellis/Alamy; **Photo 4.3:** ©McGraw-Hill Education/Photo by Lucinda Dowell; **Photo 4.4:** Courtesy of SKF, Limited.

CHAPTER 5
Opener: ©Akira Kaede/Getty Images RF; **Photo 5.1:** ©Christie's Images Ltd./SuperStock; **Photo 5.2:** ©C Squared Studios/Getty Images RF; **Photo 5.3:** ©Michel de Leeuw/Getty Images RF; **Photo 5.4:** ©maurice joseph/Alamy; **Fig. 5.18:** ©North Light Images/agefotostock; **Photo 5.5:** NASA/Carla Thomas.

CHAPTER 6
Opener: ©Lee Rentz/Photoshot; **Photo 6.1a:** ©Datacraft Co Ltd/Getty Images RF; **Photo 6.1b:** ©Fuse/Getty Images RF; **Photo 6.1c:** ©Design Pics/Ken Welsh RF; **Photo 6.2:** Courtesy Godden Collection. National Information Service for Earthquake Engineering, University of California, Berkeley; **Photo 6.3:** Courtesy of Ferdinand Beer; **Photo 6.4:** ©McGraw-Hill Education/Photo by Sabina Dowell; **Photo 6.5:** ©James Hardy/PhotoAlto RF; **Photo 6.6:** ©Mark Thiessen/National Geographic Society/Corbis; **Photo 6.7:** ©Getty Images RF.

CHAPTER 7
Opener: ©Jonatan Martin/Getty Images RF; **Photo 7.1:** ©McGraw-Hill Education/Photo by Sabina Dowell; **Fig. 7.6(Continuous beam):** ©Ross Chandler/Getty Images RF; **Fig. 7.6(Overhanging beam):** ©Goodshoot/Getty Images RF; **Fig. 7.6(Cantilever beam):** ©Ange/Alamy; **Photo 7.2:** ©Alan Thornton/Getty Images; **Photo 7.3:** ©Bob Edme/AP Photo; **Photo 7.4:** ©Thinkstock/Getty Images RF; **Photo 7.5a:** ©Ingram Publishing/Newscom; **Photo 7.5b:** ©Karl Weatherly/Getty Images RF; **Photo 7.5c:** ©Eric Nathan/Alamy.

CHAPTER 8
Opener: ©Bicot (Marc Caya); **Photo 8.2:** ©Tomohiro Ohsumi/Bloomberg/Getty Images; **Photo 8.3:** ©Leslie Miller/agefotostock; **Photo 8.4:** Courtesy of REMPCO Inc.; **Photo 8.5:** ©Bart Sadowski/Getty Images RF; **Photo 8.6:** ©Fuse/Getty Images RF.

CHAPTER 9
Opener: ©ConstructionPhotography.com/Photoshot; **Photo 9.1:** ©Barry Willis/Getty Images; **Photo 9.2:** ©loraks/Getty Images RF.

CHAPTER 10
Opener: ©Tom Brakefield/SuperStock; **Photo 10.1:** Courtesy Brian Miller; **Photo 10.2:** Courtesy of Altec, Inc.; **Photo 10.3:** Courtesy of DE-STA-CO.

Index

		$\bar{x}$	$\bar{y}$	Area
Triangular area			$\dfrac{h}{3}$	$\dfrac{bh}{2}$
Quarter-circular area		$\dfrac{4r}{3\pi}$	$\dfrac{4r}{3\pi}$	$\dfrac{\pi r^2}{4}$
Semicircular area		0	$\dfrac{4r}{3\pi}$	$\dfrac{\pi r^2}{2}$
Semiparabolic area		$\dfrac{3a}{8}$	$\dfrac{3h}{5}$	$\dfrac{2ah}{3}$
Parabolic area		0	$\dfrac{3h}{5}$	$\dfrac{4ah}{3}$
Parabolic spandrel		$\dfrac{3a}{4}$	$\dfrac{3h}{10}$	$\dfrac{ah}{3}$
Circular sector		$\dfrac{2r\sin\alpha}{3\alpha}$	0	αr^2
Quarter-circular arc		$\dfrac{2r}{\pi}$	$\dfrac{2r}{\pi}$	$\dfrac{\pi r}{2}$
Semicircular arc		0	$\dfrac{2r}{\pi}$	πr
Arc of circle		$\dfrac{r\sin\alpha}{\alpha}$	0	$2\alpha r$